Compendium of Plant Genomes

Series editor

Chittaranjan Kole, Raja Ramanna Fellow, Department of Atomic Energy, Government of India, Kalyani, West Bengal, India

Whole-genome sequencing is at the cutting edge of life sciences in the new millennium. Since the first genome sequencing of the model plant Arabidopsis thaliana in 2000, whole genomes of about 70 plant species have been sequenced and genome sequences of several other plants are in the pipeline. Research publications on these genome initiatives are scattered on dedicated web sites and in journals with all too brief descriptions. The individual volumes elucidate the background history of the national and international genome initiatives; public and private partners involved; strategies and genomic resources and tools utilized; enumeration on the sequences and their assembly; repetitive sequences; gene annotation and genome duplication. In addition, synteny with other sequences, comparison of gene families and most importantly potential of the genome sequence information for gene pool characterization and genetic improvement of crop plants are described.

Interested in editing a volume on a crop or model plant? Please contact Dr. Kole, Series Editor, at ckole2012@gmail.com

More information about this series at http://www.springer.com/series/11805

Nils Stein · Gary J. Muehlbauer
Editors

The Barley Genome

Springer

Editors
Nils Stein
Leibniz Institute of Plant Genetics
and Crop Plant Research (IPK)
Seeland, Sachsen-Anhalt
Germany

Gary J. Muehlbauer
Department of Agronomy
and Plant Genetics
University of Minnesota
St. Paul, MN, USA

ISSN 2199-4781 ISSN 2199-479X (electronic)
Compendium of Plant Genomes
ISBN 978-3-030-06453-2 ISBN 978-3-319-92528-8 (eBook)
https://doi.org/10.1007/978-3-319-92528-8

This Springer imprint is published by the registered company Springer Nature Switzerland AG
The registered company address is: Gewerbestrasse 11, 6330 Cham, Switzerland

This book series is dedicated to
my wife Phullara, and our children
Sourav, and Devleena

Chittaranjan Kole

In Memory

Dr. Patrick Schweizer

(1959–2018)

Preface to the Series

Genome sequencing has emerged as the leading discipline in the plant sciences coinciding with the start of the new century. For much of the twentieth century, plant geneticists were only successful in delineating putative chromosomal location, function, and changes in genes indirectly through the use of a number of 'markers' physically linked to them. These included visible or morphological, cytological, protein, and molecular or DNA markers. Among them, the first DNA marker, the RFLPs, introduced a revolutionary change in plant genetics and breeding in the mid-1980s, mainly because of their infinite number and thus potential to cover maximum chromosomal regions, phenotypic neutrality, absence of epistasis, and codominant nature. An array of other hybridization-based markers, PCR-based markers, and markers based on both facilitated construction of genetic linkage maps, mapping of genes controlling simply inherited traits, and even gene clusters (QTLs) controlling polygenic traits in a large number of model and crop plants. During this period, a number of new mapping populations beyond F2 were utilized and a number of computer programs were developed for map construction, mapping of genes, and for mapping of polygenic clusters or QTLs. Molecular markers were also used in studies of evolution and phylogenetic relationship, genetic diversity, DNA-fingerprinting, and map-based cloning. Markers tightly linked to the genes were used in crop improvement employing the so-called marker-assisted selection. These strategies of molecular genetic mapping and molecular breeding made a spectacular impact during the last one and a half decades of the twentieth century. But still they remained 'indirect' approaches for elucidation and utilization of plant genomes since much of the chromosomes remained unknown and the complete chemical depiction of them was yet to be unraveled.

Physical mapping of genomes was the obvious consequence that facilitated development of the 'genomic resources' including BAC and YAC libraries to develop physical maps in some plant genomes. Subsequently, integrated genetic–physical maps were also developed in many plants. This led to the concept of structural genomics. Later on, emphasis was laid on EST and transcriptome analysis to decipher the function of the active gene sequences leading to another concept defined as functional genomics. The advent of techniques of bacteriophage gene and DNA sequencing in the 1970s was extended to facilitate sequencing of these genomic resources in the last decade of the twentieth century.

As expected, sequencing of chromosomal regions would have led to too much data to store, characterize, and utilize with the-then available computer software could handle. But development of information technology made the life of biologists easier by leading to a swift and sweet marriage of biology and informatics, and a new subject was born—bioinformatics.

Thus, evolution of the concepts, strategies, and tools of sequencing and bioinformatics reinforced the subject of genomics—structural and functional. Today, genome sequencing has traveled much beyond biology and involves biophysics, biochemistry, and bioinformatics!

Thanks to the efforts of both public and private agencies, genome sequencing strategies are evolving very fast, leading to cheaper, quicker, and automated techniques right from clone-by-clone and whole-genome shotgun approaches to a succession of second generation sequencing methods. Development of software of different generations facilitated this genome sequencing. At the same time, newer concepts and strategies were emerging to handle sequencing of the complex genomes, particularly the polyploids.

It became a reality to chemically—and so directly—define plant genomes, popularly called whole-genome sequencing or simply genome sequencing.

The history of plant genome sequencing will always cite the sequencing of the genome of the model plant Arabidopsis thaliana in 2000 that was followed by sequencing the genome of the crop and model plant rice in 2002. Since then, the number of sequenced genomes of higher plants has been increasing exponentially, mainly due to the development of cheaper and quicker genomic techniques and, most importantly, development of collaborative platforms such as national and international consortia involving partners from public and/or private agencies.

As I write this preface for the first volume of the new series 'Compendium of Plant Genomes,' a net search tells me that complete or nearly complete whole-genome sequencing of 45 crop plants, eight crop and model plants, eight model plants, 15 crop progenitors and relatives, and three basal plants is accomplished, the majority of which are in the public domain. This means that we nowadays know many of our model and crop plants chemically, i.e., directly, and we may depict them and utilize them precisely better than ever. Genome sequencing has covered all groups of crop plants. Hence, information on the precise depiction of plant genomes and the scope of their utilization is growing rapidly every day. However, the information is scattered in research articles and review papers in journals and dedicated Web pages of the consortia and databases. There is no compilation of plant genomes and the opportunity of using the information in sequence-assisted breeding or further genomic studies. This is the underlying rationale for starting this book series, with each volume dedicated to a particular plant.

Plant genome science has emerged as an important subject in academia, and the present compendium of plant genomes will be highly useful both to students and teaching faculties. Most importantly, research scientists involved in genomics research will have access to systematic deliberations on the plant genomes of their interest. Elucidation of plant genomes is of interest not only for the geneticists and breeders, but also for practitioners of an array of plant science disciplines, such as taxonomy, evolution, cytology, physiology, pathology, entomology, nematology, crop production, biochemistry, and obviously bioinformatics. It must be mentioned that information regarding each plant genome is ever-growing. The contents of the volumes of this compendium are therefore focusing on the basic aspects of the genomes and their utility. They include information on the academic and/ or economic importance of the plants, description of their genomes from a molecular genetic and cytogenetic point of view, and the genomic resources developed. Detailed deliberations focus on the background history of the national and international genome initiatives, public and private partners involved, strategies and genomic resources and tools utilized, enumeration on the sequences and their assembly, repetitive sequences, gene annotation, and genome duplication. In addition, synteny with other sequences, comparison of gene families, and, most importantly, potential of the genome sequence information for gene pool characterization through genotyping by sequencing (GBS) and genetic improvement of crop plants have been described. As expected, there is a lot of variation of these topics in the volumes based on the information available on the crop, model, or reference plants.

I must confess that as the series editor, it has been a daunting task for me to work on such a huge and broad knowledge base that spans so many diverse plant species. However, pioneering scientists with lifetime experience and expertise on the particular crops did excellent jobs editing the respective volumes. I myself have been a small science worker on plant genomes since the mid-1980s and that provided me the opportunity to personally know several stalwarts of plant genomics from all over the globe. Most, if not all, of the volume editors are my longtime friends and colleagues. It has been highly comfortable and enriching for me to work with them on this book series. To be honest, while working on this series I have been and will remain a student first, a science worker second, and a series editor last. And I must express my gratitude to the volume editors and the chapter authors for providing me the opportunity to work with them on this compendium.

I also wish to mention here my thanks and gratitude to the Springer staff, Dr. Christina Eckey and Dr. Jutta Lindenborn in particular, for all their constant and cordial support right from the inception of the idea.

I always had to set aside additional hours to edit books besides my professional and personal commitments—hours I could and should have given to my wife, Phullara, and our kids, Sourav, and Devleena. I must mention that they not only allowed me the freedom to take away those hours from them but also offered their support in the editing job itself. I am really not sure whether my dedication of this compendium to them will suffice to do justice to their sacrifices for the interest of science and the science community.

Kalyani, India Chittaranjan Kole

Preface

Barley (*Hordeum vulgare* L.) was selected by early humans in the Fertile Crescent around 10,000–12,000 years ago and is likely one of the first domesticated plants. Subsequently, barley became a foundation for early human civilization and due to its adaptability is now grown in all temperate regions of the world. Currently, it is the fourth most important cereal crop behind maize, rice, and wheat. Barley is primarily used for animal feed, and malting and brewing, with a small percentage devoted to food. Due to the economic and agronomic importance of barley as a crop, it has been the subject of numerous genetic and genomics studies.

Barley has a large and highly repetitive 5.1 Gb genome, which presented significant obstacles to developing of a high-quality genome sequence. However, in 2006, a small group of barley geneticists had the foresight to form the International Barley Sequencing Consortium (IBSC), resulting in the coordination and data sharing to develop and release of a draft sequence in 2012. Subsequent effort by the IBSC resulted in releasing a high-quality genome sequence in 2017. Thus, this book is timely in that it describes the current status of barley genetics, breeding, and biology, and sets the stage for increased genome sequence-enabled understanding and improvement of this ancient crop. This volume covers aspects of the barley genome (sequencing and assembly approaches, gene prediction, chromosomal genomics, sequence diversity and structural variation, and variation in the secondary and tertiary gene pools), taxonomy, domestication, development (vegetative and inflorescence), genome characteristics (cytogenetics, repetitive sequences,), biotic and abiotic stress responses, organellar genomes, proteomics, gene cloning and expression, and genomics-enabled improvement.

It has been a great privilege to work with members of the barley research community on this book. We are indebted to all of the authors for their expertise and time. Experts in the field reviewed each chapter, and thus we are thankful for their efforts to improve the quality of this compilation. We hope that this volume will serve as a reference for those new to barley and to experienced barley researchers.

Seeland, Germany — Nils Stein
St. Paul, USA — Gary J. Muehlbauer

Acknowledgements

We are grateful to our peer reviewers who have invested their time to help improving the content of the chapters of this volume: Roland v. Bothmer, Andrea Bräutigam, Hans Peter Braun, Robin Buell, Mario Caccamo, Fred Choulet, Tim J. Close, Dave Edwards, Justin Faris, Bernd Friebe, Stephan Greiner, Perry Gustafson, Mats Hansson, David Jackson, Kostya Kanyuka, Benjamin Kilian, Takao Komatsuda, Sergio Lucretti, Tim March, Harvey Millar, Michele Morgante, Peter Morrell, Richard Pickering, Jochen C. Reif, Stuart Roy, Paula McSteen, Michael J. Scanlon, Margret Scholz, Ingo Schubert, Ian Small, Mark E. Sorrells, Nathan Springer, Jan Svensson, David Swarbreck, Cristobal Uauy, and Robbie Waugh.

Contents

Economic and Academic Importance of Barley

1

Peter Langridge

Abstract

Barley has had an interesting history. It is thought to be the first crop domesticated and developed as the staple food for the earliest farmers. It has remained an important food in many regions but its main uses now are as an animal feed and for beer production. While production for the other major cereal crops, maize, rice and wheat, has continued to grow, barley production has stagnated over the past two decades. Nevertheless, over the last century, barley has been an important crop model for a wide range of studies on genetics, biochemistry and developmental biology, particularly for barley's close relative, wheat. Many key concepts and tools in modern crop research can be traced back to early studies on barley. As techniques for genetic and genome analysis improve, and genomic research in wheat becomes more tractable, the role of barley as a model is likely to shift. However, there are several aspects of barley that are likely to keep it as an important crop for study.

P. Langridge (✉)
School of Agriculture, Food and Wine, University of Adelaide, Adelaide, SA, Australia
e-mail: peter.langridge@adelaide.edu.au

1.1 Background

Barley (*Hordeum vulgare*) is the fourth major cereal in terms of production after maize, rice and wheat. Barley with nonshattering rachises has been found at the oldest archaeological sites dated at just around 11,000 years ago. These are likely to represent the first plants morphologically modified by human selection and just predated the first domesticated wheats—diploid or einkorn wheat (*Triticum monococcum*) (Allard 1999). Wheat and barley are closely related, and it is possible to produce fertile hybrids between the two crops. However, barley is often seen as being an inferior food staple compared to wheat and has been described as the 'poor man's bread'. Despite this limitation, barley is usually hardier than wheat and has been an important crop in many regions where wheat might struggle to yield and this characteristic has ensured barley cultivation from domestication to the present day (Zohary and Hopf 1988).

1.2 World Barley Production

Barley has profited from the changes that have occurred in breeding strategies and in farming practices resulting in a steady rate of yield increases. Today, barley is grown across the temperate regions of both the northern and southern hemispheres. Figure 1.1 shows the

N. Stein and G. J. Muehlbauer (eds.), *The Barley Genome*, Compendium of Plant Genomes,
https://doi.org/10.1007/978-3-319-92528-8_1

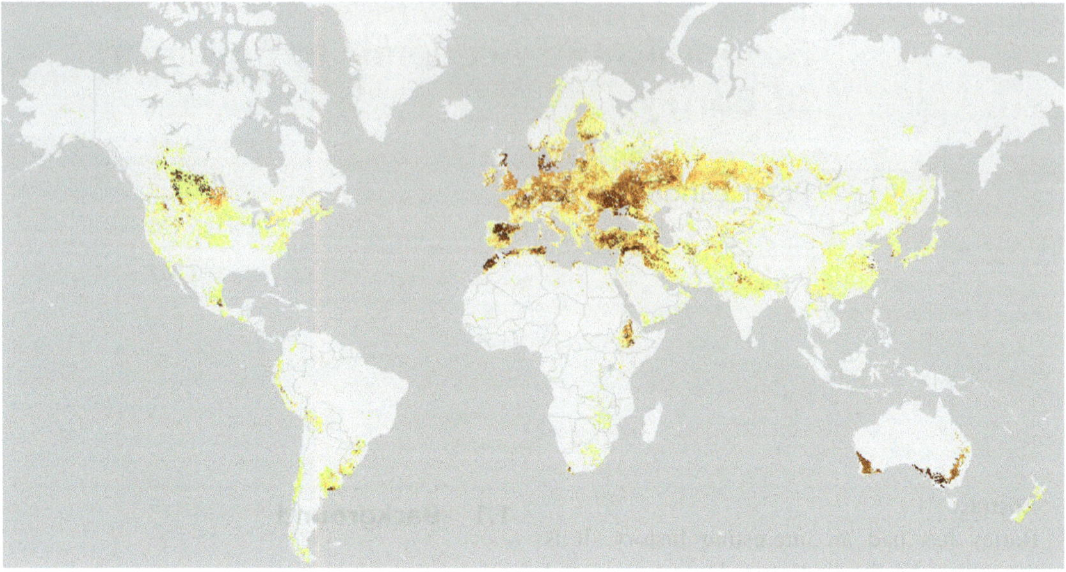

Fig. 1.1 Distribution of barley production globally (You et al. 2014 MapSpam). Colour intensity relates to the proportion of land devoted to barley production

distribution of barley production across the world. Europe and the Russian Federation account for around 65% of global production but barley has remained an important food crop in parts of North Africa, Asia and South America.

The ten biggest barley producers are shown in Fig. 1.2. The Russian Federation is not only the largest producer by quantity but also has the largest area sown to barley. It can also be seen from Fig. 1.2 that yields in Russia and Australia are quite low at around 2.5 tonne/ha compared to France and Germany where yields are usually well over 6 tonne/ha. In 2014, almost 150 million tonne of barley were produced on almost 50 million ha giving an average global yield of around 3 tonne/ha.

Fig. 1.2 The world's ten major barley producers for 2014. Data from FAOSTAT (2017)

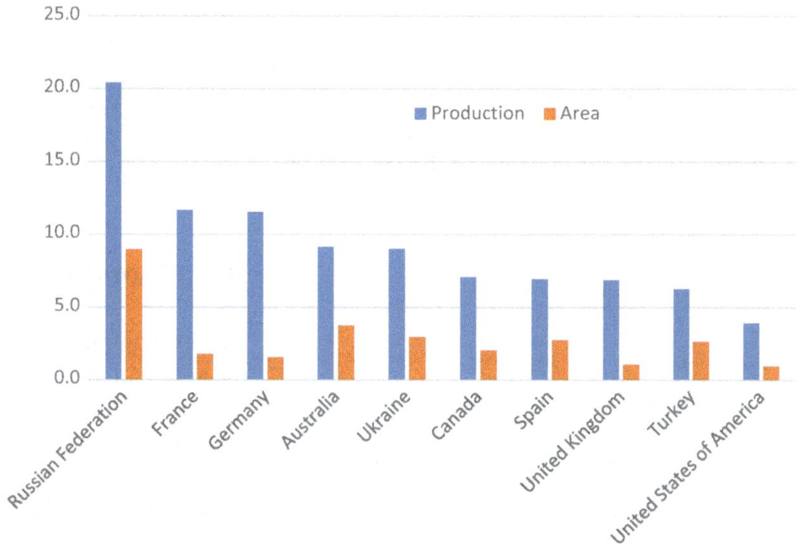

In 2013, barley exports were valued at over US$8.5 billion (31 million tonne) with the biggest exporting countries France, Australia, Argentina, Germany and Ukraine. Conversely, the value of barley imports globally was just over US$9.4 billion with Saudi Arabia (10.5 million tonne) by far the biggest importer accounting for almost one-third of the total global barley imports.

Figure 1.3 shows the global changes that have occurred in the area sown to barley and the total production since 1961. Over this period, the yields of barley have risen from an average of 1.3 tonne/ha to over 2.5 tonne. However, it is interesting to note that since the mid-1980s the area sown to barley has been declining. This is probably related to the increasing success of new maize hybrids and soybean cultivars in the USA and to the higher value of wheat in many areas. Brassica crops have also tended to replace barley as an alternative to wheat in many regions. A second factor is the reduction of barley as a traded staple. While barley yields showed rapid increase from the early 1960s until the mid-1980s, there has been little real improvement in yields for the past 15 years. This is probably related to barley being pushed out of some of the more productive cropping regions and moving further to low rainfall, stressed environments where it can outperform wheat. It is possible that the small investment in barley improvement, relative to wheat and particularly maize, is a contributing factor to the slow yield gains.

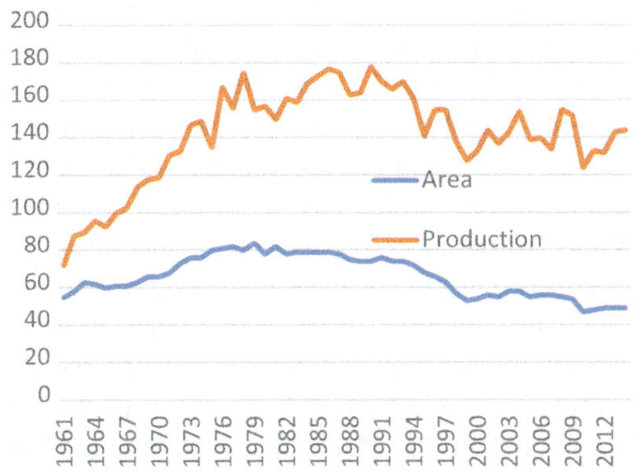

Fig. 1.3 Global barley production and area sown. Area (blue) is given in million hectares and production (orange) in million tonne. Data from FAOSTAT (2017)

1.3 Barley End Uses

Since 1960, the major use of barley is as animal feed which accounts for between 61 and 77% of barley use. However, malting represents the high-value use for barley with malting barley commanding substantial premiums compared to feed. Over the same period, between 9 and 22% of barley production goes to malting. Although barley was likely to have been originally domesticated for human food and has remained an important food source for people in many regions, currently, food consumption accounts for only around 5% of barley end use (FAOSTAT 2017). The most obvious trend in barley end use has been an increase in barley going to malting and a decrease in human consumption of barley (Fig. 1.4). In the 1960s, just over 10% of barley was used for malting and over 15% for humans. Now the situation is reversed with over 20% of barley production going to malting in some years while human consumption has remained around 5% since the 1980s (Fig. 1.4).

Barley malt is a key raw material for beer brewing and whiskey production—about 130 g of malt is used to produce a litre of beer. There are many different characteristics of barley that are important for malting. These characteristics are primarily related to the speed and consistency of germination, the breakdown of the endosperm cell walls and the degradation of starch into fermentable sugars. High protein is undesirable for malting, and efficient degradation of endogenous proteins is also important for producing high-quality malt (Fox et al. 2003). The importance of the malting process has meant that grain structure, development and germination have been intensively studied in barley and this now represents one of the best-studied cereal grains (Schulte et al. 2009).

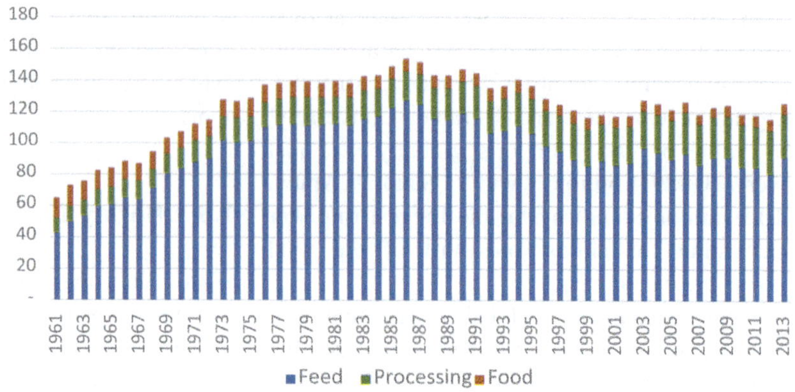

Fig. 1.4 End uses of barley grain. The major end uses, in million tonne, are for animal feed (blue), processing or malting (green) and human food (red). Data from FAOSTAT (2017)

Fig. 1.5 Beer from barley in million tonne. Data from FAOSTAT (2017)

Despite the decline in barley production over the past three decades, beer production from barley has shown steady growth (Fig. 1.5) with worldwide beer production at well over a billion hectolitres annually. This is reflected in the rising proportion of barley that has been used for malting, ranging from around 10% prior to the mid-1980s to around 20% more recently.

Food uses have been important historically in many regions including the Middle East, North Africa and northern and eastern Europe and Asia (reviewed in Baik and Ullrich 2008). Barley flour is usually prepared from pearled barley and can be incorporated into wheat-based foods (Newman and Newman 1991). However, wheat and rice provide a better quality product and better mouthfeel than barley which led to a decline in barley consumption over the past 200 years (Newman and Newman 2006). Consequently, breeders have largely ignored food quality in barley improvement.

1.3.1 New Trends in Barley End Uses

In recent years, there has been a trend away from the major breweries to small or independent brewing facilities. The craft brewing industry is made up of brewpubs, microbreweries, regional craft breweries and contract brewers. It has been estimated that in 2015 there are over 10,000 craft breweries globally with the vast majority in Europe (4486) and North America (4483) (http://ag.alltech.com/en/blog/2015-craft-brewery-count). While overall beer production in the USA has remained constant or even declined, craft breweries and imported beers have been growing rapidly (up by 6.2 and 6.8% in 2016, respectively) and the craft brewing market in the US represented 12.3% of total consumption in 2016. In 2016, the craft brewing industry in the USA was valued at US$67.8 billion and employed over 456,000 people (https://www.brewersassociation.org/statistics/economic-impact-data/). Importantly, the growth of the craft brewing industry has been accelerating with an increase in the number of craft breweries in the USA of over 16% between 2015 and 2016 (https://www.brewersassociation.org/statistics/number-of-breweries/). These numbers are significant for barley production not only because of their rapid growth in the craft brewing industry but also because they are heavy users of high-quality barley malt in contrast to many of the major breweries who use large amounts of adjunct from non-barley sources.

In addition to the rise of the craft brewing industries, opportunities also exist for increasing the use of barley for human food. Several studies indicate that barley is one of the healthiest cereals for the human diet due to high levels of some important nutrients (Table 1.1). The overall nutritional value of cereal grains varies greatly depending on how the grain is processed and consumed. There is also considerable variation between accessions (Shewry et al. 2013). However, some barley accessions have high levels of dietary fibre, particularly beta-glucans, and good levels of other bioactive compounds and minerals, such as iron and zinc (Shewry et al. 2013; HealthGrain forum https://healthgrain.org/). The high overall levels and extensive genetic variation were used to select for lines with particularly high concentration of fibre and resistant starch (https://www.thehealthygrain.com/). High fibre barleys were developed by CSIRO in Australia and have been commercialised as BARLEYmax^{TM}.

Despite the clear benefits of barley as a human food, its use for food remains low relative to other cereals and there has been no indication that its use will grow. In 2016, the global per capita food use of barley was only 1 kg/person compared to the 67 kg for wheat, 17 kg for maize and 54 kg for rice (FAO 2016). The highest per capita consumption is in North Africa, particularly Morocco (41 kg/person in 2016), Ethiopia (15 kg) and Syria (15 kg).

1.4 Academic Importance of Barley

Since the mid-1800s, there have been over 47,000 scientific publications on barley (based on a Scopus search using ‘barley’ as keyword). The number of barley and *Hordeum* publications since 1950 is shown in Fig. 1.6. The 47,000 barley publications contrast to over 150,000 publications on wheat, and around 92,000 for maize over the same period. Importantly, there were almost 14,000 publications where both wheat and barley were listed as keywords.

Barley has been an important model for wheat but as resources and technologies have advanced, wheat researchers have become increasingly independent. For the period from 1950 to around 1970, there were about twice as many

Table 1.1 Nutritional composition of major cereals

Nutritional value (/100 g raw)	Units	Whole grain wheat flour	Oats	Brown rice	Whole grain barley
Energy	kcal	340	389	357	334
Protein	g	13.2	16.9	8.3	10.6
Total fat	g	2.5	6.9	2.6	2.1
Carbohydrates	g	61.3	55.7	73.5	60.8
Fibre	g	10.7	10.6	3	14.8
Calcium	mg	34	54	12	50
Iron	mg	3.6	4.7	1.3	6
Zinc	mg	2.6	4	0.8	3.3

Data adapted from https://ec.europa.eu/jrc/en/health-knowledge-gateway/promotion-prevention/nutrition/whole-grain#_Toc479239823

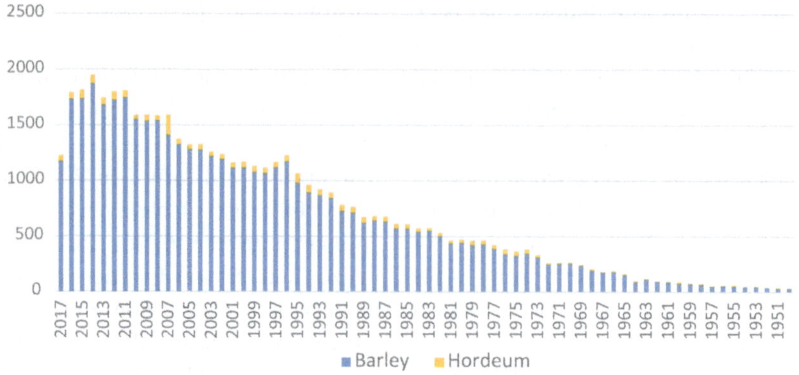

Fig. 1.6 **Numbers of publications giving 'barley' or '*Hordeum*' as keywords since 1950**. Publications listing only 'barley' as a keyword are shown in blue while those with only '*Hordeum*' are in orange. Data from Scopus

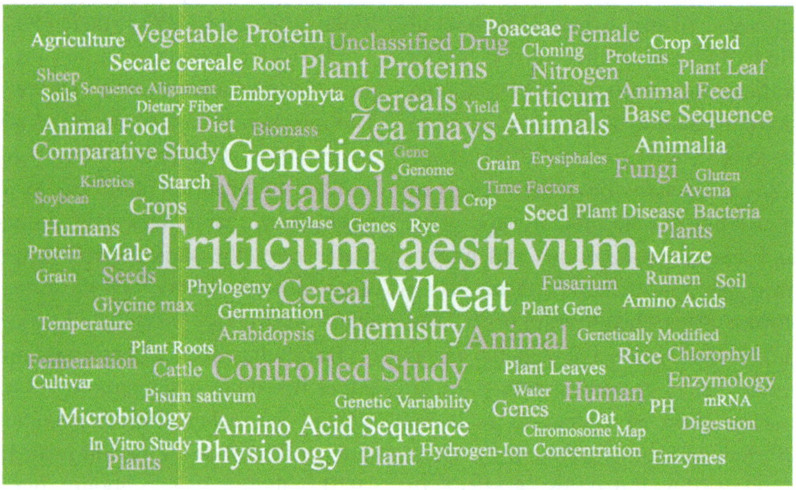

Fig. 1.7 **Word cloud based on keywords used in the 47,436 publications that also included barley as a keyword**. The size of the word in the image reflects the frequency of this word amongst the keywords. Raw data from Scopus and word cloud generated with WordItOut (https://worditout.com/)

publications on wheat compared to barley. However, since the mid-1990s, publications on wheat have exceeded barley fourfold.

The close relationship between barley and wheat research is further emphasised in the Word cloud shown in Fig. 1.7. The most frequently associated keywords are '*Triticum aestivum*' and 'wheat'. The word cloud also shows the importance of barley in research on metabolism, genetics and physiology.

1.4.1 Examples of the Broad Impact of Barley Research

1.4.1.1 Domestication

There are many aspects of barley domestication, physiology, biochemistry and developmental biology that have been critical for crop science. Many of our ideas about crop domestication have come from studies of diversity in barley, and as noted above, barley is believed to have been the first crop domesticated. The concept of geographical centres of origin for our modern crops fits well with early ideas of barley domestication in the Fertile Crescent and in the close proximity to wild barley, *Hordeum spontaneum*. However, the more recent discovery of two different rachis mutations in barley (Pourkheirandish et al. 2015) provides strong evidence for two separate domestication events (Morrell and Clegg 2007). This diversity was reinforced with new archaeological evidence for at least two separate domestication events in the Middle East (Riehl et al. 2013). These discoveries are now raising questions about the whole concept of single origin for our major crops and have stimulated a reanalysis of diversity in other species (Allaby 2015).

1.4.1.2 Disease Resistance

The *Mlo* gene was cloned from barley about 20 years ago (Büschges et al. 1997). This gene was first found in an Ethiopian landrace and became a central tool in the control of barley powdery mildew in Europe. However, powdery mildew, caused by over 650 fungal species, is a disease of around 10,000 plant species. It now seems that the *Mlo* resistance mechanism found in barley could have broad application for control of the disease in many other species. This has led to the description of *mlo* as a possible 'universal weapon to defeat powdery mildew disease' (Kusch and Panstruga 2017). This strategy was adopted to generate mutations at all three homoeoloci of *Mlo* in wheat through genome editing to provide broad-spectrum resistance to powdery mildew (Wang et al. 2014).

1.4.1.3 Mutation Research

Mutation research in barley goes back to the very start of mutation work in crop plants with the early work of Stadler, Nilson-Ehle and Gustafsson (Lundqvist 2014). Indeed, barley has been used as a model for the application of mutations to the study of pathogen resistance, physiological, biochemical and developmental processes and to the production of novel commercial varieties based on specific mutations. For example, a gamma-ray-induced mutant of the cultivar 'Valticky' was produced in 1965 and released as the variety 'Diamant'. Diamant was about 15 cm shorter than its parent variety and showed about 12% higher yield. This variety resulted in over 150 new varieties in Europe, North America and Asia (Ahloowalia et al. 2004). Another gamma ray mutant, Golden Promise derived from Maythorpe, has been a mainstay of the Scottish whiskey industry. It was originally selected for its short stature, stiff straw and good malting properties but the mutant also proved to be more salt tolerant than its parent variety (Wei et al. 2003). These characteristics are likely to have been important in two Australian varieties, Hindmarsh and La Trobe, which both have Golden Promise sister lines in their pedigree.

More recently, lipoxygenase-deficient mutants have been adopted by the malting and brewing industry for their improved effects on beer stability (Skadhauge et al. 2011).

The development of genomics resources has led to a revitalisation of barley mutant research and renewal of interest in the extensive series of development mutants identified in the 1950s to 1970s. These mutants have been used to elucidate a range of developmental and metabolic pathways in plants (Druka et al. 2011). In several cases, the work on barley has provided critical starting points for equivalent work in other cereals, notably wheat. Some key examples are the work on flowering time control through the isolation of vernalisation and photoperiod response genes (Fu et al. 2005; Turner et al. 2005; Cockram et al. 2007; Beales et al. 2007) and the analysis of floral morphology with the work on floret fertility (two-row vs. six-row barley) and the hulless traits (Komatsuda et al. 2007; Taketa et al. 2008).

1.4.1.4 Grain Development and Germination

Although more barley is used for animal feed than for beer production, malting is the high-value product of barley and there has been considerable work to understand the characteristics and properties of barley that have made it so important for beer production. Grain development determines the composition and properties of the grain, while germination is critical for the malting process and providing the sugars needed for beer fermentation. Consequently, barley has provided an important model for a range of physiological and biochemical studies around grain development and germination in the cereals. For example, isolated aleurone layers of barley were used to study the effects of phytohormones gibberellic acid and abscisic acid. Enzymes secreted by the aleurone in response to hormone treatment could be readily isolated and characterised (Chrispeels and Varner 1967; Jacobsen and Varner 1967; Slakeski and Fincher 1992; Gómez-Cadenas et al. 2001). The early work on enzyme isolation and characterisation also meant that barley grain enzymes were amongst the first plant proteins to be crystallised and with solved structures (Varghese et al. 1994; Kadziola et al. 1994).

The barley aleurone was also important in early studies of the control of gene expression in response to hormonal signals (Chandler et al. 1984) and the characterisation of gene promoters (Lanahan et al. 1992; Gubler et al. 1995).

The long history of research on barley grains has made this species a valuable model for applying new techniques for studying different aspects of seed development. For example, transcript dynamics have been measured during both grain development (Zhang et al. 2016) and germination (Betts et al. 2017), and a detailed analysis and three-dimensional reconstruction of grain development has also been produced from careful histological study (Gubatz et al. 2007). These new resources will mean that barley grain remains an important system for plant research.

1.5 Conclusions

The recent trends suggest that barley is in decline as a crop and as a research tool. Global production of barley has been falling and publications on barley are not growing as rapidly as for wheat, maize and rice. Barley was long seen as a good diploid model for hexaploid wheat and although the genome size of barley is large, it is still only a third of the size of wheat. However, wheat has now been sequenced and other genomic resources are rapidly accumulating. Gene discovery is now far less dependent on genome size and structure than only a few years ago.

Does barley still have something to offer for crop research? There are two key features of barley that are likely to ensure its continued importance as a model species. First, it is one of the hardiest of the cereal crops. Consequently, barley has been proposed as a good model for studying adaptation to climate change (Dawson et al. 2015). The second important feature of barley is diversity. There is increasing awareness that expanding the germplasm base for our crops will be critical in continuing to advance yields in the face of a wide range of societal and environmental challenges (McCouch et al. 2013). Wild barley, *H. spontaneum*, can be directly crossed to cultivated barley (primary gene pool) and resources for accessing diversity in the secondary gene pool of barley are developing rapidly (Wendler et al. 2014). There are over 400,000 accessions of barley in gene banks (Knüpfer 2009), so the scope to use barley to develop efficient strategies for exploit genetic resources is enormous. Therefore, barley is likely to remain an important model but the prime opportunities can be expected to lie in the use of barley to understand and enhance adaptation to environmental instability and the development of tools and techniques for enhancing the utilisation of diversity in landraces and wild relatives of our crops.

References

Ahloowalia BS, Maluszynski M, Nichterlein K (2004) Global impact of mutation-derived varieties. Euphytica 135:187–204

Allaby RG (2015) Barley domestication: the end of a central dogma? Genome Biol 16:176–178

Allard RW (1999) History of plant population genetics. Annu Rev Genet 33:1–27

Baik B-K, Ullrich SE (2008) Barley for food: characteristics, improvement and renewed interest. J Cereal Sci 48:233–242

Beales J, Turner A, Griffiths S, Snape J, Laurie D (2007) A pseudo-response regulator is missexpressed in the photoperiod insensitive *Ppd-D1a* mutant of wheat

(*Triticum aestivum* L.). Theor Appl Genet 115:721–733

Betts NS, Berkowitz O, Liu RJ, Collins HM, Skadhauge B, Dockter C, Burton RA, Whelan J, Fincher GB (2017) Isolation of tissues and preservation of RNA from intact, germinated barley grain. Plant J 91:754–765

Büschges R, Hollricher K, Panstruga R, Simons G, Wolter M, Frijters A, van Daelen R, van der Lee T, Diergarde P, Groenendijk J, Töpsch S, Vos P, Salamini F, Schulze-Lefert P (1997) The barley *Mlo* gene: a novel control element of plant pathogen resistance. Cell 88:695–705

Chandler PM, Zwar JA, Jacobsen JV, Higgins TJ, Inglis AS (1984) The effects of gibberellic acid and abscisic acid on α-amylase mRNA levels in barley aleurone layers studies using an α-amylase cDNA clone. Plant Mol Biol 3:407–418

Chrispeels MJ, Varner J (1967) Gibberellic acid-enhanced synthesis and release of α-amylase and ribonuclease by isolated barley and aleurone layers. Plant Physiol 42:398–406

Cockram J, Jones H, Leigh FJ, O'Sullivan D, Powell W, Laurie DA, Greenland AJ (2007) Control of flowering time in temperate cereals: genes, domestication, and sustainable productivity. J Exp Bot 58:1231–1244

Dawson IK, Russell J, Powell W, Steffensen B, Thomas WTB, Waugh R (2015) Barley: a translational model for adaption to climate change. New Phytol 206:913–931

Druka A, Franckowiak J, Lundqvist U, Bonar N, Alexander J, Houston K, Radovic S, Shahinnia F, Vendramin V, Morgante M, Stein N, Waugh R (2011) Genetic dissection of barley morphology and development. Plant Physiol 155:617–627

FAO (2016) Food outlook: biannual report on global food markets. ISSN 1560-8182, http://www.fao.org/3/a-i5703e.pdf

FAOSTAT (2017) http://www.fao.org/faostat/en/#data

Fox GP, Panozzo JF, Li CD, Lance RCM, Inkerman PA, Henry RJ (2003) Molecular basis of barley quality. Aust J Agric Res 54:1081–1101

Fu D, Szűcs P, Yan L, Helguera M, Skinner JS, von Zitzewitz J, Hayes PM, Dubcovsky J (2005) Large deletions within the first intron in VRN-1 are associated with spring growth habit in barley and wheat. Mol Genet Genomics 273:54–65

Gómez-Cadenas A, Zentella R, Walker-Simmons MK, Ho T-HD (2001) Gibberellin/abscisic acid antagonism in barley aleurone cells: site of action of the protein kinase PKABA1 in relation to gibberellin signaling molecules. Plant Cell 13:667–679

Gubatz S, Dercksen VJ, Bruess C, Weschke W, Wobus U (2007) Analysis of barley (*Hordeum vulgare*) grain development using three-dimensional digital models. Plant J 52:779–790

Gubler F, Kalla R, Roberts JK, Jacobsen JV (1995) Gibberellin-regulated expression of a myb gene in barley aleurone cells: evidence for Myb transactivation of a high-pI alpha-amylase gene promoter. Plant Cell 7:1879–1891

Jacobsen JV, Varner J (1967) Gibberellic acid-induced synthesis of protease by isolated aleurone layers of barley. Plant Physiol 42:1596–1600

Kadziola A, Abe J, Svensson B, Haser R (1994) Crystal and molecular structure of barley α-amylase. J Mol Biol 239:104–121

Knüpfer R (2009) Triticeae genetic resources in ex situ genebank collections. In Muehlbauer GJ, Feuillet C (eds) Genetic and genomics of the triticeae. Plant genetics and genomics: crops and models, vol 7. Springer Science, New York, USA, pp 31–80

Komatsuda T, Pourkheirandish M, He C, Azhaguvel P, Kanamori H, Perovic D, Stein N, Graner A, Wicker T, Tagiri A, Lundqvist U, Fujimura T, Matsuoka M, Matsumoto T, Yano M (2007) Six-rowed barley originated from a mutation in a homeodomain-leucine zipper I-class homeobox gene. Proc Natl Acad Sci USA 104:1424–1429

Kusch S, Panstruga R (2017) Mlo-based resistance: an apparently universal "weapon" to defeat powdery mildew disease. MPMI 30:179–189

Lanahan MB, Ho T, Rogers SW, Rogers JC (1992) A gibberellin response complex in cereal alpha-amylase gene promoters. Plant Cell 4:203–211

Lundqvist U (2014) Scandinavian mutation research in barley—a historical review. Hereditas 151:123–131

McCouch S, Baute GP, Bradeen J, Bramel P, Bretting PK, Buckler E, Burke JM, Charest D, Cloutier S, Cole G, Dempewolf H, Dingkuhn M, Feuillet C, Gepts P, Grattapaglia D, Guarino L, Jackson S, Knapp S, Langridge P, Lawton-Rauh A, Lijua Q, Lusty C, Michael T, Myles S, Naito K, Nelson RL, Pontarollo R, Richards CM, Rieseberg L, Ross-Ibarra J, Rounsley S, Sackville Hamilton R, Schurr U, Stein N, Tomooka N, van der Knaap E, van Tassel D, Toll J, Valls J, Varshney RK, Ward J, Waugh R, Wenzl P, Zamir D (2013) Agriculture: feeding the future. Nature 499:23–24

Morrell PL, Clegg MT (2007) Genetic evidence for a second domestication of barley (*Hordeum vulgare*) east of the Fertile Crescent. Proc Natl Acad Sci USA 104:3289–3294

Newman RK, Newman CW (1991) Barley as a food grain. Cereal Foods World 36:800–805

Newman CW, Newman RK (2006) A brief history of barley foods. Cereal Foods World 51:4–7

Pourkheirandish M, Hensel G, Kilian B, Senthil N, Chen G, Sameri M, Azhaguvel P, Sakuma S, Dhanagond S, Sharma R, Mascher M, Himmelbach A, Gottwald S, Nair SK, Tagiri A, Yukuhiro F, Nagamura Y, Kanamori H, Matsumoto T, Willcox G, Middleton CP, Wicker T, Walther A, Waugh R, Fincher GB, Stein N, Kumlehn J, Sato K, Komatsuda T (2015) Evolution of the grain dispersal system in barley. Cell 162:527–539

Riehl S, Zeidi M, Conard NJ (2013) Emergence of agriculture in the foothills of the Zagros mountains of Iran. Science 341:65–67

Schulte D, Close TJ, Graner A, Langridge P, Matsumoto T, Muehlbauer G, Sato K, Schulman AH, Waugh R, Wise RP, Stein N (2009) The international barley sequencing consortium (IBSC)—at the threshold of efficient access to the barley genome. Plant Physiol 149:142–147

Shewry PR, Hawkesford MJ, Piironen V, Lampi A-M, Gebruers K, Boros D, Andersson AAM, Aman P, Rakszegi M, Bedo Z (2013) Natural variation in grain composition of wheat and related cereals. J Agric Food Chem 61:8295–8303

Skadhauge B, Lok F, Breddam K, Olsen O, Bech M, Knudsen S (2011) Barley with reduced lipoxygenase activity and beverage prepared therefrom. US Patent 0318469A1

Slakeski N, Fincher GB (1992) Developmental regulation of (1 → 3, 1 → 4)-β-glucanase gene expression in barley tissue-specific expression of individual isoenzymes. Plant Physiol 99:1226–1231

Taketa S, Amano S, Tsujino Y, Sato T, Saisho D, Kakeda K, Nomura M, Suzuki T, Matsumoto T, Sato K, Kanamori H, Kawasaki S, Takeda K (2008) Barley grain with adhering hulls is controlled by an ERF family transcription factor gene regulating a lipid biosynthesis pathway. Proc Natl Acad Sci USA 105:4062–4067

Turner A, Beales J, Faure S, Dunford R, Laurie D (2005) The pseudo-response regulator Ppd-H1 provides adaptation to photoperiod in barley. Science 310:1031–1034

Varghese JN, Garrett TPJ, Colman PM, Chen L, Høj P, Fincher GB (1994) The three-dimensional structures of two plant β-glucan endohydrolases with distinct substrate specificities. Proc Natl Acad Sci USA 91:2785–2789

Wang YP, Cheng X, Shan QW, Zhang Y, Liu JX, Gao CX, Qiu JL (2014) Simultaneous editing of three homoeoalleles in hexaploid bread wheat confers heritable resistance to powdery mildew. Nat Biotechnol 32:947–951

Wei W, Bilsborrow PE, Hooley P, Fincham DA, Lombi E, Forster BP (2003) Salinity induced differences in growth, ion distribution and partitioning in barley between the cultivar Maythorpe and its derived mutant Golden Promise. Plant Soil 250:183–191

Wendler N, Mascher M, Noh C, Himmelbach A, Scholz U, Ruge-Wehling B, Stein N (2014) Unlocking the secondary gene-pool of barley with next generations sequencing. Plant Biotechnol J 12:1122–1131

You L, Wood-Sichra U, Fritz S, Guo Z, See L, Koo J (2014) Spatial production allocation model (SPAM) 2005 v2.0., 10 Jan 2017. Available from http://mapspam.info

Zhang RX, Tucker MR, Burton RA, Shirley NJ, Little A, Morris J, Milne L, Houston K, Hedley P, Waugh R, Fincher GB (2016) The dynamics of transcript abundance during cellularization of developing barley endosperm. Plant Physiol 170:1549–1565

Zohary D, Hopf M (1988) Domestication of plants in the old world. Clarendon Press, Oxford, England

2 Taxonomy of the Genus *Hordeum* and Barley (*Hordeum vulgare*)

Frank R. Blattner

Abstract

Barley refers to the cereal *Hordeum vulgare* subsp. *vulgare* but also more generally to the barley genus *Hordeum* that, apart from cultivated barley, comprises more than 30 wild grass species distributed in temperate and arid regions of the world. Like wheat and rye, *Hordeum* belongs to the Triticeae tribe of grasses, most conspicuously characterized by their inflorescence that is a spike instead of the panicle that occurs in most other grasses. The wild progenitor of the cereal is *H. vulgare* subsp. *spontaneum* from Southwest Asia. Together with bulbous barley (*Hordeum bulbosum*), the closest relative of the crop, and wall barley (*Hordeum murinum*) these species are grouped within subgenus *Hordeum*, while all other species belong to subgenus *Hordeastrum*. The crop is easily crossable with its wild progenitor (forming the primary gene pool of barley), while hybrids between cultivated and bulbous barley (secondary gene pool) exhibit low fertility. All other species belong to the tertiary gene pool, resulting in sterile hybrids that can only be established through embryo rescue techniques. However, barley's tertiary gene pool holds traits for pathogen resistances and adaptations to extreme environmental conditions, which are of high value if they can be transferred into cultivated barley or other cereals. Taxonomic and nomenclatural issues are discussed here in the light of recent findings in molecular systematics and gene function.

2.1 Taxonomic Principles

The field of taxonomy has three subareas, which in an ideal world would be integrated into a single consecutive workflow consisting of (i) the analysis of the evolutionary history of organisms (phylogenetics), (ii) circumscribing evolutionary meaningful categories (systematics), and (iii) providing names for such categories (nomenclature). Thus, taxonomic units like species, genera, families, etc. would all be defined through their unique evolutionary history and relationships among each other. However, since the advent of DNA-based phylogenetic analysis about 30 years ago, it became clear that many historically defined and still used taxonomic categories did not represent natural units, i.e., they are not monophyletic. Monophyly is defined as describing a group of organisms derived from the most recent common ancestor that is different from the ancestor of other such lineages (Fig. 2.1a). Imposing the monophyly criterion on systematics should

F. R. Blattner (✉)
Leibniz Institute of Plant Genetics and Crop Plant Research (IPK), Gatersleben, Germany
e-mail: blattner@ipk-gatersleben.de

N. Stein and G. J. Muehlbauer (eds.), *The Barley Genome*, Compendium of Plant Genomes,
https://doi.org/10.1007/978-3-319-92528-8_2

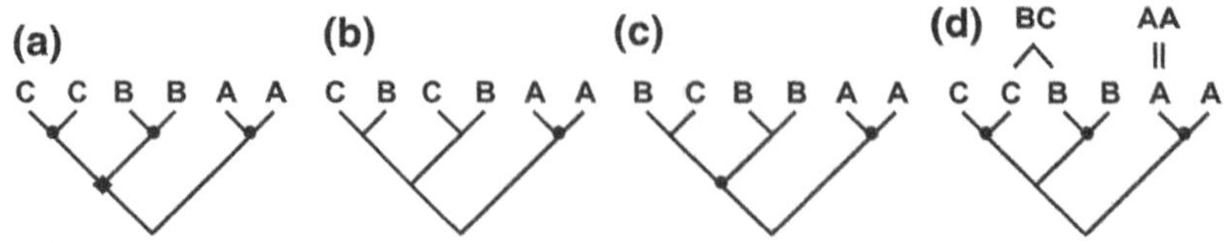

Fig. 2.1 Explanations for terms describing phylogenetic relationships. **a** Taxa A, B, and C are all monophyletic units, each reaches back to its own most recent common ancestor (•) and all clade members within A, B, and C share the same name. Taxa B and C could alternatively also be unified within a single taxon, as both go back to a common ancestor (♦). **b** Taxa B and C are both polyphyletic, i.e., they originated multiple times independently but share the same name. Such groups are taxonomically preposterous, as they are not defined through a common evolutionary history. **c** Taxon B is paraphyletic, as not all descendants (C) of its most recent common ancestor (•) carry the same name. This reflects ongoing evolution, i.e., a population starts to diverge clearly from other conspecific populations, and paraphyletic groups might therefore in some cases be tolerable taxonomic units—although defining monophyletic groups should, if possible, be preferred. **d** Through whole-genome duplication in A an autopolyploid originated, while BC is an example for an allopolyploid taxon, combining the genomes of its parents B and C

automatically result in natural units (clades), where members are more closely related to each other than to members of other units. Such clades are defined through phylogenetic analyses of morphological or molecular characters and most often the relationships of taxa are depicted in phylogenetic trees like in Fig. 2.1. As clades are the result of the evolutionary process, they are solidly fixed through their common history. A system based on this principle will automatically result in long-term stability of the names of organismic units, which hierarchically reflect gradual relationships, and has a certain predictive value (i.e., closely related organisms should share more traits than more distant relatives). Although this system cannot account for all mechanisms that drive evolution (for example, taxon relationships cannot always be represented by bifurcating trees but might involve also reticulations resulting in organisms belonging to two or more clades), and determination of such clades might still change with improving methods of phylogenetic analysis, taxonomists now consider the identification of clades the best way to come up with meaningful taxonomic units for the majority of higher plant taxa on Earth, although it might not always be possible or desirable (Brummitt 2006) to avoid paraphyletic groups (Fig. 2.1c). And also the circumscription of clades regarding how wide or narrow a taxon should be defined (Fig. 2.1a) could still be a matter of discussion.

To name taxonomic units, nomenclatural rules were specified, including the priority principle, meaning the oldest validly published name for a taxon has to be used, and that a description of the organism has to be given that at least defines the differences to the most similar other organism. For a long time, these descriptions had to be in Latin but recently also English descriptions became valid. For plants, the rules were fixed in different editions of the International Code of Botanical Nomenclature (ICN, last version: McNeill et al. 2012). This code determines, however, only *how* the naming has to be done and not the criteria that define systematic entities like species, genera, families, etc. Thus, depending on authors and the species and/or genus concepts they follow, different *correct* scientific names might exist in parallel for the same species. Hence, Löve (1984) split *Hordeum* into two genera resulting, for example, in the valid names *H. murinum* L. and *Critesion murinum* (L.) Á.Löve for wall barley. *H. murinum* L. means that this species was first described by Linnaeus (1753), while *C. murinum* (L.) Á.Löve refers to the older Linnean name, the authority now put into brackets, that was sorted into a new genus by Löve (1984). In cases where the meaning of a taxon name is explicit, giving the authority for a taxon can be omitted. In other cases, it might help to make clear to what organisms a name is referring by providing the authority together with a taxon name (Barkworth and von Bothmer 2009).

2.2 *Hordeum* and Triticeae

Hordeum is a medium-sized genus within the grass tribe Triticeae. The tribe comprises about 350 species (Barkworth and von Bothmer 2009); among them the important cereals are wheat (*Triticum* spp.), rye (*Secale cereale*) and triticale (x*Triticosecale*; an artificial wheat x rye hybrid), many forage grasses (*Elymus* and *Thinopyrum*), and ecologically important taxa of temperate grasslands (*Aegilops*, *Agropyron*, *Elymus*, *Hordeum*, *Pseudoroegneria,* and others). All Triticeae have chromosome numbers based on $x = 7$, with di-, tetra-, hexa-, and octoploid taxa. Sometimes, even higher ploidy levels can be found. The Triticeae taxa are characterized by their inflorescence that is a spike, the open leaf sheath with membranous ligules, and the hairy top of the developing grain.

Among taxonomists, disagreements exist about the generic concept to be used within the tribe (Bernhardt 2015). An extreme view is that of Stebbins (1956) who argued that the weak hybridization barriers among the different taxa allow to subsume all Triticeae species within a single genus *Triticum*. Others grouped species into different genera according to similar morphological features and life history traits (Linnaeus 1753; Bentham 1882; Nevski 1934; Hitchcock 1951; Tzevelev 1976) or according to the cytogenetic data, defining different so-called genome groups through meiotic crossing-over frequencies in interspecific hybrids (Kihara 1930; Dewey 1984; Löve 1984). Thus, Löve (1984) recognized 37 genera in Triticeae, 13 of them belonging to traditional *Aegilops* (van Slageren 1994; Yen et al. 2005; Barkworth and von Bothmer 2009). As phylogenetic relationships among the genera and species in Triticeae are currently not finally resolved (Escobar et al. 2011; Bernhardt 2015; Bernhardt et al. 2017; and references therein), a rational basis for a solid generic concept of Triticeae is still missing.

2.3 The Genus *Hordeum* and Subgeneric Units Within

In *Hordeum*, about 33 annual and perennial species are currently recognized (Blattner 2009). As some of them are divided into several subspecies, about 45 different taxa belong to the genus. They are distributed in temperate and arid parts of all continents except Australasia. *Hordeum* originated approximately 14–10 million years ago (Mya) in an area that became today's Southwest Asia and the Mediterranean and started to diversify 9 Mya (Brassac and Blattner 2015) afterward colonizing Asia, the Americas, and South Africa involving multiple intercontinental dispersals (Blattner 2006). The highest species numbers are found in southern South America, where about 16 species evolved during the last 1.5 million years, more than one-third of them being allopolyploids. As in many Triticeae and grasses generally, allopolyp loidization is an important mechanism in *Hordeum* contributing to the generation of biodiversity (Kellogg 2015, 2016). In *Hordeum*, diploid ($2n = 2\times = 14$), tetraploid ($2n = 4\times = 28$), and hexaploid ($2n = 6\times = 42$) taxa exist. Except two autopolyploid cytotypes (in *H. bulbosum* and *H. brevisubulatum*), all polyploids are allopolyploids (Jakob et al. 2004; Brassac and Blattner 2015).

Allopolyploids originate through interspecific hybridization followed by a genome duplication that stabilizes the karyotype by allowing chromosome pairing and an orderly distribution of chromosomes during meiosis. Due to the initial hybridization, allopolyploids create problems in taxonomy, as such organisms evolve from multiple parental species (within *Hordeum*) or even different genera (within Triticeae), which means they reach back to two (or more) most recent common ancestors. To account for this mechanism in the Triticeae, where the majority of species are allopolyploids, genera were defined according to the combined parental genomes/genera (Dewey 1984; Löve 1984; Barkworth and von Bothmer 2009). To name just a few examples, the allopolyploid genus *Douglasdeweya* obtained a genome each from *Agropyron* and *Pseudoroegneria*, while *Stenostachys* is characterized by the possession of an *Australopyrum* and a *Hordeum* genome, and the combination of genomes from *Pseudoroegneria* and *Hordeum* results in *Elymus*. Although this system is artificial and not consistently used throughout the tribe (Bernhardt 2015), it is the convention that most grass taxonomists currently agree on. In *Hordeum*, taxonomic problems are less pronounced, as allopolyploids evolving from within the genus are treated as new *Hordeum* species. Although the taxonomy is still not completely consistent regarding the rank and status of *Hordeum* polyploids, this will be solved in the frame of the future monograph of the genus (Blattner, in prep.).

For *Hordeum*, different taxonomic treatments exist, regarding the genus, subgeneric entities (like subgenera, sections, and series), and species or subspecific units (like subspecies or varieties). In contrast to the genera closely related to wheat, the monophyly of the taxa belonging to *Hordeum* was nearly never disputed. No matter if unified into one genus or split into two, it was clear that all species evolved from a most recent common ancestor that was different from the ancestors of other lineages within Triticeae. This is due to the unique inflorescence structure of *Hordeum*, where the spike consists of three single-flowered spikelets at each rachis node (named triplets) making *Hordeum* taxa easily recognizable. Monophyly was later also confirmed by molecular methods (below) so that this genus seems somehow exceptional within Triticeae, as it is less burdened by multiple contradicting taxonomic treatments in comparison to many other genera of the tribe.

Still, the most important changes in the systematics of the genus were the ones proposed by Dewey (1984) and Löve (1984). Based on the analysis of pairing behavior of meiotic metaphase I chromosomes in hybrids, four different genomes were recognized in *Hordeum* (von Bothmer et al. 1995). Löve (1984) therefore split *Hordeum* into *Hordeum* L. s.str., consisting only of *H. vulgare* and *Critesion* Raf., comprising all other species of the genus. Dewey (1984) arrived at a similar solution, although he added *H. bulbosum* in his *Hordeum* s.str. instead of *Critesion*.

Few taxonomists followed this approach, probably due to the clear morphological characters unifying *Hordeum* and *Critesion*, making them easily recognizable as 'belonging together'. Later, molecular systematic analyses of nuclear loci (Petersen and Seberg 2003; Blattner 2004; Petersen et al. 2011; Brassac and Blattner 2015) showed that neither Dewey's nor Löve's treatment provides monophyletic units. As both *H. vulgare* and *H. bulbosum* are nested within the *Critesion* lineage (Fig. 2.2), *Hordeum* s.str. would indeed be in both cases monophyletic. *Critesion*, however, is a paraphyletic genus, as not all species derived from the most recent common ancestor of *Critesion* would be included in this taxon. Only the transfer of *H. murinum* from *Critesion* either into *Hordeum* or a genus of its own would make *Critesion* monophyletic. Keeping all species within a single genus named *Hordeum* provided a relatively stable and intuitive solution, and it prevents botanists from learning more than 30 new taxon names.

For a long time, the *Hordeum* species were also grouped into units below the genus level, mostly sections and series that were erected to harbor species with similar morphology or certain life history traits. *Hordeum vulgare* was placed in sect. *Crithe* Doell or sect. *Cerealia* Anders., all the other annual species in sect. *Hordeastrum* Doell, the perennials with rather long awns in sect. *Critesion* (Raf.) Nevski, the short-awned species from South America into sect. *Anisolepis* Nevski, the remaining species from North America, Asia, and Europe in sect. *Stenostachys* Nevski, and *H. bulbosum* in sect. *Bulbohordeum* Nevski (Nevski 1941). Bothmer and Jacobsen (1985) recognized only the four sections *Anisolepis*, *Critesion*, *Hordeum,* and *Stenostachys*. In a later monograph of the genus, von Bothmer et al. (1995) already expressed their doubts about these sections being natural units but deterred to erect a new classification system, as they found the evidence from then emerging molecular data not strong enough to base far-reaching taxonomic changes on. Petersen and Seberg (2003) undertook an approach toward a

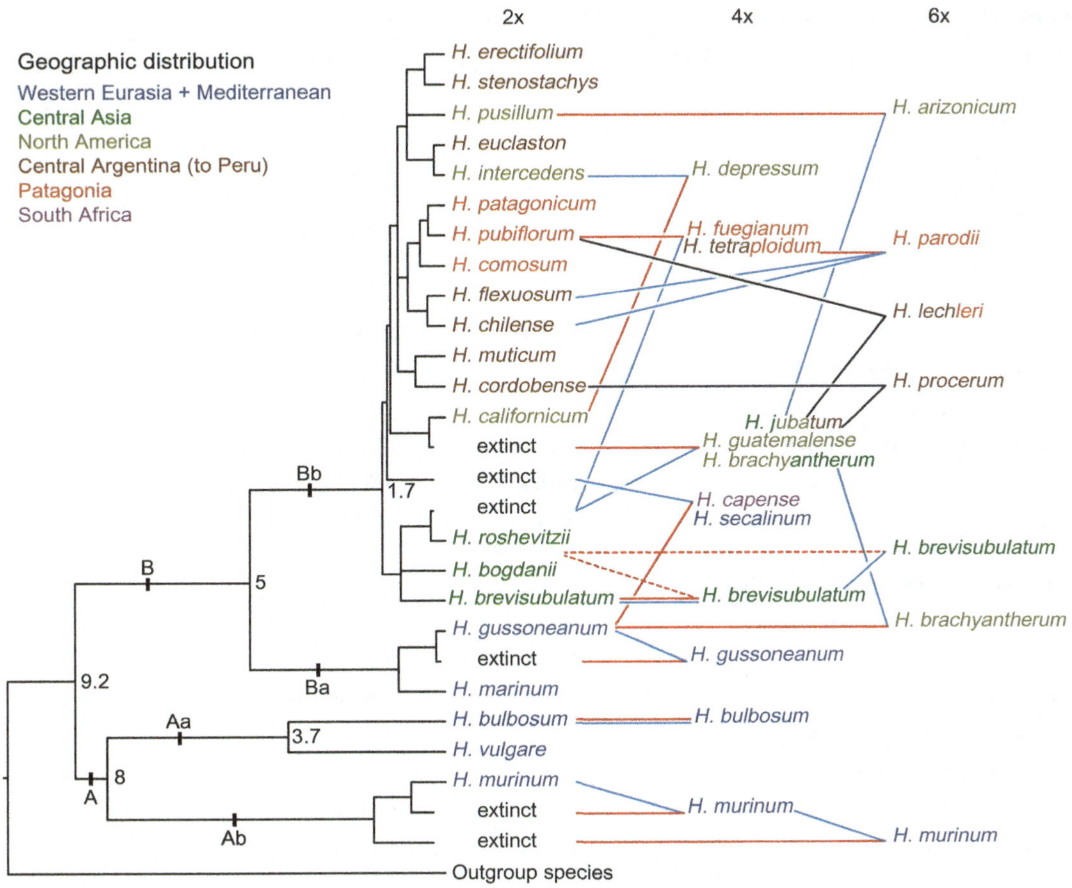

Fig. 2.2 Phylogenetic relationships of *Hordeum* species. The tree is based on the combined analysis of DNA sequences from one chloroplast and 12 nuclear single-copy genes. Diploid species are provided at the tips of the tree, polyploid species (4×, 6×) are connected through lines with their ancestral di- or polyploid progenitors. Extinct taxa/genotypes were inferred from the presence of gene copies (homeologs) in polyploids, which do not occur any more in extant diploid taxa. Numbers at major nodes in the tree provide clade ages (in million years). A = subg. *Hordeum*, Aa = sect. *Hordeum*, Ab = sect. *Trichostachys*, B = subg. *Hordeastrum*, Ba = sect. *Marinae*, Bb = sect. *Stenostachys*. The figure is modified from Brassac and Blattner (2015)

new system for *Hordeum,* based on phylogenetic data of sequences of two nuclear loci plus characters derived from the chloroplast genome. They proposed four sections *Hordeum*, *Critesion*, *Sibirica,* and *Stenostachys*. Through time, accumulating phylogenetic data (Komatsuda et al. 1999; Blattner 2004; Petersen et al. 2011; Wang et al. 2011; Brassac et al. 2012; Brassac and Blattner 2015) proved, however, that apart from sect. *Hordeum* the other sections were again not monophyletic when used in the sense of Petersen and Seberg (2003).

A new system (Blattner 2009), which tried to include all evidence available to be strictly based on natural units, now divides *Hordeum* in two subgenera (subg. *Hordeum* and *Hordeastrum*), each with two sections conforming the four genome groups occurring within *Hordeum* (von Bothmer et al. 1995), plus one section comprising three intersectional allopolyploid hybrid species of subg. *Hordeastrum* (for more details see Table 2.1). Blattner (2009), and Yen and Yang (2009) independently proposed to base *Hordeum* sections onto natural units or genomes

Table 2.1 Taxa of *Hordeum* L. (modified from Blattner 2009)

Taxon	Ploidy	Haploid genome[2]	Distribution area
Subgenus *Hordeum*			
Section *Hordeum*			
H. vulgare L.			
subsp. *vulgare*	2×	**H**	Cultivated
subsp. *spontaneum* (K. Koch) Thell.	2×	**H**	SW to C Asia
H. bulbosum L.	2×, 4×	**H, HH**	Mediterranean to C Asia
Section *Trichostachys* Dum.			
H. murinum L.			
subsp. *glaucum* (Steud.) Tzvel.	2×	**Xu**	Mediterranean to C Asia
subsp. *murinum*	4×	**XuXu**	NW Europe to Caucasus
subsp. *leporinum* (Link) Arc.	4×, 6×	**XuXu, XuXuXu**	Mediterranean to C Asia
Subgenus *Hordeastrum* (Doell) Rouy			
Section *Marinae* (Nevski) Jaaska			
H. gussoneanum Parl.	2×, 4×	**Xa, XaXa**	Mediterranean to C Asia
H. marinum Huds.	2×	**Xa**	Mediterranean
Section *Stenostachys* Nevski			
Series *Sibirica* Nevski			
H. bogdanii Will.	2×	**I**	C Asia
H. brevisubulatum (Trin.) Link[1]	2×, 4×, 6×	**I, II, III**	C Asia
H. roshevitzii Bowden	2×	**I**	C Asia
Series *Critesion* (Raf.) Blattner			
H. californicum Covas & Stebb.	2×	**I**	SW North America
H. chilense Roem. & Schult.	2×	**I**	Chile and W Argentina
H. comosum Presl	2×	**I**	S Argentina
*H. cordobense*Bothmer et al.	2×	**I**	C Argentina
H. erectifolium Bothmer et al.	2×	**I**	C Argentina
H. euclaston Steud.	2×	**I**	C Argentina, Uruguay
H. flexuosum Steud.	2×	**I**	E + C Argentina
H. intercendens Nevski	2×	**I**	SW USA, NW Mexico
H. muticum Presl	2×	**I**	C to N Andes
H. patagonicum (Haum.) Covas[1]	2×	**I**	S Argentina
H. pubiflorum Hook.f.[1]	2×	**I**	S Argentina
H. pusillum Nutt.	2×	**I**	C + E USA
H. stenostachys Godr.	2×	**I**	C Argentina
H. depressum (Scribn. & Sm.) Rydb.	4×	**II**	W USA

(continued)

Table 2.1 (continued)

Taxon	Ploidy	Haploid genome[2]	Distribution area
Interserial allopolyploids of series *Critesion* (all combining genomes of an American species with an extinct relative of *H. roshevitzii*)			
H. brachyantherum Nevski	4×	**II**	W North America, Kamchatka, Newfoundland
H. fuegianum Bothmer et al.	4×	**II**	S Argentina, S Chile
H. guatemalense Bothmer et al.	4×	**II**	Guatemala, S Mexico
H. jubatum L.	4×	**II**	NE Asia, W North America, C Argentina
H. tetraploidum Covas	4×	**II**	C Argentina
H. arizonicum Covas	6×	**III**	SW USA
H. lechleri (Steud.) Schenk	6×	**III**	C + S Argentina
H. parodii Covas	6×	**III**	C Argentina
H. procerum Nevski	6×	**III**	S Argentina
Section *Nodosa* (Nevski) Blattner			
H. brachyantherum Nevski	6×	**IIXa**	C California
H. capense Thunb.	4×	**IXa**	S Africa
H. secalinum Schreb.	4×	**IXa**	Mediterranean, C Europe

[1]Species with subspecies not further detailed here; [2]for details regarding *Hordeum* genomes, see Blattner (2009)

but unfortunately came up with partly different names for their sections. It will be a matter of time to see which system finally reaches wider acceptance within the community of grass and *Hordeum* researchers.

2.4 The Gene Pool Concept in Barley

Based on the possibility to use related species for utilization in crop breeding, the gene pool concept was introduced by Harlan and de Wet (1971), classifying crop wild relatives into three levels depending on the absence or presence and strength of crossing barriers with the crop species. For barley breeding, the primary gene pool refers to plants where no crossing barrier toward barley is present, i.e., the wild progenitor of cultivated barley *H. vulgare* subsp. *spontaneum* (see Schmid et al., Chap. 17). The secondary gene pool consists of *H. bulbosum* where crossing with barley is possible, although either fertility of the hybrid is strongly reduced or the *H. bulbosum* chromosomes are completely eliminated at very early developmental stages resulting in haploid plants (see Wendler, Chap. 18). All other species of *Hordeum* belong to the tertiary gene pool of barley, which means that crossing is only possible if embryo rescue techniques are employed and the resulting hybrids are sterile. While the transfer of favorable traits from the primary and secondary gene pool into barley can be achieved relatively easy, often characters are cotransferred that are non-desirable and have to be eliminated through series of backcrosses to the cultivar. The transfer of genes from the tertiary gene pool by traditional breeding methods is nearly impossible. New methods of genome editing might provide an easy and fast way for crop improvement when the genetic basis of certain traits of the wild species will finally be understood.

In *Hordeum,* interbreeding and the definition of gene pools are very closely correlated with the ages of the different groups. The populations of the primary gene pool are still seen as belonging to the same species as cultivated barley, which diverged under human influence only during the last 20,000 years (Weiss et al. 2008) from its wild progenitor and, even in nature, is still interbreeding when wild and cultivated stands are in close proximity (Russell et al. 2011; Jakob et al. 2014; Mascher et al. 2016). The *H. bulbosum* and the *H. vulgare* lineages split approximately 3.7 Mya (Fig. 2.2), which is still young enough for a certain amount of interfertility. In contrast, *H. murinum*, the sister species of *H. vulgare* and *H. bulbosum*, separated from them by 8 Mya, all other *Hordeum* species groups diverged 9.2 Mya (Brassac and Blattner 2015), which is well beyond the time frame where crossing is possible. These ages are much higher than the divergence ages found in, for example, *Aegilops/Triticum* (Bernhardt et al. 2017) and might explain why it is easier to utilize wild species in the wheat group in comparison to the wild *Hordeum* species. Another element is that the cultivated barley is a diploid taxon that seems to be much more sensitive toward introgression of foreign genomic material than the polyploid wheat species, where redundancy through homeologous loci might buffer against the detrimental genomic effects of hybridization (von Bothmer et al. 1995).

2.5 The Taxonomy of Barley in the Light of Its Evolution

Linnaeus (1753) in his *Species Plantarum* listed eight species belonging to his genus *Hordeum*, five of them referring to cultivated barley types (*H. vulgare*), two to bulbous barley (*H. bulbosum*), plus wall barley (*H. murinum*). These species are the most common *Hordeum* taxa of Central and Northern Europe, and therefore were easily accessible for the Swedish botanist at that time. The example of the five names Linné created for what we today understand to be a single species illustrates two major pitfalls for the taxonomy of cultivated plants. (i) Cultivars often have a much more diverse appearance than wild plants due to the breeding process that tries to generate diversity, and in addition humans select conspicuously mutated forms and propagate mutants if they might possess an advantage to them or rise their curiosity. In natural populations, these mutants would only very rarely reach a noticeable frequency or become fixed. (ii) The economic importance of crops ensures them the attention of many taxonomists, which naturally have diverse perceptions of how to classify and name the taxa in a most favorable way. For barley that has a close mutualistic relationship with humans for more than 10,000 years, both processes contributed to generate diverse systems and a plethora of taxonomic names. Here, I will provide a suitable way to taxonomically harness the diversity that is present in the crop and its wild progenitor. My conclusions are very much in accord with the treatment of *H. vulgare* by von Bothmer et al. (1995) in the last monograph of *Hordeum*, which became more and more widely accepted during the last decades.

It is important to note the difference in the objective between the taxonomy of wild and cultivated taxa in general. While for wild taxa the aim of botanists is to put names on natural units so that the meaningful names exist for comparable categories throughout the plant kingdom, in cultivated plants this is not equally straightforward, as cultivar taxonomy should fulfill different requirements. These depend very much on the users of the taxonomic system. Breeders and seed bank managers might need fine categories, which describe their plant stock quite precisely (Mansfeld 1950), but as soon as it comes to marketing of a certain variety, easy and catchy names are needed. This results in rather different taxonomic approaches. So, on the one hand there is a name like *Hordeum vulgare* L. subsp. *vulgare* convar. *distichon* (L.) Alef. var. *nutans* (Rode) Alef. The other kind of naming could be *Hordeum vulgare* 'Golden Promise' or even barley 'Golden Promise', both providing valid names for barley cultivars under the Code of Nomenclature for Cultivated Plants (CNCP; Brickell et al. 2009). The first example describes that we deal with a cultivated two-rowed, naked barley variety. The second example is a hulled spring barley cultivar of the UK breeding company Miln Marsters from the mid-1960s. It is a mutant originating through X-ray treatment from the cultivar 'Maythorpe' and was widely used in Scotland to produce ale and whisky, as it had excellent malting quality. Only the last property might be deduced from the cultivar name, if at all, every other information has to be obtained from the descriptions of the breeder. Although the first example looks like a real taxonomic name, it is without any meaning regarding monophyly of the two-rowed and naked barleys appointed with this name. Many of the morphological traits important in cultivated cereals are based on single or very few genes, which means that the specific phenotypes might originate over and over again. For example, six-rowed barley evolved several times independently through loss of the function of the *Vrs1* gene that, if functional, results in the normally two-rowed *Hordeum* spike type (Komatsuda et al. 2007). This makes the six-rowed type polyphyletic and the two-rowed type paraphyletic. Therefore, these are not taxonomically valid units no matter if referring to the assumed wild six-rowed [convar. *agriochriton* (Åberg) Bowd.] or cultivated six-rowed materials [var. *hexastichon* (L.) Aschers.]. It has to be understood that this highly scientifically looking nomenclature of cultivars is completely artificial. Taking this into account, I consider naming a cultivar *Hordeum vulgare* 'Golden Promise' or barley 'Morex' more appropriate than the system above, as it does not feign to be based on scientifically defined taxonomic categories. With its reference to *H. vulgare* or even barley, the cultivars are connected to a botanical species and therefore rooted in the botanical nomenclatural system (ICN). The cultivar name, provided in single quotation marks (CNCP), then refers to a certain entity produced by a breeder. However, as landraces and many ancient cultivars have no trademark name, both systems of naming cultivated barley will still be used in parallel for a long time depending on context.

In the scientific naming system for barley, the crop and its wild progenitor are assumed to be conspecific and subsumed under the species name *Hordeum vulgare* L. (von Bothmer et al. 1995). All cultivated types are placed in *H. vulgare* subsp. *vulgare* and the wild form in *H. vulgare* subsp. *spontaneum* (K. Koch) Thell. This treatment reflects the current knowledge that all cultivated barleys are domesticated types derived from wild barley and still group within the wild barley clade in phylogenetic analyses. There is some weak evidence that the crop could be derived from a gene pool that is different from extant wild barley and is no longer present in the wild (Jakob et al. 2014). If this should be proven, it would be valid to treat both forms as two different species: *H. vulgare* L. for the cultivars and *H. spontaneum* K. Koch for the wild populations. However, I think it requires stronger evidence than currently available to decide about the presence of one or two species. The often encountered name *H. spontaneum* in scientific publications of the last years (Dai et al. 2012, 2014 but also many others) is most probably not a statement of the authors regarding this being an independent species from cultivated barley but seems due to the effort to keep articles within the word limits of journals. Correct would, however, be to mention first the correct taxonomic name and only afterward use the abbreviated forms *H. spontaneum* or subsp. *spontaneum*.

The most important morphological character that is consistently different between both types is the brittle rachis in the wild plants versus the tough rachis in the cultivars. At maturity, the spike disarticulates in the wild type at each rachis node, releasing the triplets as dispersal units (Sakuma et al. 2011). In contrast, in cultivated forms, the tough rachis does not disarticulate and the spike can be harvested as an entire unit. This trait, as most other traits referred to as domestication syndrome (Hammer 1984), is under very strong selection in cultivated plants, as shattering immediately results in the exclusion of the respective seeds/genotypes from next year's sowing. Vice versa it is also under selection in the wild populations (Zohary 1959), as a tough rachis reduces the dispersal ability for the grains. Natural selection here is, however, much less severe than the human-enforced selection in cultivated stands. Thus, tough rachis types occur sometimes in the wild, which initially allowed their selection during the domestication process.

The rachis type is controlled by two adjacent genes (*Btr1*, *Btr2*), which both can induce the loss of the brittle rachis by recessive mutations (Pourkheirandish et al. 2015). The geographical distribution of the respective alleles indicates that two independent domestications of barley occurred (Azhaguvel and Komatsuda 2007; Dai et al. 2012; Pourkheirandish et al. 2015), one initially based on the mutated *Btr1* (that is *btr1*) in the Southern Levant, and a second one that resulted from selecting the *Btr2* mutation (that is *btr2*) in today's northern Syria and adjacent Turkey (Pourkheirandish et al. 2015). Cultivars with the mutated *Btr1* gene are today mostly distributed in Southwest Asia and Europe (from where they were introduced into American and Australian cultivars), while the *Btr2* mutants occur mainly between northern Africa and the Central Mediterranean in the West, and Central and East Asia in the East. Pourkheirandish et al. (2015) explain this distribution either with differences in ecological conditions between the two areas, favoring one or the other genotype, or that early trade or migration of farming could have resulted in the preferential distribution of the respective ancient stocks of extant cultivars. Taxonomically, it would be correct to provide different names for the independently domesticated variants of subsp. *vulgare*. However, as they cannot be distinguished without a still time-consuming screening of sequence differences at the *Btr* loci, this might provide practical obstacles for the use of such taxonomic entities.

There are regular claims of Tibet as an additional and independent domestication center for East Asian barley (Dai et al. 2012, 2014; Wang et al. 2015; and references therein) that then would also earn the status of a separate taxon. However, currently I am not finally convinced by this hypothesis. My uncertainty is caused by two not very well explained matters regarding (i) wild barley occurring under very diverse ecological conditions and (ii) the interpretation of the phylogenetic data that, to me, does not really support an independent domestication of barley in Tibet.

'Tibetan wild barley' is ecologically clearly distinct from the populations occurring in the Fertile Crescent, as they occur under much harsher conditions at elevations of 3500–4500 m on the Tibetan Plateau in comparison to the Southwest Asian stands that normally are found well below 1500 m elevation (von Bothmer et al. 1995; Zohary and Hopf 2000). It is not easy, at least to me, to envision a scenario for the Tibetan Plateau that was glaciated till 8400 years ago and influenced by a minor glaciation cycle about 4400 years ago (Kuhle 2005), to be (re-)colonized by wild barley, (re-)colonized by Neolithic humans, them starting immediately the domestication of local wild barley, and afterward spread their local high-elevation crop through trade to the vast areas in East Asia, where barley is cultivated at much lower elevation and with a very different temperature regime. Although not impossible, this seems a rather elaborate hypothesis.

How the new findings regarding the origin and domestication of the *Btr* types can contribute to the discussion about Tibetan barley has yet to be evaluated. The preferential presence of *btr2* in East Asian cultivated barley (Azhaguvel and Komatsuda 2007) was used to argue for an independent domestication of Asian barley in Tibet or China, as six-rowed brittle rachis barleys occur in Tibet and are more closely related to East Asian barley cultivars than to barley from western Eurasia (Dai et al. 2012; Wang et al. 2015). However, the resolution of this topic completely hinges around the identity of wild barley on the Tibetan Plateau. If Tibetan barley really would represent an autochthonous wild population, then Tibet has to be acknowledged as an independent domestication area. If wild barley there is merely a feral and weedy population derived from the locally adapted cultivated forms, a completely different interpretation results. von Bothmer et al. (1990) linked the observation of the locally high frequency of brittle rachis types to the specific cultivation praxis in the Tibetan highland. Due to in some years rather short vegetation periods at elevations above 3500 m, barley is often harvested in a premature, sometimes even green stage and dried in the homes of the farmers afterward. If harvesting is done before disarticulation of the spike starts, the purifying selection against brittle rachis types is no longer severe and reversals to the dominant wild-type genotypes can spread in the population. These weedy types might then accumulate in disturbed habitats in the farmed areas during long summers, when the brittle spikes disarticulate already in the field before harvesting. Classifying such secondary 'wild barley' as native wild barley would, of course, result afterward in a close relationship between them and their local cultivated progenitors when analyzed for genetic similarity. With the recent discovery of function and structure of *Btr* (Pourkheirandish et al. 2015), a route to clarifying the nature of Tibetan wild barley is opening. Thus, it would be helpful to see if Tibetan cultivated barley possesses a *btr2* allele identical to the Southwest Asian one or if a different mutation rendered *Btr2* nonfunctional. While true wild barley of Tibet should have the *Btr2* allele, feral brittle rachis types must be characterized either by a compensatory mutation reversing *btr2* of cultivated Asian material back into a functional *Btr2* in feral plants or by recombination between *Btr1/btr2* and *btr1/Btr2* types recreating a *Btr1/Btr2* brittle rachis type. The analysis of the respective gene regions in Tibetan wild barley and comparison with the materials from Southwest Asia could show if they share the Southwest Asian wild-type allele or if there are indications for a reversal to a secondary-functional brittle rachis *Btr2*. This most probably should be different, as it is unlikely that the same frameshift mutation happened twice in Southwest Asia and Tibet or was compensated by an identical reversal mutation from Tibetan cultivar's *btr2* toward the Southwest Asian-derived *Btr2* allele. Producing *Btr1/Btr2* brittle rachis barley by hybridization of *Btr1/btr2* and *btr1/Btr2* domestic barleys followed by recombination between the two loci is possible, as demonstrated experimentally (Pourkheirandish et al. 2015).

2.6 Conclusions

During the last decades, taxonomists agreed more and more that monophyletic groups (= clades) should be the principle units for nomenclatural categories, as they would—through time—result in long-term stable systematics. Although there are inherent problems in some organism groups, where monophyly is not always a useful criterion (hybrids, allopolyploids, asexual organisms, etc.) and paraphyletic groups also reflect the course of evolution, for the majority of organisms this clade-based nomenclature works satisfactorily. All species of *Hordeum* form a clade and can therefore be unified in a single genus. Earlier proposals to split *Hordeum* into two genera (*Hordeum* and *Critesion*) were mostly not accepted by the Triticeae research community, as all *Hordeum* species are easily recognized as forming one unit.

Cultivated barley and its progenitor wild barley are thought to still belong to a single species, as they are freely interbreeding and form a clade in phylogenetic analyses. To account for both forms, they are differentiated at the subspecies level with *H. vulgare* subsp. *vulgare* denoting the cultivar and *H. vulgare* subsp. *spontaneum*, the wild progenitor. While this treatment is unambiguous, within the subspecies different naming approaches exist. According to the Code of Nomenclature for Cultivated Plants, cultivar names are given in single quotation marks after a taxon name (*Hordeum* 'Morex'). However, there was and still partly is also an artificial system of naming in place that looks very much like valid taxonomic categories, although these categories do mostly not depict monophyletic units. Thus, six-rowed barley was/is named *H. agriocrithon*, implying that this is a species of its own, different from *H. vulgare*. But it could be also found as *H. vulgare* subsp. *agriocrithon* or *H. vulgare* subsp. *vulgare* convar. *agriocrithon*, assuming that it is either a subspecies equal to subsp. *vulgare* and subsp. *spontaneum* or a convariety within the cultivated barleys. As six-rowed barley originated multiple times, this character does not qualify as a valid taxonomic unit (as it does not describe a clade) but names a morphological type of barley. Using this kind of names might be useful in certain fields of agronomy or for seed bank management. When using them it should just be understood that such categories are completely artificial and very different from real taxonomic groups defined through their evolutionary history.

Recent research in barley domestication shows that assuming monophyly for cultivated barley might also not describe the history and groups completely correct. There are strong indications that the cultivated barley was domesticated at least twice from slightly different stocks of the Southwest Asian subsp. *spontaneum* population. This means that cultivated barley is polyphyletic and the two independently domesticated lineages rightfully should have two different names. However, as it is only possible to discern these lineages through analysis of the tough rachis genes and interbreeding partly merges the two gene pools, labeling these units with different scientific names seems currently impractical and essentially not necessary. Everyone working with these plants should have in mind that there are at least two cryptic lineages within cultivated barley germplasm going back to the slightly different genotypes that were initially selected by Neolithic humans during the domestication process from the same (meta) population of wild barley.

References

Azhaguvel P, Komtasuda T (2007) A phylogenetic analysis based on nucleotide sequence of a marker linked to the brittle rachis locus indicates a diphyletic origin of barley. Ann Bot 100:1009–1015

Barkworth ME, von Bothmer R (2009) Scientific names in the Triticeae. In: Muehlbauer GJ, Feuillet C (eds) Genetics and genomics of Triticeae. Springer, New York, pp 3–30

Bentham G (1882) Notes on Gramineae. J Linn Soc London, Bot 19:14–134

Bernhard N (2015) Taxonomic treatment of Triticeae and the wheat genus *Triticum*. In: Molnár-Láng M, Ceoloni C, Doležel J (eds) Alien introgression in wheat: cytogenetics, molecular biology, and genomics. Springer Int Publ, Switzerland, pp 1–19

Bernhardt N, Brassac J, Kilian B, Blattner FR (2017) Dated tribe-wide whole chloroplast genome phylogeny indicates recurrent hybridizations within Triticeae. BMC Evol Biol 17:141

Blattner FR (2004) Phylogenetic analysis of *Hordeum* (Poaceae) as inferred by nuclear rDNA ITS sequences. Mol Phylogenet Evol 33:289–299

Blattner FR (2006) Multiple intercontinental dispersals shaped the distribution area of *Hordeum* (Poaceae). New Phytol 169:603–614

Blattner FR (2009) Progress in phylogenetic analysis and a new infrageneric classification of the barley genus *Hordeum* (Poaceae: Triticeae). Breed Sci 59:471–480

Brassac J, Blattner FR (2015) Species level phylogeny and polyploid relationships in *Hordeum* (Poaceae) inferred by next-generation sequencing and *in silico* cloning of multiple nuclear loci. Syst Biol 64:792–808

Brassac J, Jakob SS, Blattner FR (2012) Progenitor-derivative relationships of *Hordeum* polyploids (Poaceae, Triticeae) inferred from sequences of TOPO6, a nuclear low-copy gene region. PLoS ONE 7:e33808

Brickell CD, Alexander C, David JC et al (2009) International code of nomenclature for cultivated plants, 8th ed. Scripta Hort 10

Brummitt RK (2006) Am I a bony fish? Taxon 55:268–269

Dai F, Nevo E, Wu D et al (2012) Tibet is one of the centers of domestication of cultivated barley. Proc Natl Acad Sci USA 109:16969–16973

Dai F, Chen ZH, Wand Y et al (2014) Transcriptome profiling reveals mosaic genomic origins of modern cultivated barley. Proc Natl Acad Sci USA 111:13403–13408

Dewey DR (1984) The genomic system of classification as a guide to intergeneric hybridization with the perennial Triticeae. In: Gustafson JP (ed) Gene manipulation in plant improvement. Plenum Publishing Corporation, New York, pp 209–279

Escobar JS, Scornavacca C, Cenci A et al (2011) Multigenic phylogeny and analysis of tree incongruences in Triticeae (Poaceae). BMC Evol Biol 11:181

Hammer K (1984) Das Domestikationssyndrom. Kulturpflanze 32:11–34

Harlan JR, de Wet JMJ (1971) Toward a rational classification of cultivated plants. Taxon 20:509–517

Hitchcock AS (1951). In: A. Chase (ed) Manual of the grasses of the United States, 2nd edn. U.S. Department of Agriculture Miscellaneous Publication No. 200. U.S. Government Printing Office, Washington, D.C.

Jakob SS, Meister A, Blattner FR (2004) The considerable genome size variation of *Hordeum* species (Poaceae) is linked to phylogeny, life form, ecology, and speciation rates. Mol Biol Evol 21:860–869

Jakob SS, Rödder D, Engler JO, Shaaf S, Özkan H, Blattner FR, Kilian B (2014) Evolutionary history of wild barley (*Hordeum vulgare* subsp. *spontaneum*) analyzed using multilocus sequence data and paleodistribution modeling. Genome Biol Evol 6:685–702

Kellogg EA (2015) Flowering plants—monocots—Poaceae. In: The families and genera of vascular plants, vol 13. Springer, Heidelberg

Kellogg EA (2016) Has the connection between polyploidy and diversification actually been tested? Curr Opin Plant Biol 30:25–32

Kihara H (1930) Genomanalyse bei Triticum und Aegilops. Cytologia (Tokyo) 1:263–284

Komatsuda T, Tanno K, Salomon B, Bryngelsson T, von Bothmer R (1999) Phylogeny in the genus *Hordeum* based on nucleotide sequences closely linked to the *vrs1* locus (row number of spikelets). Genome 42:973–981

Komatsuda T, Pourkheirandish M, He CF et al (2007) Six-rowed barley originated from a mutation in a homeodomain-leucine zipper I-class homeobox gene. Proc Natl Acad Sci USA 104:1424–1429

Kuhle M (2005) Glacial geomorphology and ice ages in Tibet and the surrounding mountains. Island Arc 14:346–367

Linnaeus C (1753) Species plantarum. Impensis Laurentii Salvii, Stockholm

Löve A (1984) Conspectus of the Triticeae. Feddes Repert 95:425–521

Mansfeld R (1950) Das morphologische System der Saatgerste, *Hordeum vulgare* L. s.l. Züchter 20:8–24

Mascher M, Schuenemann VJ, Davidovich U et al (2016) Genomic analysis of 6,000-year-old cultivated grain illuminates the domestication history of barley. Nature Genet 48:1089–1093

McNeill J, Barrie FR, Buck WR et al (2012) International code of nomenclature for algae, fungi, and plants (Melbourne Code). Regnum Veg 154

Nevski SA (1934) Agrostological studies IV. The systematics of the tribe Hordeeae. Trudy Bot Inst Akad Nauk SSSR. Flora Sist Vysshikh Rastenii 1: 9–32

Nevski SA (1941) Beiträge zur Kenntnis der wildwachsenden Gersten in Zusammenhang mit der Frage über den Ursprung von *Hordeum vulgare* L. und *Hordeum distichon* L. (Versuch einer Monographie der Gattung *Hordeum*). Trudy Bot Inst Akad Nauk SSSR, ser 1, 5:64–255

Petersen G, Seberg O (2003) Phylogenetic analyses of the diploid species of *Hordeum* (Poaceae) and a revised classification of the genus. Syst Bot 28:293–306

Petersen G, Aagesen L, Seberg O, Larsen IH (2011) When is enough, enough in phylogenetics? A case in point from *Hordeum* (Poaceae). Cladistics 27:1–19

Pourkheirandish M, Hensel G, Kilian B et al (2015) Evolution of the grain dispersal system in barley. Cell 162:527–539

Russell J, Dawson IK, Flavell AJ et al (2011) Analysis of >1000 single nucleotide polymorphisms in geographically matched samples of landrace and wild barley indicates secondary contact and chromosome-level differences in diversity around domestication genes. New Phytol 191:564–578

Sakuma S, Salomon B, Komatsuda T (2011) The domestication syndrome genes responsible for the major changes in plant form in the Triticeae crops. Plant Cell Physiol 52:738–749

Stebbins GL (1956) Taxonomy and the evolution of genera, with special reference to the family Gramineae. Evolution 10:235–245

Tzvelev NN (1976) Zlaki SSSR—grasses of the Soviet Union. Leningrad. [Engl. Ed. 1983 Transl Sharma BR]

van Slageren MW (1994) Wild wheats: a monograph of *Aegilops* L. and *Amblyopyrum* (Jaub. & Spach) Eig (Poaceae). Wageningen Agricultural University Papers

von Bothmer R, Jacobsen N (1985) Origin, taxonomy, and related species. In: Rasmuss DC (ed) Barley. Agronomy Monographs No. 26. Madison, American Society Agronomy, pp 19–56

von Bothmer R, Chi Y, Yang J (1990) Does wild, six-rowed barley, *Hordeum agriocrithon*, really exist? Plant Genet Res Newsl 77:17–19

von Bothmer R, Jacobsen N, Baden C, Jørgensen R, Linde-Laursen I (1995) An ecogeographical study of the genus *Hordeum*, 2nd edn. IPGRI, Rome

Wang H, Sun D, Sun G (2011) Molecular phylogeny of diploid *Hordeum* species and incongruence between chloroplast and nuclear datasets. Genome 54:986–992

Wang Y, Ren X, Sun D, Sun G (2015) Origin of worldwide cultivated barley revealed by *NAM-1* gene and grain protein content. Front Plant Sci 6:803

Weiss E, Kislev ME, Simchoni O, Nadel D, Tschauner H (2008) Plant-food preparation area on an Upper Paleolithic brush hut floor at Ohalo II, Israel. J Archaeol Sci 35:2400–2414

Yen C, Yang JL (2009) Historical review and prospect of taxonomy of tribe Triticeae Dumortier (Poaceae). Breed Sci 59:513–518

Yen C, Yang JL, Yen Y (2005) Hitoshi Kihara, Áskell Löve and the modern genetic concept of the genera in the tribe Triticeae (Poaceae). Acta Phytotax Sin 43:82–93

Zohary D (1959) Is *Hordeum agriocrithon* the ancestor of six-rowed cultivated barley? Evolution 13:279–280

Zohary D, Hopf M (2000) Domestication of plants in the Old World, 3rd edn. Oxford Univ Press, New York

3 Cytogenetics and Genetic Stocks for Physical Mapping and Sequencing

Andreas Houben, Lala Aliyeva-Schnorr, Giri Prasad Joshi and Takashi R. Endo

Abstract

Barley is a model for other Triticeae genomes and is frequently used for cytological studies. In this chapter, we focus on cytogenetic approaches helping to improve physical mapping of the barley genome and describe the previous utilization of repetitive and low-copy probes successfully applied to barley. We demonstrate how the analysis of barley centromeres provided a better understanding of the process of uniparental chromosome elimination resulting in haploid plants. In addition, we describe how the downsizing of barley chromosomes was achieved by the production of aneuploids, and the application of the gametocidal system and telomere seeding.

A. Houben (✉) · L. Aliyeva-Schnorr
Leibniz Institute of Plant Genetics and Crop Plant Research (IPK), Gatersleben, 06466 Stadt Seeland, Germany
e-mail: houben@ipk-gatersleben.de

L. Aliyeva-Schnorr
e-mail: lalaalieva@hotmail.com

G. P. Joshi
Central Department of Botany, Tribhuvan University, GPO 8975, EPC 2266 Kathmandu, Nepal
e-mail: gp.joshi@cdbtu.edu.np

T. R. Endo
Faculty of Agriculture, Ryukoku University, 1-5 Yokotani, Seta Oe-cho, Otsu, Shiga 520-2194, Japan
e-mail: endo.takashi.2e@gmail.com

3.1 Identification of Barley Chromosomes

The cytogenetics of *Hordeum vulgare* L. ssp. *vulgare* (cultivated barley) and its wild progenitor, *H. vulgare* L. spp. *spontaneum* (C. Koch) Thell. has been reviewed extensively (Taketa et al. 2003; Costa and Singh 2006; Houben and Pickering 2009). Cultivated barley is diploid ($2n = 14$) and possesses relatively large ($\sim$9 µm long), mitotic metaphase chromosomes, which is typical across the genus *Hordeum*. Since the successful introduction of chromosome banding technique in the tribe Triticeae species (Gill and Kimber 1974), various Giemsa C- and N-banding techniques have been applied to barley (Linde-Laursen 1975; Islam 1980; Kakeda et al. 1991). Their characteristic banding patterns allow for the identification of all barley chromosomes (Fig. 3.1). The protocols of Endo (2011) are probably the simplest methods for C-banding and N-banding in barley. Although the two banding patterns are generally similar to each other, a difference exists between them at the centromeric sites on all chromosomes. Every centromeric site consisted of N-banding positive and C-banding negative heterochromatin in every cultivar examined. Some bands are different between the two banding techniques and among cultivars

N. Stein and G. J. Muehlbauer (eds.), *The Barley Genome*, Compendium of Plant Genomes,
https://doi.org/10.1007/978-3-319-92528-8_3

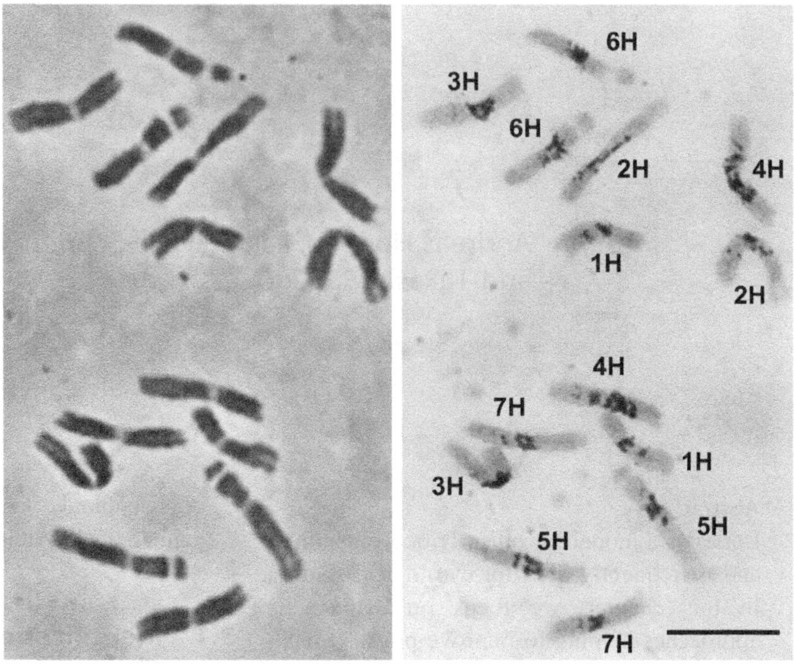

Fig. 3.1 Phase-contrast (left) and N-banding (right) images of barley, *Hordeum vulgare*, $2n = 14$. Bar = 10 μm

(Kakeda et al. 1991; Linde-Laursen 1991). C-bands tend to replicate late in the S-phase, but the late-replicating chromosomal regions do not always correspond to C-bands (Kakeda and Yamagata 1992). The application of labelled tandem trinucleotide repeats, also known as SSR markers ($(AAG)_5$, $(AGG)_5$, $(AAC)_5$, $(ACT)_5$, $(CAT)_5$, $(GTG)_5$, $(ACG)_5$ and $(CAG)_5$) showing a distribution similar to C-banding sites (Cuadrado and Jouve 2007) (Fig. 3.2a–c), allowed the identification of barley chromosomes by fluorescence in situ hybridization (FISH). Besides, 5S rDNA, pHv-365, pHv-961 (HVT01), and $(GAA)_5$ and $(ATC)_5$ microsatellites also allowed the recognition of all chromosomes in different barley cultivars (Kato 2011; Fukui et al. 1994) (Fig. 3.2d, e). Additional repetitive sequences and chromosome-specific low- and single-copy sequences that have been applied to barley FISH are listed in Tables 3.1 and 3.2. Barley chromosomes added to another genome like wheat (*Triticum* ssp.) (wheat–barley addition lines) can be identified by genomic in situ hybridization (GISH) using labelled barley genomic DNA as a probe (Mukai and Gill 1991).

During interphase, the Rabl-like polarized distribution patterns of centromeres and telomeres are conserved in interphase nuclei from different tissues (Dong and Jiang 1998) (Fig. 3.2f). The chromosomes of barley are predominantly metacentric and arm ratios vary from 1 to 1.5 (Marthe and Künzel 1994). Two chromosomes (5H and 6H) contain satellites and carry 45S rDNA containing nucleolus organizer regions (NORs). The short arm and the long arm of a chromosome are distinguishable from each other according to their physical lengths and are abbreviated as "S arm" and "L arm", respectively. The current nomenclature of the barley chromosomes was established based on the homoeologous relationship between the barley genome and the wheat genomes (Linde-Laursen et al. 1997).

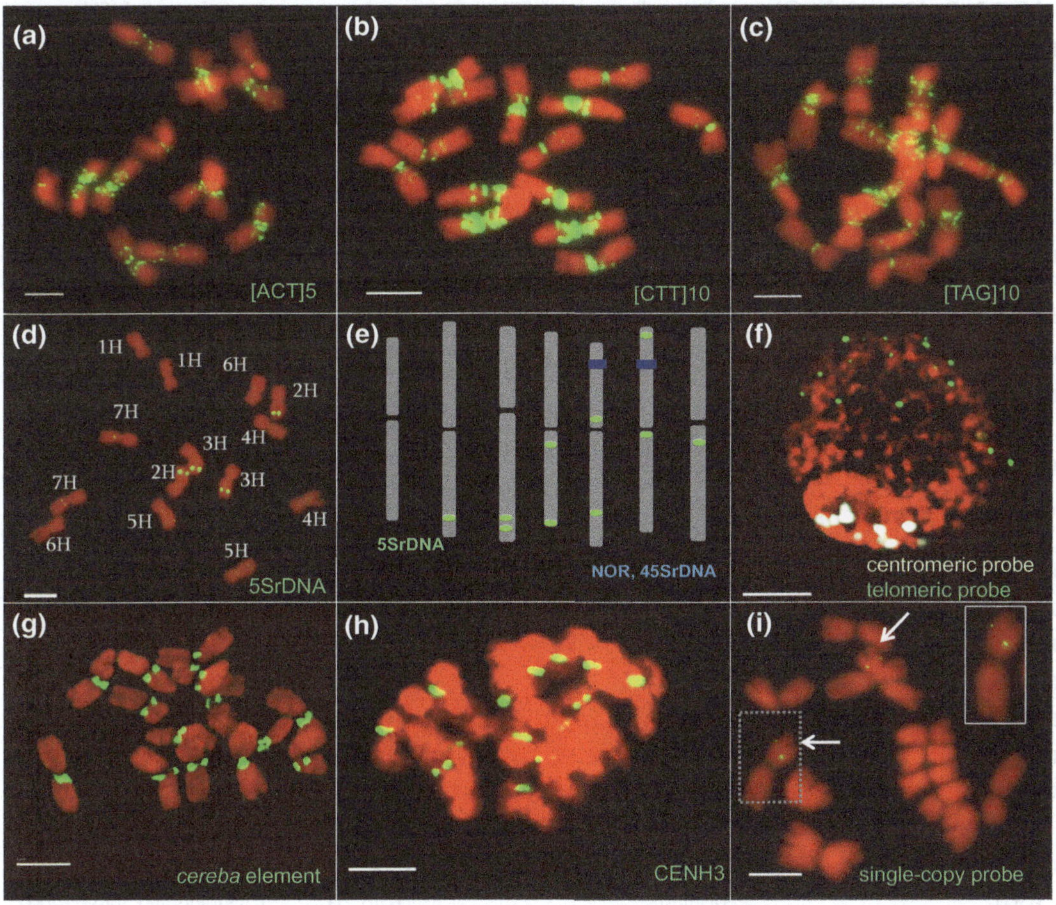

Fig. 3.2 Characterization of mitotic barley chromosomes. **a** $[ACT]_5$ trinucleotide repeat signals (green), **b** $[CTT]_{10}$ trinucleotide repeat signals (green), **c** $[TAG]_{10}$ trinucleotide repeat signals (green), **d** metaphase spread showing 5S rDNA-specific signals (green). **e** Ideogram of barley chromosomes showing the positions of the 5S rDNA (green) and two NORs (45S rDNA, red). Mitotic metaphase showing **f** interphase nucleus of barley showing the Rabl-like distribution of telomeres (green) and centromeres (white arrows), **g** centromeric *cereba* signals (green), **h** chromosomes after immunostaining with the centromere-specific CENH3 antibody (green), and **i** single-copy signals of a 3 kb long probe (green). Scale = 5 µm

Owing to a high degree of mitotic chromosome condensation, only unique sequences at least ~10 Mb apart from each other can be distinguished by FISH. To overcome this limitation, pachytene chromosomes have been used for mapping by single-copy FISH. Spatial resolution of neighbouring loci in the euchromatic region has been improved to 60–120 kb (in heterochromatic region, 0.14–1.2 Mbp) (Valarik et al. 2004). Thus, the physical distance between very closely linked sequences can be determined by using barley pachytene chromosomes (Fig. 3.3) (Aliyeva-Schnorr et al. 2015).

Table 3.1 Barley repetitive DNA probes with corresponding chromosomal positions determined by FISH

Probe	Chromosomal position	Chromosome	References
pTa71 (45S rDNA)	Intercalary	1H, 2H, 5H, 6H, 7H	Leitch and Heslop-Harrison (1992)
pTa794 (5S rDNA)	Subterminal,	2HL, 3HL, 4HL,	Leitch and Heslop-Harrison (1993)
	Intercalary	7HS	
pHvMWG2315 (AT-rich tandem repeat)	Subtelomeric	All chromosomes	Busch et al. (1995)
HvT01	Subtelomeric	1H to 7H	Schubert et al. (1998)
pHvA14 (Afa-family) pAs1	Distal	1H to 7H	Tsujimoto et al. (1997)
pAs1	Distal	1H to 7H	Brandes et al. (1995)
	Interstitial	4H	
GAA microsatellite	Pericentric	1H to 7H	Pedersen et al. (1996)
pHv-365	Pericentric	1H, 2H, 4H, 6H	Kato (2011)
pHv-177	Distal and centromeric	1HS	Kato (2011)
pHv-1112	Distal	1HL	Kato (2011)
pHv-689	Pericentric	1H to 7H	Kato (2011)
pHv-1476	Pericentric	1H to 7H	Kato (2011)
pHv-1889	Pericentric	1H to 7H	Kato (2011)
pHv-1972	Pericentric	1H to 7H	Kato (2011)
$(ATC)_5$	Polymorphic	4H, 6H, 5H	Cuadrado and Jouve (2007)
	Subtelomeric	2HL, 3HL	
$(AAG)_5$	Pericentric	1H to 7H	Cuadrado and Jouve (2007)
$(AGG)_5$	Pericentric	1H to 7H	Cuadrado and Jouve (2007)
$(AAC)_5$	Pericentric	1H to 7H	Cuadrado and Jouve (2007)
$(ACT)_5$	Intercalary/telomeric/subtelomeric	2H, 3H, 4H, 5H, 6H	Cuadrado and Jouve (2007)
$(CAT)_5$	Pericentric	4H, 5H	Cuadrado and Jouve (2007)
$(GTG)_5$	Centromeric	1H to 7H	Cuadrado and Jouve (2007)
$(ACG)_5$	Centromeric	1H to 7H	Cuadrado and Jouve (2007)
$(CAG)_5$	Centromeric	3H and 4H	Cuadrado and Jouve (2007)
14 superfamilies of transposable elements	50% of the entire genome	All chromosomes	Wicker et al. (2009)
BAC7 (cereba retroelement and GC-rich satellite)	Centromeric	All chromosomes	Presting et al. (1998), Hudakova et al. (2001)

Table 3.2 Barley-specific low- and single-copy FISH probes

Probe	Position	Chromosome	References
Alpha-amylase	Intercalary	1HL	Leitch and Heslop-Harrison (1993)
Adh	Subtelomeric	3HS, 4HS, 5HS	Stephens et al. (2004)
Ale	Intercalary	5HL	Stephens et al. (2004)
Amy 1	Distal	6HL	Stephens et al. (2004)
Amy 2	Intercalary	7HL	Stephens et al. (2004)
Brz	Subtelomeric	7HS	Stephens et al. (2004)
Chi	Subtelomeric	7HL	Stephens et al. (2004)
Chs	Distal	2HS	Stephens et al. (2004)
Dhn 6	Subtelomeric	4HS	Stephens et al. (2004)
Glb 3	Subtelomeric	3HS	Stephens et al. (2004)
Glx	Subtelomeric	7HS	Stephens et al. (2004)
His 3	Subtelomeric	4HS, 6HS, 7HS	Stephens et al. (2004)
Nar 1	Subtelomeric	6HS	Stephens et al. (2004)
Nar 7	Subtelomeric	6HL	Stephens et al. (2004)
Ubi	Subtelomeric	6HS, 7HL	Stephens et al. (2004)
BAC 0356N21	Distal	1HS	Phillips et al. (2010)
BAC 0146M01	Distal	6HL	Phillips et al. (2010)
BAC 026D09	Subtelomeric	2HL	Phillips et al. (2010)
BAC 0047K17	Distal	4HL	Phillips et al. (2010)
FLbaf140k15	Pericentric	7HS	Karafiatova et al. (2013)
FLbaf67j12	Pericentric	7HS	Karafiatova et al. (2013)
FLbaf140c21	Pericentric	7HS	Karafiatova et al. (2013)
FLbaf104j18	Pericentric	7HS	Karafiatova et al. (2013)
FLbaf151b16	Centromeric	7HS	Karafiatova et al. (2013)
FLbaf169o18	Centromeric	7HL	Karafiatova et al. (2013)
FLbaf125j04	Pericentric	7HL	Karafiatova et al. (2013)
FLbaf54a18	Pericentric	7HL	Karafiatova et al. (2013)
FLbaf148b24	Pericentric	7HL	Karafiatova et al. (2013)
FLbaf24d09	Pericentric	7HL	Karafiatova et al. (2013)
FLbaf25l12	Pericentric	7HL	Karafiatova et al. (2013)
FLbaf129g09	Pericentric	7HL	Karafiatova et al. (2013)
FLbaf89h06	Pericentric	7HL	Karafiatova et al. (2013)
FLbaf107j09	Pericentric	7HL	Karafiatova et al. (2013)
FLbaf175h04	Pericentric	7HL	Karafiatova et al. (2013)
BAC 58H2	Subtelomeric	7HS	Ma et al. (2010)
BAC 63E11	Distal	7HL	Ma et al. (2010)

(continued)

Table 3.2 (continued)

Probe	Position	Chromosome	References
FPc678	Distal	2HL	Schmutzer et al. (2014)
FPc38863	Distal	2HL	Schmutzer et al. (2014)
65 FP contigs	Pericentric intercalary	3HS and 3HL	Aliyeva-Schnorr et al. (2015)

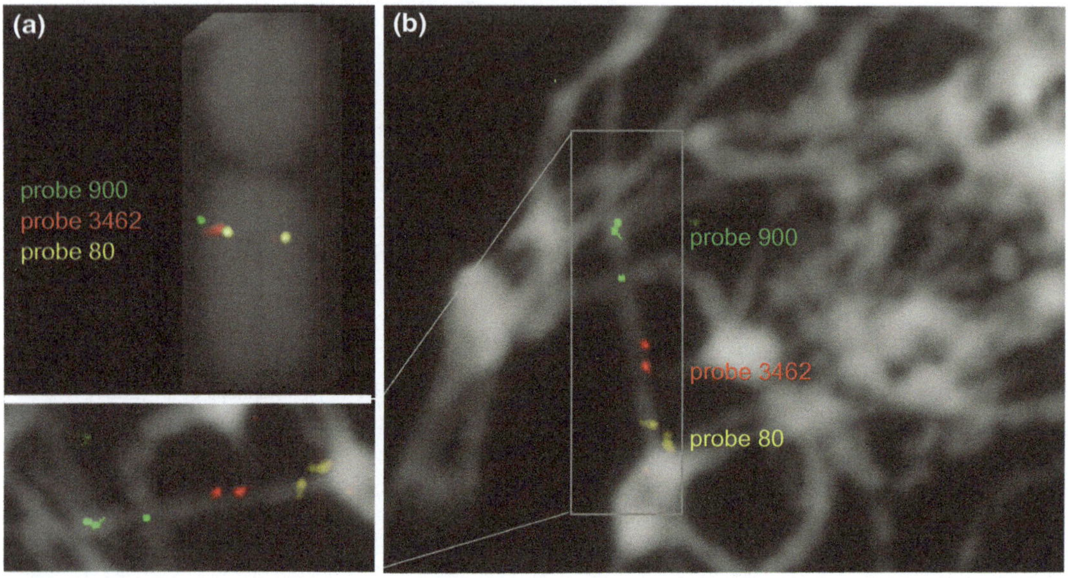

Fig. 3.3 Comparison of single-copy FISH at different resolutions. **a** FISH on metaphase chromosome 3H showed overlapping signals of probe 900, 3462 and 80. **b** In contrast, FISH using pachytene chromosomes revealed non-overlapping FISH signals

3.2 Chromatin Composition of Barley

The distribution of post-translational histone modifications has been analysed in barley at the sequence and chromosome level by chromatin immunoprecipitation DNA-sequencing (ChIP-seq) and indirect immunostaining, respectively (Baker et al. 2015; Aliyeva-Schnorr et al. 2015). One of the histone markers first studied in barley was the cell cycle-regulated phosphorylation of histone H3 at position serine 10 (Houben et al. 1999). This modification correlates with the process of chromosome segregation.

The chromosomal distribution of post-translational histone modifications, typical for eu- or heterochromatin, was determined for barley. Dimethylated lysine 9 of histone H3 (H3K9me2), a prominent mark for constitutive heterochromatin, was revealed to be uniformly distributed throughout the entire length of the chromosome, as is typical for plants with genomes larger than 500 Mbp (Houben et al. 2003; Fuchs et al. 2006). ChIP-seq data were in agreement with a cytogenetic analysis revealing an overlap between H3K9me2-marked regions with LTR retrotransposon (Baker et al. 2015). In contrast, such marks as H3K4me2 and H3K4me3, typically associated with transcriptionally potent euchromatin, were strongly enhanced in the subterminal regions of mitotic and meiotic chromosomes. Moreover, ChIP-seq analysis revealed that H3K4me2 is partially present in pericentromeric regions (Baker et al. 2015). Gene-rich subtelomeric regions are tightly covered by genic and intergenic H3K27me3, and H3K27me1 has been deleted in these regions (Baker et al. 2015). Comparable subtelomeric regions were detected with H3K27me3-specific antibodies, a mark for transcriptionally inactive gene-containing chromatin. Based on the analysis of nine histone marks by ChIP-seq, an 11-state model of the barley epigenome was proposed (Baker et al. 2015). The correlation between chromatin states and differential gene expression suggested an inverse relationship between H3K27me3 and H3K27me1.

Detection of eu- and heterochromatic histone markers revealed that the physical distribution of the 5.5 cM region anchored to the genetic centromere was correlated with chromatin containing preferentially the heterochromatin-typical post-translational histone modification H3K9me2. In contrast, antibodies recognizing the euchromatic markers H3K4me2, H3K4me3 and H3K27me3 stained mainly distal regions of mitotic metaphase chromosomes, representing 37% of the entire length of chromosome 3H (Aliyeva-Schnorr et al. 2015). A comparable distribution of histone marks was described for *Aegilops* species and rye (*Secale cereale* L.) (Oliver et al. 2013). It seems that the frequency of recombination along chromosomes correlates with a distinct chromatin structure (Higgins et al. 2014).

3.3 Telomeres and Their Application for the Generation of Barley Mini-Chromosomes

The ends of barley chromosomes are capped with ~25 kb long *Arabidopsis*-type telomere sequence arrays (Kilian et al. 1995). Besides being used for the identification of the chromosome ends, the telomere-specific repeats have been used to truncate barley chromosomes with the aim of generating a novel type of chromosomal vector system. Engineered minimal chromosomes with sufficient mitotic and meiotic stability would have an enormous potential as vectors for stacking multiple genes required for complex traits in plant biotechnology. An option to shorten chromosomes is to use telomere seeding (reviewed in Birchler et al. 2010; Mette and Houben 2015). As shown first by Farr et al. (1991), the introduction of cloned telomeric repeats into cells may randomly shorten chromosomes by the formation of new telomeres at integration sites. Such chromosome ends carrying a tandem array of telomeric repeats can serve as seeding points for the formation of telomeres. The introduction of a T-DNA construct containing a block of telomeric repeats at one end via *Agrobacterium*-mediated DNA transfer can lead to the formation of T-DNA associated de novo telomeres. Telomere seeding leading to chromosome truncation was found only in tetraploid barley, indicating that genetic redundancy facilitated the recovery of shortened chromosomes. Truncated chromosomes were transmissible in sexual reproduction but were inherited at rates lower than expected (Kapusi et al. 2012). In the future, telomere seeding in combination with genome editing tools like CRISPR/Cas should allow the targeted generation of engineered mini-chromosomes (Puchta 2015).

3.4 Analysis of the Barley Centromere and Its Role in the Process of Haploidization

The chromosomal location of the centromeric histone H3 variant CENH3 (also called CENP-A; Earnshaw et al. 2013) is the assembly site of the kinetochore complex for active centromeres (Fig. 3.2h). Any error in transcription, translation, modification or loading of CENH3 can affect its ability to assemble intact centromeric chromatin, which would result in the loss of CENH3 from the centromeres and hence in the loss of centromere identity (reviewed in Allshire and Karpen 2008). In barley, CENH3 interacts with a fraction of *cereba*, a centromeric TY3/gypsy retroelement (CR)-like repeat, and in addition with a GC-rich centromeric satellite (Houben et al. 2007). Single barley chromosome containing an average of about 200 *cereba* elements were estimated to have at least 1.4 Mb of centromeric DNA (Presting et al. 1998; Hudakova et al. 2001). The existence of non-centromeric *cereba* sites was confirmed by the identification of weak interstitial *cereba* signals along both arms of barley chromosomes in addition to accumulated FISH signal at the centromeres (Aliyeva-Schnorr et al. 2015). However, the identified barley centromeric repeats were neither sufficient nor obligatory to assemble functional centromeres as barley telocentric chromosomes derived from a 7HS isochromosome possessing neocentromeres without centromeric repeats are mitotically and meiotically stable (Nasuda et al. 2005a). Many other de novo formed centromeres have been identified in plants, yeast, flies, chicken and humans, demonstrating the epigenetic nature of the centromere (reviewed in Cuacos et al. 2015).

3-D scanning electron microscopy immunogold investigations showed that CENH3 was distributed exclusively in the interior rather than on the surface of the barley centromere (Houben et al. 2007; Schroeder-Reiter et al. 2012). Barley encodes two variants of CENH3, called αCENH3 and βCENH3 (Sanei et al. 2011). Both CENH3 variants are arranged in distinct but intermingled subdomains in the centromeres. A similar distribution of the two CENH3 variants in relationship to the spindle poles in metaphase cells suggests that both CENH3s have the same function in chromosome segregation. However, the interphase centromere composition and transcription of the CENH3 variants differ between different tissues of barley (Ishii et al. 2015). Thus, the loading of the CENH3 variants into the centromeres might vary between different tissues.

Haploid *Arabidopsis thaliana* plants were generated from crossing between wild-type and 'CENH3-tail swap' plants (Ravi and Chan 2010; Ravi et al. 2014). To elucidate whether besides the CENH3-tail swap construct, non-transgenic-induced minimal mutations in the endogenous CENH3 gene could be used for haploid induction also, an ethylmethansulfonate (EMS)-induced TILLING population of barley (Gottwald et al. 2009) was screened (Karimi-Ashtiyani et al. 2015). Assuming that either CENH3 variant can compensate for the absence of the other, viable offspring should be obtained even though it has a loss-of-function allele for either of the two CENH3 variants. AlphaCENH3 and βCENH3 signals at the centromeres were revealed in all mutants but one mutant (Hvßcenh3 L92F) showed no centromeric βCENH3 signals in mitotic, meiotic and interphase cells. The centromeric loading of the mutated βCENH3 seemed to be impaired but nonetheless the centromeres without βCENH3 have sufficient centromere functions as no obvious chromosome segregation defects were found in mitosis. The Hvβcenh3 L92F mutation was located in the CENH3 centromere targeting domain (CATD). This domain was shown to be required for centromere loading of CENH3 by Scm3/HJURP chaperons in non-plant species (Bassett et al. 2012; Foltz et al. 2009). CENH3 chaperons are highly variable among different organisms (Bai et al. 2011), and no analogue has yet been identified in plants. However, it cannot be excluded that this domain in plants also mediates interaction with a chaperon.

The potential of Hvßcenh3 L92F to act as a haploid inducer was tested. The advantage of doubled haploids for breeders is that homozygosity can be achieved in the first generation, whereas in classical breeding systems, several selfed generations are needed to obtain a high level of homozygosity (reviewed in Ishii et al. 2016). None of the F1 barley plants obtained from crosses between the wild-type and the Hvßcenh3 L92F mutant in either direction was haploid (Karimi-Ashtiyani et al. 2015). This result indicates that the Hvßcenh3 L92F mutation in the presence of native αCENH3 is not sufficient for chromosome elimination during early zygotic embryogenesis. In contrast, *A. thaliana* encodes only one variant of CENH3, and after crossing between plants with a comparable CENH3 mutation and wild-type, haploid plants were obtained. The high degree of evolutionary conservation of the identified CENH3 mutation sites offers a promising opportunity for application in a wide range of crop species where haploid technology is of interest.

The involvement of CENH3 in the process of uniparental chromosome elimination has also been shown in unstable *H. vulgare* × *Hordeum bulbosum* hybrids (Sanei et al. 2011). Probably, because of cell cycle asynchrony of the two parental genomes, CENH3 incorporation only occurs into the centromeres of *H. vulgare*. In unstable hybrids, *H. bulbosum* chromosomes lag behind *H. vulgare* chromosomes because of centromere inactivity during anaphase, subsequently forming micronuclei. Finally, micronucleated *H. bulbosum* chromatin will degrade and a haploid *H. vulgare* embryo will develop. In conclusion, a detailed knowledge of the biology of centromeres will translate into an efficient tool to speed up the process of breeding.

3.5 Barley Cytogenetic Stocks

3.5.1 Trisomics

Unlike hexaploid wheat (*Triticum aestivum* L.), hypoaneuploid seldom happens in barley due to its diploid nature. Instead, hyperaneuploid, such as trisomy, occurs spontaneously or in the progeny of autotriploids. Complete sets of trisomics were established in wild barley (*H. spontaneum*) and in a cultivated variety 'Shin Ebisu'. From those primary trisomics, secondary trisomics ($2n = 14 + 1$ isochromosome) and telotrisomics ($2n = 14 + 1$ telocentric chromosome) and other aneuploids arose spontaneously (Tsuchiya 1969). These aneuploids have been used in genetic-linkage mapping, termed trisomic analysis, in barley (Tsuchiya 1991).

3.5.2 Wheat–Barley Addition, Substitution and Translocation Lines

Common wheat or hexaploid wheat (AABBDD, $2n = 42$) can tolerate aneuploidy to some extent, thanks to its polyploid nature, and therefore many aneuploid stocks of common wheat have been established, such as monosomic, nullisomic and ditelosomic lines (Sears 1954, 1966; Sears and Sears 1978). Moreover, the hexaploid nature of common wheat enabled many alien chromosome addition lines to be established in common wheat (Jiang et al. 1994). Hybrids between the two genera *Triticum* × *Hordeum* had received little attention, presumably because the formation of these hybrids had been regarded as improbable before Kruse (1973) achieved such intergeneric hybrids, using *H. vulgare* as maternal and *T. aestivum* as paternal parents: the hybrids were vegetatively vigorous but generatively sterile. After failing to obtain self-fertile wheat–barley addition lines from F_1 hybrid between barley as the female parent and hexaploid wheat, Islam et al. (1975, 1981) attempted the reciprocal cross and obtained only one proper hybrid ($2n = 28$), forming 28 univalents at metaphase I (MI) from 3381 florets of hexaploid wheat (cultivar 'Chinese Spring') pollinated with barley (cultivar 'Betzes') pollen. After backcrossing, the 28-chromosome hybrid with 'Chinese Spring' pollen, Islam et al. (1981) obtained self-sterile 49-chromosome BC_1 plants and, by pollinating the BC_1 plants with 'Chinese Spring', subsequently produced a series of self-fertile wheat–barley addition lines for all chromosomes except for chromosome 1H. Later, Islam (1983) achieved ditelosomic additions of barley chromosome arms to common wheat, except for chromosome 1H. The addition of chromosome 1H causes extreme meiotic abnormalities leading to complete sterility (Islam et al. 1981). Islam and Shepherd (2000) finally produced a self-fertile addition line for chromosome 1H that carries a pair of heteromorphic 1H/1HS (the short arm) together with a pair of the 6H chromosomes; a disomic addition line for 1H alone was not established, because it was highly sterile. These additional lines of cv. Betzes, except for 1H, are available from the National Bioresource Project (NBRP-wheat) (http://www.shigen.nig.ac.jp/wheat/komugi/strains/aboutNbrpLgku.jsp).

Taketa and Takeda (2001) developed a complete set of wheat (Japanese common wheat cultivar 'Shinchunaga')-wild barley (*H. vulgare* ssp. *spontaneum*) whole chromosome addition lines and five telosomic addition lines. Except for 1H and 1HL, self-fertile disomic and ditelosomic addition lines were obtained. For 1H and 1HL, self-sterile monosomic and monotelosomic additions were obtained from the backcrossed progeny of a double monosomic 1H and 6H addition plant and from that of a monosomic 1HL addition plant carrying a translocation 1HS.6HL, respectively. This result showed that in wild barley, like in cultivated barley, chromosome 1H (specifically 1HL) carries a genetic factor causing sterility in the wheat genetic background and that chromosome 6H (specifically 6HL) carries a genetic factor ameliorating the sterility caused by chromosome 1H. The responsible factor for the sterility is located in the interstitial 25% region of the 1HL arm (Taketa et al. 2002). Incomplete sets of wheat–barley additions, substitutions and translocations have also been produced using other combinations of barley and wheat cultivars (see Molnár-Láng et al. (2014); Molnár-Láng and Linc (2015) for reviews). Utilization of the GISH procedure allowed for the identification of barley chromosomes in the wheat–barley addition and substitution lines and enabled precise visualization of the fusion points in wheat–barley translocation lines (Fig. 3.5).

The wheat–barley ditelosomic addition lines can be used for flow cytometry to sort individual barley chromosome arms. The chromosome arms of 2H-7H were successfully isolated (Suchánková et al. 2006). Chromosome 1H is sufficiently smaller than the other barley chromosomes to be isolated by flow sorting directly from euploid barley with 95% purity (Mayer et al. 2009). This physical isolation of chromosome arms using the wheat–barley ditelosomics is definitely a powerful tool for the genome analysis in barley.

Thus, the wheat–barley addition lines facilitate the partition of the whole barley genome into chromosomes or chromosome arms. Moreover, the presence of barley chromosome or chromosome arm-specific genes and DNA markers can be predicted from the phenotypes and marker profiles of the wheat–barley lines (Serizawa et al. 2001a).

3.5.3 Translocations Lines

Reciprocal translocations exhibit an exchange of chromosome fragments between nonhomologous chromosomes. Around 1000 lines with single translocation have been described for barley (Künzel 1992; Marthe and Künzel 1994). The distribution of translocation breakpoints induced by ionizing radiation along chromosomes is non-random. When mitosis was first evaluated after radiation, centromeres and the heterochromatin-containing proximal segments tended to be more often than expected involved in translocations, and terminal segments to be less than randomly involved in translocations. Contrary to this, small chromosomal regions in median and distal arm positions, characterized by high recombination rates and high gene density, were identified as preferred breakpoint sites of viable translocations. Apparently, the position of a translocation breakpoint has an influence on the rate of viability versus elimination of the carrier cells. Surprisingly, translocation breakpoints within centromeres and heterochromatin-containing segments seem to be more harmful for survival than those induced in gene-rich regions (Künzel et al. 2001).

Translocation lines have been used for the generation of duplicated chromosome segments (Hagberg 1995), for analysis of the position-dependent NOR activity (Schubert and Künzel 1990) and for the physical mapping of marker sequences (Sorokin et al. 1994; Künzel et al. 2000; Künzel and Waugh 2002).

Künzel et al. (2000) integrated the positions of 240 translocation breakpoints as physical landmarks into RFLP linkage maps of the seven barley chromosomes. Therefore, predefined translocation chromosomes were microdissected and used as templates for PCR with primers specific for >300 genetically mapped RFLP probes. Heterogeneous recombination rates were found along individual chromosomes. Recombination appeared to be mainly confined to a few relatively small areas spaced with large segments in which recombination was severely suppressed. The regions of higher recombination frequency (1 Mb/cM or less) corresponded to only 4.9% of the total barley genome and harboured 47.3% of the 429 markers used (Künzel et al. 2000).

3.6 Cytological Mapping of Barley Chromosomes

Different approaches of cytological mapping were applied to barley. Physical or cytological mapping by the use of chromosome deletion lines (Serizawa et al. 2001b), reciprocal translocation lines (Marthe and Künzel 1994; Sorokin et al. 1994; Künzel et al. 2000) or wheat–barley addition lines (Nasuda et al. 2005b; Islam and Shepherd 2000) has been used to physically map the barley genome. However, these strategies provide only sketchy information on the barley genome. More recently, single-copy FISH was successfully used to correlate the genetic map with the cytological map of barley chromosome 3H (Fig. 3.2i) (Aliyeva-Schnorr et al. 2015).

3.6.1 Deletion Stocks and Their Use in Cytological Mapping

Cytogenetic stocks carrying various chromosomal aberrations are valuable genetic materials to determine the location of genes. A high-resolution cytological map based on aberrant chromosomes is mainly dependent on the availability of cytogenetic stocks with variable breakpoints along the chromosomes. Cytogenetic stocks generated by radiation (*X*-rays and gamma rays) in animal cell culture lines and crop plants, known as the radiation hybrid panels (RH panels), are valuable materials for physical mapping (Kumar et al. 2014). In wheat, a unique system inducing chromosomal breakage, called the gametocidal (Gc) system, was discovered (Endo 1990, 2007). Some of the chromosomes introduced into common wheat from the genus *Aegilops* assured their existence in wheat in a selfish manner, i.e. by causing chromosomal breakage in gametes lacking themselves. The Gc system was applied as a biological mutagen inducing mild to lethal chromosomal mutations in common wheat (Endo 1988). Using this system, 436 deletion stocks were generated in common wheat (Endo and Gill 1996).

The Gc system was found to generate deletions and translocations effectively in barley chromosomes added to common wheat, as well as in wheat chromosomes (Shi and Endo 1997; 1999) (Fig. 3.4). Chromosomal aberrations involving barley chromosomes induced by the Gc system can be maintained and established in common wheat. These cytogenetic stocks carrying parts of barley chromosomes, in the form of terminal deletions or wheat–barley translocations (Fig. 3.5), are called dissection lines (Ashida et al. 2007). These dissection stocks have been used for cytological mapping using molecular markers that are assigned to specific barley chromosomes in the wheat–barley disomic addition lines, 103 AFLP markers (Serizawa et al. 2001a) and 701 EST markers (Nasuda et al. 2005b).

Based on the presence and absence of DNA makers in the dissection lines, cytological maps for 2H (Joshi et al. 2011), 3H (Sakai et al. 2009), 4H (Sakata et al. 2010) and 6H (Ishihara et al. 2014) were constructed and compared with the genetic maps of the same chromosomes constructed by Sato et al. (2009) using the same DNA markers. In general, both cytological and genetic maps show almost the same marker order in the distal regions; however, the cytological maps have better resolution in the proximal regions than the genetic maps. Moreover, it was found that the markers distributed rather evenly in the genetic maps were actually concentrated in the distal regions in the cytological maps; in other words, in the proximal regions markers are scarce and crossing over is suppressed (Fig. 3.6).

The number of dissection lines so far developed totals 433 (Table 3.3), contributing to the increase in the number of cytogenetic stocks available for cytological mapping. The number of dissection lines can infinitely be increased by the screening of the progeny of wheat–barley addition lines with monosomic Gc addition. Usually, chromosomal aberrations are cytologically identified by chromosome banding and by in situ hybridization (FISH/GISH), which is a laborious task.

Joshi et al. (2013) reported that the Gc action could be effective in post-zygotic divisions, as well as during gametogenesis after meiosis. They chose 81 plants carrying a normal-appearing 2H

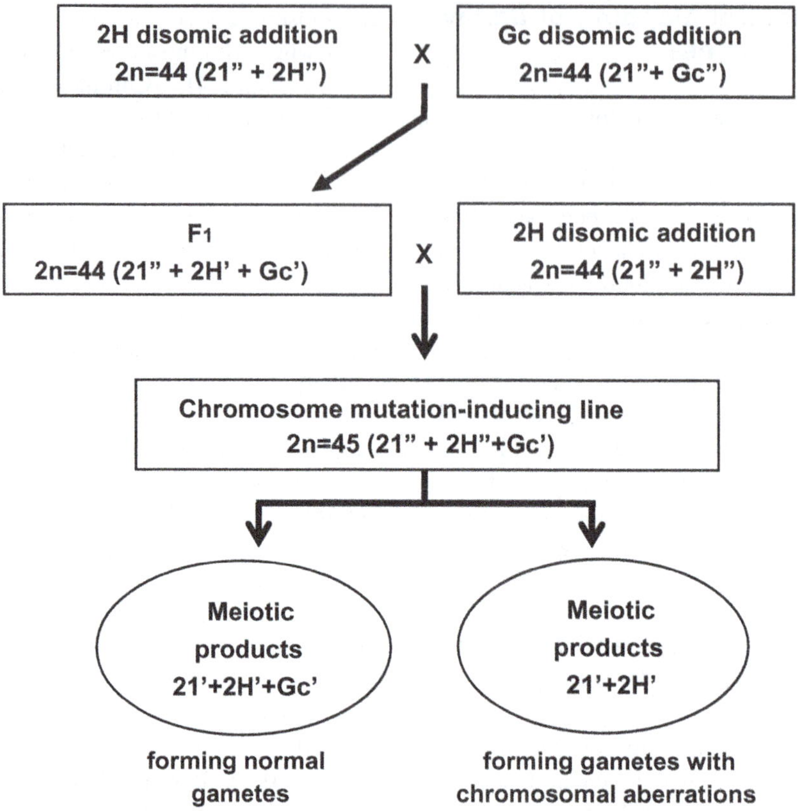

Fig. 3.4 Schematic illustration of the Gc system to induce structural changes in barley chromosome 2H added to common wheat ($2n = 42$, 21″). Gc stands for a gametocidal chromosome, 2C or $3C^{SAT}$, derived from the genus *Aegilops* (Endo and Gill 1996)

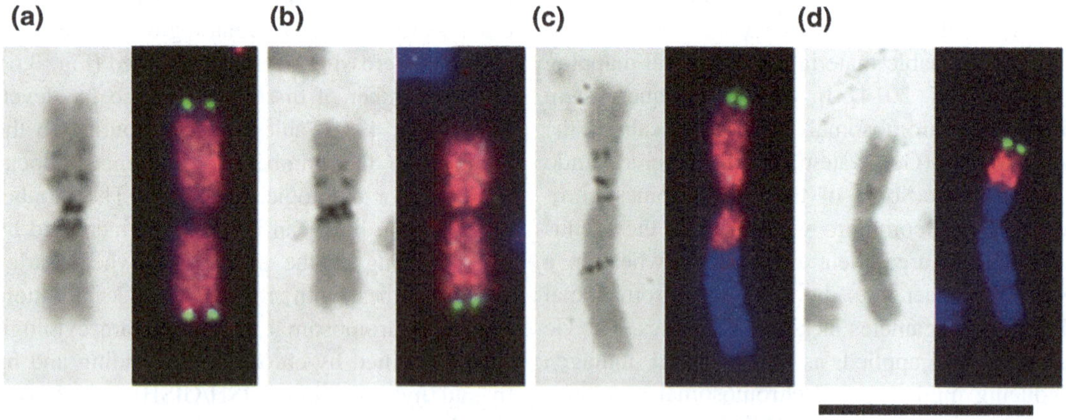

Fig. 3.5 C-band (left) and FISH/GISH (right) images of **a** normal 7H chromosome and **b**–**d** aberrant 7H chromosomes (**b**: terminal deletion, **c**: wheat–barley translocation with the barley centromere, **d**: wheat–barley translocation with the wheat centromere). Bar = 10 μm

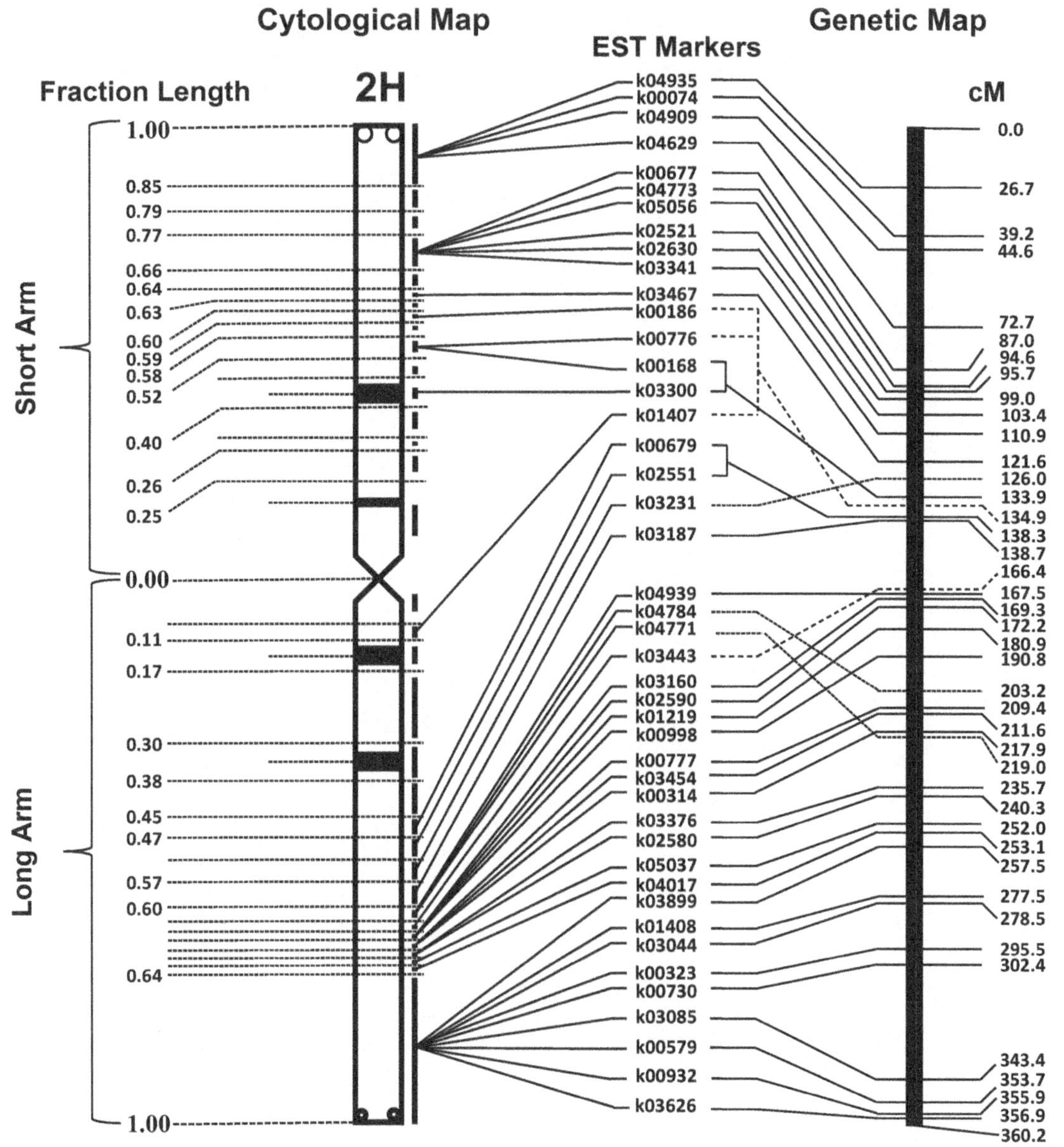

Fig. 3.6 Comparison between the cytological (left) and genetic (right) maps of chromosome 2H constructed with 44 EST markers (reproduced from Fig. 3.3 of Joshi et al. 2011). Note that the markers distributed rather evenly in the genetic map were actually concentrated in the distal regions as revealed in the cytological map

chromosome in root-tip cells by FISH/GISH from the progeny of wheat-2H plants with monosomic Gc addition, and found that six of them had 2H aberrations as revealed by the absence of some of 18 2H-specific EST sequences; all the 2H aberrations were cytologically confirmed in the next generation. This suggests that a plant carrying a normal 2H chromosome in root-tip cells (root tips are usually used for chromosomal screening) could have an aberrant 2H chromosome in the aerial parts of the same plant and vice versa. Actually, some of the aberrant chromosomes that had cytologically been identified in root-tip cells were not found in the next generation (e.g. Ashida et al. 2007). Therefore, PCR-based screening using DNA

Table 3.3 Barley dissection lines developed by the Gc system and markers used for cytological mapping

Gc chromosome	Barley chromosome								
	1H[a]	2H[b]	3H[c]	4H[d]	5H[e]	6H[f]	7H[g]		
	2C	2C and $3C^{SAT}$	2C and $3C^{SAT}$	2C and $3C^{SAT}$	$3C^{SAT}$	2C and $3C^{SAT}$	2C		
Number of dissection lines developed	5	72	50	60	123	33	90		
with deletion	5	30	30	21	30	15	43		
with translocation	0	39	18	32	83	16	46		
with other aberrations	0	3	2	7	10	2	1		
Number of breakpoints	5	82	56	68	123	36	92		
in short arm	2	33	23	29	28	6	29		
in long arm	0	30	20	15	39	15	35		
in centromere	3	19	10	24	40	6	14		
in unknown regions	0	0	3	0	16	9	14		
Number of dissection lines used for mapping	5	72	50	60	23	33	90	19	22
Number of markers used for mapping	63 (EST)	115 (EST)	36 (EST)	93 (EST)	97 (EST)	74 (EST)	51 (SSR, AFLP)	90 (EST)	45 (STS, AFLP)

[a–g]: References from which data were taken. [a]Ishihara et al. (2014), [b]Joshi et al. (2011, 2013), [c]Sakai et al. (2009), [d]Sakata et al. (2010), [e]Ashida et al. (2007), [f]Ishihara et al. (2014), [g]Serizawa et al. (2001b), Nasuda et al. (2005b) and Masoudi-Nejad et al. (2005)

markers in the terminal regions of both chromosome arms surely increases the chance to detect aberrant chromosomes. Furthermore, Joshi et al. (2013) inferred from sequencing data that the Gc system does not induce nucleotide changes, although the analysed region was small ($\sim$650 kbp). The details of the dissection lines developed so far by the Gc system for each of the barley chromosomes are summarized in Table 3.3.

Chromosome 1H Among the 75 EST markers assigned to 1H (Nasuda et al. 2005b), 15 and 48 were reassigned to the short arm and to the long arm, respectively, and the 15 markers on the short arm were divided into three regions flanked by the telomere, the centromere and two breakpoints of five dissection lines (Ishihara et al. 2014).

Chromosome 2H Joshi et al. (2011, 2013) established a total of 72 dissection lines with 82 breakpoints in total (ten aberrations had dual breakpoints). These breakpoints divided 47 EST markers on the short arm into 18 regions and 68 EST markers on the long arm into 21 regions (Joshi et al. 2013) (Fig. 3.6).

Chromosome 3H Sakai et al. (2009) generated 50 dissection lines with 53 breakpoints in total and used them for cytological mapping. These breakpoints divided 12 EST markers on the short arm and 24 EST markers on the long arm into ten regions, respectively.

Chromosome 4H Sakata et al. (2010) established 60 dissection lines with 68 breakpoints in total. These breakpoints divided 37 EST markers on the short arm into 18 regions and 56 EST markers on the long arm into eight regions.

Chromosome 5H Ashida et al. (2007) developed 123 dissection lines, from which 23 representative lines were chosen for cytological mapping. Ten breakpoints in the short arm divided 23 EST markers into nine regions and 14 breakpoints in the long arm divided 74 EST markers into 11 regions.

Chromosome 6H Ishihara et al. (2014) established 33 dissections lines with 36 breakpoints in total. These breakpoints divided 36 EST markers on the short arm into nine regions and 38 EST markers on the long arm into 13 regions.

Chromosome 7H Serizawa et al. (2001b) used 22 dissection lines for cytological mapping of 17 AFLP and 28 STS markers: Four AFLP and 21 STS markers on the short arm were localized in four and six regions, respectively; 13 AFLP and seven STS markers on the long arm were grouped into seven and three groups, respectively. Nasuda et al. (2005b) conducted cytological mapping using 19 dissection lines with a total of 23 breakpoints and assigned 49 EST markers to eight regions in the short arm and 41 EST markers to nine regions in the long arm. Using all the 90 Gc-induced 7H dissection lines with a total of 91 breakpoints, Masoudi-Nejad et al. (2005) successfully constructed an RH map of chromosome 7H consisting of 28 SSR and 23 AFLP markers.

3.7 Application of FISH to Correlate the Genetic with the Cytological Map of Barley

Until recently, cytological mapping of large-genome species by FISH remained difficult, since short sequences (<5 kb) such as cDNA were not always detectable and large sequences (BACs) may contain repeats. Stephens et al. (2004) applied a tyramide signal amplification (TSA) based on ultrasensitive FISH for the detection of single-copy cDNA probes, achieving the detection of 14 cDNA probes. Eleven of these probes were assigned to unique chromosome positions. Phillips et al. (2010) preselected by dot blot hybridization 105 BACs with a low repeat content. After FISH, seven BACs showed single chromosome signal while the remaining probes resulted in the detection of multiple hybridization loci or were not detectable. Recently, Karafiatova et al. (2013) determined the relative order of 13 chromosome-7H-specific cDNAs via FISH. In addition, during the study of synteny relationship between barley and *Brachypodium*, two out of 13 *Brachypodium* BACs showed FISH hybridization signals on barley chromosomes (Ma et al. 2010).

In order to improve our knowledge about chromosomal synteny between cultivated barley and its wild relatives, genomic single-copy and cDNA sequences specific to chromosome 3H were mapped by FISH to the chromosomes of *H. bulbosum, Hordeum marinum, Hordeum pubiflorum, Hordeum murinum* and *S. cereale* (Aliyeva-Schnorr et al. 2016). That most of the sequences showed reliable signals suggested homoeology between cultivated barley and related species. Differences in the order and position of FISH markers demonstrated the occurrence of intrachromosomal rearrangements within the genus *Hordeum* and of interchromosomal rearrangements between barley and rye. Comparison between repeat-free genomic and cDNA probes revealed that gene-containing single-copy genomic DNA (gDNA) probes perform more reliably for FISH-based analysis of synteny. All barley chromosome-specific probes used for FISH are summarized in Table 3.2.

The combination of in silico extraction of unique sequences suitable for FISH probes and a sensitive FISH method (Kato et al. 2006) provides a new tool to overcome the problem of cross-hybridization in large-genome species like barley (Schmutzer et al. 2014). The Kmasker method (http://webblast.ipk-gatersleben.de/kmasker) allows the in silico extraction of single-copy sequences. This method can be used for the integration of genetic and cytological maps in species with large and complex genomes if the information of the target sequences (e.g. BACs and sequence contigs) and a fourfold coverage of unassembled genomic sequences of the species of interest are available. By this novel approach, the chromosomal positions and order of 65 FP contigs mapped to a 5.5 cM bin around the genetic centromere of chromosome 3H were determined (Aliyeva-Schnorr et al. 2015). The study revealed that the centromeric 5.5 cM bin comprises 58% of the length of mitotic metaphase chromosome 3H (Fig. 3.7). This result is in line with earlier results that approximately 1.9 Gb corresponding to 48% of the genetically anchored physical map (3.9 Gb) is anchored to the genetic centromere (Künzel et al. 2000; International Barley Genome Sequencing et al. 2012). The unequal distribution of recombination frequency along the chromosome is the main

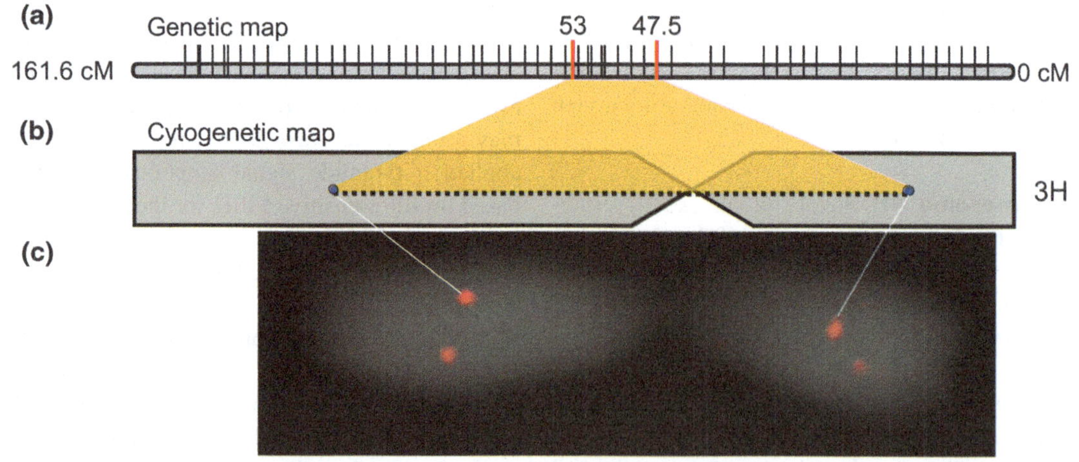

Fig. 3.7 Comparison between the genetic and cytological maps of barley chromosome 3H. **a** Genetic positions of 65 FP contigs range from 47.5 to 53.0 cM (International Barley Genome Sequencing et al. 2012). **b** Schematic representation of the cytogenetic positions determined by FISH of 70 single-copy probes deduced from in silico analysis. **c** FISH signals (in red) of bordering FP contig-derived single-copy probes on chromosome 3H

reason for non-linear translation of genetic distances into physical distances; hence, a genetic distance may translate into a shorter or a longer physical distance depending on the chromosomal region. Suppressed recombination in the centromeric region is a typical feature of Triticeae (Hordeae) species with large genomes.

References

Aliyeva-Schnorr L, Beier S, Karafiatova M et al (2015) Cytogenetic mapping with centromeric bacterial artificial chromosomes contigs shows that this recombination-poor region comprises more than half of barley chromosome 3H. Plant J 84:385–394

Aliyeva-Schnorr L, Stein N, Houben A (2016) Collinearity of homoeologous group 3 chromosomes in the genus Hordeum and Secale cereale as revealed by 3H-derived FISH analysis. Chromosome Res 24(2): 231–242

Allshire RC, Karpen GH (2008) Epigenetic regulation of centromeric chromatin: old dogs, new tricks? Nat Rev Genet 9:923–937

Ashida T, Nasuda S, Sato K, Endo TR (2007) Dissection of barley chromosome 5H in common wheat. Genes Genet Syst 82:123–133

Bai YW, Zhou Z, Feng HQ, Zhou BR (2011) Recognition of centromeric histone variant CenH3s by their chaperones Structurally conserved or not. Cell Cycle 10:3217–3218

Baker K, Dhillon T, Colas I et al (2015) Chromatin state analysis of the barley epigenome reveals a higher-order structure defined by H3K27me1 and H3K27me3 abundance. Plant J 84:111–124

Bassett EA, DeNizio J, Barnhart-Dailey MC et al (2012) HJURP uses distinct CENP-A surfaces to recognize and to stabilize CENP-A/histone H4 for centromere assembly. Dev Cell 22:749–762

Birchler JA, Krishnaswamy L, Gaeta RT et al (2010) Engineered minichromosomes in plants. Crit Rev Plant Sci 29:135–147

Brandes A, Roder MS, Ganal MW (1995) Barley telomeres are associated with two different types of satellite DNA sequences. Chromosome Res 3:315–320

Busch W, Martin R, Herrmann RG, Hohmann U (1995) Repeated DNA sequences isolated by microdissection. 1. Karyotyping of barley (*Hordeum vulgare* L.). Genome 38:1082–1090

Costa JM, Singh RJ (2006) Chromosome mapping in barley (*Hordeum vulgare* L.). In: Genetic resources, chromosome engineering, and crop improvement, vol 2, pp 257–280

Cuacos M, Franklin FCH, Heckmann S (2015) Atypical centromeres in plants—what they can tell us. Front Plant Sci 6

Cuadrado A, Jouve N (2007) The nonrandom distribution of long clusters of all possible classes of trinucleotide repeats in barley chromosomes. Chromosome Res 15:711–720

Dong FG, Jiang JM (1998) Non-rabl patterns of centromere and telomere distribution in the interphase nuclei of plant cells. Chromosome Res 6:551–558

Earnshaw WC, Allshire RC, Black BE et al (2013) Esperanto for histones: CENP-A, not CenH3, is the centromeric histone H3 variant. Chromosome Res 21:101–106

Endo TR (1988) Induction of chromosomal structural changes by a chromosome of *Aegilops cylindrical* L. in common wheat. J Hered 79:366–370

Endo TR (1990) Gametocidal chromosomes and their induction of chromosome mutations in wheat. Jpn J Genet 65:135–152

Endo TR (2007) The gametocidal chromosome as a tool for chromosome manipulation in wheat. Chromosome Res 15:67–75

Endo TR (2011) Cytological dissection of the triticeae chromosomes by the gametocidal system. Methods Mol Biol 701:247–257

Endo TR, Gill BS (1996) The deletion stocks of common wheat. J Hered 87:295–307

Farr C, Fantes J, Goodfellow P et al (1991) Functional reintroduction of human telomeres into mammalian cells. Proc Natl Acad Sci USA 88:7006–7010

Foltz DR, Jansen LET, Bailey AO et al (2009) Centromere-specific assembly of CENP-a nucleosomes is mediated by HJURP. Cell 137:472–484

Fuchs J, Demidov D, Houben A (2006) Chromosomal histone modification patterns—from conservation to diversity. Trends Plant Sci 11:199–208

Fukui K, Kamisugi Y, Sakai F (1994) Physical mapping of 5S rDNA loci by direct cloned biotinylated probes in barley chromosomes. Genome 37:105–111

Gill BS, Kimber G (1974) Giemsa C-banding and the evolution of wheat. Proc Natl Acad Sci USA 71:4086–4090

Gottwald S, Bauer P, Komatsuda T et al (2009) TILLING in the two-rowed barley cultivar 'Barke' reveals preferred sites of functional diversity in the gene HvHox1. BMC Res Notes 2:258

Hagberg A (1995) Coordinator's report: duplication of chromosome segments. Barley Genet Newsl 25:114

Higgins JD, Osman K, Jones GH et al (2014) Factors underlying restricted crossover localization in barley meiosis. Annu Rev Genet 48:29–47

Houben A, Pickering R (2009) Applying cytogenetics and genomics to wide hybridisations in the genus *Hordeum*. In: Genetics and genomics of the triticeae. Springer

Houben A, Wako T, Furushima-Shimogawara R, Fukui K (1999) The cell cycle dependent phosphorylation of histone H3 is correlated with the condensation of plant mitotic chromosomes. Plant J 18:675–679

Houben A, Demidov D, Gernand D, Meister A et al (2003) Methylation of histone H3 in euchromatin of plant chromosomes depends on basic nuclear DNA content. Plant J 33:967–973

Houben A, Schroeder-Reiter E, Nagaki K et al (2007) CENH3 interacts with the centromeric retrotransposon *cereba* and GC-rich satellites and locates to centromeric substructures in barley. Chromosoma 116:275–283

Hudakova S, Michalek W, Presting GG (2001) Sequence organization of barley centromeres. Nucleic Acids Res 29:5029–5035

International Barley Genome Sequencing C, Mayer KF, Waugh R, Brown JW et al (2012) A physical, genetic and functional sequence assembly of the barley genome. Nature 491:711–716

Ishihara A, Mizuno N, Islam AKMR et al (2014) Dissection of barley chromosomes 1H and 6H by the gametocidal system. Genes Genet Syst 89:203–214

Ishii T, Karimi-Ashtiyani R, Banaei-Moghaddam AM et al (2015) The differential loading of two barley CENH3 variants into distinct centromeric substructures is cell type- and development-specific. Chromosome Res 23:277–284

Ishii T, Karimi-Ashtiyani R, Houben A (2016) Haploidization via chromosome elimination: means and mechanisms. Annu Rev Plant Biol

Islam AKMR (1980) Identification of wheat-barley addition lines with N-banding of chromosomes. Chromosoma 76:365–373

Islam AKMR (1983) Ditelosomic additions of barley chromosomes to wheat. In: Sakamoto S (ed) Proceedings of the sixth international wheat genetics symposium, Kyoto University Press, Kyoto, Japan, 28 Nov–3 Dec 1983, pp 233–238

Islam AKMR, Shepherd KW (2000) Isolation of a fertile wheat-barley addition line carrying the entire barley chromosome 1H. Euphytica 111:145–149

Islam AKMR, Shepherd KW, Sparrow DHB (1975) Addition of individual barley chromosomes to wheat. In: Proceedings of 3rd international barley genetics symposium (Garching, BRD), pp 260–270

Islam AKMR, Shepherd KW, Sparrow DHB (1981) Isolation and characterization of euplasmic wheat-barley chromosome addition lines. Heredity 46:161–174

Jiang J, Friebe B, Gill BS (1994) Recent advances in alien gene transfer in wheat. Euphytica 73:199–212

Joshi GP, Nasuda S, Endo TR (2011) Dissection and cytological mapping of barley chromosome 2H in the genetic background of common wheat. Genes Genet Syst 86:231–248

Joshi GP, Endo TR, Nasuda S (2013) PCR and sequence analysis of barley chromosome 2H subjected to the gametocidal action of chromosome 2C. Theor Appl Genet 126:2381–2390

Kakeda K, Yamagata H (1992) Immunological analysis of chromosome replication in barley, rye and durum wheat by using anti-BrdU antibody. Hereditas 116:67–70

Kakeda K, Fukui K, Yamagata H (1991) Heterochromatic differentiation in barley chromosomes revealed by C- and N-banding techniques. Theor Appl Genet 81:144–150

Kapusi E, Ma L, Teo CH et al (2012) Telomere-mediated truncation of barley chromosomes. Chromosoma 121:181–190

Karafiatova M, Bartos J, Kopecky D et al (2013) Mapping nonrecombining regions in barley using multicolor FISH. Chromosome Res 21:739–751

Karimi-Ashtiyani R, Ishii T, Niessen M et al (2015) Point mutation impairs centromeric CENH3 loading and induces haploid plants. Proc Natl Acad Sci USA 112:11211–11216

Kato A (2011) High-density fluorescence in situ hybridization signal detection on barley (*Hordeum vulgare* L.) chromosomes with improved probe screening and reprobing procedures. Genome 54:151–159

Kato A, Albert P, Vega J et al (2006) Sensitive fluorescence in situ hybridization signal detection in maize using directly labeled probes produced by high concentration DNA polymerase nick translation. Biotech Histochem 81:71–78

Kilian A, Stiff C, Kleinhofs A (1995) Barley telomeres shorten during differentiation but grow in callus culture. Proc Natl Acad Sci USA 92:9555–9559

Kruse A (1973) *Hordeum* × *Triticum* hybrids. Hereditas 73:157–161

Kumar A, Bassi FM, de Jimenez MKM et al (2014) Radiation hybrids: a valuable tool for genetic, genomic and functional analysis of plant genomes. In: Genomics of plant genetic resources. Springer, Netherlands, pp 285–318

Künzel G (1992) Coordinator's report: translocation sand balanced tertiary trisomics. Barley Genet Newsl 22:80–102

Künzel G, Waugh R (2002) Integration of microsatellite markers into the translocation-based physical RFLP map of barley chromosome 3H. Theor Appl Genet 105:660–665

Künzel G, Korzun L, Meister A (2000) Cytologically integrated physical restriction fragment length polymorphism maps for the barley genome based on translocation breakpoints. Genetics 154:397–412

Künzel G, Gecheff KI, Schubert I (2001) Different chromosomal distribution patterns of radiation-induced interchange breakpoints in barley: first post-treatment mitosis versus viable offspring. Genome 44:128–132

Leitch IJ, Heslop-Harrison JS (1992) Physical mapping of the 18s-5.8s-26s ribosomal-rna genes in barley by in situ hybridization. Genome 35:1013–1018

Leitch IJ, Heslop-Harrison JS (1993) Physical mapping of four sites of 5S rDNA sequences and one site of the α-amylase-2 gene in barley (*Hordeum vulgare*). Genome 36:517–523

Linde-Laursen I (1975) Giemsa C-banding of the chromosomes of 'Emir' barley. Hereditas 81:285–289

Linde-Laursen L (1991) Giemsa C-banded karyotypes of cultivated and wild barley. Barley Genet Newsl 20:47–52

Linde-Laursen I, Heslop-Harrison J, Shepherd K et al (1997) The barley genome and its relationship with the wheat genomes. A survey with an internationally agreed recommendation for barley chromosome nomenclature. Hereditas 126:1–16

Ma L, Vu GTH, Schubert V et al (2010) Synteny between *Brachypodium distachyon* and *Hordeum vulgare* as revealed by FISH. Chromosome Res 18:841–850

Marthe F, Künzel G (1994) Localization of translocation breakpoints in somatic metaphase chromosomes of barley. Theor Appl Genet 89:240–248

Masoudi-Nejad A, Nasuda S, Bihoreau MT et al (2005) An alternative to radiation hybrid mapping for large-scale genome analysis in barley. Mol Genet Genomics 274:589–594

Mayer KFX, Taudien S, Maris M, Šimková et al (2009) Gene content and virtual gene order of barley chromosome 1H. Plant Physiol 151:496–505

Mette MF, Houben A (2015) Engineering of plant chromosomes. Chromosome Res 23:69–76

Molnár-Láng M, Linc G (2015) Wheat-barley hybrids and introgression lines. In: Molnár-Láng M, Ceoloni C, Doležel, J (eds) Alien introgression in wheat, pp 315–345

Molnár-Láng M, Linc G, Szakács E′ (2014) Wheat–barley hybridization: the last 40 years. Euphytica 195:315–329

Mukai Y, Gill BS (1991) Detection of barley chromatin added to wheat by genomic in situ hybridization. Genome 34:448–452

Nasuda S, Hudakova S, Schubert I et al (2005a) Stable barley chromosomes without centromeric repeats. Proc Natl Acad Sci USA 102:9842–9847

Nasuda S, Kikkawa Y, Ashida T et al (2005b) Chromosomal assignment and deletion mapping of barley EST markers. Genes Genet Syst 80:357–366

Oliver C, Pradillo M, Corredor E et al (2013) The dynamics of histone H3 modifications is species-specific in plant meiosis. Planta 238:23–33

Pedersen C, Rasmussen SK, Linde-Laursen I (1996) Genome and chromosome identification in cultivated barley and related species of the Triticeae (Poaceae) by in situ hybridization with the GAA-satellite sequence. Genome 39:93–104

Phillips D, Nibau C, Ramsay L et al (2010) Development of a molecular cytogenetic recombination assay for barley. Cytogenet Genome Res 129:154–161

Presting GG, Malysheva L, Fuchs J et al (1998) A Ty3/gypsy retrotransposon-like sequence localizes to the centromeric regions of cereal chromosomes. Plant J 16:721–728

Puchta H (2015) Using CRISPR/Cas in three dimensions: towards synthetic plant genomes, transcriptomes and epigenomes. Plant J

Ravi M, Chan SW (2010) Haploid plants produced by centromere-mediated genome elimination. Nature 464:615–618

Ravi M, Marimuthu MPA, Tan EH et al (2014) A haploid genetics toolbox for *Arabidopsis thaliana*. Nat Commun 5

Sakai K, Nasuda S, Sato K, Endo TR (2009) Dissection of barley chromosome 3H in common wheat and a comparison of 3H physical and genetic maps. Genes Genet Syst 84:25–34

Sakata M, Nasuda S, Endo TR (2010) Dissection of barley chromosome 4H in common wheat by the gametocidal system and cytological mapping of chromosome 4H with EST markers. Genes Genet Syst 85:19–29

Sanei M, Pickering R, Kumke K et al (2011) Loss of centromeric histone H3 (CENH3) from centromeres precedes uniparental chromosome elimination in interspecific barley hybrids. Proc Natl Acad Sci USA 108:E498–E505

Sato K, Nankaku N, Takeda K (2009) A high-density transcript linkage map of barley derived from a single population. Heredity 103:110–117

Schmutzer T, Ma L, Pousarebani N et al (2014) Kmasker-a tool for in silico prediction of single-copy FISH probes for the large-genome species *Hordeum vulgare*. Cytogenet Genome Res 142: 66–78

Schroeder-Reiter E, Sanei M, Houben A et al (2012) Current SEM techniques for de- and re-construction of centromeres to determine 3D CENH3 distribution in barley mitotic chromosomes. J Microsc 246:96–106

Schubert I, Künzel G (1990) Position-dependent NOR activity in barley. Chromosoma 99:352–359

Schubert I, Shi F, Fuchs J et al (1998) An efficient screening for terminal deletions and translocations of barley chromosomes added to common wheat. Plant J 14:489–495

Sears ER (1954) The aneuploids of common wheat. Missouri Agric Exp Stat Res Bull 572:1–59

Sears ER (1966) Nullisomic-tetrasomic combinations in hexaploid wheat. In: Riley R, Lewis KR (eds) Chromosome manipulations and plant genetics. Oliver & Boyd, Edinburgh, pp 29–45

Sears ER, Sears LMS (1978) The telocentric chromosomes of common wheat. In: Ramanujam RS (ed) Proceedings of 5th international wheat genetics symposium, New Delhi, India, pp 389–407

Serizawa N, Nasuda S, Endo TR (2001a) Barley chromosome addition lines of wheat for screening of AFLP markers on barley chromosomes. Genes Genet Syst 76:107–110

Serizawa N, Nasuda S, Shi F et al (2001b) Deletion-based physical mapping of barley chromosome 7H. Theor Appl Genet 103:827–834

Shi F, Endo TR (1997) Production of wheat-barley disomic addition lines possessing an *Aegilops cylindrica* gametocidal chromosome. Genes Genet Syst 72:243–248

Shi F, Endo TR (1999) Genetic induction of structural changes in barley chromosomes added to common wheat by a gametocidal chromosome derived from *Aegilops cylindrica*. Genes Genet Syst 74:49–54

Sorokin A, Marthe F, Houben A et al (1994) Polymerase chain reaction mediated localization of RFLP clones to microisolated translocation chromosomes of barley. Genome 37:550–555

Stephens JL, Brown SE, Lapitan NLV et al (2004) Physical mapping of barley genes using an ultrasensitive fluorescence in situ hybridization technique. Genome 47:179–189

Suchánková P, Kubaláková M, Kovárová P et al (2006) Dissection of the nuclear genome of barley by chromosome flow sorting. Theor Appl Genet 113:651–659

Taketa S, Takeda K (2001) Production and characterization of a complete set of wheat-wild barley (*Hordeum vulgare* ssp. *spontaneum*) chromosome addition lines. Breed Sci 51:199–206

Taketa S, Choda M, Ohashi R et al (2002) Molecular and physical mapping of a barley gene on chromosome arm 1HL that causes sterility in hybrids with wheat. Genome 45:617–625. https://doi.org/10.1139/g02-024

Taketa S, Linde-Laursen I, Künzel G (2003) Cytogenetic diversity. In: von Bothmer R, van Hintum Th, Knüppffer H, Sato K (eds) Diversity in barley (*Hordeum vulgare*). Elsevier Science BV, Amsterdam, The Netherlands, pp 97–119

Tsuchiya T (1969) Status of studies of primary trisomics and other aneuploids in barley. Genetica 40:216

Tsuchiya T (1991) Chromosome mapping by means of aneuploid analysis in barley. In: Gupta PK, Tsuchiya T (eds) Chromosome engineering in plants: genetics, breeding, evolution part A. Elsevier, pp 361–384

Tsujimoto H, Mukai Y, Akagawa K et al (1997) Identification of individual barley chromosomes based on repetitive sequences: conservative distribution of Afa-family repetitive sequences on the chromosomes of barley and wheat. Genes Genet Syst 72:303–309

Valarik M, Bartos J, Kovarova P et al (2004) High-resolution FISH on super-stretched flow-sorted plant chromosomes. Plant J 37:940–950

Wicker T, Krattinger SG, Lagudah ES et al (2009) Analysis of intraspecies diversity in wheat and barley genomes identifies breakpoints of ancient haplotypes and provides insight into the structure of diploid and hexaploid triticeae gene pools. Plant Physiol 149:258–270

Chromosomal Genomics of Barley

4

Hana Šimková, Marie Kubaláková, Jan Vrána, Petr Cápal and Jaroslav Doležel

Abstract

Barley (*Hordeum vulgare* L.) genome has been difficult to analyze, map, and sequence due to its size ($\sim$5 Gb/1C) and high repeat content (over 80%). Flow cytometric sorting of plant mitotic chromosomes, which was first reported in 1984, provided an approach for reduction of complexity to simplify genome analysis and laid the basis of a new discipline—chromosomal genomics. Chromosome flow sorting in barley was established in 1999 and since that time played an important role in genomics of this crop. Initially, flow-sorted chromosomes facilitated assignment of DNA markers to chromosomes or their regions, providing a complement and validation to genetic mapping. A protocol for multiple displacement amplification of chromosomal DNA, established for mapping on Illumina genotyping platform, was shown compatible with next-generation sequencing. This enabled identification of gene content of individual barley chromosomes and establishment of their putative order by an innovative approach, which was later followed also in wheat and rye. Preparation of high-molecular-weight DNA from flow-sorted chromosomes, while developed in barley, opened avenues to constructing chromosome-specific BAC libraries in bread wheat and, more recently, generating optical maps in several cereal species. Thus, the development of a chromosome genomics toolbox contributed to the advances in genomics of other important crops, in addition to barley. Chromosomal genomics plays an important role also in the post-sequencing era by enabling cost-efficient gene cloning, analyzing chromatin structure by chromatin conformation capture approaches, and characterizing the proteome of mitotic chromosomes.

Abbreviations

BAC	Bacterial artificial chromosome
HMW DNA	High-molecular-weight DNA
FISH	Fluorescence in situ hybridization
FISHIS	Fluorescence in situ hybridization in suspension
MDA	Multiple displacement amplification
NGS	Next-generation sequencing
OPA	Oligonucleotide Pool Assay
RFLP	Restriction fragment length polymorphism
STS	Sequence-tagged site

H. Šimková (✉) · M. Kubaláková · J. Vrána
P. Cápal · J. Doležel
Institute of Experimental Botany, Šlechtitelů 31, Olomouc, Czech Republic
e-mail: simkovah@ueb.cas.cz

N. Stein and G. J. Muehlbauer (eds.), *The Barley Genome*, Compendium of Plant Genomes,
https://doi.org/10.1007/978-3-319-92528-8_4

4.1 Development of Barley Chromosomal Genomics

Genomes of many important crops, including cereals of the tribe Triticeae, are difficult to analyze, map, and sequence due to their complexity and a high content of repetitive DNA sequences. Genome size of ~5 Gb/1C ranks barley to the largest genomes among cereal crops, challenged only by rye (*Secale cereale*, ~8 Gb/1C), pasta wheat (*Triticum durum* L., ~12 Gb/1C), and bread wheat (*Triticum aestivum* L., ~17 Gb/1C). Dissection of nuclear genomes into smaller and well-defined parts such as chromosomes or chromosome arms enabled complexity reduction by about one order of magnitude and offered an attractive approach to tackle huge genomes.

First attempts to dissect barley genome and analyze it *per partes* appeared in 1992 with applying chromosome microdissection technique (Fukui et al. 1992). The microdissected chromosomes were used for targeted development of RFLP markers (Schondelmaier et al. 1993) and probes for FISH (Busch et al. 1995) and also for mapping STS markers to physically defined chromosome segments by using DNA of microdissected translocated barley chromosomes as a template for PCR (Sorokin et al. 1994; Künzel et al. 2000). The advantage of microdissection techniques is that the operator can isolate single chromosomes of interest and even their parts. However, these techniques are not suitable for downstream applications that require high-molecular-weight DNA. In fact, even the quality of DNA amplified from microdissected chromosomes is rather low and DNA fragments obtained after amplification are typically only up to several kilobases in length (Fominaya et al. 2005; Hobza and Vyskot 2007; Zhou and Hu 2007; Datta et al. 2015). This is likely a consequence of the harsh treatments of chromosomes during the fixation step and preparation of metaphase spreads on microscope slides.

Over the years, it became obvious that flow cytometry is the method of choice for providing large quantities of pure chromosome fractions that are suitable for molecular methods demanding high DNA amount and quality (reviewed in Doležel et al. 2014). Flow cytometry is a technique based on measuring optical properties of microscopic particles (e.g., cells, cellular organelles, chromosomes, bacteria, viruses, etc.) flowing one by one through a strong light beam (usually laser). If the particles differ in optical parameters (light scatter, fluorescence intensity), they can be discriminated into individual classes (populations). Each population can be subsequently isolated, usually by means of electrostatic deflection of droplets carrying particles from the population of interest (see Fig. 4.1 for basic principles of flow cytometry).

Flow cytometric analysis and sorting of metaphase chromosomes (collectively called flow cytogenetics) was introduced to plant genomic research in 1984 (De Laat and Blaas 1984) and over the years became an indispensable tool with published protocols for more than 40 plant species (Vrána et al. 2016). Like in other plant species, there were several issues in barley to be resolved for enabling large-scale purification of mitotic chromosomes, namely, (a) obtaining sufficient amount of actively dividing cells with condensed chromosomes, (b) release of mitotic metaphase chromosomes from cells with rigid walls and (c) discrimination of individual chromosome types. The challenges in preparation of suspensions of intact mitotic chromosomes were overcome by Doležel et al. (1992) who developed a chromosome isolation method based on synchronization of root-tip meristem cells and subsequent release of metaphase chromosomes from formaldehyde-fixed root tips by tissue chopping. The manual tissue chopping was later replaced by mechanical homogenization (Gualberti et al. 1996), and the procedure was optimized for barley (*Hordeum vulgare* L.) by Lysák et al. (1999). In a barley cultivar with a standard karyotype, the authors succeeded in discriminating and sorting only the smallest chromosome (1H) (Fig. 4.2a), but up to three chromosomes could be sorted from translocation lines carrying morphologically distinct chromosomes.

To implement the flow sorting as a technique of choice for broader chromosome-based studies,

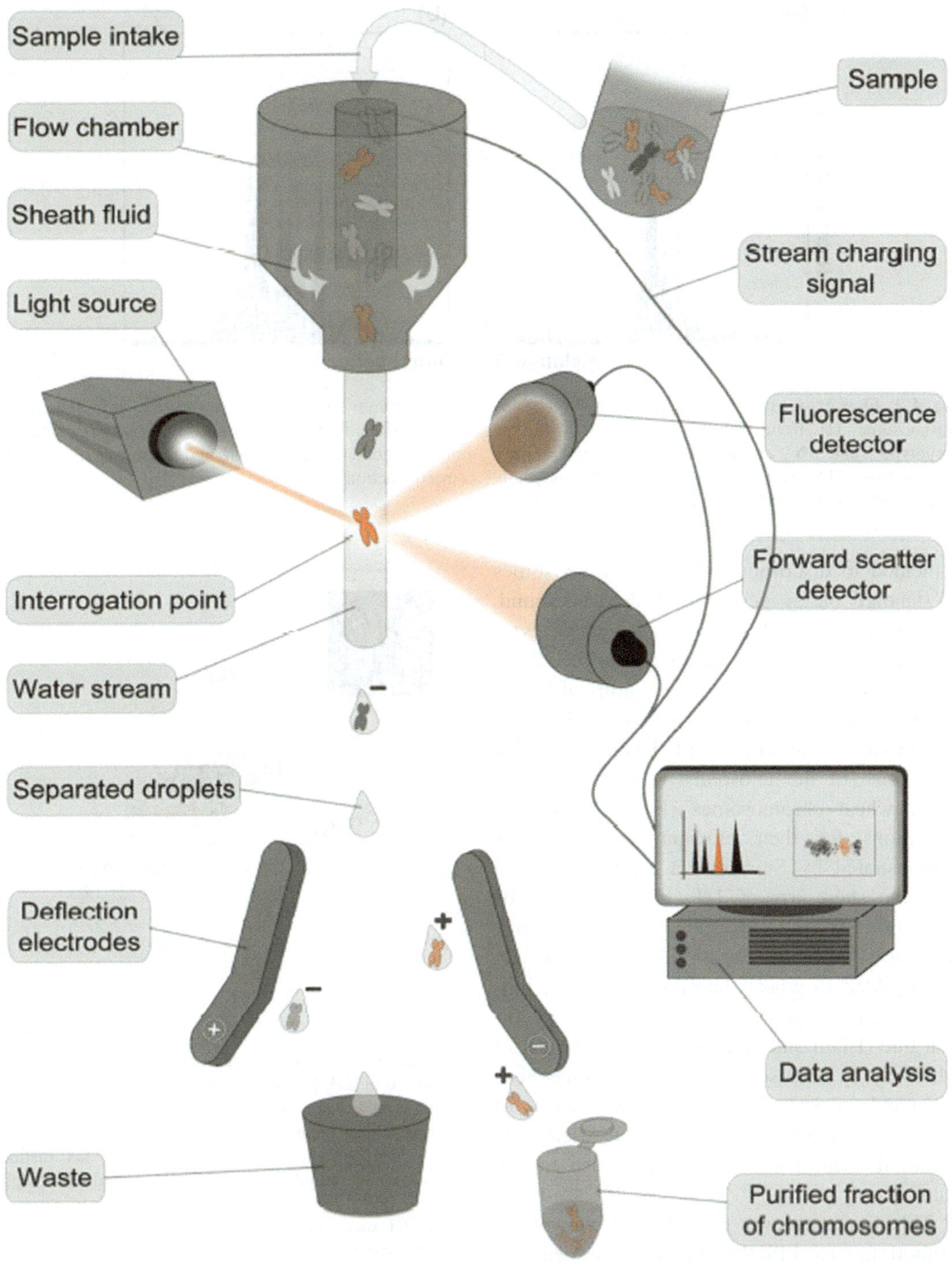

Fig. 4.1 Basic principle of flow cytometry. A suspension of intact chromosomes is introduced into the flow chamber, where the particles are arranged in a single file and at high speed intercept one-by-one a strong light beam. Optical signals generated from each chromosome (light scatter and fluorescence) are electronically processed and used to classify chromosomes into populations with similar properties. The population of interest is physically isolated from the rest by deflecting electrically charged droplets during the passage through a strong electrostatic field

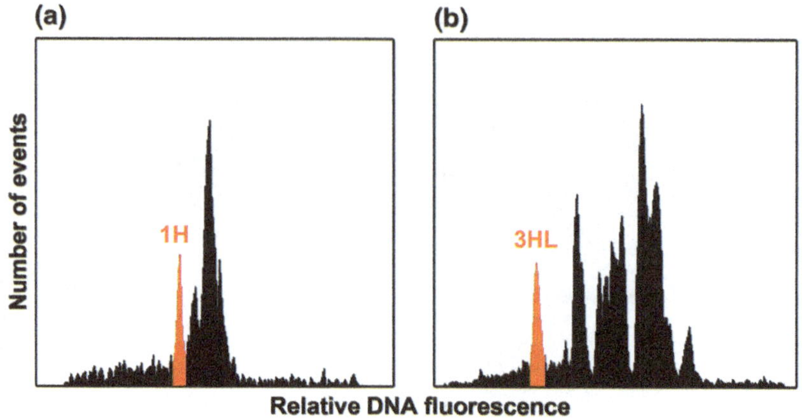

Fig. 4.2 Flow cytometric analysis and sorting of **a** barley chromosome 1H from a barley line with wild-type karyotype and **b** chromosome arm 3HL from wheat-barley addition line, in which the 3HL is stably maintained as a telosome. Significantly smaller size of the chromosome 1H and the chromosome arm 3HL, compared to the remaining chromosomes in both samples, enables good discrimination and a high purity of the sorted fraction

a method to discriminating other chromosomes in standard barley genotypes had to be found. This issue was resolved by using a set of cytogenetic stocks—wheat-barley addition lines, carrying individual chromosomal arms of barley (2HS–7HL) as telosomes in a bread wheat background (Islam et al. 1981; Islam 1983). The small size of the telosomes in comparison to bread wheat chromosomes (Fig. 4.2b) creates conditions for their efficient purification as demonstrated by Suchánková et al. (2006). In fact, the short arm of 1H (1HS) can also be flow-sorted from a corresponding addition line, in contrast to the long arm of 1H (1HL), which, when added to wheat, introduces sterility, thus impeding generation of 1HL wheat-barley addition line (Islam and Shepherd 2000). The set of wheat-barley ditelosomic addition lines has been generated for individual chromosomal arms of cv. Betzes, which for a long time restricted the use of the chromosome-based approach to that cultivar only.

This limitation was recently overcome by a method termed FISHIS (FISH In Suspension) developed by Giorgi et al. (2013), in which mitotic chromosomes are differentially labeled by hybridizing them with fluorescently-labeled oligonucleotide probes targeting specific microsatellite sequences, typically GAA

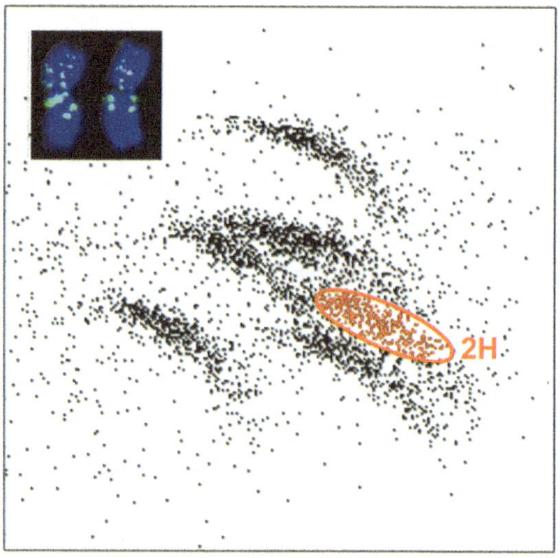

Fig. 4.3 Flow cytometric analysis and sorting of barley chromosome 2H. Suspension of metaphase chromosomes of barley cv. Foma was labeled by a probe for GAA microsatellites using FISHIS and counterstained with DAPI. Biparametric analysis discriminated six distinct chromosomal populations with 2H population highlighted. Inset: The population corresponding to chromosome 2H was sorted, and the purity was estimated by microscopic observation of the fluorescent labeling pattern. Y axis: GAA-FITC fluorescence; X axis: DAPI fluorescence

(Fig. 4.3). This approach adds another parameter for discriminating individual chromosome types and enables obtaining sorted fractions highly enriched in the desired chromosome from any cultivar of barley.

Hundred per cent purity of a particular chromosome type can then be achieved by single-chromosome sorting (Cápal et al. 2015), which is based on sorting single anonymous chromosomes that are identified only after amplifying their DNA, either by the use of chromosome-specific markers or after sequencing. Although this approach may appear similar to chromosome microdissection, the principal difference lies in the quality of DNA obtained. DNA of flow-sorted chromosomes is intact and well accessible to DNA polymerases as they are not subjected to the harsh treatments as are the chromosomes during the preparation of slides for microdissection (Hobza and Vyskot 2007).

4.2 Applications of Flow Cytogenetics

4.2.1 Physical Mapping

The first application of flow cytogenetics in barley, the PCR-based physical mapping, was inspired by the microdissection-based approach proposed by Sorokin et al. (1994) and relied on small numbers of chromosomes. Running PCR on 500 chromosomes flow-sorted from three barley translocation lines enabled assigning barley STS markers to the translocated segments (Lysák et al. 1999), similarly as in the study of Künzel et al. (2000). The flow cytogenetics approach was easy and straightforward as flow sorting hundreds of chromosomes of one type took only a few minutes (Fig. 4.4).

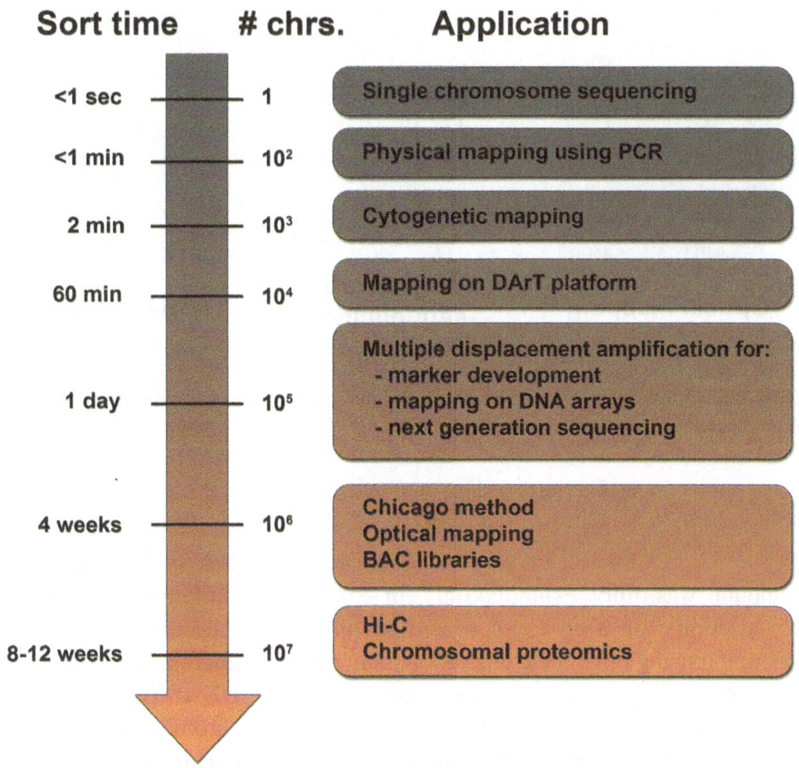

Fig. 4.4 Portfolio of chromosomal genomics applications as developed in cereals. The amounts of chromosomes needed for particular applications together with the corresponding sorting time, relating to purification of a particular chromosome type, are shown on the left side

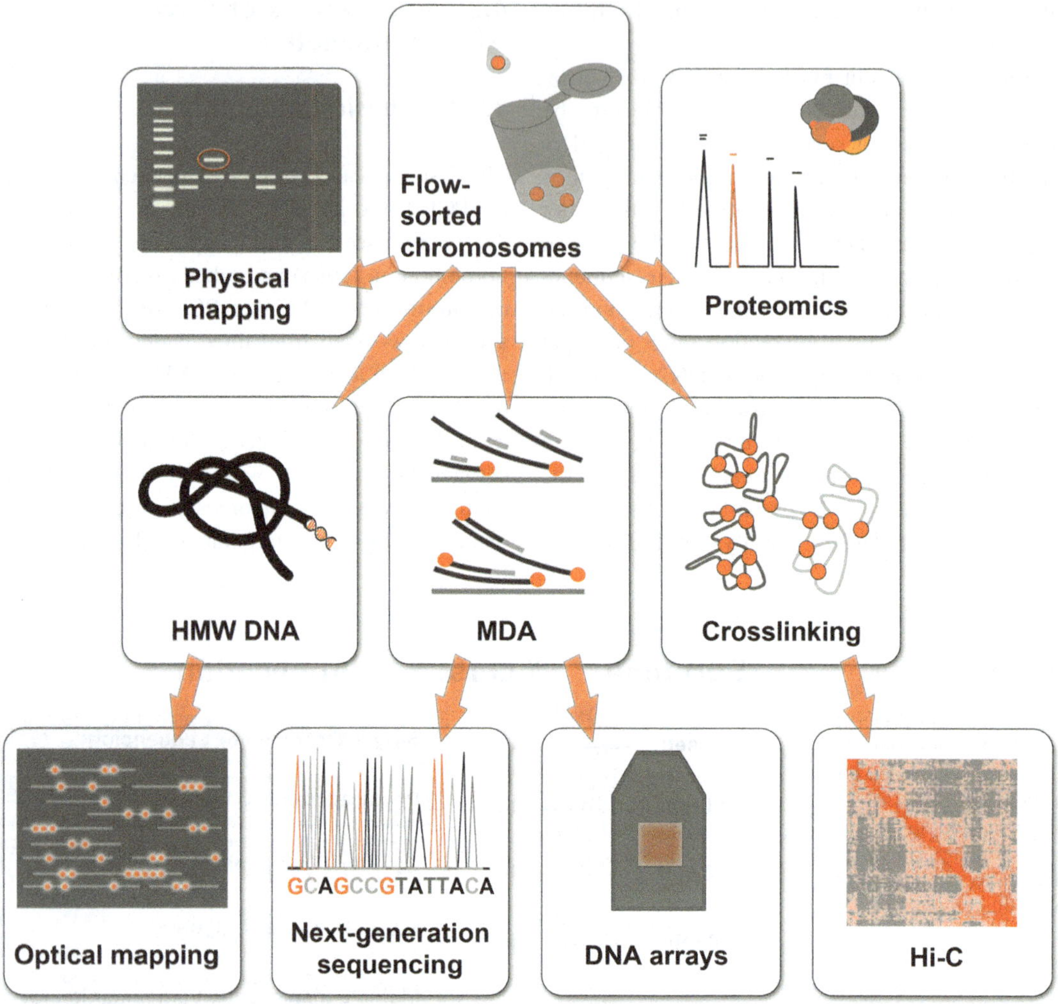

Fig. 4.5 Downstream applications of flow-sorted chromosomes established in barley. The NGS data were further used for generating blueprints of individual barley chromosomes, constructing genome zippers or gene cloning

Nevertheless, the potential of flow cytogenetics to surpass microdissection in terms of DNA amount was fully exploited only in 2008, in connection with a pilot experiment on physical mapping using oligonucleotide pool assay (OPA, Rostoks et al. 2006). Yet, the Illumina SNP genotyping platform used for the experiment required micrograms of DNA, corresponding to millions of chromosomes, which were laborious to purify even using flow cytometry (Fig. 4.4). To significantly reduce the sorting time, a DNA amplification step was proposed, and the method of choice became the isothermal multiple displacement amplification (MDA) (Fig. 4.5), exploiting the highly accurate Phi29 polymerase (Dean et al. 2002). This procedure yielded micrograms of amplified DNA from 10 ng input DNA. To test the compatibility of flow-sorted chromosomes with MDA, a protocol combining purification of chromosomal DNA with isothermal amplification by GenomiPhi DNA Amplification Kit (GE Healthcare, UK) was developed (Šimková et al. 2008) and applied to prepare 1H-specific DNA with fragment sizes ranging from 5 to 30 kb. Assaying the amplification product by quantitative PCR for 10 genes located on various chromosomes showed high purity of the sorted fraction and low quantitative amplification bias among the genes studied (maximum 2.3-fold). A whole-genome analysis was then performed by pilot OPA for interrogation of 1524 SNP loci across the barley genome. Comparison of unamplified genomic DNA of barley cv. Akcent with three replicates of amplified genomic DNA revealed >99% concordance. Besides, assaying amplified 1H-specific DNA by OPA confirmed the 1H position for all loci assigned to this chromosome by linkage mapping and mapped 40 new loci to 1H (Šimková et al. 2008). The study, showing high fidelity and representativeness of the Phi29-amplified chromosomal DNA, charted an easy way to obtaining microgram amounts of chromosome-specific DNA and opened avenues to multiple chromosomal studies not only in barley but also in other cereals, grasses, and legumes.

An identical approach and SNP genotyping platform were used in a follow-up study of Muñoz-Amatriaín et al. (2011) who aimed to produce an improved consensus linkage map for barley, combining data from 10 populations totaling 1133 individuals. The authors used DNA from barley chromosome 1H and from the arms of barley chromosomes 2H–7H. Chromosome DNA samples were amplified by an improved Illustra GenomiPhi V2 DNA Amplification Kit (GE Healthcare, UK) and the amplification products were applied on the same OPA platforms as used for the linkage mapping (Close et al. 2009). A novel application of the genotyping platform for gene detection allowed the assignment of 2930 SNP-containing genes to flow-sorted chromosomes or arms, confirmed the position of 2545 SNP-mapped loci, added chromosome or arm allocations to an additional 370 SNP loci, and delimited pericentromeric regions for chromosomes 2H–7H. The knowledge of the arm, to which each SNP mapped, resolved several recurring data clustering patterns, which previously had uncertain interpretations, and thus improved the precision of the barley consensus map.

Amplified chromosomal DNA was also shown compatible with a long-oligonucleotide array. A complete set of samples originating from individual barley chromosome arms and chromosome 1H was one-by-one hybridized on an Agilent microarray with 42,302 barley sequences representing around 30,000 nonredundant barley genes. This allowed assigning 16,804 genes from the array to individual chromosome arms at high confidence, providing a complement and confirmation to data generated by next-generation sequencing (Mayer et al. 2011).

Physical mapping using chromosome-arm-specific DNA proved a useful complement to genetic mapping also in case of the diversity arrays technology (DArT) platform (Jaccoud et al. 2001). The technology requires only a few nanograms of input DNA but relies on methylation patterns, which are not reproduced by the Phi29 polymerase. Thus, several nanograms of nonamplified DNA of individual barley chromosome/arms were used to prepare genomic representations that, after analysis on the barley DArT array, helped in validating the chromosome assignment of individual DArT markers (A. Kilian, pers. comm.).

4.2.2 Next-Generation Sequencing

A possibility to produce chromosome-(arm)-specific DNA in sufficient amount and quality stimulated attempts toward obtaining sequences of individual barley chromosomes or even arms. The advent of next-generation sequencing (NGS) technologies, featured by high throughput and relatively low cost, made this effort feasible and affordable.

The pilot study used DNA amplified from flow-sorted chromosome 1H and employed by that time the most advanced NGS technology—Roche 454, to generate sequence data representing 1.3-fold chromosome coverage (Mayer et al. 2009). Sequence comparison against the reference genomes of rice and sorghum, and against wheat and barley EST datasets, led to the estimation of 4600–5800 genes on chromosome

1H. The integration of shotgun sequencing information of barley chromosome 1H with the collinear gene order of orthologous rice and sorghum genes allowed proposing a virtual gene order map of the chromosome, including 1987 loci. The syntenic integration also allowed ordering genes in regions with limited genetic resolution such as subcentromeric and centromeric regions, which are intractable to genetic mapping.

The same approach was used later by Mayer et al. (2011) for all barley chromosomes. The low-pass 454-acquired sequences (2.2-fold average coverage) of individual barley chromosome arms and the 1H were combined with the available full-length cDNA sequences and microarray hybridization data, and the set of model grass genomes was supplemented by *Brachypodium*. This effort allowed 86% of the estimated ~32,000 barley genes to be assigned to individual chromosome arms. As a result of a conserved synteny-based modeling, a series of so-called genome zippers were constructed, which put in a putative linear order 21,766 barley genes, including 3125 (14%) genes belonging to nonrecombining genetic centromeres. The integrative approach using three model genomes in parallel helped to overcome limitations imposed by species-specific regional restructuring, made the prediction of the gene order highly reliable and laid a basis for a large-scale comparative genomics study across the whole barley genome.

Sequencing flow-sorted chromosomes combined with conserved synteny-based ordering inspired other scientific communities working on nonsequenced cereal or grass species and lead to generation of chromosomal draft sequences and construction of genome zippers for rye (Martis et al. 2013), fescue (Kopecký et al. 2013), and bread wheat (IWGSC 2014). The genome zippers proved an excellent tool for evolutionary studies, a favorable genomic resource for anchoring physical maps in poorly recombining regions, and for targeted marker development.

Illumina-sequenced DNAs of individual barley chromosomes were used in a study of Muñoz-Amatriaín et al. (2015) that described the sequencing of 15,622 gene-bearing bacterial artificial chromosome (BAC) clones aiming to characterize gene-dense regions of the barley genome. Assembled sequences of individual barley arms and chromosome 1H allocated 15,216 BAC assemblies (97.4% of those sequenced) at high confidence to chromosome 1H and arms of chromosomes 2H–7H. The number of assigned BACs per chromosome ranged from 1936 for 6H to 2439 for 2H, proportional to the molecular size of the corresponding barley chromosome arm as reported by Suchánková et al. (2006). Using the same dataset of chromosomal shotgun assemblies, Ounit et al. (2015) assigned 125 (36 at high confidence) out of 50,646 barley unigenes (Close et al. 2006) to centromeric regions of 2H–7H, which were identified as overlaps between assemblies of short and long arms of particular chromosomes.

4.2.3 Preparation of High-Molecular-Weight DNA and Its Applications

The above studies (Mayer et al. 2009, 2011; Muñoz-Amatriaín et al. 2015) demonstrated compatibility of Phi29-amplified chromosomal DNA with Roche-454 and Illumina sequencing technologies. The amplified DNA, however, showed problematic for mate-pair sequencing with insert sizes >3 kb, probably due to chimerism introduced by MDA (Belova et al. 2013). Moreover, third-generation sequencing technologies, such as single-molecule real-time (SMRT) sequencing developed by Pacific Biosciences and a nanopore-based technology commercialized by Oxford Nanopore, producing reads of tens kilobases, require input DNA exceeding 50–100 kb in length. Such high-molecular-weight (HMW) DNA cannot be obtained after MDA amplification of chromosomal DNA and must be prepared through a special procedure that includes embedding intact flow-sorted chromosomes in agarose plugs, in which chromosomal DNA is gently purified while avoiding mechanical breakage. Such procedure was developed for flow-sorted chromosomes as early as in 2003 and provided HMW DNA of hundreds to thousands of kilobases in length (Šimková et al. 2003). While optimized in barley, it was most frequently applied in wheat, where HMW DNA was used to construct the complete set of chromosome arm-specific BAC libraries (Šafář et al. 2010, http://olomouc.ueb.cas.cz/dna-libraries/cereals).

The ability to prepare HMW DNA from flow-sorted chromosomes was exploited in barley only 12 years later for optical mapping, a technology visualizing short sequence motives along DNA molecules hundreds of kilobases in length (Fig. 4.5). HMW chromosomal DNA showed to be an optimal input material for the optical mapping platform of BioNano Genomics (Staňková et al. 2016), outperforming HMW DNA preparations from nuclei purified by flow cytometry (H. Šimková, unpublished). Thus, chromosomes purified by flow sorting became the material of choice for generating barley whole-genome optical map assembly through the BioNano platform (Beier et al. 2017; Mascher et al. 2017). In this work, all barley chromosomes (1H–7H) were sorted simultaneously. Besides, chromosome 1H was also flow-sorted separately and used to construct a chromosomal optical map on the same platform. While the parameters of the whole-genome and the 1H assemblies were similar, the chromosomal assembly featured a lower level of chimerism (H. Toegelová, unpublished).

We anticipate that HMW DNA prepared from flow-sorted/flow-purified chromosomes will find a wider use in combination with new sequencing technologies and approaches that rely on long DNA. Due to laboriousness of preparing multiple micrograms of nonamplified DNA (Fig. 4.4), the flow cytogenetics couples best with technologies less demanding in DNA amount, such as nanopore sequencing of Oxford Nanopore Technologies (input ≥ 400 ng), GemCode Technology of 10× Genomics (input ≥ 20 ng), or the Chicago method of Dovetail Genomics (input ≥ 500 ng).

4.2.4 Chromosomal Genomics in the Post-sequencing Era

The advent of long-read sequencing technologies and new assembly algorithms, which can produce high-quality de novo assemblies even for very large and complex genomes, reduced the need to dissect the genome for obtaining whole-genome sequences. Yet, there are areas of research, where the reduction of complexity, enabled by purification of a specific chromosome type, simplifies the analyses and makes the studies more focused and affordable.

Purification and sequencing of a target chromosome from several mutants laid a basis of a new approach to gene isolation in cereals called MutChromSeq (Sánchez-Martín et al. 2016). The efficiency of the approach was demonstrated on a previously cloned barley *Eceriferum-q* gene for epicuticular aliphatic wax accumulation. Chromosome 2H, to which the gene had been assigned, was flow-sorted after labeling by FISHIS from the wild-type cultivar Foma (Fig. 4.3) as well as from six ethyl methanesulphonate (EMS)-derived wax-less mutants. The chromosomal DNA was Phi29-amplified and sequenced on Illumina platform. A de novo assembly of 2H from Foma was produced, and sequence reads from the six mutants were compared to the wild-type assembly, which resulted in a rapid identification of a single candidate gene based on mutation overlap. This demonstrated the power of MutChromSeq to identify induced, causal mutations without the need for positional fine mapping. The approach showed also beneficial for gene cloning in bread wheat (Sánchez-Martín et al. 2016) and is applicable in any plant species where the flow cytogenetics has been developed (Vrána et al. 2016) and the chromosome of interest can be discriminated and sorted.

The advent of chromosome conformation capture approaches such as Hi-C (Lieberman-Aiden et al. 2009) provided means to unraveling three-dimensional (3D) organization of DNA in cell nuclei and its impact on genome function and gene expression. These methods employ crosslinking of chromatin in the nucleus, enzymatic DNA digestion, proximity ligation, and massively parallel sequencing to construct spatial proximity maps of a genome (Fig. 4.5). In contrast to the advances in understanding 3D organization of cell nuclei, the way DNA molecules are arranged in condensed mitotic chromosomes remains obscure. The ability to prepare purified fractions of mitotic chromosomes provides a unique opportunity to study their architecture by chromatin conformation capture approaches. The procedure was recently established for barley chromosomes (T. Beseda, unpublished), and it is expected that it will lead to solving the mystery.

The way the nuclear genome is organized into mitotic chromosomes cannot be fully understood without characterization of their proteome. Apart from the assembly of chromosomes consisting of sister chromatids, chromosomal proteins also play a major role in their behavior during mitosis. Despite the general importance, a very small number of mitotic chromosome proteins have been identified to date. In order to characterize the proteome of barley nuclei in different phases of the cell cycle, Blavet et al. (2017) performed mass spectrometry of proteins isolated from nuclei purified by flow sorting. The work is in progress to characterize the proteome of barley mitotic chromosomes purified by flow sorting, which provides an elegant approach to avoid contamination by cytoplasmic proteins (Petrovská et al., unpublished).

Acknowledgements We thank our colleagues Helena Toegelová, Tomáš Beseda, and Beáta Petrovská for sharing their unpublished results. This work was supported by the Czech Ministry of Education, Youth and Sports (award LO1204 from the National Program of Sustainability I).

References

Beier S, Himmelbach A, Colmsee C, Zhang X-Q, Barrero RA, Zhang Q, Li L, Bayer M, Bolser D, Taudien S, Groth M, Felder M, Hastie A, Šimková H, Staňková H, Vrána J, Chan S, Muñoz-Amatriaín M, Ounit R, Wanamaker S, Schmutzer T, Aliyeva-Schnorr L, Grasso S, Tanskanen J, Sampath D, Heavens D, Cao S, Chapman B, Dai F, Han Y, Li H, Li X, Lin C, McCooke JK, Tan C, Wang S, Yin S, Zhou G, Poland JA, Bellgard MI, Houben A, Doležel J, Ayling S, Lonardi S, Langridge P, Muehlbauer GJ, Kersey P, Clark MD, Caccamo M, Schulman AH, Platzer M, Close TJ, Hansson M, Zhang G, Braumann I, Li C, Waugh R, Scholz U, Stein N, Mascher M (2017) Construction of a map-based reference genome sequence for barley, *Hordeum vulgare* L. Sci Data 4:170044

Belova T, Zhan BJ, Wright J, Caccamo M, Asp T, Šimková H, Kent M, Bendixen C, Panitz F, Lien S, Doležel J, Olsen OA, Sandve SR (2013) Integration of mate pair sequences to improve shotgun assemblies of flow-sorted chromosome arms of hexaploid wheat. BMC Genomics 14:222

Blavet N, Uřinovská J, Jeřábková H, Chamrád I, Vrána J, Lenobel R, Beinhauer J, Šebela M, Doležel J, Petrovská B (2017) UNcleProt (Universal Nuclear Protein database of barley): the first nuclear protein database that distinguishes proteins from different phases of the cell cycle. Nucleus 8:70–80

Busch W, Martin R, Herrmann RG, Hohmann U (1995) Repeated DNA sequences isolated by microdissection. I. Karyotyping of barley (*Hordeum vulgare* L.). Genome 38:1082–1090

Cápal P, Blavet N, Vrána J, Kubaláková M, Doležel J (2015) Multiple displacement amplification of the DNA from single flow-sorted plant chromosome. Plant J 84:838–844

Close TJ, Wanamaker S, Roose ML, Lyon M (2006) HarvEST. Methods Mol Biol 406:161–177

Close TJ, Bhat PR, Lonardi S, Wu YH, Rostoks N, Ramsay L, Druka A, Stein N, Svensson JT, Wanamaker S, Bozdag S, Roose ML, Moscou MJ, Chao SAM, Varshney RK, Szucs P, Sato K, Hayes PM, Matthews DE, Kleinhofs A, Muehlbauer GJ, DeYoung J, Marshall DF, Madishetty K, Fenton RD, Condamine P, Graner A, Waugh R (2009) Development and implementation of high-throughput SNP genotyping in barley. BMC Genomics 10:582

Datta S, Malhotra L, Dickerson R, Chaffee S, Sen CK et al (2015) Laser capture microdissection: big data from small samples. Histol Histopathol 30:1255–1269

De Laat AMM, Blaas J (1984) Flow cytometric characterization and sorting of plant chromosomes. Theor Appl Genet 67:463–467

Dean FB, Hosono S, Fang L, Wu X, Faruqi AF, Bray-Ward P, Sun Z, Zong Q, Du Y, Du J, Driscoll M, Song W, Kingsmore SF, Egholm M, Lasken RS (2002) Comprehensive human genome amplification using multiple displacement amplification. Proc Natl Acad Sci USA 99:5261–5266

Doležel J, Číhalíková J, Lucretti S (1992) A high-yield procedure for isolation of metaphase chromosomes from root tips of *Vicia faba* L. Planta 188:93–98

Doležel J, Vrána J, Cápal P, Kubaláková M, Burešová V, Šimková H (2014) Advances in plant chromosome genomics. Biotechnol Adv 32:122–136

Fominaya A, Linares C, Loarce Y, Ferrer E (2005) Microdissection and microcloning of plant chromosomes. Cytogenet Genome Res 109:8–14

Fukui K, Minezawa M, Kamisugi Y, Ishikawa M, Ohmido N, Yanagisawa T, Fujishita M, Sakai F (1992) Microdissection of plant chromosomes by argon-ion laser-beam. Theor Appl Genet 84:787–791

Giorgi D, Farina A, Grosso V, Gennaro A, Ceoloni C, Lucretti S (2013) FISHIS: fluorescence *in situ* hybridization in suspension and chromosome flow sorting made easy. PLoS One 8:e57994

Gualberti G, Doležel J, Macas J, Lucretti S (1996) Preparation of pea (*Pisum sativum* L.) chromosome and nucleus suspensions from single root tips. Theor Appl Genet 92:744–751

Hobza R, Vyskot B (2007) Laser microdissection-based analysis of plant sex chromosomes. Methods Cell Biol 82:433–453

International Wheat Genome Sequencing Consortium (2014) A chromosome-based draft sequence of the hexaploid wheat (*Triticum aestivum*) genome. Science 345:1251788

Islam AKMR (1983) Ditelosomic additions of barley chromosomes to wheat. In: Sakamoto S (ed) Proceedings of 6th international wheat genetics symposium, Kyoto University Press, Kyoto, Japan, pp 233–238

Islam AKMR, Shepherd KW (2000) Isolation of a fertile wheat–barley addition line carrying the entire barley chromosome 1H. Euphytica 111:145–149

Islam AKMR, Shepherd KW, Sparrow DHB (1981) Isolation and characterization of euplasmic wheat–barley chromosome addition lines. Heredity 46:161–174

Jaccoud D, Peng K, Feinstein D, Kilian A (2001) Diversity arrays: a solid state technology for sequence information independent genotyping. Nucl Acids Res 29:e25

Kopecký D, Martis M, Číhalíková J, Hřibová E, Vrána J, Bartoš J, Kopecká J, Cattonaro F, Stočes Š, Novák P, Neumann P, Macas J, Šimková H, Studer B, Asp T, Baird JH, Navrátil P, Karafiátová M, Kubaláková M, Šafář J, Mayer KFX, Doležel J (2013) Flow sorting and sequencing meadow fescue chromosome 4F. Plant Physiol 163:1323–1337

Künzel G, Korzun L, Meister A (2000) Cytologically integrated physical restriction fragment length polymorphism maps for the barley genome based on translocation breakpoints. Genetics 154:397–412

Lieberman-Aiden E, van Berkum NL, Williams L, Imakaev M, Ragoczy T, Telling A, Amit I, Lajoie BR, Sabo PJ, Dorschner MO, Sandstrom R, Bernstein B, Bender MA, Groudine M, Gnirke A, Stamatoyannopoulos J, Mirny LA, Lander ES, Dekker J (2009) Comprehensive mapping of long-range interactions reveals folding principles of the human genome. Science 326:289–293

Lysák MA, Číhalíková J, Kubaláková M, Šimková H, Künzel G, Doležel J (1999) Flow karyotyping and sorting of mitotic chromosomes of barley (*Hordeum vulgare* L.). Chromosome Res 7:431–444

Martis MM, Zhou R, Haseneyer G, Schmutzer T, Vrána J, Kubaláková M, König S, Kugler KG, Scholz U, Hackauf B, Korzun V, Schön CC, Doležel J, Bauer E, Mayer KFX, Stein N (2013) Reticulate evolution of the rye genome. Plant Cell 25:3685–3698

Mascher M, Gundlach H, Himmelbach A, Beier S, Twardziok SO, Wicker T, Radchuk V, Dockter C, Hedley PE, Russell J, Bayer M, Ramsay L, Liu H, Haberer G, Zhang X-Q, Zhang Q, Barrero RA, Li L, Taudien S, Groth M, Felder M, Hastie A, Šimková H, Staňková H, Vrána J, Chan S, Muñoz-Amatriaín M, Ounit R, Wanamaker S, Bolser D, Colmsee C, Schmutzer T, Aliyeva-Schnorr L, Grasso S, Tanskanen J, Chailyan A, Sampath D, Heavens D, Clissold L, Cao S, Chapman B, Dai F, Han Y, Li H, Li X, Lin C, McCooke JK, Tan C, Wang P, Wang S, Yin S, Zhou G, Poland JA, Bellgard MI, Borisjuk L, Houben A, Doležel J, Ayling S, Lonardi S, Kersey P, Langridge P, Muehlbauer GJ, Clark MD, Caccamo M, Schulman AH, Mayer KFX, Platzer M, Close TJ, Scholz U, Hansson M, Zhang G, Braumann I, Spannagl M, Li C, Waugh R, Stein N (2017) A chromosome conformation capture ordered sequence of the barley genome. Nature 544:427–433

Mayer KFX, Taudien S, Martis M, Šimková H, Suchánková P, Gundlach H, Wicker T, Petzold A, Felder M, Steuernagel B, Scholz U, Graner A, Platzer M, Doležel J, Stein N (2009) Gene content and virtual gene order of barley chromosome 1H. Plant Physiol 151:496–505

Mayer KFX, Martis M, Hedley PE, Šimková H, Liu H, Morris JA, Steuernagel B, Taudien S, Roessner S, Gundlach H, Kubaláková M, Suchánková P, Murat F, Felder M, Nussbaumer T, Graner A, Salse J, Endo TR, Sakai H, Tanaka T, Itoh T, Sato K, Platzer M, Matsumoto T, Scholz U, Doležel J, Waugh R, Stein N (2011) Unlocking the barley genome by chromosomal and comparative genomics. Plant Cell 23:1249–1263

Muñoz-Amatriaín M, Moscou MJ, Bhat PR, Svensson JT, Bartoš J, Suchánková P, Šimková H, Endo TR, Fenton RD, Lonardi S, Castillo AM, Chao S, Cistué L, Cuesta-Marcos A, Forrest KL, Hayden MJ, Hayes PM, Horsley RD, Makoto K, Moody D, Sato K, Vallés MP, Wulff BBH, Muehlbauer GJ, Doležel J, Close TJ (2011) An improved consensus linkage map of barley based on flow-sorted chromosomes and single nucleotide polymorphism markers. Plant Genome 4:238–249

Muñoz-Amatriaín M, Lonardi S, Luo M-C, Madishetty K, Svensson JT, Moscou MJ, Wanamaker S, Jiang T, Kleinhofs A, Muehlbauer GJ, Wise RP, Stein N, Ma Y, Rodriguez E, Kudrna D, Bhat PR, Chao S, Condamine P, Heinen S, Resnik J, Wing R, Witt HN, Alpert M, Beccuti M, Bozdag S, Cordero F, Mirebrahim H, Ounit R, Wu Y, You F, Zheng J, Šimková H, Doležel J, Grimwood J, Schmutz J, Duma D, Altschmied L, Blake T, Bregitzer P, Cooper L, Dilbirligi M, Falk A, Feiz L, Graner A, Gustafson P, Hayes PM, Lemaux P, Mammadov J, Close TJ (2015) Sequencing of 15 622 gene-bearing BACs clarifies the gene-dense regions of the barley genome. Plant J 84:216–227

Ounit R, Wanamaker S, Close TJ, Lonardi S (2015) CLARK: fast and accurate classification of metagenomic and genomic sequences using discriminative k-mers. BMC Genomics 16:236

Rostoks N, Ramsay L, Mackenzie K, Cardle L, Bhat PR, Roose ML, Svensson JT, Stein N, Varshney RK, Marshall DF, Graner A, Close TJ, Waugh R (2006) Recent history of artificial outcrossing facilitate whole-genome association mapping in elite inbred crop varieties. Proc Natl Acad Sci USA 103:18656–18661

Šafář J, Šimková H, Kubaláková M, Číhalíková J, Suchánková P, Bartoš J, Doležel J (2010) Development of chromosome-specific BAC resources for genomics of bread wheat. Cytogenet Genome Res 129:211–223

Sánchez-Martín J, Steuernagel B, Ghosh S, Herren G, Hurni S, Adamski N, Vrána J, Kubaláková M, Krattinger SG, Wicker T, Doležel J, Keller B, Wulff BBH (2016) Rapid gene isolation in barley and wheat by mutant chromosome sequencing. Genome Biol 17:221

Schondelmaier J, Martin R, Jahoor A, Houben A, Graner A, Koop HU, Herrmann RH, Jung C (1993) Microdissection and microcloning of the barley (*Hordeum vulgare* L) chromosome 1HS. Theor Appl Genet 86:629–636

Šimková H, Číhalíková J, Vrána J, Lysák MA, Doležel J (2003) Preparation of HMW DNA from plant nuclei and chromosomes isolated from root tips. Biol Plant 46:369–373

Šimková H, Svensson JT, Condamine P, Hřibová E, Suchánková P, Bhat PR, Bartoš J, Šafář J, Close TJ, Doležel J (2008) Coupling amplified DNA from flow-sorted chromosomes to high-density SNP mapping in barley. BMC Genomics 9:294

Sorokin A, Marthe F, Houben A, Pich U, Graner A, Kunzel G (1994) Polymerase chain-reaction mediated localization of RFLP clones to microisolated translocation chromosomes of barley. Genome 37:550–555

Staňková H, Hastie AR, Chan S, Vrána J, Tulpová Z, Kubaláková M, Visendi P, Hayashi S, Luo M, Batley J, Edwards D, Doležel J, Šimková H (2016) BioNano genome mapping of individual chromosomes supports physical mapping and sequence assembly in complex plant genomes. Plant Biotechnol J 14:1523–1531

Suchánková P, Kubaláková M, Kovářová P, Bartoš J, Číhalíková J, Molnár-Láng M, Endo TR, Doležel J (2006) Dissection of the nuclear genome of barley by chromosome flow sorting. Theor Appl Genet 113:651–659

Vrána J, Cápal P, Šimková H, Karafiátová M, Čížková J, Doležel J (2016) Flow analysis and sorting of plant chromosomes. Curr Protoc Cytom 78:5.3.1–5.3.43

Zhou R-N, Hu Z-M (2007) The development of chromosome microdissection and microcloning technique and its applications in genomic research. Curr Genomics 8:67–72

Barley Genome Sequencing and Assembly—A First Version Reference Sequence

5

Nils Stein and Martin Mascher

Abstract

A reference sequence of the barley genome—some 12 years ago—this goal seemed unrealistic to achieve based on the available technology. Still, a group of international barley scientists developed a vision of how such a resource could be developed by joint efforts and by revisiting once established strategies against new possibilities facilitated through technological innovation. This chapter provides an overview of the main steps taken toward the publication of a first reference sequence of the barley genome in 2017 and how this represents the beginning rather than the end of genome research in barley.

5.1 Introduction

About 15 years ago, complete sequencing and assembly of large plant genomes was considered to be unfeasible. This assumption was mainly based on the prohibitive costs of using Sanger sequencing technology. To reduce costs, researchers at that time developed strategies to focus only on the gene space of a genome by filtering (Barbazuk et al. 2005; Whitelaw et al. 2003) or enriching for gene sequences (Madishetty et al. 2007) before sequencing. Even without any indication of the feasibility for developing a whole-genome sequence, however, the barley research community prepared itself toward this goal and established the International Barley genome Sequencing Consortium (IBSC) in 2006. A whitepaper (www.barleygenome.org) was published providing an outline of the envisioned feasibility within a decade of sequencing at least the barley gene space on the basis of a BAC-based physical map. A whole-genome physical map was targeted first because of its immediate value to the community as a necessary general resource for map-based cloning of genes underlying relevant barley traits (Schulte et al. 2009). Soon after, the introduction of second- and next-generation sequencing technology (for review: Shendure et al. 2004, 2017; Shendure and Ji 2008) opened a new perspective for whole-genome sequencing of multi-gigabase sized genomes mainly through cutting overall sequencing costs and the required time for sequencing. Thus, IBSC adopted these new possibilities and almost exactly 10 years after publishing the research strategy of sequencing the barley gene space, IBSC exceeded its goal by establishing a physical map-based, first version of a barley reference sequence (Mascher et al. 2017). The present chapter summarizes the overall characteristics of this first version of a

N. Stein (✉) · M. Mascher
Leibniz Institute of Plant Genetics and Crop Plant Research (IPK), Gatersleben, 06466 Seeland, Germany
e-mail: stein@ipk-gatersleben.de

N. Stein and G. J. Muehlbauer (eds.), *The Barley Genome*, Compendium of Plant Genomes,
https://doi.org/10.1007/978-3-319-92528-8_5

barley reference genome sequence, existing limitations, and the potential for further research advances. The main intermediate achievements toward whole-genome sequencing are also summarized in brief. More details on these aspects have been covered in previous reviews (Eversole et al. 2009; Stein 2014; Stein and Steuernagel 2014).

5.2 Whole-Genome Sequencing

5.2.1 Sequencing Strategy and Assembly

The sequencing strategy for the barley genome defined by the IBSC required the development of a genome-wide physical map. Such a map would be constituted of overlapping Bacterial Artificial Chromosome (BAC) clones which then provide the units to be sequenced (Schulte et al. 2009). Initial attempts of developing a physical map focused on the gene space, a strategy based on the observation that genes are not randomly distributed in the barley genome. In a large effort, 83,831 genes containing BAC clones were selected from the first available barley BAC library (Yu et al. 2000) by gene-specific overgo probe hybridization to high-density colony filter membranes (Madishetty et al. 2007). Out of these, 72,052 BACs were fingerprinted to define a Minimal Tiling Path (MTP) of 15,711 overlapping clones (Muñoz-Amatriaín et al. 2015). In parallel, a larger effort was initiated to build a generic genome-wide physical map taking advantage of a set of six independent BAC libraries derived by partial digestion with different enzymes or random shearing of high-molecular-weight genomic DNA in order to maximize for cloned genome representation (Schulte et al. 2011). 690,912 BACs were analyzed by High-Information-Content Fingerprinting (HICF) (Luo et al. 2003) and 570,007 passed the filter criteria to be included in the physical map (Ariyadasa et al. 2014). This was equivalent to approximately 14-fold haploid genome coverage and produced a map comprising 9265 BAC contigs with an average size of 538 kb, covering about 4.9 Gb of the estimated 5.1 Gb barley genome (Ariyadasa et al. 2014). A genome-wide MTP defined from the physical map comprised of 68,047 individual BAC clones. While this effort produced the basis for a hierarchical clone-by-clone sequencing strategy, IBSC had to evaluate newly emerging sequencing technologies in order to decide the most cost- and time-efficient way forward, ruling out the need of Sanger sequencing, which came at a given cost of USD ~ 1000 per clone.

Initial attempts of next-generation sequencing of barley BAC clones used the early Roche/454 GS20 platform producing 100-nucleotide (nt)-long sequence reads, which were considered at that time as too short for producing good assemblies in genomes with long stretches of repetitive DNA. Meaningful assemblies were obtained only after hybrid assembly with low coverage of ABI Sanger sequence reads, however, the Roche/454 sequencing sufficed for gene detection in de novo sequenced BAC clones (Wicker et al. 2006). With the introduction of longer sequence reads on the advanced Roche/454 GS FLX platform and the possibility of multiplexing BAC clones in the same sequencing run enabled by the use of Multiplex Identifiers (MID), also commonly called barcode adaptor sequences, provided a first encouraging perspective for BAC-based barley genome sequencing. It was possible to sequence individual clones in 91 multiplexes and assemble to almost finished quality, comparable to Sanger sequenced BACs (Steuernagel et al. 2009). Assembly quality improved as read length improved on the further advanced Roche/454 Titanium platform and in combination with short read mate-pair sequence information generated from larger BAC pools on the Solexa/Illumina platform, scaffolding of 454 Titanium contig assemblies appeared to be feasible (Taudien et al. 2011). Based on this strategy, two teams at IPK Gatersleben and the Fritz Lipmann Institute in Jena, Germany, initiated the sequencing of the MTP of the first barley chromosome, 3H—selected for the reason of the perspective for comparative genomic analysis to the homoeologous wheat chromosome 3B, which was in progress for sequencing in France at the same time (Choulet et al. 2014). Soon after, the approach was further optimized in order to take advantage of the higher sequencing output of the new Illumina Hiseq 2000 platform. This was feasible by establishing a much higher level of multiplexing on the basis of 668 barcode adaptors and sequencing of such a complex pool of BAC clones on a single lane in the 2 × 100 nt high-output mode of Hiseq 2000. In combination with mate-pair sequence data from an even more complex non-barcoded BAC pool generated on the Miseq (2 × 250 nt) and the Hiseq (2 × 100 nt), platforms produced similar quality assemblies like those sequenced on the Roche/454 platform (Beier et al. 2016). With 667 BACs per sequencing lane of one flow cell, the capacity of a single Hiseq 2000 run (8 lanes, 2-3 Tbp) enabled the sequencing of all MTP clones of a single-barley chromosome [MTP clone numbers in: (Beier et al. 2017)} within 1.5–2 complete Hiseq 2000 runs. After this breakthrough, sequencing all barley chromosomes seemed feasible and the task of sequencing all MTP BAC clones was distributed among members of IBSC. Chromosomes 1H, 3H, and 4H were sequenced at IPK Gatersleben/FLI Jena, Germany; chromosomes 2H and unanchored BACs (0H) were sequenced at Earlham Institute (EI), UK, formerly known as The Genome Analysis Center (TGAC) under the lead of the James Hutton Institute (JHI), UK; chromosomes 5H, 6H, and 7H were sequenced at the Beijing Genomics Institute (BGI), China, under the lead of Murdoch University, Australia/Zhejiang University China, (5H, 7H), and the Carlsberg Laboratory (6H), Denmark. The different sequencing centers followed similar but not identical protocols and methods for multiplexing, sequencing, and assembly/scaffolding, however, with very comparable success and results (Beier et al. 2017).

At the same time as sequencing the chromosomal MTPs were carried on, also the gene space MTP consisting of 15,661 gene-bearing BACs was sequenced and assembled (Muñoz-Amatriaín et al. 2015). This relied on a completely different strategy, omitting the laborious steps of labeling the DNA of individual BACs by specific barcode adaptors. Here, the BAC clones were pooled in a multidimensional combinatorial design, which allowed, before assembly, the deconvolution of reads belonging to individual clones on the basis of the position of each clone in the combinatorial design. Both MTP sequence datasets were integrated into one final reference sequence assembly (Beier et al. 2017) comprising 4.8 Gb of nonredundant sequence (Mascher et al. 2017). This combined the sequence data of 87,105 sequenced BAC clones (Table 5.1). Based on a statistical kmer-based representation analysis, the IBSC RefSeq v1.0 is very close to complete representation of the barley genome (Fig. 5.1). However, the fact that major repeat clusters (rDNA, Cereba centromeric repeats, telomeres) are still not represented in the assembly (Mascher et al. 2017) indicates that the difference between estimated genome size and assembled sequence size is real and some parts of the barley genome are still not represented in the reference sequence, although a conservative BAC MTP-based sequencing and assembly strategy was followed. Underrepresentation of some genomic regions in the physical map might be explained by incompatibility or toxicity of the cloned eukaryotic DNA when introduced into *E. coli*.

IBSC and other barley researchers continuously evaluated also in parallel the possibility of using next-generation sequencing technology for de novo Whole-Genome Shotgun (WGS) sequencing and assembly of the barley genome. In 2012, IBSC published the sequence-enriched physical map of the "Morex" genome (The International Barley Genome Sequencing Consortium (IBSC) 2012) and this study included a de novo WGS assembly, which was generated based on Illumina Hiseq 2000 paired-end (PE) 100 shotgun sequencing. In the same study, the genomes of two other cultivars "Barke" and "Bowman" were sequenced and de novo assembled (Table 5.1). These early shotgun assemblies were largely incomplete with only about 37% of the estimated genome size in the assemblies. Also, they were highly fragmented with between 2 and 3 million assembled contigs and N50 contig length between 1.5 and 2 kb.

Table 5.1 Overview barley genome assembly statistics

Genotype/Cultivar	Morex[a]	Barke[a]	Bowman[a]	Haruna Nijo[b]	Lasa Goumang[c]	Zangqing320[d]	Morex[e]
Total size of assembled scaffolds (Gbp)	1.9	2.0	1.8	2.0	3.9	4.8	4.8
Total sequence length anchored to chromosomes (Gbp)	1.4				3.5	4.6	4.6
Number of scaffolds					37,441 (>2 kb)	91,879	6347
Longest scaffold (Mbp)					3.1	2.7	14.0
N50 length, scaffolds (kb)					242	171.1	1900
Total size of assembled contigs (Gbp)	1.9	2.0	1.8	2.0	3.6	3.3	4.8
Number of contigs	2,670,738	2,742,077	2,077,901	1,712,236	287,049 (>2 kb)	632,382	466,070
Longest contig (kb)	36.1	38.4	37.4		277	298	297
Contig N50 (kb)	1.4	1.4	2.0	3.5	18.1		79
GC content (%)	44			45	44	44	44–45
Repeat content (%)	84			61	80	n.a.	81
Number of annotated gene loci	26,159			30,606	36,151	46,787	39,734

[a]The International Barley Genome Sequencing Consortium (IBSC) (2012)
[b]Sato et al. (2015)
[c]Zeng et al. (2015)
[d]Dai et al. (2018)
[e]Mascher et al. (2017)

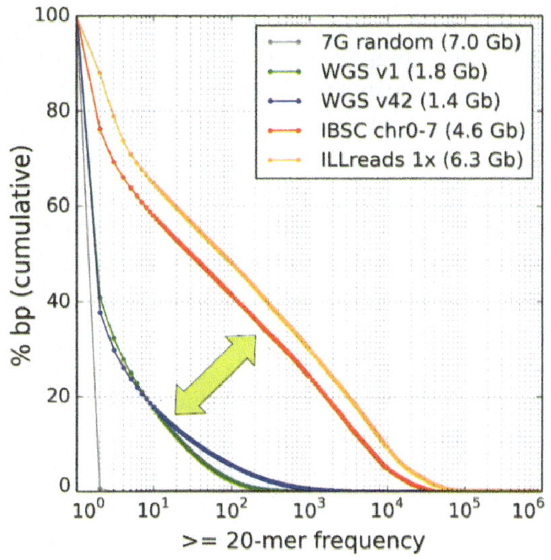

Fig. 5.1 Analysis of genome representation in barley genome assemblies based on kmer frequencies. The representation of the barley RefSeq v1.0 was benchmarked on the basis of kmer-20 frequencies. The gray line represents kmer frequencies in a random sequence sample from NCBI Genbank of the size similar to the haploid barley genome. Green and blue lines show kmer frequencies in draft WGS assemblies (The International Barley Genome Sequencing Consortium (IBSC) 2012). The orange line indicates kmer frequencies in a haploid barley genome equivalent of random barley DNA Illumina short sequence reads. The red line indicates the kmer frequency profile of barley RefSeq v1.0 which follows the same dynamics as random barley sequence indication very good representation of the barley genome in contrast to previous draft assemblies (yellow double-headed arrow)

However, comprehensive gene sets could be annotated for these assemblies and integrated to the physical map on the basis of sequence overlap to BAC end sequences, which were available for 50% of all clones in the physical map assembly, as well as to several thousand randomly selected BAC clones (The International Barley Genome Sequencing Consortium (IBSC) 2012). A fourth genotype ("Haruna Nijo") was sequenced and assembled by a mixed strategy using Hiseq 2000 paired-end data for assembly and Roche/454 long mate-pair sequence information for scaffolding (Sato et al. 2015). Again, the assembly represented only 2 Gb of the estimated 5.1 Gb barley genome, however, the N50 contig length increased to over 3.5 kb which was an advantage for improved gene prediction (Table 5.1).

Two additional attempts were made to shotgun sequence and assemble Tibetan barley cultivars (Table 5.1). First, the genome of a landrace "Lasa Guomang" was assembled from 178-fold sequence coverage generated from different paired-end and mate-pair libraries with insert sizes ranging from 250 bp up to 40 kb (Zeng et al. 2015). Almost 3.9 Gb of sequence was assembled, and the contig N50 was significantly increased to over 18 kb. This assembly profited from the work of the IBSC, because the sequence-enriched physical map of "Morex" provided the physical framework for assigning the large scaffolds of the "Lasa Guomang" assembly to chromosomal bins. A second shotgun assembly of the Tibetan cultivar "Zangqing320" became available very recently (Dai et al. 2018). This is the first nearly complete barley de novo shotgun assembly with a total length of 4.84 Gb (Table 5.1). The assembly was based again on high-coverage short read sequencing but profited from the Hiseq 2500 platform. Overlapping reads (pseudo-long reads) were generated from small insert (250–300 nt) paired-end libraries, providing a better start for de novo assembly. Furthermore, long-read real-time single-molecule (SMRT) sequencing on the Pacific Biosystems platform was used for closing gaps between contigs/scaffolds. The majority of this assembly was ordered and assigned to chromosomes by relating it to the information provided by the "Morex" reference sequence v1.0.

Shortly after the "Morex" BAC MTP-based reference sequence was published, a first high-quality reference sequence of the tetraploid wild emmer wheat (*Triticum turgidum* subsp. *dicoccoides*) genome was published. This assembly was based purely on short read WGS sequencing (Avni et al. 2017). In comparison to the early studies of de novo shotgun sequencing based on short reads, this study emphasized the importance of generating very high-quality (PCR-free) sequencing libraries and high-coverage of sequence data from different but precisely sized paired-end (for assembly) and mate-pair (for scaffolding) libraries. Furthermore, longer pseudo-long reads, generated by Illumina Hiseq 2500 sequencing (rapid run mode 2 × 250 nt) of 450 bp insert paired-end libraries, were used and proved to be crucially important. Wild emmer wheat is a tetraploid species with two homoeologous sub-genomes, each of the sizes, and complexity comparable to the barley genome. Given the progress in assembly quality that can now be obtained based on Illumina short read sequencing, it is almost ironic that after reaching the goal of an MTP-based barley reference sequence within the predicted time frame, this strategy can now be considered to be outdated and will probably no longer be applied to large genome sequencing in the future.

5.2.2 Genetic and Physical Ordering—Chromosome-Like Assemblies

Sequencing of overlapping BAC clones alone did not lead immediately to chromosome-scale assemblies of the barley genome. The BAC-based physical map was composed of 9265 individual BAC contigs. This was an average of ~1300 contigs per barley chromosome. Only 4556 of the contigs, or 3.9 Gb of the cumulative length of the physical map, could be anchored to the genetic map initially (The International Barley Genome Sequencing Consortium [IBSC] 2012). Although this represented 76% of the barley genome, still the contigs were assigned to 1501 distinct genetic bins, thus a correct linear ordering of the genome for regions without recombination (50% of the physical length of all barley chromosomes) required additional tools and sources of information.

Chromosome Survey Sequencing (CSS)

The first breakthrough for physical assignment of sequence resources to individual barley chromosomes was made by new possibilities provided by the chromosomal genomics toolbox developed by Jaroslav Dolezel and his team at Institute of Experimental Biology, Olomouc, Czech Republic (Doležel et al. 2014), (Chap. 4 of this book). After the first proof of concept study on shotgun sequencing of flow-sorted barley chromosome 1H (Mayer et al. 2009), the gene content and gene order of all seven barley chromosomes were computed mainly by exploiting synteny information to related grass genomes with annotated reference sequence information available (Mayer et al. 2011). This approach provided very helpful intermediate gene order information before whole-genome sequencing was realistically achievable. These results triggered similar studies in wheat (The International Wheat Genome Sequencing Consortium 2014) and rye (Martis et al. 2013). Furthermore, the shotgun sequence data from individual chromosomes (Chromosome Survey Sequences, CSS) could be mapped to non-repetitive sequences in contigs of the de novo WGS assemblies or from individually sequenced BAC clones. This generated a highly accurate predictor for the genomic origin of the respective contig sequence (The International Barley Genome Sequencing Consortium (IBSC) 2012; The International Wheat Genome Sequencing Consortium 2014). The CSS data proved also to be very useful for the identification of mis-assemblies or chimeric scaffolds during quality assessment of the barley Refseq v1.0 (Beier et al. 2017; Mascher et al. 2017).

Population Sequencing (POPSEQ)

CSS-based chromosomal assignment of sequence resources is a very valuable first step toward sorting the genomic origin of otherwise anonymous DNA assemblies. This method, however, does not provide linear ordering information and thus is really powerful only in combination with other sources of information (e.g., synteny, dense genetic marker maps, etc.). Genetic maps used to be the main tool for anchoring and ordering of physical maps and sequence assemblies. This "genetic" anchoring process used to be limited first by the number of available molecular markers and the density of the existing experimental genetic maps and second by the fact that large parts of the genomes do not show meiotic crossovers, which are the basis of measuring genetic distances. In the absence of alternatives for physical mapping, genetic markers and maps still provided the best information for anchoring and ordering of BAC-based physical maps and sequences. In collaboration with the US Department of Energy–Joint Genome Institute (DOE–JGI), the IBSC generated two genetic maps of unprecedented density in barley by sequencing to onefold haploid genome coverage each of the 90 and 82 individuals of a recombinant inbred line (RIL) and a doubled haploid (DH) mapping population, respectively. The two datasets provided 5.1 and 6.5 million markers for genetic anchoring of the "Morex" de novo shotgun assembly (Mascher et al. 2013) and two million nonredundant markers could be used to revise the genetic anchoring of the barley physical map (Ariyadasa et al. 2014).

Three-Dimensional (3D) Chromosome Conformation Capture Sequencing (Hi-C)

Physical ordering of BAC contigs or sequence scaffolds in non-recombining regions of the genome remained an insurmountable obstacle toward the goal of a fully anchored and ordered chromosome-scale reference quality sequence assembly in barley. Additional methods for physical mapping like deletion bin mapping (Sourdille et al. 2004), HAPPY mapping (Dear and Cook 1993), radiation hybrid mapping (Tiwari et al. 2016), or Fluorescent In situ Hybridisation (FISH) (Aliyeva-Schnorr et al. 2015, 2016) were long seen as the only possibilities of solving the linear order of physical map contigs or sequence scaffolds that were allocated to the non-recombining region of barley chromosomes. In parallel to genome sequencing and assembly in many organisms, however, major progress was made by research on chromatin conformation analysis in yeast and human. Several technologies were established that enabled testing 3D physical proximity of chromosomal domains that are located at varying distances on the linear chromosome (Dekker et al. 2013). One method, Hi-C, was developed to test in parallel any genome-wide interaction of chromatin domains by high throughput sequencing (Lieberman-Aiden et al. 2009). While this method has stimulated a fascinating new area of research on the impact of 3D chromatin structure on nucleus organization and gene regulation (Dekker et al. 2013), it was also demonstrated that Hi-C sequence data can be exploited for scaffolding WGS assemblies of individual human chromosomes (Burton et al. 2013). After adapting the existing protocols from human genome research to barley (Mascher et al. 2017), Hi-C libraries were produced and used to bring all sequence scaffolds of Morex RefSeq v1.0 into linear order, thus producing the first chromosome-scale assemblies in barley (Fig. 5.2). The method was further streamlined to work in different cereal species and Hi-C datasets helped subsequently to produce chromosome-scale assemblies in wild emmer wheat (Avni et al. 2017), durum wheat (Maccaferri et al. 2018), common wheat (The International Wheat Genome Sequencing Consortium (IWGSC) 2018), and rye (our own unpublished data). Furthermore, the method enabled detection of large megabase-scale structural variation between different barley genotypes (Himmelbach et al. 2018). Recently, Hi-C for genome scaffolding has been established also on the basis of in vitro reconstituted chromatin ("Chicago", Putnam et al. 2016) and both methods have now become standard tools in genome sequencing and assembly.

Optical Mapping by Nanofluidics-Supported Analysis of Megabase Sized DNA Molecules

Physical mapping of genomes by fingerprinting of long DNA molecules by the support of image analysis tools ("optical mapping") has long been suggested as a powerful alternative for developing genome maps in large genome species and difficult-to-access genomic regions like extensive tandem repeats or non-recombining regions (Aston et al. 1999). The power of the method for validating genome assemblies was demonstrated in large genome crop species like maize (Schnable et al. 2009); however, this challenging technique was not applied routinely until commercial solutions for automated analysis of large DNA molecules became available. The introduction of nanofluidics solutions in

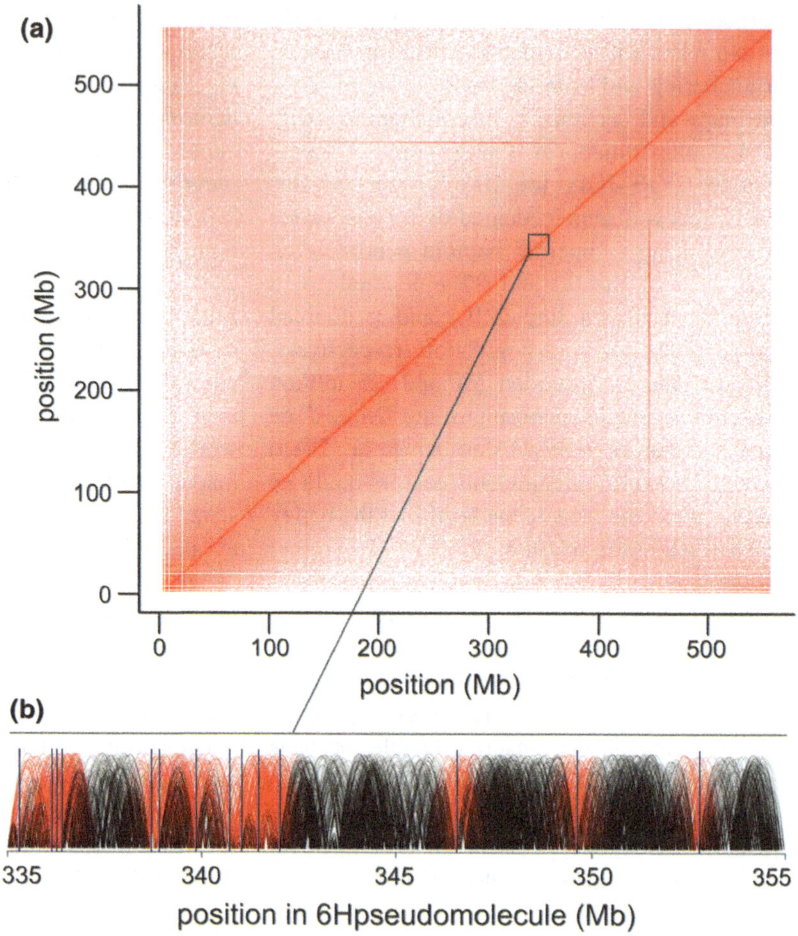

Fig. 5.2 Hi-C sequencing for scaffolding. **a** Hi-C interaction matrix is shown for barley chromosome 6H. **b** Zoom into a region of the Hi-C interaction matrix at Mb position 335 to 355 of chromosome 6H. Blue vertical lines indicate gaps between scaffolds concatenated on the basis of the Hi-C interaction matrix. Red connector lines illustrate linking Hi-C sequence pairs between scaffolds, whereas black connectors illustrate links of Hi-C sequence pairs within the same scaffold

optical mapping provided the necessary technical breakthrough (Dimalanta et al. 2004; Tegenfeldt et al. 2004). In combination with fluorescent labeling of large DNA molecules by sequence-specific enzymes, the method of optical mapping became more broadly applicable and suited for automation (Das et al. 2010). Since then it has evolved to become an important component of de novo genome assembly strategies, especially as part of super-scaffolding and for validating highly contiguous sequence assemblies as shown for different animal and plant species (Howe and Wood 2015; Jarvis et al. 2017; Mostovoy et al. 2016). When Alan Schulman, University Helsinki, Finland, hit the jackpot of Bionano Genomics' raffle among participants of their workshop at the Plant and Animal Genome Conference in January 2014, it provided the exciting possibility to generate a Bionano-based optical map of the reference barley genotype Morex. An equivalent of 4.3 Gb of the barley genome was assembled into optical molecule maps of which 3.9 Gb (85%) could be aligned to the BAC-based sequence clusters (Mascher et al. 2017). It turned out that the optical map of Morex was of great value as an independent source of information for validating the chromosome-scale assemblies obtained after Hi-C based scaffolding (Fig. 5.3).

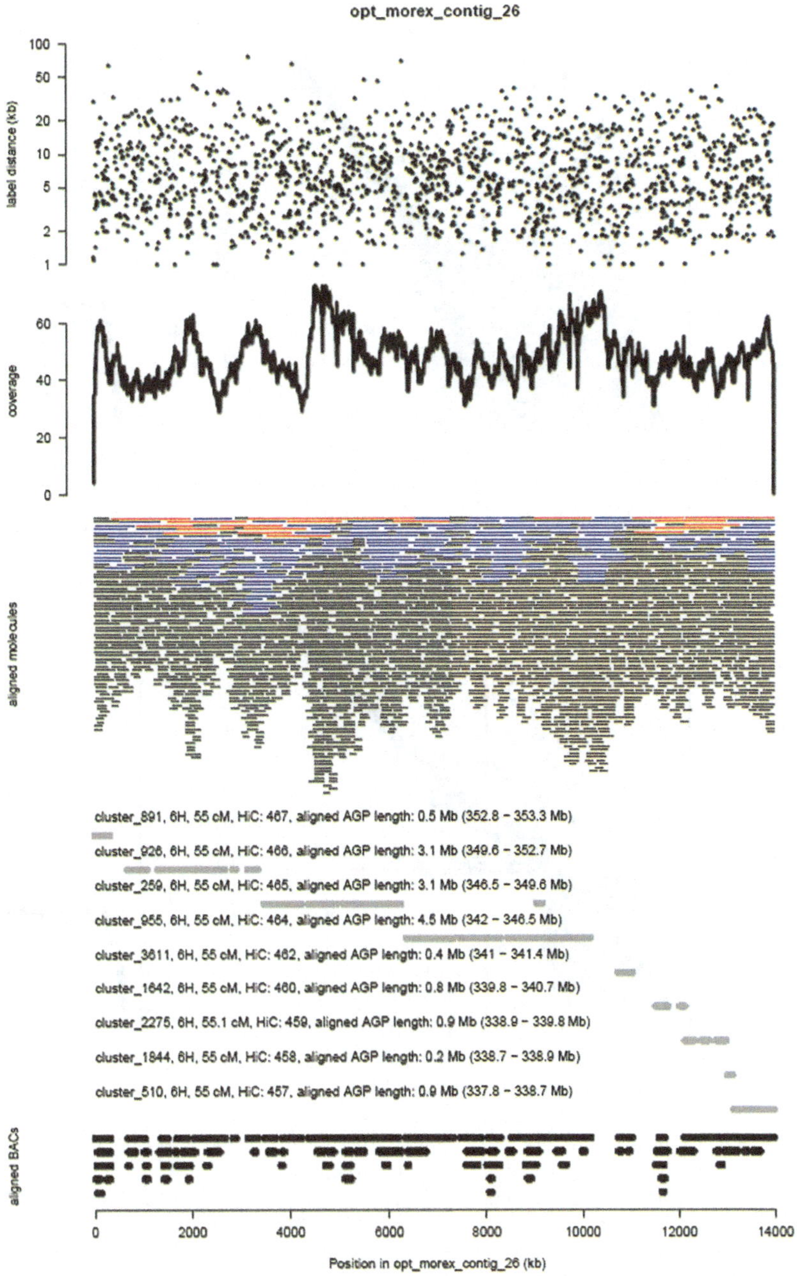

Fig. 5.3 Validation of Hi-C-based scaffolding by alignment to a Bionano optical molecule map of barley cultivar "Morex". The upper three panels show label density distribution, molecule coverage, and aligned molecule distribution in the optical molecule map contig 26 of the Bionano "Morex" genome map. Below, a successive array of sequence scaffolds of barley RefSeq v1.0, ordered by Hi-C based scaffolding, is shown, reproducing exactly the optical molecule map. The lowest panel illustrates the position and alignment of the individual BAC clones of the physical map underlying the reference sequence assembly

5.2.3 Barley RefSeq v1.0—Practical Considerations

The international effort of IBSC delivered a first reference sequence of the barley genome within a decade (Mascher et al. 2017). The genome was fully annotated on the basis of extensive transcribed gene evidences (see Chaps. 6 and 7), and all underlying data resources have been made publicly available and searchable (Beier et al. 2017; Colmsee et al. 2015; Mascher et al. 2017; Chap. 21) and these resources should be consulted further if more details on the barley genome are of interest to researchers. Here,

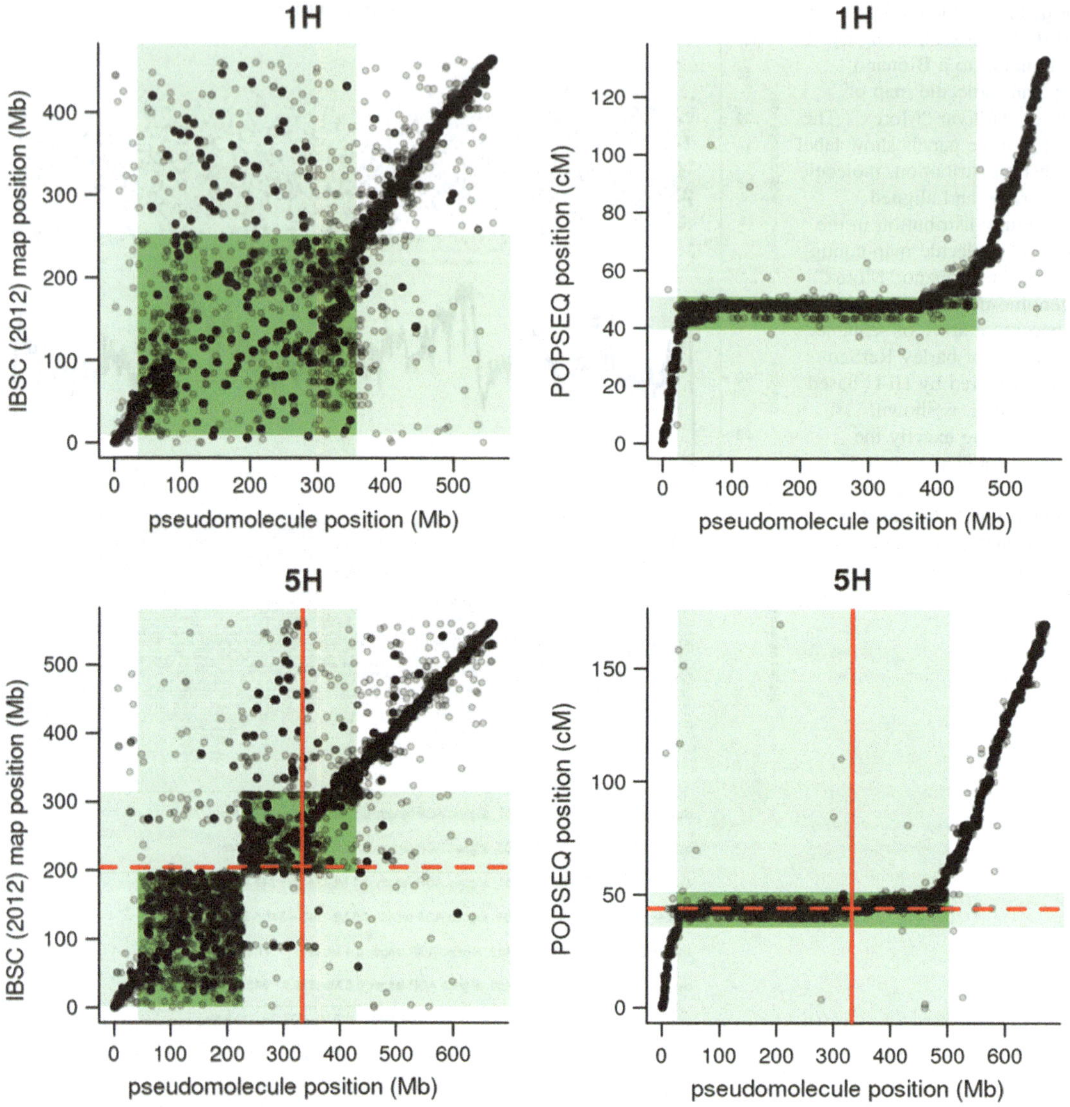

Fig. 5.4 Comparison of the barley draft (2012, 2014) and reference sequence assembly. The sequence information of two chromosomes 1H (upper two panels) and 5H (lower two panels) from the draft genome sequence before (left panels, The International Barley Genome Sequencing Consortium (IBSC) 2012) and after high-density POPSEQ-based genetic anchoring (right panels, Ariyadasa et al. 2014) is shown in comparison to the respective pseudomolecules (chromosome-scale assemblies) of barley RefSeq v1.0. Centromeric and peri-centromeric regions in the respective sequence resources are shown in light green shading. This highlights that genome ordering information was almost random for 50% of the length of the chromosome in the 2012 resource (dark green area in left panels). Higher density genetic anchoring masked this problem only by binning all this information into a 5–10 cM bin (dark green shading in the right panels). The massively improved physical resolution of genomic sequence scaffolding is illustrated for chromosome 5H for the position of the gene *HvLAX* (Jost et al. 2016) (red lines). While the approximated location of the gene remained in a region of large uncertainty in the draft genome sequence versions (2012, 2014), which complicated cloning of the gene, its physical position was completely resolved in the correct context in RefSeq v1.0, highlighting the highly increased value of RefSeq v1.0 for gene cloning in barley

instead, we draw the attention to the changes and improvements that were introduced and delivered since the publication of the draft genome sequence (The International Barley Genome Sequencing Consortium (IBSC) 2012). The most important difference from previous barley genomic resources is that sequence clusters/super-scaffolds are ordered throughout the non-recombining peri-centromeric and centromeric regions of the seven barley chromosomes (Fig. 5.4). In the 2012 draft, genome information (BAC contigs and WGS contigs) were placed into a single (or few) centromeric genetic bin in nearly random order. This genetic interval equalled about 50% of the physical extension of the barley chromosomes. In the new reference sequence, these regions were ordered by physical information generated through Hi-C sequencing and validated by comparison to the optical map, thus providing the correct sequence context for any gene in these regions.

The second important aspect about the Morex RefSeq v1.0 for users to consider is that this first version reference sequence is not perfect; there is still room for improvement, as future genetic/physical information that may conflict with or contradict parts of the assembly should always be considered. It needs to be kept in mind that the reference sequence was assembled from individually sequenced BAC clones. The minority of these BACs were assembled into one contig/scaffold sequence, but the majority were represented by a number of large and small contigs and scaffolding/ordering was not always finalized for these BACs. Thus, within the map position of each individual BAC clone ($\sim$100 kb bin), the order and orientation of sequence information may be subject to change as additional evidence may emerge. As an example, only 3.6 Gb (85%) of the 4.3 Gb optical map assembly could be aligned to the BAC-based sequence assembly. This can be attributed in part to limited information content of the sequence-specific labeling pattern of the optical molecule maps. However, the random orientation of contigs within scaffolds and super-scaffolds must also contribute to the observed limitation in sequence/optical molecule map alignment.

5.3 More Barley Genomes: The Barley Pan-Genome

A major milestone for barley research and crop improvement was achieved with the availability of the barley reference sequence. While this is a major advance, however, one should not ignore the fact that access to a single high-quality-reference sequence reveals only part of the total genome composition of a given species. Originating from the area of microbial genomics (Tettelin et al. 2008), the concept of "pan-genome" analysis has firmly taken hold also in plant sciences, where an increasing number of studies address the problem of characterizing the "core" and "dispensable" components also of crop species genomes (Hirsch et al. 2014; Montenegro et al. 2017; Zhao et al. 2018). The "core" genome refers to parts of the genome (especially genes) found in every haplotype, whereas the "dispensable" genome refers to genomic components found in fewer haplotypes of a given species. Structural variation, including presence/absence variation, was shown to be high (Muñoz-Amatriaín et al. 2013 and Chap. 8), thus indicating the importance of pan-genome analysis also for barley. Efforts are underway to characterize genomic diversity of the global domesticated barley population by reduced representation re-sequencing of more than 20,000 barley accessions of the largest ex situ seed bank of the EU28 (BRIDGE project: http://bridge.ipk-gatersleben.de). Based on this information, an initial set of five diverse genotypes including also a wild barley accession was selected for de novo WGS following the approach used for sequencing the wild emmer and the *Aegilops tauschii* genomes (Avni et al. 2017; Zhao et al. 2017). Thus, shortly after publishing the first reference sequence of the barley ("Morex") genome, access to more complex barley genome information on the basis of a first pan-genome analysis is already in sight.

Acknowledgements We kindly acknowledge Drs. Heidrun Gundlach and Manuel Spannagl for conceiving Fig. 5.1.

References

Aliyeva-Schnorr L, Beier S, Karafiátová M, Schmutzer T, Scholz U, Dolezel J, Stein N, Houben A (2015) Cytogenetic mapping of BAC contigs assigned to the genetic centromere of barley chromosome 3H. Plant J 84:385–394

Aliyeva-Schnorr L, Stein N, Houben A (2016) Collinearity of homoeologous group 3 chromosomes in the genus *Hordeum* and *Secale cereale* as revealed by 3H-derived FISH analysis. Chromosom Res 24: 231–242

Ariyadasa R, Mascher M, Nussbaumer T, Schulte D, Frenkel Z, Poursarebani N, Zhou R, Steuernagel B, Gundlach H, Taudien S, Felder M, Platzer M, Himmelbach A, Schmutzer T, Hedley P, Muehlbauer G, Scholz U, Korol A, Mayer K, Waugh R, Langridge P, Graner A, Stein N (2014) A sequence-ready physical map of barley anchored genetically by two million SNPs. Plant Physiol 164:412–423

Aston C, Mishra B, Schwartz DC (1999) Optical mapping and its potential for large-scale sequencing projects. Trends Biotechnol 17:297

Avni R, Nave M, Barad O, Baruch K, Twardziok SO, Gundlach H, Hale I, Mascher M, Spannagl M, Wiebe K, Jordan KW, Golan G, Deek J, Ben-Zvi B, Ben-Zvi G, Himmelbach A, MacLachlan RP, Sharpe AG, Fritz A, Ben-David R, Budak H, Fahima T, Korol A, Faris JD, Hernandez A, Mikel MA, Levy AA, Steffenson B, Maccaferri M, Tuberosa R, Cattivelli L, Faccioli P, Ceriotti A, Kashkush K, Pourkheirandish M, Komatsuda T, Eilam T, Sela H, Sharon A, Ohad N, Chamovitz DA, Mayer KFX, Stein N, Ronen G, Peleg Z, Pozniak CJ, Akhunov ED, Distelfeld A (2017) Wild emmer genome architecture and diversity elucidate wheat evolution and domestication. Science 357:93–97

Barbazuk WB, Bedell JA, Rabinowicz PD (2005) Reduced representation sequencing: a success in maize and a promise for other plant genomes. BioEssays 27:839–848

Beier S, Himmelbach A, Schmutzer T, Felder M, Taudien S, Mayer K, Platzer M, Stein N, Scholz U, Mascher M (2016) Multiplex sequencing of bacterial artificial chromosomes for assembling complex plant genomes. Plant Biotechnol J 14:1511–1522

Beier S, Himmelbach A, Colmsee C, Zhang X-Q, Barrero RA, Zhang Q, Li L, Bayer M, Bolser D, Taudien S, Groth M, Felder M, Hastie A, Šimková H, Staňková H, Vrána J, Chan S, Muñoz-Amatriaín M, Ounit R, Wanamaker S, Schmutzer T, Aliyeva-Schnorr L, Grasso S, Tanskanen J, Sampath D, Heavens D, Cao S, Chapman B, Dai F, Han Y, Li H, Li X, Lin C, McCooke JK, Tan C, Wang S, Yin S, Zhou G, Poland JA, Bellgard MI, Houben A, Doležel J, Ayling S, Lonardi S, Langridge P, Muehlbauer GJ, Kersey P, Clark MD, Caccamo M, Schulman AH, Platzer M, Close TJ, Hansson M, Zhang G, Braumann I, Li C, Waugh R, Scholz U, Stein N, Mascher M (2017) Construction of a map-based reference genome sequence for barley, *Hordeum vulgare* L. Sci Data 4:170044

Burton JN, Adey A, Patwardhan RP, Qiu R, Kitzman JO, Shendure J (2013) Chromosome-scale scaffolding of de novo genome assemblies based on chromatin interactions. Nat Biotech. https://doi.org/10.1038/nbt.2727

Choulet F, Alberti A, Theil S, Glover N, Barbe V, Daron J, Pingault L, Sourdille P, Couloux A, Paux E, Leroy P, Mangenot S, Guilhot N, Le Gouis J, Balfourier F, Alaux M, Jamilloux V, Poulain J, Durand C, Bellec A, Gaspin C, Safar J, Dolezel J, Rogers J, Vandepoele K, Aury J-M, Mayer K, Berges H, Quesneville H, Wincker P, Feuillet C (2014) Structural and functional partitioning of bread wheat chromosome 3B. Science 345. https://doi.org/10.1126/science.1249721

Colmsee C, Beier S, Himmelbach A, Schmutzer T, Stein N, Scholz U, Mascher M (2015) BARLEX—the barley draft genome explorer. Mol Plant. https://doi.org/10.1016/j.molp.2015.03.009

Dai F, Wang X, Zhang X-Q, Chen Z, Nevo E, Jin G, Wu D, Li C, Zhang G (2018) Assembly and analysis of a qingke reference genome demonstrate its close genetic relation to modern cultivated barley. Plant Biotechnol J 16:760–770

Das SK, Austin MD, Akana MC, Deshpande P, Cao H, Xiao M (2010) Single molecule linear analysis of DNA in nano-channel labeled with sequence specific fluorescent probes. Nucleic Acids Res 38:e177–e177

Dear PH, Cook PR (1993) Happy mapping: linkage mapping using a physical analogue of meiosis. Nucl Acids Res 21:13–20

Dekker J, Marti-Renom MA, Mirny LA (2013) Exploring the three-dimensional organization of genomes: interpreting chromatin interaction data. Nat Rev Genet 14:390–403

Dimalanta ET, Lim A, Runnheim R, Lamers C, Churas C, Forrest DK, dePablo JJ, Graham MD, Coppersmith SN, Goldstein S, Schwartz DC (2004) A microfluidic system for large DNA molecule arrays. Anal Chem 76:5293–5301

Doležel J, Vrána J, Cápal P, Kubaláková M, Burešová V, Šimková H (2014) Advances in plant chromosome genomics. Biotechnol Adv 32:122–136

Eversole K, Graner A, Stein N (2009) Wheat and barley genome sequencing. In: Feuillet C, Muehlbauer GJ (eds) Genetics and genomics of the Triticeae. Springer, Berlin, pp 713–742

Himmelbach A, Walde I, Šimková H, Doležel J, Hastie A, Stein N, Mascher M (2018) Discovery of large polymorphic inversions by chromosome conformation capture sequencing. Plant J (under review)

Hirsch CN, Foerster JM, Johnson JM, Sekhon RS, Muttoni G, Vaillancourt B, Peñagaricano F, Lindquist E, Pedraza MA, Barry K, de Leon N, Kaeppler SM, Buell CR (2014) Insights into the Maize pan-genome and pan-transcriptome. Plant Cell 26:121–135

Howe K, Wood JMD (2015) Using optical mapping data for the improvement of vertebrate genome assemblies. GigaScience 4:10

Jarvis DE, Ho YS, Lightfoot DJ, Schmöckel SM, Li B, Borm TJA, Ohyanagi H, Mineta K, Michell CT, Saber N, Kharbatia NM, Rupper RR, Sharp AR, Dally N, Boughton BA, Woo YH, Gao G, Schijlen EGWM, Guo X, Momin AA, Negrão S, Al-Babili S, Gehring C, Roessner U, Jung C, Murphy K, Arold ST, Gojobori T, Linden CGvd, van Loo EN, Jellen EN, Maughan PJ, Tester M (2017) The genome of *Chenopodium quinoa*. Nature (advance online publication), pp 1–6

Jost M, Taketa S, Mascher M, Himmelbach A, Yuo T, Shahinnia F, Rutten T, Druka A, Schmutzer T, Steuernagel B, Beier S, Taudien S, Scholz U, Morgante M, Waugh R, Stein N (2016) A homolog of blade-on-petiole 1 and 2 (BOP1/2) controls internode length and homeotic changes of the barley inflorescence. Plant Physiol 171:1113–1127

Lieberman-Aiden E, van Berkum NL, Williams L, Imakaev M, Ragoczy T, Telling A, Amit I, Lajoie BR, Sabo PJ, Dorschner MO, Sandstrom R, Bernstein B, Bender MA, Groudine M, Gnirke A, Stamatoyannopoulos J, Mirny LA, Lander ES, Dekker J (2009) Comprehensive mapping of long-range interactions reveals folding principles of the human genome. Science 326:289–293

Luo M-C, Thomas C, You FM, Hsiao J, Ouyang S, Buell CR, Malandro M, McGuire PE, Anderson OD, Dvorak J (2003) High-throughput fingerprinting of bacterial artificial chromosomes using the snapshot labeling kit and sizing of restriction fragments by capillary electrophoresis. Genomics 82:378

Maccaferri M, Harris NS, Twardziok SO, Gundlach H, Spannagl M, Ormanbekova D, Lux T, Prade V, Milner S, Himmelbach A, Mascher M, Bagnaresi P, Faccioli P, Cozzi P, Lauria M, Lazzari B, Stella A, Manconi A, Gnocchi M, Avni R, Deek J, Biyiklioglu S, Frascaroli E, Corneti S, Salvi S, Tuberosa R, Sonnante G, Desiderio F, Marè C, Crosatti C, Mica E, Ozkan H, Vita PD, Marone D, Joukhadar R, Pasam RK, Mazzucotelli E, Nigro D, Gadaleta A, Chao S, Faris J, Melo ATO, Pumphrey M, Pecchioni N, Milanesi L, Wiebe K, MacLachlan RP, Clarke JM, Sharpe AG, Koh K, Liang KYH, Taylor GJ, Knox R, Budak H, Mastrangelo AM, Xu SS, Stein N, Hale I, Distelfeld A, Hayden MJ, Walkowiak S, Mayer KFX, Ceriotti A, Pozniak CJ, Cattivelli L (2018) Durum wheat genome reveals past domestication signatures and future improvement targets (under review)

Madishetty K, Condamine P, Svensson JT, Rodriguez E, Close TJ (2007) An improved method to identify BAC clones using pooled overgos. Nucl Acids Res 35:e5

Martis MM, Zhou R, Haseneyer G, Schmutzer T, Vrána J, Kubaláková M, König S, Kugler KG, Scholz U, Hackauf B, Korzun V, Schön C-C, Doležel J, Bauer E, Mayer KFX, Stein N (2013) Reticulate evolution of the rye genome. Plant Cell 25:3685–3698

Mascher M, Muehlbauer GJ, Rokhsar DS, Chapman J, Schmutz J, Barry K, Muñoz-Amatriaín M, Close TJ, Wise RP, Schulman AH, Himmelbach A, Mayer KFX, Scholz U, Poland JA, Stein N, Waugh R (2013) Anchoring and ordering NGS contig assemblies by population sequencing (POPSEQ). Plant J 76:718–727

Mascher M, Gundlach H, Himmelbach A, Beier S, Twardziok SO, Wicker T, Radchuk V, Dockter C, Hedley PE, Russell J, Bayer M, Ramsay L, Liu H, Haberer G, Zhang X-Q, Zhang Q, Barrero RA, Li L, Taudien S, Groth M, Felder M, Hastie A, Šimková H, Staňková H, Vrána J, Chan S, Muñoz-Amatriaín M, Ounit R, Wanamaker S, Bolser D, Colmsee C, Schmutzer T, Aliyeva-Schnorr L, Grasso S, Tanskanen J, Chailyan A, Sampath D, Heavens D, Clissold L, Cao S, Chapman B, Dai F, Han Y, Li H, Li X, Lin C, McCooke JK, Tan C, Wang P, Wang S, Yin S, Zhou G, Poland JA, Bellgard MI, Borisjuk L, Houben A, Doležel J, Ayling S, Lonardi S, Kersey P, Langridge P, Muehlbauer GJ, Clark MD, Caccamo M, Schulman AH, Mayer KFX, Platzer M, Close TJ, Scholz U, Hansson M, Zhang G, Braumann I, Spannagl M, Li C, Waugh R, Stein N (2017) A chromosome conformation capture ordered sequence of the barley genome. Nature 544:427–433

Mayer KFX, Taudien S, Martis M, Simkova H, Suchankova P, Gundlach H, Wicker T, Petzold A, Felder M, Steuernagel B, Scholz U, Graner A, Platzer M, Dolezel J, Stein N (2009) Gene content and virtual gene order of barley chromosome 1H. Plant Physiol (Rockv) 151:496–505

Mayer KFX, Martis M, Hedley P, Šimková H, Liu H, Morris JA, Steuernagel B, Taudien S, Roessner S, Gundlach H, Kubaláková M, Suchánková P, Murat F, Felder M, Nussbaumer T, Graner A, Salse J, Endo T, Sakai H, Tanaka T, Itoh T, Sato K, Platzer M, Matsumoto T, Scholz U, Doležel J, Waugh R, Stein N (2011) Unlocking the barley genome by chromosomal and comparative genomics. Plant Cell 23:1249–1263

Montenegro JD, Golicz AA, Bayer PE, Hurgobin B, Lee H, Chan C-KK, Visendi P, Lai K, Doležel J, Batley J, Edwards D (2017) The pangenome of hexaploid bread wheat. Plant J 90:1007–1013

Mostovoy Y, Levy-Sakin M, Lam J, Lam ET, Hastie AR, Marks P, Lee J, Chu C, Lin C, Džakula Ž, Cao H, Schlebusch SA, Giorda K, Schnall-Levin M, Wall JD, Kwok P-Y (2016) A hybrid approach for de novo human genome sequence assembly and phasing. Nat Meth 13:587–590

Muñoz-Amatriaín M, Eichten SR, Wicker T, Mascher M, Richmond TA, Jeddeloh JA, Steuernagel B, Taudien S, Platzer M, Scholz U, Ariyadasa R, Spannagl M, Nussbaumer T, Mayer KFX, Springer NM, Muehlbauer GJ, Stein N (2013) Distribution, functional impact and origin mechanisms of copy number variation in the barley genome. Genome Biol 14:R58

Muñoz-Amatriaín M, Lonardi S, Luo M, Madishetty K, Svensson JT, Moscou MJ, Wanamaker S, Jiang T, Kleinhofs A, Muehlbauer GJ, Wise RP, Stein N, Ma Y, Rodriguez E, Kudrna D, Bhat PR, Chao S, Condamine P, Heinen S, Resnik J, Wing R, Witt HN, Alpert M, Beccuti M, Bozdag S, Cordero F, Mirebrahim H, Ounit R, Wu Y, You F, Zheng J, Simková H, Dolezel J, Grimwood J, Schmutz J, Duma D, Altschmied L, Blake T, Bregitzer P, Cooper L, Dilbirligi M, Falk A, Feiz L, Graner A, Gustafson P, Hayes PM, Lemaux P, Mammadov J, Close TJ (2015) Sequencing of 15622 gene-bearing BACs clarifies the gene-dense regions of the barley genome. Plant J 84:216–227

Putnam NH, O'Connell BL, Stites JC, Rice BJ, Blanchette M, Calef R, Troll CJ, Fields A, Hartley PD, Sugnet CW, Haussler D, Rokhsar DS, Green RE (2016) Chromosome-scale shotgun assembly using an in vitro method for long-range linkage. Genome Res 26:342–350

Sato K, Tanaka T, Shigenobu S, Motoi Y, Wu J, Itoh T (2015) Improvement of barley genome annotations by deciphering the Haruna Nijo genome. DNA Res

Schnable PS, Ware D, Fulton RS, Stein JC, Wei F, Pasternak S, Liang C, Zhang J, Fulton L, Graves TA, Minx P, Reily AD, Courtney L, Kruchowski SS, Tomlinson C, Strong C, Delehaunty K, Fronick C, Courtney B, Rock SM, Belter E, Du F, Kim K, Abbott RM, Cotton M, Levy A, Marchetto P, Ochoa K, Jackson SM, Gillam B, Chen W, Yan L, Higginbotham J, Cardenas M, Waligorski J, Applebaum E, Phelps L, Falcone J, Kanchi K, Thane T, Scimone A, Thane N, Henke J, Wang T, Ruppert J, Shah N, Rotter K, Hodges J, Ingenthron E, Cordes M, Kohlberg S, Sgro J, Delgado B, Mead K, Chinwalla A, Leonard S, Crouse K, Collura K, Kudrna D, Currie J, He R, Angelova A, Rajasekar S, Mueller T, Lomeli R, Scara G, Ko A, Delaney K, Wissotski M, Lopez G, Campos D, Braidotti M, Ashley E, Golser W, Kim H, Lee S, Lin J, Dujmic Z, Kim W, Talag J, Zuccolo A, Fan C, Sebastian A, Kramer M, Spiegel L, Nascimento L, Zutavern T, Miller B, Ambroise C, Muller S, Spooner W, Narechania A, Ren L, Wei S, Kumari S, Faga B, Levy MJ, McMahan L, Van Buren P, Vaughn MW, Ying K, Yeh C-T, Emrich SJ, Jia Y, Kalyanaraman A, Hsia A-P, Barbazuk WB, Baucom RS, Brutnell TP, Carpita NC, Chaparro C, Chia J-M, Deragon J-M, Estill JC, Fu Y, Jeddeloh JA, Han Y, Lee H, Li P, Lisch DR, Liu S, Liu Z, Nagel DH, McCann MC, SanMiguel P, Myers AM, Nettleton D, Nguyen J, Penning BW, Ponnala L, Schneider KL, Schwartz DC, Sharma A, Soderlund C, Springer NM, Sun Q, Wang H, Waterman M, Westerman R, Wolfgruber TK, Yang L, Yu Y, Zhang L, Zhou S, Zhu Q, Bennetzen JL, Dawe RK, Jiang J, Jiang N, Presting GG, Wessler SR, Aluru S, Martienssen RA, Clifton SW, McCombie WR, Wing RA, Wilson RK (2009) The B73 maize genome: complexity, diversity, and dynamics. Science 326:1112–1115

Schulte D, Close TJ, Graner A, Langridge P, Matsumoto T, Muehlbauer G, Sato K, Schulman AH, Waugh R, Wise RP, Stein N (2009) The international barley sequencing consortium—at the threshold of efficient access to the barley genome. Plant Physiol (Rockv) 149:142–147

Schulte D, Ariyadasa R, Shi B, Fleury D, Langridge P, Collins N, Saski C, Atkins M, deJong P, Graner A, Stein N (2011) BAC library resources for map-based cloning and physical map construction in barley (*Hordeum vulgare* L.). BMC Genom 12:247

Shendure J, Ji H (2008) Next-generation DNA sequencing. Nat Biotechnol 26:1135–1145

Shendure J, Mitra RD, Varma C, Church GM (2004) Advanced sequencing technologies: methods and goals. Nat Rev Genet 5:335–343

Shendure J, Balasubramanian S, Church GM, Gilbert W, Rogers J, Schloss JA, Waterston RH (2017) DNA sequencing at 40: past, present and future. Nature 550:345

Sourdille P, Singh S, Cadalen T, Brown-Guedira G, Gay G, Qi L, Gill B, Dufour P, Murigneux A, Bernard M (2004) Microsatellite-based deletion bin system for the establishment of genetic-physical map relationships in wheat (*Triticum aestivum* L.). Funct Integr Genomics 4:12–25

Stein N (2014) Development of sequence resources. In: Kumlehn J, Stein N (eds) Biotechnological approaches to barley improvement. Springer, Berlin

Stein N, Steuernagel B (2014) Advances in sequencing the barley genome. In: Tuberosa R, Frison E, Graner A (eds) Advances in genomics of plant genetic resources. Springer Netherlands, Dordrecht, pp 391–403

Steuernagel B, Taudien S, Gundlach H, Seidel M, Ariyadasa R, Schulte D, Petzold A, Felder M, Graner A, Scholz U, Mayer KFX, Platzer M, Stein N (2009) De novo 454 sequencing of barcoded BAC pools for comprehensive gene survey and genome analysis in the complex genome of barley. BMC Genom 10:547

Taudien S, Steuernagel B, Ariyadasa R, Schulte D, Schmutzer T, Groth M, Felder M, Petzold A, Scholz U, Mayer KF, Stein N, Platzer M (2011) Sequencing of BAC pools by different next generation sequencing platforms and strategies. BMC Res Notes 4:411

Tegenfeldt JO, Prinz C, Cao H, Huang RL, Austin RH, Chou SY, Cox EC, Sturm JC (2004) Micro- and nanofluidics for DNA analysis. Anal Bioanal Chem 378:1678–1692

Tettelin H, Riley D, Cattuto C, Medini D (2008) Comparative genomics: the bacterial pan-genome. Curr Opin Microbiol 11:472–477

The International Barley Genome Sequencing Consortium (IBSC), Mayer KFX, Waugh R, Langridge P, Close TJ, Wise RP, Graner A, Matsumoto T, Sato K, Schulman A, Muehlbauer GJ, Stein N (2012) A physical, genetical and functional sequence assembly of the barley genome. Nature (Lond) 491:711–716

The International Wheat Genome Sequencing Consortium (2014) A chromosome-based draft sequence of the hexaploid bread wheat (*Triticum aestivum*) genome. Science 345. https://doi.org/10.1126/science.1251788

The International Wheat Genome Sequencing Consortium (IWGSC) (2018) Shifting the limits in wheat research and breeding through a fully annotated and anchored reference genome sequence (under review)

Tiwari VK, Heesacker A, Riera-Lizarazu O, Gunn H, Wang S, Wang Y, Gu YQ, Paux E, Koo D-H, Kumar A, Luo M-C, Lazo G, Zemetra R, Akhunov E, Friebe B, Poland J, Gill BS, Kianian S, Leonard JM (2016) A whole-genome, radiation hybrid mapping resource of hexaploid wheat. Plant J 86:195–207

Whitelaw CA, Barbazuk WB, Pertea G, Chan AP, Cheung F, Lee Y, Zheng L, van Heeringen S, Karamycheva S, Bennetzen JL, SanMiguel P, Lakey N, Bedell J, Yuan Y, Budiman MA, Resnick A, Van Aken S, Utterback T, Riedmuller S, Williams M, Feldblyum T, Schubert K, Beachy R, Fraser CM, Quackenbush J (2003) Enrichment of gene-coding sequences in maize by genome filtration. Science 302:2118–2120

Wicker T, Schlagenhauf E, Graner A, Close TJ, Keller B, Stein N (2006) 454 sequencing put to the test using the complex genome of barley. BMC Genom 7:275

Yu Y, Tomkins JP, Waugh R, Frisch DA, Kudrna D, Kleinhofs A, Brueggeman RS, Muehlbauer GJ, Wise RP, Wing RA (2000) A bacterial artificial chromosome library for barley (*Hordeum vulgare* L.) and the identification of clones containing putative resistance genes. Theor Appl Genet 101:1093–1099

Zeng X, Long H, Wang Z, Zhao S, Tang Y, Huang Z, Wang Y, Xu Q, Mao L, Deng G, Yao X, Li X, Bai L, Yuan H, Pan Z, Liu R, Chen X, WangMu Q, Chen M, Yu L, Liang J, DunZhu D, Zheng Y, Yu S, LuoBu Z, Guang X, Li J, Deng C, Hu W, Chen C, TaBa X, Gao L, Lv X, Abu YB, Fang X, Nevo E, Yu M, Wang J, Tashi N (2015) The draft genome of Tibetan hulless barley reveals adaptive patterns to the high stressful Tibetan Plateau. Proc Natl Acad Sci 112:1095–1100

Zhao G, Zou C, Li K, Wang K, Li T, Gao L, Zhang X, Wang H, Yang Z, Liu X, Jiang W, Mao L, Kong X, Jiao Y, Jia J (2017) The Aegilops tauschii genome reveals multiple impacts of transposons. Nature Plants. https://doi.org/10.1038/s41477-017-0067-8

Zhao Q, Feng Q, Lu H, Li Y, Wang A, Tian Q, Zhan Q, Lu Y, Zhang L, Huang T, Wang Y, Fan D, Zhao Y, Wang Z, Zhou C, Chen J, Zhu C, Li W, Weng Q, Xu Q, Wang Z-X, Wei X, Han B, Huang X (2018) Pan-genome analysis highlights the extent of genomic variation in cultivated and wild rice. Nat Genet. https://doi.org/10.1038/s41588-018-0041-z

6 Gene Prediction in the Barley Genome

Sven O. Twardziok, Heidrun Gundlach, Thomas Lux, Pete E. Hedley, Micha Bayer, Matthias Platzer, Marco Groth, Robbie Waugh and Manuel Spannagl

Abstract

Gene prediction in large and highly repetitive grass genomes like barley is complicated by large numbers of transposable elements (TEs), pseudogenes and often incomplete or un-/miss-oriented genomic sequence. In this chapter, we describe the automated gene prediction and annotation pipeline used for the latest barley reference genome sequence, as well as the genomic evidence used to predict gene models. Additional topics cover the (automated) functional annotation, the evaluation of the gene models, and a comprehensive discussion about shortcomings of the current annotation and ways to improve it further.

S. O. Twardziok · H. Gundlach · T. Lux
M. Spannagl (✉)
Plant Genome and Systems Biology (PGSB), Helmholtz Zentrum München, 85764 Neuherberg, Germany
e-mail: manuel.spannagl@helmholtz-muenchen.de

P. E. Hedley · M. Bayer · R. Waugh
James Hutton Institute, Dundee DD2 5DA, UK

M. Platzer · M. Groth
Leibniz-Institut für Alternsforschung, Fritz-Lipmann-Institut e.V. (FLI), Beutenbergstraße 11, 07745 Jena, Germany

6.1 Introduction

Until recently, the identification of barley's full gene inventory was limited by the absence of a reference quality genome sequence. An automated gene annotation is usually one of the first steps after sequencing of a genome especially if a complete and validated set of protein sequences is not available. The predicted gene structures allow for targeted gene analyses, comparative genomic studies as well as the application of transcriptomics approaches. Gene prediction on the 2012 barley genome resources identified 26,159 high-confidence gene models and another 53,220 low-confidence gene loci (International Barley Sequencing Consortium 2012). The gene calls presented in the 2012 study were derived by the integration of gene models predicted on Whole-Genome Shotgun (WGS)-assembled sequences and a nonredundant set of publicly available fl-cDNAs that were not, or only partially identified on the WGS contigs.

With the availability of the latest barley reference genome sequence in 2017 (Mascher et al. 2017), both gene prediction quality and completeness increased substantially, along with the quality and contiguity/completeness of the genome assembly. As a result, 39,734 high-confidence gene models and 41,949 low-confidence gene loci have been predicted on the latest barley reference genome sequence. This book chapter describes the genomic

N. Stein and G. J. Muehlbauer (eds.), *The Barley Genome*, Compendium of Plant Genomes,
https://doi.org/10.1007/978-3-319-92528-8_6

evidence used during automated gene prediction in barley, outlines the automated gene prediction procedure (structural gene prediction) including confidence classification and filtering of transposable elements and summarizes the evaluation of the barley gene calls. An additional section deals with the functional annotation of gene models and last but not least strategies to improve the barley gene annotation in the near future are discussed.

6.2 Evidence for Automated Gene Prediction

To assist the automated calling of gene structures, different types of genomic "evidence" proved to be useful. These include, most prominently, transcriptome/gene expression data and mapping of homologous gene models from related reference species. This chapter describes the genomic evidence used in the automated gene prediction on the latest barley reference genome sequence (Mascher et al. 2017).

Transcript resources

To assist with gene model identification, transcript datasets were generated using both Illumina RNA-Seq and Pacific Biosciences Iso-Seq. A single barley cultivar, Morex, was used throughout, and was the basis of the current reference (Mascher et al. 2017) and previous draft (International Barley Sequencing Consortium 2012) genome sequences. This ensured robust and accurate mapping of the RNA-Seq and Iso-Seq reads, with no expected DNA polymorphism, enabling generation of high-confidence predicted gene models.

Illumina RNA-Seq

Plant material was grown under controlled conditions as individual plants or pooled dissected tissues where the material was limiting. In total, 16 barley tissues were isolated:

Vegetative tissues: Leaf tissue (LEA) was sampled from seedlings at 17 days after planting (dap), and root tissue at both 17 dap (ROO1) and 28 dap (ROO2). The third stem internode (NOD) was dissected at 42 dap, and senescing leaf (SEN) isolated at 56 dap. Seedlings were grown to 10 dap in the dark to isolate etiolated leaf (ETI) material. Epidermal strips (EPI) were made from 28 dap plant leaf tissue.

Inflorescence tissues: Whole developing inflorescence tissue was sampled at 30 dap (INF1) and 50 dap (INF2). From plants 42 dap, inflorescences were dissected to isolate lodicule (LOD), lemma (LEM) and palea (PAL). Rachis (RAC) was isolated from plants 35 dap.

Developing grain tissues: Whole developing grain (caryopsis) was sampled at 5 days post-anthesis (dpa; CAR5) and 15 dpa (CAR15). Only the central caryopses were sampled from each spike.

Germinating grain tissues: Mature grain was germinated and after 4 days, embryonic tissue (including mesocotyl and seminal roots; EMB) was dissected.

RNA extraction and RNA-Seq methods are described in Mascher et al. (2017).

In addition to using this dataset to assist gene annotation of the barley genome, it provides an extensive gene transcript atlas to study differential gene expression profiles and isoform analysis (Flores et al., in prep).

PacBio Iso-Seq

Pacific Biosciences (PacBio) SMRT (Single Molecule, Real-Time) sequencing technology has the potential to generate reads over 60 kbp (http://www.pacb.com/smrt-science/smrt-sequencing/read-lengths/). This has the potential for sequencing entire mRNA transcripts in a single read, thus obviating the need for de novo assembly. PacBio's approach to this, Iso-Seq, includes both wet-lab protocols and bioinformatics analysis (http://www.pacb.com/blog/intro-to-iso-seq-method-full-leng/).

Iso-Seq was performed on RNA from germinating grain tissues (EMB), as above for the Illumina RNA-Seq.

With PacBio's RSII sequencer short sequences can be overrepresented, so libraries of different sizes (1–2, 2–4, and 3–8 kbp) were sequenced. ICE/Quiver, the consensus calling tool in the SMRTPipe software (https://github.com/

PacificBiosciences/SMRT-Analysis/wiki/SMRT-Pipe-Reference-Guide-v2.1) was used to generate consensus sequences from multiple reads-of-insert (ROI) representing the same transcript. A ROI itself represents a consensus sequence of the multiple subreads generated through repeat passes of the polymerase over the double-stranded DNA insert (http://files.pacb.com/software/smrtanalysis/2.2.0/doc/smrtportal/help/!SSL!/Webhelp/Portal_PacBio_Glossary.htm). ROIs are grouped into isoform clusters using the ICE algorithm, and Quiver is used to recruit non-full-length reads to these clusters and compute a more accurate consensus sequence based on these ("polishing"). Based on the level of confidence in the consensus calls, Quiver annotates its output as either high or low quality. Polished high-and low-quality read-of-insert sequences from all runs were merged into a single FASTA file of 122,190 sequences.

Reads of this type may still contain errors at this stage, the prevalent type being small indels (Ross et al. 2013). This is in contrast to Illumina sequencing, where substitutions are the predominant error type, albeit at much lower error rates than for raw PacBio output. PacBio and Illumina data, thus, complement each other, in that PacBio reads provide long reads with relatively high error rates, while Illumina data provide higher accuracy at the base level. The two technologies can be combined to provide reads that are both long *and* accurate, using either hybrid assembly or error correction. We opted for an error correction approach, where we corrected the PacBio sequences with Illumina data generated from the same RNA sample from germinating grain tissues, using proovread v.2.13.8 (Hackl et al. 2014). This resulted in 123,875 corrected sequences—more than in the input dataset, a result of splitting chimeric reads. Figure 6.1 shows the trimodal frequency distribution for the length of the trimmed, error-corrected Iso-Seq reads, reflecting merging of the three libraries.

The Illumina data was mapped back to both the uncorrected and the corrected PacBio data to quantify the effect of the error correction, using Bowtie2 (Langmead and Salzberg 2012).

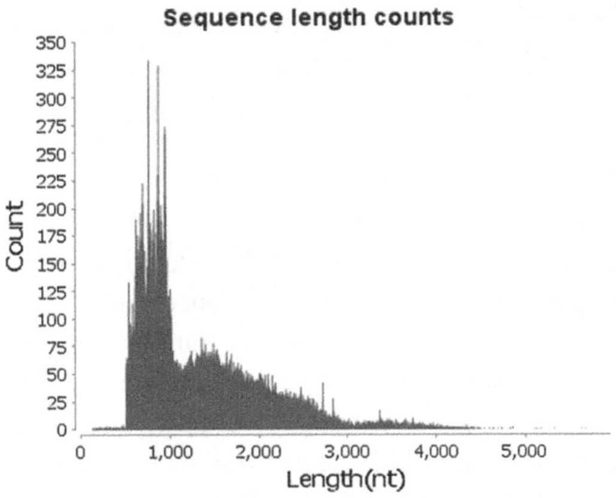

Fig. 6.1 Frequency distribution of sequence lengths for the trimmed, error-corrected PacBio Iso-Seq reads

The resulting mappings were then subjected to coverage analysis with the GATK's DepthOfCoverage tool, and the percentage of reference bases that had no Illumina support (zero coverage) was calculated from the output. This decreased from 18.3 to 2.9% in the corrected reference, a clear improvement resulting from the error correction. We also ran the VarScan variant caller v2.3.7 (Koboldt et al. 2012) over both mappings to establish the number of indels before and after error correction. The error correction resulted in a dramatic reduction of the number of indels from 59,765 (for the whole dataset of 122,190 uncorrected sequences) to only 127 in the corrected sequences.

Transcriptome data availability

Transcriptome datasets were deposited in the European Nucleotide Archive (ENA) and are available from the following accession numbers (Table 6.1).

Additional transcriptome data

For barley, a comprehensive, publicly available and high-quality collection of full-length cDNA sequences exist (Matsumoto et al. 2011). These sequences were generated from the *Haruna Nijo* barley variety, whereas the genome sequence is from the *Morex* variety. This resource includes a total of 28,592 sequences that were further utilized during gene prediction.

Table 6.1 Transcriptome data availability

Transcriptome sequencing		
Illumina RNA-Seq reads	ENA	PRJEB3149, PRJEB14349
PacBio Iso-Seq reads	ENA	PRJEB14446

Mapping of homologous reference gene models Predicted gene structures from related reference organisms can be used as evidence during gene annotation, given there is substantial sequence conservation between the reference sequence and the homologous/orthologous copy in the barley genome. Previously predicted barley protein sequences (International Barley Sequencing Consortium 2012) were selected, as well as gene predictions from the closely related grass species *Brachypodium distachyon* (Vogel et al. 2010), rice (*O. sativa*) (Matsumoto et al. 2005), and sorghum (*S. bicolor*) (Paterson et al. 2009).

- Barley 2012 gene models (PGSB PlantsDB): **26,159** genes
- *Sorghum bicolor* v2.1 (Phytozome): **33,032** genes
- *Brachypodium distachyon* v2.1 (Phytozome): **31,694** genes
- Rice MSU7.0 (MSU): **39,049** genes.

6.3 Automated Gene Prediction Pipeline

To predict gene structures on whole-genome sequence assemblies from complex genomes such as barley, automated gene prediction pipelines are used. Gene-supporting evidence (as introduced and described in the section before) typically constitute the starting point of such automated gene prediction pipelines and are processed in different ways as the first step (Fig. 6.2).

Plant genomes are usually repeat masked prior to gene detection for two reasons:

(i) to reduce computing times, as transposable elements cover 50 to over 80% of the sequences and generate high hit numbers in homology searches due to their repetitive nature;
(ii) to avoid picking up all the transposon-related gene models, which outnumber canonical genes and can distort downstream data analyses. The drawback of repeat-masking is the danger of "overmasking" which means that parts of genes are masked by gene fragments that might have been picked up as parts of transposons and as such find their way into the transposon libraries.

Due to such minor but clearly observable overmasking issues for barley, we directly annotated the genes on the unmasked, whole-genome sequence assembly. As a consequence, a vast number of transposon-related genes, were sorted out in a later step by an elaborate keyword search on the functional gene descriptions (see "Confidence classification"). We decided for two reasons for an entirely evidence-driven gene prediction process rather than incorporating de novo gene finders such as Augustus: (a) de novo gene finders typically show high error rates in predicting many false positive predictions that need to be dealt with in a subsequent consolidation step and (b) given the great amount of transcriptome and homology mapping data for barley at this stage we expect most bonafide genes to be supported by at least one of these data types, rather than being identified by de novo gene finders alone.

In the first step of the pipeline, predicted protein sequences from barley and closely related grass species brachypodium, rice, and sorghum, as well as predicted open reading frames from full-length cDNA sequences from barley, were aligned against the unmasked chromosome sequences using the splice-aware alignment software GenomeThreader (Gremme et al. 2005) (version 1.6.2; parameters used:—species

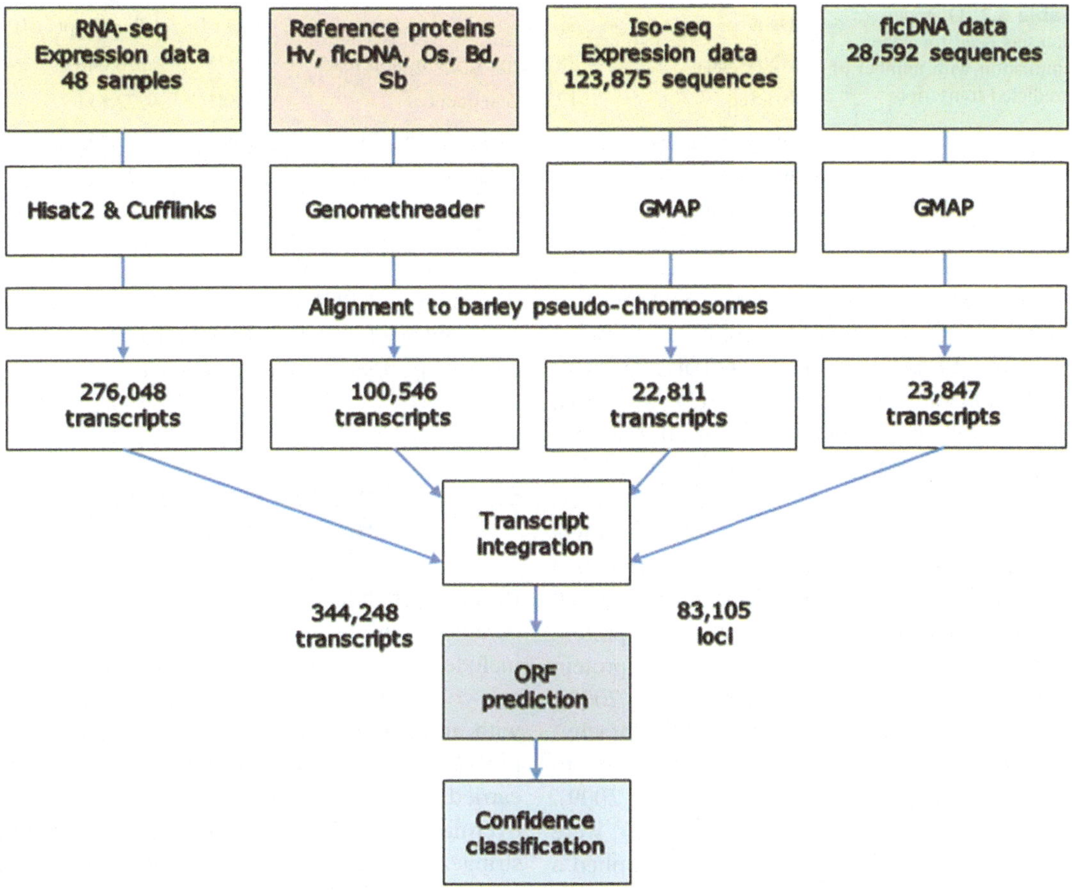

Fig. 6.2 Overview of the gene prediction pipeline, which combined information from four evidence sources in order to predict gene structures on the barley genome assembly

rice—gcmincoverage 30—prseedlength 7—prhdist 4—force). Resulting transcript structure predictions were then merged using Cuffcompare from the Cufflinks package (Trapnell et al. 2012). RNA-Seq reads were mapped against the chromosomes using Hisat2 (Kim et al. 2015) (default parameters). Resulting alignment files were then assembled into transcript structures by using the Cufflinks software package and Iso-Seq reads were aligned against the chromosome sequences using GMAP (Wu and Watanabe 2005) (default parameters). A Python script from the PacBio repository (*collapse_isoforms_by_sam.py*) was used to predict transcript structures and to remove redundant transcripts. For the alignments of fl-cDNA sequences to chromosomes, the GMAP software (Wu and Watanabe 2005) was used once again. RNA-Seq-based transcript structures were clustered with the reference-based gene model predictions, the structure information from Iso-Seq alignments as well as with structure information from fl-cDNA sequence alignments and a consensus transcript set was generated by using Cuffcompare from the Cufflinks software package (Trapnell et al. 2012).

6.4 ORF Prediction and Selection

Merging of transcript sequences from RNA-Seq, reference proteins, fl-cDNAs and Iso-Seq sequences resulted in 344,248 transcript sequences, which were clustered into 83,105 potential gene loci (Table 6.2).

Table 6.2 Overview of datasets used for gene annotation with number of predicted transcripts

Data	Size	Loci (#)	Transcripts (#)
RNA-Seq	48 samples, 1.9 billion reads	63,075	276,048
Iso-Seq	123,875 sequences	15,881	22,811
Reference proteins	Hv, HvflcDNA, Os, Bd, Sb	56,682	100,546
fl-cDNA	28,592 sequences	19,721	23,847
Combination		83,105	344,248

To determine putative open reading frames, as well as corresponding peptide sequences including PFAM domain predictions, Transdecoder software (Haas et al. 2013) was used (version rel_16Jan; parameters: -m 30—retain_long_orfs 90—search_pfam pfam.AB.hmm.bin). This resulted in several alternative predicted peptides for each transcript. To select a single best translation per transcript ("representative gene model"), we used BLASTP to compare all predicted peptides with a comprehensive protein database including high-confidence protein sequences from arabidopsis (Kaul et al. 2000), *Zea mays* L. (Schnable et al. 2009), brachypodium (Vogel et al. 2010), rice (Matsumoto et al. 2005) and sorghum (Paterson et al. 2009). BLASTP hits with an e-value below 10^{-5} were considered as significant hits. We then applied a sequential filtering strategy to identify a single best translation for each transcript. In each filtering step, all remaining peptide sequences were sorted according to a specific attribute/category and those translations within the highest category (or with the highest value) remained for the next filtering step. The six filtering steps were as follows:

1. Three categories: (i) Neither homology support nor PFAM domains, (ii) without homology support and with PFAM domains, and (iii) with homology support;
2. Total length of translation;
3. Four categories: (i) CDS without start and without stop codon, (ii) CDS without start and with stop codon, (iii) CDS with start codon and no stop codon, 4, and (iv) CDS with start and with stop codon;
4. Number of PFAM domains;
5. Number of significant BLAST hits;
6. Start position on the chromosome.

6.5 Confidence Classification of Gene Predictions

Besides bonafide genes, the candidate gene set obtained from the steps outlined before typically includes transposons, pseudogenes, long non-coding RNAs, and other genetic elements with an open reading frame. This is even more pronounced here where no repeat masking was carried out before the gene calling to avoid overmasking (see section before). As a result, stringent classification (along with appropriate criteria) of predicted genetic elements is needed to indicate their putative role. The need for such a classification process reflects the difficulty and limitations of annotating large plant genomes, in general, and is not a standard procedure for any plant annotation strategy.

The classification scheme was developed to discriminate between loci representing high-confidence (HC) protein-coding genes and less reliable low-confidence (LC) genes. LC genes typically consist of gene fragments, putative pseudogenes, transposable elements and non (-protein)-coding transcripts, whereas the HC gene complement should mainly contain bonafide protein-coding gene models.

In the first step, gene models with significant homology to a repeat elements library are classified as low-confidence genes. Second, sequence homology searches against the protein datasets of barley, brachypodium, rice, sorghum, arabidopsis, and predicted protein sequences from the barley fl-cDNAs define significantly conserved gene models which are promoted to the HC gene set as a consequence.

We predicted putative function for all genes (high confidence and low confidence) using the AHRD pipeline (https://github.com/groupschoof/AHRD) on the basis of one representative protein sequence for each gene (see details in section "Functional annotation"). To discriminate between TE-genes and genes, we searched the functional description for typical transposon-related keywords, like retrotransposon, GAG/Pol, or integrase. Low-confidence genes with predicted functions other than these TE domains and unknown function were moved to the high-confidence gene set and high confidence genes that were annotated as transposable element were downgraded to the set of low-confidence genes.

Based on functional annotation and final refinement, we defined ten confidence subclasses:

- HC_G: high-confidence genes with predicted function
- HC_TE: high-confidence genes which might be transposable element due to conflicting information
- HC_u: high-confidence gene without function assignment
- HC_U: high-confidence gene with annotation of unknown function (based on homology to other proteins with unknown function)
- LC_M: genes that were manually removed from the set of high-confidence genes
- LC_nof: genes without open reading frame

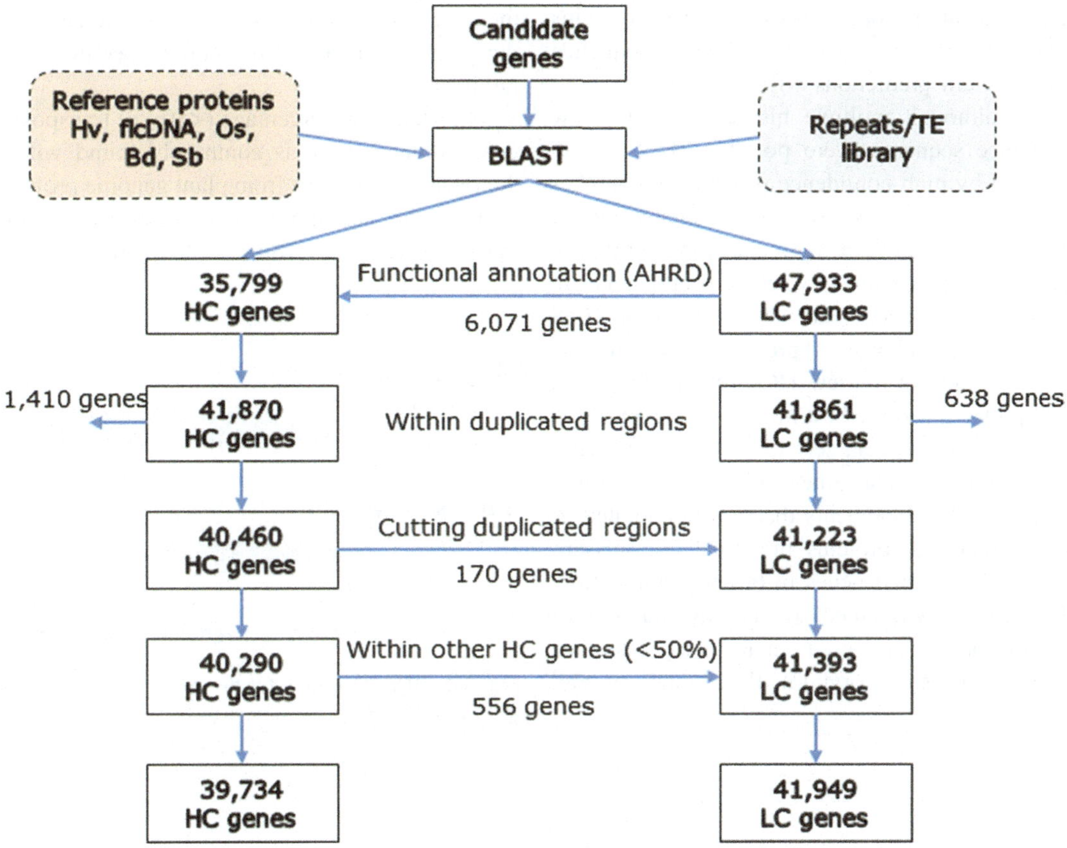

Fig. 6.3 Final confidence assignment pipeline and manual refinement

- LC_TE: genes that are annotated as transposable elements
- LC_u: genes without functional annotation
- LC_U: genes with annotation of unknown function (based on homology to other proteins with unknown function).

An overview of the gene classification pipeline is provided in Fig. 6.3.

6.6 Evaluation of Gene Calls

6.6.1 Comparison to Gene Models in the WGS Assembly of Barley cv. Morex

We used BLAST to compare the representation of previously annotated barley proteins (International Barley Sequencing Consortium 2012) with the latest gene annotation (Mascher et al. 2017). A large proportion of the earlier proteins (96.4%) were represented by reference sequence v1.0 protein predictions with at least 75% coverage, although multiple hits to a single, new reference sequence were possible. 92.2% were covered by high-confidence proteins and another 28.3% were covered by low-confidence genes. This overlap between high- and low-confidence genes can be explained by sets of pseudogenes in the low-confidence gene set. A total of 1058 "2012 proteins" were represented by new LC proteins and not by new HC proteins. Of these, 572 proteins had the best hit to a low-confidence protein from subclasses LC_TE or LC_TE?, which indicates that some of these proteins might be transposable elements that were annotated as high-confidence proteins previously or missannotated TE-related genes in the new annotation. This can be explained by the low number of transposon genes present in the used reference protein sets from Sorghum, Brachypodium, and rice.

6.6.2 Benchmarking Universal Single-Copy Orthologs (BUSCO)

A frequently used method to validate the completeness of genome assemblies as the basis for gene annotation is BUSCO (Simao et al. 2015). BUSCO consists of a set of conserved gene models across a wide range of plant taxa and the abundance and conservation of these genes serves as a proxy for genome assembly and annotation completeness. For barley, BUSCO gene abundance measures in the 2012 predictions and the latest IBSC 2017 predictions show almost 98% completeness in the combined HC and LC gene set of 2017 (Fig. 6.4), as compared with 95% in 2012. These numbers compare well with similar plant genomes sequenced to a reference standard and are typically significantly higher than for draft WGS assemblies. Due to its setup, however, BUSCO does not provide a measure for the correctness or completeness of the more Triticeae- or barley-specific gene families.

A certain low percentage (<5%) of transposon gene contamination is commonly found within the released gene sets from plant genome projects even if the annotation was performed on the masked sequences. Some of them are very

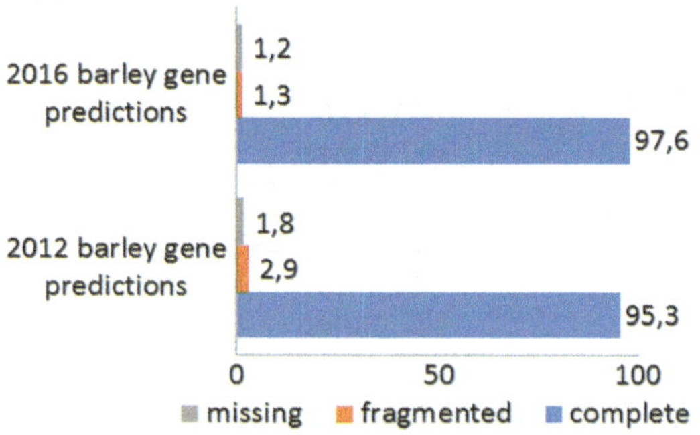

Fig. 6.4 BUSCO results for the barley gene sets as of 2012 and 2017 (%)

obvious and can easily be extracted by a keyword search for transposon categories and components in the functional description as well as the Pfam and InterPro designations.

6.7 (Automated) Functional Annotation of Gene Models in Barley

In contrast to the structural annotation of gene models outlined before, the main objective of functional annotation is to assign a (potential) biological function to individual gene models. While manual (expert) curation of gene functions is considered to be the gold standard with respect to reliability and accuracy, automated functional annotation aims to assign potential biological functions to as many gene predictions as possible. Both approaches should not be considered mutually exclusive, in fact both strategies successfully complement each other, especially in genome projects dealing with reference organisms or mature studies with a larger user community (examples from the plant world are arabidopsis and rice). While manual curation of gene models involves significant amounts of expert knowledge for different areas/gene families and is very labor intensive, this approach usually cannot be applied to complete gene sets. To be able to assign potential biological functions to a larger number of gene models nevertheless, automatic computational approaches have been developed and often used in whole-genome sequencing projects since then.

In general, these approaches utilize different criteria for assigning potential biological functions to gene models. Most prominently, sequence homology/conservation of the yet uncharacterized gene model to known reference gene sequences, ideally manually curated, is used to infer and transfer potential function. While this has been proven valid and useful in many cases (Conesa and Gotz 2008), sequence homology or orthology does not necessarily infer conserved biological function. In principle, this method works best when high-quality (manually curated) gene annotations are available from a closely related reference organism. Other criteria often used to infer functional descriptions for uncharacterized gene models are structural sequence features such as domains and conserved (functional) motifs. These can be automatically derived from InterproScan (Jones et al. 2014) computations with the protein sequences of all gene predictions as an input.

The AHRD (Automated Human Readable Descriptions) (https://github.com/groupschoof/AHRD) method has been applied to the (automated) barley gene predictions to determine functional descriptions ("human-readable descriptions") for as many gene models as possible. A short discussion about the issues of functional (as well as structural) annotation in barley, in particular, as well as an outlook, can be found later in this chapter.

AHRD assigns functional descriptions on the basis of sequence homology searches to reference proteins as well as by domain signatures identified on the protein sequence. The method is fully adjustable, meaning users can select the protein databases searched, based on the availability of expert/validated functional annotations in species with a position close to the subject in the taxonomic tree. For barley, searches against the gene complement of arabidopsis (Kaul et al. 2000) as well as against UniProt90 were performed. The BLAST hit descriptions are tokenized into informative words, and a lexical analysis scores these tokens according to their frequency and the quality of the blast hits they occur in. Finally, the best scoring description is assigned. This approach is especially useful for Triticeae genes, where the UniProt databases contain large amounts of uncharacterized genes with often wordy nondescriptive headers, like "Chromosome 3B, genomic scaffold, cultivar Chinese Spring" or simply "barley Uncharacterized protein". Here, a pure best BLAST hit approach would lead to about 30% of 'unknown' descriptions. In addition, AHRD makes use of protein domain signatures such as PFAM which were computed using the InterproScan software package (Jones et al. 2014). AHRD scores blast hits taken from searches against different databases on the basis of the trust put into these databases and the alignment quality. Details on the results for the 2017 barley annotation are given in the next section.

6.8 Characteristics and Number of Predicted Gene Models

We identified 39,734 high- and 41,949 low-confidence gene models in the barley genome assembly as of 2017 (Mascher et al. 2017). A potential function was assigned to genes based on their representative proteins (see section above). Using this procedure, we were able to assign a potential function to the majority (n = 31,899) of high-confidence genes (subclass HC_G). Some high-confidence genes were likely candidates for transposable elements (subclass HC_TE?) but remained in the set of high-confidence genes due to significant homology with high-confidence genes from other species or from previous gene annotation. Most low-confidence genes (n = 27,922) had no predicted function (subclass LC_u). A great number of low-confidence genes (n = 8975) were annotated as transposable elements (subclass LC_TE) and 948 low-confidence genes were likely candidates for being transposable elements (subclass LC_TE?). A small number of low-confidence genes (n = 403) had no open reading frame of at least 150 bp length (subclass LC_nof). The number of HC genes per chromosome varied between 4380 for chromosome 4H and 6518 for chromosome 2H and we predicted 2157 high-confidence genes on unanchored scaffolds (U). Basic gene model statistics are given in Table 6.3 and compared to gene model statistics of other plant species.

6.9 Data Availability and Download

The barley gene annotation can be accessed and downloaded from the following sites and DOIs (Table 6.4):

IPK Gatersleben: http://webblast.ipk-gatersleben.de/barley_ibsc/downloads/

PGSB PlantsDB: http://pgsb.helmholtz-muenchen.de/plant/barley/index.jsp

Ensembl Plants: http://plants.ensembl.org/Hordeum_vulgare/Info/Index.

6.10 Annotation Shortcomings and Ways to Improve the Annotation

The barley gene annotation described here, both structural and functional, represents a fully automated process with little to no manual curation of gene models or gene families. The final annotation result directly reflects the choices made in each step of the annotation process. It is noteworthy that changes of software components or the logic of the workflow have direct consequences on the gene prediction results, thus different annotation pipelines can produce substantially different gene sets.

Gene predictions mainly rely on genomic evidence such as transcriptome data and homologous gene mappings from related reference species. Although both of these sources of evidence and the

Table 6.3 Basic gene model statistics for the 2017 barley annotation, in comparison to other gene predictions

Species	Number of genes	Mean gene length	Mean transcript length	Mean CDS length	Mean exon length	Mean exons/transcript
Rice	42,189	2870	3173	1112	316	4
Wheat TGAC	104,089	4072	4482	1244	362	5
Arabidopsis	27,416	2205	2334	1218	260	5
Barley 2017	39,734	6010	5857	968	277	6
Maize	40,557	5992	5992	1061	251	5

Table 6.4 DOIs for accessing the 2017 barley gene annotation

Annotation of transcribed regions		
High-confidence (HC) gene set GTF file	IPK	https://doi.org/10.5447/ipk/2016/38
Low-confidence (LC) gene set GTF file	IPK	https://doi.org/10.5447/ipk/2016/46
HC gene set CDS (all and representative only)	IPK	https://doi.org/10.5447/ipk/2016/40, https://doi.org/10.5447/ipk/2016/43
LC gene set CDS (all and representative only)	IPK	https://doi.org/10.5447/ipk/2016/48, https://doi.org/10.5447/ipk/2016/51
HC gene set transcripts (all and repr. only)	IPK	https://doi.org/10.5447/ipk/2016/39, https://doi.org/10.5447/ipk/2016/42
LC gene set transcripts (all and repr. only)	IPK	https://doi.org/10.5447/ipk/2016/47, https://doi.org/10.5447/ipk/2016/50
HC gene set proteins (all and repr. only)	IPK	https://doi.org/10.5447/ipk/2016/41, https://doi.org/10.5447/ipk/2016/44
LC gene set proteins (all and repr. only)	IPK	https://doi.org/10.5447/ipk/2016/49, https://doi.org/10.5447/ipk/2016/52
HC gene set functional descriptions	IPK	https://doi.org/10.5447/ipk/2016/45
LC gene set functional descriptions	IPK	https://doi.org/10.5447/ipk/2016/53
Long non-coding RNAs	IPK	https://doi.org/10.5447/ipk/2016/18
microRNA loci	IPK	https://doi.org/10.5447/ipk/2016/15

automated gene calling pipeline have been evaluated and adapted specifically for the challenges involved with the large and complex barley genome (e.g., annotation on unmasked sequence, Iso-Seq data etc.), all automated gene annotations suffer from intrinsic problems that become obvious once working with the gene models. These problems include incorrect structural gene predictions such as gene fusions, splits, fragmented gene models (missing exons), and wrong exon–intron junctions. Some of those "obviously wrong" gene predictions were supported by genomic evidence such as transcriptome data, while others are the result of incorrect mappings and alignment artifacts of transcriptome data to the genome assembly sequence. On a number of occasions, errors or missassemblies in the genomic sequence template introduce termination codons or other features causing gene prediction to fail in predicting the correct structure, or a gene feature at all. This issue is also closely linked with the definition and prediction of pseudogenes, where the boundary between a "true" pseudogene and annotation of an erroneous gene model is often blurred. Especially in large and highly repetitive genomes, such as the barley genome, transposable elements (TEs) can be a major source of incorrect gene predictions as some of them contain gene-like features and can also be transcribed. A high-quality, comprehensive and species-specific library of TEs can help gene prediction tremendously to differentiate between TEs and protein-coding genes, and the availability of these for barley has been the precondition for calling gene structures on the unmasked genome sequence.

Some of the issues and incorrect gene predictions described before can be detected, although not fixed, in an automated fashion. This has been tried in the barley gene annotation both during confidence classification and filtering of gene models by their functional description assigned. As a result, barley genes with conflicting information supporting both TE and protein-coding gene classification could be assigned to respective categories within the main confidence classes HC and LC (see chapter for details). These categories can be the main target for manual/expert curation/inspection and yield an efficient improvement of the overall annotation.

Another shortcoming of automated gene predictions is the annotation accuracy of individual gene families. This becomes more evident the larger/more active the user community is and the more important the gene family is for the respective organism. The accuracy of automated gene predictions with respect to individual gene families both translates into structural and functional features. On the structural side, it is desirable to have all gene family members annotated, preferably with their correct and experimentally proven (where available) structures. On the functional aspect, meaningful descriptions of the gene function, established in the literature, are on the wish list. As discussed before, automated gene predictions will only be able to deliver part of this and even more and higher quality genomic evidence will not be able to overcome all of these issues. As a consequence, many (plant and non-plant) genome-sequencing projects initiated systems and strategies to add a level of manual curation to their automated annotations, either for selected gene families only or for the full gene complement, at least as an objective. These strategies include the development of platforms for the manual curation of gene models by the community (also as a long-term initiative) as well as collecting input from experts and expert groups for particular gene families (also as part of gene annotation releases).

A prominent example of a user community platform for the improvement of a plant whole-genome annotation is the TAIR database (Lamesch et al. 2012), where both user input and literature knowledge were incorporated into regular annotation update releases. Such community annotation platforms can, therefore, help to improve both the structural and functional annotation, e.g., by resolving gene fusions and incorporating up-to-date experimental findings. Main requirements for any user annotation platform are identifying (and accepting) a single authority for the annotation as well as providing the technical infrastructure and database. In addition, user annotation platforms usually require dedicated funding schemes as at least minimal database curation is needed in addition to the user annotations to avoid inconsistencies. Thus, for barley, a strong user community commitment would be needed to identify and realize a funding scheme/authority to establish such a community annotation platform. On the other hand, the quality of community annotation efforts strongly depends on the establishment and use of consistent criteria for gene prediction and curation across all participating groups. Here, many community efforts tend to fall short and would/will require additional curator oversight. As a result, an alternative strategy proven useful for arabidopsis is the automated identification of potential annotation issues (both structural and functional) by computational approaches, then resolved by manual curation from experienced curators. Besides more and better data supporting gene prediction/annotation and expert curation, improved and/or additional/different computational approaches can also yield significant improvements of gene annotation quality and completeness. This, for instance, includes specific approaches to annotate ln-cRNA or pseudogenes with the existing data, improved transcript assembly or identification of alignment artifacts.

Another strategy to especially improve the functional annotation of gene calls, in general, is the integration of literature knowledge (experimental data) with (automated) whole-genome data. In the case of gene predictions, the automated functional descriptions are hereby replaced by expert-curated functional descriptions derived from experimental studies on individual genes, gene families, or even larger gene sets. This approach also involves the transfer of gene names (or gene symbols) established by experts or experimental studies to the corresponding gene model in the automated gene annotation. An example from barley would be the alpha-amylase gene family with its gene symbols *amy1–amy4,* which can be associated with the corresponding gene IDs in the automated annotation and become searchable for both query terms in genome databases as a consequence.

A comparable effort is currently ongoing within a project of the Wheat Initiative, where annotations from the Wheat Gene Catalogue

(Mcintosh et al. 2003) are going to be integrated with the (automated/whole-genome) genomics resources currently and recently generated in several international consortia. The catalogue of Gene Symbols for Wheat contains more than 400 pages of curated wheat genetic information collected over almost 50 years now. This includes symbols for loci and alleles controlling quantitative characters, as well as SNPs, but also guidelines and rules for Gene Symbolization and Nomenclature in wheat. Much of the data describes genetic features with no or little genomic sequence associated. This data is notoriously hard to integrate with genome sequence data, at least in an automated way. The features associated with sequences, however, are a worthwhile target for integration with genomic data such as the automated annotation. The possible result of such an integration will be the mutual exchange of gene symbol/name and (automated) gene identifier in each of the records, extending the additional information into the respective domain such as genome databases and gene catalogue.

In this context, it has been discussed whether a single, uniform gene name/symbol could be established for a gene model over the Triticeae, or at least a subgroup of the Triticeae (such as barley and wheat). While this should be intuitive and helpful for many users working with the data, arguments related to different gene functions in different Triticeae species and established species-specific terminologies were raised.

While no such "gene catalogue" exists for barley to our knowledge, plenty of comparable genetic information is available or has been published for barley over decades. Although not centrally organized or collected yet, integrating this existing genetic data with the genomic information would make an important contribution to efficiently use either of the datasets in the future. As a starting point, discussions with identified barley gene family experts with barley genome database providers (such as Ensembl-Plants (Bolser et al. 2017)) have already taken place and should be broadened and intensified as more and more genomic data is generated.

For barley, unlike for many other high-throughput genome-sequencing projects, both expert groups with a focus on specific gene families as well as the literature on many genes exist. This makes the improvement of the current

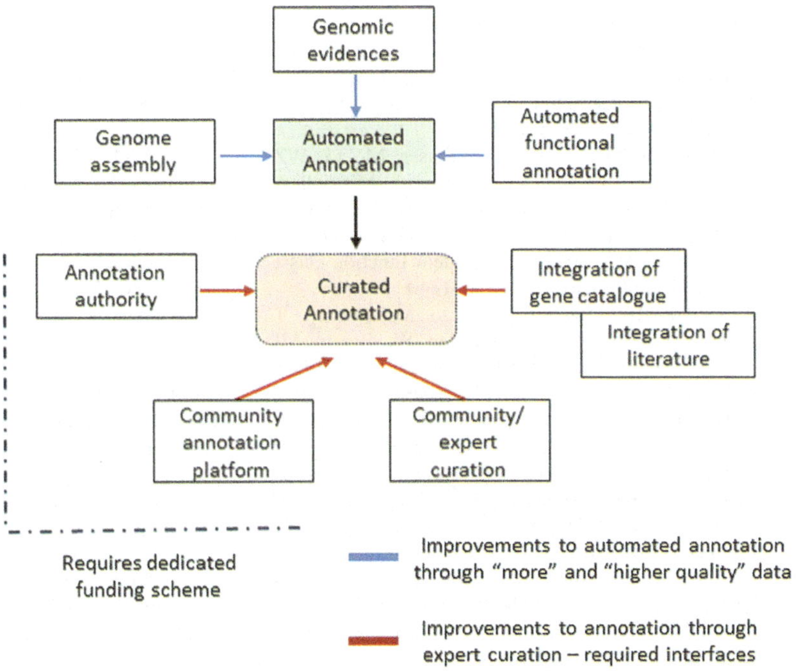

Fig. 6.5 Schematic and simplified illustration of the different components and interfaces for the automated and curated gene annotation in barley

barley annotation possible at all in the first place, but also valuable and desirable to invest in. Figure 6.5 summarizes the steps and options towards an improved barley gene annotation and highlights the need for a dedicated funding scheme for a curated annotation.

References

Bolser DM, Staines DM, Perry E, Kersey PJ (2017) Ensembl plants: integrating tools for visualizing, mining, and analyzing plant genomic data. Methods Mol Biol 1533:1–31

Conesa A, Gotz S (2008) Blast2GO: a comprehensive suite for functional analysis in plant genomics. Int J Plant Genomics 2008:619832

Gremme G, Brendel V, Sparks ME, Kurtz S (2005) Engineering a software tool for gene structure prediction in higher organisms. Inf Softw Technol 47:965–978

Haas BJ, Papanicolaou A, Yassour M, Grabherr M, Blood PD, Bowden J, Couger MB, Eccles D, Li B, Lieber M, MacManes MD, Ott M, Orvis J, Pochet N, Strozzi F, Weeks N, Westerman R, William T, Dewey CN, Henschel R, Leduc RD, Friedman N, Regev A (2013) De novo transcript sequence reconstruction from RNA-seq using the Trinity platform for reference generation and analysis. Nat Protoc 8:1494–1512

Hackl T, Hedrich R, Schultz J, Forster F (2014) Proovread: large-scale high-accuracy PacBio correction through iterative short read consensus. Bioinformatics 30:3004–3011

International Barley Sequencing Consortium (2012) A physical, genetic and functional sequence assembly of the barley genome. Nature 491:711

Jones P, Binns D, Chang HY, Fraser M, Li WZ, McAnulla C, McWilliam H, Maslen J, Mitchell A, Nuka G, Pesseat S, Quinn AF, Sangrador-Vegas A, Scheremetjew M, Yong SY, Lopez R, Hunter S (2014) InterProScan 5: genome-scale protein function classification. Bioinformatics 30:1236–1240

Kaul S, Koo HL, Jenkins J, Rizzo M, Rooney T, Tallon LJ, Feldblyum T, Nierman W, Benito MI, Lin XY, Town CD, Venter JC, Fraser CM, Tabata S, Nakamura Y, Kaneko T, Sato S, Asamizu E, Kato T, Kotani H, Sasamoto S, Ecker JR, Theologis A, Federspiel NA, Palm CJ, Osborne BI, Shinn P, Conway AB, Vysotskaia VS, Dewar K, Conn L, Lenz CA, Kim CJ, Hansen NF, Liu SX, Buehler E, Altafi H, Sakano H, Dunn P, Lam B, Pham PK, Chao Q, Nguyen M, Yu GX, Chen HM, Southwick A, Lee JM, Miranda M, Toriumi MJ, Davis RW, Wambutt R, Murphy G, Dusterhoft A, Stiekema W, Pohl T, Entian KD, Terryn N, Volckaert G, Salanoubat M, Choisne N, Rieger M, Ansorge W, Unseld M, Fartmann B, Valle G, Artiguenave F, Weissenbach J, Quetier F, Wilson RK, de la Bastide M, Sekhon M, Huang E, Spiegel L, Gnoj L, Pepin K, Murray J, Johnson D, Habermann K, Dedhia N, Parnell L, Preston R, Hillier L, Chen E, Marra M, Martienssen R, McCombie WR, Mayer K, White O, Bevan M, Lemcke K, Creasy TH, Bielke C, Haas B, Haase D, Maiti R, Rudd S, Peterson J, Schoof H, Frishman D, Morgenstern B, Zaccaria P, Ermolaeva M, Pertea M, Quackenbush J, Volfovsky N, Wu DY, Lowe TM, Salzberg SL, Mewes HW, Rounsley S, Bush D, Subramaniam S, Levin I, Norris S, Schmidt R, Acarkan A, Bancroft I, Quetier F, Brennicke A, Eisen JA, Bureau T, Legault BA, Le QH, Agrawal N, Yu Z, Martienssen R, Copenhaver GP, Luo S, Pikaard CS, Preuss D, Paulsen IT, Sussman M, Britt AB, Selinger DA, Pandey R, Mount DW, Chandler VL, Jorgensen RA, Pikaard C, Juergens G, Meyerowitz EM, Theologis A, Dangl J, Jones JDG, Chen M, Chory J, Somerville MC, In AG (2000) Analysis of the genome sequence of the flowering plant Arabidopsis thaliana. Nature 408:796–815

Kim D, Landmead B, Salzberg SL (2015) HISAT: a fast spliced aligner with low memory requirements. Nat Methods 12:357–U121

Koboldt DC, Zhang QY, Larson DE, Shen D, McLellan MD, Lin L, Miller CA, Mardis ER, Ding L, Wilson RK (2012) VarScan 2: somatic mutation and copy number alteration discovery in cancer by exome sequencing. Genome Res 22:568–576

Lamesch P, Berardini TZ, Li DH, Swarbreck D, Wilks C, Sasidharan R, Muller R, Dreher K, Alexander DL, Garcia-Hernandez M, Karthikeyan AS, Lee CH, Nelson WD, Ploetz L, Singh S, Wensel A, Huala E (2012) The Arabidopsis Information Resource (TAIR): improved gene annotation and new tools. Nucleic Acids Res 40:D1202–D1210

Langmead B, Salzberg SL (2012) Fast gapped-read alignment with Bowtie 2. Nat Methods 9:357–U354

Mascher M, Gundlach H, Himmelbach A, Beier S, Twardziok SO, Wicker T, Radchuk V, Dockter C, Hedley PE, Russell J, Bayer M, Ramsay L, Liu H, Haberer G, Zhang XQ, Zhang Q, Barrero RA, Li L, Taudien S, Groth M, Felder M, Hastie A, Simkova H, Stankova H, Vrana J, Chan S, Munoz-Amatriain M, Ounit R, Wanamaker S, Bolser D, Colmsee C, Schmutzer T, Aliyeva-Schnorr L, Grasso S, Tanskanen J, Chailyan A, Sampath D, Heavens D, Clissold L, Cao S, Chapman B, Dai F, Han Y, Li H, Li X, Lin C, McCooke JK, Tan C, Wang P, Wang S, Yin S, Zhou G, Poland JA, Bellgard MI, Borisjuk L, Houben A, Dolezel J, Ayling S, Lonardi S, Kersey P, Langridge P, Muehlbauer GJ, Clark MD, Caccamo M, Schulman AH, Mayer KFX, Platzer M, Close TJ, Scholz U, Hansson M, Zhang G, Braumann I, Spannagl M, Li C, Waugh R, Stein N (2017) A chromosome conformation capture ordered sequence of the barley genome. Nature 544:427–433

Matsumoto T, Tanaka T, Sakai H, Amano N, Kanamori H, Kurita K, Kikuta A, Kamiya K,

Yamamoto M, Ikawa H, Fujii N, Hori K, Itoh T, Sato K (2011) Comprehensive sequence analysis of 24,783 barley full-length cDNAs derived from 12 clone libraries. Plant Physiol 156:20–28

Matsumoto T, Wu JZ, Kanamori H, Katayose Y, Fujisawa M, Namiki N, Mizuno H, Yamamoto K, Antonio BA, Baba T, Sakata K, Nagamura Y, Aoki H, Arikawa K, Arita K, Bito T, Chiden Y, Fujitsuka N, Fukunaka R, Hamada M, Harada C, Hayashi A, Hijishita S, Honda M, Hosokawa S, Ichikawa Y, Idonuma A, Iijima M, Ikeda M, Ikeno M, Ito K, Ito S, Ito T, Ito Y, Ito Y, Iwabuchi A, Kamiya K, Karasawa W, Kurita K, Katagiri S, Kikuta A, Kobayashi H, Kobayashi N, Machita K, Maehara T, Masukawa M, Mizubayashi T, Mukai Y, Nagasaki H, Nagata Y, Naito S, Nakashima M, Nakama Y, Nakamichi Y, Nakamura M, Meguro A, Negishi M, Ohta I, Ohta T, Okamoto M, Ono N, Saji S, Sakaguchi M, Sakai K, Shibata M, Shimokawa T, Song JY, Takazaki Y, Terasawa K, Tsugane M, Tsuji K, Ueda S, Waki K, Yamagata H, Yamamoto M, Yamamoto S, Yamane H, Yoshiki S, Yoshihara R, Yukawa K, Zhong HS, Yano M, Sasaki T, Yuan QP, Shu OT, Liu J, Jones KM, Gansberger K, Moffat K, Hill J, Bera J, Fadrosh D, Jin SH, Johri S, Kim M, Overton L, Reardon M, Tsitrin T, Vuong H, Weaver B, Ciecko A, Tallon L, Jackson J, Pai G, Van Aken S, Utterback T, Reidmuller S, Feldblyum T, Hsiao J, Zismann V, Iobst S, de Vazeille AR, Buell CR, Ying K, Li Y, Lu TT, Huang YC, Zhao Q, Feng Q, Zhang L, Zhu JJ, Weng QJ, Mu J, Lu YQ, Fan DL, Liu YL, Guan JP, Zhang YJ, Yu SL, Liu XH, Zhang Y, Hong GF, Han B, Choisne N, Demange N, Orjeda G, Samain S, Cattolico L, Pelletier E, Couloux A, Segurens B, Wincker P, D'Hont A, Scarpelli C, Weissenbach J, Salanoubat M, Quetier F, Yu Y, Kim HR, Rambo T, Currie J, Collura K, Luo MZ, Yang TJ, Ammiraju JSS, Engler F, Soderlund C, Wing RA, Palmer LE, de la Bastide M, Spiegel L, Nascimento L, Zutavern T, O'Shaughnessy A, Dike S, Dedhia N, Preston R, Balija V, McCombie WR, Chow TY, Chen HH, Chung MC, Chen CS, Shaw JF, Wu HP, Hsiao KJ, Chao YT, Chu MK, Cheng CH, Hour AL, Lee PF, Lin SJ, Lin YC, Liou JY, Liu SM, Hsing YI, Raghuvanshi S, Mohanty A, Bharti AK, Gaur A, Gupta V, Kumar D, Ravi V, Vij S, Kapur A, Khurana P, Khurana P, Khurana JP, Tyagi AK, Gaikwad K, Singh A, Dalal V, Srivastava S, Dixit A, Pal AK, Ghazi IA, Yadav M, Pandit A, Bhargava A, Sureshbabu K, Batra K, Sharma TR, Mohapatra T, Singh NK, Messing J, Nelson AB, Fuks G, Kavchok S, Keizer G, Llaca ELV, Song RT, Tanyolac B, Young S, Il KH, Hahn JH, Sangsakoo G, Vanavichit A, de Mattos LAT, Zimmer PD, Malone G, Dellagostin O, de Oliveira AC, Bevan M, Bancroft I, Minx P, Cordum H, Wilson R, Cheng ZK, Jin WW, Jiang JM, Leong SA, Iwama H, Gojobori T, Itoh T, Niimura Y, Fujii Y, Habara T, Sakai H, Sato Y, Wilson G, Kumar K, McCouch S, Juretic N, Hoen D, Wright S, Bruskiewich R, Bureau T, Miyao A, Hirochika H, Nishikawa T, Kadowaki K, Sugiura M, Project IRGS (2005) The map-based sequence of the rice genome. Nature 436:793–800

Mcintosh RA, Yamazaki Y, Devos KM, Dubcovsky J, Rogers WJ, Appels R (2003) Catalogue of gene symbols for wheat

Paterson AH, Bowers JE, Bruggmann R, Dubchak I, Grimwood J, Gundlach H, Haberer G, Hellsten U, Mitros T, Poliakov A, Schmutz J, Spannagl M, Tang HB, Wang XY, Wicker T, Bharti AK, Chapman J, Feltus FA, Gowik U, Grigoriev IV, Lyons E, Maher CA, Martis M, Narechania A, Otillar RP, Penning BW, Salamov AA, Wang Y, Zhang LF, Carpita NC, Freeling M, Gingle AR, Hash CT, Keller B, Klein P, Kresovich S, McCann MC, Ming R, Peterson DG, Mehboob-ur-Rahman Ware D, Westhoff P, Mayer KFX, Messing J, Rokhsar DS (2009) The Sorghum bicolor genome and the diversification of grasses. Nature 457:551–556

Ross MG, Russ C, Costello M, Hollinger A, Lennon NJ, Hegarty R, Nusbaum C, Jaffe DB (2013) Characterizing and measuring bias in sequence data. Genome Biology 14

Schnable PS, Ware D, Fulton RS, Stein JC, Wei FS, Pasternak S, Liang CZ, Zhang JW, Fulton L, Graves TA, Minx P, Reily AD, Courtney L, Kruchowski SS, Tomlinson C, Strong C, Delehaunty K, Fronick C, Courtney B, Rock SM, Belter E, Du FY, Kim K, Abbott RM, Cotton M, Levy A, Marchetto P, Ochoa K, Jackson SM, Gillam B, Chen WZ, Yan L, Higginbotham J, Cardenas M, Waligorski J, Applebaum E, Phelps L, Falcone J, Kanchi K, Thane T, Scimone A, Thane N, Henke J, Wang T, Ruppert J, Shah N, Rotter K, Hodges J, Ingenthron E, Cordes M, Kohlberg S, Sgro J, Delgado B, Mead K, Chinwalla A, Leonard S, Crouse K, Collura K, Kudrna D, Currie J, He RF, Angelova A, Rajasekar S, Mueller T, Lomeli R, Scara G, Ko A, Delaney K, Wissotski M, Lopez G, Campos D, Braidotti M, Ashley E, Golser W, Kim H, Lee S, Lin JK, Dujmic Z, Kim W, Talag J, Zuccolo A, Fan C, Sebastian A, Kramer M, Spiegel L, Nascimento L, Zutavern T, Miller B, Ambroise C, Muller S, Spooner W, Narechania A, Ren LY, Wei S, Kumari S, Faga B, Levy MJ, McMahan L, Van Buren P, Vaughn MW, Ying K, Yeh CT, Emrich SJ, Jia Y, Kalyanaraman A, Hsia AP, Barbazuk WB, Baucom RS, Brutnell TP, Carpita NC, Chaparro C, Chia JM, Deragon JM, Estill JC, Fu Y, Jeddeloh JA, Han YJ, Lee H, Li PH, Lisch DR, Liu SZ, Liu ZJ, Nagel DH, McCann MC, SanMiguel P, Myers AM, Nettleton D, Nguyen J, Penning BW, Ponnala L, Schneider KL, Schwartz DC, Sharma A, Soderlund C, Springer NM, Sun Q, Wang H, Waterman M, Westerman R, Wolfgruber TK, Yang LX, Yu Y, Zhang LF, Zhou SG, Zhu Q, Bennetzen JL, Dawe RK, Jiang JM, Jiang N, Presting GG, Wessler SR, Aluru S, Martienssen RA, Clifton SW, McCombie WR, Wing RA, Wilson RK (2009) The B73 maize genome: complexity, diversity, and dynamics. Science 326:1112–1115

Simao FA, Waterhouse RM, Ioannidis P, Kriventseva EV, Zdobnov EM (2015) BUSCO: assessing genome assembly and annotation completeness with single-copy orthologs. Bioinformatics 31:3210–3212

Trapnell C, Roberts A, Goff L, Pertea G, Kim D, Kelley DR, Pimentel H, Salzberg SL, Rinn JL, Pachter L (2012) Differential gene and transcript expression analysis of RNA-seq experiments with TopHat and Cufflinks. Nat Protoc 7:562–578

Vogel JP, Garvin DF, Mockler TC, Schmutz J, Rokhsar D, Bevan MW, Barry K, Lucas S, Harmon-Smith M, Lail K, Tice H, Grimwood J, McKenzie N, Huo NX, Gu YQ, Lazo GR, Anderson OD, You FM, Luo MC, Dvorak J, Wright J, Febrer M, Idziak D, Hasterok R, Lindquist E, Wang M, Fox SE, Priest HD, Filichkin SA, Givan SA, Bryant DW, Chang JH, Wu HY, Wu W, Hsia AP, Schnable PS, Kalyanaraman A, Barbazuk B, Michael TP, Hazen SP, Bragg JN, Laudencia-Chingcuanco D, Weng YQ, Haberer G, Spannagl M, Mayer K, Rattei T, Mitros T, Lee SJ, Rose JKC, Mueller LA, York TL, Wicker T, Buchmann JP, Tanskanen J, Schulman AH, Gundlach H, de Oliveira AC, Maia LD, Belknap W, Jiang N, Lai JS, Zhu LC, Ma JX, Sun C, Pritham E, Salse J, Murat F, Abrouk M, Bruggmann R, Messing J, Fahlgren N, Sullivan CM, Carrington JC, Chapman EJ, May GD, Zhai JX, Ganssmann M, Gurazada SGR, German M, Meyers BC, Green PJ, Tyler L, Wu JJ, Thomson J, Chen S, Scheller HV, Harholt J, Ulvskov P, Kimbrel JA, Bartley LE, Cao PJ, Jung KH, Sharma MK, Vega-Sanchez M, Ronald P, Dardick CD, De Bodt S, Verelst W, Inze D, Heese M, Schnittger A, Yang XH, Kalluri UC, Tuskan GA, Hua ZH, Vierstra RD, Cui Y, Ouyang SH, Sun QX, Liu ZY, Yilmaz A, Grotewold E, Sibout R, Hematy K, Mouille G, Hofte H, Pelloux J, O'Connor D, Schnable J, Rowe S, Harmon F, Cass CL, Sedbrook JC, Byrne ME, Walsh S, Higgins J, Li PH, Brutnell T, Unver T, Budak H, Belcram H, Charles M, Chalhoub B, Baxter I, Initiative IB (2010) Genome sequencing and analysis of the model grass Brachypodium distachyon. Nature 463:763–768

Wu TD, Watanabe CK (2005) GMAP: a genomic mapping and alignment program for mRNA and EST sequences. Bioinformatics 21:1859–1875

7 The Expressed Portion of the Barley Genome

Craig G. Simpson, Roberto A. Barrero, Micha Bayer, Andrew Flavell, Paulo Rapazote-Flores, Matthew I. Bellgard, Pete E. Hedley, Runxuan Zhang and Robbie Waugh

Abstract

In this chapter, we refer to the expressed portion of the barley genome as the relatively small fraction of the total cellular DNA that either contains the genes that ultimately produce proteins, or that directly/indirectly controls the level, location and/or timing of when these genes are expressed and proteins are produced. We start by describing the dynamics of tissue and time-dependent gene expression and how common patterns across multiple samples can provide clues about gene networks involved in common biological processes. We then describe some of the complexities of how a single mRNA template can be differentially processed by alternative splicing to generate multiple different proteins or provide a mechanism to regulate the amount of functional gene product in a cell at a given point in time. We extend our analysis, using a number of biological examples, to address how diverse families of small non-coding microRNAs specifically regulate gene expression, and complete our appraisal by looking at the physical/molecular environment around genes that can result in either the promotion or repression of gene expression. We conclude by assessing some of the issues that remain around our ability to fully exploit the depth and power of current approaches for analysing gene expression and propose improvements that could be made using new but available sequencing and bioinformatics technologies.

C. G. Simpson · M. Bayer · P. Rapazote-Flores
P. E. Hedley · R. Zhang · R. Waugh (✉)
James Hutton Institute, Dundee, UK
e-mail: robbie.waugh@hutton.ac.uk

R. A. Barrero · M. I. Bellgard
Murdoch University, Perth, Australia

A. Flavell · R. Waugh
Division of Plant Sciences, University of Dundee, Dundee, UK

7.1 Introduction

The expressed portion of the barley genome as defined here refers to the part that is transcribed by an RNA polymerase into either a pre-messenger RNA that is processed and ultimately translated into a protein and the major class of endogenous small RNAs (mainly microRNAs (miRNAs)) that mediate both genetic and epigenetic processes. In relation to gene expression, the path from gene to protein involves layers of regulation that ultimately determine where, when and how much of a protein is present within a cell. These layers include temporal accessibility to DNA sequences upon which the transcriptional machinery assembles and copies DNA into pre-mRNA, the nature of the specific regulatory sequences in and

N. Stein and G. J. Muehlbauer (eds.), *The Barley Genome*, Compendium of Plant Genomes,
https://doi.org/10.1007/978-3-319-92528-8_7

around the genes themselves that control transcription and then a host of post-transcriptional processes, including interactions with specific miRNAs, that determine if and how the mRNA is translated into its designated protein product. Investigating the steps between gene and protein has been enabled by the development of molecular technologies that allow genome-wide assessment of structural features of the genome, the transcriptome and, while not covered here, the proteome itself. From a very recent historical perspective, due to its relative ease of interpretation, the transcriptome has proved most amenable to analysis. Starting less than 20 years ago with the development and popularisation of microarray technology, and now more commonly RNA-Seq and other parallel mRNA quantification approaches (e.g. nanostring technology—https://www.nanostring.com/), many studies use transcriptomics to investigate complex biological processes. These generally come in a restricted number of flavours that include descriptive tissue-level comparisons (e.g. 'transcriptional atlas'), time-series, case-control and genetical experiments. Most aim at illuminating wide and complex interactions and can be highly informative when exploring the mechanisms underpinning how temporal, developmental, biochemical or environmental variables condition a biological response (e.g. van Esse et al. 2017; Bull et al. 2017). In the microarray era, the barley research community was one of the first crop plant communities to develop a commercial transcriptomics platform that was openly available to all through an alliance with the Affymetrix company (Close et al. 2004). It was similarly pioneering in establishing a central repository, BarleyBase, for storing and querying barley transcriptomic experimental data (Shen et al. 2005), promoting sharing and community-wide reuse of expensive-to-generate data. While these have now largely been superseded by newer online resources that allow sequence-based transcriptomic data to be queried (e.g. BARLEX, https://apex.ipk-gatersleben.de/apex/f?p=284:10), they retain intrinsic value and were an early demonstration of an ethos that has been important for the small barley community to prosper. In this chapter, we summarise some of the ways in which intricacies of the barley transcriptome have been explored, starting with the transcriptome itself, describing new observations on post-transcriptional variation and regulation, miRNA discovery and diversity, and ending with the epigenome that we now know influences if, where and when transcription can occur.

7.2 Dynamics of Gene Expression in Barley

7.2.1 Organ Specificity and Temporal Expression

Gene expression responses between cells and organs to different environmental conditions are determined by the coordinated regulation of different groups of genes. Previously, barley full-length cDNAs and RNA-Seq reads from eight stages of barley development (highlighted in Fig. 7.1) were mapped to the first sequence assembly of the barley genome (IBGSC 2012), and it was found that 36–55% of the high-confidence barley gene sets were differentially regulated between these barley organs. Only 10% of the genes were uniquely expressed in any one of the tissues, and 65% of the genes showed expression in all organs tested (IBGSC 2012). The new chromosome-scale assembly of the barley genome utilised RNA-Seq from the same eight developmental tissues used to characterise gene models in the original assembly and an additional eight tissues from different organs and stages of development (Mascher et al. 2017). Using data from these 16 tissues in combination with the improved genome assembly identified 39,734 high-confidence protein-coding genes encoding 248,270 different transcripts (also see Chap. 6 (this volume)).

A transcriptome study of the 16 tissues found common expression of genes in related tissues (Fig. 7.1). Both whole inflorescences at 30 and 50 days after planting, and developing grain at 5 and 10 days post anthesis, showed commonly expressed gene sets between the two organs. However, distinct tissues of the inflorescence

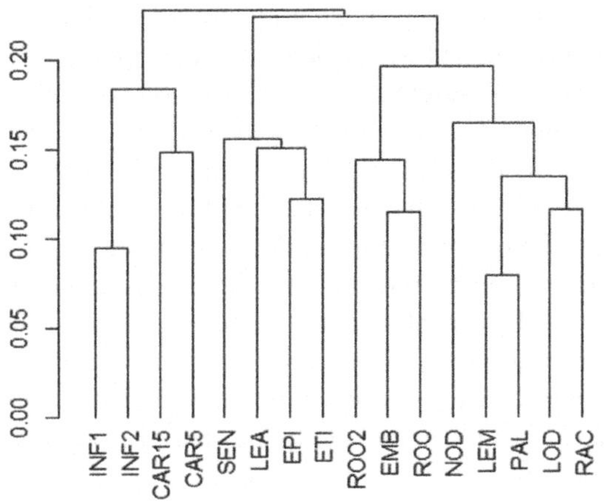

Fig. 7.1 Hierarchical clustering analysis of the global transcriptional profiles between different tissues and developmental stages using Jensen–Shannon distances between conditions. The bar height shows linkage cluster distance ($\log^2$ FPKM (fragments per kilobase of exon per million fragments mapped)). EMB*, Embryonic tissue, mesocotyl and seminal roots, 4 days after germination; ETI, Etiolated leaf, 10 days after planting; LEA*, Leaf, 17 days after planting; EPI, Epidermal strips, from leaf, 28 days after planting; ROO*, Root, 17 days after planting; ROO2, Root, 28 days after planting; NOD*, Third stem internode, 42 days after planting; RAC, Rachis, 35 days after planting; INF1*, Whole developing inflorescence tissue, 30 days after planting; INF2*, Whole developing inflorescence tissue, 50 days after planting; CAR5*, Developing grain (caryopsis), 5 days post anthesis; CAR15*, Developing grain (caryopsis), 15 days post anthesis; LEM, Lemma dissected from inflorescences, 42 days after planting; LOD, Lodicule dissected from inflorescences, 42 days after planting; PAL, Palea dissected from inflorescences, 42 days after planting; SEN, Senescing leaf, 56 days after planting. *Eight samples from the original study (IBGSC 2012)

architecture (rachis, lemma, lodicule and palea) showed expression of related genes that were discrete from the whole inflorescences, but overlapped with those of internode tissue. Leaf material that also included epidermal strips, etiolated leaf and senescing leaf tissue formed an additional clade of genes expression. Genes expressed in root were related to those expressed in germinating embryonic tissue, that includes both mesocotyl and seminal root tissues (Fig. 7.1). The ability to distinguish multiple tissues by their overall patterns of transcript accumulation indicates strongly synchronised gene expression. More detailed transcriptome profiling in barley tissues is needed to distinguish the cellular changes that occur in gene expression at this scale (Giacomello et al. 2017).

Some genes are under strong regulation and show distinct organ-specific expression, for example those genes which encode UDP-Glycosyltransferases, which are part of a large family of enzymes that catalyse the addition of a glycosyl group from a UTP-sugar to a protein. Over-expression of a barley UDP-Glycosyltransferase gene in wheat showed resistance to *Fusarium* by detoxification of the fungal mycotoxin and virulence factor deoxynivalenol (Li et al. 2015). In the barley genome, 230 genes are annotated as UDP-Glycosyltransferases, and members of this family show distinct regulated patterns of expression in specific organs, thereby suggesting distinct tissue-specific roles. The barley UDP-Glycosyltransferase HORVU1Hr1G007720 is highly expressed in germinating embryo with little expression in other tissues tested (Fig. 7.2a). However, other members of the gene family show more general expression profiles, but have preferential expression in other tissues: HORVU5Hr1G047150 shows highest levels of expression in leaf shoots (Fig. 7.2b), contrasting with HORVU3Hr1G078850 which is highest in senescing leaf tissue (Fig. 7.2c). Other gene family members show constitutive expression through all the tissues indicating a more general role (data not shown).

7.2.2 Gene Expression Clustering and Co-regulation

Based on the gene expression patterns of regulatory or functional genes, a co-expression network can be created by pairing the selected genes with genes showing a similar expression profile across different samples or conditions. Previously, a global barley co-expression network was produced from 45 different microarray experiments (Mochida et al. 2011). The resulting clusters clearly identified gene sets involved in drought stress response and cellulose biogenesis. Another study generated a gene co-expression

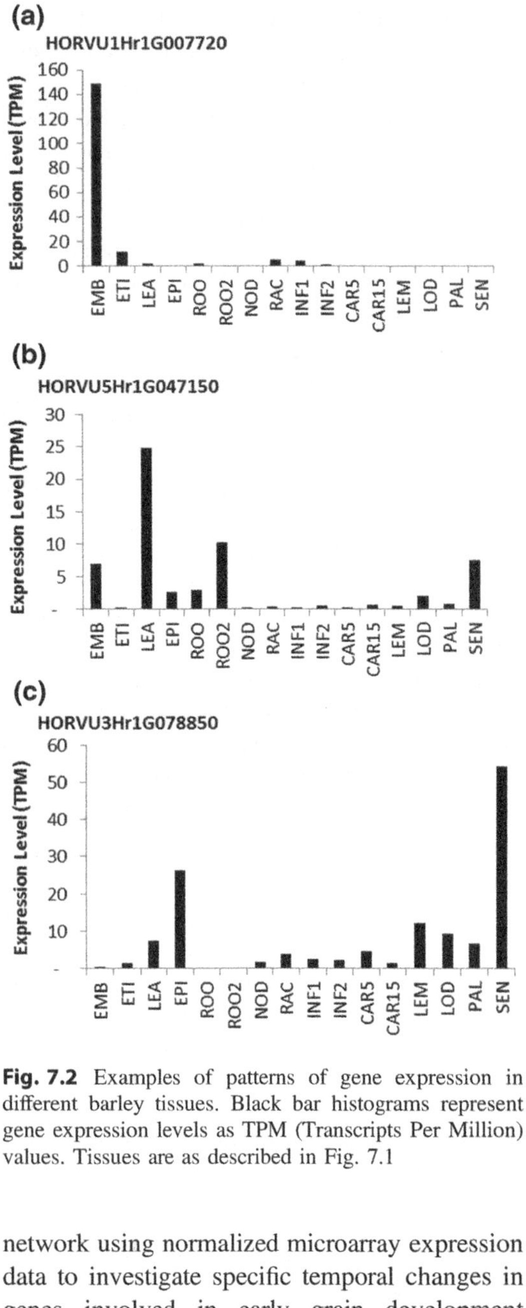

Fig. 7.2 Examples of patterns of gene expression in different barley tissues. Black bar histograms represent gene expression levels as TPM (Transcripts Per Million) values. Tissues are as described in Fig. 7.1

network using normalized microarray expression data to investigate specific temporal changes in genes involved in early grain development (Zhang et al. 2016a, b, c). Between 3 and 8 days after pollination (DAP), 47 gene expression modules were discovered, clustered into 10 groups. For example, during the early stages of endosperm development (3–4 DAP), a cluster of genes encoding expansins, arabinogalactan proteins and hydrolytic enzymes were co-expressed. Transcription and post-transcriptional regulatory factors are key regulatory hubs involved in induction and regulation of many genes. The spatial representation of 57 co-expressed transcription factors with 89 cell wall-related genes identified 4 clusters of genes that peaked at different DAP (Zhang et al. 2016a, b, c). Applying co-expression and network analysis to RNA-Seq datasets in barley, including those mentioned in this Chapter, should be readily achievable and this will provide invaluable information for cereal researchers, for example on gene interactions and identification of tissue-specific promoter elements.

7.2.3 Summary

The accumulation and analysis of transcriptome data will find new associations, improve gene networks and promote gene discovery in barley. This will require a coordinated effort from the barley research community to assemble, database and analyse the vast quantities of RNA-Seq data which has already been generated, and will be established in the future. It will also be critical to ensure that associated metadata detailing sample provenance and experimental conditions are recorded and linked to the RNA-Seq data. This will ensure that the gene expression data can be exploited fully in the future. In addition, the advent of single-cell transcriptome sequencing will profile gene expression to unheralded new levels of cellular definition, thereby enhancing candidate gene selection (Efroni and Birnbaum 2016). Although not covered in this Chapter, such approaches would require very careful planning, sampling and high levels of replication, to enable accurate analysis of single-cell data to be performed.

7.3 Transcript Isoforms in Barley

7.3.1 Barley Pre-mRNA Intron Splicing Characteristics

Barley genes, as in all higher plants, are interrupted with intron sequences that are removed after transcription by a process called pre-mRNA splicing. The expressed transcriptome of the 16 barley organs assembled against the latest ordered barley genome sequence identified 344,248 transcript sequences, clustered into 83,105 potential gene positions that include high- and low-confidence protein-coding genes, non-coding RNAs, pseudogenes and transposons (Mascher et al. 2017). A set of high-confidence gene sequences (see Table 7.1 for selection criteria) show that around 80% of the genes encoding full-length proteins were intron-containing, multi-exonic genes with on an average ~11 exons per gene (Table 7.1). The most complex intron–exon gene structure found contains 51 exons in a 22.8 kb gene annotated as a Glucan Synthase-Like family member (HORVU7Hr1G003460). An average of approximately 121,000 different introns are spliced out in individual barley tissues, supporting the widespread nature of splicing and its significance to barley gene expression.

Table 7.1 Intron splicing characteristics of a high-confidence gene set in barley

	Barley
Number of genes	27,309
Single exon	5,943 (22%)
Multi-exon	21,366 (78%)
Number of distinct exons	306,172
Mean number of distinct exons per gene	11.2
Number of multi-exon genes with alternative transcripts	10,318 (48%)
Total number of predicted transcripts	52,847
Mean number of transcripts per gene	1.94
Mean gene locus size (first to last exon) (bp)	3,249
Mean transcript size (UTR and CDS) (nt)	2,310
Mean exon size (nt)	398

A set of barley high-confidence gene sequences (IBGSC 2012) were selected by comparison to protein sequences from different grass species *Brachypodium distachyon, Oryza sativa, Sorghum bicolor* and other *Poaceae* (Uniprot database selection of complete proteins from maize rye, oat, wheat and others). Transcripts matching more than 50% of the length of a protein (identity > 50%) were selected. In addition, the 22,651 barley full-length (fl) cDNAs (Matsumoto et al. 2011) were mapped to the transcripts. The comparisons to the reference protein and fl-cDNA sequences resulted in 27,309 genes

A scan of the splice sites identified that 96% of the introns have the expected canonical 5′ splice site GT and 3′ splice site AG dinucleotides at their borders. A total of 2.6% introns show the rarer GC-AG dinucleotide borders and 1.3% introns show the U12 dependent splicing AT-AC dinucleotide borders. Alternative acceptor splices 3 bp apart from each other (NAGNAG or tandem acceptor sites) are commonly found in human genomes (>20% of alternatively spliced genes) and may be subject to regulation (Busch and Hertel 2012; Shi et al. 2014). Around 10% of barley alternative 3′ splice sites showed the NAGNAG alternative 3′ splice site duplication. Exons can vary in size from a single base (At3g05870, Guo and Liu 2015) to thousands of bases long. A clear example of a mini- or micro-exon 5 bp long within a cysteine proteinase superfamily protein (HORVU4Hr1G010390) is conserved in Arabidopsis (At1g02305). It is, however, possible that there may be shorter exon sequences in RNA-Seq reads that have not been assembled into the final transcript. Evidence shows that the characteristics and the prevalence of splicing in barley are comparable with other higher plants (Reddy 2007; Marquez et al. 2012; Zhiguo et al. 2013; Min et al. 2015).

7.3.2 Barley Alternative Splicing

Alternative selection of splice sites produces multiple mRNA transcripts derived from the same gene, increasing the protein-coding capacity of eukaryote genomes. It has an important role in regulation and fine-tuning of gene expression. It can occur throughout the pre-mRNA transcript, influencing not only the protein-coding sequence but also the length of 5′ and 3′ UTRs, which affects transcript stability (nonsense-mediated decay; NMD) and the occurrence of upstream open-reading frames (uORFs) (Kalyna et al. 2012).

Barley genes that show alternative splicing are involved in a wide range of biological processes and functions, including: abiotic and biotic responses; water and ion homeostasis; starch synthesis; metabolite formation; transcription factors; seed and flower development; and pre-harvest sprouting (Thorbjørnsen et al. 1996; Schmitz et al. 2000; Halterman et al. 2003; Xue and Loveridge 2004; Rodríguez-Suárez et al. 2011; Hove et al. 2015; Shahzad et al. 2015; Zhang et al. 2016b). Allelic differences, natural and induced mutations further contribute to the alternatively spliced transcript landscape between varieties and landraces, through elimination and activation of splice sites and splicing signals (McKibben et al. 2002; Rodríguez-Suárez et al. 2011).

Alternative splicing is common in plant genes ranging from 6% in *Brachypodium* (Walters et al. 2013) to 61% in the more widely studied Arabidopsis (Marquez et al. 2012) and 75% in wheat (Pingault et al. 2015). The majority of plant alternative splicing events use an alternative selection of 5′ and 3′ splice sites, exon skipping, intron retention and combinations of these alternative selections within transcripts (Reddy et al. 2013). In barley, a screen of barley expressed sequence tags and RNA-Seq analysis of eight barley organs show that at least 51–55% of barley genes were alternatively spliced (Panahi et al. 2015; IBGSC 2012). Comparison of the original eight barley organs shows that 17% of the genes shared the same alternative splicing event between samples and 17–27% showed organ-specific alternative splicing. This evidence indicates a pronounced role for pre-mRNA splicing and alternative splicing in the regulation of barley gene expression. In the latest genome assembly, the range of alternative splicing across 16 different barley organs has not yet been determined, but it is expected to be similar to the previous analysis.

To illustrate the complex nature of alternative splicing, HORVU7Hr1G044850 represents a WW domain-containing protein that has a complex alternative splicing pattern represented by 23 alternative RNA-Seq transcripts (Fig. 7.3a). Exon 2 shows an alternative 5′ splice site, intron 5 contains an alternative exon, and between exons 6 and 7 there are transcripts that retain the intron, an alternative exon and an alternative 5′ splice site. Different combinations of these alternative splicing events lead to the complex multi-transcript splicing pattern for this gene. High-resolution RT-PCR (HR-RT-PCR) analysis of the exon 2 region (Primer pair Hv78) and exons 6–7 region (Primer pair Hv79) identified accurately all the alternative splicing events found in the RNA-Seq assembly and found a further 3 additional minor products (Fig. 7.3b), which may indicate an underestimation of alternative splicing in barley genes. The expected products that produce a fully spliced (FS) transcript are the main splicing event and will produce a full-length protein. The other major alternative splicing events (alternative 5′ splice sites and alternative exon) change the protein codon reading frame and will result in premature termination (Fig. 7.3b). Isoform sequencing (e.g. Isoseq on PacBio Sequel instruments) will allow us to accurately determine how these different alternative splice site selections combine to produce individual transcripts.

Variation in alternative splicing is often tissue specific and the pattern of alternative splicing can be developmentally and environmentally regulated (Staiger and Brown 2013). To illustrate regulated alternative splicing between different barley organs, HR-RT-PCR analysis of an alternative 5′ splice site in intron 3 of HORVU5Hr1G027080, shows changing proportions of alternative transcripts between barley inflorescence, leaf, root, embryo and internodal tissues. Selection of the proximal 5′ splice site leads to the presence of a stop codon, which would terminate translation and lead to a truncated protein (Fig. 7.4). HR-RT-PCR analysis indicates large differences in both transcription and alternative transcript levels (Fig. 7.4). Alternative splicing changes particularly between the early and later stage inflorescence and embryo organs. This result shows the discrete differences in alternative transcript levels that occur between different organs.

In many cases, changes in alternative splicing may be temporal or dynamic as the plant responds to different stresses and times in the

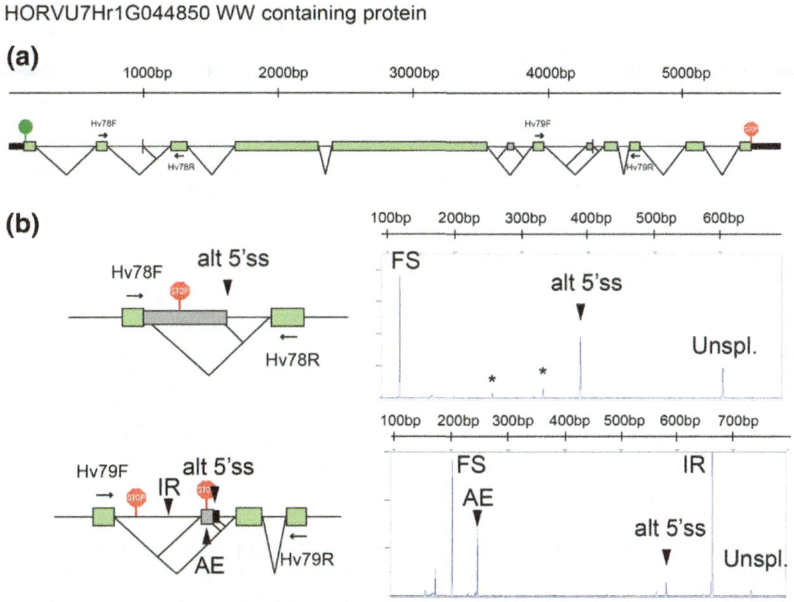

Fig. 7.3 Alternative splicing in a WW domain-containing protein. **a** Gene model for HORVU7Hr1G044850. Green boxes represent protein-coding exons, black boxes show the 5′ and 3′ UTRs, grey boxes show alternative exons and alternative splice sites are indicated as bars. Green lollipop indicates start codon and stop and premature stop codons are shown as red lollipops. Introns are shown as a line between the boxes and the splicing events shown as a triangular line between the exons. The length of the gene (bp) is indicated by the bar at the top of the figure, and the HR-RT-PCR primers used are indicated as arrows. **b** Alternatively spliced regions of HORVU7Hr1G044850 validated by HR-RT-PCR. The alternatively spliced regions are shown in the gene model schematics for primer pairs Hv78 and Hv79. Electropherogram output from the sequencer show the HR RT-PCR products, the X-axis shows the HR RT-PCR product length (bp) and the Y-axis the relative fluorescence units. The products expected from RNA-Seq are indicated as: FS, Fully spliced; AE, Alternative exon; alt 5′ ss, Alternative 5′ splice site; IR, intron retention and; Unspl., Unspliced. * indicates unknown alternative transcripts

day. In germinating seed of four different barley varieties, alternative splicing was found in 14–20% of intron-containing genes in a range of different genes, including those involved in cell wall polysaccharide metabolism (Zhang et al. 2016a, b). Over half of the alternative transcripts were replaced with new alternative transcripts in a 24 h period of germination, highlighting the highly dynamic changes in transcript populations that occur over short periods of time.

7.3.3 Nonsense-Mediated Decay

Alternative splicing is closely associated with mRNA stability through the action of nonsense-mediated decay (NMD). NMD targets aberrantly spliced transcripts for degradation that have altered upstream open reading frames (uORFs), alternative polyadenylation sites or transcripts that contain premature termination codons (PTCs) (Kalyna et al. 2012). In yeast and animal systems, the efficiency and regulation of NMD are altered by stress through the expression of miRNAs, phosphorylation of eIF2α and through auto- and cross-regulation mechanisms (Karam et al. 2013). In Arabidopsis, mutant NMD alleles change gene expression and alternative splicing profiles, increase salicylic acid levels, induce expression of Pathogenesis-related (PR) genes and stimulate a pathogen hypersensitive defence response (Gloggnitzer et al. 2014).

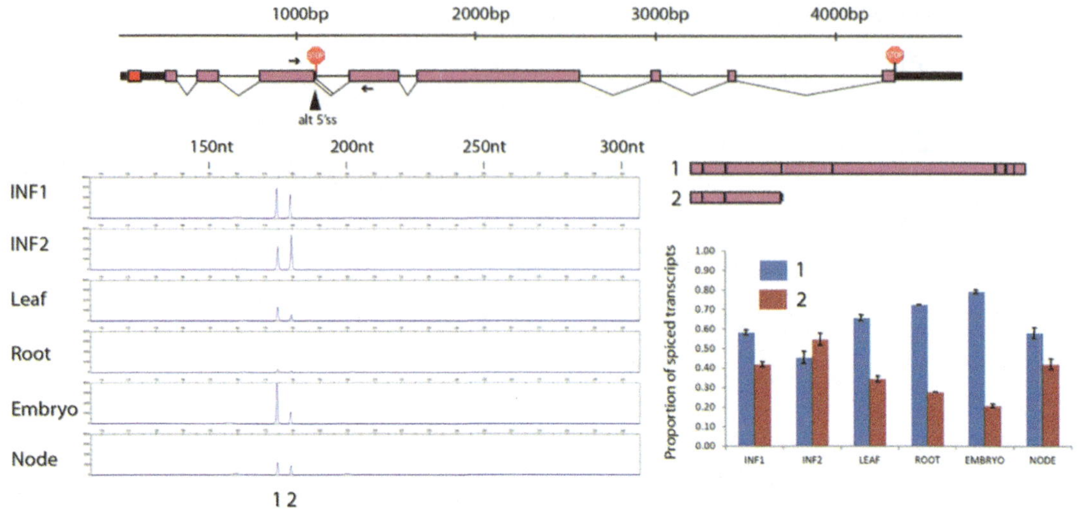

Fig. 7.4 Regulated alternative splicing. Gene model for HORVU5Hr1G027080. Pink boxes represent protein-coding regions, black boxes show the 5′ and 3′ UTRs, red box shows a putative uORF. Introns are shown as a line between the boxes and the alternative 5′ splice site is indicated by a triangle. The length of the gene (bp) is indicated by the bar at the top of the figure, and the HR-RT-PCR primers used are indicated as arrows. The alternative protein-coding sequences produced by alternative splicing are shown as 1 and 2 and the alternative transcripts that produce these alternative proteins are also labelled 1 and 2. Alternatively spliced regions of HORVU5Hr1G027080 are validated by HR-RT-PCR. The electropherogram output from the sequencer show the HR RT-PCR products produced for this gene, the X-axis shows the HR RT-PCR product length (nt) and the Y-axis the relative fluorescence units. The bar graph shows the proportion of alternative transcripts produced in barley inflorescence stages 1 and 2; leaf, root, embryo and internode organs (IBGSC 2012)

In barley, low abundance PTC containing transcripts may be the result of NMD (Gadjieva et al. 2004). Cold stress leads to an increase in abundance of alternatively spliced transcripts in circadian clock genes, many of which contain PTCs. Transcripts with the PTC containing alternative exon 6a of the key circadian clock gene *LHY* are sensitive to cycloheximide, a translational inhibitor and thereby NMD inhibitor, suggesting that these transcripts are sensitive to NMD (Calixto et al. 2016). It is not clear what role this transcript may have at the lower temperature or whether the increased abundance of this alternative exon in *LHY* transcripts is the result of regulated alternative splicing or a change in NMD efficiency at the lower temperature. But this example highlights the interplay that can occur between these post-transcriptional processes to allow barley to adapt to changing conditions.

7.3.4 Summary

Alternative splicing and NMD are fundamental to plant gene expression and the development of next-generation sequencing technologies allows us to investigate regulation of barley-alternative transcripts in unprecedented detail. The formation of alternative transcripts by alternative splicing and transcript turnover by NMD over multiple genes will result in co-ordinated alternative transcript patterns particular to different developmental organs and as a network of response to varying environmental conditions. We can now begin to identify important gene transcripts and determine their roles in gene regulation and ultimately plant traits.

7.4 Barley MicroRNAs

7.4.1 Background

Plant genomes encode small RNAs that mediate genetic or epigenetic silencing of central biological processes including development, reproduction and genome reprogramming. The large variety of small RNA pathways in plants is likely to contribute to their phenotypic plasticity (Borges and Martienssen 2015). One of the major classes of endogenous small RNAs in plants is microRNAs (miRNAs), which are involved in post-transcriptional gene silencing via transcript cleavage or translational repression. Since the discovery of plant miRNAs (Reinhart et al. 2002), a large number of these small functional RNAs have been reported across a wide range of species.

Screening EST datasets initially found barley miRNAs for stem-loop structures (Colaiacovo et al. 2010; Curaba et al. 2012; Hackenberg et al. 2013; Kantar et al. 2010). Subsequently, next-generation sequencing technologies enabled mapping of small RNA sequences onto related species genomic sequences such as known miRNA precursors (Lv et al. 2012; Schreiber et al. 2011), and barley BAC sequences (Schreiber et al. 2011). The advent of the first barley draft genome sequence (International Barley Genome Sequencing et al. 2012) enabled the discovery of novel barley miRNAs (Wu et al. 2014). The miRBase (Release 21) contains 69 barley precursors of miRNAs (pre-miRNAs) and 71 mature miRNA sequences (Kozomara and Griffiths-Jones 2014). More recently, 792 miRNA loci representing 312 putative miRNA families were found in the barley genome (Mascher et al. 2017).

In barley, dehydration, boron, phosphate-deficiency, salinity and drought stresses trigger differential expression of barley miRNAs (Deng et al. 2015; Hackenberg et al. 2013, 2015; Kantar et al. 2010; Ozhuner et al. 2013). Functional genomics studies to uncover the role of miRNAs in barley are still in its infancy. This section presents the progress made in three areas along with a discussion on the necessity for a unified miRNA annotation framework to minimise false-positive predictions and maximise the use of community resources to progress the understanding of barley miRNAs in related crop species.

7.4.2 miRNA Regulation During Early Seed Development

The early development of the grain is controlled by a complex interaction of signaling and gene regulation networks to allow the proper expansion and specialization of the different tissues that will constitute the mature grain (Curaba et al. 2012). miRNAs participate in gene expression regulation and phytohormone crosstalk during seed development and germination (Dass et al. 2015; Bai et al. 2017a). For example, miR393 modulates auxin signaling via targeting of two auxin receptors TIR1/AFBs. Gain-of-function of the miR393 increases length–width ratio of grains, whereas its loss-of-function decreases the 1000-grain weight of seeds (Bai et al. 2017b). Moreover, misexpression of miR393 impacts gibberellic acid and abscisic acid homeostasis during seed development and early seedling growth necessary for proper growth (Bai et al. 2017a).

7.4.3 miRNAs Contribute to Accession-Specific Pathogen Resistance

Plants possess two major classes of immune receptors for pathogen detection and defense response (Sreenivasulu et al. 2008). Pattern recognition receptors (PRRs) distinguish conserved pathogen-associated molecular patterns (PAMP) leading to a plant PAMP-triggered immunity (PTI) response (Nakabayashi et al. 2005), while intracellular leucine-rich repeat (NLR) receptors recognize isolate-specific pathogen effectors and trigger effector-triggered immunity (ETI), often accompanied by a hypersensitive reaction (Locascio et al. 2014; Silva et al. 2016). Increasing evidence supports the involvement of small RNAs in the regulation of PTI and ETI responses against bacterial and fungal pathogens (Jones-Rhoades et al. 2006; Meng et al. 2013). In barley, expression of miR398 is modulated by the *Mla* NLR immune receptor, influencing ETI response to powdery mildew fungus (Xu et al. 2014). More recently, the miR9863 family was shown to regulate distinct *Mla* alleles to attenuate NLR receptor-triggered disease resistance and cell-death signaling upon infection by *Blumeria graminis* f. sp. *hordei*. The miR9863 specific regulation of a subset of *Mla* alleles confers race-specific resistance to powdery mildew fungus (Bai et al. 2017b).

7.4.4 miRNAs Modulate Tolerance to Abiotic Stresses

In plants, several miRNAs families were shown to be involved in response to nutrient deficiencies as well as drought, salinity, heat and other abiotic stresses (Koroban et al. 2016; Li et al. 2017). Recently, comparison of a diploid and autotetraploid *Hordeum bulbosum* identified differentially expressed miRNAs in response to salt stress, including miR399, miR169, miR159-3p, miR319, miR156, miR157 and miR399. Additionally, four miRNAs (miR171, miR479, miR5048-5p and miR6196) were not only found to be involved in the salt-stress response but also shown to have increased expression levels owing to genome duplication (Liu et al. 2017). Analysis of drought tolerant and susceptible barley cultivars found 20 miRNAs differentially expressed between the two varieties including miR397, miR393, miR156, miR167 and miR396 (Fard et al. 2017). Recently, miR393-mediated auxin signaling regulation was shown to be involved in root elongation inhibition in response to toxic levels of aluminum (Bai et al. 2017a). Breeding programs aiming to increase the copy number of beneficial miRNA genes may enable the development of abiotic stress resilient cultivars.

7.4.5 To Be or not to Be a miRNA Locus?

Over the past decade, there has been a notable increase in the number of miRNA genes (miRBase rel. 21) facilitated by access to more affordable NGS technologies and the availability of a range of computational tools. However, recent reports suggest that only approximately 16% of metazoan miRBase entries are robustly supported as miRNA genes (Fromm et al. 2015). Typically, time-consuming human intervention is required to curate and validate automated predictions. For metazoan miRNAs, four filtering criteria find robust miRNA loci: (i) Two 20–26 nt long reads expressed from each of the two arms derived from a hairpin precursor with 2 nt offsets between the 5p and 3p arms; (ii) 5′-end homogeneity of expression; (iii) At least 16 nt complementarity between the two arm expressed sequences; and (iv) the loop sequence is at least 8 nt in length (Fromm et al. 2015). For example, application of these criteria to the 792 loci miRNA barley (Mascher et al. 2017) and exclusion of low expressed miRNA loci yielded 45 miRNA loci with robust miRNA support (Fig. 7.5a). Degradome data collected during the early seed development (Curaba et al. 2012) supported downstream targets for eight barley miRNAs including three novel miRNAs (Fig. 7.5b) not found in miRBase (rel 21) nor in the wheat (iwgsc_refseqv1.0), *Aegilops tauschii* (ASM34733v1) and *Oryza sativa* (IRGSP-1.0) genomes. The novel-hvu-miRNA-00273-k (Mascher et al. 2017) is predicted to target HORVU5HG0061600.2 and HORVU7HG0393700.1 transcripts encoding a *syntaxin-51*-like and an RNA recognition domain-containing gene, respectively (Fig. 7.6). The validation of barley-specific miRNAs using a human-mediated curation approach highlights the need to implement a unified annotation framework to curate public databases.

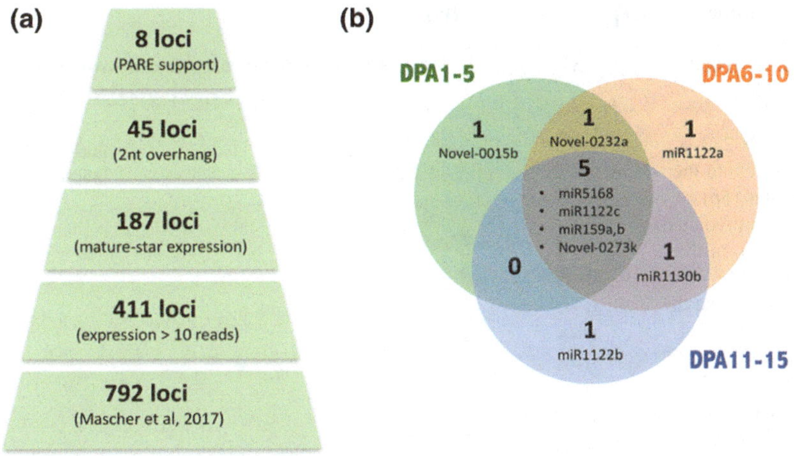

Fig. 7.5 Human-mediated curation of bioinformatic-predicted miRNA data. Nowadays, standard processes are employed to find genome-wide miRNA. However, most computed outputs are not typically subjected to human curation to verify prediction algorithms work as anticipated in all scenarios. **a** Curation of 792 barley miRNA loci found using miRDeep-P (Yang and Li 2011) filtered for plant miRNA signatures and additional filtering steps including removal of candidate miRNAs found in a single NGS dataset (Mascher et al. 2017). The number of retained miRNA loci upon application of additional curation steps is shown, including manual inspection of 2 nt-overhangs between the mature and star miRNA sequences. PARE = Parallel Analysis of RNA Ends method catalogs miRNA-mediated cleave sites of mRNAs. **b** Overlap of barley miRNAs with PARE-validated targets during early seed development (Curaba et al. 2012). DPA = days post-anthesis

7.4.6 Summary

Many studies have shown differential expression of barley miRNAs in response to environmental stresses. Barley miRNAs play a role in seed development and germination as well as resistance to fungal diseases. A consistent annotation framework is likely to promote the discovery of genuine lineage-specific miRNAs, Triticeae-specific miRNAs or grass-specific miRNAs that may play a unique role in monocotyledonous plant species.

7.5 The Barley Epigenome

7.5.1 Background

The epigenome comprises the covalent modifications to the genome and its associated chromatin proteins. Eukaryotic DNA is methylated at a subset of its cytosine bases and the histones comprising the nucleosome are modified by the addition of a variety of chemical groups that affect the affinity of nucleosomes for both each other and for other proteins. This, in turn, modulates the wrapping of DNA into highly condensed heterochromatin and less-condensed euchromatin, which facilitates the establishment and maintenance of accessible chromatin compartments for gene expression and inaccessible heterochromatin regions for suppression of transposable elements (Berger 2007; Dorn and Cook 2011).

7.5.2 The Barley Epigenomic Profile

Little or no studies of barley genome DNA methylation have been published to date and this summary is confined to histone modification. The barley epigenome received little study prior to the work of Baker et al. (2015), who mapped the genomic and genic distributions of nine

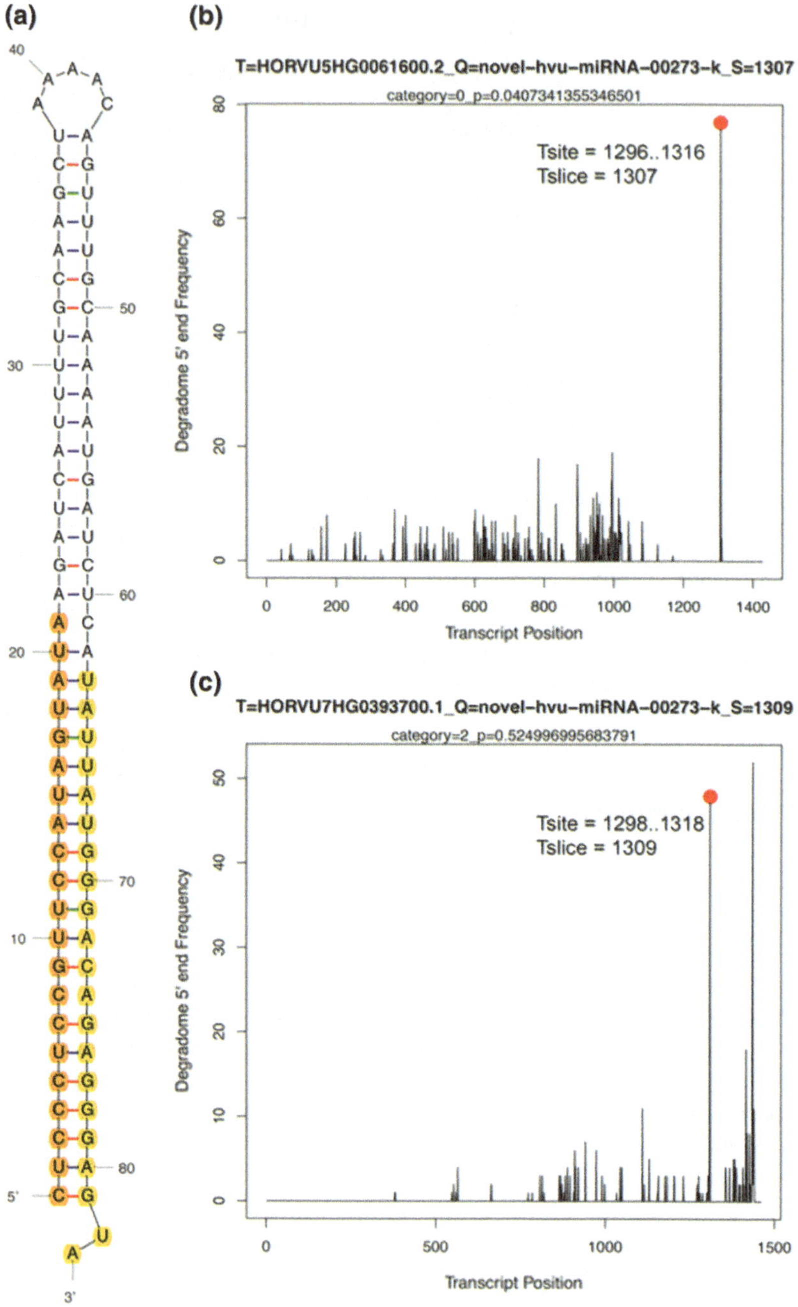

Fig. 7.6 Barley-specific miRNA structure and predicted targets. **a** Predicted MFOLD RNA-folding structure of the novel-hvu-miRNA-00273-k. Highlighted fonts denote the mature (3p) and star (5p) sequences. **b** Shows predicted cleavage site (Tslice) on the HORVU5HG0061600.2 transcript. Tsite = indicate the start and end coordinates of the alignment of the novel-hvu-miRNA-00273-k. **c** Predicted miRNA target HORVU7HG0393700.1 transcript

covalent histone modifications known to play important roles in the genic and intergenic environments of other plants (Berger 2007; Dorn and Cook 2011; Roudier et al. 2011). Mappings were made to a pseudogenome based upon the genome data of IBGSC (2012) and subsequent re-mapping to the updated genome assembly (Mascher et al. 2017) has revealed essentially the same profiles (M. Mascher, personal communication).

The chromosomal profiles for these modifications fall into four classes (Fig. 7.7b–e). The

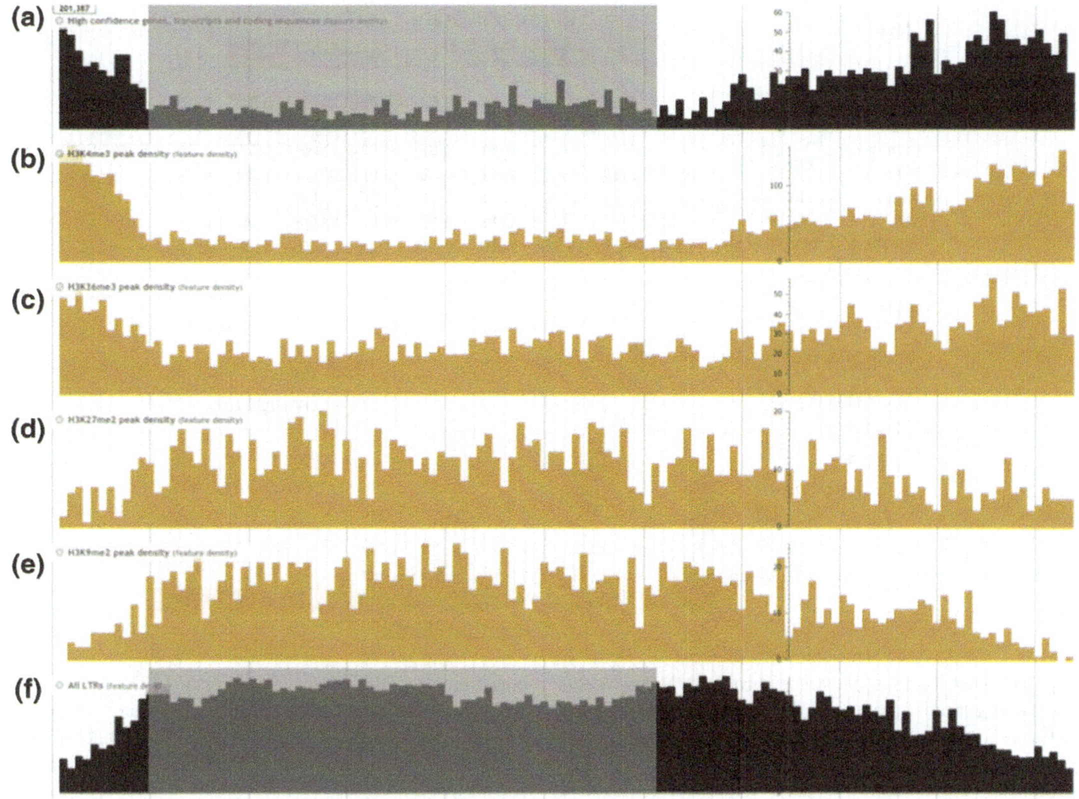

Fig. 7.7 Densities of four classes of histone modification across barley chromosome 1H. Local peak densities (peaks/1Mbp bin) for **a** High-confidence genes; **b** H3K4me3 **c** H3K36me3 **d** H3K27me2; **e** H3K9me2; **f** LTR retrotransposons are plotted against pseudo-physical position on barley chromosome 1H in JBrowse (Baker et al. 2015). The location of the low-recombining pericentromeric region (Baker et al. 2014) is in grey

profiles for H3K4me3 (b) and H3K9me2 (e) align closely with the frequencies of genes and LTR retrotransposons, respectively, and the other two class profiles (c–d) are intermediate between these extremes (Fig. 7.7). Genes and LTR retrotransposons show inverse chromosomal distributions in barley and the above data are consistent with the model that the genic and LTR retrotransposon-rich intergenic genomic spaces of barley are marked by corresponding disparate repertoires of modified histones.

7.5.3 Barley Chromatin States

Combinations of differentially modified histones associate with the different genic and genomic features to regulate the functioning of the epigenome, leading to the concept of 'chromatin states', each of which has a particular combination of different histones occupying the same region of DNA (Jenuwein and Allis 2001; Roudier et al. 2011). Baker et al. (2015) used peak sharing between barley histone modifications to define ten chromatin states (Figs. 7.7, 7.8 and 7.9), together with an eleventh 'zero' state lacking histone modification, which describe the different local epigenetic environments in the barley epigenome. For example, H3K4me3 (Fig. 7.7b) contributes in differing proportions to states 7–3 (Fig. 7.7a), whereas H3K36me3 (Fig. 7.7c) is mostly involved in only states 5 and 6. By retrieving the sequence annotations associated with these states Baker et al. were able

(a)

Chromatin State

	7	5	6	4	3	2	1	8	9	10	11
Gene	4.23	10.23	5.27	3.53	3.77	0.45	0.07	0.05	0.03	0.02	0.08
TSS	5.34	7.26	5.21	4.38	4.12	0.49	0.09	0.09	0.04	0.05	0.13
Coding seq	4.96	9.53	4.54	3.76	4.95	0.41	0.05	0.03	0.02	0.02	0.06
Exon	5.12	8.82	5.00	3.87	4.32	0.42	0.06	0.05	0.03	0.03	0.09
Intron	3.00	12.43	5.82	2.99	2.70	0.44	0.08	0.04	0.02	0.01	0.06
TES	5.25	7.27	5.23	4.56	4.05	0.48	0.07	0.07	0.05	0.06	0.13
LTR Retro	0.31	0.26	0.62	0.53	0.20	0.46	1.16	1.21	1.23	1.24	1.17

(b)

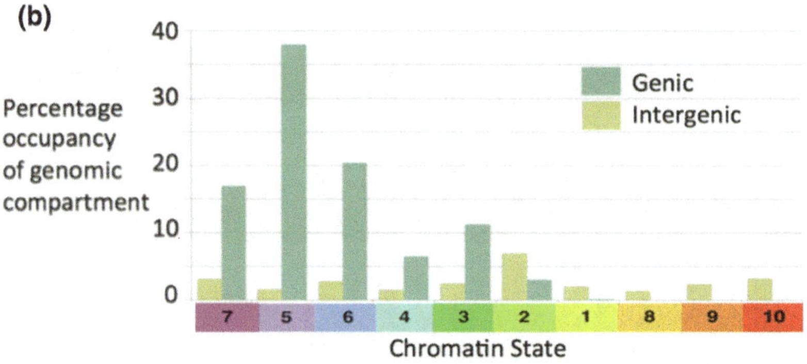

Fig. 7.8 Distributions of barley chromatin states versus annotated genomic features. **a** Enrichments of chromatin states in annotated genomic features. Cell values are colour-coded from dark green (highly enriched) through white (no enrichment) to dark red (strong negative enrichment). TSS, Transcription start site; TES, Transcription end site; LTR Retro, LTR retrotransposon. **b** Total percentage occupancies for each state in total annotated genic (green) and intergenic (khaki) spaces are shown

to explore the properties of these environments (Fig. 7.8) and order them in terms of their involvement with active vs inactive chromatin. Five 'active' states (states 3–7) are all strongly genic, with high enrichments for gene-associated annotations and weak overlaps with LTR retrotransposons (Fig. 7.8a). Four 'inactive' states (states 1, 8, 9, 10) are rare in gene annotations and enriched for LTR retrotransposons. The tenth State (state 2), transitions between the genic and intergenic environments, with similar annotation frequencies for both genes and retrotransposons.

When the total genic and intergenic annotated spaces were explored (Fig. 7.8b), almost all of the gene space (93%) was shown to be occupied by the five active genic states 3–7. State 3 superimposes H3K27me3 upon other gene-associated marks, suggesting that it is acting in barley to control tissue-specific gene expression. State 2 occupies both genic (3%) and intergenic (7%) spaces but as the latter is 20-fold larger than genic space in barley (IBGSC 2012), state 2 is overwhelmingly intergenic. The four inactive repressive chromatin states (1, 8, 9, 10) are almost absent from the genic compartment.

7.5.4 An Unexpected High-Order Structure for the Barley Epigenome

Mapping the chromatin states at the genome level revealed a striking high-order ultrastructure for the barley epigenome (Baker et al. 2015; Fig. 7.9). Telomere-proximal (TP) regions contain high densities of states 2 and 3, gene-rich interior (GRI) regions interior to these are characterized by gene-associated states 4–7 and the gene-poor LR-PC interior regions are dominated by the intergenic retrotransposon-associated states 1, 8, 9 and 10, with occasional gene islands decorated with the corresponding genic chromatin states. These three regions (TP, GRI, LR-PC) correspond to Zones 1–3 (distal,

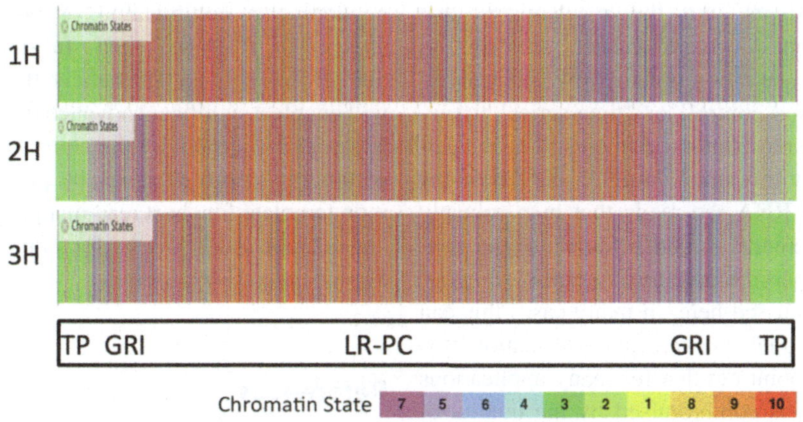

Fig. 7.9 High-order epigenomic structure of barley chromosomes. Densities in 1 Mbp bins for all chromatin states are plotted on chromosomes 1H–3H, respectively. Chromosome lengths are normalized against each other. Telomere-proximal (TP), Chromatin states are colour coded as shown in the key and the three epigenomic regions described in the text (TP, GRI and LR-PC are shown schematically in the lower image)

interstitial, proximal) respectively, defined by Mascher et al. (2017) on the basis of median 20-mer frequencies.

States 2 and 3, which define the TP/Zone 1 region, are both characterized by high levels of H3K27me3 (Baker et al. 2015), which represses transcription in plants, fungi and animals (Jenuwein and Allis 2001; Roudier et al. 2011). The rarity of the retrotransposon-associated repressive states in the TP/Zone 1 region (Baker et al. 2015) implies that intergenic state 2 is promoting the maintenance of facultative heterochromatin in this region to repress transcription. GRI/Zone 2 regions have low levels of states 2–3 and increasing amounts of the retrotransposon-associated intergenic repressive states 1, 8, 9 and 10, suggesting that the GRI/Zone 2 intergenic heterochromatic environment is maintained by these states. This situation is much more apparent within the LR-PC/Zone 3 regions, which are dominated by heterochromatic marks (Baker et al. 2015).

7.5.5 Summary

The barley chromatin epigenome displays an unexpected higher order chromatin structure comprising three epigenomic regions defined by the relative abundances of H3K27me3, other active chromatin marks and the repressive modifications which tag LTR retrotransposons. This ultrastructure has also been found on the basis of DNA sequence repeat distribution and we suggest that the barley epigenome responds to both the regional accumulation of LTR retrotransposons and the imperative to define and preserve active and inactive chromatin for gene expression and repression across this fluctuating background.

7.6 Conclusions

The control of gene expression in barley, as with all organisms, is the result of a complex layer of interactions that starts with accessibility of the genome to the transcriptional machinery, sequence and structural elements in the promoters and gene bodies of individual genes that control levels and specificity of transcription and post-transcriptional processing events that modify the abundance of mature and intact mRNAs (e.g. by miRNA-directed cleavage) that are ultimately translated into proteins. Currently, groups across the barley research community are using high throughput short-read RNA-sequencing to explore a wide range of biological processes. For example, insights into the mechanisms that underpin plant development or response to external cues are being obtained by the analysis of 'case versus control' or 'time-series' tissue-level gene expression experiments. These analyses generally rely upon counting the frequency of individual RNA-seq reads that map robustly against a reference sequence template (the genome) and then comparing amongst samples. While, as discussed here, in many cases this can provide a good level of relevant information, it is important to point out that for many applications this approach is compromised by the quality of the template upon which the RNA-seq data is mapped (e.g. a poorly assembled and annotated genome sequence) and/or the computational algorithms used in the mapping process. In addition, little or no consideration is generally given to widely occurring post-transcriptional processes that will, in many cases, significantly skew the quantification of individual transcript abundances.

So how can we make the best out of RNA-seq datasets? We have proposed that one way to improve RNA-seq data analysis would be to generate a community-wide Reference Transcript Dataset (RTD) comprised of all transcript isoforms (i.e. not simply full length genes) detected in the transcriptome of a target species and against which short read RNA-seq data could then be mapped. Such a resource should allow new and emerging sensitive methods in bioinformatics and computational biology to be applied to the analysis of RNA-seq data, allowing a better understanding of how gene expression is regulated at both the gene and individual transcript level. Being able to analyse gene expression at the transcript level will enhance all dependent downstream analyses including co-expression and regulatory network analyses, and building the mathematical models that promote our understanding of developmental or environmental responses. As many of these will underpin future crop improvement strategies, the rationale for generating an RTD for barley is strong as it will enable, improve and unify the analysis of RNA-seq datasets throughout the barley research community. We have recently started to do this by generating a combination of deep paired-end short-read RNA-seq and full-length PacBio Sequel Iso-seq data. We believe that the outcome will enable a significant advance in transcriptomic analyses carried out by the entire barley research community, unify analyses by providing a common template for short read mapping and provide considerable added-value for improving annotation of the barley genome.

References

Bai B, Bian H, Zeng Z, Hou N, Shi B, Wang J, Zhu M, Han N (2017a) miR393-mediated auxin signaling regulation is involved in root elongation inhibition in response to toxic aluminum stress in barley. Plant Cell Physiol 58:426–439

Bai B, Shi B, Hou N, Cao Y, Meng Y, Bian H, Zhu M, Han N (2017b) miRNAs participate in gene expression regulation and phytohormone cross-talk in barley embryo during seed development and germination. BMC Plant Biol 17(1):150

Baker K, Dhillon T, Colas I, Cook N, Milne I, Cardle L, Bayer M, Flavell AJ (2015) Chromatin state analysis of the barley epigenome reveals a higher order structure defined by H3K27me1 and H3K27me3 abundance. Plant J. 84:111–124

Baker K, Bayer M, Cook N, Dreißig S, Dhillon T, Russell J, Hedley PE, Morris J, Ramsay L, Colas I, Waugh R, Steffenson B, Milne I, Stephen G, Marshall D, Flavell AJ(2014) The low-recombining pericentromeric region of barley restricts gene diversity and evolution but not gene expression. Plant J 79 (6):981–992

Berger SL (2007) The complex language of chromatin regulation during transcription. Nature 447:407–412

Borges F, Martienssen RA (2015) The expanding world of small RNAs in plants. Nat Rev Mol Cell Bio 16:727–741

Bull H, Casao MC, Zwirek M, Flavell AJ, Thomas WTB, Guo W, Zhang R, Rapazote-Flores P, Kyriakidis S, Russell J, Druka A, McKim SM, Waugh R(2017) Barley SIX-ROWED SPIKE3 encodes a putative Jumonji C-type H3K9me2/me3 demethylase that represses lateral spikelet fertility. Nat Commun 8 (1)

Busch A, Hertel KJ (2012) Extensive regulation of NAGNAG alternative splicing: new tricks for the spliceosome? Genome Biol 13:143

Calixto CP, Simpson CG, Waugh R, Brown JWS (2016) Alternative splicing of barley clock genes in response to low temperature. PLoS ONE 11:e0168028

Close TJ, Wanamaker SI, Caldo RA, Turner SM, Ashlock DA, Dickerson JA, Wing RA, Muehlbauer GJ, Kleinhofs A, Wise RP (2004) A new resource for cereal genomics: 22K barley GeneChip comes of age. Plant Physiol 134:960–968

Colaiacovo M, Subacchi A, Bagnaresi P, Lamontanara A, Cattivelli L, Faccioli P (2010) A computational-based update on microRNAs and their targets in barley (Hordeum vulgare L.). BMC Genomics 11. https://doi.org/10.1186/1471-2164-11-595

Curaba J, Spriggs A, Taylor J, Li Z, Helliwell C (2012) miRNA regulation in the early development of barley seed. BMC Plant Biol 12. https://doi.org/10.1186/1471-2229-12-120

Dass SS, Karmakar P, Nandi AK, Sanan-Minshra N (2015) Small RNA mediated regulation of seed germination. Front Plant Sci 6:828

Deng PC, Wang L, Cui LC, Feng KW, Liu FY, Du XH, Tong W, Nie XJ, Ji WQ, Weining S (2015) Global identification of microRNAs and their targets in barley under salinity stress. Plos One 10. https://doi.org/10.1371/journal.pone.0137990

Dorn ES, Cook JG (2011) Nucleosomes in the neighborhood: new roles for chromatin modifications in replication origin control. Epigenetics 6:552–559

Efroni I, Birnbaum KD (2016) The potential of single-cell profiling in plants. Genome Biol 17:65. https://doi.org/10.1186/s13059-016-0931-2

Fard EM, Bakhshi B, Keshavarznia R, Nikpay N, Shahbazi M, Salekdeh GH (2017) Drought responsive microRNAs in two barley cultivars differing in their level of sensitivity to drought stress. Plant Physiol Biochem 118:121–129

Fromm B, Billipp T, Peck LE, Johansen M, Tarver JE, King BL, Newcomb JM, Sempere LF, Flatmark K, Hovig E et al (2015) A uniform system for the annotation of vertebrate microRNA genes and the evolution of the human microRNAome. Annu Rev Genet 49:213–242

Gadjieva R, Axelsson E, Olsson U, Vallon-Christersson J, Hansson M (2004) Nonsense-mediated mRNA decay in barley mutants allows the cloning of mutated genes by a microarray approach. Plant Physiol Biochem 42:681–685

Giacomello S, Salmén F, Terebieniec BK, Vickovic S, Navarro JF, Alexeyenko A, Reimegård J, McKee LS, Mannapperuma C, Bulone V, Ståhl PL, Sundström JF, Street NR, Lundeberg J (2017) Spatially resolved transcriptome profiling in model plant species. Nat Plants 3:17061

Gloggnitzer J, Akimcheva S, Srinivasan A, Kusenda B, Riehs N, Stampfl H, Bautor J, Dekrout B, Jonak C, Jiménez-Gómez JM, Parker JE, Riha K (2014) Nonsense-mediated mRNA decay modulates immune receptor levels to regulate plant antibacterial defense. Cell Host Microbe 16:376–390

Guo L, Liu C-M (2016) A single-nucleotide exon found in Arabidopsis. Sci Rep 5 (1)

Hackenberg M, Gustafson P, Langridge P, Shi BJ (2015) Differential expression of microRNAs and other small RNAs in barley between water and drought conditions. Plant Biotechnol J 13:2–13

Hackenberg M, Huang PJ, Huang CY, Shi BJ, Gustafson P, Langridge P (2013) A comprehensive expression profile of MicroRNAs and other classes of non-coding small RNAs in barley under phosphorous-deficient and -sufficient conditions. DNA Res 20:109–125

Halterman DA, Wei F, Wise RP (2003) Powdery mildew-induced Mla mRNAs are alternatively spliced and contain multiple upstream open reading frames. Plant Physiol 131:558–567

Hove RM, Ziemann M, Bhave M (2015) Identification and expression analysis of the barley (Hordeum vulgare L.) aquaporin gene family. PLoS ONE 10:e0128025

International Barley Genome Sequencing Consortium (IBGSC) (2012) A physical, genetic and functional sequence assembly of the barley genome. Nature 491:711–716

Jenuwein T, Allis CD (2001) Translating the histone code. Science 293:1074–1080

Jones-Rhoades MW, Bartel DP, Bartel B (2006) MicroRNAS and their regulatory roles in plants. Annu Rev Plant Biol 57:19–53

Kalyna M, Simpson CG, Syed NH, Lewandowska D, Marquez Y, Kusenda B, Marshall J, Fuller J, Cardle L, McNicol J, Dinh HQ, Barta A, Brown JWS (2012) Alternative splicing and nonsense-mediated decay modulate expression of important regulatory genes in Arabidopsis. Nucleic Acids Res 40:2454–2469

Kantar M, Unver T, Budak H (2010) Regulation of barley miRNAs upon dehydration stress correlated with target gene expression. Funct Integr Genomic 10:493–507

Karam R, Wengrod J, Gardner LB, Wilkinson MF (2013) Regulation of nonsense-mediated mRNA decay: implications for physiology and disease. Biochim Biophys Acta 1829:624–633

Koroban NV, Kudryavtseva AV, Krasnov GS, Sadritdinova AF, Fedorova MS, Snezhkina AV, Bolsheva NL, Muravenko OV, Dmitriev AA, Melnikova NV (2016) The role of microRNA in abiotic stress response in plants. Mol Biol 50:337–343

Kozomara A, Griffiths-Jones S (2014) miRBase: annotating high confidence miRNAs using deep sequencing data. Nucleic Acids Res 42:D68–D73

Li S, Castillo-González C, Yu B, Zhang X (2017) The functions of plant small RNAs in development and in stress responses. Plant J 90:654–670

Li X, Shin S, Heinen S, Dill-Macky R, Berthiller F, Nersesian N, Clemente T, McCormick S, Muehlbauer GJ (2015) Transgenic wheat expressing a barley UDP-glucosyltransferase detoxifies deoxynivalenol and provides high levels of resistance to *Fusarium graminearum*. Mol Plant Microbe Interact 28:1237–1246

Locascio A, Roig-Villanova I, Bernardi J, Varotto S (2014) Current perspectives on the hormonal control of seed development in Arabidopsis and maize: a focus on auxin. Front Plant Sci 5:412

Lv SZ, Nie XJ, Wang L, Du XH, Biradar SS, Jia XO, Weining S (2012) Identification and characterization of microRNAs from barley (Hordeum vulgare L.) by high-throughput sequencing. Int J Mol Sci 13:2973–2984

Marquez Y, Brown JWS, Simpson CG, Barta A, Kalyna M (2012) Transcriptome survey reveals increased complexity of the alternative splicing landscape in Arabidopsis. Genome Res 22:1184–1195

Mascher M, Gundlach H, Himmelbach A, Beier S, Twardziok SO, Wicker T, Radchuk V, Dockter C, Hedley PE, Russell J, Bayer M, Ramsay L, Liu H, Haberer G, Zhang XQ, Zhang Q, Barrero RA, Li L, Taudien S, Groth M, Felder M, Hastie A, Šimková H, Staňková H, Vrána J, Chan S, Muñoz-Amatriaín M, Ounit R, Wanamaker S, Bolser D, Colmsee C, Schmutzer T, Aliyeva-Schnorr L, Grasso S, Tanskanen J, Chailyan A, Sampath D, Heavens D, Clissold L, Cao S, Chapman B, Dai F, Han Y, Li H, Li X, Lin C, McCooke JK, Tan C, Wang P, Wang S, Yin S, Zhou G, Poland JA, Bellgard MI, Borisjuk L, Houben A, Doležel J, Ayling S, Lonardi S, Kersey P, Langridge P, Muehlbauer GJ, Clark MD, Caccamo M, Schulman AH, Mayer KFX, Platzer M, Close TJ, Scholz U, Hansson M, Zhang G, Braumann I, Spannagl M, Li C, Waugh R, Stein N (2017) A chromosome conformation capture ordered sequence of the barley genome. Nature 544:427–433

Matsumoto T, Tanaka T, Sakai H, Amano N, Kanamori H, Kurita K, Kikuta A, Kamiya K, Yamamoto M, Ikawa H, Fujii N, Hori K, Itoh T, Sato K (2011) Comprehensive sequence analysis of 24,783 barley full-length cDNAs derived from 12 clone libraries. Plant Physiol 156:20–28

McKibbin RS, Wilkinson MD, Bailey PC, Flintham JE, Andrew LM, Lazzeri PA, Gale MD, Lenton JR, Holdsworth MJ (2002) Transcripts of Vp-1 homeologues are misspliced in modern wheat and ancestral species. Proc Natl Acad Sci USA 99:10203–10208

Meng F, Liu H, Wang K, Liu L, Wang S, Zhao Y, Yin J, Li Y (2013) Development-associated microRNAs in grains of wheat (Triticum aestivum L.). BMC Plant Biol 13:140

Min XJ, Powell B, Braessler J, Meinken J, Yu F, Sablok G (2015) Genome-wide cataloging and analysis of alternatively spliced genes in cereal crops. BMC Genomics 16(1):721

Mochida K, Uehara-Yamaguchi Y, Yoshida T, Sakurai T, Shinozaki K (2011) Global landscape of a co-expressed gene network in barley and its application to gene discovery in Triticeae crops. Plant Cell Physiol 52:785–803

Nakabayashi K, Okamoto M, Koshiba T, Kamiya Y, Nambara E (2005) Genome-wide profiling of stored mRNA in Arabidopsis thaliana seed germination: epigenetic and genetic regulation of transcription in seed. Plant J 41:697–709

Ozhuner E, Eldem V, Ipek A, Okay S, Sakcali S, Zhang BH, Boke H, Unver T (2013) Boron stress responsive microRNAs and their targets in barley. Plos One 8. https://doi.org/10.1371/journal.pone.0059543

Panahi B, Mohammadi SA, Ebrahimi Khaksefidi R, Fallah Mehrabadi J, Ebrahimie E (2015) Genome-wide analysis of alternative splicing events in Hordeum vulgare: highlighting retention of intron-based splicing and its possible function through network analysis. FEBS Lett 589:3564–3575

Pingault L, Choulet F, Alberti A, Glover N, Wincker P, Feuillet C, Paux E (2015) Deep transcriptome sequencing provides new insights into the structural and functional organization of the wheat genome. Genome Biol 16:29

Reddy AS (2007) Alternative splicing of pre-messenger RNAs in plants in the genomic era. Annu Rev Plant Biol 58:267–294

Reddy AS, Marquez Y, Kalyna M, Barta A (2013) Complexity of the alternative splicing landscape in plants. Plant Cell 25:3657–3683

Reinhart BJ, Weinstein EG, Rhoades MW, Bartel B, Bartel DP (2002) MicroRNAs in plants (vol 16, pg 1616, 2002). Gene Dev 16:2313

Rodríguez-Suárez C, Atienza SG, Pistón F (2011) Allelic variation, alternative splicing and expression analysis of Psy1 gene in Hordeum chilense Roem. et Schult. PLoS ONE 6:e19885

Roudier F, Ahmed I, Bérard C et al (2011) Integrative epigenomic mapping defines four main chromatin states in Arabidopsis. EMBO J 30:1928–1938

Schmitz J, Franzen R, Ngyuen TH, Garcia-Maroto F, Pozzi C, Salamini F, Rohde W (2000) Cloning, mapping and expression analysis of barley MADS-box genes. Plant Mol Biol 42:899–913

Schreiber AW, Shi BJ, Huang CY, Langridge P, Baumann U (2011) Discovery of barley miRNAs through deep sequencing of short reads. BMC Genomics 12. https://doi.org/10.1186/1471-2164-12-129

Shahzad K, Rauf M, Ahmed M, Malik ZA, Habib I, Ahmed Z, Mahmood K, Ali R, Masmoudi K, Lemtiri-Chlieh F, Gehring C, Berkowitz GA, Saeed NA (2015) Functional characterisation of an intron retaining K(+) transporter of barley reveals intron-mediated alternate splicing. Plant Biol (Stuttg) 17:840–851

Shen L, Gong J, Caldo RA, Nettleton D, Cook D, Wise RP, Dickerson JA (2005) BarleyBase—an expression profiling database for plant genomics. Nucleic Acids Res 1:33

Shi Y, Sha G, Sun X (2014) Genome-wide study of NAGNAG alternative splicing in Arabidopsis. Planta 239:127–138

Silva AT, Ribone PA, Chan RL, Ligterink W, Hilhorst HW (2016) A predictive coexpression network identifies novel genes controlling the seed-to-seedling phase transition in arabidopsis thaliana. Plant Physiol 170:2218–2231

Sreenivasulu N, Usadel B, Winter A, Radchuk V, Scholz U, Stein N, Weschke W, Strickert M, Close TJ, Stitt M et al (2008) Barley grain maturation and germination: metabolic pathway and regulatory network commonalities and differences highlighted by new MapMan/PageMan profiling tools. Plant Physiol 146:1738–1758

Staiger D, Brown JWS (2013) Alternative splicing at the intersection of biological timing, development, and stress responses. Plant Cell 25:3640–3656

Thorbjørnsen T, Villand P, Kleczkowski LA, Olsen OA (1996) A single gene encodes two different transcripts for the ADP-glucose pyrophosphorylase small subunit from barley (Hordeum vulgare). Biochem J 313:149–154

van Esse GW, Walla A, Finke A, Koornneef M, Pecinka A, von Korff M, (2017) Six-Rowed Spike3 (VRS3) is a histone demethylase that controls lateral spikelet development in barley. Plant Physiol 174 (4):2397–2408

Walters B, Lum G, Sablok G, Min XJ (2013) Genome-wide landscape of alternative splicing events in Brachypodium distachyon. DNA Res 20:163–171

Wu XM, Hornyik C, Bayer M, Marshall D, Waugh R, Zhang RX (2014) In silico identification and characterization of conserved plant microRNAs in barley. Cent Eur J Biol 9:841–852

Xu W, Meng Y, Wise RP (2014) Mla- and Rom1-mediated control of microRNA398 and chloroplast copper/zinc superoxide dismutase regulates cell death in response to the barley powdery mildew fungus. New Phytol 201:1396–1412

Xue GP, Loveridge CW (2004) HvDRF1 is involved in abscisic acid-mediated gene regulation in barley and produces two forms of AP2 transcriptional activators, interacting preferably with a CT-rich element. Plant J 37:326–339

Yang X, Li L (2011) miRDeep-P: a computational tool for analyzing the microRNA transcriptome in plants. Bioinformatics 27:2614–2615

Zhang Q, Zhang X, Pettolino F, Zhou G, Li C (2016a) Changes in cell wall polysaccharide composition, gene transcription and alternative splicing in germinating barley embryos. J Plant Physiol 191:127–139

Zhang Q, Zhang X, Wang S, Tan C, Zhou G, Li C (2016b) Involvement of alternative splicing in barley seed germination. PLoS ONE 11:e0152824

Zhang R, Tucker MR, Burton RA, Shirley NJ, Little A, Morris J, Milne L, Houston K, Hedley PE, Waugh R, Fincher GB (2016c) The dynamics of transcript abundance during cellularization of developing barley endosperm. Plant Physiol 170:1549–1565

Zhiguo E, Wang L, Zhou J (2013) Splicing and a lternative splicing in rice and humans. BMB Rep 46:439–447

8 Sequence Diversity and Structural Variation

María Muñoz-Amatriaín and Martin Mascher

Abstract

Barley is a phenotypically and genetically diverse crop that has adapted successfully to many agricultural environments. Moreover, its wild progenitor species is still abundant in Western Asia and provides a large untapped reservoir of allelic variation. Deriving a complete inventory of sequence variants of the barley genome for targeted utilization of natural diversity in crop improvement has long been a key goal of barley research. While the assessment of genetic diversity has traditionally focused on a few selected—mostly genic—loci, recent technology advances have made it possible to simultaneously obtain genome-wide sequence information for genes and noncoding regions alike. In addition to small-scale sequence changes, larger scale structural variation such as presence–absence or copy number variants are widespread throughout the barley genome. These types of variation can affect large genomic regions, possibly containing multiple genes or adjacent regulatory regions. Here, we review the recent progress on assaying genetic variants, large and small, in barley, understanding the mutational processes underlying them, and their relationship to phenotype and ultimately crop performance.

M. Muñoz-Amatriaín (✉)
Department of Botany and Plant Sciences, University of California Riverside, Riverside, CA 92521, USA
e-mail: maria.munoz-amatriain@ucr.edu

M. Mascher
Leibniz Institute of Plant Genetic and Crop Plant Research (IPK) Gatersleben, 06466 Seeland, Germany

M. Mascher
German Centre for Integrative Biodiversity Research (iDiv) Halle-Jena-Leipzig, 04103 Leipzig, Germany

8.1 Introduction

Since barley was domesticated about 10,000 years ago (Badr et al. 2000), human populations have used natural diversity for barley improvement by selecting for desirable traits and/or specific adaptation and, more recently, through modern plant breeding. It is well known that domestication and breeding create genetic bottlenecks, making current elite varieties less diverse and, hence, more vulnerable to environmental changes than local landraces and wild relatives (Tanksley and McCouch 1997; Yamasaki et al. 2005; Russell et al. 2016). Like other crop scientists, barley researchers are facing additional challenges related to human population growth, variation in weather patterns, and degradation and erosion of agricultural land, among others. Solutions to these many challenges largely lay in the efficient exploitation of barley natural diversity.

N. Stein and G. J. Muehlbauer (eds.), *The Barley Genome*, Compendium of Plant Genomes,
https://doi.org/10.1007/978-3-319-92528-8_8

Barley is one of the most highly adaptable crops, growing from the Arctic Circle to high mountain regions and to the edge of the deserts. Its adaptability to this wide range of environments may reside in the species' genetic variation. Computational and technological advances are allowing opportunities to unveil the complexity of the barley genome, to gain access to its natural diversity, and to further correlate genetic differences with phenotypic variations (reviewed in Stein and Steuernagel 2014; Muñoz-Amatriaín et al. 2014b). In this endeavor, the availability of a reference genome sequence (International Barley Genome Sequencing Consortium 2012; Mascher et al. 2017) as well as tools to assess variation (e.g., Comadran et al. 2012; Poland et al. 2012; Mascher et al. 2013b) is as relevant as having access to rich barley germplasm resources. The barley community is fortunate to possess diverse germplasm collections of cultivated (*H. vulgare* ssp. *vulgare*) and wild barley (*H. vulgare* ssp. *spontaneum*) (Bockelman and Valkoun 2010). It is estimated that nearly 300,000 accessions of barley are conserved in ex situ genebank collections worldwide, with almost half of those accessions being landraces and around 15% being wild relatives (Bockelman and Valkoun 2010). The genetic characterization of these materials is useful not only for allele mining for biotic and abiotic stress tolerance and improving overall yield (e.g., Pasam et al. 2012; Ames et al. 2015; Tamang et al. 2015) but also for the preservation of barley germplasm diversity for present and future generations. In the near future, genome editing technologies such as CRISPR/Cas9 (clustered regularly interspaced short palindromic repeats/CRISPR-associated 9) could allow to directly transfer any beneficial allele into elite germplasm (Wang et al. 2017a). This would boost the characterization and use of genebank accessions (especially of wild crop relatives) in crop improvement as limitations associated with the use of exotic germplasm including cross-incompatibility and linkage drag will be eliminated (Dempewolf et al. 2017).

An important factor affecting both the generation and deployment of genetic variation is recombination. Progress in characterizing, understanding and manipulating the patterns of recombination is being made in barley (Aliyeva-Schnorr et al. 2015; Muñoz-Amatriaín et al. 2015; Rey et al. 2015; Mascher et al. 2017). In addition, targeted mitotic recombination via CRISPR-Cas9 technology, which was first demonstrated in yeast (Sadhu et al. 2016), is now achievable in crop plants (Filler Hayut et al. 2017) and has the potential to considerably increase genetic gains (Bernardo 2017).

In the following sections, we review current knowledge on the patterns of genetic variation in the barley genome, and revise the known mechanisms generating genetic variation—including single-nucleotide polymorphisms (SNPs) and structural variants—and the successes and challenges in identifying and using favorable variants for barley improvement.

8.2 Landscape of Genetic Diversity in the Barley Genome

Genetic variation can be generally divided into SNPs, small insertions and deletions (INDELs), and structural variations (SVs). SVs encompass balanced variants, such as inversions and translocations, and unbalanced changes, including copy number variations (CNVs), of at least 50 bp in size (Alkan et al. 2011; Weischenfeldt et al. 2013). Presence/absence variations (PAVs) are an extreme type of CNV, in which a particular sequence is completely missing in one or more individuals while present in others (Springer et al. 2009).

8.2.1 Single-Nucleotide and Insertion/Deletion Polymorphisms

Because of their abundance in the genome, their amenability for high-throughput genotyping, and accessibility through short-read sequencing technologies, SNPs have been the most studied type of genetic variation in barley. Until recently, the patterns of sequence diversity in the barley genome have been studied only by resequencing a small number of loci of interest (Jakob et al. 2014) or by using comparatively low-density marker assays with several thousands of markers (Russell et al. 2011; Muñoz-Amatriaín et al. 2014a). The recent development of integrated sequence and mapping resources (International Barley Genome Sequencing Consortium 2012; Mascher et al. 2013a) has enabled detailed studies of the distribution of sequence variation along the chromosomes. Concomitant with the development of advanced genomics resources, the progress in high-throughput sequencing technology has made it possible to resequence large parts or even the entirety of the genomes of different individuals of a species. The short-read sequences obtained by these technologies are aligned to a reference genome and then sequence variants are detected and genotyped by computational methods (reviewed by Nielsen et al. 2011).

Using this approach, several studies enquiring about the genetic diversity in barley have found that sequence variation is not distributed uniformly along the barley genome. The International Barley Genome Sequencing Consortium (2012) detected more than 15 million SNPs in whole-genome shotgun resequencing data from four barley cultivars and one accession of barley's wild progenitor *H. vulgare* ssp. *spontaneum*. They reported that SNP density decreased towards proximal regions of all barley chromosomes. This finding was confirmed on a larger scale by Russell et al. (2016), who resequenced the gene space of 267 diverse accessions from a worldwide diversity panel of wild and domesticated barleys by means of exome sequencing. The barley exome capture assay targets approximately 60 Mb of mRNA-coding exons and adjacent intronic regions, enabling a cost-effective selective resequencing of genic regions in the large and repeat-rich barley genome (Mascher et al. 2013b). The survey of Russell et al. (2016) detected ~1.7 million SNPs and 140,000 INDELs. Eighty-two percent of these variant were assigned to chromosomal positions in the POPSEQ genetic map (Mascher et al. 2013a). About two-thirds of variants were located in annotated gene models. A comparison of sequence diversity between wild and domesticated barleys highlighted regions on several chromosomes that had a substantially lower diversity in landraces than in *H. vulgare* ssp. *spontaneum*. Some of these regions were collocated with genes selected during initial domestication or subsequent crop improvements such as the *BRITTLE RACHIS* genes on chromosome 3H (Pourkheirandish et al. 2015) or the *NUDUM* gene on chromosome 7H (Taketa et al. 2008).

However, the most prominent pattern in both wild and domesticated barley was a sharp decrease of nucleotide diversity in peri-centromeric regions that was accompanied by a lower recombination rate on an evolutionary timescale (Fig. 8.1). This observation is related to the nonuniform distribution of recombination rate along the chromosomes. Cytogenetic mapping has revealed that crossovers predominantly occur in the distal regions of the chromosomes and that vast regions of the chromosomes show very low recombination frequencies (Künzel 1982; Künzel et al. 2000). These regions are commonly referred by as the "genetic centromeres". The genetic centromeres of a chromosome encompass not only the cytogenetically defined centromeres, i.e., the sites of kinetochore attachment, and the peri-centromeric regions but also large tracts of interstitial regions. Thus, the genetic centromeres make up about two-thirds of the physical length of a barley chromosome.

A similar concentration of crossovers in gene-rich, distal regions has also been reported in wheat (Gill et al. 1996) and maize (Gore et al. 2009). More generally, a positive correlation between nucleotide diversity and recombination rate has been first observed in Drosophila (Begun and Aquadro 1992). Maynard Smith and Haigh (1974) have put forward the hypothesis that the action of linked selection decreases neutral sequence diversity. Thus, a selective sweep acting on a single gene may reduce levels of sequence polymorphisms also in linked chromosomal regions. This effect would be strongest in low-recombining regions where haplotype blocks can persist uninterrupted for a longer period of time. Another possible explanation is

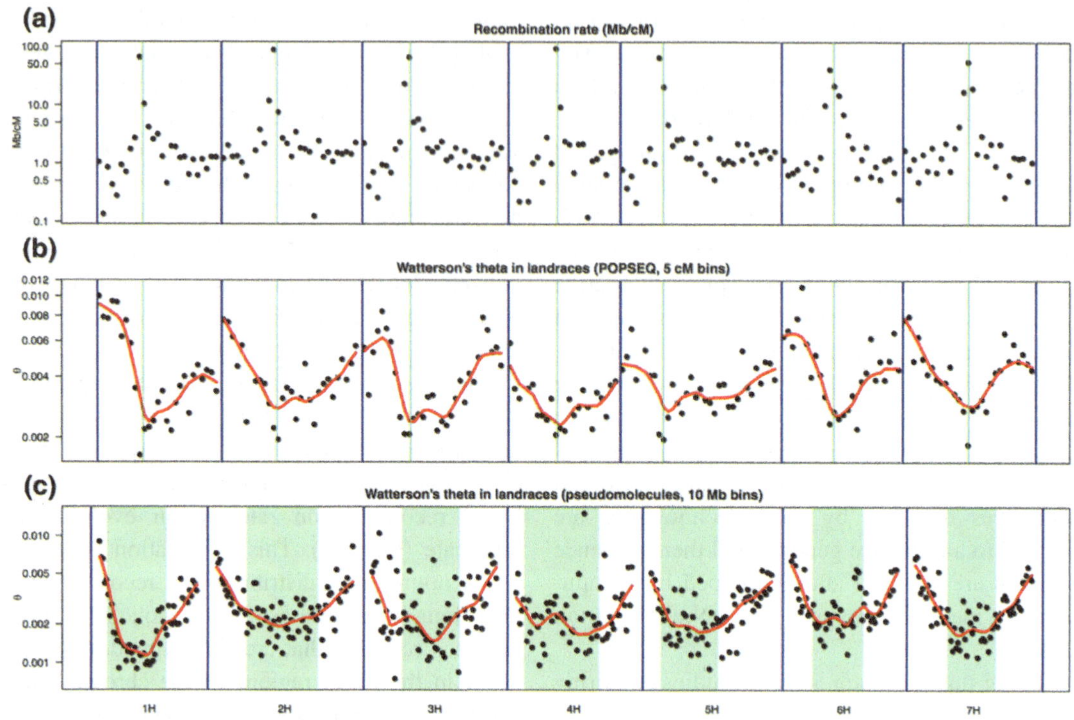

Fig. 8.1 **Relationship between sequence diversity and recombination rate in barley**. Recombination rates (**a**) and sequence diversity (**b**, **c**) were estimated using population sequencing of a biparental population (Mascher et al. 2013a) and exome sequencing data of 137 barley landraces (Russell et al. 2016) and plotted along the genome. The *x*-axes of panels **a** and **b** use the POPSEQ genetic map (Mascher et al. 2013a), while panel **c** uses the map-based reference sequence assembly of Mascher et al. (2017). The red line is a smoothed Loess curve. Blue lines separate chromosomes. Green shading marks the low-recombining peri-centromeric regions, which collapse to singles points on a genetic scale (**a**, **b**), but encompass 30–50% of the physical length of the chromosomes **c**

background selection against deleterious variants (Charlesworth et al. 1993; Hudson and Kaplan 1995) that reduces neutral diversity in genomic regions with low recombination rates. Lastly, the differences between sequence diversity between distal and proximal regions of the barley genome may be explained by variation in mutation rate as has been observed in mammalian genomes (Hodgkinson and Eyre-Walker 2011).

8.2.2 Structural Variations

Structural variation—particularly CNV—has been extensively studied in humans due to its impact in disease susceptibility (Weischenfeldt et al. 2013). Although our knowledge on SV in plant genomes is more limited, barley is one of the few species where CNV (including PAV) has been explored at a genome-wide scale (Muñoz-Amatriaín et al. 2013). Array comparative genomic hybridization (aCGH) on a panel of eight cultivars and six wild barleys identified CNV affecting 14.9% of all sequences—and 9.5% of the coding regions—represented on the array. The barley CGH array targeted ~50 Mb of genome sequence from cv. Morex, the reference genome (Mascher et al. 2017). CNV was not equally distributed along the barley chromosomes, with frequencies increasing towards their distal ends (Fig. 8.2), which correlated with recombination rate (Muñoz-Amatriaín et al. 2013). The only exception was chromosome 4H, in which CNV rates in the peri-centromeric region were as high as in the distal ends (Fig. 8.2c). However, most of those variants were lost during domestication and breeding, as it is shown by the pronounced differences between domesticated and wild barley on the peri-centromeric region of 4H (Fig. 8.2; Muñoz-Amatriaín et al. 2013). A lower SNP frequency had been previously observed on 4H (International Barley Genome Sequencing Consortium 2012). The lower recombination rates on this chromosome (the shortest genetically but not physically) could be limiting the emergence of structural variation (via nonallelic homologous recombination) as well as reducing CNV diversity through the effect of background selection. Even though a lower number of wild barleys were used in the study, almost half of all CNVs identified were present only in *H. vulgare* ssp. *spontaneum*, indicating that domestication and breeding bottlenecks also affected CNV diversity. Still, that study likely underestimated the number of CNVs present in wild barley considering that sequences not present in cv. Morex were not surveyed (Muñoz-Amatriaín et al. 2013). There are some limitations associated with aCGH, including the inability to detect changes in copy number for sequences not present in the reference, the lack of knowledge about the location of duplicated sequences, and the difficulties in defining breakpoints if those occur in an area not covered by the array (Marroni et al. 2014). Next-generation sequencing technologies can provide solutions to some of these limitations.

Although more rare in nature, some large spontaneous inversions and translocations have been detected in barley using cytological techniques (Ramage et al. 1961; Konishi and Linde-Laursen 1988). The study of Konishi and Linde-Laursen (1988) provided the most comprehensive report of this type of chromosomal rearrangements by investigating over a thousand cultivated accessions and 120 wild barleys. Four of the cultivated accessions (all from Ethiopia) carried a reciprocal translocation involving chromosomes 2H and 4H. More events were identified in wild barley, with three wild accessions carrying translocations between 2H and 4H, 3H and 5H, and 3H and 6H. In a more recent study of wild barley, Fang et al. (2014) identified, using SNP markers, a potential chromosomal structural variant on a similar region of chromosome 2H, and another on 5H.

New long-read sequencing technologies such as Single Molecule Real Time (SMRT) sequencing have the potential to revolutionize the study of SV (Chaisson et al. 2015). Still, very large structural variants and those residing in highly repetitive regions are difficult to resolve with any existing sequencing method as they may be larger than the individual reads. Additional non-sequence-based technologies such optical mapping could provide access to unexplored SVs (Lam et al. 2012; Tang et al. 2015). In barley, an optical map of the genome of the reference cultivar Morex has been developed using BioNano Genomics' Irys platform (Beier et al. 2017). Its comparison with optical maps from other accessions would likely reveal complex SVs present in barley genomes.

8.3 SNP Variation and Genome-Wide Association Studies

In the last years, barley has experienced a rapid increase in the number of marker-trait associations determined by genome-wide association studies (GWAS). This has been enabled by the availability of high-throughput genotyping technologies and their application to assembled panels of diverse germplasm. Part of the success has been, however, due to large collaborative projects such as the "Barley CAP" and "Triticeae CAP (TCAP)", which provided a framework for exploiting diverse germplasm resources for barley improvement (reviewed in Muñoz-Amatriaín et al. 2014b). The great potential of GWAS for unveiling the genetic architecture of complex traits in barley became evident when Von Zitzewitz et al. (2011), using a panel of just 148 cultivars breeding lines and ~3000 SNPs, identified the same winterhardiness loci that had been characterized in over 20 years of biparental mapping. Also, using contemporary US breeding

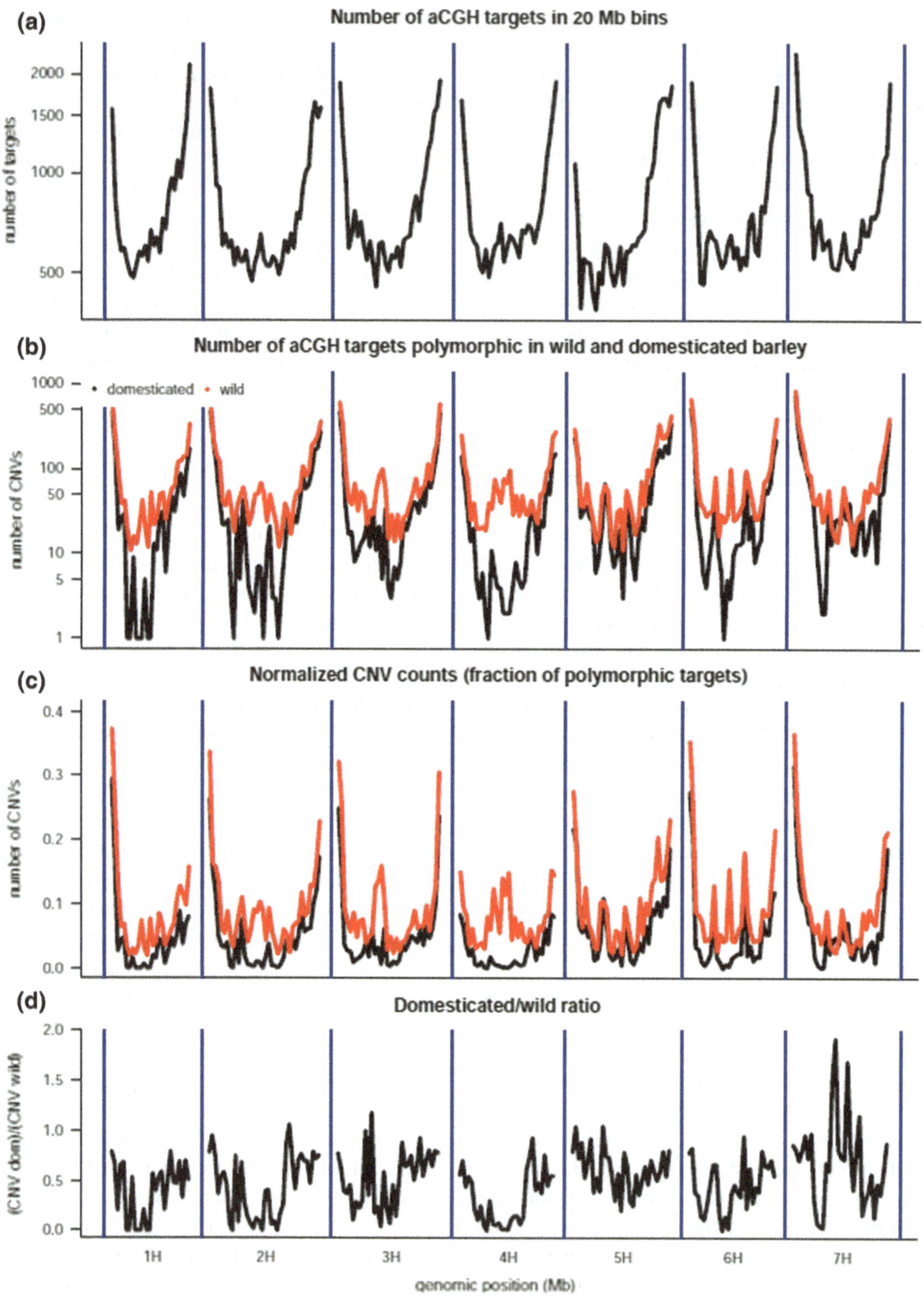

Fig. 8.2 Variation in copy number along the barley genome assessed by aCGH. a Number of sequence fragments targeted by the barley CGH array across each barley chromosome. Distribution (**b**) and frequency (**c**) of CNVs across the chromosomes for six wild (red) and eight domesticated (black) barley accessions. CNV frequencies in panel **c** are expressed as the number of variants per number of targets. **d** Ratio between cultivated and wild CNVs. Genomic positions for all panels are based on the reference sequence assembly of Mascher et al. (2017) and the window size was set to 20 Mb

germplasm, Massman et al. (2011) mapped previously reported QTLs for Fusarium head blight (FHB) resistance at a higher resolution, Gutiérrez et al. (2011) identified QTLs for key traits associated with malting quality, some which were novel, while Berger et al. (2013) identified both novel and previously described QTLs for agronomic, grain quality, and disease resistance traits. The larger extent of linkage disequilibrium in barley breeding populations made possible the application of GWAS to relatively small populations genotyped with only a few thousand markers.

As higher density SNP platforms became available, more diverse germplasm could be evaluated. One of the main objectives of the TCAP was to access diversity within the USDA National Small Grains Core Collection containing 2415 locally adapted landraces and breeding materials from all over the world. Its genetic characterization using the 9K iSelect array (Comadran et al. 2012) allowed an understanding of genetic structure and linkage disequilibrium within the population. This collection was used to successfully conduct GWAS of a few morphological and agronomic traits, and to develop smaller mini-core sets capturing most of the allelic diversity present in the core collection (Muñoz-Amatriaín et al. 2014a). This genetic resource is being extensively used for identifying beneficial alleles for several abiotic and biotic stresses. There are already examples of success including the identification of novel QTLs for spot form net blotch (Neupane et al. 2015; Tamang et al. 2015), net form net blotch (Richards et al. 2017), Russian wheat aphid (Dahleen et al. 2015) and spot blotch (Wang et al. 2017b) resistance. QTLs for grain quality including tocochromanol (vitamin E) content (Graebner et al. 2015) have also been identified in the USDA Core collection.

Array-based genotyping has also been extensively used for association genetic studies by the European barley community. For example, Visioni et al. (2013) genotyped 184 barley genotypes from the Mediterranean basin with the Barley Oligo Pooled Array 1 (BOPA1, 1536 markers, Close et al. (2009)). They found a small number of SNPs associated with frost tolerance, two of which were close to regions previously implicated in cold tolerance. Using the 9K iSelect array on 216 cultivars representing European two-row spring barley diversity in the twentieth century, Tondelli et al. (2013) found markers associated with plant height and straw strength characters such as lodging and leaning. They also observed regional reductions of genetic diversity, a likely consequence of strong selection by breeders. Neumann et al. (2017) combined 9K SNP chip data with biomass measurements collected through nondestructive automatic high-throughput phenotyping. This system made it possible to conduct dynamic association scans across plant development, pinpointing different QTLs for early and late biomass formation. The recent decrease in per-basepair costs has unlocked the potential of high-throughput sequencing for association genetics. For example, Looseley et al. (2017) performed exome sequencing on a panel of 48 winter and spring from the United Kingdom that differed for diastatic power (DP). This panel was divided into four bulks (winter/spring × low/high DP) to identify markers with contrasting allele frequencies in high and low DP pools. The marker-trait associations were validated in a wider panel using Kompetitive Allele Specific PCR (KASP) assays.

Collections of wild barley germplasm are also utilized for identifying new sources of disease resistance. For example, Roy et al. (2010) found new spot blotch resistance loci using the Wild Barley Diversity Collection of over 300 *Hordeum vulgare* subsp. *spontaneum* accessions (Steffenson et al. 2007), while Ames et al. (2015) mined the same collection for powdery mildew resistance. More recently, 25 accessions from the Wild Barley Diversity Collection have been crossed to a barley cultivar to develop an advanced backcross-nested association mapping (AB-NAM) population (Nice et al. 2017). This resource will permit easier identification and mapping of exotic alleles from diverse wild germplasm at a high resolution.

8.4 Contribution of Structural Variation to Phenotypes

Structural variations can significantly impact phenotypes by changing gene structure and dosage, or altering regulatory elements. In the barley genome, over 9% of sequences overlapping protein-coding regions were found to be affected by CNV (Muñoz-Amatriaín et al. 2013). Barley genes affected by CNV were enriched for annotations related to stress and plant defense response, predominantly nucleotide-binding site leucine-rich repeat (NBS-LRR) and other classes of R proteins, which has been also shown in other plant species such as rice (Yu et al. 2011) and soybean (McHale et al. 2012). In a more recent study, Cantalapiedra et al. (2016) showed that PAVs in a cluster of NBS-LRRs at a powdery mildew resistance locus identified in a Spanish barley landrace could determine resistance/susceptibility to a wide range of fungal isolates.

There are also several examples of structural variations affecting other important traits such as flowering time. Large deletions/insertions within the first intron of *VRN-H1* have been associated with higher levels of gene expression and earlier flowering time without vernalization (Fu et al. 2005; von Zitzewitz et al. 2005; Cockram et al. 2007), while the complete deletion of *VRN-H2*, a repressor of *VRN-H1*, results in spring growth habit (von Zitzewitz et al. 2005; Szűcs et al. 2007). Also, increased copy numbers of *VRN-H3* (*HvFT1*), ortholog of *FLOWERING LOCUS T*, have been associated with earlier transcriptional upregulation of the gene and earlier flowering time in the barley genetic stock BGS213 (Nitcher et al. 2013). Later, Loscos et al. (2014) confirmed the presence of CNV at *HvFT1* in 49 European spring-type barleys, although its effects appeared to be dependent on the genetic background.

Structural variation has also been observed at *Frost resistance-2 (Fr-H2)*, a major low-temperature tolerance locus coinciding with a cluster of 13 different *C-Repeat Binding Factor (CBF)* genes (Francia et al. 2004; Pasquariello et al. 2014). Knox et al. (2010) showed that the winter-hardy accession "Nure" had more copies of the *CBF* genes *CBF2* and *CBF4* than the frost-sensitive spring-type "Tremois", which suggested that multiple copies of these two *CBF* genes could account for the phenotypic effect of *Fr-H2*. A higher accumulation of *CBF2* and *CBF4* transcripts had been previously associated with the "Nure" allele at *Fr-H2* (Stockinger et al. 2007). In a more recent study, Francia et al. (2016) examined CNV of specific barley CBF genes at *FR-H2* in a panel of 41 accessions. These authors found that higher copies of the proximal gene cluster *HvCBF2-HvCBF4* were carried by the most frost-tolerant accessions, together with lower copies of *HvCBF3*, which is located in the distal cluster of *FR-H2*. Interestingly, in the large-scale aCGH survey of Muñoz-Amatriaín et al. (2013), the winter cultivars "Igri" and "Franka" exhibited lower copy numbers of *CBF3* than the spring types.

Also, structural variation in barley has been associated with the acquisition of tolerance to an excess of certain minerals in the soil. In 2007, Sutton et al. (2007) reported that an increased copy number of the boron transporter gene *Bot1* in the Algerian landrace "Sahara" conferred boron-toxicity tolerance. More recently, Fujii et al. (2012) showed that aluminum-tolerant accessions have a 1-kb insertion in the upstream of *HvAACT1*, which encodes a citrate transporter. This insertion causes an increase in the expression of the gene in the whole root but more markedly in the root tips, hence increasing the amount of citrate secretion and protecting those root tips from aluminum toxicity (Fujii et al. 2012). This indicated that the aforementioned insertion is responsible for the acquisition of aluminum tolerance in barley.

8.5 Molecular Mechanisms Generating Genetic Variation in Barley

Major insights into the molecular mechanisms generating structural variation come from human studies, in an attempt to understand the molecular basis of genetic disorders (reviewed in Hastings et al. 2009; Carvalho and Lupski 2016). Structural changes can arise from homologous and nonhomologous recombination mechanisms. When homologous recombination (HR) between sister chromatids or homologs is used for DNA repair, no structural changes will occur unless there is an incorrect pairing in regions of extended homology. This is called nonallelic homologous recombination (NAHR) and it has been identified as a major source of recurrent structural variation. However, any DNA repair mechanism that uses sequences with no homology or very short microhomologies has the potential to change the genome structure. Nonhomologous mechanisms including nonhomologous end joining (NHEJ) and replicative mechanisms such as replication slippage, fork stalling and template switching (FoSTeS) and microhomology-mediated break-induced replication (MMBIR) are the basis of most nonrecurrent rearrangements. Single-strand annealing (SSA), a mechanism that repairs double-strand breaks (DSBs) occurring between two flanking repeats, also generates small deletions (Hastings et al. 2009).

Transposable element (TE) activity is also a major source of structural variation in both animal and plant genomes (Kidwell and Lisch 1997). Class I elements or retrotransposons transpose via RNA intermediates, while Class II TEs or DNA transposons use a cut-and-paste transposition mechanism. Retrotransposons, particularly long terminal repeat (LTR) retrotransposons, are the most abundant in plant genomes and a major cause of TE-driven structural variations (Wang and Dooner 2006; Morgante et al. 2007; Pinosio et al. 2016). LTRs are largely responsible for changes in plant genome size: their replication is a major contributor to genome size increase, while mechanisms to purge these proliferating LTRs can lead to genome contraction (Bennetzen and Wang 2014; Lee and Kim 2014). Removal mechanisms including unequal homologous recombination between LTRs and NHEJ after DSBs can produce chromosomal rearrangements (Bennetzen and Wang 2014).

In barley, sequence comparisons at the *Rph7* leaf rust disease resistance locus from two cultivars showed high intraspecific haplotype variability primarily due to differences in LTR retrotransposons (Scherrer et al. 2005). Additional insights into the mechanisms that can generate structural variation were provided by sequence-based comparisons between two barley cultivars. In particular, WGS sequence data from cv. Barke were compared to those from cv. Morex in genomic regions containing CNVs/PAVs identified by aCGH (Muñoz-Amatriaín et al. 2013). Due to the nature of the array design, most variants analyzed (299) were relatively small, averaging 492 bp, although some spanned over 7 kb. In over 40% of them, short sequence motifs bordering the breakpoint and repeated at the other end inside the deleted region were found, indicating double-strand breaks repaired via SSA. Those sequence signatures, previously attributed to "illegitimate recombination", have been found in maize flanking small deletions occurring during the process of fractionation, where genes are preferentially removed from one of the two homeologs after an ancient tetraploidy event (Woodhouse et al. 2010). Although SSA seems to be a minor player of CNV formation in the human genome (Hastings et al. 2009), results from this study suggest that this may be a frequent mechanism generating small deletions in diploid plants. Also, sequence signatures indicating DSB repair through synthesis-dependent strand annealing (SDSA; Agmon et al. 2009) pathway of HR were found in almost 13% of the structural variants analyzed. In those cases, segments of nonhomologous sequences are used as "filler" DNA to repair the DSB. Finally, template slippage was identified as the mechanism responsible for >15% of small insertions/deletions (1–6 bp). In those cases, the complete sequence of the insertion/deletion is repeated perfectly in the immediate flanking region. The study could not attribute any mechanism for 30.4% of analyzed variants (Muñoz-Amatriaín et al. 2013). One of the limitations of the barley CGH array is that it did not allow for an exploration of large genomic regions with high sequence similarity. Therefore, major mechanisms contributing to CNV formation such as NAHR could not be assessed. Similarly, TE-mediated structural variants could not be investigated as unique probes were chosen for the aCGH design. It is possible, however, that some single-, low-copy-, or unidentified TEs could be responsible for some of the variants identified in the study. Future availability of high-quality and highly contiguous assemblies from additional barley accessions would provide the breakpoint resolution needed to further understand the mutational mechanisms generating structural variation in barley.

8.6 Conclusions

Advances in the throughput and information content of sequence data have enabled exciting advances in assessing the extant genetic variation at all levels, from single-nucleotide polymorphisms to complex structural variants. The next 5 years will likely see a further accumulation of genome variation datasets at the population scale. Among other things, such data can provide new insights in mutational dynamics, selection pressure, and phenotypic relevance of structural variation in barley. A key challenge is matching the growing amount of genotypic information with phenotypic data of similar breadth and quality to underpin GWAS of genetically complex, agronomically important traits such as yield, yield stability, and durable pathogen resistance. Other avenues for future research are the development of computational platforms to efficiently store large amounts of data, to integrate various data domains (genetic variation, phenotypes, gene expression), and to provide easy-to-use tools for rapid access and interactive analysis.

References

Agmon N, Pur S, Liefshitz B, Kupiec M (2009) Analysis of repair mechanism choice during homologous recombination. Nucleic Acids Res 37:5081–5092

Aliyeva-Schnorr L, Beier S, Karafiatova M, Schmutzer T, Scholz U, Dolezel J, Stein N, Houben A (2015) Cytogenetic mapping with centromeric bacterial artificial chromosomes contigs shows that this recombination-poor region comprises more than half of barley chromosome 3H. Plant J 84:385–394

Alkan C, Coe BP, Eichler EE (2011) Genome structural variation discovery and genotyping. Nat Rev Genet 12:363–376

Ames N, Dreiseitl A, Steffenson BJ, Muehlbauer GJ (2015) Mining wild barley for powdery mildew resistance. Plant Pathol 64:1396–1406

Badr A, Sch R, El Rabey H, Effgen S, Ibrahim H, Pozzi C, Rohde W, Salamini F (2000) On the origin and domestication history of barley (*Hordeum vulgare*). Mol Biol Evol 17:499–510

Begun DJ, Aquadro CF (1992) Levels of naturally occurring DNA polymorphism correlate with recombination rates in D. melanogaster. Nature 356:519–520

Beier S, Himmelbach A, Colmsee C, Zhang X-Q, Barrero RA, Zhang Q, Li L, Bayer M, Bolser D, Taudien S (2017) Construction of a map-based reference genome sequence for barley, *Hordeum vulgare* L. Sci Data 4

Bennetzen JL, Wang H (2014) The contributions of transposable elements to the structure, function, and evolution of plant genomes. Annu Rev Plant Biol 65:505–530

Berger GL, Liu S, Hall MD, Brooks WS, Chao S, Muehlbauer GJ, Baik B-K, Steffenson B, Griffey CA (2013) Marker-trait associations in Virginia Tech winter barley identified using genome-wide mapping. Theor Appl Genet 126:693–710

Bernardo R (2017) Prospective targeted recombination and genetic gains for quantitative traits in maize. Plant Genome 10

Bockelman HE, Valkoun J (2010) Barley germplasm conservation and resources. Barley: improvement, production, and uses. Wiley-Blackwell, Oxford, UK, pp 144–159

Cantalapiedra CP, Contreras-Moreira B, Silvar C, Perovic D, Ordon F, Gracia MP, Igartua E, Casas AM (2016) A cluster of nucleotide-binding site–leucine-rich repeat genes resides in a barley powdery mildew resistance quantitative trait loci on 7HL. Plant Genome 9

Carvalho CM, Lupski JR (2016) Mechanisms underlying structural variant formation in genomic disorders. Nat Rev Genet 17:224

Chaisson MJ, Huddleston J, Dennis MY, Sudmant PH, Malig M, Hormozdiari F, Antonacci F, Surti U, Sandstrom R, Boitano M (2015) Resolving the complexity of the human genome using single-molecule sequencing. Nature 517:608–611

Charlesworth B, Morgan MT, Charlesworth D (1993) The effect of deleterious mutations on neutral molecular variation. Genetics 134:1289–1303

Close TJ, Bhat PR, Lonardi S, Wu Y, Rostoks N, Ramsay L, Druka A, Stein N, Svensson JT, Wanamaker S, Bozdag S, Roose ML, Moscou MJ, Chao S, Varshney RK, Szucs P, Sato K, Hayes PM, Matthews DE, Kleinhofs A, Muehlbauer GJ, Deyoung J, Marshall DF, Madishetty K, Fenton RD, Condamine P, Graner A, Waugh R (2009) Development and implementation of high-throughput SNP genotyping in barley. BMC Genomics 10:582

Cockram J, Jones H, Leigh FJ, O'Sullivan D, Powell W, Laurie DA, Greenland AJ (2007) Control of flowering time in temperate cereals: genes, domestication, and sustainable productivity. J Exp Bot 58:1231–1244

Comadran J, Kilian B, Russell J, Ramsay L, Stein N, Ganal M, Shaw P, Bayer M, Thomas W, Marshall D (2012) Natural variation in a homolog of Antirrhinum CENTRORADIALIS contributed to spring growth habit and environmental adaptation in cultivated barley. Nat Genet 44:1388–1392

Dahleen LS, Bregitzer P, Mornhinweg D, Klos KE (2015) Genetic diversity for Russian wheat aphid resistance as determined by genome-wide association mapping and inheritance in progeny. Crop Sci 55:1925–1933

Dempewolf H, Baute G, Anderson J, Kilian B, Smith C, Guarino L (2017) Past and future of wild relatives in crop breeding. Crop Sci 57:1070–1082

Fang Z, Gonzales AM, Clegg MT, Smith KP, Muehlbauer GJ, Steffenson BJ, Morrell PL (2014) Two genomic regions contribute disproportionately to geographic differentiation in wild barley. G3-Genes Genom Genet 4:1193–1203

Filler Hayut S, Melamed Bessudo C, Levy AA (2017) Targeted recombination between homologous chromosomes for precise breeding in tomato. Nat Commun 8:15605

Francia E, Rizza F, Cattivelli L, Stanca A, Galiba G, Toth B, Hayes P, Skinner J, Pecchioni N (2004) Two loci on chromosome 5H determine low-temperature tolerance in a 'Nure' (winter) × 'Tremois' (spring) barley map. Theor Appl Genet 108:670–680

Francia E, Morcia C, Pasquariello M, Mazzamurro V, Milc JA, Rizza F, Terzi V, Pecchioni N (2016) Copy number variation at the HvCBF4–HvCBF2 genomic segment is a major component of frost resistance in barley. Plant Mol Biol 92:161–175

Fu D, Szűcs P, Yan L, Helguera M, Skinner JS, von Zitzewitz J, Hayes PM, Dubcovsky J (2005) Large deletions within the first intron in VRN-1 are associated with spring growth habit in barley and wheat. Mol Genet Genomics 273:54–65

Fujii M, Yokosho K, Yamaji N, Saisho D, Yamane M, Takahashi H, Sato K, Nakazono M, Ma JF (2012) Acquisition of aluminium tolerance by modification of a single gene in barley. Nat Commun 3:713

Gill KS, Gill BS, Endo TR, Taylor T (1996) Identification and high-density mapping of gene-rich regions in chromosome group 1 of wheat. Genetics 144:1883–1891

Gore MA, Chia JM, Elshire RJ, Sun Q, Ersoz ES, Hurwitz BL, Peiffer JA, McMullen MD, Grills GS, Ross-Ibarra J, Ware DH (2009) A first-generation haplotype map of maize. Science 326:1115–1117

Graebner RC, Wise M, Cuesta-Marcos A, Geniza M, Blake T, Blake VC, Butler J, Chao S, Hole DJ, Horsley R (2015) Quantitative trait loci associated with the tocochromanol (vitamin E) pathway in barley. PLoS One 10:e0133767

Gutiérrez L, Cuesta-Marcos A, Castro AJ, von Zitzewitz J, Schmitt M, Hayes PM (2011) Association mapping of malting quality quantitative trait loci in winter barley: positive signals from small germplasm arrays. Plant Genome 4:256–272

Hastings PJ, Lupski JR, Rosenberg SM, Ira G (2009) Mechanisms of change in gene copy number. Nat Rev Genet 10:551

Hodgkinson A, Eyre-Walker A (2011) Variation in the mutation rate across mammalian genomes. Nat Rev Genet 12:756

Hudson RR, Kaplan NL (1995) Deleterious background selection with recombination. Genetics 141:1605–1617

International Barley Genome Sequencing Consortium (2012) A physical, genetic and functional sequence assembly of the barley genome. Nature 491:711–716

Jakob SS, Rodder D, Engler JO, Shaaf S, Ozkan H, Blattner FR, Kilian B (2014) Evolutionary history of wild barley (*Hordeum vulgare* subsp. *spontaneum*) analyzed using multilocus sequence data and paleodistribution modeling. Genome Biol Evol 6:685–702

Kidwell MG, Lisch D (1997) Transposable elements as sources of variation in animals and plants. Proc Natl Acad Sci USA 94:7704–7711

Knox AK, Dhillon T, Cheng H, Tondelli A, Pecchioni N, Stockinger EJ (2010) CBF gene copy number variation at Frost Resistance-2 is associated with levels of freezing tolerance in temperate-climate cereals. Theor Appl Genet 121:21–35

Konishi T, Linde-Laursen I (1988) Spontaneous chromosomal rearrangements in cultivated and wild barleys. Theor Appl Genet 75:237–243

Künzel G (1982) Differences between genetic and physical centromere distances in the case of two genes for male sterility in barley. Theor Appl Genet 64:25–29

Künzel G, Korzun L, Meister A (2000) Cytologically integrated physical restriction fragment length polymorphism maps for the barley genome based on translocation breakpoints. Genetics 154:397–412

Lam ET, Hastie A, Lin C, Ehrlich D, Das SK, Austin MD, Deshpande P, Cao H, Nagarajan N, Xiao M (2012) Genome mapping on nanochannel arrays for structural variation analysis and sequence assembly. Nat Biotechnol 30:771–776

Lee S-I, Kim N-S (2014) Transposable elements and genome size gariations in plants. Genomics Inform 12:87–97

Looseley ME, Bayer M, Bull H, Ramsay L, Thomas W, Booth A, De La Fuente Canto C, Morris J, Hedley PE, Russell J (2017) Association mapping of diastatic power in UK winter and spring barley by exome sequencing of phenotypically contrasting variety sets. Front Plant Sci 8

Loscos J, Igartua E, Contreras-Moreira B, Gracia MP, Casas AM (2014) HvFT1 polymorphism and effect—survey of barley germplasm and expression analysis. Front Plant Sci 5:251

Marroni F, Pinosio S, Morgante M (2014) Structural variation and genome complexity: is dispensable really dispensable? Curr Opin Plant Biol 18:31–36

Mascher M, Muehlbauer GJ, Rokhsar DS, Chapman J, Schmutz J, Barry K, Muñoz-Amatriaín M, Close TJ,

Wise RP, Schulman AH, Himmelbach A, Mayer KFX, Scholz U, Poland JA, Stein N, Waugh R (2013a) Anchoring and ordering NGS contig assemblies by population sequencing (POPSEQ). Plant J 76:718–727

Mascher M, Richmond TA, Gerhardt DJ, Himmelbach A, Clissold L, Sampath D, Ayling S, Steuernagel B, Pfeifer M, D'Ascenzo M, Akhunov ED, Hedley PE, Gonzales AM, Morrell PL, Kilian B, Blattner FR, Scholz U, Mayer KF, Flavell AJ, Muehlbauer GJ, Waugh R, Jeddeloh JA, Stein N (2013b) Barley whole exome capture: a tool for genomic research in the genus Hordeum and beyond. Plant J 76:494–505

Mascher M, Gundlach H, Himmelbach A, Beier S, Twardziok SO, Wicker T, Radchuk V, Dockter C, Hedley PE, Russell J, Bayer M, Ramsay L, Liu H, Haberer G, Zhang XQ, Zhang Q, Barrero RA, Li L, Taudien S, Groth M, Felder M, Hastie A, Simkova H, Stankova H, Vrana J, Chan S, Munoz-Amatriain M, Ounit R, Wanamaker S, Bolser D, Colmsee C, Schmutzer T, Aliyeva-Schnorr L, Grasso S, Tanskanen J, Chailyan A, Sampath D, Heavens D, Clissold L, Cao S, Chapman B, Dai F, Han Y, Li H, Li X, Lin C, McCooke JK, Tan C, Wang P, Wang S, Yin S, Zhou G, Poland JA, Bellgard MI, Borisjuk L, Houben A, Dolezel J, Ayling S, Lonardi S, Kersey P, Langridge P, Muehlbauer GJ, Clark MD, Caccamo M, Schulman AH, Mayer KFX, Platzer M, Close TJ, Scholz U, Hansson M, Zhang G, Braumann I, Spannagl M, Li C, Waugh R, Stein N (2017) A chromosome conformation capture ordered sequence of the barley genome. Nature 544:427–433

Massman J, Cooper B, Horsley R, Neate S, Dill-Macky R, Chao S, Dong Y, Schwarz P, Muehlbauer G, Smith K (2011) Genome-wide association mapping of Fusarium head blight resistance in contemporary barley breeding germplasm. Mol Breed 27:439–454

Maynard Smith J, Haigh J (1974) The hitch-hiking effect of a favourable gene. Genet Res 23:23–35

McHale LK, Haun WJ, Xu WW, Bhaskar PB, Anderson JE, Hyten DL, Gerhardt DJ, Jeddeloh JA, Stupar RM (2012) Structural variants in the soybean genome localize to clusters of biotic stress-response genes. Plant Physiol 159:1295–1308

Morgante M, De Paoli M, Slobodanka R (2007) Transposable elements and the plant pan-genomes. Curr Opin Plant Biol 10:149–155

Muñoz-Amatriaín M, Eichten SR, Wicker T, Richmond TA, Mascher M, Steuernagel B, Scholz U, Ariyadasa R, Spannagl M, Nussbaumer T, Mayer KFX, Taudien S, Platzer M, Jeddeloh JA, Springer NM, Muehlbauer GJ, Stein N (2013) Distribution, functional impact, and origin mechanisms of copy number variation in the barley genome. Genome Biol 14:R58

Muñoz-Amatriaín M, Cuesta-Marcos A, Endelman JB, Comadran J, Bonman JM, Bockelman HE, Chao S, Russell J, Waugh R, Hayes PM (2014a) The USDA barley core collection: genetic diversity, population structure, and potential for genome-wide association studies. PLoS One 9:e94688

Muñoz-Amatriaín M, Cuesta-Marcos A, Hayes PM, Muehlbauer GJ (2014b) Barley genetic variation: implications for crop improvement. Brief Funct Genomics 13:341–350

Muñoz-Amatriaín M, Lonardi S, Luo M, Madishetty K, Svensson JT, Moscou MJ, Wanamaker S, Jiang T, Kleinhofs A, Muehlbauer GJ, Wise RP, Stein N, Ma Y, Rodriguez E, Kudrna D, Bhat PR, Chao S, Condamine P, Heinen S, Resnik J, Wing R, Witt HN, Alpert M, Beccuti M, Bozdag S, Cordero F, Mirebrahim H, Ounit R, Wu Y, You F, Zheng J, Šimková H, Doležel J, Grimwood J, Schmutz J, Duma D, Altschmied L, Blake T, Bregitzer P, Cooper L, Dilbirligi M, Falk A, Feiz L, Graner A, Gustafson P, Hayes PM, Lemaux P, Mammadov J, Close TJ (2015) Sequencing of 15 622 gene-bearing BACs clarifies the gene-dense regions of the barley genome. Plant J 84:216–227

Neumann K, Zhao Y, Chu J, Keilwagen J, Reif JC, Kilian B, Graner A (2017) Genetic architecture and temporal patterns of biomass accumulation in spring barley revealed by image analysis. BMC Plant Biol 17:137

Neupane A, Tamang P, Brueggeman R, Friesen T (2015) Evaluation of a barley core collection for spot form net blotch reaction reveals distinct genotype-specific pathogen virulence and host susceptibility. Phytopathology 105:509–517

Nice LM, Steffenson BJ, Blake TK, Horsley RD, Smith KP, Muehlbauer GJ (2017) Mapping agronomic traits in a wild barley advanced backcross–nested association mapping population. Crop Sci

Nielsen R, Paul JS, Albrechtsen A, Song YS (2011) Genotype and SNP calling from next-generation sequencing data. Nat Rev Genet 12:443–451

Nitcher R, Distelfeld A, Tan C, Yan L, Dubcovsky J (2013) Increased copy number at the HvFT1 locus is associated with accelerated flowering time in barley. Mol Genet Genomics 288:261–275

Pasam RK, Sharma R, Malosetti M, van Eeuwijk FA, Haseneyer G, Kilian B, Graner A (2012) Genome-wide association studies for agronomical traits in a world wide spring barley collection. BMC Plant Biol 12:16

Pasquariello M, Barabaschi D, Himmelbach A, Steuernagel B, Ariyadasa R, Stein N, Gandolfi F, Tenedini E, Bernardis I, Tagliafico E (2014) The barley Frost resistance-H2 locus. Funct Integr Genomics 14:85–100

Pinosio S, Giacomello S, Faivre-Rampant P, Taylor G, Jorge V, le Paslier MC, Zaina G, Bastien C, Cattonaro F, Marroni F, Morgante M (2016) Characterization of the poplar pan-genome by genome-wide identification of structural variation. Mol Biol Evol 33:2706–2719

Poland JA, Brown PJ, Sorrells ME, Jannink J-L (2012) Development of high-density genetic maps for barley and wheat using a novel two-enzyme genotyping-by-sequencing approach. PLoS One 7: e32253

Pourkheirandish M, Hensel G, Kilian B, Senthil N, Chen G, Sameri M, Azhaguvel P, Sakuma S, Dhanagond S, Sharma R, Mascher M, Himmelbach A, Gottwald S, Nair SK, Tagiri A, Yukuhiro F, Nagamura Y, Kanamori H, Matsumoto T, Willcox G, Middleton CP, Wicker T, Walther A, Waugh R, Fincher GB, Stein N, Kumlehn J, Sato K, Komatsuda T (2015) Evolution of the grain dispersal system in barley. Cell 162:527–539

Ramage R, Burnham C, Hagberg A (1961) A summary of translocation studies in barley. Crop Sci 1:277–279

Rey M-D, Calderón MC, Prieto P (2015) The use of the ph1b mutant to induce recombination between the chromosomes of wheat and barley. Front Plant Sci 6:160

Richards JK, Friesen TL, Brueggeman RS (2017) Association mapping utilizing diverse barley lines reveals net form net blotch seedling resistance/susceptibility loci. Theor Appl Genet 130:915–927

Roy JK, Smith KP, Muehlbauer GJ, Chao S, Close TJ, Steffenson BJ (2010) Association mapping of spot blotch resistance in wild barley. Mol Breed 26:243–256

Russell J, Dawson IK, Flavell AJ, Steffenson B, Weltzien E, Booth A, Ceccarelli S, Grando S, Waugh R (2011) Analysis of >1000 single nucleotide polymorphisms in geographically matched samples of landrace and wild barley indicates secondary contact and chromosome-level differences in diversity around domestication genes. New Phytol 191:564–578

Russell J, Mascher M, Dawson IK, Kyriakidis S, Calixto C, Freund F, Bayer M, Milne I, Marshall-Griffiths T, Heinen S (2016) Exome sequencing of geographically diverse barley landraces and wild relatives gives insights into environmental adaptation. Nat Genet 48:1024–1030

Sadhu MJ, Bloom JS, Day L, Kruglyak L (2016) CRISPR-directed mitotic recombination enables genetic mapping without crosses. Science 352:1113–1116

Scherrer B, Isidore E, Klein P, Kim JS, Bellec A, Chalhoub B, Keller B, Feuillet C (2005) Large intraspecific haplotype variability at the *Rph7* locus results from rapid and recent divergence in the barley genome. Plant Cell 17:361–374

Springer NM, Ying K, Fu Y, Ji T, Yeh C-T, Jia Y, Wu W, Richmond T, Kitzman J, Rosenbaum H (2009) Maize inbreds exhibit high levels of copy number variation (CNV) and presence/absence variation (PAV) in genome content. PLoS Genet 5:e1000734

Steffenson BJ, Olivera P, Roy JK, Jin Y, Smith KP, Muehlbauer GJ (2007) A walk on the wild side: mining wild wheat and barley collections for rust resistance genes. Crop Pasture Sci 58:532–544

Stein N, Steuernagel B (2014) Advances in sequencing the barley genome. In: Genomics of plant genetic resources. Springer

Stockinger EJ, Skinner JS, Gardner KG, Francia E, Pecchioni N (2007) Expression levels of barley Cbf genes at the Frost resistance-H2 locus are dependent upon alleles at Fr-H1 and Fr-H2. Plant J 51:308–321

Sutton T, Baumann U, Hayes J, Collins NC, Shi B-J, Schnurbusch T, Hay A, Mayo G, Pallotta M, Tester M (2007) Boron-toxicity tolerance in barley arising from efflux transporter amplification. Science 318:1446–1449

Szűcs P, Skinner JS, Karsai I, Cuesta-Marcos A, Haggard KG, Corey AE, Chen TH, Hayes PM (2007) Validation of the VRN-H2/VRN-H1 epistatic model in barley reveals that intron length variation in VRN-H1 may account for a continuum of vernalization sensitivity. Mol Genet Genomics 277:249–261

Taketa S, Amano S, Tsujino Y, Sato T, Saisho D, Kakeda K, Nomura M, Suzuki T, Matsumoto T, Sato K, Kanamori H, Kawasaki S, Takeda K (2008) Barley grain with adhering hulls is controlled by an ERF family transcription factor gene regulating a lipid biosynthesis pathway. Proc Natl Acad Sci USA 105:4062–4067

Tamang P, Neupane A, Mamidi S, Friesen T, Brueggeman R (2015) Association mapping of seedling resistance to spot form net blotch in a worldwide collection of barley. Phytopathology 105:500–508

Tang H, Lyons E, Town CD (2015) Optical mapping in plant comparative genomics. GigaScience 4:3

Tanksley SD, McCouch SR (1997) Seed banks and molecular maps: unlocking genetic potential from the wild. Science 277:1063–1066

Tondelli A, Xu X, Moragues M, Sharma R, Schnaithmann F, Ingvardsen C, Manninen O, Comadran J, Russell J, Waugh R, Schulman AH, Pillen K, Rasmussen SK, Kilian B, Cattivelli L, Thomas WTB, Flavell AJ (2013) Structural and temporal variation in genetic diversity of European spring two-row barley cultivars and association mapping of quantitative traits. Plant Genome 6

Visioni A, Tondelli A, Francia E, Pswarayi A, Malosetti M, Russell J, Thomas W, Waugh R, Pecchioni N, Romagosa I, Comadran J (2013) Genome-wide association mapping of frost tolerance in barley (*Hordeum vulgare* L.). BMC Genomics 14:424

von Zitzewitz J, Szűcs P, Dubcovsky J, Yan L, Francia E, Pecchioni N, Casas A, Chen TH, Hayes PM, Skinner JS (2005) Molecular and structural characterization of barley vernalization genes. Plant Mol Biol 59:449–467

von Zitzewitz J, Cuesta-Marcos A, Condon F, Castro AJ, Chao S, Corey A, Filichkin T, Fisk SP, Gutierrez L, Haggard K (2011) The genetics of winterhardiness in barley: perspectives from genome-wide association mapping. Plant Genome 4:76–91

Wang Q, Dooner HK (2006) Remarkable variation in maize genome structure inferred from haplotype diversity at the *bz* locus. Proc Natl Acad Sci USA 103:17644–17649

Wang C, Hu S, Gardner C, Lubberstedt T (2017a) Emerging avenues for utilization of exotic germplasm. Trends Plant Sci 22:624–637

Wang R, Leng Y, Ali S, Wang M, Zhong S (2017b) Genome-wide association mapping of spot blotch resistance to three different pathotypes of Cochliobolus sativus in the USDA barley core collection. Mol Breed 37:44

Weischenfeldt J, Symmons O, Spitz F, Korbel JO (2013) Phenotypic impact of genomic structural variation: insights from and for human disease. Nat Rev Genet 14:125–138

Woodhouse MR, Schnable JC, Pedersen BS, Lyons E, Lisch D, Subramaniam S, Freeling M (2010) Following tetraploidy in maize, a short deletion mechanism removed genes preferentially from one of the two homeologs. PLoS Biol 8:e1000409

Yamasaki M, Tenaillon MI, Bi IV, Schroeder SG, Sanchez-Villeda H, Doebley JF, Gaut BS, McMullen MD (2005) A large-scale screen for artificial selection in maize identifies candidate agronomic loci for domestication and crop improvement. Plant Cell 17:2859–2872

Yu P, Wang C, Xu Q, Feng Y, Yuan X, Yu H, Wang Y, Tang S (2011) Detection of copy number variations in rice using array-based comparative genomic hybridization. BMC Genomics 12:372

The Repetitive Landscape of the Barley Genome

9

Thomas Wicker, Heidrun Gundlach and Alan H. Schulman

Abstract

While transposable elements (TEs) comprise the bulk of plant genomic DNA, how they contribute to genome structure and organization is still poorly understood. Especially, in large genomes where TEs make the majority of genomic DNA, it is still unclear whether TEs target specific chromosomal regions or whether they simply accumulate where they are best tolerated. The barley genome with its vast repetitive fraction is an ideal system to study chromosomal organization and evolution of TEs. Genes make only about 2% of the genome, while over 80% is derived from TEs. The TE fraction is composed of at least 350 different families. However, 50% of the genome is comprised of only 15 high-copy TE families, while all other TE families are present in moderate or low-copy numbers. The barley genome is highly compartmentalized with different types of TEs occupying different chromosomal "niches", such as distal, interstitial or proximal regions of chromosome arms. Furthermore, gene space represents its own distinct genomic compartment that is enriched in small non-autonomous DNA transposons, suggesting that these TEs specifically target promoters and downstream regions. Some TE families also show a strong preference to insert in specific sequence motifs which may, in part, explain their distribution. The family-specific distribution patterns result in distinct TE compositions of different chromosomal compartments.

T. Wicker (✉)
Department of Plant and Microbial Biology, University of Zurich, Zurich, Switzerland
e-mail: wicker@botinst.uzh.ch

H. Gundlach
PGSB—Plant Genome and Systems Biology, Helmholtz Center Munich, German Research Center for Environmental Health, Neuherberg, Germany

A. H. Schulman
Institute of Biotechnology and Viikki Plant Science Centre, University of Helsinki, Helsinki, Finland

A. H. Schulman
Green Technology, Natural Resources Institute Finland (Luke), Helsinki, Finland

9.1 Introduction

Transposable elements (TEs) are small genetic units with the ability to move around in the genome, make copies of themselves, or both. They range in size from a few dozen bp to tens of kb. TEs are found in practically all known organisms, which suggest that they are of very ancient evolutionary origin. It is generally believed that they evolved from cellular enzymes that modify or synthesize DNA (Malik and Eickbush 2001; Gladyshev and Arkhipova 2011.). TEs can be divided into two main classes

N. Stein and G. J. Muehlbauer (eds.), *The Barley Genome*, Compendium of Plant Genomes,
https://doi.org/10.1007/978-3-319-92528-8_9

based on their mode of replication (Wicker et al. 2007). These two main classes can be further differentiated into at least 29 superfamilies, ancient lineages which are found in most eukaryotes (Wicker et al. 2007).

Class I elements (retrotransposons) transpose via reverse transcription of an mRNA intermediate that is transcribed from a cellular DNA copy. Autonomous retrotransposons all encode a reverse transcriptase-RNAse (RT-RH), which produces the double-stranded DNA from the mRNA template. The long terminal repeat (LTR) retrotransposons, which are evolutionarily younger than the long-interspersed elements (LINE) retrotransposons (Malik and Eickbush 2001), encode an integrase (IN) as well, which integrates the dsDNA made by RT-RH back into the genome. The RT-RH and IN are generally expressed as a polyprotein, which is cleaved into functional subunits by the aspartic proteinase that is also part of the polyprotein (Schulman 2013). The LTR retrotransposons can reach very high-copy numbers because each replication cycle from a single mRNA transcript can produce a new copy in the genome, which can, in turn, generate new copies. Due to their large size ($\sim$9 kb) and replicative capacity, LTR retrotransposons are the most dominant elements that determine the size of most plant genomes (Paterson et al. 2009; Schnable et al. 2009; International Brachypodium Initiative 2010; Mascher et al. 2017), while in mammals the LINE elements and not the LTR retrotransposons dominate (Chalopin et al. 2015).

Class II elements (DNA transposons) have the ability to excise from, and insert into, the genome by a "cut and paste" mechanism involving a transposase enzyme encoded by autonomous forms of the elements (Kempken and Windhofer 2001). A typical DNA transposon is flanked by terminal inverted repeats (TIRs), which act as recognition sites for the transposase. Five different superfamilies of DNA transposons have been described in most plant genomes: *Harbinger*, *Mariner*, *hAT*, *CACTA* and *Mutator* (Wicker et al. 2007). Occasionally, entire superfamilies go extinct. For example, the banana genome apparently has lost all *CACTA* elements (d'Hont et al. 2012). In addition to the typical TIR DNA transposons, plant genomes contain considerable amounts of *Helitrons*, which belong to a curious sub-class of DNA transposons that do not have transposase genes but instead encode helicase enzymes. It is, therefore, assumed that they replicate via a rolling-circle mechanism (Kapitonov and Jurka 2007). *Helitrons* are very abundant in some plant genomes (e.g. maize; Yang and Bennetzen 2009) and their actual contribution to plant genomes is probably under-reported because they are extremely diverse and difficult to identify.

Autonomous TEs are defined as transposable elements that possess all genes and regulatory sequences needed to make copies of themselves and/or move around in the genome. Often, these autonomous elements give rise to large populations of non-autonomous derivatives, which have lost some or all their genes and which depend on enzymes encoded by autonomous elements for their transposition. The most extreme cases are the so-called miniature inverted-repeat transposable elements (MITEs; Bureau and Wessler 1994a, b) which range in size from roughly 80 to 500 bp. Most plant MITEs are derived from elements of the *Mariner* and *Harbinger* superfamilies. In grasses, MITEs can vastly outnumber their autonomous partners. Indeed, the *Brachypodium distachyon* genome contains over 20,000 *Mariner* MITEs but only a few dozen potentially autonomous elements. Non-autonomous elements from other superfamilies are usually longer. For example, the highly abundant *Helitron*, *Mothra* from rice, is over 1200 bp long (Roffler and Wicker 2015). Similarly, all *Triticeae* species contain very large non-autonomous *CACTA* elements, which encode only partial genes or no genes at all (Wicker et al. 2003).

Whereas classification of TEs into superfamilies is relatively simple, it is at the family level where most TE diversity is found. TE families are usually defined as groups of TE sequences that can be aligned as DNA over most of their sequence (>80% sequence identity over >80% of the entire TE length; Wicker et al. 2007). This definition of TE family is somewhat controversial, but nevertheless, it has proven useful and practical. Most plant genomes contain hundreds of different TE families. In the relatively small *B. distachyon* genome of 275 Mbp, over 170 different TE families have been described (International Brachypodium Initiative 2010). Curiously, the number of TE families is similar in the much larger sorghum and maize genomes (Paterson et al. 2009; Schnable et al. 2009). Thus, it is not the number of different TE families that defines the genome size, but the copy numbers within individual TE families.

The barley genome is among the largest plant genomes sequenced and well assembled so far (Mascher et al. 2017). With 5100 Mb, it is close to the average of the plant genome sizes estimated to date (Wicker et al. 2017). Nevertheless, the distribution of plant genome sizes has a mode (i.e. peak) at approximately 587 Mbp, with a long tail towards very large genomes. Thus, it appears that there is some selection for genome sizes in the range of 100–1000 Mb and an apparent "typical" size of approximately 600 Mbp. The smallest plant genome sequenced so far is that of the carnivorous *Genlisea aurea,* which has a size of only 63 Mb (Leushkin et al. 2013). Interestingly, there seems to be no clear upper limit for genome sizes; many plants tolerate very large genomes with no phenotypic effect. The largest plant genome described so far is that of the lily *Fritillaria assyriaca*, which has a size of 120,000 Mbp (Leitch et al. 2007; Kelly et al. 2015). All angiosperms have very similar numbers of genes in their basic (monoploid) chromosome set; 32,000 genes of 3.5 kb each comprise together only about 112 Mbp of DNA. Hence, genome size is determined almost exclusively by the amount of TE-derived sequences. The barley genome is much larger than the mode of the genome size distribution for angiosperms and much larger than well-studied genomes until now, but close to the average of genome sizes that have been estimated. Thus, it can show us what to expect when even larger plant genomes will be sequenced in the future.

9.2 The Repetitive Fraction of the Barley Genome Is Dominated by a Small Number of High-Copy TE Families

Early on, it became obvious that the barley genome contains a few TE families that are present in extremely high-copy numbers (Vicient et al. 1999; Middleton et al. 2012). The completion of the barley genome sequence (Mascher et al. 2017) revealed that ten *Gypsy*, three *Copia*, and two *CACTA* families together comprise over 50% of the whole genome (Fig. 9.1). How many copies each of these families have in the genome is difficult to say, because many copies are fragmented by deletions or nested insertions of other TEs, or reduced to solo LTRs through intra-element recombination. Copy numbers of individual TE families can be estimated by dividing the total number of annotated base pairs by the length of the reference sequence for the respective TE. Using this approach, it was estimated that the 10 most abundant TE families together represent approximately 230,000 individual copies (Table 9.1, Wicker et al. 2017). The rest of the repetitive landscape is comprised of at least 350 TE families with moderate or low-copy numbers (Mascher et al. 2017; Wicker et al. 2017. As described above for sorghum and maize, also in barley the number of families is similar to that found in smaller plant genomes. Indeed, the relatively small *B. distachyon* genome (275 Mbp) was annotated in detail, leading to the identification of over 170 different TE families (International Brachypodium Initiative 2010). By comparison, barley has less than twice as many TE families, although the barley genome is almost 20 times larger than the *B. distachyon* genome. Thus, the factor that determines genome size is the copy numbers of the most abundant families.

Almost 81% of the barley genome was classified as derived from TEs (Mascher et al. 2017). Considering that gene space contributes only 2–3% to the barley genome, approximately ~16% remains un-annotated. This proportion of un-annotated sequence is comparable to that in other genomes. In maize,

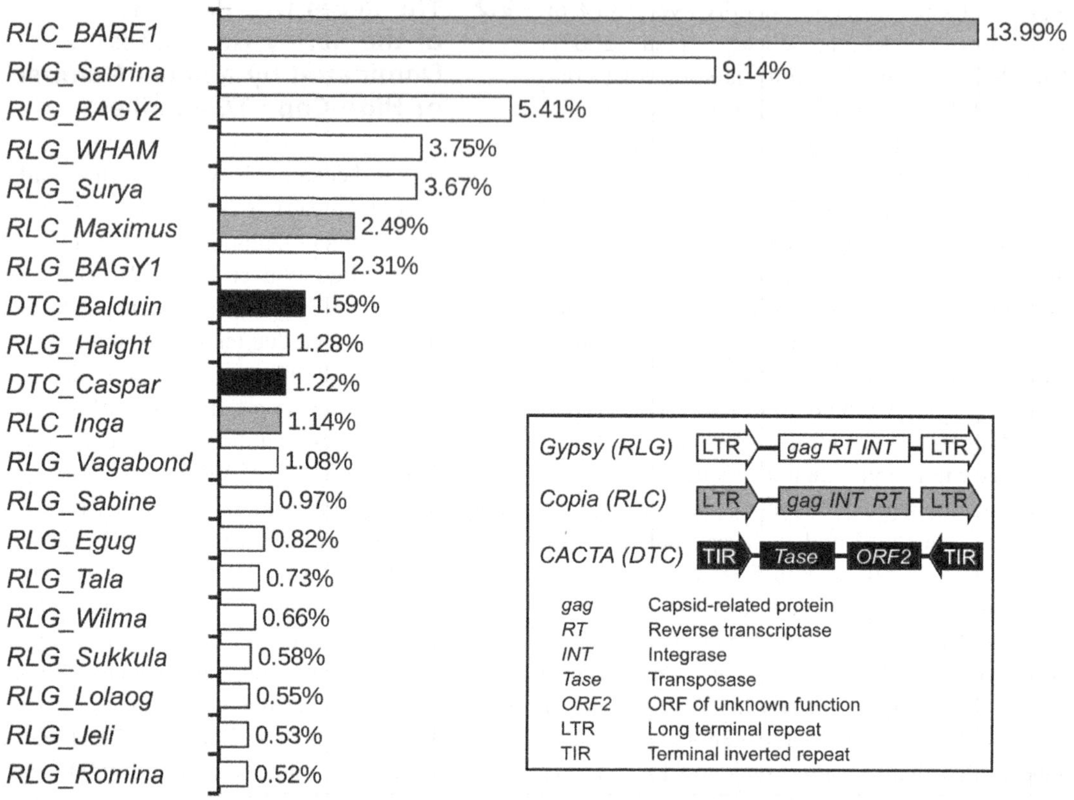

Fig. 9.1 Contribution to total genome sequence of the top 20 TE families in the barley genome. Most broadly represented is the Gypsy superfamily to which 15 of the top 20 TE families belong (name prefix "*RLG_*"). The *Copia* superfamily is represented by three families (prefix "*RLC_*"). The only Class II superfamily represented in the top 20 is the *CACTA* superfamily (prefix "*DTC_*"). The inset shows the schematic sequence organization of these three superfamilies

approximately 12% remained un-annotated (Schnable et al. 2009), while in *B. distachyon*, un-annotated sequences comprise approximately 25% of the genome (International Brachypodium Initiative 2010). It is assumed that the un-annotated portions of these genomes contain additional, yet uncharacterized, TE families (Wicker et al. 2017). These could be highly degenerated TEs, or exotic TE types that have very low-copy numbers and thus escape detection. Future efforts will be needed to further characterize the un-annotated fractions in various genomes, but it is safe to say that the actual complexity of the repetitive fraction of plant genomes has likely been under-estimated.

9.3 *BARE1*—The Most Abundant TE Family in the Barley Genome

As previously described (Vicient et al. 1999; Chang and Schulman 2008; Middleton et al. 2012), the *Copia* family *RLC_BARE1* is the most abundant in terms of copy numbers (>72,000) as well as absolute contribution to the genome (>14%; Fig. 9.1; Table 9.1), Together with other *Copia* RTNs, it is preferentially localized in the gene-rich distal regions of chromosomes (Mascher et al. 2017). *BARE1* is among the best characterized TE families in plants. Autonomous copies of *BARE1* contain a canonical *Copia*

Table 9.1 Copy number estimates of the most abundant Class 1 and Class 2 element families in the barley genome

TE family	Total kb[a]	Length[b]	Copy number[c]
RLC_BARE1	623,043	8630	72,195
RLG_Sabrina	407,047	8030	50,691
RLG_BAGY2	240,798	8630	27,902
RLG_WHAM	167,138	9450	17,687
RLG_Surya	163,300	14,470	11,285
RLC_Maximus	110,928	14,400	7703
RLG_BAGY1	102,843	14,400	7142
DTC_Balduin	70,688	11,740	6021
RLG_Haight	57,185	13,080	4372
DTC_Caspar	54,465	11,568	4708
Total	1,997,435		209,707
DTT_Thalos	2865	163	17,574
DTT_Pan	716	123	5822
DTT_Athos	394	81	4868
DTT_Icarus	555	117	4747
DTT_Hades	392	108	3627
DTT_SAF	177	85	2087
DTT_Eos	506	326	1552
DTT_Oleus	231	150	1540
DTT_Pluto	328	274	1197
DTT_Stolos	205	274	749
Total	6369		43,763
DTH_Thorne	716	273	2624
DTH_Kerberos	594	285	2086
DTH_Xumet	591	376	1571
DTH_Rong	1218	1227	993
DTT_Marimom	2024	2129	951
DTH_Orpheus	183	272	674
DTH_Xenon	203	312	650
DTH_Xian	650	1161	560
DTH_Kong	489	2119	231
DTH_Tibone	187	1037	180
DTH_Zong	278	2396	116
Total	7133		10,634

Superfamilies are indicated by a three-letter code preceding the family name
DTC *CACTA*, DTH *Harbinger*, DTT *Mariner*, RLC *Copia*, RLG *Gypsy*
[a]Total kb annotated as respective family-specific
[b]Length of the reference TE that was used for annotation
[c]Copy number estimate based on total kb occupied by the TE family and length of it consensus sequence

coding domain between the two LTRs that encodes, in the direction of transcription, the capsid protein Gag, integrase (INT), aspartic proteinase (AP), and the reverse transcriptase-RNAse H complex (RT-RH, Manninen and Schulman 1993; Suoniemi et al. 1996). *BARE1* is not only actively transcribed, but also translated, and forms virus-like particles (VLPs) (Jääskeläinen et al. 2013, 1999). *BARE1* produces two groups of transcripts, one can replicate via reverse transcription, but is not capped, polyadenylated, or translated (Chang et al. 2013). The other set is capped, polyadenylated, and translated, but not replicated. The second set of transcripts is also differentially spliced in response to stress to make more Gag protein for the formation of VLPs. *BARE1* is not only actively transcribed, but is also translated, and forms virus-like particles (VLPs) (Jääskeläinen et al. 2013, 1999).

Non-autonomous retrotransposons, lacking one or more functional coding domain(s), are commonly encountered and perhaps are the dominant form in plant genomes (Sabot and Schulman 2006). The inability of many of non-autonomous elements to carry out a full retrotransposon life cycle of transcription, translation, reverse transcription, packaging and integration (Fig. 9.2) may be complemented by other, autonomous retrotransposons with which they share signals needed for these steps. The *BARE1* element has a non-autonomous form, *BARE2* (Tanskanen et al. 2007). *BARE2* elements cannot synthesize their own Gag due to a deletion of the initiating ATG in the *gag* ORF, which is conserved among *BARE2* elements. Nevertheless, *BARE2* contains the key *BARE1* signals for replication, including the PBS (Primer Binding Site) for reverse transcription, the DIS (DImerization Signal) for the association of the

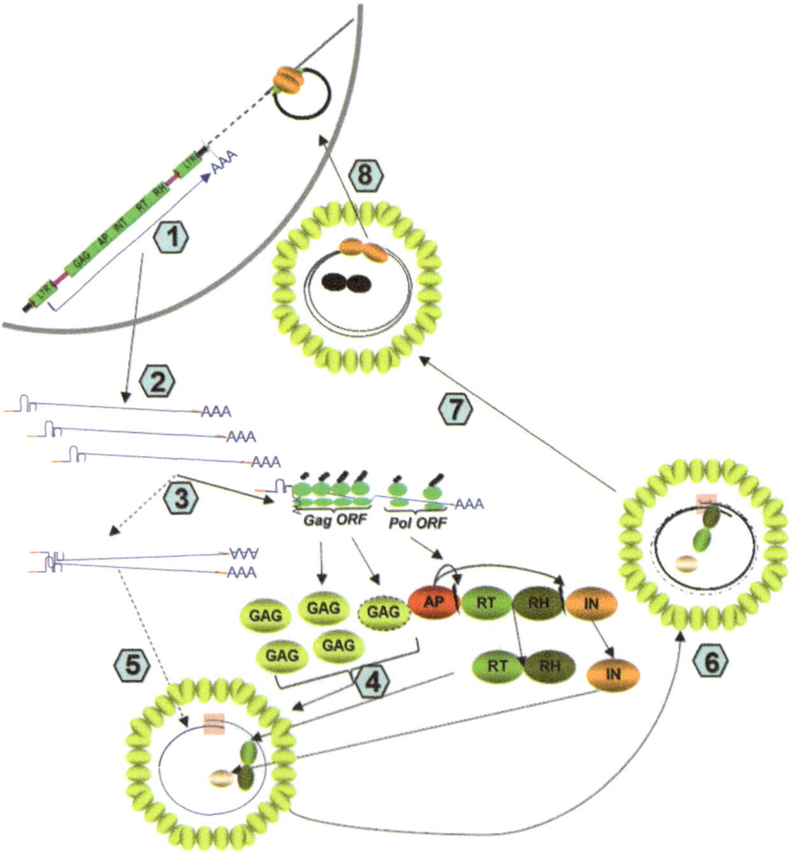

Fig. 9.2 Retrotransposon *BARE1* life cycle. The steps of *BARE* replication are schematically depicted: 1 transcription; 2 nuclear export; 3 alternatively translation or packaging; 4 formation of virus-like particles (VLPs); 5 packaging; 6 reverse transcription; 7 nuclear localization; 8 integration. The grey curve represents the nuclear envelope. GAG capsid protein; AP aspartic proteinase; RT reverse transcriptase; RH RNaseH; IN integrase

two RNAs to be packaged, and the PSI (Packaging SIgnal) for packaging into VLPs. Indeed, *BARE2* economizes by not synthesizing the sub-genomic *gag* RNA but does transcribe the replication-competent RNAs (Chang et al. 2013). These are packaged into *BARE1* VLPs. The success of the *BARE2* strategy is indicated by its outnumbering *BARE1* by about 2:1 in the genomes of cultivated and wild barley, respectively, *Hordeum vulgare* and *H. spontaneum*.

The ability of new RTN insertions to be inherited and drive genome size growth critically depends on where in the plant replication occurs. Immunostaining with anti-Gag antibodies and in situ hybridization have shown that *BARE* protein and transcript localization strongly varies from tissue to tissue (Jääskeläinen et al. 2013). Gag is strongly localized to provascular tissues and to companion cells in mature vascular tissues. Gag and *BARE* RNA appears in the developing floral spike, following transition to flowering. The localization of Gag in the floral meristems suggests that newly replicated copies there can be passed to the next generation. The visualized expression patterns are consistent with those expected from the response elements that have been identified in the *BARE* promoter.

9.4 The Barley Genome Contains Large Populations of Non-autonomous Retrotransposons

Beyond *BARE2* described above, three of the five most abundant TE families seem to be non-autonomous (*RLG_Sabrina*, *RLG_WHAM*, and *RLG_Surya*), because they have none or only fragments of the genes that are typically found in autonomous elements (Fig. 9.3). Thus, it is assumed that they rely on enzymes encoded by other TEs for their proliferation. The *Gypsy* family *RLG_Surya*, is possibly cross-mobilized by the much less abundant *RLG_Sukkula* family. Indications for this are a similar chromosomal distribution (Fig. 9.3a) and strong sequence homology in the *RLG_Surya* and *RLG_Sukkula* LTRs. LTRs contain regulatory regions and serve as start points for replication. Such cross-mobilization has been described previously for *BARE2* elements (see above).

For the *Gypsy* families *RLG_Sabrina* and *RLG_WHAM*, no putative autonomous elements have been identified so far. Both *RLG_Sabrina* and *RLG_WHAM* can be subdivided into different subfamilies, some of them contain no coding sequences at all, i.e. the LTRs flank an internal domain of a few kb, which has no coding capacity. These are reminiscent of the widely distributed LARD elements, which are of full length ($\sim$9 kb), but lack any coding capacity (Kalendar et al. 2004). Additionally, *RLG_Sabrina* and *RLG_WHAM* have subfamilies that contain a gene that probably encodes a Gag-like protein and a partial reverse transcriptase, similar to *Morgane* elements in wheat and its near relatives (Sabot et al. 2006). Sequence similarity of these partial proteins suggests that their autonomous master elements are homologs of the *Athila* retrotransposon from Arabidopsis (*Athila* clade, Fig. 9.3a). Moreover, the *Copia* family *RLC_Giselle* likely depends upon closely related autonomous *RLC_Inga* family elements for transposition, because *RLC_Giselle* does not have *rt* and *int* genes whereas *RLC_Inga* does (Fig. 9.3b). These observations indicate that non-autonomous retrotransposons mobilized by a relatively small number of autonomous elements contribute substantially to barley genome size.

In addition to the large *Gypsy*, *Copia* and *CACTA* elements, which can range in size from roughly 2 kb to over 30 kb, the barley genome also contains approximately 54,000 small non-autonomous DNA transposons of the *Mariner* and *Harbinger* superfamily (i.e. MITEs; Table 9.1). Most dominant is the *Mariner* superfamily, which is represented by at least 36 families. The 10 most abundant *Mariner* families are all small non-autonomous elements ranging in size from 81 bp (*DTT_Athos*) to 274 bp (*DTT_Stolos* and *DTT_Pluto*; Table 9.1). Such small *Mariner* elements are also referred to as *Stowaway* MITEs (Bureau and Wessler 1994b). The most abundant *Mariner* family, *DTT_Thalos*, is present in more than 17,000 copies. Interestingly, only about 150 potentially

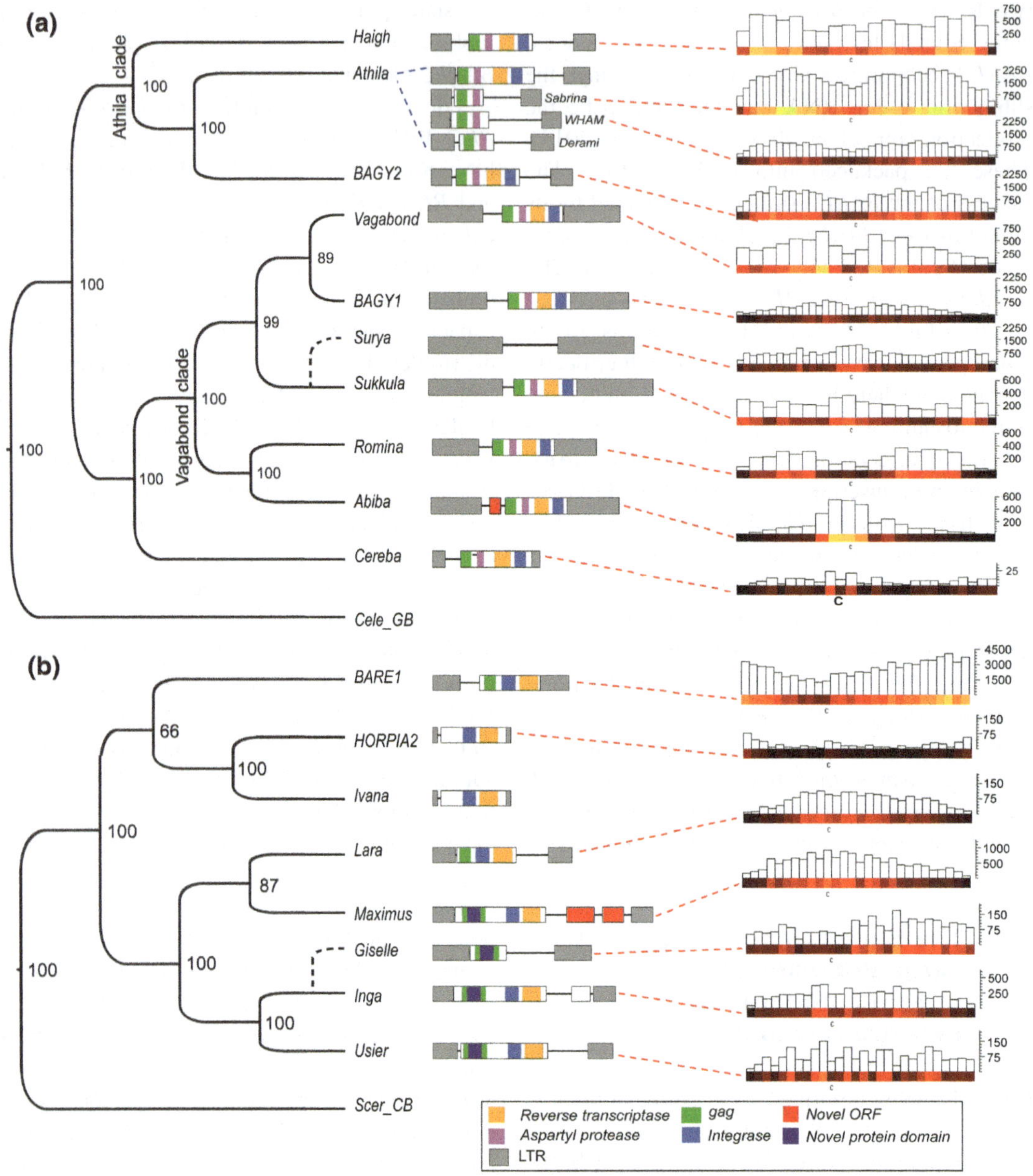

Fig. 9.3 Distribution of *Gypsy* and *Copia* retrotransposons along barley chromosomes. Phylogenetic relationships of a selection of abundant families are shown at the left. Retrotransposon structure and gene content are shown in the center. Chromosomal distributions are shown at the right in bins of 20–40 Mb (depending on the copy number) as heat maps and bar plots to indicate absolute numbers. The y-axis indicates the total number of kb that is occupied by the TE family in each bin (Note that scales differ between families). Retrotransposon families with different evolutionary histories show different chromosomal distribution patterns. **a** Distribution of *Gypsy* elements on chromosome 2. **b** Distribution of *Copia* elements on chromosome 1

functional, autonomous *Mariner* elements have been identified in the barley genome (Wicker et al. 2017). Thus, a vast number of non-autonomous DNA transposons apparently rely on a very small number of functional master elements for their potential mobilization. The situation is similar for non-autonomous *Harbinger* transposons, although these elements are about four times less abundant (Table 9.1). Despite their enormously high-copy numbers, the contribution of *Harbinger* and *Mariner* to barley genome size is negligible because of their shortness. Interestingly, these non-autonomous TEs are present in copy numbers similar to that in smaller genomes. Both rice and *B. distachyon* contain roughly 25,000 MITEs, whereas the barley genome contains approximately 54,000—only about twice as many—despite having an over tenfold larger genome. We assume that this is related to MITEs being enriched near genes in all grasses (Bureau and Wessler 1994a, b; Buchmann et al. 2012; Roffler and Wicker 2015), including barley (Wicker et al. 2016) and gene number being very similar in the monoploid set of chromosomes in all plant genomes.

9.5 Individual TE Lineages Occupy Distinct Chromosomal 'Niches'

Gypsy and *Copia* LTR retrotransposons are found throughout the genome, resulting in a more or less even distribution of coding sequences for reverse transcriptase and integrase along the chromosomes. However, at the individual family level, distributions vary strongly. For example, the *Copia* element *RLC_BARE1* is enriched in the distal regions of chromosome arms, as is the closely related but far less abundant *RLC_HORPIA2* (Fig. 9.3b). In contrast, *RLC_Lara* and *RLC_Maximus* show a clear preference for proximal (peri-centromeric) chromosomal regions (Fig. 9.3b). Retrotransposon families of the *Gyspy* superfamily occupy complementary genomic niches: the interstitial regions of chromosome arms are dominated by families from the *Athila* clade (*RLG_Sabrina, RLG_WHAM* and *RLG_Derami*, Fig. 9.3a), while *RLG_Surya* and *RLG_Sukkula* are enriched in the proximal and distal regions. Generally, closely related families tend to have similar distribution patterns. Indeed, there is a good congruence between the phylogenetic tree of LTR retrotransposons and their chromosomal location (Fig. 9.3). An interesting exception is the *RLG_Abiba* family, which is highly enriched in peri-centromeric regions, while its closest relative *RLG_Romina* shows a virtually inverse chromosomal distribution (Fig. 9.3b).

DNA (Class II) transposon families also show distinct individual distribution patterns. The superfamily of the *CACTA* transposons is highly abundant and represented by at least 20 families in the barley genome. In total, they contribute at least 5% to the genome as a whole (Mascher et al. 2017). Among them, the proximal (centromeric and peri-centromeric) regions are preferably occupied by the high-copy *CACTA* family *DTC_Balduin* (Fig. 9.4). In contrast, families of the *Caspar* clade are strongly enriched in distal regions. Indeed, over 75% of *DTC_Caspar* elements are located in the terminal 20% of chromosome arms, the strongest niche enrichment we found for any TE group (Fig. 9.4). For less abundant Class II superfamilies, such as *Mutator*, *Mariner* or *Harbinger*, we observed the familiar pattern of enrichment in distal regions that was found in other plant genomes (Paterson et al. 2009; International Brachypodium Initiative 2010; Han et al. 2013).

It is important to mention that compartmentalization into chromosomal niches was only observed for the distribution of large TEs such as LTR retrotransposons and *CACTA* elements. The very extensive populations of short non-autonomous elements (i.e. MITEs) tend to cluster near genes (Bureau and Wessler 1994a, b; Paterson et al. 2009; International Brachypodium Initiative 2010; Han et al. 2013; see below), making their overall distribution largely congruent with that of genes.

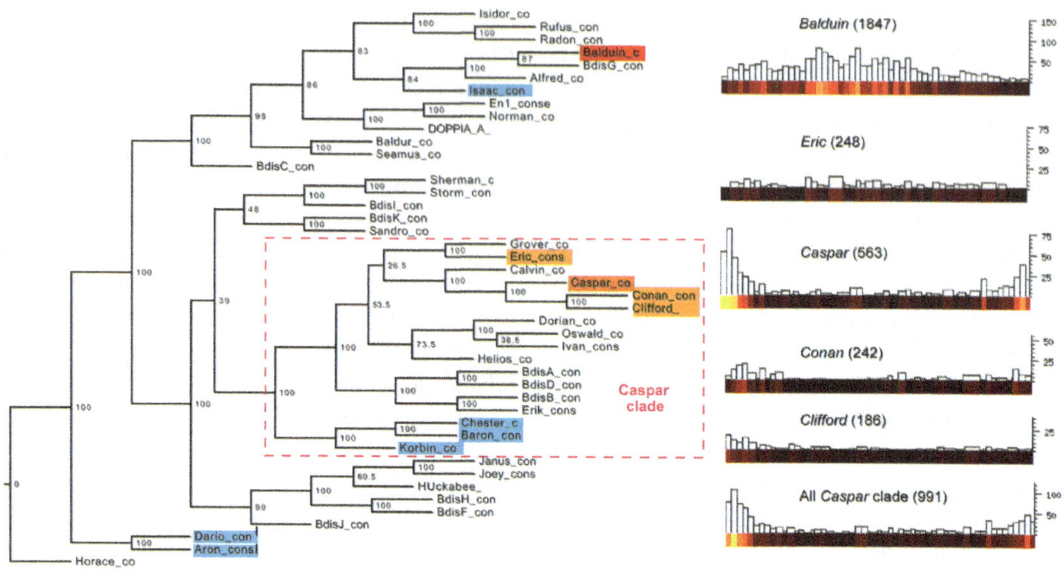

Fig. 9.4 Phylogeny and distribution of *CACTA* elements in the barley genome. For the phylogenetic tree, 44 predicted *CACTA* transposase proteins deposited at TREP were used. *CACTA* sequences come from *Brachypodium distachyon*, Sorghum, rice, *Arabidopsis thaliana* and the Triticeae. High-copy elements from the Triticeae are highlighted in orange and red, while Triticeae low-copy families are highlighted in blue. Chromosomal distributions (shown is chromosome 1H as the representative for all barley chromosomes) are shown at the right in windows of 10 Mbp. Total copy number on 1H are given in parentheses next to the family name. *DTC_Balduin* dominates in centromeric regions, while elements of the *DTC_Caspar* clade occupy telomeres. It is not certain whether the preference for different chromosomal regions is evolutionarily conserved, as similar analyses have not been done yet in other grasses. However, it is clear that *DTC_Caspar* and *DTC_Balduin* represent ancient lineages that were present already in the common ancestor of the grasses (Buchmann et al. 2014)

9.6 The Space Surrounding Genes Is a Distinct Genomic Compartment

In addition to large-scale gradients in the distribution of TE families, the barley gene space represents its own unique genomic compartment with its own TE 'environment'. Genes tend to be enriched in the distal chromosomal regions in of barley, with gene density forming an exponential gradient from centromeres to telomeres (Mascher et al. 2017). In addition to this gradient along chromosomes, genes are distributed non-randomly. They are found mostly in clusters of two–seven genes, (here, genes that are separated by less than 20 kb were defined as belonging to the same cluster). Individual clusters usually are separated by long stretches (hundreds of kb) comprising exclusively TE sequences.

Notably, the TE landscape close to genes differs strongly from that of intergenic regions (here, we arbitrarily define "intergenic regions" as stretches of at least 200 kb that do not contain genes). This particular gene space environment is strictly local, including the gene and a few kilobases upstream and downstream, and largely independent of the gene's particular location along the chromosome. Of particular interest are insertions of LTR retrotransposons and *CACTA*s that are very near genes, because these generally large elements have the potential to influence the function of the gene. Retrotransposon composition changes drastically near genes: starting approximately 10 kb upstream and downstream of genes, the frequency of LTR retrotransposons

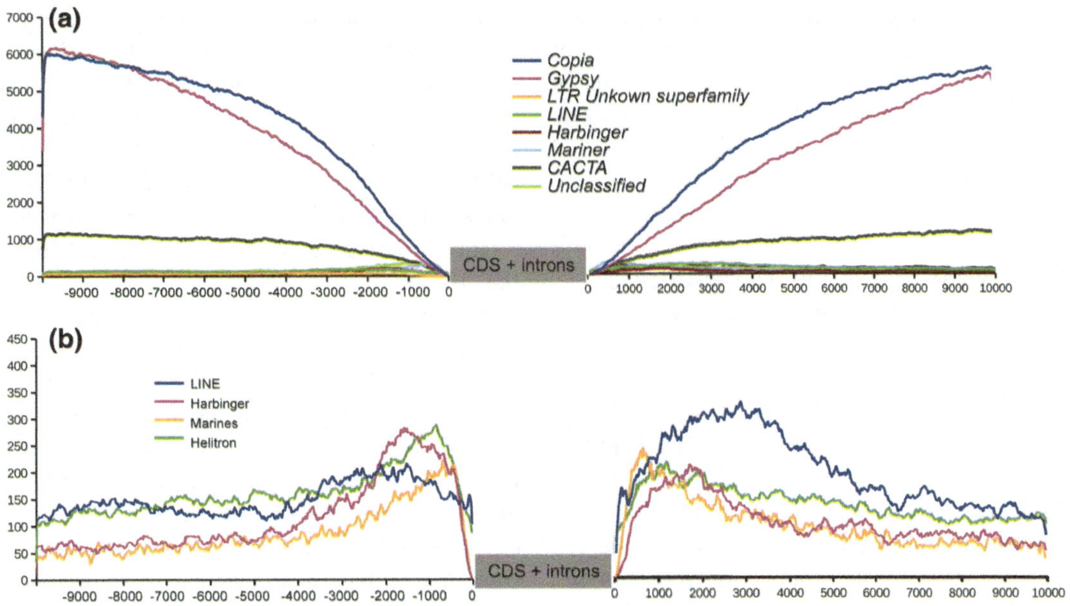

Fig. 9.5 TE composition upstream and downstream of genes. The CDS of 28,316 high-confidence genes were used as anchor points. TE composition was determined 10 kb upstream and downstream of the gene. The *x*-axis indicates the position relative to the gene while the *y*-axis indicates how many genes had a TE of the respective superfamily at the particular position up- or downstream. Close to genes, Class 2 and LINE elements dominate. *Helitrons* and Harbinger elements have a clear preference for promoter regions while *LINE* elements are found preferentially downstream of genes. **a.** Distribution of all major TE superfamilies around genes. With increasing distance from genes, *Gypsy* (RLG) and *Copia* (RLC) elements completely dominate genomic sequences, reflecting the overall composition of the barley genome. **b.** Zoom-in of the graph from (**a**), displaying only the TE superfamilies that are enriched near genes. The *y*-axis was adjusted for better visibility of the less abundant superfamilies and the most abundant ones (*Gypsy*, *Copia* and *CACTA*) were omitted

(i.e. *Gypsy* and *Copia* elements) and *CACTA* elements drops sharply (Fig. 9.5a).

As mentioned above, close to genes, we find mostly small, non-autonomous DNA transposons. More than a third (36%) of *Mariner* and 25.7% of *Harbinger* transposons are found within 5 kb of genes, a highly significant enrichment. Within 10 kb, this enrichment increases to almost 50% of *Mariner* and over 40% of *Harbinger* elements (Fig. 9.5b). Interestingly, different types of elements also occupy different niches in the proximity of genes: *Mariner* transposons preferably reside immediately upstream and downstream of the coding regions of genes (Fig. 9.5c), whereas *Harbinger* transposons are found further away. The observed distribution of different TE types around genes may reflect selective pressures that allow the smallest elements (*Mariners*) to be tolerated closest to genes. Most interestingly, *Helitrons* (particularly the high-copy families *DHH_Walter* and *DHH_Xobar*), as well as elements of the *Harbinger* superfamily, have a clear preference for promoter regions and are less abundant in downstream regions (Fig. 9.5a). This asymmetric distribution suggests that their presence in promoters may be advantageous. However, it could also be that DNA transposons are preferably inserted near genes because chromatin is most accessible in transcriptionally active regions. Curiously, also LINEs show an asymmetric distribution around genes, and are found preferentially in downstream regions (Fig. 9.5b). One explanation for the higher frequency of *LINEs* downstream of genes is that insertion of these relatively large elements in promoters might be deleterious, while they are tolerated more easily in downstream regions.

9.7 What Molecular Mechanisms Drive Genomic TE Niche Specificity?

It is still unclear how TE families "find" their chromosomal niches. Niche specificity could be driven by a preference of the respective transposase or integrase enzymes to bind to specific sequence motifs. Analysis of the insertion sites of several high-copy TEs, including *RLC_BARE1* and *RLG_Sabrina*, as well as of multiple families of *Mariner, Harbinger* and *Helitron* elements, revealed pronounced differences in target site preference (Fig. 9.6). Class II elements target very specific motifs: *Mariner* elements prefer A/T-rich targets with having the consensus [T/A][T/A]nnT-Ann[T/A][T/A], where the dash represents the insertion site (Fig. 9.6a); whereas *Harbinger* transposons prefer a short TAA motif (Fig. 9.6b). AT-rich motifs (e.g. TATA boxes) are enriched in gene promoters of the barley genome (Table 9.2). This target preference could in part explain their preference for promoter sequences, or it could be the result of selection for promoters as their preferred site of insertion. However, it is also possible, as mentioned above, that these elements might simply target open chromatin (i.e. transcriptionally active) regions during transposition and establish themselves close to genes, because their small size does not disrupt promoter function.

Interestingly, *Helitrons* have a preference for an asymmetric target, requiring an AAA triplet starting 8 bp downstream of an A-T insertion site (Fig. 9.6c). Previous studies reported the preference of *Helitrons* for a 5′-AT-3′ insertion site (Wicker et al. 2007) and for generally A/T-rich sequences (Yang and Bennetzen 2009). However, preference for an asymmetric target has, to our knowledge, not been reported previously for any type of TE. The asymmetric sequence composition of the target site suggests that the helicase/recombinase protein of *Helitrons* binds the target DNA at the insertion site as well as one rotational period away in the DNA double-helix (i.e. 10 bp).

In contrast to DNA transposons, LTR retrotransposons do not show obvious target site preferences: the high-copy LTR retrotransposon *RLC_BARE1* has a weak preference for G/C 7–8 bp away from the insertion site, while *RLG_Sabrina* has a slight preference for GGG motif 3–4 bp upstream of the insertion site and a CC motif 4 bp downstream. These findings are consistent with previous studies that showed very little target site preference for LTR elements (Abe and Gemmell 2014). Nevertheless, different LTR retrotransposon families show very distinct chromosomal distributions. This suggests that their integrase enzymes target epigenetic patterns, such as histone modifications, rather than DNA sequence motifs. Previous studies reported that *RLG_Cereba* retrotransposons are particularly enriched in peri-centromeric regions (Hudakova et al. 2001), as are its homologs (the CRM elements) in maize, rice, and *B. distachyon* (Schnable et al. 2009, International Rice Genome Sequencing Project 2005; International Brachypodium Initiative 2010). However, for barley, we could not confirm such enrichment (Fig. 9.3a). Instead, we found that the *RLG_Abiba* family has taken over the proximal (peri-centromeric) "niche" in barley. We speculate that its unique preference for centromeric regions may be due to the product encoded by an ORF that is not found in any other retrotransposon family (Fig. 9.3a). This protein might have novel properties that enable *Abiba* elements to specifically target centromeric regions, potentially similar to the chromodomains in the integrase proteins of CRM elements that likely target centromere-specific histone modifications (Neumann et al. 2011).

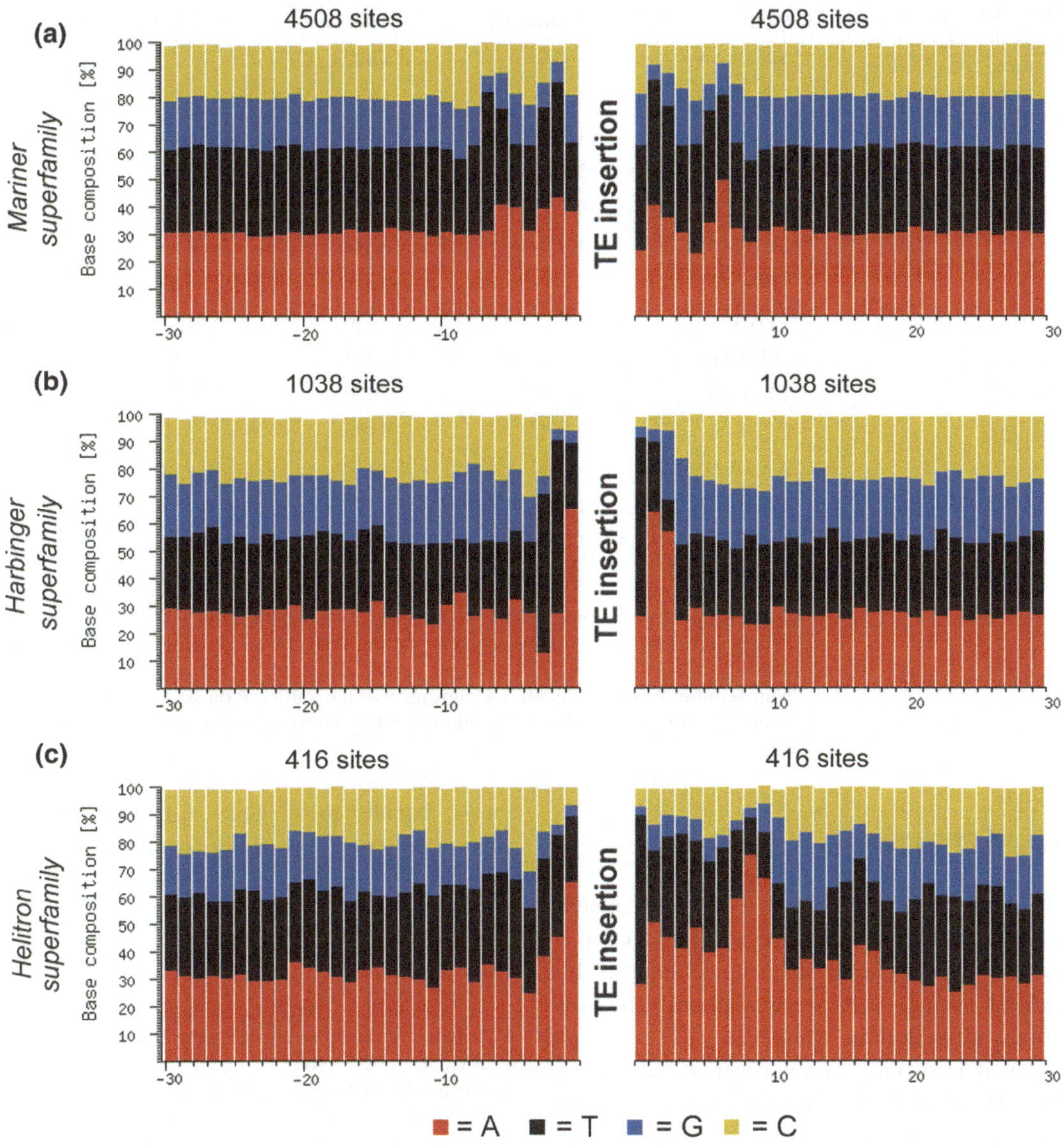

Fig. 9.6 Target site preferences of high-copy Class 2 transposons from barley. For the plots, the 30 bp flanking complete (i.e. not truncated) elements on both sides were collected. Then the different nucleotides at each position were counted across all insertion sites of a given TE type. The *x*-axis is the bp position relative to the TE insertion site, while the *y*-axis shows the relative nucleotide composition for each position. **a** *Mariner* elements have a strong preference for A/T dinucleotides 2 and 5 bp away from the insertion site, while (**b**) *Harbinger* elements almost invariably prefer 3 bp A/T-rich motifs. **c** Notably, *Helitrons* have a preference for an asymmetric target, strongly preferring an AAA motif 8 bp downstream of the insertion site

Table 9.2 Sequence motifs that are enriched in promoter compared to intergenic sequences (i.e. 10 kb upstream of genes)

Motif	Promoter[a]	Intergenic[b]	Enrichment
AAAAAA	5554	2492	2.2
CCGCCG	1333	585	2.2
CGCCGC	1380	619	2.2
GCCGCC	1378	666	2
TTTTTT	4874	2427	2
GCGGCG	1165	611	1.9
TAAAAA	2197	1219	1.8
TTAAAA	1549	850	1.8
CGGCGG	1056	617	1.7
TTTTAA	1573	920	1.7
TTTTTA	2110	1193	1.7
AAAAAG	1791	1112	1.6
TAAAAT	1481	875	1.6
TATAAA	1258	784	1.6
TTTAAA	1567	922	1.6

Copy numbers were compiled for 2000 gene loci comparing the 1 kb upstream of the transcription start site with the segment starting at 10 kb and ending at 9 kb upstream of it

[a]Number of motifs found in the 1 kb upstream of the transcription start site of 2000 genes

[b]Number of motifs found in the window 9–10 kb upstream of the transcription start site

9.8 Conclusions

The 5100 Mb barley genome provides insight into the repetitive landscape of plant genomes that are large by current standards of analysis, but near the average size for angiosperms. The most striking characteristic of the barley TE landscape is that it is strongly compartmentalized. As we have described in this chapter, there are several hints as to what could drive this chromosomal niche specificity. The TEs surrounding genes are of particular interest because insertions in or near genes can alter the expression and function of genes (Hirsch and Springer 2016). However, it is still not clear whether TEs play an active role in driving barley genome evolution by providing long-term fitness advantages, or whether they are merely present because they are selfish elements that have evolved successful strategies to amplify within the genome without causing deleterious effects. These two views of selectionist and neutralist, respectively, need not apply, of course, equally to all TEs in the genome, but rather represent a framework for future research.

References

Abe H, Gemmell NJ (2014) Abundance, arrangement, and function of sequence motifs in the chicken promoters. BMC Genom 15:900

Buchmann JP, Matsumoto T, Stein N, Keller B, Wicker T (2012) Interspecies sequence comparison of Brachypodium reveals how transposon activity corrodes genome colinearity. Plant J 488:213–217

Buchmann JP, Löytynoja A, Wicker T, Schulman AH (2014) Analysis of CACTA transposases reveals intron loss as major factor influencing their exon/intron structure in monocotyledonous and eudicotyledonous hosts. MobDNA 5(1):24. https://doi.org/10.1186/1759-8753-5-24

Bureau T, Wessler SR (1994a) Mobile inverted-repeat elements of the Tourist family are associated with the genes of many cereal grasses. Proc Natl Acad Sci USA 9:1411–1415

Bureau T, Wessler SR (1994b) Stowaway: a new family of inverted repeat elements associated with the genes of both monocotyledonous and dicotyledonous plants. Proc Natl Acad Sci USA 9:907–916

Chalopin D, Naville M, Plard F, Galiana D, Volff JN (2015) Comparative analysis of transposable elements highlights mobilome diversity and evolution in vertebrates. Genome Biol Evol 7:567–580

Chang W, Schulman AH (2008) BARE retrotransposons produce multiple groups of rarely polyadenylated transcripts from two differentially regulated promoters. Plant J 56:40–50

Chang W et al (2013) *BARE* Retrotransposons are translated and replicated via distinct RNA pools. PLoS ONE 8:e72270

d'Hont A, Denoeud F, Aury JM, Baurens FC, Carreel F, Garsmeur O, Noel B, Bocs S, Droc G, Rouard M et al (2012) The banana (*Musa acuminata*) genome and the evolution of monocotyledonous plants. Nature 488:213–217

Gladyshev EA, Arkhipova IR (2011) A widespread class of reverse transcriptase-related cellular genes. Proc Natl Acad Sci USA 108:20311–20316

Han Y, Qin S, Wessler SR (2013) Comparison of class 2 transposable elements at superfamily resolution reveals conserved and distinct features in cereal grass genomes. BMC Genom 14:71

Hirsch CD, Springer NM (2016) Transposable element influences on gene expression in plants. Biochim Biophys Acta S1874–9399:30100–30106

Hudakova S, Michalek W, Presting GG, ten Hoopen R, dos Santos K, Jasencakova Z, Schubert I (2001) Sequence organization of barley centromeres. Nucleic Acids Res 29:5029–5035

International Brachypodium Initiative (2010) Genome sequencing and analysis of the model grass *Brachypodium distachyon*. Nature 463:763–768

International Rice Genome Sequencing Project (2005) The map-based sequence of the rice genome. Nature 436:793–800

Jääskeläinen M et al (1999) Retrotransposon *BARE*-1: expression of encoded proteins and formation of virus-like particles in barley cells. Plant J 20:413–422

Jääskeläinen M et al (2013) Retrotransposon *BARE* displays strong tissue-specific differences in expression. New Phytol 200:1000–1008

Kalendar R, Vicient CM, Peleg O, Anamthawat-Jonsson K, Bolshoy A, Schulman AH (2004) LARD retroelements: novel, non-autonomous components of barley and related genomes. Genetics 166:1437–1450

Kapitonov VV, Jurka J (2007) Helitrons on a roll: eukaryotic rolling-circle transposons. Trends Genet 23:521–529

Kelly LJ, Renny-Byfield S, Pellicer J, Macas J, Novák P, Neumann P et al (2015) Analysis of the giant genomes of *Fritillaria* (Liliaceae) indicates that a lack of DNA removal characterizes extreme expansions in genome size. New Phytol 208:596–607

Kempken F, Windhofer F (2001) The hAT family: a versatile transposon group common to plants, fungi, animals, and man. Chromosoma 110:1–9

Leitch IJ, Beaulieu JM, Cheung K, Hanson L, Lysak MA, Fay MF (2007) Punctuated genome size evolution in Liliaceae. J Evol Biol 20:2296–2308

Leushkin EV, Sutormin RA, Nabieva ER, Penin AA, Kondrashov AS, Logacheva MD (2013) The miniature genome of a carnivorous plant *Genlisea aurea* contains a low number of genes and short non-coding sequences. BMC Genom 14:476

Malik HS, Eickbush TH (2001) Phylogenetic analysis of ribonuclease H domains suggests a late, chimeric origin of LTR retrotransposable elements and retroviruses. Genome Res 11:1187–1197

Manninen I, Schulman AH (1993) *BARE*-1, a *copia*-like retroelement in barley (*Hordeum vulgare* L.). Plant Mol Biol 22:829–846

Mascher M, Gundlach H, Himmelbach A, Beier S, Twardziok SO, Wicker T et al (2017) A chromosome conformation capture ordered sequence of the barley genome. Nature 544:427–433

Middleton CP, Stein N, Keller B, Kilian B, Wicker T (2012) Comparative analysis of genome composition in Triticeae reveals strong variation in transposable element dynamics and nucleotide diversity. Plant J. 73:347–356

Neumann P, Navrátilová A, Koblížková A, Kejnovský E, Hřibová E, Hobza R, Widmer A, Doležel J, Macas J (2011) Plant centromeric retrotransposons: a structural and cytogenetic perspective. Mob DNA. 2:4

Paterson AH, Bowers JE, Bruggmann R, Dubchak I, Grimwood J, Gundlach H, Haberer G, Hellsten U, Mitros T, Poliakov A, Schmutz J, Spannagl M, Tang H, Wang X, Wicker T, Bharti AK, Chapman J, Feltus FA, Gowik U, Grigoriev IV, Lyons E, Maher CA, Martis M, Narechania A, Otillar RP, Penning BW, Salamov AA, Wang Y, Zhang L, Carpita NC, Freeling M, Gingle AR, Hash CT, Keller B, Klein P, Kresovich S, McCann MC, Ming R, Peterson DG, Mehboob-ur-Rahman, Ware D, Westhoff P, Mayer KF, Messing J, Rokhsar DS (2009) The *Sorghum bicolor* genome and the diversification of grasses. Nature 457:551–556. https://doi.org/10.1038/nature07723

Roffler S, Wicker T (2015) Genome-wide comparison of Asian and African rice reveals high recent activity of DNA transposons. Mob DNA. 6:8

Sabot F, Schulman AH (2006) Parasitism and the retrotransposon life cycle in plants: a hitchhiker's guide to the genome. Heredity 97:381–388

Sabot F, Sourdille P, Chantret N, Bernard M (2006) *Morgane*, a new LTR retrotransposon group, and its subfamilies in wheats. Genetica 128:439–447

Schnable PS, Ware D, Fulton RS, Stein JC, Wei F, Pasternak S, Liang C, Zhang J, Fulton L, Graves TA, Minx P, Reily AD, Courtney L, Kruchowski SS, Tomlinson C, Strong C, Delehaunty K, Fronick C, Courtney B, Rock SM, Belter E, Du F, Kim K, Abbott RM, Cotton M, Levy A, Marchetto P, Ochoa K, Jackson SM, Gillam B, Chen W, Yan L, Higginbotham J, Cardenas M, Waligorski J, Applebaum E, Phelps L, Falcone J, Kanchi K, Thane T, Scimone A, Thane N, Henke J, Wang T, Ruppert J, Shah N, Rotter K, Hodges J, Ingenthron E, Cordes M, Kohlberg S, Sgro J, Delgado B, Mead K, Chinwalla A, Leonard S, Crouse K, Collura K, Kudrna D, Currie J,

He R, Angelova A, Rajasekar S, Mueller T, Lomeli R, Scara G, Ko A, Delaney K, Wissotski M, Lopez G, Campos D, Braidotti M, Ashley E, Golser W, Kim H, Lee S, Lin J, Dujmic Z, Kim W, Talag J, Zuccolo A, Fan C, Sebastian A, Kramer M, Spiegel L, Nascimento L, Zutavern T, Miller B, Ambroise C, Muller S, Spooner W, Narechania A, Ren L, Wei S, Kumari S, Faga B, Levy MJ, McMahan L, Van Buren P, Vaughn MW, Ying K, Yeh CT, Emrich SJ, Jia Y, Kalyanaraman A, Hsia AP, Barbazuk WB, Baucom RS, Brutnell TP, Carpita NC, Chaparro C, Chia JM, Deragon JM, Estill JC, Fu Y, Jeddeloh JA, Han Y, Lee H, Li P, Lisch DR, Liu S, Liu Z, Nagel DH, McCann MC, SanMiguel P, Myers AM, Nettleton D, Nguyen J, Penning BW, Ponnala L, Schneider KL, Schwartz DC, Sharma A, Soderlund C, Springer NM, Sun Q, Wang H, Waterman M, Westerman R, Wolfgruber TK, Yang L, Yu Y, Zhang L, Zhou S, Zhu Q, Bennetzen JL, Dawe RK, Jiang J, Jiang N, Presting GG, Wessler SR, Aluru S, Martienssen RA, Clifton SW, McCombie WR, Wing RA, Wilson RK (2009) The B73 maize genome: complexity, diversity, and dynamics. Science 326:1112–1115. https://doi.org/10.1126/science.1178534

Schulman AH (2013) Retrotransposon replication in plants. Curr Opin Virol 3:604–614

Suoniemi A et al (1996) The *BARE*-1 retrotransposon is transcribed in barley from an LTR promoter active in transient assays. Plant Mol Biol 31:295–306

Tanskanen JA et al (2007) Life without GAG: The *BARE*-2 retrotransposon as a parasite's parasite. Gene 390:166–174

Vicient CM, Suoniemi A, Anamthawat-Jonsson K, Tanskanen J, Beharav A, Nevo E, Schulman AH (1999) Retrotransposon BARE-1 and Its role in genome evolution in the genus Hordeum. Plant Cell 11:1769–1784

Wicker T, Guyot R, Yahiaoui N, Keller B (2003) *CACTA* transposons in Triticeae. A diverse family of high-copy repetitive elements. Plant Physiol 132:52–63

Wicker T, Sabot F, Hua-Van A, Bennetzen JL, Capy P, Chalhoub B et al (2007) A unified classification system for eukaryotic transposable elements. Nat Rev Genet 8:973–982

Wicker T, Yu Y, Haberer G, Mayer KFX, Marri PR, Rounsley S et al (2016) DNA transposon activity is associated with increased mutation rates in genes of rice and other grasses. Nat Commun 7:12790

Wicker T, Schulman AH, Tanskanen J, Spannagl M, Twardziok S, Mascher M, Springer NM, Li Q, Waugh R, Li C, Zhang G, Stein N, Mayer KFX, Gundlach H (2017) The repetitive landscape of the 5,100 Mbp barley genome, Mob. DNA 8:22

Yang L, Bennetzen JL (2009) Distribution, diversity, evolution, and survival of Helitrons in the maize genome. Proc Natl Acad Sci USA 106:19922–19927

10 Molecular Mapping and Cloning of Genes and QTLs

Mats Hansson, Takao Komatsuda, Nils Stein and Gary J. Muehlbauer

Abstract

The barley genome is comprised of more than 39,000 high-confidence genes, which represent many valuable targets for breeders as well as plant researchers trying to understand the genetic network controlling the various grass species, especially members of the Triticeae tribe including barley, wheat, and rye. The present chapter provides an overview of how past activities with barley mutants, markers, and genetic maps have laid the foundation for the present physical map based on the barley genome. We also describe how this new genome sequence resource can be integrated with mapping approaches to facilitate the cloning of genes and quantitative trait loci (QTL). Although the cost of genomic sequencing is likely to decrease, we assume that mapping of genes deficient in mutants will remain an important approach for gene identification. We present a comprehensive list of barley genes identified up to 2017.

M. Hansson (✉)
Department of Biology, Lund University, 22361 Lund, Sweden
e-mail: mats.hansson@biol.lu.se

T. Komatsuda
National Institute of Agrobiological Sciences, 305-8602 Tsukuba, Japan

N. Stein
Leibniz Institute of Plant Genetics and Crop Plant Research (IPK), OT Gatersleben, 06466 Stadt Seeland, Germany

G. J. Muehlbauer
Department of Plant and Microbial Biology, University of Minnesota, St. Paul, MN 55108, USA

G. J. Muehlbauer
Department of Agronomy and Plant Genetics, University of Minnesota, St. Paul, MN 55108, USA

10.1 Introduction

Barley (*Hordeum vulgare* L.) is one of the oldest crop plant species that was domesticated in the Fertile Crescent approximately 10,000 years ago (Badr et al. 2000). Barley also has a long history as a model organism for plant genetic analyses. This is obvious from the rich collection of various barley mutants (Fig. 10.1) that are available at various seed stock deposits (e.g., the Nordic Genetic Resource Center at Alnarp (Sweden), the Genebank at IPK (Gatersleben, Germany), and the National Small Grain Collection at the USDA-ARS (Aberdeen, ID, USA)). Past work with mutants has resulted in a long list of mapped traits of immediate interest for plant breeding activities (Franckowiak 1996). The identification of genes underlying these traits will facilitate our understanding of basic physiological and molecular processes in plants, which, in turn, will generate novel ideas for how crop plants can be further developed to serve society. In parallel,

N. Stein and G. J. Muehlbauer (eds.), *The Barley Genome*, Compendium of Plant Genomes,
https://doi.org/10.1007/978-3-319-92528-8_10

Fig. 10.1 Three examples of barley mutants, all exhibiting spike phenotypes. The *uzu1.a* mutant is semidwarf with short awns and a compact spike. *Lks1.a* (*Awnless*) is an awnless mutant. *Kap1.a* (*Hooded lemma*) is characterized by an appendage to the lemma, which develops as a trifurcate structure consisting of a deformed floret at its center with two triangular leaf-like projections called lemma wings. *Lks1.a* and *Kap1.a* are dominant mutations in contrast to the recessive *uzu1.a* mutation (www.nordgen.org/bgs)

accumulated knowledge concerning the linkage of genetic markers to valuable traits can be directly used in plant breeding. Such genetic markers are suitable for high-throughput screening, resulting in accelerated cultivar development.

Mapping genes and traits in barley have been performed for decades (Hor 1924; Miyake and Imai 1922; Nilsson-Ehle 1922). However, the step from mapping to the actual identification of the gene through map-based cloning has been a major challenge. This is mainly due to the large genome size (5.3 Gbp) and the high content of repetitive DNA (84%) like retrotransposons (International Barley Genome Sequencing et al. 2012; Mascher et al. 2017). The first gene that was successfully identified in barley through map-based cloning was *Mlo*, which provides resistance against powdery mildew based on the absence of a functional *Mlo* gene (Büschges et al. 1997). Initially, successful examples of genes identified through map-based cloning were dominated by genes conferring resistance to fungal and viral pathogens—*Rar1* (Shirasu et al. 1999), *Rpg1* (Brueggeman et al. 2002), *Mla6* (Halterman et al. 2001), *Ror2* (Collins et al. 2003), and *rym4*/*rym5* (Stein et al. 2005). More recently, there are also examples of genes related to morphological and biochemical traits that have been isolated including: spike row-type (*Vrs1*, *Vrs2*, *Vrs4,* and *Int-c*; Komatsuda et al. 2007; Koppolu et al. 2013; Ramsay et al. 2011; Youssef et al. 2017), kernel morphology (*Nud*; Taketa et al. 2008), early flowering (*Ppd-H1*, *Eam8*, *Mat-a*; Turner et al. 2005; Zakhrabekova et al. 2012), spike morphology (*Ert-m*, *Lax-a*; Jost et al. 2016; Zakhrabekova et al. 2015), seed dormancy (*Qsd1*; Nakamura et al. 2016; Sato et al. 2016), closed flowering (*Cly1*; Nair et al. 2010), epicuticular waxes (*Cer-cqu*; Schneider et al. 2016), chlorophyll biosynthesis (*Fch2*; Mueller et al. 2012), shoot architecture (*Cul4*; Tavakol et al. 2015), leaf development (*Blf1*; Jöst et al. 2016), and culm length (*Ari-e*, *Ari-o*, *Ari-u*, *Brh13*, *Ert-u*, *Uzu1*; Chono et al. 2003; Dockter et al. 2014; Wendt et al. 2016). Table 10.1 provides a comprehensive list of the cloned genes regardless of used method.

Table 10.1 Cloned genes in barley

Locus symbol	Locus name	Gene/gene product	Cloning strategy	References
Architecture				
ari-e	*breviaristatum-e, HvDEP1*	Heterotrimeric G-protein γ-subunit	Candidate gene	Wendt et al. (2016)
ari-m, brh1	*brachytic 1, breviaristatum-m, brachytic 1*	Alpha subunit of heterotrimeric G-protein	Positional cloning/candidate gene	Braumann and Hansson (2012), Ito et al. (2017)
ari-o, brh. af, brh14, brh16, ert-u, ert-zd	*HvDIM*	*DIMINUTO*	Candidate gene	Dockter et al. (2014)
ari-u, brh3, ert-t	*HvBRD*	*BRASSINOSTEROID-6-OXIDASE*	Candidate gene	Dockter et al. (2014)
blf1	*broad leaf 1*	INDETERMINATE DOMAIN protein transcriptional regulator	Positional cloning/candidate gene	Jöst et al. (2016)
brh13, brh18	*HvCPD*	*CONSTITUTIVE PHOTOMORPHOGENIC DWARF*	Candidate gene	Dockter et al. (2014)
btr1	*nonbrittle rachis 1*	Novel	Positional cloning	Pourkheirandish et al. (2015)
btr2	*nonbrittle rachis 2*	Novel	Positional cloning	Pourkheirandish et al. (2015)
com2	*compositum 2*	AP2-domain transcription factor	Positional cloning	Poursarebani et al. (2015)
cul4	*uniculme4*	Blade-on-petiole transcription factor homolog	Positional cloning	Tavakol et al. (2015)
Eli-a	*Eligulum-a*	Unknown protein	Transcriptional cloning	Okagaki et al. (2018)
ert-m	*erectoides-m*	Leucine-rich repeat receptor-like kinase	Mapping/candidate gene	Zakhrabekova et al. (2015)
Ert-r, Zeo1	*Zeocrithon 1, Erectoides-r*	HvAP2	GWAS	Houston et al. (2013)
fst2	*fragile stem 2*	HvCesA4	Transcriptional cloning	Burton et al. (2010)
int-a, vrs3	*six-rowed spike 3*	Histone demethylase	Cloning-by-sequencing Positional cloning/candidate gene	Bull et al. (2017), van Esse et al. (2017)
int-c	*intermedium-C*	TEOSINTE BRANCHED 1 homolog	GWAS, mutant analysis	Ramsay et al. (2011)
Kap1	*Hooded*	*Knox3*	Candidate gene	Müller et al. (1995)
lax-a	*laxatum-a*	Blade-on-petiole transcription factor homolog	Positional cloning/cloning-by-sequencing	Jost et al. (2016)

(continued)

Table 10.1 (continued)

Locus symbol	Locus name	Gene/gene product	Cloning strategy	References
lks	*short awn*	SHI-family transcription factor	Positional cloning	Yuo et al. (2012)
Lnt1	*Low number of tillers 1*	Jubel 2	Positional cloning/expression	Dabbert et al. (2010)
mnd6	*many-noded dwarf*	Cytochrome P450 oxidase	Cloning-by-sequencing	Mascher et al. (2014)
nld1	*narrow leafed dwarf1*	WUSCHEL-RELATED HOMEOBOX 3 (WOX3)	Positional cloning	Yoshikawa et al. (2016)
nud	*Naked caryopsis*	Ethylene response factor (ERF) family transcription factor	Positional cloning	Taketa et al. (2008)
sdw1	*semidwarf 1*	Gibberellin 20-oxidase	Candidate gene	Xu et al. (2017)
trd1	*third outer glume 1*	GATA zinc finger	Mapping/synteny	Houston et al. (2012)
uzu1	*HvBRI1*	BR-insensitive 1 (BRI1)	Synteny	Chono et al. (2003)
vrs1	*six-rowed spike 1*	Homeodomain-leucine zipper I-class homeobox gene	Positional cloning	Komatsuda et al. (2007)
vrs2	*six-rowed spike 2*	SHORT INTERNODES (SHI)	Positional cloning	Youssef et al. (2017)
vrs4	*six-rowed spike 4*	RAMOSA 2 homolog	Positional cloning	Koppolu et al. (2013)
Biotic stress resistance				
Mla	*Mla*	NBS-LRR	Positional cloning	Wei et al. (1999)
mlo	*MLO*	Seven transmembrane domain calmodulin binding protein	Positional cloning	Büschges et al. (1997)
Rar1	*RAR1*	CHORD domain zink binding protein	Positional cloning	Shirasu et al. (1999)
Ror2	*ROR2*	SNARE	Positional cloning/synteny	Collins et al. (2003)
Rpg1	*Rpg1*	NBS-LRR	Positional cloning/synteny	Brueggeman et al. (2002)
Rdg2a	*Resistance to Drechslera graminea 2a*	CC-NB-LRR type resistance gene	Positional cloning	Bulgarelli et al. (2010)
rpg4	*rpg4*	Actin depolymerizing factor-like protein	Positional cloning	Brueggeman et al. (2008)
Rpg5	*Rpg5*	NBS-LRR-Kinase	Positional cloning	Brueggeman et al. (2008)
rym4/5	*rym4*	eIF4E	Positional cloning	Stein et al. (2005)
rym11	*Rym11*	PDIL5-1	Positional cloning	Yang et al. (2014)

(continued)

Table 10.1 (continued)

Locus symbol	Locus name	Gene/gene product	Cloning strategy	References
Flower development				
cly1	*cleistogamy 1*	AP2-domain transcription factor	Positional cloning	Nair et al. (2010)
Flowering time				
eam5	*early maturity 5*	Phytochrome C	Cloning-by-sequencing	Pankin et al. (2014)
eam8, mat-a	*praematurum-a*	Circadian clock regulator, Early Flowering 3	Mapping/synteny/candidate gene	Faure et al. (2012), Zakhrabekova et al. (2012)
eam10	*early maturity 10*	LUX ARRHYTHMO homolog	Mapping/candidate gene	Campoli et al. (2013)
mat-c	*HvCEN*	CENTRORADIALIS homolog	GWAS, mutant analysis	Comadran et al. (2012)
Ppd-H1	*photo period dependent 1*	Pseudo-response regulator	Positional cloning	Turner et al. (2005)
vrn-H1	*vernalization response H1*	Apetala-1-like	Synteny/candidate gene	Fu et al. (2005)
vrn-H2	*vernalization response H2*	ZCCT-H	Synteny/candidate gene	von Zitzewitz et al. (2005)
vrn-H3	*vernalization response H3*	FLOWERING LOCUS T-like	Candidate gene/mapping	Faure et al. (2007)
Kernel morphology				
sex6	*shrunken endosperm xenia 6*	Starch synthase IIa	Candidate gene	Morell et al. (2003)
Photosynthesis				
fch2	*chlorina seedling 2*	Chlorophyllide a oxygenase	Candidate gene	Mueller et al. (2012)
tig-d	*tigrina-d*	*FLU*	Candidate gene	Lee et al. (2003)
xan-f	*xantha-f*	Magnesium chelatase large subunit	Candidate gene	Jensen et al. (1996), Olsson et al. (2004)
xan-g	*xantha-g*	Magnesium chelatase middle-size subunit	Candidate gene	Axelsson et al. (2006)
xan-h	*xantha-h*	Magnesium chelatase small subunit	Candidate gene	Hansson et al. (1999), Jensen et al. (1996)
xan-l	*xantha-l*	Mg-protoporphyrin IX monomethyl ester cyclase	Candidate gene	Rzeznicka et al. (2005)
Physiology				
bot1	*boron toxicity tolerance 1*	BOR1 efflux transporter	Synteny/mapping	Sutton et al. (2007)
cer-c	*eceriferum-c*	CHS-like synthase	Positional cloning/candidate gene	Schneider et al. (2016)

(continued)

Table 10.1 (continued)

Locus symbol	Locus name	Gene/gene product	Cloning strategy	References
cer-q	*eceriferum-q*	Lipase/carboxyl transferase	Positional cloning/candidate gene	Schneider et al. (2016)
cer-u	*eceriferum-u*	Cytochrome P450 protein	Positional cloning/candidate gene	Schneider et al. (2016)
Qsd1	*seed dormancy 1*		Positional cloning	Sato et al. (2016)
Qsd2	*seed dormancy 2*	Mitogen-activated Protein Kinase 3	Positional cloning	Nakamura et al. (2016)
Pigmentation				
ant1	*anthocyanin-less 1*	R2R3 MYB transcription factor	Mapping/candidate gene	Himi and Taketa (2015b), Zakhrabekova et al. (2015)
ant2	*anthocyanin-less 2*	Basic helix-loop-helix (bHLH) transcription factor	GWAS/mapping/candidate gene	Cockram et al. (2010)
ant17	*proanthocyanidin-free 17*	Flavanone-3-hydroxylase	Candidate gene	Himi and Taketa (2015a)
ant30	*proanthocyanidin-free 30*	Chalcone isomerase	Candidate gene	Druka et al. (2003)
Sterility				
des10	*desynapsis 10*	MutL-Homolog 3	Positional cloning/candidate gene	Colas et al. (2016)

The given examples are all based on studies with mutants. This demonstrates the immediate importance of mutants for gene identification activities and link modern genetic analyses to studies over the past hundred years. The chlorophyll mutants *xantha-m.3* and *albina-c.7*, and the short-culm mutant *uzu1.a*, where all isolated before 1922 (Gustafsson 1938, 1940; Miyake and Imai 1922). These early spontaneous mutants were quickly outnumbered when induced mutagenesis by physical treatment (X-rays) was initiated in the 1920s (Stadler 1928) and chemical treatment (mustard gas) was initiated in the 1940s (Gustafsson and Mac Key 1948). The induced mutants exhibited very distinct phenotypic characters like the absence of epicuticular waxes (e.g., *eceriferum*, *glossy sheath*) or divergent spike morphology (e.g., *erectoides*, *laxatum*) (www.nordgen.org/bgs).

10.2 Mutants, Markers, and Linkage Maps

Early cytological studies of barley chromosomes and genetic studies of phenotypic barley mutants (Smith 1951) laid the foundation for the later construction of various genetic linkage maps and cloning of genes. Cytological studies of condensed chromosomes made it possible to distinguish the seven pairs of barley chromosomes from each other (Löve and Löve 1961). In addition, microscopy was also used to follow translocations of large chromosomal segments (Hagberg and Hagberg 1968). Further, the translocation lines could be used in genetic crosses to barley mutants with distinct phenotypic characters. The linkage between the translocation breakpoints and the mutations made it possible to assign genes to specific chromosomes. The dense spike genes *Ert-a*, *Ert-b*, *Ert-c*, *Ert-d*, *Ert-e*, *Ert-i*, *Ert-g*, *Ert-m*, and *Ert-n* were mapped in this manner (Persson and Hagberg 1969). Therefore, the translocation lines functioned as a very important bridge linking mutated genes to certain chromosomes. Once the chromosomal locations had been established for mutants like the nine *Ert* genes just mentioned, the mutants could be used in further crosses to establish the chromosomal location of other genes. Later, wheat lines carrying an arm of one of the seven barley chromosomes (Islam et al. 1981) were also used to determine the chromosomal location of barley genes (Bilgic et al. 2007; Cho et al. 2006; Hansson et al. 1998; Rzeznicka et al. 2005). Over the years, data from various studies were assembled into genetic maps comprising more than 150 phenotypic markers distributed in seven linkage groups corresponding to the haploid chromosome number of barley (Franckowiak 1996).

However, the number of mapped genes is dwarfed by several thousand barley mutants that have been isolated of which more than 20,000 are stored in genetic seed stock centers. To map genes represented by the thousands of barley mutants, linkage maps based on DNA markers were developed. Different marker types, dominated by restriction fragment length polymorphism (RFLP) and simple sequence repeats (SSR), emerged in the late 1980s–late 1990s and were used to develop genetic maps (Graner et al. 1991; Heun et al. 1991; Kleinhofs et al. 1993; Ramsay et al. 2000; Shin et al. 1990). These studies provided up to 568 markers distributed on the seven barley chromosomes. The maps were consistent in chromosome size and covered a total map length ranging from 1173 to 1453 cM (Graner et al. 1991; Kleinhofs et al. 1993; Ramsay et al. 2000). The sizes of the seven chromosomes were reported in these studies and they exhibited an average of slightly less than 200 cM. A decade later, the number of identified markers had increased and several high density maps based on 1000–3500 markers were published (Hearnden et al. 2007; Marcel et al. 2007; Potokina et al. 2008; Rostoks et al. 2005; Sato et al. 2009; Stein et al. 2007; Szücs et al. 2009; Varshney et al. 2007; Wenzl et al. 2006).

In 2009, Close and colleagues explored more than a half million of EST sequences available at the public dbEST database, which had been obtained through Sanger sequencing from several barley cultivars (Close et al. 2009). Complemented with sequenced PCR amplicons derived from genomic sequences, approximately 22,000 SNPs were identified, of which 3072 were selected for two production-scale Illumina GoldenGate oligonucleotide pool assays named BOPA1 and BOPA2. A consensus genetic map was constructed based on four double haploid populations—Morex × Barke, Oregon Wolfe Barley, Steptoe × Morex, and Haruna Nijo × OHU602. The consensus map involved 2943 SNP markers with 975 unique positions over a total map length of 1099 cM. An improved genetic map was developed with the same SNP platform on 10 mapping populations and 2994 SNP loci were mapped to 1163 unique positions with a total length of 1137 cM (Muñoz-Amatriaín et al. 2011). More recently, population sequencing (POPSEQ) of the Morex × Barke recombinant inbred line and Oregon Wolfe doubled haploid populations resulted in identification and mapping of over 11 million SNPs (Mascher et al. 2013a). SNP markers are suitable for the development of high-throughput screening and are transferable across mapping populations. Thus, SNPs can serve as anchors to link mapping populations, and additionally can anchor genetic maps to the barley reference genome.

10.3 Identification of Genes Through Map-Based Cloning

10.3.1 Genetic Mapping

Map-based cloning is a strategy to clone genes of interest without prior knowledge of the gene product (Krattinger et al. 2009). The often tedious process consists of three major steps—construction of a genetic map that indicates the location of the gene relative to the genotyped markers, translation of the genetic map to a physical map that reveals candidate genes between the closest markers, and finally DNA sequencing of candidate genes. A more detailed scheme for the construction of a genetic map, which is the initial step in map-based cloning, could look like the following:

1. Construction of a mapping population. Typically, a mutant carrying a recessive mutation is crossed to a cultivar ("wild type"). To obtain markers covering the genome, it is important that the cultivar has a different genetic background compared to the mutant strain. To ensure adequate marker polymorphisms, it is also advantageous to make crosses of the mutant strains to several different genetic backgrounds. Flowers are first emasculated in the mother plant to avoid self-pollination and after 2–4 days anthers are transferred from the father plant. If the mutant is used as mother plant, unsuccessful emasculation (which causes self-pollination) will be revealed by the resulting F1 plant which will display a mutant phenotype. 50 F1 seeds are expected to generate approximately 4000 F2 seeds, which is a standard size of a mapping population.
2. Cultivation of the F2 population for low-resolution mapping. The F2 seeds are planted and followed as individual plants. In the first round, a smaller population of approximately 100 F2 plants are analyzed. The obtained resolution is normally above 1 cM.
3. Phenotyping. The F2 population is phenotyped. Generally, any population derived from the cross of a recessive mutant and a "wild-type" cultivar should display 25% mutant and 75% wild-type phenotypes.
4. Low-resolution genotyping. Leaf material is collected and genotyped with genetic markers. The relatively few individuals used in this initial low-resolution mapping are analyzed with many markers representing the range from each of the maps from all seven chromosomes. The markers have been identified through comparisons between the original mutant and the "wild-type" cultivar. Thus, it is known which markers represent the mutant and which represent the wild-type cultivar. Most markers will occur with equal frequencies in plants with a mutant or a wild-type phenotype, i.e., they are not linked to the mutation and far away from the gene to be identified. The closer a marker is located to the mutation, the fewer recombination events are likely to occur between the marker and the mutated gene, i.e., the marker and the mutation are linked. For example, following 25 plants expressing the mutant phenotype, a marker showing a wild-type genotype in only one plant is closer to the mutation than a marker showing a wild-type genotype in two plants. The two markers, which are flanking the mutation are selected for high-resolution mapping. Presently, next-generation sequencing technology is utilized for genotyping (genotyping-by-sequencing), which often reveals markers close to the introgression sites of the analyzed plants as well as a large set of markers distributed over the genomes which can be used for further high-resolution mapping. Genotyping-by-sequencing can allow both DNA and RNA to be sequenced and mined for SNPs, which can be converted into genetic markers. In restriction-site associated DNA sequencing (RAD-Seq) single- or double-digested DNA fragments are sequenced (Baird et al. 2008). To reduce the complexity of the barley genome, RNA sequencing can be performed although the plant material does not need to be collected from tissues specifically displaying the mutant phenotype. This is because SNPs in all genes located near the gene of interest and which are expressed in the samples can be used as markers for the mapping (Li et al. 2013; Liu et al. 2012). In the so-called bulked segregant RNA-Seq (BSR-Seq) analyses, samples from mutant and non-mutant plants of a segregating F2 population were combined into two separate pools and subjected to RNA-Seq, which provided accurate mapping of a maize *glossy3* gene encoding an R2R3-type MYB

transcription factor regulating genes involved in synthesis of epicuticular waxes (Li et al. 2013). Another approach to reduce the complexity of the genome is to use the barley exome capture platform, which can capture 61.6 Mb of exon space (Mascher et al. 2013b). Exome capture sequencing of phenotypic bulks has been conducted to map and isolate *many-noded dwarf* (Mascher et al. 2014).

5. Cultivation of the F2 population for high-resolution mapping. F2 seeds are planted again and followed as individual plants. More than 2000 F2 plants have often been required to obtain a high-resolution genetic map of less than 0.5 cM. The larger the mapping population the better are the chances of obtaining a high-resolution map.
6. Genotyping. All plants are screened with the two flanking markers, which were identified as the closest markers through the low-resolution mapping above. This will result in a small number of plants which show recombination between the two markers. These plants are further analyzed with additional molecular markers, which are located between the two markers from the low-resolution mapping.
7. Phenotyping. The recombinants identified in the high-resolution mapping are phenotyped as above. The mapping is said to have reached saturation when flanking markers show 100% linkage to the phenotype. These markers will be located closest to the mutated gene and explored in the physical map.

10.3.2 Physical Mapping

In the second major step of the process of map-based cloning, the results from genetic mapping are translated to the nucleotide sequence of the genome, i.e., the physical map. The units of the genetic map are given in cM, reflecting recombination frequency, whereas the physical map is measured in base pairs. Since recombination in barley occurs more frequently in telomeric regions of a chromosome compared to its centromeric and pericentromeric regions a cM unit can correspond to less than a million base pairs near the telomere but several million base pairs near the centromere. Nevertheless, in the process of map-based cloning, the physical stretch of DNA spanning the interval between the two flanking genetic markers needs to be identified. Specific examples are instructive here. In search for the two closely linked genes, *Btr1* and *Btr2*, regulating brittleness of the rachis (the node of the central floral axis to which the kernel is attached), a 0.19 cM region was identified on chromosome *3H* (Pourkheirandish et al. 2015). This distance on the genetic map corresponded to a 403 kb region on the physical map. Similarly, a 0.07 cM region on chromosome *2H* was identified containing the six-rowed spike gene *Vrs1* (Komatsuda et al. 2007). The region corresponded to 518 kb in the physical map. During the cloning of the powdery mildew resistant gene, *Mlo*, a 0.64 cM region was identified on chromosome *4H*, which corresponded to approximately 200 kbp (Büschges et al. 1997).

The physical map of the barley genome (Beier et al. 2017; Mascher et al. 2017) is a game changer that will significantly facilitate the process of map-based cloning. Identified flanking markers from the genetic maps will now be easily found on the physical map, which will also reveal the genes between the markers. Previously, the identified flanking markers had to be identified in BAC clones or YAC clones containing genomic barley DNA. A BAC clone typically contains 80–150 kb of DNA (International Barley Genome Sequencing et al. 2012). Thus, several BAC clones had to be linked in order to bridge between the flanking markers. First, the BAC clones containing the identified markers were sequenced. The sequence information made it possible to identify new BAC clones partially overlapping with the initially sequenced BAC clones. Repeating this chromosomal-walk procedure made it possible to construct a consensus sequence of the region between the flanking markers. Six and 20 BAC clones covered the *Vrs1* and *Nud* region, respectively (Komatsuda et al. 2007; Taketa et al.

2008), demonstrating the major efforts of this approach to identify genes of interest.

10.3.3 Candidate Gene Sequencing

The identified region in the physical map flanked by the two markers contains genes of which one is the gene of interest. To identify the correct gene, the candidate genes are normally sequenced. Since the F2 mapping population is based on a genetic difference between a mutant and a wild type, the correct gene should have a point mutation, a deletion or any kind of mutation that is likely to cause the observed mutant phenotype. In the process of identifying the correct gene, it is very helpful to have access to a set of allelic mutants since they all should have mutations in the same gene, which provides strong evidence that the correct gene has been found. For example, more than 50 allelic mutations in each of three genes were used to identify *Cer-c*, *Cer-q*, and *Cer-u* required for accumulation of epicuticular aliphatic waxes (Schneider et al. 2016).

In studies involving bacteria and yeast, a complementation test of the mutant is a common approach to verify that the correct gene has been identified. In such tests, the candidate gene is typically cloned in a plasmid from where it can be expressed. The plasmid is transformed into the mutant strain and if the mutant can be rescued from its deficiency this provides very strong evidence that the correct gene has been found. Barley can be transformed but compared to many unicellular organisms it is a major effort (Mrizova et al. 2014). In addition, not all genotypes are easily transformed. Therefore, this approach is still not routinely used in barley research to verify candidate genes. One early successful example concerns the verification of *RPG1* as the resistance gene to the stem rust fungus *Puccinia graminis* f. sp. *tritici* (Horvath et al. 2003). *RPG1* encodes a putative receptor kinase and is present in the stem rust resistant cultivar Morex but not the susceptive cultivar Golden Promise (Brueggeman et al. 2002). Transformation of *RPG1* into Golden Promise transferred the resistance to the transgenic plant (Horvath et al. 2003). A more recent example is provided by a complementation test of a line carrying the long dormancy *qsd1* allele of wild barley H602 in a genetic background of Golden Promise with a construct carrying the cloned Haruna Nijo short dormancy *Qsd1* allele (Sato et al. 2016).

10.4 Exploring Near-Isogenic Lines

A collection of near-isogenic lines is also an important resource for gene identification. To generate the near-isogenic lines, 881 historic barley mutants were recurrently backcrossed (often 6 times) to the cultivar Bowman with phenotyping in each F2 cycle (Druka et al. 2011). Through backcrossing, the regions close to the mutations were fixed regardless of whether one or several genes were involved in the formation of the phenotypic character. In addition, the near-isogenic lines were genotyped with 3072 SNP markers, which revealed the chromosomal location of the involved genes as well as provided initial markers for further high-resolution mapping (Druka et al. 2011). Also, without further mapping, the near-isogenic lines have been successfully used to identify mutated genes. The approach was to compare the location of candidate genes to the location of introgression regions in near-isogenic lines displaying relevant phenotypes. In this way, three genes involved in brassinosteroid biosynthesis were identified (Dockter et al. 2014). Interrupted brassinosteroid biosynthesis causes a semidwarf phenotype in several plant species (Kim and Wang 2010; Vriet et al. 2012). Dockter and colleagues found that barley genes orthologous to *BRASSINOSTEROID-6-OXIDASE*, *CONSTITUTIVE PHOTOMORPHOGENIC DWARF*, and *DIMINUTO*, encoding three enzymes of the brassinosteroid biosynthetic pathway in Arabidopsis, were located on barley chromosome *2H*, *5H*, and *7H*, respectively (Dockter et al. 2014). The location of the three genes overlapped with the introgression regions of near-isogenic lines with semidwarf phenotypes. Sequencing of the three genes from the near-isogenic lines as well as other allelic mutants revealed mutations. Thus, genotype and phenotype could be connected and the mutated genes were identified.

10.5 Identification of Genes by Genomic Sequencing of Allelic Mutants

Whole-genome sequencing of mutant lines to identify the mutated gene is a viable option if several allelic mutants are available. However, with current sequencing techniques, it is an expensive option given the 5.3 Gbp size of the barley genome. Mutant lines are likely to be very polygenic especially if they have been isolated through mutagenesis with physical or chemical treatment, which introduces hundreds or thousands of mutations randomly in the genome. To succeed with this approach, it is, therefore, essential to have access to allelic mutants since random mutations in each line can be filtered away and relevant mutations in the causal gene identified.

Today, sequencing the barley genome requires significant resources and it is beyond the economic means of most research groups to sequence the genomic DNA from multiple mutants and wild-type individuals. Obviously, a significant drop in the costs for genomic sequencing is required before this is a realistic option. In the meantime, sequencing of the exome, the transcriptome, or single chromosomes or pieces of chromosomes are attractive alternatives to reduce sequence complexity of the barley genome. An in-solution hybridization-based sequence capture platform has been developed for selective enrichment of coding DNA sequences (Mascher et al. 2013b). The capture targets exome DNA and is based on predicted genes from the genome of the barley cultivar Morex, publically available full-length cDNA sequences and RNA-Seq consensus sequence contigs. The capture is estimated to represent at least 86% of the barley genes. Approximately 80–95% of all captured reads could be mapped to these genes (Mascher et al. 2013b). The exome capture approach has the advantage that genomic DNA from the lines to be analyzed is used. Thus, the growing conditions of the lines to be analyzed are less important, which must be strictly considered if transcriptome analyses are to be performed. In RNA-Seq analyses, the transcriptome at a certain time point or a certain growth phase is sequenced. It is therefore important that mRNA is isolated from the correct tissues at the correct time point when the mutant phenotype is displayed. Despite these concerns, there might be many instances when it would be advantageous to perform RNA-Seq analyses since the barley transcriptome is much less than the genome. RNA-Seq analyses might, therefore, be a cost-efficient approach to identify a mutated gene in a group of allelic mutants. Recently, flow-sorted chromosomes were sequenced from allelic *cer-q* wax-less mutants (Sanchez-Martin et al. 2016). With this approach, named MutChromSeq, sequence complexity was significantly reduced and sequencing of five chemically induced mutants where enough to confidently identify the *Cer-q* gene.

10.6 Conclusions and Future Perspectives

The publication of the barley genome (Beier et al. 2017; Mascher et al. 2017) accelerated barley research into the post-genomic era. In this era, scientific activities are often focused on the elucidation of the identified open reading frames and gene models. Most genes encoding enzymes of metabolic pathways will be easy to annotate due to their sequence similarity to genes of other organisms which gene products have been thoroughly studied through molecular and biochemical analyses. Other gene products will be possible to group according to general function based on characteristic sequence motifs but it might be a challenge to determine in which biological process they are involved in. This can be exemplified by genes obviously encoding kinases, proteases, and various transcription factors. Gene editing techniques are indeed emerging and TILLING populations exist (Caldwell et al. 2004; Gottwald et al. 2009; Szarejko et al. 2016; Talame et al. 2008), which make it possible to use a reverse genetic approach where a gene of unknown function can be inactivated and the phenotypic effect can be studied to reveal the function of the gene. However, forward genetic approaches are more straightforward in revealing the genes of the many existing barley cultivars with desired and important traits, as well as the rich collection of induced barley mutants. In work involving forward genetic approaches, the now published physical map will be of immense value since less time will be spent on translating genetic maps to physical maps. Still, classic methods like the mapping of genes and QTLs will remain an important tool in gene hunting projects, since mapping makes it possible to isolate and characterize a certain locus, and thereby exclude many other possible gene candidates.

References

Axelsson E, Lundqvist J, Sawicki A et al (2006) Recessiveness and dominance in barley mutants deficient in Mg-chelatase subunit D, an AAA protein involved in chlorophyll biosynthesis. Plant Cell 18:3606–3616

Badr A, Muller K, Schafer-Pregl R et al (2000) On the origin and domestication history of barley (*Hordeum vulgare*). Mol Biol Evol 17:499–510

Baird NA, Etter PD, Atwood TS et al (2008) Rapid SNP discovery and genetic mapping using sequenced RAD markers. PLoS ONE 3:e3376

Beier S, Himmelbach A, Colmsee C et al (2017) Construction of a map-based reference genome sequence for barley *Hordeum vulgare* L. Sci Data 4:170044

Bilgic H, Cho S, Garvin DF, Muehlbauer GJ (2007) Mapping barley genes to chromosome arms by transcript profiling of wheat-barley ditelosomic chromosome addition lines. Genome 50:898–906

Braumann I, Hansson M (2012) Barley plants with short culm. In: Danish Patent and Trademark Office, Denmark

Brueggeman R, Druka A, Nirmala J et al (2008) The stem rust resistance gene *Rpg5* encodes a protein with nucleotide-binding-site, leucine-rich, and protein kinase domains. Proc Natl Acad Sci USA 105:14970–14975

Brueggeman R, Rostoks N, Kudrna D et al (2002) The barley stem rust-resistance gene *Rpg1* is a novel disease-resistance gene with homology to receptor kinases. Proc Natl Acad Sci USA 99:9328–9333

Bulgarelli D, Biselli C, Collins NC et al (2010) The CC-NB-LRR-type *Rdg2a* resistance gene confers immunity to the seed-borne barley leaf stripe pathogen in the absence of hypersensitive cell death. PLoS ONE 5

Bull H, Casao CM, Zwirek M et al (2017) Barley *SIX-ROWED SPIKE3* encodes a putative Jumonji C-type H3K9me2/me3 demethylase that represses lateral spikelet fertility. Nature Commun 8:936

Burton RA, Ma G, Baumann U et al (2010) A customized gene expression microarray reveals that the brittle stem phenotype *fs2* of barley is attributable to a retroelement in the *HvCesA4* cellulose synthase gene. Plant Physiol 153:1716–1728

Büschges R, Hollricher K, Panstruga R et al (1997) The barley *Mlo* gene: a novel control element of plant pathogen resistance. Cell 88:695–705

Caldwell DG, McCallum N, Shaw P, Muehlbauer GJ, Marshall DF, Waugh R (2004) A structured mutant population for forward and reverse genetics in Barley (*Hordeum vulgare* L.). Plant J 40:143–150

Campoli C, Pankin A, Drosse B, Casao CM, Davis SJ, von Korff M (2013) *HvLUX1* is a candidate gene underlying the *early maturity 10* locus in barley: phylogeny, diversity, and interactions with the circadian clock and photoperiodic pathways. New Phytol 199:1045–1059

Cho S, Garvin DF, Muehlbauer GJ (2006) Transcriptome analysis and physical mapping of barley genes in wheat-barley chromosome addition lines. Genetics 172:1277–1285

Chono M, Honda I, Zeniya H et al (2003) A semidwarf phenotype of barley uzu results from a nucleotide substitution in the gene encoding a putative brassinosteroid receptor. Plant Physiol 133:1209–1219

Close TJ, Bhat PR, Lonardi S et al (2009) Development and implementation of high-throughput SNP genotyping in barley. BMC Genom 10:582

Cockram J, White J, Zuluaga DL et al (2010) Genome-wide association mapping to candidate polymorphism resolution in the unsequenced barley genome. Proc Natl Acad Sci U S A 107:21611–21616

Colas I, Macaulay M, Higgins JD et al (2016) A spontaneous mutation in MutL-Homolog 3 (HvMLH3) affects synapsis and crossover resolution in the barley desynaptic mutant des10. New Phytol 212:693–707

Collins NC, Thordal-Christensen H, Lipka V et al (2003) SNARE-protein-mediated disease resistance at the plant cell wall. Nature 425:973–977

Comadran J, Kilian B, Russell J et al (2012) Natural variation in a homolog of *Antirrhinum CENTRORADIALIS* contributed to spring growth habit and environmental adaptation in cultivated barley. Nat Genet 44:1388–1392

Dabbert T, Okagaki RJ, Cho S, Heinen S, Boddu J, Muehlbauer GJ (2010) The genetics of barley low-tillering mutants: *low number of tillers-1* (*lnt1*). Theor Appl Genet 121:705–715

Dockter C, Gruszka D, Braumann I et al (2014) Induced variations in brassinosteroid genes define barley height and sturdiness, and expand the green revolution genetic toolkit. Plant Physiol 166:1912–1927

Druka A, Franckowiak J, Lundqvist U et al (2011) Genetic dissection of barley morphology and development. Plant Physiol 155:617–627

Druka A, Kudrna D, Rostoks N, Brueggeman R, von Wettstein D, Kleinhofs A (2003) Chalcone isomerase gene from rice (*Oryza sativa*) and barley (*Hordeum vulgare*): physical, genetic and mutation mapping. Gene 302:171–178

Faure S, Higgins J, Turner A, Laurie DA (2007) The *FLOWERING LOCUS T*-like gene family in barley (*Hordeum vulgare*). Genetics 176:599–609

Faure S, Turner AS, Gruszka D et al (2012) Mutation at the circadian clock gene *EARLY MATURITY 8* adapts domesticated barley (*Hordeum vulgare*) to short growing seasons. Proc Natl Acad Sci USA 109:8328–8333

Franckowiak J (1996) Revised linkage maps for morphological markers in barley, *Hordeum vulgare*. Barley Genet Newsl 26:9–21

Fu D, Szucs P, Yan L et al (2005) Large deletions within the first intron in VRN-1 are associated with spring growth habit in barley and wheat. Mol Genet Genomics 273:54–65

Gottwald S, Bauer P, Komatsuda T, Lundqvist U, Stein N (2009) TILLING in the two-rowed barley cultivar 'Barke' reveals preferred sites of functional diversity in the gene *HvHox1*. BMC Res Notes 2:258

Graner A, Jahoor A, Schondelmaier J et al (1991) Construction of an RFLP map of barley. Theor Appl Genet 83:250–256

Gustafsson Å (1938) Studies on the genetic basis of chlorophyll formation and the mechanism of induced mutating. Hereditas 24:33–93

Gustafsson Å (1940) The mutation system of the chlorophyll apparatus. Lunds Univ Årsskr NF 36:1–40

Gustafsson Å, Mac Key J (1948) The genetical effects of mustard gas substances and neutrons. Hereditas 34:371–386

Hagberg A, Hagberg G Induced traslocations in barley. In: Proceedings of the mutations in plant breeding II, 1968. International Atomic Energy Agency, Vienna, Austria

Halterman DA, Zhou FS, Wei FS, Wise RP, Schultze-Lefert P (2001) The Mla6 coiled-coil, NBS-LRR protein confers AvrMla6-dependent resistance specific to *Blumeria graminis* f. sp. *hordei* in barley and wheat. Plant J 25:335–348

Hansson A, Kannangara CG, von Wettstein D, Hansson M (1999) Molecular basis for semidominance of missense mutations in the XANTHA-H (42-kDa) subunit of magnesium chelatase. Proc Natl Acad Sci USA 96:1744–1749

Hansson M, Gough SP, Kannangara CG, von Wettstein D (1998) Chromosomal location of six barley genes encoding enzymes of chlorophyll and heme biosynthesis and the sequence of the ferrochelatase gene identify two regulatory genes. Plant Physiol Biochem 36:545–554

Hearnden PR, Eckermann PJ, McMichael GL, Hayden MJ, Eglinton JK, Chalmers KJ (2007) A genetic map of 1,000 SSR and DArT markers in a wide barley cross. Theor Appl Genet 115:383–391

Heun M, Kennedy AE, Anderson JA, Lapitan NLV, Sorrells ME, Tanksley SD (1991) Construction of a restriction fragment length polymorphism map for barley (*Hordeum vulgare*). Genome 34:437–447

Himi E, Taketa S (2015a) Barley Ant17, encoding *flavanone 3-hydroxylase* (*F3H*), is a promising target locus for attaining anthocyanin/proanthocyanidin-free plants without pleiotropic reduction of grain dormancy. Genome 58:43–53

Himi E, Taketa S (2015b) Isolation of candidate genes for the barley *Ant1* and wheat *Rc* genes controlling anthocyanin pigmentation in different vegetative tissues. Mol Genet Genomics 290:1287–1298

Hor KS (1924) Interrelations of genetic factors in barley. Genetics 9:151–180

Horvath H, Rostoks N, Brueggeman R, Steffenson B, von Wettstein D, Kleinhofs A (2003) Genetically engineered stem rust resistance in barley using the *Rpg1* gene. Proc Natl Acad Sci USA 100:364–369

Houston K, Druka A, Bonar N et al (2012) Analysis of the barley bract suppression gene *Trd1*. Theor Appl Genet 125:33–45

Houston K, McKim SM, Comadran J et al (2013) Variation in the interaction between alleles of *HvAPETALA2* and microRNA172 determines the density of grains on the barley inflorescence. Proc Natl Acad Sci USA 110:16675–16680

International Barley Genome Sequencing Consortium, Mayer KF, Waugh R et al (2012) A physical, genetic and functional sequence assembly of the barley genome. Nature 491:711–716

Islam AKMR, Shephard KW, Sparrow DHB (1981) Isolation and characterization of euplamisc wheat-barley chromosome addition lines. Heredity 46:161–174

Ito A, Yasuda A, Yamaoka K et al (2017) *Brachytic 1* of barley (*Hordeum vulgare* L.) encodes the alpha subunit of heterotrimeric G protein. J Plant Physiol 213:209–215

Jensen PE, Willows RD, Petersen BL et al (1996) Structural genes for Mg-chelatase subunits in barley: *Xantha-f*, *-g* and *-h*. Mol Gen Genet 250:383–394

Jost M, Taketa S, Mascher M et al (2016) A homolog of *Blade-On-Petiole 1* and *2* (*BOP1/2*) controls internode length and homeotic changes of the barley inflorescence. Plant Physiol 171:1113–1127

Jöst M, Hensel G, Kappel C et al (2016) The INDETERMINATE DOMAIN protein BROAD LEAF1 limits barley leaf width by restricting lateral proliferation. Curr Biol 26:903–909

Kim TW, Wang ZY (2010) Brassinosteroid signal transduction from receptor kinases to transcription factors. Annu Rev Plant Biol 61:681–704

Kleinhofs A, Kilian A, Saghai Maroof MA et al (1993) A molecular, isozyme and morphological map of the barley (*Hordeum vulgare*) genome. Theor Appl Genet 86:705–712

Komatsuda T, Pourkheirandish M, He C et al (2007) Six-rowed barley originated from a mutation in a homeodomain-leucine zipper I-class homeobox gene. Proc Natl Acad Sci USA 104:1424–1429

Koppolu R, Anwar N, Sakuma S et al (2013) *Six-rowed spike4* (*Vrs4*) controls spikelet determinacy and row-type in barley. Proc Natl Acad Sci USA 110:13198–13203

Krattinger S, Wicker T, Keller B (2009) Map-based cloning of genes in Triticeae (wheat and barley). In: Feulliet CJ, Muehlbauer G (eds) Genetics and genomics of the Triticeae, plant genetics and genomics: crops and models, vol 7, pp 337–357

Lee KP, Kim C, Lee DW, Apel K (2003) *TIGRINA d*, required for regulating the biosynthesis of tetrapyrroles in barley, is an ortholog of the *FLU* gene of *Arabidopsis thaliana*. FEBS Lett 553:119–124

Li L, Li D, Liu S et al (2013) The maize *glossy13* gene, cloned via BSR-Seq and Seq-walking encodes a putative ABC transporter required for the normal accumulation of epicuticular waxes. PLoS ONE 8: e82333

Liu S, Yeh CT, Tang HM, Nettleton D, Schnable PS (2012) Gene mapping via bulked segregant RNA-Seq (BSR-Seq). PLoS ONE 7:e36406

Löve Á, Löve D (1961) Chromosome numbers of central and northwest European plant species. Opera Bot 5:1–581

Marcel TC, Varshney RK, Barbieri M et al (2007) A high-density consensus map of barley to compare the distribution of QTLs for partial resistance to *Puccinia hordei* and of defence gene homologues. Theor Appl Genet 114:487–500

Mascher M, Gundlach H, Himmelbach A et al (2017) A chromosome conformation capture ordered sequence of the barley genome. Nature 544:427–433

Mascher M, Jost M, Kuon JE et al (2014) Mapping-by-sequencing accelerates forward genetics in barley. Genome Biol 15:R78

Mascher M, Muehlbauer GJ, Rokhsar DS et al (2013a) Anchoring and ordering NGS contig assemblies by population sequencing (POPSEQ). Plant J 76:718–727

Mascher M, Richmond TA, Gerhardt DJ et al (2013b) Barley whole exome capture: a tool for genomic research in the genus *Hordeum* and beyond. Plant J 76:494–505

Miyake K, Imai Y (1922) Genetic studies in barley I. Bot Mag Tokyo 36:25–38

Morell MK, Kosar-Hashemi B, Cmiel M et al (2003) Barley *sex6* mutants lack starch synthase IIa activity and contain a starch with novel properties. Plant J 34:173–185

Mrizova K, Holaskova E, Oz MT, Jiskrova E, Frebort I, Galuszka P (2014) Transgenic barley: a prospective tool for biotechnology and agriculture. Biotechnol Adv 32:137–157

Mueller AH, Dockter C, Gough SP, Lundqvist U, von Wettstein D, Hansson M (2012) Characterization of mutations in barley *fch2* encoding chlorophyllide *a* oxygenase. Plant Cell Physiol 53:1232–1246

Muñoz-Amatriaín M, Moscou MJ, Bhat PR et al (2011) An improved consensus linkage map of barley based on flow-sorted chromosomes and single nucleotide polymorphism markers. Plant Genome 4:238–249

Müller KJ, Romano N, Gerstner O et al (1995) The barley *Hooded* mutation caused by a duplication in a homeobox gene intron. Nature 374:727–730

Nair SK, Wang N, Turuspekov Y et al (2010) Cleistogamous flowering in barley arises from the suppression of microRNA-guided *HvAP2* mRNA cleavage. Proc Natl Acad Sci USA 107:490–495

Nakamura S, Pourkheirandish M, Morishige H et al (2016) *Mitogen-Activated Protein Kinase Kinase 3* regulates seed dormancy in barley. Curr Biol 26:775–781

Nilsson-Ehle H (1922) Über freie Kombination und Koppelung verschiedener Chlorophyllerbeinheiten bei Gerste. Hereditas 3:191–199

Okagaki RJ, Haaning A, Bilgic H, Heinen S, Druka A, Bayer M, Waugh R, Muehlbauer GJ (2018) *ELIGULUM-A* regulates lateral branch and leaf development in barley. Plant Physiol. https://doi.org/10.1104/pp.17.01459

Olsson U, Sirijovski N, Hansson M (2004) Characterization of eight barley *xantha-f* mutants deficient in magnesium chelatase. Plant Physiol Biochem 42:557–564

Pankin A, Campoli C, Dong X et al (2014) Mapping-by-sequencing identifies *HvPHYTOCHROME C* as a candidate gene for the *early maturity 5* locus modulating the circadian clock and photoperiodic flowering in barley. Genetics 198:383–396

Persson G, Hagberg A (1969) Induced variation in a quantitative character in Barley. Morphology and cytogenetics of *erectoides* mutants. Hereditas 61:115–178

Potokina E, Druka A, Luo Z, Wise R, Waugh R, Kearsey M (2008) Gene expression quantitative trait locus analysis of 16000 barley genes reveals a complex pattern of genome-wide transcriptional regulation. Plant J 53:90–101

Pourkheirandish M, Hensel G, Kilian B et al (2015) Evolution of the grain dispersal system in barley. Cell 162:527–539

Poursarebani N, Seidensticker T, Koppolu R et al (2015) The genetic basis of composite spike form in barley and 'miracle-wheat'. Genetics 201:155–165

Ramsay L, Comadran J, Druka A et al (2011) *INTERMEDIUM-C*, a modifier of lateral spikelet fertility in barley, is an ortholog of the maize domestication gene *TEOSINTE BRANCHED 1*. Nat Genet 43:169–172

Ramsay L, Macaulay M, degli Ivanissevich S et al (2000) A simple sequence repeat-based linkage map of barley. Genetics 156:1997–2005

Rostoks N, Mudie S, Cardle L et al (2005) Genome-wide SNP discovery and linkage analysis in barley based on genes responsive to abiotic stress. Mol Genet Genomics 274:515–527

Rzeznicka K, Walker CJ, Westergren T et al (2005) *Xantha-l* encodes a membrane subunit of the aerobic Mg-protoporphyrin IX monomethyl ester cyclase involved in chlorophyll biosynthesis. Proc Natl Acad Sci USA 102:5886–5891

Sanchez-Martin J, Steuernagel B, Ghosh S et al (2016) Rapid gene isolation in barley and wheat by mutant chromosome sequencing. Genome Biol 17:221

Sato K, Nankaku N, Takeda K (2009) A high-density transcript linkage map of barley derived from a single population. Heredity (Edinb) 103:110–117

Sato K, Yamane M, Yamaji N et al (2016) Alanine aminotransferase controls seed dormancy in barley. Nat Commun 7:11625

Schneider LM, Adamski NM, Christensen CE et al (2016) The *Cer-cqu* gene cluster determines three key players in a beta-diketone synthase polyketide pathway synthesizing aliphatics in epicuticular waxes. J Exp Bot 67:2715–2730

Shin JS, Chao S, Corpuz L, Blake T (1990) A partial map of the barley genome incorporating restriction fragment length polymorphism, polymerase chain reaction, isozyme, and morphological marker loci. Genome 33:803–810

Shirasu K, Lahaye T, Tan MW, Zhou F, Azevedo C, Schulze-Lefert P (1999) A novel class of eukaryotic zinc-binding proteins is required for disease resistance signaling in barley and development in *C. elegans*. Cell 99:355–366

Smith L (1951) Cytology and genetics of barley. Bot Rev 17:1–51

Stadler LJ (1928) The rate of induced mutation in relation to dormancy, temperature and dosage. Anatomical Record 41

Stein N, Perovic D, Kumlehn J et al (2005) The eukaryotic translation initiation factor 4E confers multiallelic recessive *Bymovirus* resistance in *Hordeum vulgare* (L.). Plant J 42:912–922

Stein N, Prasad M, Scholz U et al (2007) A 1,000-loci transcript map of the barley genome: new anchoring points for integrative grass genomics. Theor Appl Genet 114:823–839

Sutton T, Baumann U, Hayes J et al (2007) Boron-toxicity tolerance in barley arising from efflux transporter amplification. Science 318:1446–1449

Szarejko I, Szurman-Zubrzycka M, Nawrot M et al (2016) Creation of a TILLING population in barley after chemical mutagensesis with sodium azide and MNU. In: Jankowicz-Cieslak J, Tai TH, Kumlehn J, Till BJ (eds) Biotechnologies for plant mutation breeding. Springer International Publication, Berlin, pp 91–111

Szücs P, Blake VC, Bhat PR et al (2009) An integrated resource for barley lingage map and malting quality QTL alignment. Plant Genome 2:134–140

Taketa S, Amano S, Tsujino Y et al (2008) Barley grain with adhering hulls is controlled by an ERF family transcription factor gene regulating a lipid biosynthesis pathway. Proc Natl Acad Sci USA 105:4062–4067

Talame V, Bovina R, Sanguineti MC, Tuberosa R, Lundqvist U, Salvi S (2008) TILLMore, a resource for the discovery of chemically induced mutants in barley. Plant Biotechnol J 6:477–485

Tavakol E, Okagaki R, Verderio G et al (2015) The barley *Uniculme4* gene encodes a BLADE-ON-PETIOLE-like protein that controls tillering and leaf patterning. Plant Physiol 168:164–174

Turner A, Beales J, Faure S, Dunford RP, Laurie DA (2005) The pseudo-response regulator *Ppd-H1* provides adaptation to photoperiod in barley. Science 310:1031–1034

van Esse GW, Walla A, Finke A, Koornneef M, Pecinka A, von Korff M (2017) Six-rowed spike3 (VRS3) is a histone demethylase that controls lateral spikelet development in barley. Plant Physiol 174:2397–2408

Varshney RK, Marcel TC, Ramsay L et al (2007) A high density barley microsatellite consensus map with 775 SSR loci. Theor Appl Genet 114:1091–1103

Wei F, Gobelman-Werner K, Morroll SM et al (1999) The *Mla* (powdery mildew) resistance cluster is associated with three NBS-LRR gene families and suppressed recombination within a 240-kb DNA interval on chromosome 5S (1HS) of barley. Genetics 153:1929–1948

Wendt T, Holme I, Dockter C et al (2016) *HvDep1* is a positive regulator of culm elongation and grain size in barley and impacts yield in an environment-dependent manner. PLoS ONE 11:e0168924

Wenzl P, Li H, Carling J et al (2006) A high-density consensus map of barley linking DArT markers to SSR, RFLP and STS loci and agricultural traits. BMC Genom 7:206

von Zitzewitz J, Szucs P, Dubcovsky J et al (2005) Molecular and structural characterization of barley vernalization genes. Plant Mol Biol 59:449–467

Vriet C, Russinova E, Reuzeau C (2012) Boosting crop yields with plant steroids. Plant Cell 24:842–857

Xu Y, Jia Q, Zhou G et al (2017) Characterization of the *sdw1* semi-dwarf gene in barley. BMC Plant Biol 17:11

Yang P, Lupken T, Habekuss A et al (2014) *PROTEIN DISULFIDE ISOMERASE LIKE 5-1* is a susceptibility factor to plant viruses. Proc Natl Acad Sci U S A 111:2104–2109

Yoshikawa T, Tanaka SY, Masumoto Y et al (2016) Barley *NARROW LEAFED DWARF1* encoding a WUSCHEL-RELATED HOMEOBOX 3 (WOX3) regulates the marginal development of lateral organs. Breed Sci 66:416–424

Youssef HM, Eggert K, Koppolu R et al (2017) VRS2 regulates hormone-mediated inflorescence patterning in barley. Nat Genet 49:157–161

Yuo T, Yamashita Y, Kanamori H et al (2012) A *SHORT INTERNODES* (*SHI*) family transcription factor gene regulates awn elongation and pistil morphology in barley. J Exp Bot 63:5223–5232

Zakhrabekova S, Dockter C, Ahmann K et al (2015) Genetic linkage facilitates cloning of *Ert-m* regulating plant architecture in barley and identified a strong candidate of *Ant1* involved in anthocyanin biosynthesis. Plant Mol Biol 88:609–626

Zakhrabekova S, Gough SP, Braumann I et al (2012) Induced mutations in circadian clock regulator *Mat-a* facilitated short-season adaptation and range extension in cultivated barley. Proc Natl Acad Sci USA 109:4326–4331

11 Genomics Approaches to Mining Barley Germplasm Collections

Martin Mascher, Kazuhiro Sato and Brian Steffenson

Abstract

Barley has been diversified during the long process of evolution. The genetic diversity that was lost during domestication and crop improvement can be introduced from various collections with extant wild relatives of barley being a particularly rich source. Thousands of diverse accessions of cultivated and wild barley have been collected, preserved in ex situ collections, phenotyped for various traits, genotyped with molecular markers, and catalogued in databases. Such attributes make these collections readily accessible for germplasm mining. High-throughput sequencing methods for assessing intraspecific diversity have become available recently through the implementation of exome sequencing and genotype-by-sequencing in barley. These methods enable the systematic collection of molecular passport data of entire collections to inform genebank management decisions. They can also guide the selection of core collections for further in-depth studies linking phenotype and genotype. Finally, the joint analysis of genetic data and information on collection sites of accession can give insights about the population structure, dispersal, and evolutionary history of the crop.

11.1 Introduction

The genus *Hordeum* belongs to the tribe Triticeae, which includes a number of important cereal crops such as barley (*Hordeum vulgare*), bread wheat (*Triticum aestivum*), and rye (*Secale cereale*). These genera share a common wild ancestor from 8 to 9 million years ago (Middleton et al. 2014), but evolved independently to the diverse present-day species. Natural mutations and genome arrangements occurred during this evolutionary period. Unlike bread wheat, an allohexaploid crop with three genome donors [*Aegilops tauschii* (DD), *Triticum urartu* (AA), and likely *Aegilops speltoides* (SS ~ BB)], barley is a simple diploid crop domesticated from its wild progenitor *H. vulgare* ssp. *spontaneum* (see also Chap. 17 for barley domestication). Barley was one of the first crops domesticated by humans in the Near East (Zohary et al. 2012).

M. Mascher
Leibniz Institute of Plant Genetics and Crop Plant Research (IPK) Gatersleben, 06466 Seeland, Germany
e-mail: mascher@ipk-gatersleben.de

K. Sato
Institute of Plant Science and Resources, Okayama University, Kurashiki 710-0046, Japan
e-mail: kazsato@rib.okayama-u.ac.jp

B. Steffenson (✉)
Department of Plant Pathology, University of Minnesota, St. Paul, MN 55108, USA
e-mail: bsteffen@umn.edu

N. Stein and G. J. Muehlbauer (eds.), *The Barley Genome*, Compendium of Plant Genomes,
https://doi.org/10.1007/978-3-319-92528-8_11

After domestication, it was distributed and adapted across a wide diversity of environments, surviving and, in some cases, thriving in environments that are too cool, too wet, too infertile, or too saline for wheat (Dawson et al. 2015). Barley is an important crop for producers across a range of agro-systems: from the arid subsistence lands of some developing countries to the high input, intensively managed systems of developed countries (Newton et al. 2011). As such, barley is vulnerable to a number of biotic and abiotic stresses. The crop has several different important end uses, including malt for production of various alcoholic beverages (e.g., beer, whiskey, etc.), animal feed, and human food. These different end use products mandate different target traits in breeding and sometimes caused serious bottlenecks (e.g., high level of susceptibility to *Barley Yellow Mosaic Virus* in malting barley germplasm). Genetic diversity is essential for the survival of all plant species on earth. Plant breeding is predicated on genetic diversity for crop improvement. This makes the allelic diversity contained in germplasm collections the essential element for sustaining and enhancing a crop species. Where wheat has three genome donors and a number of wild relatives in which to source genetic diversity, barley has three genepools but only its wild progenitor (primary genepool) and *Hordeum bulbosum* (secondary genepool) can be used for recombination with cultivated barley. The objective of this chapter is to review the genomic approaches available for exploiting various germplasm collections for enhancing barley and our basic knowledge about the crop.

11.2 Sources of Diversity and Germplasm Mining

11.2.1 *The Genus* Hordeum *and Germplasm Collections*

The genus *Hordeum* is comprised of over 30 species including cultivated barley, *H. vulgare* ssp. *vulgare* (see also Chap. 2 for detailed descriptions of these species). The wild ancestral form of cultivated barley is the subspecies *H. vulgare* ssp. *spontaneum*. This taxon is a rich source of new alleles for barley breeding because it is genetically diverse and can readily hybridize with the cultivated form (von Bothmer et al. 2003). Other wild *Hordeum* species are more distantly related to cultivated barley and have reproductive barriers limiting hybridization (see also Chaps. 2 and 18). Barley germplasm has been extensively collected because of its economic importance in agricultural production and end uses for malting and animal feed (Ullrich 2011). Thus, most of the accessions in ex situ (genebank) collections are comprised of landraces of *H. vulgare* ssp. *vulgare*. The number of barley accessions housed in genebanks around the world is extensive and second only to wheat (FAO 2010). According to Knüpffer (2009), there are over 485,000 accessions of *Hordeum* residing in over 200 different genebanks around the world. These collections include 299,165 accessions of *H. vulgare* ssp. *vulgare* (primarily new and old cultivars and landraces), 32,385 accessions of *H. vulgare* ssp. *spontaneum* as well as 4,681 accessions of wild species, and substantial numbers of genetic stocks, breeding lines, and mapping populations. Many accessions are duplicated among genebanks for safety or as part of germplasm exchanges.

11.2.2 Barley Core and Other Germplasm Collections

Since the numbers of barley accessions held in genebanks are far too numerous and expensive to fully characterize, an international barley core collection was developed that consists of a reduced number of accessions. This condensed germplasm set was selected to represent the greatest genetic diversity possible, but still be manageable for researchers to conduct various phenotyping studies in the laboratory, greenhouse, or field. In the international barley core collection (BCC) consisting of about 1500 accessions, there are eight groups of accessions: (1) 285 landraces and cultivars from West Asia and North Africa selected by the International

Center for Agricultural Research in the Dry Areas (ICARDA); (2) 380 landraces and cultivars from South and East Asia selected by Okayama University in Japan; (3) 320 landraces and cultivars from Europe selected by the Leibniz Institute of Plant Genetics and Crop Plant Research (IPK) in Germany; (4) 155 landraces and cultivars from the Americas selected by the United States Department of Agriculture–Agriculture Research Service (USDA–ARS) National Small Grains Collection (NSGC) in the United States; (5) 11 cultivars from Oceania and other parts of the world selected by researchers at the Australian Winter Cereals Collection (AWCC) in Australia; (6) 150 *H. vulgare* ssp. *spontaneum* accessions selected by ICARDA; (7) 45 accessions of 22 different wild *Hordeum* species (excluding *H. vulgare* ssp. *spontaneum*) selected by the Nordic Genetic Resource Center (NORDGEN) in Sweden; and (8) about 200 genetic stocks selected by the USDA–ARS NSGC (Knüpffer and van Hintum 2003). Subsets of the BCC have been used to investigate the genetic architecture of diverse traits such as row type, heading date, height, grain weight, grain starch content, grain protein content, and grain cadmium content (Pasam et al. 2012; Wu et al. 2015).

In addition to the international barley core collection, other sets of diverse cultivated core collections are available at several genebanks. For example, the USDA established the "NSGC Barley Core" consisting of about 10% of the >33,000 barley accessions in the genebank (Muñoz-Amatriaín et al. 2014). This core consists of 2417 accessions, but SNP (single nucleotide polymorphism) genotyping revealed 172 cases where two or more accessions appeared identical. Altogether, these cases involved 520 accessions. In the final analysis, 82 accessions were retained from this group, and 438 were discarded from the original starting number of 2298 accessions passing quality control for SNP calls (Muñoz-Amatriaín et al. 2014). That over 19% of this rigorously curated collection was discarded due to duplication, which indicates that significant savings and efficiencies can be made with other collections housed in genebanks worldwide. The final "informative" Core or iCore collection of the NSGC contains 1860 accessions from 94 different countries. At the IPK in Germany, Pasam et al. (2014) established a diverse panel of landrace germplasm in order to capture wide adaptation to different climates. This Landrace Collection (LRC1485) consists of 1485 spring landraces selected from more than >22,000 barley accessions based on various taxonomic traits (growth habit, row type, seed color, etc.), available descriptions of the original collection sites, and passport data. LRC1485 includes accessions from a wide area (from 5.63° to 62.47° north latitude and from 16.62° to 71.5° east longitude) of Europe, West and Central Asia, and also North and East Africa. Two-rowed types comprise 47.7% (708) and six-rowed types comprise 52.3% (777) of this collection. To reduce the LRC to a more manageable number for various phenotyping and molecular studies, the M strategy implemented in MSTRAT (Gouesnard 2001) was used, resulting in a reduced core set of 648 landraces. This panel can be reduced by about half again by selecting only two- or six-rowed types depending on the research objective (Pasam et al. 2014).

Two large association genetics studies were made on elite barley breeding germplasm in the United Kingdom and the United States. The Association Genetics of United Kingdom Elite Barley (AGOUEB) project was based on over 1000 lines and cultivars (both spring and winter types) evaluated in the national trials and utilized SNP markers. In addition to resolving basic questions about the level and extent of linkage disequilibrium (LD) and population structure in this elite germplasm (Rostoks et al. 2006), AGOUEB contributed valuable data on the mapping of both qualitative and quantitative traits important for barley production and the environment in the UK. The Barley Coordinated Agricultural Project (BCAP) was a parallel and concurrent association genetics project in the United States. It focused on developing the tools for GWAS with 3840 breeding lines from 10 programs (Hamblin et al. 2010; Zhou et al. 2012). Using 3072 SNP markers, many marker–trait associations were discovered for yield components (Pauli et al. 2014), agronomic traits (Cuesta-Marcos et al. 2010; Pauli et al. 2014), end use quality (Berger et al. 2013; Gutiérrez et al. 2011), and disease resistance (Berger et al. 2013; Massman et al. 2011; Zhou et al. 2014; Zhou and Steffenson 2013a, b). With the advanced breeding status of these BCAP lines (at least to F_4) and publically available genotype and phenotype datasets, breeders can readily identify suitable parents from other programs to introgress various traits into their germplasm. In this sense, the BCAP unified 10 different breeding programs into one.

In Europe, a genomics-based project (EXBARDIV) was launched to exploit useful genetic variation in landrace and wild barley for cultivated barley improvement (Tondelli et al. 2013). For this project, an incremental association mapping approach was used based on different population types for the discovery of new alleles in wild and landrace barley. One specific objective of EXBARDIV was to determine the efficiencies of identification and extraction of useful alleles from these exotic sources in barley breeding programs, especially with wild barley in advanced backcross programs.

To capture the rich genetic diversity present in the progenitor of cultivated barley, the Wild Barley Diversity Collection (WBDC) was established by Steffenson et al. (2007). This collection is an expansion of the *H. vulgare* ssp. *spontaneum* set contained in BCC mentioned above and consists of 318 accessions selected on the basis of various ecogeographic characters such as longitude/latitude, elevation, temperature, rainfall, soil type, disease resistance, etc. Since wild barley originated in the Fertile Crescent and is still home to large populations today, most of the accessions selected for the WBDC were from this region ($\sim$64.5%). However, the habitat range of *H. vulgare* ssp. *spontaneum* is far more extensive, so representative samples were also included from North Africa, the Caucasus region, Central Asia, and South Asia. The WDBC has been genotyped with several marker systems [e.g., Diversity Arrays Technology sequencing (DArTseq) and SNP markers generated from oligonucleotide pooled assays and genotype-by-sequencing (GBS)] and phenotyped for over 30 traits related to yield, agronomics, morphology, quality, and disease resistance (Ames et al. 2015; Roy et al. 2010; Sallam et al. 2017; Steffenson et al. 2007). Comprehensive GWAS are currently in progress to elucidate the genetic basis of these traits and their genomic architecture. Another wild barley core collection was established by Fu and Horbach (2012) from the 3782 accessions of *H. vulgare* ssp. *spontaneum* held at the Plant Gene Resources of Canada (PGRC). This wild barley core collection was comprised of 269 accessions selected by random sampling stratified with respect to country of origin.

Since raw *H. vulgare* ssp. *spontaneum* accessions are difficult to work with in terms of phenotyping in the field and assessing the true breeding value of their alleles, researchers have developed backcross introgression populations (also called Recombinant Chromosome Substitution Lines or RCSLs) having every chromosome segment of a particular wild barley accession transferred into the adapted background of a barley cultivar (von Korff et al. 2004; Pillen et al. 2004; Sato and Takeda 2009; Schnaithmann et al. 2014; Schmalenbach et al. 2008). These populations are developed with the aid of molecular markers and provide an opportunity to assess wild barley alleles in an adapted genetic background as proposed by Tanksley and McCouch (1997). Sato and Takeda (2009) used this type of population to separate QTL controlling seed dormancy into a single locus and cloned the major effect QTL *Qsd1* (Sato et al. 2016). Such backcross introgression populations are possible for *H. vulgare* ssp. *spontaneum*, but not members of the secondary or tertiary genepools of *Hordeum* due to crossing barriers. Still, given the recent progress in identifying cultivated barley with introgressions of *H. bulbosum* chromatin via exome capture re-sequencing data, the rich diversity of this sole member of the secondary genepool is now more accessible for breeders (Wendler et al. 2015).

Nested association mapping (NAM) populations are superior to other methods (i.e., conventional biparental linkage mapping and association mapping) for QTL detection in plants because they have low marker density requirements, high allelic richness, high mapping resolution, and enhanced statistical power—all without the drawbacks of the other two methods (Yu et al. 2008). They are constructed by crossing a series of different exotic parents to one common parent. As such, they capture both historic (i.e., occurring throughout the life history of the exotic and common parents) and recent (i.e., occurring after crossing the exotic parents with the common parent) recombination events. In NAM populations, the linkage between the common parent and exotic parents is detected by different haplotype combinations. Recently, NAM populations were constructed with different sets of diverse *H. vulgare* ssp. *spontaneum* accessions by research groups in Germany (with cultivar Barke) (Maurer et al. 2015; Saade et al. 2016) and in the United States (with cultivar Rasmusson) (Nice et al. 2016). These populations offer the opportunity to assess the value and genetic architecture of multiple allelic variants of wild germplasm in an adapted cultivated barley background.

11.3 Access to Germplasm Data

11.3.1 Databases

Comprehensive and accurate databases on germplasm collections are crucial for many types of studies regarding evolution, natural mutation, and germplasm mining. Most of the pertinent information regarding *Hordeum* germplasm collections is available online via the respective genebank databases (see Table 11.1). Additionally, a comprehensive global database that includes records on *Hordeum* accessions

Table 11.1 Databases of barley germplasm holdings at major genebanks around the world

Genebank	URL (Cited 18 Nov 2016)
Plant Gene Resources of Canada	http://pgrc3.agr.ca/search_grinca-recherche_rirgc_e.html
National Small Grains Germplasm Research Facility, USA	http://www.ars-grin.gov/npgs
International Centre for Agricultural Research in the Dry Areas	https://www.genesys-pgr.org/welcome
Department of Applied Genetics, John Innes Centre, UK	http://www.jic.ac.uk/GERMPLAS/bbsrc_ce/index.htm
Leibniz Institute of Plant Genetics and Crop Plant Research, Germany	http://gbis.ipk-gatersleben.de
Institute of Crop Germplasm Resources, Chinese Academy of Agricultural Sciences	http://icgr.caas.net.cn/cgris_english.html
N. I. Vavilov All-Russian Scientific Research Institute of Plant Industry	http://91.151.189.38/virdb/
Okayama University, Japan	www.shigen.nig.ac.jp/barley/
Nordic Genetic Resources Centre, Sweden	www.nordgen.org/ngb/
Genetic Resources Center, NARO, Japan	http://www.gene.affrc.go.jp/plant/
Lieberman Germplasm Bank, Tel-Aviv University, Israel	http://www2.tau.ac.il/ICCI/default.asp
Institute for Plant Genetic Resources, Bulgaria	http://eurisco.ecpgr.org
Centre for Genetic Resources, the Netherlands	http://www.cgn.wur.nl/UK/
Israel Gene Bank for Agricultural Crops, Israel	http://igb.agri.gov.il/
Plant Genetic Resources Documentation in the Czech Republic	http://genbank.vurv.cz/genetic/resources/
Research Institute of Plant Production Piestany, Slovakia	http://eurisco.ecpgr.org
Suceava Genebank, Romania	http://www.svgenebank.ro/index.htm

Fig. 11.1 Size of barley collections in genebanks available through Genesys (https://www.genesys-pgr.org/welcome)

collected from other genebanks is available through Genesys at https://www.genesys-pgr.org/welcome (Fig. 11.1). Other special databases supporting the characterization of various collections and breeding germplasm are also available. For example, two large association genetics projects focusing on United States breeding germplasm and the NSGC barley core collection were recently completed with all genotype and phenotype data being deposited in "The Triticeae Toolbox" (T3) (https://triticeaetoolbox.org/). The German Genebank Information System (GBIS) is a multilayered system that (i) provides passport information and an ordering system accessible by the general public, and (ii) maintains records of seed and line management as well as observational data for internal use by genebank managers (Oppermann et al. 2015). The Global Biodiversity Information Facility (GBIF) (http://www.gbif.org/) is an open data infrastructure to access data about all types of life on earth, including barley. Accessions have to be submitted to GBIF in order to gain access to the data. A search using the term "*H. vulgare* L." revealed the mapped locations of accessions submitted to the GBIF (Fig. 11.2). The European Search Catalogue for Plant Genetic Resources (EURISCO: eurisco.ecpgr.org) provides information about 1.9 million accessions of crop plants and their wild relatives, preserved ex situ by almost 400 institutes. It is based on a network of national inventories of 43 member countries and represents an important effort for the preservation of the world's agrobiological diversity by providing information about the large genetic diversity kept by the collaborating institutions. EURISCO contains both passport data and phenotypic data. DivSeek (http://www.divseek.org/) develops and shares methodologies, open-source software tools, and practices to facilitate the generation, integration, and sharing of data and information related to plant genetic resources.

11.3.2 Use of Collection Site Data for Germplasm Mining

Passport data and the ecogeographic parameters associated with them are important for germplasm mining. However, passport data for some accessions, especially cultivated landraces, may be of limited value due to the widespread movement of germplasm via the seed trade or

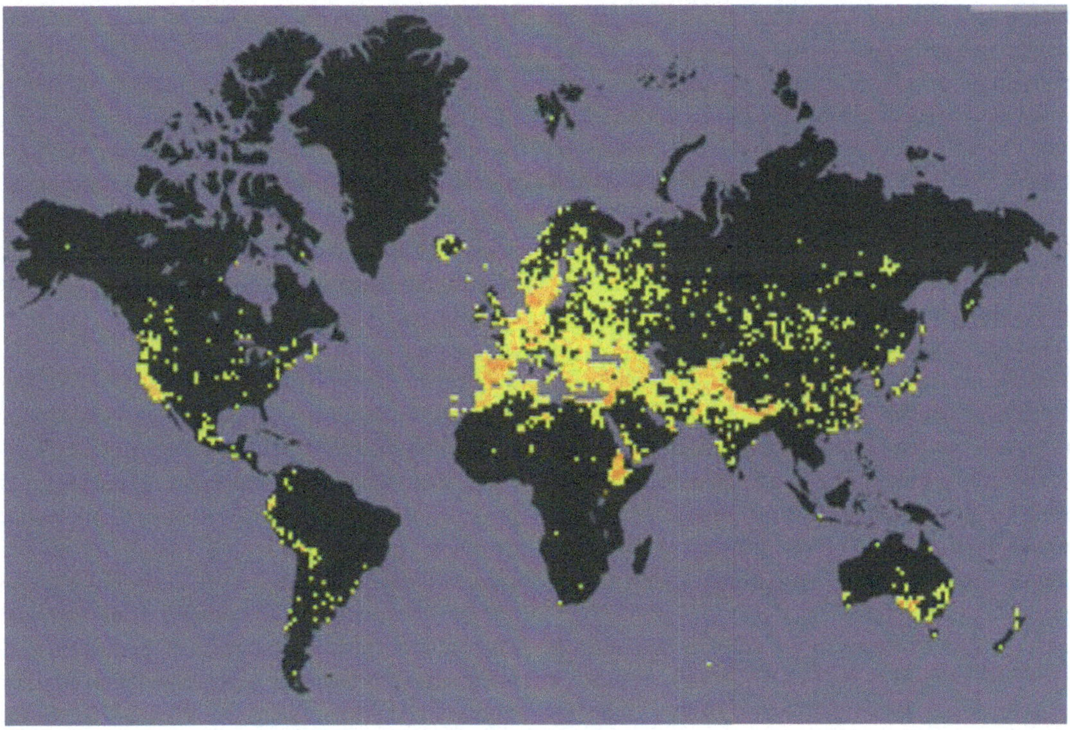

Fig. 11.2 Locations of accessions by the search "*Hordeum vulgare* L." in Global Biodiversity Information Facility (www.gbif.org)

incomplete information concerning the original collection site. Typically, one would expect greater correspondence between the ecogeographic data of collection sites and adaptive evolution of wild barley accessions, which presumably coevolved for thousands of years in these sites. Still, movement of these wild species also occurred via humans and animals, extending the taxon's range far beyond their center of origin. Moreover, gene flow can occur within ex situ *H. vulgare* ssp. *spontaneum* collections when accessions are grown in close proximity to other wild or cultivated accessions during regeneration in genebanks (Jakob et al. 2014). In a recent investigation, Russell et al. (2016a) utilized the exome sequences of geo-referenced landrace and wild barley accessions and demonstrated the strong influence of geography on the observed patterns of variation. Moreover, individual genes controlling major traits such as heading date and height showed variants by environment associations.

The ecogeographical data associated with accession collection sites can be extremely important for allele mining. For example, if a target trait is strongly related to environmental conditions that favor continual challenge by a pathogen or insect, the Focused Identification of Germplasm Strategy (FIGS) (Mackay and Street 2004; http://figs.icarda.org/) can be employed to achieve better targeting of a markedly reduced number of accessions that likely carry the desired allele(s). This technique has been successfully used in barley to identify, within a much reduced germplasm sample, additional resistance to the net blotch pathogen (Endresen et al. 2011) and also in wheat to identify resistance to the Russian wheat aphid, powdery mildew pathogen, and stem rust pathogen, specifically the widely virulent race TTKSK (isolate synonym Ug99) from Africa (Endresen et al. 2011). Some of the identified resistances in these studies were novel and found in accessions from environmentally similar sites. In addition to ecogeographic factors, the presence of other plant species can have a marked effect on the evolution of disease resistance in wild barley. Two cases in point involve leaf (caused by *Puccinia hordei*) and stem rust (*Puccinia graminis* f. sp. *tritici*) of barley. From the evaluation of the WBDC to leaf rust, all of the resistant accessions originated from the Fertile Crescent, a region where the alternate host of the pathogen (Star of Bethlehem lily: *Ornithogalum umbellatum*) is present (Dawson et al. 2015). Although the number of WBDC accessions from Central Asia was considerably smaller than those from the Fertile Crescent, no leaf rust resistant accessions were found in this region. In contrast, the frequency of stem rust resistance from Central Asia was markedly higher than those accessions from the Fertile Crescent. This may be attributed to the presence of common barberry (*Berberis vulgaris*), the alternate host of *P. graminis* f. sp. *tritici* (Dawson et al. 2015). These examples clearly demonstrate that other factors must be considered when implementing the strategy of FIGS.

11.4 Sequencing Methods for Barley Diversity

11.4.1 High-Throughput Sequencing Methods for Assessing Intraspecific Diversity

A high-quality reference genome sequence is often only assembled for one or a very few number of individuals of a species. Reference genome projects are often followed by construction of haplotype maps (HapMaps) for assessing intraspecific diversity at a genome-wide scale, taking advantage of the newly developed sequence resources. Prominent examples in crop plants include the HapMaps of maize (Gore et al. 2009), rice (Huang et al. 2010), sorghum (Morris et al. 2013), and wheat (Jordan et al. 2015). The pan-genome and core genome sizes of wheat are $140{,}500 \pm 102$ and $81{,}070 \pm 1631$ genes, respectively, with an average of 128,656 genes in each cultivar (Montenegro et al. 2017). While HapMap projects in species with small genomes can be performed by whole genome sequencing (WGS), large genome species require the reduction of genome complexity prior to sequencing for cost reasons. The various means of complexity reduction (Davey et al. 2011) differ in the genome coverage and marker density they achieve. Panel size and sequencing depth need to be well balanced in large genome species, where complexity reduction is necessary prior to sequencing in order to obtain sufficient sequencing depth for reliable variant calling (Nielsen et al. 2011). In barley, whole genome sequencing is still impractical for large panels. Once genotypic data have been collected, HapMaps can provide insights into the distribution of nucleotide diversity, recombination rate, decay of linkage disequilibrium, and genetic relationship patterns among accessions. Additionally, HapMaps can also underpin the connection of genotype with phenotype by means of genome-wide association studies (GWAS) (Huang and Han 2013). In the following subsections, we will summarize recent results from exome capture re-sequencing, GBS genotyping of diversity panels, and entire genebank collections.

11.4.2 Exome Sequencing to Assess Genetic Variation of Diversity Panels

Whole exome sequencing is a method of sequencing that utilizes oligonucleotide probes to selectively enrich sequencing libraries for fragments from genic regions (Hodges et al. 2007). As the design of an exome capture assay requires prior sequence information about the genomic sequence of target regions to design oligonucleotide baits, exome sequencing usually requires a reference genome sequence. Exome sequencing was first used in human genomics (Ng et al. 2009) and has since been applied to several crop species such as wheat (Jordan et al. 2015), maize (Yang et al. 2015), and switchgrass (Evans et al. 2014). An exome capture liquid array is also available for re-sequencing of the barley gene space (Mascher et al. 2013a). The barley exome capture probe set was designed mainly from the gene models predicted based on the whole genome sequencing of cv. Morex (International Barley Genome Sequencing Consortium 2012).

A proof-of-principle study with 13 barley cultivars and barley wild relatives revealed that the assay reliably captures ~60 Mb of mRNA-coding sequences in diverse germplasm (Mascher et al. 2013a). Subsequent to these pilot experiments, the exome capture platform was used to explore patterns of genetic diversity in a representative collection of 228 geo-referenced landraces and 91 accessions of *H. vulgare* ssp. *spontaneum* (Russell et al. 2016a). This study reported that the nucleotide diversity was, on average, reduced by 27% in domesticated barley relative to *H. vulgare* ssp. *spontaneum*. Some regions with reduced exonic diversity in cultivated barley were colocated with known domestication traits such as the *BRITTLE RACHIS* (*Btr1* and *Brt2*) genes on chromosome 3H (Pourkheirandish et al. 2015) and the *NUDUM* (*nud*) gene on chromosome 7H (Taketa et al. 2008). Moreover, wild barley harbored more private alleles and exhibited a more rapid decay of linkage disequilibrium than domesticated barley. Russell et al. (2016b) also tested for associations between genetic markers and environmental variables and found that geo-climatic variables associated with temperature and dryness were driving the adaptive responses as barley dispersed from the site of initial domestication in the Fertile Crescent across temperate regions of the world to more northern latitudes (Russell et al. 2016b). An in-depth analysis of haplotype structure in known flowering time genes revealed strong geographical structuring. The exome capture dataset of Russell et al. (2016b) will be an important community resource for barley, affording geo-referenced information on species-wide diversity linked for the majority of barley genes. The first example of how exome capture data can be used was provided by Mascher et al. (2016) who exploited it for understanding the relationship between extant landraces and 6000-year-old barley. During the excavation of caves in the Judean desert, Israeli archaeologists found well-preserved barley grains from the Chalcolithic period. The excellent preservation of the grains under the ambient conditions at the site made it possible to extract and sequence DNA fragments from them using state-of-the-art ancient DNA protocols (Knapp and Hofreiter 2010). The sequence analysis of ancient DNA sequences in the context of the extant diversity panel of present-day landraces from the Levant showed that the 6000-year-old barley sample was likely of the two-rowed type and carried domesticated alleles of the *BRITTLE RACHIS* genes. Moreover, the study of Mascher et al. (2016) provided further evidence for repeated gene flow between wild and domesticated barley in regions where they occur sympatrically.

11.4.3 Genotyping-by-Sequencing for Characterizing Entire Germplasm Collections

Genotyping entire genebank collections will generate comprehensive and coherent information on genetic diversity and population structure within the stored accessions and will facilitate the systematic valorization of genomics data. The long-term prospect is the evolution of genebanks from service centers managing the collection and maintenance of plant genetic resources into integrated biological and digital repositories for enhancing crop diversity and breeding. This transition will require the development of an innovative and intuitive diversity informatics and data warehouse infrastructure to accommodate and analyze big datasets (McCouch et al. 2013). A first step toward this aim is collection of genome-wide genotypic data for all accessions in a genebank collection. The per-sample costs and labor associated with the construction and sequencing of exome capture or whole genome shotgun libraries still prohibit the analysis of entire genomes and exomes of all accessions in a genebank collection. Thus, a technique for greater complexity reduction and easier parallel sample processing is required.

GBS employs digestion with restriction enzymes for complexity reduction prior to high-throughput sequencing (Elshire et al. 2011). In contrast to exome capture, GBS does not require prior knowledge of target sequences or the cost-intensive design and synthesis of bait sequences. The GBS protocol commonly used in large genome cereal species such as barley and wheat (Poland et al. 2012) uses two different restriction enzymes (*Pst*I and *Msp*I) for complexity reduction. The methylation-sensitive *Pst*I effects a selection for fragments from gene-rich, distal regions of the chromosomes (Wenzl et al. 2004). Coverage calculation in available GBS datasets (Mascher et al. 2013b) indicated that ~3 Mb of the barley genome associated with *Pst*I and *Msp*I restriction sites are reliably targeted by GBS, i.e., represented in ≥90% of samples with twofold read coverage. The main uses of GBS in agricultural genomics are genetic mapping (in biparental populations and GWAS panels) and genomic prediction in breeding programs (Poland and Rife 2012). GBS is a highly streamlined, multiplexed wet lab protocol that relies only on standard laboratory techniques and reagents (Elshire et al. 2011). These attributes have made it the method of choice for the cost-effective, sequenced-based characterization of very large germplasm collections encompassing thousands of individuals (Romay et al. 2013; Lu et al. 2015).

Researchers at IPK in Germany are currently using GBS to collect genotypic data for all ~22,000 accessions in the ex situ barley collection. Although GBS may be a more laborious procedure than large-scale SNP arrays, it can provide valuable data on genetic polymorphisms without ascertainment bias (Moragues et al. 2010). The first results on ~3000 winter barley accessions indicate that GBS can yield ~35000 SNP markers with genotype calls in at least 90% of accessions. A comparison of published datasets with both exome capture (Russell et al. 2016b) and GBS data (Pourkheirandish et al. 2015) for the same panel of accessions has shown a very high concordance between genotype calls (>99.5%) at SNP positions present in both sets. It is important to note in this context that some genebank accessions panel may harbor intra-accession diversity. Thus, it will be worthwhile to study the heterogeneity of accessions to derive a better estimate of the predicted total diversity in the collection. Nevertheless, the GBS data generated by IPK will provide a valuable estimate of the genetic diversity between accessions at a global scale. Moreover, preliminary analyses have revealed a substantial number of duplicates: at least 10% of sequenced individuals share marker profiles with one or more other individuals. How to handle potentially duplicated accessions will require careful decisions by genebank managers in the future.

In addition to underpinning population genetic studies, GBS affords in silico fingerprints that can be used as molecular passport data to (i) identify duplicated accessions that share identical genotypic data, potentially allowing for a significant savings in germplasm maintenance costs (see Sect. 5.2); (ii) establish core collections that provide the benefits of increasing the efficiency of evaluations, while at the same time reducing management and evaluation expenses (Diwan et al. 1995; Escribano et al. 2008); and (iii) highlight potential mix-ups where the origins of accessions reported in the genebank information system (Oppermann et al. 2015) are inconsistent with the patterns of genetic diversity. With respect to the latter point, it will often be difficult to trace back the sources of errors, e.g., seed exchange, incorrect recording of passport data, and mistakes during the GBS workflow.

The costs for library construction and deep sequencing associated with both exome capture and whole genome sequencing are currently prohibitive for genotyping thousands of accessions with these methods. The 9K SNP array (Comadran et al. 2012) suffers from ascertainment bias. How well the recently developed 50K SNP array (Bayer et al. 2017) performs for assessing genetic diversity in diverse germplasm including wild barley remains yet to be seen. Even though more powerful marker platform may become available in the future, the comparatively low-density GBS data can be used for the informed selection of a manageable number of accessions for core collections from the thousands contained in genebanks. These core collections can be used for high-coverage re-sequencing, allele mining in known genes and in-depth phenotypic characterization to underpin GWAS linking phenotype and genotype.

11.5 Conclusions and Outlook

The advent of high-throughput genomics technology has revolutionized the study of barley genetics. Completion of a reference genome for barley, (http://webblast.ipk-gatersleben.de/barley_ibsc/) together with genome-wide re-sequencing datasets for assessing the diversity of *Hordeum*, has provided important new knowledge on the demographic history of barley. Key findings include regions of the genome selected during domestication of *H. vulgare* ssp. *spontaneum*, adaptive evolution, and other aspects important to our understanding of the biology, ecology, and utilization of this important crop species. With respect to crop improvement, genome-wide re-sequencing has provided a powerful platform for relating many economically important traits to genotypes by GWAS. Such studies can provide an initial starting point for the cloning of both qualitative and quantitative acting genes (Comadran et al. 2012; Sato et al. 2016). The germplasm resources for *Hordeum* are extensive and held within many genebanks across the world. While several useful databases exist for barley online, there is a strong need for a fully integrated database that includes all collected data (phenotype, genotype, etc.) on the *Hordeum* genetic resources available worldwide. Given the limited resources available for germplasm maintenance, characterization, and utilization, as well as the problem of duplication in genebank collections, it is essential that the barley community come together to genetically characterize as many accessions as possible across genebanks and establish a new comprehensive core collection based on re-sequencing data that captures the maximum amount of genetic diversity while avoiding any redundancies. Toward this goal, the IPK is genotyping all of their 22,000 *H. vulgare* holdings by GBS. Once other major genebanks conduct similar characterizations, the comprehensive core collection can be assembled and increased in sufficient seed quantities for all to utilize in the barley community. Considering the limitations many research groups have for various phenotyping and molecular assays, the comprehensive core collection could be divided into manageable subsets, similar to that done for barley landraces by Pasam et al. (2014). Other non-duplicated *H. vulgare* germplasm not included in the comprehensive core collection will serve as an additional reservoir of allelic diversity mined via SNP diversity analysis or even FIGS. Exploiting members of the secondary and tertiary genepools of *Hordeum* remains a challenge (Dempewolf et al. 2017). However, for certain important classes of genes, i.e., disease resistance or R genes, some major advances could be made. Most of the plant R genes characterized to date encode proteins that contain Nucleotide-binding and Leucine-Rich repeat (NLR) motifs. Based on these attributes, Arora et al. (2018) developed a robust and rapid protocol for identifying and cloning R genes from a diversity panel of the wild wheat species *A. tauschii* ssp. *strangulata*. This protocol combines Association genetics with R gene enrichment Sequencing (AgRenSeq) to exploit pan-genome variation in germplasm panels and is especially applicable to crop wild relatives. With AgRenSeq, NLR gene complements could be efficiently isolated from various wild *Hordeum* species, providing many targets for R gene pyramiding in barley cultivars. This could be achieved through the transformation of multigene cassettes, which should extend the effectiveness of resistance against pathogens with high evolutionary potential. Additional advances in such technologies and the cost-effectiveness of re-sequencing throughput will lead to many exciting discoveries in barley genomics.

References

Ames N, Dreiseitl A, Steffenson BJ, Muehlbauer GJ (2015) Mining wild barley for powdery mildew resistance. Plant Pathol 64:1396–1406

Arora S, Steuernagel B, Chandramohan S, Long Y, Matny O, Johnson R, Enk J, Periyannan S, Md Hatta M, Athiyannan N, Cheema J, Yu G, Kangara N, Ghosh S, Szabo LJ, Poland J, Bariana H, Jones JDG, Bentley AR, Ayliffe M, Olson E, Xu SS, Steffenson BJ, Lagudah E, Wulff BBH (2018) Resistance gene discovery and cloning by sequence capture and association genetics. bioRxiv. https://doi.org/10.1101/248146

Bayer MM, Rapazote-Flores P, Ganal M, Hedley PE, Macaulay M, Plieske J, Ramsay L, Russell J, Shaw PD, Thomas W, Waugh R (2017) Development and evaluation of a Barley 50k iSelect SNP Array. Front Plant Sci. https://doi.org/10.3389/fpls.2017.01792

Berger GL, Liu S, Hall MD, Brooks WS, Chao S, Muehlbauer GJ et al (2013) Marker-trait associations in Virginia Tech winter barley identified using genome-wide mapping. Theor Appl Genet 126:693–710. Available at: http://www.ncbi.nlm.nih.gov/pubmed/23139143

Comadran J, Kilian Benjamin, Russell Joanne, Ramsay Luke, Stein Nils, Ganal Martin, Shaw Paul et al (2012) Natural variation in a homolog of Antirrhinum CENTRORADIALIS contributed to spring growth habit and environmental adaptation in cultivated barley. Nat Genet 44(12):1388–1392

Cuesta-Marcos A, Szucs P, Close TJ, Filichkin T, Muehlbauer GJ, Smith KP et al (2010) Genome-wide SNPs and re-sequencing of growth habit and inflorescence genes in barley: implications for association mapping in germplasm arrays varying in size and structure. BMC Genom 11:707

Davey JW, Hohenlohe PA, Etter PD, Boone JQ, Catchen JM, Blaxter ML (2011) Genome-wide genetic marker discovery and genotyping using next-generation sequencing. Nat Rev Genet 12 (7):499–510. https://doi.org/10.1038/nrg3012

Dawson IK, Russell J, Powell W, Steffenson B, Thomas WTB, Waugh R (2015) Barley: a translational model for adaptation to climate change. New Phytol 206:913–931

Dempewolf H, Baute G, Anderson J, Kilian B, Smith C, Guarino L (2017) Past and future use of wild relatives in crop breeding. Crop Sci 57:1070–1082

Diwan N, McIntosh MS, Bauchan GR (1995) Methods of developing a core collection of annual *Medicago* species. Theor Appl Genet 90:755–761

Elshire RJ, Glaubitz JC, Sun Q, Poland JA, Kawamoto K, Buckler ES, Mitchell SE (2011) A robust, simple genotyping-by-sequencing (GBS) approach for high diversity species. PLoS ONE 6(5):e19379. https://doi.org/10.1371/journal.pone.0019379

Endresen DTF, Street K, Mackay M, Bari A, de Pauw E (2011) Predictive association between biotic stress traits and eco-geographic data for wheat and barley landraces. Crop Sci 51:2036–2055

Escribano P, Viruel MA, Hormaza JI (2008) Comparison of different methods to construct a core germplasm collection in woody perennial species with simple sequence repeat markers. A case study in cherimoya (*Annona cherimola*, Annonaceae), an underutilized subtropical fruit tree species. Ann Appl Biol 153:25–32

Evans J, Kim J, Childs KL, Vaillancourt B, Crisovan E, Nandety A, Gerhardt DJ, Richmond TA, Jeddeloh JA, Kaeppler SM, Casler MD, Buell CR (2014) Nucleotide polymorphism and copy number variant detection using exome capture and next-generation sequencing in the polyploid grass *Panicum virgatum*. Plant J Cell Mol Biol 79(6):993–1008. https://doi.org/10.1111/tpj.12601

FAO (2010) The second report on the state of the world's plant genetic resources for food and agriculture. Commission on Genetic Resources for Food and Agriculture, Food and Agriculture Organization of the United Nations, Rome, 399 p. http://www.fao.org/docrep/013/i1500e/i1500e00.htm

Fu YB, Horbach C (2012) Genetic diversity in a core subset of wild barley germplasm. Diversity 4:239–257

Gore MA, Chia JM, Elshire RJ, Sun Q, Ersoz ES, Hurwitz BL, Peiffer JA, McMullen MD, Grills GS, Ross-Ibarra J, Ware DH, Buckler ES (2009) A first-generation haplotype map of maize. Science 326 (5956):1115–1117. https://doi.org/10.1126/science.1177837

Gouesnard B (2001) MSTRAT: an algorithm for building germplasm core collections by maximizing allelic or phenotypic richness. J Hered 92:93–94

Gutiérrez L, Cuesta-Marcos A, Castro AJ, von Zitzewitz J, Schmitt M, Hayes PM (2011) Association mapping of malting quality quantitative trait loci in winter barley: positive signals from small germplasm arrays. Plant Genome J 4:256

Hamblin MT, Close TJ, Bhat PR, Chao S, Kling JG, Abraham KJ et al (2010) Population structure and linkage disequilibrium in U.S. barley germplasm: implications for association mapping. Crop Sci 50:556–566

Hodges E, Xuan Z, Balija V, Kramer M, Molla MN, Smith SW, Middle CM, Rodesch MJ, Albert TJ, Hannon GJ, McCombie WR (2007) Genome-wide in situ exon capture for selective resequencing. Nat Genet 39(12):1522–1527. https://doi.org/10.1038/ng.2007.42

Huang X, Han B (2013) Natural variations and genome-wide association studies in crop plants. Annu Rev Plant Biol. https://doi.org/10.1146/annurev-arplant-050213-035715

Huang X, Wei X, Sang T, Zhao Q, Feng Q, Zhao Y, Li C, Zhu C, Lu T, Zhang Z, Li M, Fan D, Guo Y, Wang A, Wang L, Deng L, Li W, Lu Y, Weng Q, Liu K, Huang T, Zhou T, Jing Y, Li W, Lin Z, Buckler ES, Qian Q, Zhang QF, Li J, Han B (2010) Genome-wide association studies of 14 agronomic traits in rice landraces. Nat Genet 42(11):961–967. https://doi.org/10.1038/ng.695

International Barley Genome Sequencing Consortium (2012) A physical, genetic and functional sequence assembly of the barley genome. Nature 491 (7426):711–716. https://doi.org/10.1038/nature11543

Jakob SS, Rödder D, Engler JO, Shaaf S, Ozkan H, Blattner FR et al (2014) Evolutionary history of wild barley (*Hordeum vulgare* subsp. *spontaneum*) analyzed using multilocus sequence data and paleodistribution modeling. Genome Biol Evol 6:685–702

Jordan KW, Wang S, Lun Y, Gardiner LJ, MacLachlan R, Hucl P, Wiebe K, Wong D, Forrest KL, Consortium I, Sharpe AG, Sidebottom CH, Hall N, Toomajian C, Close T, Dubcovsky J, Akhunova A, Talbert L, Bansal UK, Bariana HS, Hayden MJ, Pozniak C, Jeddeloh JA, Hall A, Akhunov E (2015) A haplotype map of allohexaploid wheat reveals distinct patterns of selection on homoeoloe genomes. Genome Biol 16:48. https://doi.org/10.1186/s13059-015-0606-4

Knapp M, Hofreiter M (2010) Next generation sequencing of ancient DNA: requirements, strategies and perspectives. Genes (Basel) 1(2):227–243. https://doi.org/10.3390/genes1020227

Knüpffer H (2009) Triticeae genetic resources in ex situ Genebank Collections. In: Feuillet C, Muehlbauer GJ (eds) Genetics and genomics of the triticeae. Springer, New York, pp 31–79

Knüpffer H, van Hintum THJL (2003) Summarised diversity—the barley core collection. In: von Bothmer R, van Hintum TH, Knüpffer H, Sato K (eds) Diversity in barley (*Hordeum vulgare*). Elsevier Science B.V., Amsterdam, The Netherlands, pp 250–258

Lu F, Romay MC, Glaubitz JC, Bradbury PJ, Elshire RJ, Wang T, Li Y, Li Y, Semagn K, Zhang X, Hernandez AG, Mikel MA, Soifer I, Barad O, Buckler ES (2015) High-resolution genetic mapping of maize pan-genome sequence anchors. Nat Commun 6:6914. https://doi.org/10.1038/ncomms7914

Mackay MC, Street KS (2004) Focused identification of germplasm strategy—FIGS. In: Rebetzke GJ, Black CK, Panozzo JK (eds) Proceedings of the 54th Australian cereal chemistry conference and the 11th wheat breeders' assembly. Royal Australian Chemical Institute, Melbourne, Australia, pp 138–141

Mascher M, Richmond TA, Gerhardt DJ, Himmelbach A, Clissold L, Sampath D, Ayling S, Steuernagel B, Pfeifer M, D'Ascenzo M, Akhunov ED, Hedley PE, Gonzales AM, Morrell PL, Kilian B, Blattner FR, Scholz U, Mayer KF, Flavell AJ, Muehlbauer GJ, Waugh R, Jeddeloh JA, Stein N (2013a) Barley whole exome capture: a tool for genomic research in the genus Hordeum and beyond. Plant J Cell Mol Biol 76 (3):494–505. https://doi.org/10.1111/tpj.12294

Mascher M, Wu S, Amand PS, Stein N, Poland J (2013b) Application of genotyping-by-sequencing on semiconductor sequencing platforms: a comparison of genetic and reference-based marker ordering in barley. PLoS ONE 8(10):e76925. https://doi.org/10.1371/journal.pone.0076925

Mascher M, Schuenemann VJ, Davidovich U, Marom N, Himmelbach A, Hubner S, Korol A, David M, Reiter E, Riehl S, Schreiber M, Vohr SH, Green RE, Dawson IK, Russell J, Kilian B, Muehlbauer GJ, Waugh R, Fahima T, Krause J, Weiss E, Stein N (2016) Genomic analysis of 6,000-year-old cultivated grain illuminates the domestication history of barley. Nat Genet 48(9):1089–1093. https://doi.org/10.1038/ng.3611

Massman J, Cooper B, Horsley R, Neate S, Dill-Macky R, Chao S et al (2011) Genome-wide association mapping of Fusarium head blight resistance in contemporary barley breeding germplasm. Mol Breed 27:439–454

Maurer A, Draba V, Jiang Y, Schnaithmann F, Sharma R, Schumann E, Kilian B, Reif JC, Pillen K (2015) Modelling the genetic architecture of flowering time control in barley through nested association mapping. BMC Genomic 16:290

McCouch S, Baute GJ, Bradeen J, Bramel P, Bretting PK, Buckler E, Burke JM, Charest D, Cloutier S, Cole G, Dempewolf H, Dingkuhn M, Feuillet C, Gepts P, Grattapaglia D, Guarino L, Jackson S, Knapp S, Langridge P, Lawton-Rauh A, Lijua Q, Lusty C, Michael T, Myles S, Naito K, Nelson RL, Pontarollo R, Richards CM, Rieseberg L, Ross-Ibarra J, Rounsley S, Hamilton RS, Schurr U, Stein N, Tomooka N, van der Knaap E, van Tassel D, Toll J, Valls J, Varshney RK, Ward J, Waugh R, Wenzl P, Zamir D (2013) Agriculture: feeding the future. Nature 499(7456):23–24. https://doi.org/10.1038/499023a

Middleton CP, Senerchia N, Stein N, Akhunov ED, Keller B, Wicker T et al (2014) Sequencing of chloroplast genomes from wheat, barley, rye and their relatives provides a detailed insight into the evolution of the Triticeae tribe. PLoS ONE 9:e85761

Montenegro JD, Golicz AA, Bayer PE, Hurgobin B, Lee H, Chan CKK, Visendi P, Lai K, Doležel J, Batley J, Edwards D (2017) The pangenome of hexaploid bread wheat. Plant J 90:1007–1013. https://doi.org/10.1111/tpj.13515

Moragues M, Comadran J, Waugh R, Milne I, Flavell AJ, Russell JR (2010) Effects of ascertainment bias and marker number on estimations of barley diversity from high throughput SNP genotype data Theor. Appl. Genet. 120:1525–1534

Morris GP, Ramu P, Deshpande SP, Hash CT, Shah T, Upadhyaya HD, Riera-Lizarazu O, Brown PJ, Acharya CB, Mitchell SE, Harriman J, Glaubitz JC, Buckler ES, Kresovich S (2013) Population genomic and genome-wide association studies of agroclimatic traits in sorghum. Proc Natl Acad Sci USA 110 (2):453–458. https://doi.org/10.1073/pnas.1215985110

Muñoz-Amatriaín M, Cuesta-Marcos A, Endelman JB, Comadran J, Bonman JM, Bockelman HE et al (2014) The USDA barley core collection: genetic diversity, population structure, and potential for genome-wide association studies. PLoS ONE 9:e94688

Newton AC, Flavell AJ, George TS, Leat P, Mullholland B, Ramsay L et al (2011) Crops that feed the world 4. Barley: a resilient crop? Strengths and weaknesses in the context of food security. Food Secur 3:141–178. Available at: http://link.springer.com/10.1007/s12571-011-0126-3. Accessed 23 Mar 2014

Ng SB, Turner EH, Robertson PD, Flygare SD, Bigham AW, Lee C, Shaffer T, Wong M, Bhattacharjee A, Eichler EE, Bamshad M, Nickerson DA, Shendure J (2009) Targeted capture and massively parallel sequencing of 12 human exomes. Nature 461 (7261):272–276. https://doi.org/10.1038/nature08250

Nice LM, Steffenson BJ, Brown-Guedira GL, Akhunov ED, Liu C, Kono TJY et al (2016) Development and genetic characterization of an advanced backcross-nested association mapping (AB-NAM) population of wild x cultivated barley. Genetics 203:1453–1467

Nielsen R, Paul JS, Albrechtsen A, Song YS (2011) Genotype and SNP calling from next-generation sequencing data. Nat Rev Genet 12(6):443–451. https://doi.org/10.1038/nrg2986

Oppermann M, Weise S, Dittmann C, Knupffer H (2015) GBIS: the information system of the German Genebank. Database (Oxford) 2015:bav021. https://doi.org/10.1093/database/bav021

Pasam RK, Sharma R, Malosetti M, van Eeuwijk FA, Haseneyer G et al (2012) Genome-wide association studies for agronomical traits in a world-wide spring barley collection. BMC Plant Biol 12:16

Pasam RK, Sharma R, Walther A, Özkan H, Graner A, Kilian B (2014) Genetic diversity and population structure in a legacy collection of spring barley landraces adapted to a wide range of climates. PLoS ONE 9:e116164

Pauli D, Muehlbauer G, Smith K, Cooper B (2014) Association mapping of agronomic QTLs in US Spring barley breeding germplasm. Plant Genome 7. https://doi.org/10.3835/plantgenome2013.11.0037

Pillen K, Zacharias A, Leon J (2004) Comparative AB-QTL analysis in barley using a single exotic donor of *Hordeum vulgare* ssp. *spontaneum*. Theor Appl Genet 108:1591–1601

Poland JA, Rife TW (2012) Genotyping-by-sequencing for plant breeding and genetics. Plant Genome 5 (3):92–102. https://doi.org/10.3835/plantgenome2012.05.0005

Poland JA, Brown PJ, Sorrells ME, Jannink J-L (2012) Development of high-density genetic maps for barley and wheat using a novel two-enzyme genotyping-by-sequencing approach. PLoS ONE 7 (2):e32253

Pourkheirandish M, Hensel G, Kilian B, Senthil N, Chen G, Sameri M, Azhaguvel P, Sakuma S, Dhanagond S, Sharma R, Mascher M, Himmelbach A, Gottwald S, Nair SK, Tagiri A, Yukuhiro F, Nagamura Y, Kanamori H, Matsumoto T, Willcox G, Middleton CP, Wicker T, Walther A, Waugh R, Fincher GB, Stein N, Kumlehn J, Sato K, Komatsuda T (2015) Evolution of the grain dispersal system in barley. Cell 162(3):527–539. https://doi.org/10.1016/j.cell.2015.07.002

Romay MC, Millard MJ, Glaubitz JC, Peiffer JA, Swarts KL, Casstevens TM, Elshire RJ, Acharya CB, Mitchell SE, Flint-Garcia SA, McMullen MD, Holland JB, Buckler ES, Gardner CA (2013) Comprehensive genotyping of the USA national maize inbred seed bank. Genome Biol 14(6):R55. https://doi.org/10.1186/gb-2013-14-6-r55

Rostoks N, Ramsay L, MacKenzie K, Cardle L, Bhat PR, Roose ML et al (2006) Recent history of artificial outcrossing facilitates whole-genome association mapping in elite inbred crop varieties. Proc Natl Acad Sci USA 103:18656–18661

Roy JK, Smith KP, Muehlbauer GJ, Chao S, Close TJ, Steffenson BJ (2010) Association mapping of spot blotch resistance in wild barley. Mol Breed 26:243–256

Russell J, Mascher M, Dawson IK, Kyriakidis S, Calixto C, Freund F, Bayer M, Milne I, Marshall-Griffiths T, Heinen S (2016a) Exome sequencing of geographically diverse barley landraces and wild relatives gives insights into environmental adaptation. Nat Genet 48:1024–1030

Russell J, Mascher M, Dawson IK, Kyriakidis S, Calixto C, Freund F, Bayer M, Milne I, Marshall-Griffiths T, Heinen S, Hofstad A, Sharma R, Himmelbach A, Knauft M, van Zonneveld M, Brown JWS, Schmid KJ, Kilian B, Muehlbauer G, Stein N, Waugh R (2016b) Adaptation of barley to different environments revealed in the exomes of a range-wide collection of landraces and wild relatives. Nat Genetics

Saade S, Maurer A, Shahid M, Oakey H, Schmöckel SM, Negrão S et al (2016) Yield-related salinity tolerance traits identified in a nested association mapping (NAM) population of wild barley. Sci Rep 6:1–9

Sallam AH, Tyagi P, Gina Brown-Guedira G, Muehlbauer GJ, Hulse A, Steffenson BJ (2017) Genome-wide association mapping of stem rust resistance in *Hordeum vulgare* subsp. *spontaneum*. G3 Genes Genomes Genet 7:3491–3507

Sato K, Takeda K (2009) An application of high-throughput SNP genotyping for barley genome mapping and characterization of recombinant chromosome substitution lines. Theor Appl Genet 119:613–619

Sato K, Yamane M, Yamaji N, Kanamori H, Tagiri A, Schwerdt JG et al (2016) Alanine aminotransferase controls seed dormancy in barley. Nat Commun 7:11625

Schmalenbach I, Korber N, Pillen K (2008) Selecting a set of wild barley introgression lines and verification of QTL effects for resistance to powdery mildew and leaf rust. Theor Appl Genet 117:1093–1106

Schnaithmann F, Kopahnke D, Pillen K (2014) A first step toward the development of a barley NAM population and its utilization to detect qtls conferring leaf rust seedling resistance. Theor Appl Genet 127:1513–1525

Steffenson BJ, Olivera P, Roy JK, Jin Y, Smith KP, Muehlbauer GJ (2007) A walk on the wild side:

mining wild wheat and barley collections for rust resistance genes. Aust J Agric Res 58:532–534

Taketa S, Amano S, Tsujino Y, Sato T, Saisho D, Kakeda K, Nomura M, Suzuki T, Matsumoto T, Sato K, Kanamori H, Kawasaki S, Takeda K (2008) Barley grain with adhering hulls is controlled by an ERF family transcription factor gene regulating a lipid biosynthesis pathway. Proc Natl Acad Sci USA 105 (10):4062–4067. https://doi.org/10.1073/pnas.0711034105

Tanksley SD, McCouch SR (1997) Seed banks and molecular maps: unlocking genetic potential from the wild. Science 277:1063–1066

Tondelli A, Xu X, Moragues M et al (2013) Structural and temporal variation in genetic diversity of European spring two-row barley cultivars and association mapping of quantitative traits. Plant Genome 6:0. https://doi.org/10.3835/plantgenome2013.03.0007

Ullrich SE (2011) Significance, adaptation, production and trade of barley. In: Steven E. Ullrich (ed) Barley: production, improvement, and uses. Wiley-Blackwell, Ames, IA, pp 3–13

von Bothmer R, Sato K, Komatsuda T, Yasuda S, Fischbeck G (2003) The domestication of cultivated barley. In: von Bothmer R, van Hintum Th, Knüpffer H, Sato K (eds) Diversity in Barley (*Hordeum vulgare*). Elsevier Science B.V., Amsterdam, The Netherlands, pp 9–27

von Korff M, Wang H, Leon J, Pillen K (2004) Development of candidate introgression lines using an exotic barley accession (*Hordeum vulgare* ssp. *spontaneum*) as donor. Theor Appl Genet 109:1736–1745

Wendler N, Mascher M, Himmelbach A, Johnston P, Pickering R, Stein N (2015) Bulbosum to go: a toolbox to utilize *Hordeum vulgare*/bulbosum introgressions for breeding and beyond. Mol Plant 8:1507–1519

Wenzl P, Carling J, Kudrna D, Jaccoud D, Huttner E, Kleinhofs A, Kilian A (2004) Diversity arrays technology (DArT) for whole-genome profiling of barley. Proc Natl Acad Sci USA 101(26):9915–9920. https://doi.org/10.1073/pnas.0401076101

Wu D, Sato K, Ma JF (2015) Genome-wide association mapping of cadmium accumulation in different organs of barley. New Phytol 208:817–829. https://doi.org/10.1111/nph.13512

Yang J, Jiang H, Yeh CT, Yu J, Jeddeloh JA, Nettleton D, Schnable PS (2015) Extreme-phenotype genome-wide association study (XP-GWAS): a method for identifying trait-associated variants by sequencing pools of individuals selected from a diversity panel. The Plant journal: for cell and molecular biology 84(3):587–596. https://doi.org/10.1111/tpj.13029

Yu J, Holland JB, McMullen MD, Buckler ES (2008) Genetic design and statistical power of nested association mapping in maize. Genetics 178:539–551. Available at: http://www.pubmedcentral.nih.gov/articlerender.fcgi?artid=2206100&tool=pmcentrez&rendertype=abstract. Accessed 10 May 2015

Zhou H, Steffenson B (2013a) Genome-wide association mapping reveals genetic architecture of durable spot blotch resistance in US barley breeding germplasm. Mol Breed 32:139–154

Zhou H, Steffenson BJ (2013b) Association mapping of Septoria speckled leaf blotch resistance in U.S. barley breeding germplasm. Phytopathology 103:600–9. Available at: http://www.ncbi.nlm.nih.gov/pubmed/23342987

Zhou H, Muehlbauer G, Steffenson B (2012) Population structure and linkage disequilibrium in elite barley breeding germplasm from the United States. J Zhejiang Univ Sci B 13:438–451. Available at: http://www.springerlink.com/index/10.1631/jzus.B1200003. Accessed 23 Mar 2014

Zhou H, Steffenson BJ, Muehlbauer G, Wanyera R, Njau P, Ndeda S (2014) Association mapping of stem rust race TTKSK resistance in US barley breeding germplasm. Theor Appl Genet 127:1293–1304

Zohary D, Hoph M, Weiss E (2012) Domestication of plants in the old world: the origin and spread of domesticated plants in Southwest Asia, Europe, and the Mediterranean Basin, 4th edn. Oxford University Press Inc., New York

12 Barley Inflorescence Architecture

Sarah M. McKim, Ravi Koppolu and Thorsten Schnurbusch

Abstract

Cultivated barley, *Hordeum vulgare* ssp. *vulgare*, is the fourth most abundantly grown cereal in the world (www.fao.org/faostat) and is long associated with human civilisations. Although most barley grain grown today is destined for animal feed and malting, barley remains an important source of primary calories in many parts of the world. Increasing barley yield in the face of challenges posed by increasing world population and climate change is a major goal of current research efforts. Grain is the ultimate product of inflorescence development and maturation. As such, understanding the genetics underlying inflorescence architecture in barley and then learning how to apply this knowledge to manipulate inflorescence development are important steps towards improving yield. The barley reference genome sequence represents an invaluable resource to support the identification and functional characterization of genes controlling inflorescence architecture. Resolving the relationships between gene and inflorescence traits are critical to support breeding as well as to provide insight about fundamental questions in cereal developmental biology. In this chapter, we first provide an overview of inflorescence development in cereals, highlighting the transitions in meristem identity associated with species-specific architectures. From here, we describe the development of key morphological features associated with the barley spike, spikelet, floret and grain, while discussing the identification and functions of genes which regulate their development. We also discuss those genes whose variation contributed to architectural changes during domestication and those with yield potential. Lastly, we describe environmental control of inflorescence development, with special attention to flowering time and the agronomic environment.

S. M. McKim (✉)
University of Dundee at the James Hutton Institute, Errol Road, Invergowrie, DD2 5DA Dundee, Scotland, UK
e-mail: s.mckim@dundee.ac.uk

R. Koppolu · T. Schnurbusch
HEISENBERG-Research Group Plant Architecture, Leibniz Institute of Plant Genetics and Crop Plant Research (IPK), Corrensstr. 3, OT Gatersleben, 06466 Seeland, Germany

T. Schnurbusch
Faculty of Natural Sciences III, Institute of Agricultural and Nutritional Sciences, Martin Luther University Halle-Wittenberg, Von-Seckendorff-Platz 3, Halle, 06108, Germany

12.1 Background

Over the course of evolution, flowering plants rapidly radiated into new species with dramatically diverse flowering forms which now

N. Stein and G. J. Muehlbauer (eds.), *The Barley Genome*, Compendium of Plant Genomes,
https://doi.org/10.1007/978-3-319-92528-8_12

dominate almost every terrestrial niche (Scotland and Wortley 2003; Endress 2011; Magallón et al. 2015). While beautifying our world, this variation in form also laid fertile ground for early humans to domesticate grain crops—events that founded early agrarian societies and further civilizations (Harlan and Zohary 1966). Now, cultivated cereals, including wheat, rice, barley, oat and maize, provide the majority of calories to the human diet.

12.1.1 Inflorescences in Cereals

Cereal grain yield is inevitably linked to flowering and the formation of **spikelets**, the basic reproductive units of grasses (Malcomber et al. 2006). A single spikelet can give rise to one or more flowers or **florets** whose fertilization leads to **grain**. The total arrangement of spikelets and florets on the shoot axis is called an **inflorescence**.

Inflorescences provide a protective substrate for flowering and pollination, deliver photosynthate to fuel growth and grain set, and facilitate grain dispersal. While grain crops are morphologically similar during vegetative growth, their inflorescences show remarkable species-specific diversity, depending in part on whether the **inflorescence meristem** at the shoot tip makes **spikelet meristems** directly or **branches** into further inflorescence axes first (Fig. 12.1; Bommert et al. 2005; Doust 2007). For instance, rice and oat inflorescences develop spikelets along highly branched and elongated structures known as panicles, while in barley, rye and wheat rows of spikelets line a single, jointed vertical axis or rachis, an inflorescence type called a spike (Fig. 12.2a). The nature of the spikelet meristems also influences the number of **floret meristems** developed from each spikelet: the wheat spikelet, for example, produces an indeterminate number of florets, while the determinate barley spikelet forms a single floret (Fig. 12.1). Thus, a species' inflorescence architecture reflects both the shifting identity and activity of meristems from inflorescence to spikelet to floret and their indeterminate versus determinate nature. The idea of shifting meristem identities appears most favorable in explaining the observed diversity of grass inflorescences (McSteen et al. 2000; Laudencia-Chingcuanco and Hake 2002), including their extensive branching patterns (Whipple 2017). Another concept, however, suggesting a "transient model" for meristem determinacy in *Arabidopsis thaliana* and other species, mainly based on the antagonistic actions of only two proteins (Prusinkiewicz et al. 2007), may be too simplistic to explain complex grass inflorescences.

Each floret contains outer, leaf-like organs that encircle inner sexual organs. Following fertilization, a small embryo develops surrounded by endosperm and pericarp, which collectively make up the grain. In cereals, the grain kernel, or caryopsis, is covered by an additional pericarp layer and may be further enclosed by a hull. Grain endosperm is the major energy source for the germinating seedling, but early humans, like other animals, discovered that endosperm is also a dense packet of easily stored and consumed calories. However, harvesting and processing grain was challenging due to architectural features of wild cereals, especially the fall of grain from the mother plant (shatter), as well as long, lodging stems and tough, protective organs coating the grain. Thus, our ancestors cultivated inflorescence variants which minimized these traits and enhanced others to improve yield, selecting for non-shattering yet easily harvested grain, increased floret number and grain size, modified floral branching patterns, and increased compaction (Sakuma et al. 2011). Over a relatively short time, these selection pressures transformed grain into our major food supply. Moreover, as cereal cultivation spread globally, farmers also modified flowering time sensitivity to photoperiod and temperature to grow crops over wider climates and latitudes (Cockram et al. 2007).

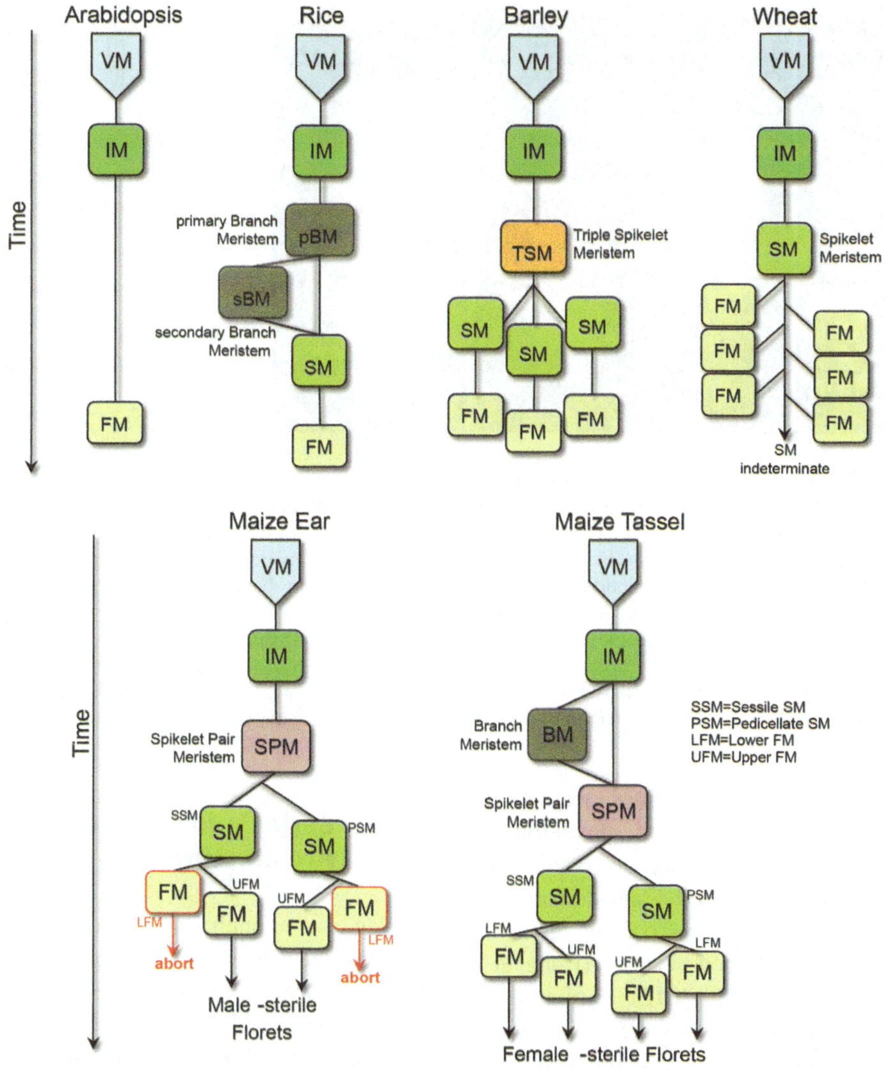

Fig. 12.1 Shifting meristem identities in crops. Meristem differentiation in Arabidopsis: Vegetative Meristem (VM) transforms into an inflorescence meristem (IM) and IM differentiates into Floret Meristems (FMs). Meristem differentiation in rice: VM transforms into IM, which further differentiates primary Branch Meristems (pBMs) and pBM differentiates into secondary Branch Meristems (sBMs). Spikelet Meristems (SMs) are produced on both pBMs and sBMs. A single floret is produced in each SM. Meristem differentiation in barley: VM transforms into an IM and IM produces barley-specific Triple Spikelet Meristems (TSMs). TSM differentiates into three SMs (one central and two lateral). Each SM produces one FM on a determinate rachilla meristem (determinate SM). Meristem differentiation in wheat: VM transforms into an IM and IM directly produces SMs. Each wheat SM produces up to 10 FMs on an indeterminate rachilla meristem (Indeterminate SM). Meristem differentiation in the maize ear: VM transforms into an IM and IM directly produces one SPM, which further differentiates into two SMs, Sessile Spikelet Meristem (SSM) and Pedicellate Spikelet Meristem (PSM). Each SM produces two FMs, the Lower FM, (LFM) and Upper FM (UFM), and each LFM aborts, resulting in two male-sterile UFs. Meristem differentiation in the maize tassel: VM transforms into an IM and IM directly produces one SPM, which further differentiates into two SMs (SSM and PSM), while each SM produces two FMs (LFM and UFM) all of which produce female-sterile florets. VM: Vegetative Meristem; IM: Inflorescence Meristem; PBM: Primary Branch Meristem; SBM: Secondary Branch Meristem; TSM: Triple Spikelet Meristem; SPM: Spikelet Pair Meristem; SM: Spikelet Meristem; FM: Floret Meristem; SSM: Sessile Spikelet Meristem; PSM: Pedicellate Spikelet Meristem; UFM: Upper Floret Meristem; LFM: Lower Floret Meristem

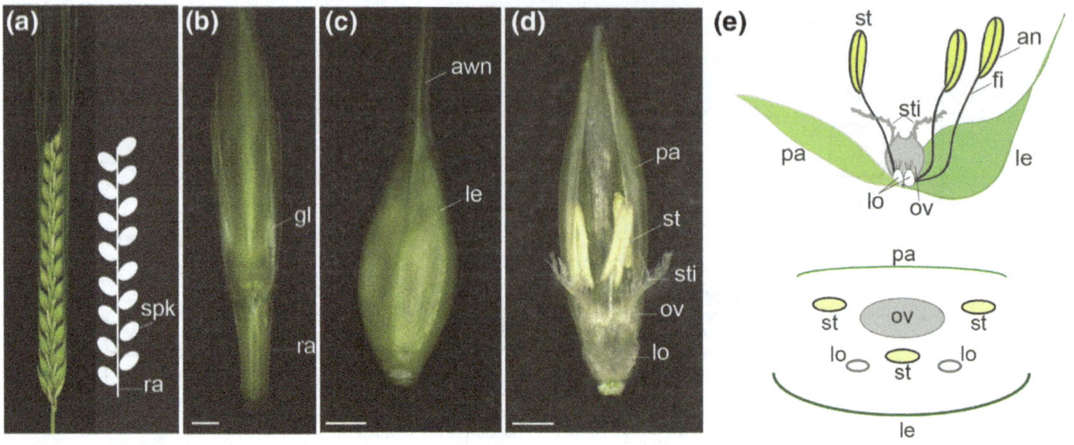

Fig. 12.2 Morphology of the barley spike, spikelet and floret. **a** Two-rowed barley spike, cultivar (*cv.*) Bowman. Diagram on right shows alternating spikelet nodes along the unbranched axis or rachis. **b** Barley reproductive phytomer. Rachis internode with a single spikelet flanked by a pair of subtending glumes. **c** Abaxial face of floret with lemma and awn. **d** Floret with lemma removed to show internal floret organs: lodicules, stamens, pistil (single carpelovary and stigma). Adaxial palea remains attached. **e** Floret diagrams showing external and internal organs. A longitudinal profile (adaxial, left and abaxial, right) shown in top diagram and a bird's eye view in bottom diagram. The lemma and palea organs encircle the inner floret organs. Two abaxial lodicules form between the lemma and ovary. Three stamens develop at the base of the ovary, each comprising a long filament tipped by an anther. The ovary is tipped with stigma. Scale bars (b, c, d) 1 mm. spk: spikelet; ra: rachis; gl: glume; le: lemma; lo: lodicules; st: stamen; ov: ovary; sti: stigma; pa: palea; fi: filament; an: anther

Modern breeding continues to modify inflorescence architecture and developmental timing in cereals to improve yield. Nonetheless, projected rates of increased grain demand coupled with climate change present enormous challenges to global food security (Kang et al. 2009; Lobell and Gourdji 2012). Exploiting and generating novel changes in cereal reproductive architecture is a key avenue to improve yield, yet we still know relatively little about the underlying mechanisms regulating variation in reproductive architecture, both within and between species (Sreenivasulu and Schnurbusch 2012; Schrager-Lavelle et al. 2017). The explosion of plant genomic resources, including those for traditionally recalcitrant cereals such as barley, holds exciting potential (Ford-Lloyd et al. 2014). Yet to fulfill this promise, we must unravel the genetic control underlying reproductive architecture and its variation in cereals. Barley, as a diploid plant and dominant global crop with plentiful mutant collections and the widest climatic range of cultivated cereals is an ideal organism on which to focus this effort.

12.2 Barley's Inflorescence Architecture

Barley is uniquely positioned to reveal the genetic underpinnings of inflorescence architecture in temperate cereals. Here, we first describe the morphological events and features associated with spike development in barley and then discuss the species-specific and shared functions of genes known to control barley inflorescence architecture, noting those genes whose variation contributed to architectural changes during domestication and those with yield potential (Table 12.1). Throughout, we mention the genomic tools and approaches used to identify these genes and understand their function (please see Appendix for Genetic and Genomic Resources for Gene Cloning and Functional Analysis), and highlight certain inflorescence architecture loci yet to be cloned. Lastly, we describe environmental control of inflorescence development, with special attention to flowering time and the agronomic environment.

Table 12.1 Loci associated with inflorescence architecture in barley

Locus/loci	Inheritance	Gene name	Chromosome	Function	Approach	References
Eam5	Dominant/Semi-dominant	*HvPHYC*	5HL	Flowering time	Exome capture/Mapping-by-sequencing	Pankin et al. (2014)
Eam6/Eps2/Mat-c	Dominant	*HvCEN*	2H	Flowering time	Genome-Wide Association Study	Comadran et al. (2012)
eam8	Recessive	*HvELF3*	1HL	Flowering time	Map-based cloning	Faure et al. (2012), Zakhrabekova et al. (2012)
eam10	Recessive	*HvLUX1*	3HL	Flowering time	Mapping and mutant analysis	Campoli et al. (2013), Gawroński et al. (2014)
Ppd–H1	Recessive	*HvPRR*	2HS	Flowering time	Map-based cloning	Turner et al. (2005)
Ppd-H2	–	*HvFT3*	1HL	Flowering time	Candidate gene approach	Faure et al. (2007), Kikuchi et al. (2007)
		HvFT2	3HS	Flowering time		
VRN–H1		*HvFUL1*	5HL	Growth habit	Homology based on wheat	Yan et al. (2003)
VRN–H2		*HvZCCT*	5HL	Growth habit	Homology based on wheat	Yan et al. (2004)
VRN–H3		*HvFT1*	7HS	Growth habit	Homology based on wheat	Yan et al. (2006)
vrs1	Recessive	*HvHOX1*	2HL	Lateral spikelet development	Map-based cloning	Komatsuda et al. (2007)
vrs2	Recessive	*HvSHI*	5HL	Lateral spikelet development	Map-based cloning	Youssef et al. (2017a)
vrs3	Recessive	*HvJMJC*	1H	Lateral spikelet development	Mapping and mutant analysis	Bull et al. (2017), van Esse et al. (2017)
vrs4	Recessive	*HvRA2*	3HS	Lateral spikelet development/spike-branching/floret determinacy	Mapping and mutant analysis	Koppolu et al. (2013)

(continued)

Table 12.1 (continued)

Locus/loci	Inheritance	Gene name	Chromosome	Function	Approach	References
int–c	Recessive	*HvTB1*	4HS	Lateral spikelet development	Genome-Wide Association Study	Ramsay et al. (2011)
lab	Recessive	–	2H/6HL	Spikelet development	–	Youssef et al. (2014)
deficiens	Recessive	–	2HL	Lateral spikelet development	–	Sakuma et al. (2017)
sls1	Recessive	–	1H	Lateral spikelet development	–	Druka et al. (2011), Franckowiak (2010a)
als	Recessive	–	3HL	Lateral spikelet development	–	Franckowiak (2013)
btr1	Recessive		3H	Spike rachis development and grain shattering	Map-based cloning	Pourkheirandish et al. (2016)
btr2	Recessive		3H	Spike rachis development and grain shattering	Map-based cloning	Pourkheirandish et al. (2016)
com1	Recessive	–		Spike-branching/floret determinacy		Lundqvist and Franckowiak (2010a)
com2	Recessive	*HvFZP*	2HS	Spike-branching/floret determinacy	Map-based cloning	Poursarebani et al. (2015)
rtt1.a	Recessive	–	2H	Spike-branching/floret determinacy	–	Franckowiak (1997c)
lnt1	Recessive	*HvJUBEL2*	3HL	Spike-branching/floret determinacy	Genetic mapping and Expression analysis	Dabbert et al. (2010)
mul2	Semi-dominant	–		floret determinacy	–	Franckowiak (1997d)
flo	Recessive	–		Spikelet determinacy	–	Lundqvist and Franckowiak (2015)

(continued)

Table 12.1 (continued)

Locus/loci	Inheritance	Gene name	Chromosome	Function	Approach	References
mov1	Recessive	–	7HL	Ovary development	–	Kleinhofs and Franckowiak (2013)
mov2	Recessive	–	3HS	Ovary development	–	Kleinhofs and Franckowiak (2013)
mov3	Recessive	–	1H	Ovary development	–	Konishi and Franckowiak (2002)
mov4	Recessive	–	–	Ovary development	–	Kleinhofs and Franckowiak (2013)
mov5	Recessive	–	–	Ovary development	–	Kleinhofs and Franckowiak (2013)
ovl1	Recessive/very weak co-dominance	–	4H	Ovary development	–	Franckowiak (2005)
ovl2	Recessive	–	–	Ovary development	–	Kleinhofs (2013)
ovl3	Recessive	–	–	Ovary development	–	Kleinhofs (2013)
Lks1	Dominant	–	2HL	Awn length	–	Franckowiak (2011)
lks2	Recessive	*HvSHI*	7HL	Awn length	Map-based cloning	You et al. (2012)
lks5	Recessive	–	4HL	Awn length	–	Franckowiak and Lundqvist (2011a)
lks6	Recessive	–	1H/5H/6H	Awn length	–	Kleinhofs and Franckowiak (2013)
lel1	Recessive	–	1HL	lemma development	–	Franckowiak (2002)
lga1	Dominant	–	7HS	Awn development	–	Franckowiak (2014a)

(continued)

Table 12.1 (continued)

Locus/loci	Inheritance	Gene name	Chromosome	Function	Approach	References
Kap	Dominant	*HvKN3*	4HS	Awn development	Map-based cloning	Muller et al. (1995)
suK		–		Awn development	–	Roig et al. (2004), Osnato et al. (2010)
triple awned lemma (trp1)	Recessive	–	4HL	Lemma/awn development	–	Franckowiak (2011)
Smooth awn 1 (raw1)	Recessive	–	5HL	Awn length	–	Franckowiak (1997b)
raw2	Recessive	–	5HL	Awn length	–	Franckowiak (1997a)
raw6	Recessive	–	6HL	Awn length	–	Franckowiak (2014a)
raw5	Recessive	–	5HL	Awn length	–	Franckowiak (2014b)
cal-b	Recessive	–	5HL	Awn development	–	Lundqvist and Franckowiak (2014)
cal-c	Dominant	–	5HL	Awn development	–	Franckowiak and Lundqvist (2011b)
cal-d	Recessive	–	3H	Awn development	–	Lundqvist and Franckowiak (2010b)
cal-e	Recessive	–	5HS	Awn development	–	Lundqvist and Franckowiak (2002)
trd1	Recessive	*GATA TF*	1HL	Glume development	Map-based cloning	Whipple et al. (2010), Houston et al. (2012)
eog1	Recessive	–	2HL	Glume development	–	Franckowiak and Lundqvist (2013)

(continued)

Table 12.1 (continued)

Locus/loci	Inheritance	Gene name	Chromosome	Function	Approach	References
uzu	Recessive	*HvBRI1*	3HL	Rachis internode elongation	Sequence homology with rice	Chono et al. (2003)
Zeo1	Dominant	*HvAP2*	2HL	Rachis internode elongation	Genome-Wide Association Study and Map-based cloning	Houston et al. (2013)
Zeo2	Dominant	–	2HL	Rachis internode elongation	–	Franckowiak (2011)
Zeo3	Dominant	–	4HS	Rachis internode elongation	–	Franckowiak (2002)
ari-e	Recessive	*HvDEP1*	5HL	Rachis internode elongation		Wendt et al. (2016)
cly1	Recessive	*HvAP2*	2HL	Lodicule development	–	Nair et al. (2010)
nud1	Recessive	*ERF TF*	7HL	Caryopsis development	Map-based cloning	Taketa et al. (2008)
lax.a	Recessive	*HvLAX.A*	5H	Rachis internode elongation/stamen development	Map-based cloning	Jost et al. (2016)
accordian rachis1	Recessive	–	2HL	Spike rachis development	–	Franckowiak (2010a)
accordian rachis2	Recessive	–	4HL	Spike rachis development	–	Franckowiak (2010b)
accordian rachis3	Recessive	–	1HL	Spike rachis development	–	Franckowiak (2010c)
accordian rachis4	Recessive	–	2H/6HL	Spike rachis development	–	Lundqvist and Franckowiak (2011)

12.2.1 Barley Spike Ontogeny

The barley spike (Fig. 12.2a) emerges from the flag leaf sheath just prior to anthesis. Known as "**heading**", this event is a common proxy for flowering time in cereals but flowering actually occurs earlier in most spring barleys (Alqudah and Schnurbusch 2017). In fact, the transition to reproductive development happens when the shoot apex identity changes from vegetative to reproductive, which can occur several weeks before heading (Kirby and Appleyard 1987). During vegetative development, the shoot apex generates phytomers, basic plant body building blocks, typically comprising a node, a lateral organ (leaf), an axillary bud (tiller meristem formed in the leaf's axil), and an intervening stem internode. At the reproductive transition, the shoot apex in barley ceases making vegetative phytomers, and switches to making floral phytomers, each made up of a rachis internode, rachis node and axillary bud which generates a spikelet meristem (Fig. 12.2b). Each spikelet meristem matures into an unifloretted barley spikelet which holds a single floret flanked by two small and thin glumes (Fig. 12.2b, c). The floret is made up of the enveloping sterile lemma and palea surrounding two lodicules (grass "petals"), three stamens and central carpel housing a single ovary (Fig. 12.2c–e).

The floral transition is marked by dramatic lengthening of the stubby barley shoot apex into an elongated apical dome (Fig. 12.3a). The apex generates successive, alternating pairs of transverse **double ridges** (**DR**s). Each DR marks a reproductive phytomer and consists of a large, lower bract primordium ridge cupping a smaller, rounder lateral meristem ridge in its axil. The meristem ridge enlarges into a wide meristematic band called a **triple spikelet meristem** (**TSM**) or triple mound which splits into three spikelet meristems (Fig. 12.3b): a larger **central spikelet meristem** (**CSM**) bordered by two slightly basal, smaller **lateral spikelet meristems** (**LSM**s) (Fig. 12.3b, c). The lowermost node's bract ridge forms a small collar, separating the inflorescence from the vegetative stem, while all other bract ridges arrest. Indeterminate apex activity continually displaces nodes towards the base, leading to a young spike lined with triplet spikelet nodes that mature along a bottom-up or acropetal gradient (Fig. 12.3c). In many species, flowers are connected to the main inflorescence axis by a stalk or pedicel. In spikes, the spikelet pedicel is greatly reduced to absent, such that the spikelet base essentially fuses to the glumes and rachis, rendering the spikelets sessile. This attachment has implications for shattering behavior in wild versus domesticated forms (discussed below).

Commitment to spikelet fate is marked by initiation of two **glumes** on the lower flanks of each SM, first noticeable in the CSM whose development is advanced compared to the LSMs (Fig. 12.3c). After glume primordia formation, the SM transitions to a **floret meristem** (**FM**; Fig. 12.3c) which is subtended by the greatly reduced spikelet axis or "rachilla". Unlike rye and wheat whose indeterminate spikelets generate a variable number of FMs along the rachilla, barley spikelets produce only one FM on a rudimentary rachilla. Each FM forms an abaxial buttress primordium which differentiates to form a large, outer (abaxial) **lemma** that folds around the floret (Fig. 12.3d, e). Later on, an adaxial primordium ridge forms a thinner, inner (adaxial) palea (Fig. 12.1c, d). The lemma rapidly grows over the inner floret dome, while its tip develops an **awn primordium** (AP) that develops into a long, photosynthetically-active projection or **awn** (Figs. 12.3e and 12.1c). Although the apex continues to form SMs, the number of spikelets is determined around the AP stage, since additional SMs rarely develop completely (Alqudah et al. 2014).

The spike broadly maintains an acropetal pattern of maturation; however, the lower two nodes are usually delayed compared to the central nodes (Bonnett 1935; Fig. 12.3c, d). Within the inner floret whorls, the **stamen** primordia arise first, initiating adaxial to the lemma and palea (Fig. 12.3d, e), and each differentiate into a distal anther and proximal filament (Figs. 12.3f and 12.1c, d). Barley is self- or wind pollinated and does not have large, showy petals; rather, it develops lodicules, the highly-derived petals of grasses (Whipple et al. 2007; Kellogg 2015), which form as two heavily vascularised, small,

Fig. 12.3 Early and late stages of barley reproductive development. a The lengthening shoot apex starts to form double ridges. **b** Triple spikelet meristems cleave to form a central spikelet meristem and two lateral spikelet meristems. **c** Each spikelet meristem forms a single floret meristem flanked by two glume primordia. Lemma primordium form on the abaxial side of the floret meristem. **d** Awn primordia develop on the extending lemmas as stamen primordia emerge from the floret meristem. **e** Awns extend from the expanding lemmas. **f** Stamens form green anthers on filaments. Lodicules swell. **g** Dissected swollen lodicules. **h** Yellow anthers dehisce. Lodicules somewhat deflated. **i** Carpel with single ovary and biforked stigma. **j** Expanding caryopsis. **k** Dried grain with adherent hull. Scale bars (a, b) 100 μm; (c) 200 μm (d–f) 500 μm; (g–k) 1 mm. TSM: triple spikelet meristem; CSM: central spiklet meristem; LSM: lateral spikelet meristem; FM: floret meristem; GP: glume primordium; LP: lemma primordium; SP: stamen primordium; AP: awn primordium. *cv* Bowman for all images. *Micrograph/Photo credits* Sarah McKim

fringed sacs between the base of the stamens and the lemma (Figs. 12.3f–h and 12.2d, e). The central single-carpel pistil consumes the remaining FM, developing into a single **ovary** whose two styles project a biforked, feathery **stigma** (Figs. 12.3i and 12.2d, e). Lodicules swell late in floret development, generally during anthesis or pollen shed, to push the lemma outward from the floret thereby facilitating pollen exchange (Fig. 12.3g, h). Lengthening stamen filaments project the anthers upward where they dehisce pollen onto the stigma (Figs. 12.3h and 12.2d). The fertilized ovary enlarges as the **embryo** and **endosperm** differentiate to form the mature caryopsis enclosed by the adherent lemma and palea **hull** (Fig. 12.3j, k).

12.3 Row-Type and Lateral Spikelet Fertility

The triple spikelet form is unique to the *Hordeum* species and gives barley spikes an identity known as "row-type". In wild barley and many cultivated varieties, each spikelet node develops only one central fertile spikelet and one grain, bordered by rudimentary, female-sterile lateral spikelets whose floral organ development is

Fig. 12.4 Barley spike morphological mutants. **a** Two-rowed barley spike (*Six-rowed spike 1, Vrs1*) with sterile lateral spikelets and fertile central spikelets. **b** Six-rowed barley spike (*vrs1*) with fertile central and lateral spikelets. **c** *vrs2* mutant spike showing enlarged lateral spikelets and additional spikelets at spike base. **d** *vrs3* mutant spike. **e** *vrs4* mutant spike with complete restoration of lateral spikelet fertility and spike-branching (arrow heads). **f** *vrs5* mutant spike. **g** *labile* (*lab*) barley spike showing random fertility and grain set in lateral spikelets. **h–j** *compositum1* (*com1*, h), *compositum2* (*com2*, i), and *low number of tillers-1* (*lnt1*, j) mutant spikes showing non-canonical spike-branching phenotype. **k** *rattail spike 1* (*rtt1*) mutant showing highly branched spike without grain set. **l** *extrafloret* (*flo*) mutant spike showing formation of extra spikelets (arrow heads) from the adaxial base of central spikelets. **m** *mulitflorous 2.b* (*mul2.b*) mutant spike with formation of extra florets (due to extended rachilla) only in the lateral spikelets (arrow heads)

inhibited. Since adjacent spikelet nodes are oppositely positioned up the rachis, spikes have two rows of central grain (Figs. 12.4a and 12.2a), the "**two-rowed**" ancestral state (Zohary and Hopf 2000). Most wild barleys (*Hordeum vulgare* ssp. *spontaneum*), the progenitors of cultivated barley, are two-rowed with a brittle rachis, which enables disarticulation of arrowhead-like triple spikelet for efficient grain dispersal (Zohary 1963). Sterile lateral spikelets are thought to act as barbs to facilitate animal dispersal and/or grain burial following spike shatter (Sakuma et al. 2011). Soon after domestication, lines were selected which developed **fertile lateral spikelets** leading to three grains per node and six rows of grain (Fig. 12.4b). "**Six-rowed**" barleys are still cultivated today, yet their yield advantage is minimal since these varieties develop fewer tillers (Kirby and Riggs 1978; Lundqvist et al. 1997). Moreover, grains from two-rowed varieties are more uniform and thus preferred by the malting industry compared to six-rowed varieties, which as a result are usually destined for animal feed.

Understanding genetic control of row-type and its relationship to lateral spikelet fertility, tillering, and grain set is a focus of intense research and has a long history of study (Engeldow 1920, 1921, 1924; Leonard 1942; Nečas 1963; Takeda and Saito 1988). Quantitative variation observed in lateral spikelet growth in segregating two- by six-rowed progenies suggested multiple loci associate with row-type, while mutation screens of two-rowed barley showed that lateral floret fertility is modified by at least 11 loci, many of which cause an intermediate phenotype between the two- and six-rowed forms (Lundqvist and Lundqvist 1988). There are at least five known genetic loci that independently convert two-rowed barley to six-rowed: ***Six-rowed spike 1*** (*vrs1*, formerly for *vulgare row-type spike 1*), *vrs2, vrs3, vrs4*, and *Intermedium spike-c* (*Int-c*, or *vrs5*). These mutants display a variable degree of lateral spikelet fertility (Koppolu et al. 2013) (Fig. 12.4b–g). The development of six-rowed spikes is predominantly controlled by a single recessive gene *Vrs1*. These mutants do not persist in wild populations since they lack the adaptive dispersal advantage of the barbed spikelet; however, six-rowed barleys were cultivated preferentially by early farmers due to more kernels per spike. The earliest domesticated barley (9500–8400 yBP) had two-rowed spikes whereas cultivation of six-rowed barley started later, approximately from 8800 to 8000 yBP (Helback 1959; Zohary and Hopf 2000). The gene underlying *Vrs1* was identified by map-based cloning as encoding a homeodomain-leucine zipper class I transcription factor (Komatsuda et al. 2007), which shows homology to the maize *grassy tillers 1* which inhibits tiller outgrowth (Whipple et al. 2011). *Vrs1* is an inhibitor of lateral spikelet fertility, so that in a functional state it promotes two-rowed condition (*Vrs1.b* alleles), while non-functional *vrs1.a* alleles promote lateral spikelet development and are present in majority of the cultivated six-rowed germplasm (Fig. 12.4a, b). *Vrs1* was recently shown to negatively influence leaf primordium size, leaf vein number and also leaf area (Thirulogachandar et al. 2017). *Vrs1* also pleiotropically influences tiller number in barley (Liller et al. 2015; Alqudah et al. 2016).

Along with *vrs1.a*, a majority of the six-rowed cultivars carry the *Int-c.a*, a dominant missense allele of *VRS5*, which contributes to increased grain fullness in commercial six-rowed cultivars (Fig. 12.4f). *vrs5*, cloned by a combination of map-based cloning and association genetics followed by allele resequencing, encodes a TEOSINTE BRANCHED 1/CYCLOIDEA/PCF1 (TCP) TCP transcription factor with homology to the *teosinte branched 1* (*tb1*) domestication gene in maize (Ramsay et al. 2011). Higher expression of *tb1* in modern corn led to growth arrest of axillary meristems (which otherwise give rise to tillers), and thereby increasing apical dominance (Doebley et al. 1997), highlighting a striking conservation of lateral growth suppression between maize and barley, although deployed in different architectural contexts. Interestingly, *tb1* promotes *gt1* expression, especially in unfavorable environments (Whipple et al. 2011); however it is unclear if a similar relationship exists in barley between VRS5 and VRS1.

There are several six-rowed-like barleys (*vrs1* look-alikes), which do not harbor either a mutation in *Vrs1* gene or an altered *Vrs1* expression (Komatsuda et al. 2007; Youssef et al. 2017b). These findings suggest that the suppression of lateral spikelet fertility can be completely abolished even in the presence of fully functional VRS1 protein and hints that there is potentially more naturally occurring variation for row-type apart from variation in *Vrs1* and *Vrs5*. From mutational studies it is clear that other loci contribute to lateral spikelet infertility. For example, *Vrs2* encodes a member of the plant-specific SHORT INTERNODES (SHI) transcription factor family protein, which are known to be involved in floral development, auxin biosynthesis, gibberellin (GA) response, and leaf development (Youssef et al. 2017a) (Fig. 12.4c). Mutants at the *vrs4* locus show a complete recovery of lateral spikelet fertility as seen in *vrs1* mutants (Fig. 12.4e). The gene responsible for the *vrs4* phenotype was genetically mapped to chromosome 3H and shown to encode a barley ortholog of the maize inflorescence architecture gene *RAMOSA2* (*HvRA2*), which encodes a LATERAL ORGAN BOUNDARY (LOB) domain transcription factor (Koppolu et al. 2013). In maize, *RA2* (*ZmRA2*) functions as a central regulator of the *ramosa* pathway, important for inflorescence branch formation (Bortiri et al. 2006). Genetics and gene expression analyses suggest that HvRA2 regulates the barley row-type pathway by promoting the transcription of *Vrs1*; when *vrs4* is mutated, transcription of *Vrs1* is lowered resulting in six-rowed spikes (Koppolu et al. 2013). Another row-type locus isolated through mutant screens and analysis of natural quantitative variation is *Vrs3* (*Int-a*) located on chromosome 1H (Fig. 12.4d). Cloning *vrs3*, through a combination of mapping and synteny approaches followed by extensive mutant resequencing, revealed that, unlike other cloned row-type genes that encode DNA-binding transcription factors, *Vrs3* encodes a putative Jumonji C-type (JMJC) H3K9me2/3 histone demethylase (Bull et al. 2017; van Esse et al. 2017), orthologous to the rice gene *OsJMJ706* known to regulate rice spikelet morphology (Sun and Zhou 2008). Based on RNA-seq and qPCR, *Vrs3* promotes the expression of all known row-type genes (Bull et al. 2017; van Esse et al. 2017), and modulates the expression of genes involved in hormone and sugar metabolism as well as stress signaling (Bull et al. 2017).

Labile or **irregular barley** (*H. vulgare* L. convar. *labile* (Schiem.) Mansf.) represents a unique spike-type endemic to highlands of Ethiopia, Eritrea (Åberg and Wiebe 1945) and districts of North India and Pakistan (Takeda and Saito 1988). Spikes of *labile* are characterized by random or irregular lateral spikelet fertility (Fig. 12.4g). For example within a single plant, in one spike the lateral spikelets can be completely reduced to a deficiens phenotype (extreme lateral spikelet abortion, for phenotype see Sakuma et al. 2017), whereas other spikes show various degrees of lateral spikelet fertility (Youssef et al. 2012). Moreover, there is a continuous variation in the number of fertile lateral spikelets from genotype to genotype (Djalali 1970). Recent phenotypic and molecular evidence linked the occurrences of labile barleys mainly to higher altitudes above 2800 m average sea level [subpopulation T6; (Tanto Hadado et al. 2010)]. Youssef et al. (2012) showed that a majority of labile genotypes carry *vrs1.a* (non-functional) and *Int-c.a* alleles at the respective loci. Apart from these loci, the lateral spikelet fertility and development is also severely affected in mutants like *small lateral spikelets1* (*sls1*), *deficiens* (Sakuma et al. 2017), *semi-deficiens* and *angustifolium* barley mutants, where the organs of lateral spikelets are highly reduced and rudimentary. The identification of the underlying genes in these mutants and the uncharacterized row-type mutants (Youssef et al. 2017b) and their functional understanding together with the known row-type genes may unravel the mechanism of row-type regulation unique to barley.

12.4 Determinacy

12.4.1 The "Unbranched" Spike

Inflorescence branching is dependent on the developmental fate of the axillary shoot meristems (Ward and Leyser 2004). The compound spike may represent the ancestral form from which different inflorescence types have evolved (Endress 2010; Kellogg et al. 2013; Remizowa et al. 2013), and among those the non-ramified spike represents the most derived inflorescence form (Vegetti and Anton 1995). In panicles, the spikelets are produced after one or more rounds of branching; in contrast, the sessile spikelets in barley are produced directly on the branchless rachis. Inflorescence branch formation in barley is thus genetically suppressed, a prominent feature characteristic of barley and other Triticeae species. Classical induced mutagenesis approaches identified a group of barley mutants called *compositum* (*com*) which show **non-canonical spike-branching** (Franckowiak and Lundqvist 2011a, b). The mutant loci in this group include *compositum 1* (5HL) and *compositum 2* (2HS) (Bossinger et al. 1992), while wild-type alleles actively suppress spike-branching by specifying SM identity and determinacy in all spikelet primordia. In *com1* and *com2* mutants the spikelet meristems lose their identity and either produce branches, which develop into miniature spikes, or supernumerary spikelets or florets (Fig. 12.4h, i). By map-based cloning and mutant resequencing, Poursarebani et al. (2015) identified *COM2* as an AP2-ERF transcription factor, barley ortholog of maize *branched silkless 1* (*bd1*) (Chuck et al. 2002), rice *FRIZZY PANICLE/BRANCHED FLORETLESS 1* (*FZP/BFL1*) (Komatsu et al. 2003; Zhu et al. 2003), and *Brachypodium distachyon MORE SPIKELETS 1* (*MOS1*) (Derbyshire and Byrne 2013). Analogous loss-of-function mutant phenotypes in maize, rice, and Brachypodium indicate a conserved function for this gene in regulation of inflorescence branching across grass species. The gene responsible for the *com1* locus is not yet known.

Apart from *compositum* mutants, several other barley mutants such as *vrs4*, *low number of tillers 1* (*lnt1*), and spike-lethal *rattail* (*rtt1.a*) also display variable levels of spike-branching (Fig. 12.4e, j, k). *Lnt1* encodes a BELL-like homeodomain transcription factor *JuBel2* on chromosome 3HL (Dabbert et al. 2010). Although *lnt1.a* mutants show reduced shoot-branching (tillering), the spike-branching of *lnt1* mutants is comparable to that of *com1* and *com2* (RK pers. obs.) suggesting that they function in similar pathways. In case of the *rtt1.a* mutant, the floret meristems reiteratively produce several spike-like meristems which ultimately fail to form floret meristems (Fig. 12.4k). Impaired HvRA2 function in the *vrs4* mutants also leads to severe spike-branching as seen in *compositum* mutants (Fig. 12.4e). The defects in *vrs4* mutants include formation of extra spikelets/florets in addition to the basic triple spikelet and also extension of the CSM into a miniature spike (Koppolu et al. 2013; Poursarebani et al. 2015). Poursarebani et al. (2015) showed that expression of *COM2* is decreased in *vrs4* mutants, suggesting a possible mechanism explaining increased spike-branching. Lastly, a role for microRNAs in spike branch suppression was demonstrated by Brown and Bregitzer (2011), where a Ds transposon insertion in miR172 gene led to spike-branching, suggesting that miR172 targets may promote spike-branching and loss of SM identity.

12.4.2 Determinacy of Triple Spikelet

The triplet spikelet node originating from the TSM is a unique feature of the barley inflorescence. Although wheat spikes also develop through a DR stage, the meristematic ridge does not cleave, limiting wheat spike nodes to one spikelet (Kirby and Appleyard 1984). While certain genes influencing branch formation and row-type also regulate spike-branching (described above) and spikelet determinacy (described below), mutant phenotypes suggest that they do not affect the transition from DR to TSM and are not required for TSM formation per se. Similarly, the *ramosa* pathway controlling inflorescence branching in maize is not responsible for the maize-specific spikelet paired meristem formation in the first instance, but for the determinacy imposed on meristems with spikelet fate (Eveland et al. 2014). Formation of boundaries within the meristematic ridge to generate three separate spikelet meristems likely involves a canalized program of boundary gene expression; however, the upstream drivers of this event are unknown. Mutant collections do not include spikes without triple spikelet node architecture, suggesting that genes controlling TSM architecture may be important for viability, or are redundant.

Beyond suppressing lateral spikelet development, multiple row-type genes, such as *Vrs4*, *Vrs3,* and *Vrs2*, also have roles in **preventing additional spikelets** and **florets** developing from the spike once the spikelet triplet pattern is established. For instance, loss of VRS4 function not only causes spike-branching but also leads to ectopic meristematic bulges on either side of the LSM which develop into either glume-flanked spikelets or solitary florets, probably due to the disturbance in boundary establishment on the lateral edges of the LSM (Koppolu et al. 2013). Based on microarray analysis, gene expression in *vrs4* spikes showed signatures of enhanced cytokinin and downregulation of trehalose phosphate phosphatase (TPP), an enzyme which neutralizes the growth-promoting trehalose 6-phophate signal (Smeekens et al. 2010), consistent with increased growth compared to wild-type (Koppolu et al. 2013). *vrs4* ectopic spikelets are more common in middle to basal nodes, while *vrs2* mutants show supernumerary spikelet formation around the two lowest nodes of the spike (Youssef et al. 2017a), suggesting that both *Vrs4* and *Vrs2* are also necessary to maintain triple spikelet architecture in the spike by regulating boundary formation and proper auxin: cytokinin (CK) homeostasis along the spike, respectively (Youssef et al. 2017a). Both these studies emphasize the contribution of hormone and sugar metabolism to the mechanistic explanation of branching and lateral fertility in the barley spike. The development of in vivo reporters for hormone regulators and sugar signaling would significantly advance our understanding of the regulatory network controlling meristem initiation and outgrowth, and of the parallels with other systems.

Fig. 12.5 **Barley multifloret mutants. a** Barley spikelet triplet with canonical unifloreted spikelets produced on a determinate rachilla (arrow heads). **b** Wheat spikelet with four florets (F1–F4) produced on an indeterminate rachilla. **c–h** Spikelets of barley multifloret mutants *multiflorus 2* (*mul2*, c), *compositum 1* (*com1*, d), *compositum 2* (*com2*, e), *low number of tillers 1* (*lnt1*, f), and *Six-rowed spike 4* (*vrs4*, g) showing more than one floret per spikelet produced on an indeterminate rachilla. Arrow heads indicate rachilla extension into extra florets. F: floret; CS: central spikelet; LS: lateral spikelet

12.4.3 Spikelet Determinacy

The number of florets produced in a spikelet ranges in Triticeae species from one to >2 (Sakuma et al. 2011) and depends on the meristematic activity and growth of the rachilla (Forster et al. 2007). In barley, the rachilla becomes vestigial after producing one floret, and shows highly reduced internode elongation and meristematic activity, leading to a single floret per barley spikelet (**determinate rachilla**; Fig. 12.5a). However, in wheat, where up to 10 florets are produced per spikelet, the rachilla assumes an indeterminate developmental mode due to the continued development of rachilla phytomers and meristematic activity (**indeterminate rachilla**; Fig. 12.5b). Such a profound developmental difference suggests a possible species-specific regulation or expression pattern of a genetic determinant that may underlie rachilla determinacy. Several barley mutants deviate from the set rule "determinate rachilla", and show more than one floret per spikelet. In the barley mutant *multiflorus 2.b* (*mul2.b*), the rachillae of lateral spikelets consistently elongate to produce up to three florets per spikelet; yet fascinatingly, rachillae of central spikelets remain determinate and bear just one floret per spikelet, suggesting that control of rachilla elongation in central versus lateral spikelets can be genetically separated (Figs. 12.5c and 12.4m). Further, an additional floret phenotype is also observed in the lateral spikelets of all spike-branching mutants discussed so far (*com1*, *com2*, *Int2* and *vrs4*; Fig. 12.5d–f). However, in the spike-branching mutants, the rachillae of central spikelets assume an inflorescence identity and produce a miniature spike-like structure, which is not seen in *mul2* mutants. By this observation it is plausible to infer that the gene underlying the *mul2* phenotype could be a downstream target of any of the spike-branching regulators.

Spikelet determinacy is also affected in barley *extrafloret* (*flo*) mutants *flo.a*, *.b*, and *.c* (Gustafsson et al. 1969). In these mutants an extra spikelet (but not a floret as mentioned in the locus name) develops from the adaxial base of existing central spikelets (primary spikelets). The extra spikelet as well as the primary central spikelet all form spikelet organs (Fig. 12.4l). The underlying gene for this phenotype is not yet known.

12.5 Floral Organ Differentiation

Following glume initiation, the SM transitions into a single FM that generates a canalized number and arrangement of asexual and sexual floral organs (Fig. 12.2d, e). Barley mutant collections contain a whole array of floral mutants which show floral organ number (merosity) and/or identity phenotypes (Franckowiak and Lundqvist 2010; Druka et al. 2011). Across angiosperms, flower and floral organ formation is associated with the function of MADS-box transcription factors (Gramzow and Theissen 2010; Wellmer and Riechmann 2010) and disruptive variation in these genes may well underlie several floral mutants in barley (Rossini et al. 2006). Comparative transcriptomics in Digel et al. (2015) showed suites of upregulated genes associated with floral development, including multiple MADS-box genes. In addition, *BARLEY FLORICAULA/LEAFY* (*BFL*), *PANICLE PHYTOMER2* (*HvPAP2*, a MADS-box gene), and *HvCONSTANS* (*HvCO6*), homologues of key floral genes in rice and arabidopsis, were downregulated in *vrs2* mutants, consistent with *vrs2* delayed floral maturation phenotypes (Youssef et al. 2017a). Taken together, these expression patterns suggest an importance in floral fate and differentiation; however, we do not yet have corresponding mutant phenotypes. Notably, although normally expressed in the spike, ectopic overexpression of a subset of MADS-box genes, the SHORT VEGETATIVE PHASE (SVP) homologues in barley, *BM1*/*BM10*/*VRN2*, caused reversion to shoot identity in floral meristems and aberrant spike and spikelet differentiation (Trevaskis et al. 2007), suggesting they are under strict spatial and/or expression level control to permit normal floral meristem identity and floral progression.

12.5.1 Bracts

Grasses actively suppress the development of a subtending bract early in spike development, potentially to prevent removal of resources from the floret meristem (Coen and Nugent 1994; Whipple et al. 2007). Suppression is lost in barley *third outer glume* (*trd*) mutants which develop large bracts subtending the spikelet nodes and a leaf at the collar (Ivanova 1937; Houston et al. 2012; Fig. 12.6a). The subtending bract develops at DR-TSM stage (Whipple et al. 2010), before normal flanking glume differentiation, and likely represents the growth of the lower meristem ridge. Synteny and mutant allele sequencing suggested that *HvTRD* encodes a GATA-like transcription factor which has an orthologous function in rice and maize (Whipple et al. 2010), supported by the identification of 14 independent alleles through extensive resequencing of allelic mutant populations (Houston et al. 2012).

12.5.2 Glumes

The barley spikelet develops a pair of flanking glumes that differentiate into small, awn-tipped

Fig. 12.6 Selection of spike and spikelet architectural mutants. a The *third outer glume*, (*trd*) mutant shows ectopic development of a large central abaxial glume on each spikelet. **b** The *elongated outer glume* (*eog1*) mutant develops long awns on each glume. **c** The *leafy lemma 1* (*lel1*) mutant shows leaf-like, wide lemmas. **d** The *laxatum.a8* (*lax.a*) mutant lacks a distinct awn-lemma boundary. **e** The dominant *Awnless1* (*Lks1*) mutant shows loss of awn formation. **f** The *lax.a* mutant also develops stamens instead of lodicules. **g** The *Zeocriton1.b* (*Zeo1.b*) mutant lodicules fail to expand. **h** The *multiple ovary1* (*mov1.a*) forms multiple ovaries in the pistil. **i** *Zeo1.b* mutants form dense spikes due to reduced rachis internode elongation. **j** The *accordian rachis 1* (*acr1.a*) mutant develops extended rachis internodes. All mutants shown are from the Bowman near-isogenic line population. Scale bars (f, g) 1 mm. *Photo credits* (a, b, c, e, h, j) Arnis Druka; (d, f, g, h, i) Sarah McKim

bracts on either side of the floret. Genes necessary for glume identity include *VRS4* where glumes are often replaced by ectopic florets and can also remain undeveloped and/or fused (Koppolu et al. 2013). In addition, in rice *OsMADS1/LEAFY HULL STERILE* (*LHS*) expression promotes lemma over glume identity, suggesting that exclusion of *LHS*, and potentially the barley orthologue, from the glume is important for glume identity (Jeon et al. 2000; Prasad et al. 2001, 2005). Glume size in barley is controlled by the *Elongated outer glume1* (*eog1*) locus on chromosome 2HL (Druka et al. 2011; Fig. 12.6b). Depending on the *eog1* allele and genetic background, the glume width and glume awn length can be significantly increased or reduced, with some alleles converting glumes into lemma-like organs (Franckowiak and Lundqvist 2013). Glume awn length, but not glume width, is specifically regulated by the *Long glume awn 1* (*Lga1*) locus on chromosome 7HS, where the dominant allele (*Lga1.a*) confers long awns and while recessive alleles (*lga1.b*) lead to short awns. The short-awned allele is considered the mutant form originating in western two-rowed cultivars (Franckowiak 2014a). Variation at *Lga1* does not seem to affect lemma-awn length (Franckowiak 2014a) suggesting independent genetic control of elongation growth depending on the basal organ of awn attachment.

12.5.3 Lemma and Palea

Whether these protective sheaths are homologous to bracts, or represent floral organs analogous to the sepals in eudicot flowers, has been long debated (Arber 1934; Bell 1991; Kellogg 2001; Lombardo and Yoshida 2015). However, molecular and genetic studies in rice and maize indicate that developing lemma and palea express floral genes under distinct genetic control (as argued in Lombardo and Yoshida 2015). As mentioned, *LHS* homologue expression may also confer lemma identity in barley, although this remains to be tested. Two barley *leafy lemma* mutants, *lel1* (Fig. 12.6c) and *lel2* show conversion of lemmas into leaves (Bossinger et al. 1992; Pozzi et al. 2000), suggesting the underlying loci are critical for lemma identity. The *short awn 2* (*lks2*) allele is required for full expression of the *leafy lemma* phenotype, in addition to roles in lemma-awn development (see below; Pozzi et al. 2000). The *laxatum-a* (*lax.a*) locus in barley is associated with rachis internode length (see below) but *lax.a* mutants also show narrowed paleas and lemmas and lack the lemma-awn boundary (Fig. 12.6d). Using an ingenious combination of positional mapping and mapping-by-sequencing of pooled recombinants, Jost et al. (2016) showed that the *lax.a* phenotype is caused by a large deletion in the centromeric region of chromosome 5H which deletes a gene encoding a homologue of the BLADE-ON-PETIOLE (BOP) transcription factors, known to promote boundary formation, leaf margin growth and floral organ identity in arabidopsis (Hepworth et al. 2005; McKim et al. 2008).

12.5.4 Awns

Photosynthesis in the distinctive long awn extending from the lemma tip likely contributes to grain fill (Abebe et al. 2009; Taketa et al. 2011; Liller et al. 2017). Controlled by multiple loci, changes in awn identity and/or length are common phenotypes in both natural and mutant populations. Awn mutants fall into two general categories: mutants whose awn phenotype is part of a pleiotropic effect on elongation growth; and those specific to lemma-awn development (Frankowiack and Lundqvist 2012; Rossini et al. 2014). A major locus essential for awn formation is *awnless*, *Lks1*, on chromosome 2H, whose dominant *Lks1.a* alleles lead to a loss of awn development, the extent of which is environmentally variable (Druka et al. 2011; Franckowiak and Lundqvist 2011a; Fig. 12.6e).

The first developmental gene cloned in barley encodes a Class I KNotted1-like HomeobOX (KNOX) transcription factor, BARLEY KNOX3 (BKn3) whose ectopic expression, due to a duplicated intron in the *BKn3 Kap* (*K*) allele (chromosome 4HS), leads to the semi-dominant *Hooded* phenotype where the lemma develops side wings and an additional, inverted spikelet or "hood" on its tip (Stebbins and Yagil 1966; Muller et al. 1995; Richardson et al. 2016). Recently, Richardson et al. (2016) revealed that *BKn3* expression in the lemma-awn boundary leads to altered tissue orientation and growth rates via induction and reorientation of an auxin efflux transporter, Sister of PINFORMED1 (SoPIN1) in the ectopic hood meristem, which may relate to the side wing phenotype, along with higher expression of auxin importers and boundary gene expression, providing clues of the BKn3 downstream mechanism. The five *calcaroides* loci all lead to ectopic wing formation on the lemma due to reversed growth orientation but not hood formation (Pozzi et al. 2000). It will be interesting to explore the possible mechanistic links between the growth reorientation in *calcaroides* and *Hooded*.

In a screen for suppressed hood formation, Roig et al. (2004) isolated five *su*ppressors of *K* (*suK*) loci. Ethylene treatment also rescues the hooded phenotype and is associated with decreases in *BKn3* expression, potentially mediated by Barley Ethylene-Response Factor1 (BERF1), Barley Ethylene Insensitive Like1 (BEIL1), and Barley Growth Regulating Factor1 (BGRF1) proteins which directly bind the *BKN3* intron (Santi et al. 2003; Osnato et al. 2010). Based on map positions of genes, these factors are independent from the yet uncloned *suK* loci (Osnato et al. 2010). Ectopic meristem formation of BKn3 *Kap* is also suppressed by the short awn *lks2* mutant (Takahashi et al. 1953) which is also independent from the *suK* loci (Roig et al. 2004).

Short awns may confer advantages under stress in barley (Schaller and Qualset 1975), and wheat (Rebetzke et al. 2016). Natural *lks2* alleles are concentrated in the Far East, suggesting benefits under higher humidity (Takahashi 1987; Yuo et al. 2012). Awns of *lks2* are also finer and less stiff, and *lks2* florets have pistils with shorter styles (Yuo et al. 2012). The gene underlying *lks2*, roughly mapped to chromosome 7H (Taketa et al. 2008), was further fine-mapped by position and synteny with rice to a gene encoding a SHI transcription factor in the same grass-specific clade as *VRS2* (Yuo et al. 2012). In arabidopsis, SHI proteins are also involved in apical growth, including style formation (Kuusk et al. 2002), and promote auxin biosynthesis (Sohlberg et al. 2006; Eklund et al. 2010), highlighting a potentially conserved relationship between SHI proteins, apical tissues and auxin, with awns interpreted as apical tissues of the lemma. The interval underlying a major QTL for natural variation for awn length on chromosome 7H was narrowed down using fine-mapping and NIL RNA-seq to reveal a candidate gene (HORVU7Hr1G118500.1), encoding a zinc finger protein 3 which showed no homology with either *LKS2* or other genes more generally affecting elongation growth (Liller et al. 2017).

Awns of wild barley and certain cultivated varieties are lined with minute barbs leading to a rough texture which may have aided in dispersal similar to the sterile lateral barbs (von Bothmer et al. 1995). Many cultivated forms show full or partial **loss of rough awns**, since the barbs can be unpleasant in fodder for animals (Takahashi 1955). Variation at multiple loci can reduce the number of barbs, a phenotype often linked with the reduction in stigma hair formation, suggesting a shared regulatory pathway for both structures. Recessive *raw1.a* alleles at the *Smooth awn 1* locus (chromosome 5HL) lead to semi-smooth or completely smooth awns depending on the genetic background (Harlan 1920; Franckowiak 1997c). Mutant alleles of *Smooth awn 2* (*raw2*) reduce awn barb number and in combination with *raw1*, lead to a very smooth awn (Franckowiak 1997c). *Smooth awn 6* (*raw6*) and *Smooth awn 5* (*raw5*), both on 6HL, are more severe with very smooth awns and few stigma hairs (Franckowiak 1997a).

12.5.5 Lodicules

Although lodicules are morphologically very different from the typical eudicot petal, molecular evidence of "B" class gene expression supports homology between petals and lodicules in grasses (Whipple et al. 2007; Yoshida and Nagato 2011). The aforementioned *lax.a* mutants show homeotic conversion of lodicules to stamens with two rather than the normal four lodicules suggesting that HvLAX.A inhibits stamen identity in lodicules (Jost et al. 2016; Fig. 12.6f).

Swelling of lodicules at anthesis leads to "open-flowering" and possible pollen transfer, while loss of lodicule enlargement causes cleistogamy or "closed flowering" where foreign pollen is excluded from the floret (Lord 1981). Thus, cleistogamy is an attractive feature to control gene flow between populations. Nair et al. (2010) showed that variation in the *Cly1* (syn. *Zeo*, *HvAP2*) was associated with cleistogamy: SNPs in *cly1* cleistogamous alelles caused a loss of miR172-regulation of *Cly1* transcript and presumed increased CLY1 levels, suggesting that *Cly1* inhibits lodicule swelling (Fig. 12.6g). *Cly1* encodes a miR172-regulated AP2-domain containing transcription factor (Nair et al. 2010), homologous to APETALA2 in arabidopsis and SHATTERING ABORTION1 in rice which are similarly miR172 regulated and involved in petal or lodicule development (Chen 2004; Aukerman and Sakai 2003; Zhou et al. 2012). *Cly1* is also subject to epigenetic regulation at its promoter (Wang et al. 2015). Timing of anthesis to spike emergence from the flag leaf is also key for pollen exchange. Dominant alleles at the *Cly2* locus cause premature anthesis and lodicule swelling prior to spike heading, leading to a cleistogamous phenotype despite functional lodicules (Wang et al. 2013).

12.5.6 Stamens and Ovaries

Stamen identity is controlled by the *multiovary* (*mov*) loci whose alleles all cause a homeotic transformation of stamens into carpels (Franckowiak et al. 2005; Fig. 12.6h). The unique *mov1* mutant also shows transformation of lodicules into leaf-like organs (Kleinhofs 2013). In addition to homeotic conversion of stamens to carpels, the *mov4* mutant also produces extra stamens which are associated with extra grain set. Other mutants show losses of floral determinacy associated with multiple carpels per spikelet, such as *vrs4* mutants (Franckowiak et al. 2005; Koppolu et al. 2013). Two *Ovaryless* (*Ovl*) loci affect pistil development where recessive alleles result in defective ovaries: *ovl1* is mapped to chromosome 4H and confers a highly pleiotropic phenotype including loss of awns and midveins, while *ovl2* is normal with respect to awns and leaf midveins apart from rudimentary ovaries (Franckowiak 2005; Kleinhofs 2013). Unsurprisingly, mutations affecting carpel development often reduce fertility; for instance, the *gigas1.a* allele, mapped to chromosome 2H, shows elongated ovaries, lemmas, stigma with fewer papillae, and is semi-sterile in some genetic backgrounds (Harvey et al. 1968).

A recent study showed that anther development is controlled by the *MALE STERILITY1* (*MS1*) gene on chromosome 5 which encodes a PHD transcription factor (Fernández and Wilson 2014). HvMS1 was identified by reverse genetics, an approach commonly used in arabidopsis but will likely become more prevalent in barley research, whereby RNAi targeting of a barley orthologue of the rice and arabidopsis genes known to control pollen wall development, conferred male sterility (Fernández and Wilson 2014). Anther development is also controlled by several, recessive *Male-sterile genetic* (*msg*) loci (https://www.nordgen.org/bgs/index.php) where recessive alleles lead to rudimentary anthers and loss of filament elongation (Franckowiak 1997b).

12.5.7 Spike Growth and Rachis Elongation

Initially close together, spikelet nodes separate from each other during floret differentiation as rachis internodes lengthen. The extent of elongation and the number of spikelet nodes determines spike length and spike density (spikelets/rachis length), two important agronomic traits which along with peduncle length, influence heading time. Spike density variation is associated with multiple loci across the germplasm (Houston et al. 2013) and is a common phenotype catalogued across mutant populations (Druka et al. 2011).

Consistent with other plant species, gibberellin (GA) and brassinosteroid (BR) hormone metabolism and signaling regulate internode elongation throughout the barley plant, including the rachis. One of the first architectural loci cloned in barley, *uzu1*, is an orthologue of *BRASSINOSTEROID INSENSITIVE1* (*BRI1*), which encodes the receptor for BR (Chono et al. 2003). The *uzu1.a* allele used to generate semi-dwarf "*uzu* barley" in Japan, contains a single nucleotide polymorphism in *HvBRI1* which leads to an overall reduction in elongation growth in internodes, awns, glumes and the grain (Chono et al. 2003). By careful screening of mutant populations for the diagnostic BR mutant phenotype profile, which includes wavy leaf margins and erect leaf and awn angles, multiple other BR-related genes were cloned, including three genes encoding BR biosynthetic enzymes (*HvBRASSINOSTEROID-6-OXIDASE, HvCONSTITUTIVE PHOTOMORPHOGENIC DWARF, HvDIMINUTO*), and further *HvBRI1* alleles isolated (Dockter et al. 2014). Comparative analysis between *HvBRI1* alleles showed that the *uzu1.a* mutations cause temperature-dependent modifications of BRI1 function. The new *HvBRI1* alleles were more complete functional knockouts that may provide more robust semi-dwarfism under variable conditions (Dockter et al. 2014). GA has a particular relevance to cultivated cereals where Green Revolution semi-dwarfing alleles in wheat and rice are now known to reflect defects in GA signaling or biosynthesis (Hedden 2003). One of the two semi-dwarf alleles are introgressed into European cultivated barley, *denso1/semi-dwarf1.a* (*d1/sdw1.a*), was recently confirmed to result from a defective *GA20oxidase* gene, long suspected (Jia et al. 2009), and now proven through allele resequencing (Xu et al. 2017). The other major semi-dwarfing allele in European barleys, *breviaristatum-e* (*ari-e*), the *cv*. Golden Promise semi-dwarfing allele, is located to chromosome 5H (Liu et al. 2014). By exploiting mutant populations followed by transgenic complementation, *ari-e* was recently shown to encode a heterotrimeric G protein (Wendt et al. 2016).

In addition to genes involved in hormone metabolism and signaling, several developmental transcription factors control rachis internode growth in barley. The *HvAPETALA2* (*HvAP2*/*Cly1*) gene underlies the major QTL and GWAS peak on chromosome 2H linked to spike density in spring barley (Houston et al. 2013). HvAP2 inhibits rachis elongation since *HvAP2* overexpression in the miR172-resistant *Zeo* alleles leads to significant increases in spike density associated with delayed transition to rachis internode elongation (Houston et al. 2013; Fig. 12.6i). Recently, studies on the *Q* gene which encodes an HvAP2 homologue in wheat showed that variation in the miR172-targeting the *Q* transcript was likely responsible for the domestication traits linked to this locus, including spike compaction, suggesting that this role is conserved in the Triticeae (Debernardi et al. 2017; Greenwood et al. 2017). Further, loss of function *Q* alleles and overexpression of miR172 resulted in elongated spikes and rachillae, and revealed a novel function of *Q* in controlling floret number per spike (Debernardi et al. 2017; Greenwood et al. 2017). It is unknown if spikelet determinacy is also controlled by the *Q* orthologue in barley; however, its expression was reduced in the *com2* mutant, consistent with a role in meristem indeterminacy (Poursarebani et al. 2015).

Variation in other loci cause increased internode elongation. *HvLAX.A* actually derives its name from the increased rachis internode elongation in *lax.a* spikes, which gives them a "lax" or relaxed appearance, and suggests that HvLAX.A function shortens rachis internodes. Multiple other *lax* loci are known (*lax.b-n*) but are not yet molecularly identified. As the name implies, recessive alleles of the *long basal rachis internode* locus (chromosome 4HS) cause excessive elongation in the lower rachis internode (Franckowiak and Lundqvist 2014), perhaps reflecting an incomplete switch from culm to spike identity. Consistent with a role of the floral progression to the regulation of rachis internode elongation, overexpression of the aforementioned *SVP*-like genes also cause extreme rachis internode elongation, similar to the phenotype seen in the *accordian rachis1* mutant (Trevaskis et al. 2007; Franckowiak and Lundqvist 2010; Druka et al. 2011; Fig. 12.6j).

12.5.8 Hull

Domestication and cultivation also altered several spikelet and grain architectural traits relevant to the downstream uses of the barley grain. Unlike other grasses, barley's lemma and palea hull firmly adheres to the caryopsis. Complete loss of hull adherence in so-called "**naked**" barley make the harvested grain readily edible. Naked varieties are still grown throughout the world, most notably in the Tibetan plateau where barley is grown for human consumption (Takahashi 1955) while barley destined for animal feed and malting is "covered," retaining the adherent hull that protects the embryo and helps filter liquids during malting.

The distinctive covered barley grain arises from a sticky lipid "cementing layer" secreted from the pericarp epidermis which adheres the lemma and palea to the expanding grain (Harlan 1920; Gaines et al. 1985; Taketa et al. 2008). Naked barley shows a complete loss of this layer due to a single loss-of-function mutation in *NUD* on chromosome 7HL (Taketa et al. 2008; Franckowiack and Konishi 1997). *NUD* encodes an AP2-like ETHYLENE-RESPONSE FACTOR (ERF) transcription factor which is expressed in the testa or seed coat underneath the pericarp (Taketa et al. 2008). *NUD* shows homology to arabidopsis *WAX INDUCER1/ SHINE1* (*WIN1/SHN1*) which encodes a transcription factor that regulates cuticular wax biosynthesis, especially on floral organs in the arabidopsis plant (Aharoni et al. 2004; Kannangara et al. 2007). In barley, *nud* (naked) mutants do not show obvious additional wax phenotypes, suggesting that NUD is specialized for a non-redundant role in barley hull adhesion. Although the regulatory cascade originated by NUD to create the cementing layer is largely unknown, comparative transcriptome work between Morex and Tibetan Hulless landrace Dulihuang caryopses' highlighted misexpression of numerous regulatory and structural cuticle-related genes in Dulihuang (Duan et al. 2015), in support of a conserved function between NUD and WIN1. Furthermore, three *eceriferum* (*cer*) mutants impaired in wax load on aerial tissues, *cer-yl*, *cer-ym* and *cer-zv*, also present naked grain (Taketa et al. 2008), consistent with a link between wax biosynthesis and hull adhesion, which may hint that these loci are NUD targets.

In fact, naked and covered barley represent the extremes of a sticky spectrum: intermediate phenotypes where adhesion is weak rather than absent causes a phenomenon called "skinning" where the hull partially sheds (Hoad et al. 2016). Skinning is a significant problem in the malting industry worldwide, since variable hull loss reduces malt homogeneity, decreasing efficiency and output. Increasing prevalence of skinning appears to be a germplasm-dependent effect of new cultivars (Brennan et al. 2017), which can lead to entire grain lots being rejected by maltsters. No loci have yet been linked to skinning. It will be intriguing to learn if defective steps in wax metabolism or transport are implicated.

12.5.9 Brittle Rachis

Barley domestication was driven by major architectural and developmental changes that allowed efficient planting and grain harvest, improved quality of grain and increased grain yields. Grasses employ diverse modes of grain shatter based on placement of the abscission zone, a specialized boundary that marks the separation layer between leaving organ and the main plant body. In wild barley, "constriction grooves" marking abscission zones form across the rachis just above the spikelet node, weakening the rachis such that the spike disarticulates into whole wedge-shaped spikelet + rachis fragments at maturity, preventing easy grain harvest (Ubisch 1915; von et al. 2005). Tight attachment of barley grain to the rachis presented another barrier to isolate grain. Domestication of barley in the first instance involed early selection by ancient farmers for tough rachis variants whose grain readily abscised from the spike (Komatsuda et al. 2004;

Sang 2009). Variation in two linked loci, *Non-brittle rachis1* (*btr1*) and *btr2* underlie the tough rachis typical of cultivated barley which is associated with increased cell wall thickness in the constriction grooves (Pourkheirandish et al. 2015). *BTR1* and *BTR2* were shown to encode unrelated proteins, with BTR1 encoding a potentially membrane-associated protein and BTR2 showing limited similarity to INFLORESCENCE DEFICIENT in ABSCISSION (IDA), a small peptide key for organ abscission in arabidopsis (Stenvik et al. 2006).

12.6 Environmental Control of Inflorescence Architecture

We understand most about the genetically determined aspects of plant architecture. However, in response to the challenges posed by their sedentary life habit, plants have evolved remarkable architectural plasticity. Environmental inputs often converge on many of the same players involved in genetically predetermined traits. As such, plant development in the field represents an ongoing adaptive dialogue between these intrinsic genetic programs and external signals.

12.6.1 Time to Anthesis

In addition to reduced genetic diversity, modern crops also show altered plasticity for climatic cues compared to their wild relatives, due to selection during post-domestication range expansion. By far the best understood response to environmental inputs in cultivated barley is flowering time. In particular, barley cultivars grown in northern regions show insensitivity to vernalisation and photoperiod-related flowering time cues to ensure that crops flower early in spring (Comadran et al. 2012 and references therein).

Barley, like all plants, regulate the floral transition carefully, flowering only after accumulating sufficient resources to fuel development and grain fill, yet early enough to limit the window for disease, all the while ensuring the transition occurs within permissive climates for optimal pollination and fertilization. In response to these wide-ranging pressures, plants evolved exquisitely sensitive internal and external signaling pathways networked through central integrator hubs to regulate flowering. Like most other flowering plants, cereals rely on environmental cues to adapt and optimize the time to anthesis. As naturally evolved winter-annual plants of the temperate regions, cold response and a period of cold exposure (i.e., **vernalisation**) are essential prerequisites for initiating floral induction in barley and wheat. This ancestral state leads to the winter growth type, while loss of this requirement underlies the spring growth habit. Identified in either wheat or barley through positional cloning, three *spring growth habit* (*sgh*) loci mainly control the vernalization requirement and winter growth habit (Takahashi and Yasuda 1971; Laurie et al. 1995): *Vrn-H1*, a floral meristem identity MADS-box transcription factor of the APETALA1 (AP1) class that regulates the transition from the vegetative to the reproductive meristem (Laurie et al. 1995; Danyluk et al. 2003; Yan et al. 2003); *Vrn-H2*, a zinc finger CCT-domain containing repressor of floral induction (Yan et al. 2004; Chen and Dubcovsky 2012); and *Vrn-H3* (syn. *HvFT1*), the barley orthologous protein of arabidopsis FT (Flowering locus T), which possesses the typical florigenic properties for floral induction and can be considered as the downstream integrator of photoperiod (see below) and vernalization pathways (Yan et al. 2006). While all three proteins build a feedback regulatory loop in leaf cells, allelic variation around the two genuine vernalization loci *Vrn-H1* and *Vrn-H2* is sufficient to convert winter-annual barleys into spring types (von Korff and Campoli 2014). To induce the transition from vegetative to inflorescence meristem state, HvFT1 must travel from the leaf to the shoot apex where it upregulates *VRN-H1*, thereby promoting floral meristem identity (Yan et al. 2006; Hemming et al. 2008). Loss of this requirement underlies the spring growth rather than winter growth habit.

Another important environmental cue for floral transition is **day length**. Cereal crops are facultative long day (LD) plants, and experience LD conditions from around 10 hours (h) light or longer, while short days (SD) are reliably sensed from 8 h or less. Since floral induction and development has to occur under suitable temperatures, sensing day length is a sensible proxy for cereal crops to avoid frost conditions during reproductive development. One central day length-sensing genetic locus in barley, *PHOTOPERIOD RESPONSE LOCUS1* (*Ppd-H1*), encodes for a PSEUDO-RESPONSE REGULATOR (HvPRR37) protein that promotes floral transition and development under LD conditions via upregulation of HvFT1 (Turner et al. 2005; Hemming et al. 2008). Under noninductive SD photoperiods, however, barley plants were capable of initiating floral primordia but always failed to progress beyond this stage, indicating a clear requirement for *Ppd-H1* expression in photoperiod-sensitive plants to successfully complete inflorescence development (Digel et al. 2015). Importantly, the photoperiod- and *Ppd-H1*-dependent successful induction of inflorescence development correlated with the expression of both *HvFT1* and *HvFT2*, suggesting that *HvFT1* alone may not be sufficient under these conditions (Digel et al. 2015).

A less photoperiod-sensitive allele of *Ppd-H1* (i.e. *ppd-H1*) is associated with delayed heading time, an adaptation to higher latitudes selected during the domestication for higher barley yields in central European spring barleys (Turner et al. 2005; Alqudah et al. 2014). Later heading time reflects a *ppd-H1* allele-dependent decrease in *CONSTANS1* (*HvCO1*) and *HvCO2* expression linked to reduced HvFT1 induction (Turner et al. 2005; Campoli et al. 2012). Interestingly, regulation of heading time in *ppd-H1* carrying genotypes under LD conditions was predominantly associated with allelic variation at the *HvCO1* locus in a GWAS of spring barleys (Alqudah et al. 2014), while photoperiod-sensitive genotypes only showed an association with *HvCO2* (Alqudah et al. 2014). Responses to SD conditions are usually conferred by allelic variation at the *Ppd-H2* (*HvFT3*) locus (Faure et al. 2007; Kikuchi et al. 2011) or *HvCO9* (Kikuchi et al. 2011). For example, winter cultivars are often combined with a recessive *ppd-H2* in order to delay heading under SD photoperiods (Cockram et al. 2015). Moreover, alleles at *Ppd-H2* were highly associated with time to heading and pre-anthesis phase duration under LD, independent of the allelic status at *Ppd-H1* in a GWAS of spring barleys (Alqudah et al. 2014), reinforcing the notion that *Ppd-H2* is an important photoperiod regulator for barley adaptation and yield (von Korff et al. 2008; Francia et al. 2011; Tondelli et al. 2014).

Photoperiod-independent induction of floral development is usually related to the so-called *earliness* per se (*eps*) or *early maturity* (*eam*) loci in barley (Laurie et al. 1995). All of the currently known *Eam* genes are associated with the **circadian clock**, such as *Eam5* (*Pankin* et al. 2014), *Eam8* (*Early Flowering3*, *HvELF3*; (Faure et al. 2012; Zakhrabekova et al. 2012)) and *Eam10* (*LUX ARRHYTHMO*, *HvLUX1*; (Mizuno et al. 2012; Campoli et al. 2013; Gawroński et al. 2014). The functional forms of these proteins all suppress *HvPRR37* or *HvFT1* expression under noninductive conditions (SD), while loss-of-function mutants show induced *HvPRR37* and/or *HvFT1* expression, and hence floral progression, regardless of day length. Furthermore, early floral induction in the *elf3* mutant was not dependent on *HvFT1* transcript levels alone, but also on GA biosynthesis genes, especially *HvGIBBERELLIN 20 OXIDASE2*, and enhanced bioactive GA_1 levels in the leaves, highlighting for the first time that GA is an important florigenic signal in barley, as in arabidopsis (Boden et al. 2014). GA-mediated early heading in *elf3* mutants was also highly associated with increased transcript levels of essential floral identity genes, such as of *SUPPRESSOR OF OVEREXPRESSION OF CONSTANS1*

(*HvSOC1*), *FLOWERING-PROMOTING FACTOR3* (*HvFPF3*), *BFL* and *HvPAP2*, which all promote floral development (Boden et al. 2014). Based on their findings, the authors propose a model in which *Vrn-H1* expression in the apex appears pivotal for the development of a receptive meristem that is "competent" to transmit the florigenic signals GA and HvFT1, which then provoke floral development by triggering expression of floral identity genes (Boden et al. 2014). A very similar set of floral identity genes were downregulated (i.e., *BFL*, *HvPAP24*, *HvCO6*) in GA-mediated late heading of the *Six-rowed spike2* (*vrs2*) mutant (Youssef et al. 2017a). *Vrs2*, which encodes a grass-specific SHORT INTERNODES (SHI) transcriptional regulator, is involved in pre-anthesis phase duration and time to heading by regulating sucrose, another florigenic signal, and GA levels during barley spike development (Youssef et al. 2017a). GA_3 leaf treatment completely rescued the late heading phenotype of *vrs2* mutants, indicating that lower GA levels in *vrs2* mutants predominantly caused the delay in floral induction and development (Youssef et al. 2017a).

Interestingly, the proximodistal distribution of GA levels detected in young, developing spike sections of wild-type plants resembled the known proximodistal floral patterning at anthesis, i.e., florets in the middle of the spike usually develop fastest and self-fertilize first; these findings from young spike sections show that the proximodistal GA distribution, which favors central floral progression, has already been established during early spike development (Youssef et al. 2017a). While GA-mediated floral development in barley appears to be executed, at least in part, by *BFL* expression (Boden et al. 2014; Youssef et al. 2017a), it is evidentially surprising how little we know about the function of *BFL* in Triticeae. Recent GWAS results in spring barley, though, suggest that alleles at the *BFL* locus might be important for variation in pre-anthesis phase duration and time to heading, specifically in carriers of the *ppd-H1* allele (Alqudah et al. 2014). Transgenic knockdown lines of its rice ortholog, *RICE FLORICAULA/LEAFY* (*RFL*; syn. *ABERRANT PANICLE ORGANIZATION2*, *APO2*), show late or no heading at all, which was in line with lowered transcript levels, whereas overexpression of *RFL/APO2* resulted in precocious floral induction (Rao et al. 2008). Moreover, mutants of *RFL/APO2* exhibited aberrant floral organ identities and loss of floral meristem identity, which similarly suggests a central role during inflorescence formation (Ikeda-Kawakatsu et al. 2012). *RFL/APO2* seems to regulate *OsSOC1* and *FRIZZY PANICLE* (*FZP*) transcripts (Rao et al. 2008), while it controls the transition from the inflorescence meristem to the spikelet meristem, cooperatively with *APO1* (syn. rice *UNUSUAL FLORAL ORGANS*, *OsUFO*) (Ikeda-Kawakatsu et al. 2012). A similar function with similar interactions may exist for *BFL*; however, this remains untested. Pre-anthesis phase durations have implications for spike development because they influence the number of spikelets that will abort prior to anthesis. Spikelet abortion in barley mainly appears in the apical spike part and can produce substantial spikelet loss. Alqudah and Schnurbusch (2014) found that higher spikelet survival was positively associated with longer pre-anthesis phase duration regardless of growth condition or row-type.

12.6.2 Branching

In addition to new climates, elite cultivars grow within an environment far removed from their pre-domestication history. The agronomic environment is characterized by several human-derived inputs such as enhanced mineral availability due to fertilizer application and shade from dense planting. Although rarely studied directly, cultivated varieties likely show altered plasticity in response to the agricultural environment compared to their wild relatives (Carriedo et al. 2016). Branching is exquisitely environmentally sensitive as plants use cues such as light quality, soil mineral levels and sugar status to balance branching benefits, such as accessing new canopy, with its significant energetic cost (Casal and Deregibus 1986, 1988; Mason et al. 2014; Drummond et al. 2015; Kebrom and Mullet 2015; Evers and Bastiaans 2016; see Teichmann and Muhr 2015) for an in-depth review. Regulation of bud activation and outgrowth contribute to branching plasticity (Teichmann and Muhr 2015). Tiller bud outgrowth is repressed by *Tb1* in maize, which also appears to be a central hub for environmental integration of branching (Sawers et al. 2005; Kebrom et al. 2006, 2012; Mason et al. 2014; Drummond et al. 2015). Although the canonical barley spike is unbranched, breeders have long observed branching spikes in the field depending on environmental conditions (H. Bull, pers comm). The extent and form of inflorescence branching in dicot models, rice and maize also responds to environmental control and often involves many similar players as in vegetative branching plasticity including *Tb1* and hormonal homeostasis. Interestingly, *vrs3* and *vrs5* phenotype expressivity is environmentally sensitive (Bull et al. 2017) suggesting that functional VRS3 and VRS5 buffer the two-rowed phenotype from environmental change, consistent with the that certain regulators act to stabilize developmental networks in the face of environmental variation. Tb1 suppresses lateral branching in response to shade through activation of *grassy tillers 1* (*gt1*; Whipple et al. 2011), the VRS1 homologue, suggesting a potentially conserved molecular relationships between VRS5 and VRS1; however, we do not know whether Tb1-like *VRS5* is involved in responding to environmental cues or whether it follows the same regulatory pathway as Tb1. Furthermore, genetic evidence indicates that *VRS1* is transcriptionally regulated directly or indirectly by VRS4, hinting that environmental signals may also be integrated differently. In fact, the *vrs4* mutant phenotype was more severe under spring conditions, suggesting an interaction between by VRS4 and environmental control (Koppulu et al. 2013).

12.6.3 Floral Structure

Basic floral structure tends to be canalized and species-specific. However, cold and short-day conditions can rescue the *Hooded* phenotype, suggested to reflect changes in hormonal homeostasis (Yagil and Stebbins 1969), consistent with Osnato et al. (2010). Multiple reports suggest interactions between floral architecture and environmental stress; for instance, shifts to carpelloidy in stamens induced by weather were reported in English barley (Gregory and Purvis 1947). However, despite the risks posed by climate change to crop security, the plasticity of floral form in cereals is relatively understudied.

12.7 Conclusions

Cultivated barley, *H. vulgare* ssp. *vulgare*, is the fourth most abundantly grown cereal in the world (www.fao.org/faostat) and is long associated with human societies. Currently, barley grain is used primarily for animal feed and as stock for malt during brewing and distilling; nonetheless, barley remains an important primary food staple in many parts of the world, especially in cases where barley is better suited to growing conditions compared to wheat such as the Far Eastern plateaus.

With the advent of high-quality genomic resources, the barley reference genome sequence will provide an invaluable community resource for cereal genetics and genomics. It may also furnish a much better contextualization of population genomic datasets and facilitate comparative genomic analyses with other Triticeae, such as wheat, in non-recombining regions that have been inaccessible to analysis of gene collinearity until now. But perhaps most importantly it will enable positional cloning and gene identification of mutant phenotypes following various approaches (Table 12.1). Regardless of the approach used, genetic analyses of mutants and functional characterization of the underlying genes may pave the way for a much better insight into barley inflorescence development and will help resolve the relationships between gene and inflorescence trait, which can support breeding as well as answer fundamental questions in cereal developmental biology.

Appendix

Genetic and Genomic Resources for Gene Cloning and Functional Analysis

A robust genetic map is the starting point for a good physical map, which in turn is crucial for identification of physical contigs harboring genes of interest in positional cloning studies. Over the past two decades, many barley genetic linkage maps became available based on different marker systems, including RFLPs (Graner et al. 1991; Kleinhofs et al. 1993), SSRs (Varshney et al. 2007), DArTs (Wenzl et al. 2006), ESTs (Potokina et al. 2008; Stein et al. 2007), SNPs (Close et al. 2009; Rostoks et al. 2005), and a combination of different marker systems (Hearnden et al. 2007; Marcel et al. 2007). A major advance was the development of the barley genome zippers by positioning the gene-based barley 454 sequence reads in synteny with the genes of rice, *Brachypodium*, and Sorghum (Mayer et al. 2011). These syntenic 454 reads were positioned in between the gene-based SNP markers reported in the consensus map developed by Close et al. (2009). A dense genetic map-based on Morex × Barke RIL population (3973 SNP markers) was developed by Comadran et al. (2012) and (IBSC 2012). Mascher et al. (2013a) further refined the Morex × Barke RIL map by introducing a strategy based on sequencing progenies of Morex × Barke of population (POPSEQ) that allowed de novo production of a genetically anchored linear assembly (an ultra dense genetic map with 4.3 million SNPs). Ariyadasa et al. (2014) developed a genome-wide physical map of barley by map-based sequencing (4.9 Gb representing 96% of the physical length). A draft genome sequence of barley was made available by the International Barley Sequencing Consortium (IBSC 2012). Further A high-quality barley reference genome assembly was developed by chromosome conformation capture mapping (HiC) to derive the linear order of sequences across the pericentromeric regions. Such resources greatly facilitate high-resolution trait mapping, gene isolation, and comparative genome or transcriptome studies. The availability of gold standard reference genome, physical maps and marker systems paved the way for several NGS-based fast-forward genetic analysis and gene cloning strategies in barley such as mapping-by-sequencing (Liu et al. 2014; Mascher et al. 2014) exome capture (Mascher et al. 2013b; Russell et al. 2016), and MutChromSeq (Sánchez-Martín et al. 2016). Gene cloning was also accelerated due to robust genetic analyses such as Genome-Wide Association scans (GWAS) benefitting from variation across germplasm (Comadran et al. 2012; Houston et al. 2015; Houston et al. 2013; Ramsay et al. 2011; Youssef et al. 2017).

Gene expression patterns can often reveal insight about possible gene function and links to mutant phenotypes. A high-quality RNA seq data set from 16 different tissues (including six inflorescence tissues) is available for all genes in barley and the expression data can be visualized through the BARLEX database (Colmsee et al. 2015). Apart from this, developmental stage (from SAM until AP) and position specific (LSM, CSM, and IM) reference transcriptome data has been generated from Bowman (Schnurbusch et al., Unpublished), which will provide the expression dynamics of inflorescence developmental genes across time and space.

Once the gene underlying a phenotype is identified, the further functional analysis is essential to substantiate the gene function. Towards this measure, barley has a rich collection of chemical and ionizing ration induced mutants defective in their inflorescence development. All of these mutants are genetically characterized (Lundqvist 2009; Scholz and Lehmann 1961) and have been systematically maintained in geneticist and breeder collections at the National Small Grain Collection, Aberdeen, ID (http://www.ars.usda.gov/Main/docs.htm?docid=2922), IPK Genebank and the Nordic Genetic Resource

Center, Alnarp, Sweden (https://www.nordgen.org/bgs/index.php). Several of these mutants have been backcrossed to (BW001 to BW979) a near-isogenic background in the genotype Bowman (Franckowiak et al. 1985) making their functional evaluation less cumbersome. Druka et al. (2011) performed SNP genotyping of all Bowman backcrossed mutants, which delimited the genomic introgression responsible for the mutant phenotype. The historical mutant resources together with SNP introgression data would be a wonderful starting point for the functional studies and positional isolation of responsible gene/s in these mutants.

Most of the historical barley developmental mutants identified in the past have more than one mutant allele by which the underlying gene function can be validated. However, for genes being isolated by conventional bi-parental genetic analyses or GWAS, and for those mutants lacking enough number of mutant alleles, it is necessary to validate identified gene candidates using reverse genetic approaches such as TILLING (Targeting Induced Local Lesions IN Genomes) or transgenics. As of today, four different TILLING populations are available for barley. These include (i) TILLMore, a sodium azide induced six-rowed TILLING population from Morex (Talamè et al. 2008). (ii) Two-rowed EMS (Ethyl Methyl Sulfonate) induced Barke TILLING population (Gottwald et al. 2009). (iii) Two-rowed EMS-induced Optic TILLING population (Caldwell et al. 2004) The other unpublished resources include EMS-induced TILLING lines from two-rowed cultivar Haruna Nijo (Sato et al. unpublished). These reverse genetic tools shall enable functional evaluation of genes in mutants of interest.

Testing genes of interest by transgenics is a common functional genetics approach (Hensel et al. 2011). Several transgenic approaches were established for barley by which a preferred gene can be complemented, downregulated, knocked-out or over-expressed. Complementation is a straight forward approach by which a mutated gene can be complemented by the functional wild-type form of the gene. The quantitative effect of a gene can be analyzed through approaches like RNA interference (RNAi), Virus-Induced Gene Silencing (VIGS) (Hein et al. 2005; Holzberg et al. 2002), which function through homology-dependant post-transcriptional gene silencing. However, these approaches suffer from potential off-target effects (gene family members). This problem can be circumvented through a recent technology called Transcriptional Activator-Like Effector Nucleases (TALENs), which allows sequence specific *in planta* gene targeting with typically no off-target effects in the genetic background of the host genome (Boch et al. 2009; Joung and Sander 2013; Zhang et al. 2013). This technology has already been proven successful in barley (Gurushidze et al. 2014; Wendt et al. 2013). Recently, the RNA-guided Cas9 system (CRISPR/Cas) has been proven successful to knockout genes in barley (Lawrenson et al. 2015). The comprehensive set of available genetic and genomic resources discussed will be useful to isolate and characterize genetic determinants underlying various developmental processes in barley.

Bibliography

Abebe T, Wise RP, Skadsen RW (2009) Comparative transcriptional profiling established the awn as the major photosynthetic organ of the barley spike while the lemma and the palea primarily protect the seed. Plant Genome 2:247–259

Åberg E, Wiebe G (1945) Ash content of barley awns and kernels as influenced by location, season and variety. J Agron 37:583–586

Aharoni A, Dixit S, Jetter R, Thoenes E, van Arkel G et al (2004) The SHINE clade of AP2 domain transcription factors activates wax biosynthesis, alters cuticle properties, and confers drought tolerance when overexpressed in arabidopsis. Plant Cell 16:2463–2480

Alqudah AM, Schnurbusch T (2014) Awn primordium to tipping is the most decisive developmental phase for spikelet survival in barley. Funct Plant Biol 41:424–436

Alqudah AM, Sharma R, Pasam RK, Graner A, Kilian B et al (2014) Genetic dissection of photoperiod response based on GWAS of pre-anthesis phase duration in spring barley. PLoS ONE 9:e113120

Alqudah AM, Koppolu R, Wolde GM, Graner A, Schnurbusch T (2016) The genetic architecture of barley plant stature. Front Genet 7

Amanda S-L, Harry K, Amanda F, Madelaine B (2017) Grass flowers: an untapped resource for floral evo-devo. J Syst Evol 55:525–541

Arber A (1934) The Gramineae: a study of cereal, bamboo, and grass. Cambridge University Press, New York

Ariyadasa R, Mascher M, Nussbaumer T, Schulte D, Frenkel Z, Poursarebani N, Zhou R, Steuernagel B, Gundlach H, Taudien S, Felder M, Platzer M, Himmelbach A, Schmutzer T, Hedley PE, Muehlbauer GJ, Scholz U, Korol A, Mayer KFX, Waugh R, Langridge P, Graner A, Stein N (2014) A sequence-ready physical map of barley anchored genetically by two million single-nucleotide polymorphisms. Plant Physiol 164:412–423

Aukerman MJ, Sakai H (2003) Regulation of flowering time and floral organ identity by a microRNA and its *APETALA2-like* target genes. The Plant Cell 15:2730–2741

Bell AD (1991) Plant form: an illustrated guide to flowering plant morphology. Oxford University Press, Oxford

Boch J, Scholze H, Schornack S, Landgraf A, Hahn S, Kay S, Lahaye T, Nickstadt A, Bonas U (2009) Breaking the code of DNA binding specificity of TAL-type III Effectors. Science 326:1509–1512

Boden SA, Weiss D, Ross JJ, Davies NW, Trevaskis B et al (2014) *EARLY FLOWERING3* regulates flowering in spring barley by mediating gibberellin production and *FLOWERING LOCUS T* expression. Plant Cell 26:1557–1569

Bommert P, Satoh-Nagasawa N, Jackson D, Hirano H-Y (2005) Genetics and evolution of inflorescence and flower development in grasses. Plant Cell Physiol 46:69–78

Bonnett OT (1935) The development of the barley spike. J Agric Res 51:451–457

Bortiri E, Chuck G, Vollbrecht E, Rocheford T, Martienssen R et al (2006) *ramosa2* encodes a LATERAL ORGAN BOUNDARY domain protein that determines the fate of stem cells in branch meristems of maize. Plant Cell 18:574–585

Bossinger GLU, Rohde W, Salamini F (1992) Genetics of plant development in barley. In: Munck L (ed) Barley genetics VI. Munksgaard International Publishers, Copenhagen, Denmark, pp 989–1017

Brennan M, Shepherd T, Mitchell S, Topp CFE, Hoad SP (2017) Husk to caryopsis adhesion in barley is influenced by pre- and post-anthesis temperatures through changes in a cuticular cementing layer on the caryopsis. BMC Plant Biol 17:169

Brown RH, Bregitzer P (2011) A *Ds* insertional mutant of a barley *mir172* gene results in indeterminate spikelet development. Crop Sci 51:1664–1672

Bull H, Casao MC, Zwirek M, Flavell AJ, Thomas WTB et al (2017) Barley SIX-ROWED SPIKE3 encodes a putative Jumonji C-type H3K9me2/me3 demethylase that represses lateral spikelet fertility. Nat Commun 8(1):936

Caldwell DG, McCallum N, Shaw P, Muehlbauer GJ, Marshall DF, Waugh R (2004) A structured mutant population for forward and reverse genetics in Barley (*Hordeum vulgare* L.). Plant J 40:143–150

Campoli C, Drosse B, Searle I, Coupland G, von Korff M (2012) Functional characterisation of *HvCO1*, the barley (*Hordeum vulgare*) flowering time ortholog of *CONSTANS*. Plant J 69:868–880

Campoli C, Pankin A, Drosse B, Casao CM, Davis SJ et al (2013) *HvLUX1* is a candidate gene underlying the *early maturity 10* locus in barley: phylogeny, diversity, and interactions with the circadian clock and photoperiodic pathways. New Phytol 199:1045–1059

Carriedo LG, Maloof JN, Brady SM (2016) Molecular control of crop shade avoidance. Curr Opin Plant Biol 30:151–158

Casal JJ (1988) Persistent effects of changes in phytochrome status on internode growth in light-grown mustard: Occurrence, kinetics and locus of perception. Planta 175:214–220

Casal JJ, Deregibus VA (1986) The effect of plant density on tillering: the involvement of R/FR ratio and the proportion of radiation intercepted per plant. Environ Exp Bot 26:365–371

Chen X (2004) A microRNA as a translational repressor of *APETALA2* in Arabidopsis flower development. Science 303:2022–2025

Chen A, Dubcovsky J (2012) Wheat TILLING mutants show that the vernalization gene *VRN1* down-regulates the flowering repressor *VRN2* in leaves but is not essential for flowering. PLoS Genet 8:e1003134

Chono M, Honda I, Zeniya H, Yoneyama K, Saisho D et al (2003) A semidwarf phenotype of barley *uzu* results from a nucleotide substitution in the gene encoding a putative brassinosteroid receptor. Plant Physiol 133:1209–1219

Chuck G, Muszynski M, Kellogg E, Hake S, Schmidt RJ (2002) The control of spikelet meristem identity by the *branched silkless1* gene in maize. Science 298:1238–1241

Close TJ, Bhat PR, Lonardi S, Wu YH, Rostoks N, Ramsay L, Druka A, Stein N, Svensson JT, Wanamaker S, Bozdag S, Roose ML, Moscou MJ, Chao SAM, Varshney RK, Szucs P, Sato K, Hayes PM, Matthews DE, Kleinhofs A, Muehlbauer GJ, DeYoung J, Marshall DF, Madishetty K, Fenton RD, Condamine P, Graner A, Waugh R (2009) Development and implementation of high-throughput SNP genotyping in barley. BMC Genomics 10:582

Cockram J, Jones H, Leigh FJ, O'Sullivan D, Powell W et al (2007) Control of flowering time in temperate cereals: genes, domestication, and sustainable productivity. J Exp Bot 58:1231–1244

Cockram J, Horsnell R, Soh EH, Norris C, O'Sullivan DM (2015) Molecular and phenotypic characterization of the alternative seasonal growth habit and flowering

time in barley (*Hordeum vulgare* ssp. *vulgare* L.). Mol Breed 35:165
Coen ES, Nugent JM (1994) Evolution of flowers and inflorescences. Development 1994:107–116
Colmsee C, Beier S, Himmelbach A, Schmutzer T, Stein N, Scholz U, Mascher M (2015) BARLEX—the Barley draft genome explorer. Mol Plant 8:964–966
Comadran J, Kilian B, Russell J, Ramsay L, Stein N, Ganal M, Shaw P, Bayer M, Thomas W, Marshall D, Hedley P, Tondelli A, Pecchioni N, Francia E, Korzun V, Walther A, Waugh R (2012) Natural variation in a homolog of Antirrhinum *CENTRORADIALIS* contributed to spring growth habit and environmental adaptation in cultivated barley. Nat Genet 44:1388–1392
Dabbert T, Okagaki RJ, Cho S, Heinen S, Boddu J et al (2010) The genetics of barley low-tillering mutants: *low number of tillers-1* (*lnt1*). Theor Appl Genet 121:705–715
Danyluk J, Kane NA, Breton G, Limin AE, Fowler DB et al (2003) *TaVRT-1*, a putative transcription factor associated with vegetative to reproductive transition in cereals. Plant Physiol 132:1849–1860
Debernardi JM, Lin H, Chuck G, Faris JD, Dubcovsky J (2017) *microRNA172* plays a crucial role in wheat spike morphogenesis and grain threshability. Development 144:1966–1975
Derbyshire P, Byrne ME (2013) *MORE SPIKELETS1* is required for spikelet fate in the inflorescence of brachypodium. Plant Physiol 161:1291–1302
Digel B, Pankin A, von Korff M (2015) Global transcriptome profiling of developing leaf and shoot apices reveals distinct genetic and environmental control of floral transition and inflorescence development in barley. Plant Cell 27:2318–2334
Djalali M (1970) Investigations on expressivity and penetrance of labile character of barley (*Hordeum vulgare* L.). Z Pflanzenzüchtung 63:274–322
Dockter C, Gruszka D, Braumann I, Druka A, Druka I et al (2014) Induced variations in brassinosteroid genes define barley height and sturdiness, and expand the green revolution genetic toolkit. Plant Physiol 166:1912–1927
Doebley J, Stec A, Hubbard L (1997) The evolution of apical dominance in maize. Nature 386:485–488
Doust A (2007) Architectural evolution and its implications for domestication in grasses. Ann Bot 100:941–950
Druka A, Franckowiak J, Lundqvist U, Bonar N, Alexander J, Houston K, Radovic S, Shahinnia F, Vendramin V, Morgante M, Stein N, Waugh R (2011) Genetic dissection of barley morphology and development. Plant Physiol 155:617–627
Drummond RSM, Janssen BJ, Luo Z, Oplaat C, Ledger SE et al (2015) Environmental control of branching in Petunia. Plant Physiol 168:735–751
Duan R, Xiong H, Wang A, Chen G (2015) Molecular mechanisms underlying hull-caryopsis adhesion/separation revealed by comparative transcriptomic analysis of covered/naked barley (*Hordeum vulgare* l.). Int J Mol Sci 16:14181
Eklund DM, Ståldal V, Valsecchi I, Cierlik I, Eriksson C et al (2010) The *Arabidopsis thaliana* STYLISH1 protein acts as a transcriptional activator regulating auxin biosynthesis. Plant Cell 22:349–363
Endress PK (2010) Disentangling confusions in inflorescence morphology: patterns and diversity of reproductive shoot ramification in angiosperms. J Syst Evol 48:225–239
Endress PK (2011) Evolutionary diversification of the flowers in angiosperms. Am J Bot 98:370–396
Engeldow FL (1920) Inheritance in barley. I. The lateral florets and the rachilla. J Genet 10:93–108
Engledow F (1921) Inheritance in barley. II. The awn and the lateral floret. J Agric Sci 11:159–196
Engledow FL (1924) Inheritance in barley. III. The awn and the lateral floret (cont'd): fluctuation: a linkage: multiple allelomorphs. J Genet 14:49–87
Eveland AL, Goldshmidt A, Pautler M, Morohashi K, Liseron-Monfils C et al (2014) Regulatory modules controlling maize inflorescence architecture. Genome Res 24:431–443
Evers JB, Bastiaans L (2016) Quantifying the effect of crop spatial arrangement on weed suppression using functional-structural plant modelling. J Plant Res 129:339–351
Faure S, Higgins J, Turner A, Laurie DA (2007) The *FLOWERING LOCUS T*-like gene family in barley (*Hordeum vulgare*). Genetics 176:599–609
Faure S, Turner AS, Gruszka D, Christodoulou V, Davis SJ et al (2012) Mutation at the circadian clock gene *EARLY MATURITY 8* adapts domesticated barley (*Hordeum vulgare*) to short growing seasons. P Natl Acad Sci USA 109:8328–8333
Fernández GJ, Wilson ZA (2014) A barley PHD finger transcription factor that confers male sterility by affecting tapetal development. Plant Biotechnol 12:765–777
Finlayson SA, Krishnareddy SR, Kebrom TH, Casal JJ (2010) Phytochrome regulation of branching in arabidopsis. Plant Physiol 152:1914–1927
Ford-Lloyd B, Engels JMM, Jackson M (2014) Genetic resources and conservation challenges under the threat of climate change. In: Jackson M, Ford-Lloyd, B., Parry, M (eds) Plant genetic resources and climate change
Forster BP, Franckowiak JD, Lundqvist U, Lyon J, Pitkethly I et al (2007) The barley phytomer. Ann Bot 100:725–733
Francia E, Tondelli A, Rizza F, Badeck FW, Li Destri Nicosia O et al (2011) Determinants of barley grain yield in a wide range of Mediterranean environments. Field Crop Res 120:169–178
Franckowiack JD, Konishi T (1997) Naked caryopsis. Barley Genet Newslett 26:51–52
Franckowiak JD (1997a) *Smooth awn 2*. Barley Genet Newslett 26:289
Franckowiak JD (1997b) *Smooth awn 1*. Barley Genet Newslett 26:261
Franckowiak JD (1997c) *Rattail spike 1*. Barley Genet Newsl 26:87

Franckowiak JD (1997d) *Multiflorus 2*. Barley Genet Newsl 26:232
Franckowiak JD (2005) *Ovaryless 1*. Barley Genet Newslett 35:191
Franckowiak JD (2010a) *Accordian rachis 1*. Barley Genet Newsl 40:56–57
Franckowiak JD (2010b) *Accordian rachis 2*. Barley Genet Newsl 40:65–66
Franckowiak JD (2010c) *Accordian rachis 3*. Barley Genet Newsl 40:85–86
Franckowiak JD (2010d) *Small lateral spikelets 1*. Barley Genet Newsl 40:78
Franckowiak JD (2013) *Absent lower laterals*. Barley Genet Newsl 43:74–75
Franckowiak JD (2014a) *Long glume awn 1*. Barley Genet Newslett 44:183–184
Franckowiak JD (2014b) *Rough awn 5*. Barley Genet Newsl 44:112
Franckowiak J, Lundqvist U (2010) Descriptions of barley genetic stocks for 2010. Barley Genet Newslett 40:45–177
Franckowiak JD, Lundqvist U (2011a) Descriptions of barley genetic stocks for 2011. Barley Genet Newslett 40:45–177
Franckowiak JD, Lundqvist U (2011b) *Calcaroides-c*. Barley Genet Newsl 41:195–196
Franckowiak JD, Lundqvist U (2013) Barley Genet Newslett 43:64–66
Franckowiak JD, Lundqvist U (2014) Barley Genet Newslett 44:93–94
Franckowiak JD, Foster AE, Pederson VD, Pyler RE (1985) Registration of 'Bowman' barley. Crop Sci 25:883
Franckowiak JD, Kleinhofs A, Lundqvist U (2005) Descriptions of barley genetic stocks for 2005. Barley Genet Newslett 35:155–210
Gaines RL, Bechtel DB, Pomeranz Y (1985) A microscopic study on the development of a layer in barley that causes hull-caryopsis adherence. Cereal Chem 62:35–40
Gawroński P, Ariyadasa R, Himmelbach A, Poursarebani N, Kilian B et al (2014) A distorted circadian clock causes early flowering and temperature-dependent variation in spike development in the *Eps-3A*^{m} mutant of einkorn wheat. Genetics 196:1253–1261
Gottwald S, Bauer P, Komatsuda T, Lundqvist U, Stein N (2009) TILLING in the two-rowed barley cultivar 'Barke' reveals preferred sites of functional diversity in the gene *HvHox1*. BMC Res Notes 2:1–14
Gramzow L, Theissen G (2010) A hitchhiker's guide to the MADS world of plants. Genome Biol 11:214
Graner A, Jahoor A, Schondelmaier J, Siedler H, Pillen K, Fischbeck G, Wenzel G, Herrmann RG (1991) Construction of an RFLP map of barley. Theor Appl Genet 83:250–256
Greenwood JR, Finnegan EJ, Watanabe N, Trevaskis B, Swain SM (2017) New alleles of the wheat domestication gene *Q* reveal multiple roles in growth and reproductive development. Development 144:1959–1965
Gregory FG, Purvis ON (1947) Abnormal flower development in barley involving sex reversal. Nature 160:221–222
Gurushidze M, Hensel G, Hiekel S, Schedel S, Valkov V, Kumlehn J (2014) True-breeding targeted gene knock-out in barley using designer TALE-nuclease in haploid cells. PLoS ONE 9:e92046
Gustafsson ÅA, Hagberg A, Lundqvist U, Persson G (1969) A proposed system of symbols for the collection of barley mutants at Svalöv. Hereditas 62:409–414
Harlan HV (1920) Daily development of kernels of Hannchen barley from flowering to maturity, at Aberdeen Idaho. J Agric Res 19:393–429
Harlan JR, Zohary D (1966) Distribution of wild wheats and barley. Science 153:1074–1080
Harvey BL, Reinbergs E, Somaroo BH (1968) Inheritance of female sterility in barley. Canad J Plant Science 48:417–418
Hearnden PR, Eckermann PJ, McMichael GL, Hayden MJ, Eglinton JK, Chalmers KJ (2007) A genetic map of 1,000 SSR and DArT markers in a wide barley cross. Theor Appl Genet 115:383–391
Hedden P (2003) The genes of the green revolution. Trends Genet 19:5–9
Hein I, Pacak MB, Hrubikova K, Williamson S, Dinesen M, Soenderby IE, Sundar S, Jarmolowski A, Shirasu K, Lacomme C (2005) Virus-induced gene silencing-based functional characterization of genes associated with powdery mildew resistance in barley. Plant Physiol 138:2155–2164
Helback H (1959) Domestication of food plants in the old world. Joint efforts by botanists and archeologists illuminate the obscure history of plant domestication. Science 130:365–372
Hemming MN, Peacock WJ, Dennis ES, Trevaskis B (2008) Low-temperature and daylength cues are integrated to regulate *FLOWERING LOCUS T* in barley. Plant Physiol 147:355–366
Hensel G, Himmelbach A, Chen W, Douchkov DK, Kumlehn J (2011) Transgene expression systems in the Triticeae cereals. J Plant Physiol 168:30–44
Hepworth SR, Zhang Y, McKim S, Li X, Haughn GW (2005) BLADE-ON-PETIOLE–dependent signaling controls leaf and floral patterning in arabidopsis. Plant Cell 17:1434–1448
Hoad SP, Brennan M, Wilson GW, Cochrane PM (2016) Hull to caryopsis adhesion and grain skinning in malting barley: identification of key growth stages in the adhesion process. J Cereal Sci 68:8–15
Holzberg S, Brosio P, Gross C, Pogue GP (2002) Barley stripe mosaic virus-induced gene silencing in a monocot plant. Plant J 30:315–327
Houston K, Druka A, Bonar N, Macaulay M, Lundqvist U et al (2012) Analysis of the barley bract suppression gene *Trd1*. Theor Appl Genet 125:33–45
Houston K, McKim SM, Comadran J, Bonar N, Druka I, Uzrek N, Cirillo E, Guzy-Wrobelska J, Collins NC, Halpin C, Hansson M, Dockter C, Druka A, Waugh R (2013) Variation in the interaction between alleles of

HvAPETALA2 and *microRNA172* determines the density of grains on the barley inflorescence. P Natl Acad Sci USA 110:16675–16680
Houston K, Burton RA, Sznajder B, Rafalski AJ, Dhugga KS, Mather DE, Taylor J, Steffenson BJ, Waugh R, Fincher GB (2015) A Genome-wide association study for culm cellulose content in barley reveals candidate genes co-expressed with members of the CELLULOSE SYNTHASE a gene family. PLoS ONE 10:e0130890
IBSC (2012) A physical, genetic and functional sequence assembly of the barley genome. Nature 491:711–716
Ikeda-Kawakatsu K, Maekawa M, Izawa T, Itoh J-I, Nagato Y (2012) *ABERRANT PANICLE ORGANIZATION 2/RFL*, the rice ortholog of arabidopsis *LEAFY*, suppresses the transition from inflorescence meristem to floral meristem through interaction with *APO1*. Plant J 69:168–180
Ivanova K (1937) A new character in barley, "third outer glume": its inheritance and linkage with the colour of the flowering glumes. Bull Appl Bot Genet Plant Breed. (Russia) Series II 7:339–353
Jeon J-S, Jang S, Lee S, Nam J, Kim C et al (2000) *leafy hull sterile1* is a homeoticmutation in a rice MADS box gene affecting rice flower development. Plant Cell 12:871–884
Jia Q, Zhang J, Westcott S, Zhang X-Q, Bellgard M et al (2009) GA-20 oxidase as a candidate for the semidwarf gene *sdw1/denso* in barley. Funct Integr Genomic 9:255–262
Jost M, Taketa S, Mascher M, Himmelbach A, Yuo T et al (2016) A homolog of *Blade-On-Petiole 1* and *2* (*BOP1/2*) controls internode length and homeotic changes of the barley inflorescence. Plant Physiol 171:1113–1127
Joung JK, Sander JD (2013) INNOVATION TALENs: a widely applicable technology for targeted genome editing. Nat Rev Mol Cell Biol 14:49–55
Kang Y, Khan S, Ma X (2009) Climate change impacts on crop yield, crop water productivity and food security—a review. Prog Nat Sci 19:1665–1674
Kannangara R, Branigan C, Liu Y, Penfield T, Rao V et al (2007) The transcription factor *WIN1/SHN1* regulates cutin biosynthesis in *Arabidopsis thaliana*. Plant Cell 19:1278–1294
Kebrom TH, Mullet JE (2015) Photosynthetic leaf area modulates tiller bud outgrowth in sorghum. Plant Cell Environ 38:1471–1478
Kebrom TH, Burson BL, Finlayson SA (2006) Phytochrome B represses *Teosinte Branched1* expression and induces sorghum axillary bud outgrowth in response to light signals. Plant Physiol 140:1109–1117
Kebrom TH, Chandler PM, Swain SM, King RW, Richards RA et al (2012) Inhibition of tiller bud outgrowth in the *tin* mutant of wheat is associated with precocious internode development. Plant Physiol 160:308–318
Kellogg EA (2001) Evolutionary history of the grasses. Plant Physiol 125:1198–1205
Kellogg EA (2015) Flowering plants. Monocots Poaceae. In: Kubitzky K (ed) The families and genera of vascular plants, p 408
Kellogg E, Camara P, Rudall P, Ladd P, Malcomber S et al (2013) Early inflorescence development in the grasses (Poaceae). Front Plant Sci 4
Kikuchi R, Kawahigashi H, Oshima M, Ando T, Handa H (2011) The differential expression of *HvCO9*, a member of the *CONSTANS*-like gene family, contributes to the control of flowering under short-day conditions in barley. J Exp Bot 63:773–784
Kirby EJM, Appleyard M (1987) Cereal development guide, 2nd edn. Arable Unit, National Agricultural Centre, Warwickshire, Kenilworth, England
Kirby EJM, Riggs TJ (1978) Developmental consequences of two-row and six-row ear type in spring barley: 2. Shoot apex, leaf and tiller development. J Agric Sci 91:207–216
Kleinhofs A (2013) *Ovaryless 2*. Barley Genet Newslett 43:169
Kleinhofs A, Franckowiak JD (2013) *Multiovary 1*. Barley Genet Newslett 43:59–60
Kleinhofs A, Kilian A, Saghai Maroof MA, Biyashev RM, Hayes P, Chen FQ, Lapitan N, Fenwick A, Blake TK, Kanazin V, Ananiev E, Dahleen L, Kudrna D, Bollinger J, Knapp SJ, Liu B, Sorrells M, Heun M, Franckowiak JD, Hoffman D, Skadsen R, Steffenson BJ (1993) A molecular, isozyme and morphological map of the barley (Hordeum vulgare) genome. Theor Appl Genet 86:705–712
Komatsu M, Chujo A, Nagato Y, Shimamoto K, Kyozuka J (2003) *FRIZZY PANICLE* is required to prevent the formation of axillary meristems and to establish floral meristem identity in rice spikelets. Development 130:3841–3850
Komatsuda T, Maxim P, Senthil N, Mano Y (2004) High-density AFLP map of *nonbrittle rachis 1* (*btr1*) and *2* (*btr2*) genes in barley (*Hordeum vulgare* L.). Theor Appl Genet 109:986–995
Komatsuda T, Pourkheirandish M, He C, Azhaguvel P, Kanamori H et al (2007) Six-rowed barley originated from a mutation in a homeodomain-leucine zipper I-class homeobox gene. P Natl Acad Sci USA 104:1424–1429
Konishi T, Franckowiak JD (2002) *Multiovary 3*. 32:101
Koppolu R, Anwar N, Sakuma S, Tagiri A, Lundqvist U et al (2013) Six-rowed spike4 (*Vrs4*) controls spikelet determinacy and row-type in barley. P Natl Acad Sci USA 110:13198–13203
Kuusk S, Sohlberg JJ, Long JA, Fridborg I, Sundberg E (2002) *STY1* and *STY2* promote the formation of apical tissues during arabidopsis gynoecium development. Development 129:4707–4717
Laudencia-Chingcuanco D, Hake S (2002) The *indeterminate floral apex1* gene regulates meristem determinacy and identity in the maize inflorescence. Development 129:2629–2638
Laurie DA, Pratchett N, Snape JW, Bezant JH (1995) RFLP mapping of five major genes and eight quantitative trait loci controlling flowering time in a

winter × spring barley (*Hordeum vulgare* L.) cross. Genome 38:575–585

Lawrenson T, Shorinola O, Stacey N, Li C, Østergaard L, Patron N, Uauy C, Harwood W (2015) Induction of targeted, heritable mutations in barley and *Brassica oleracea* using RNA-guided Cas9 nuclease. Genome Biol 16:258

Leonard WH (1942) Inheritance of reduced lateral spikelet appendages in the Nudihaxtoni variety of barley. J Am Soc Agron 34:211–221

Leyser O (2009) The control of shoot branching: an example of plant information processing. Plant Cell Environ 32:694–703

Liller CB, Neuhaus R, von Korff M, Koornneef M, van Esse W (2015) Mutations in barley row type genes have pleiotropic effects on shoot branching. PLoS ONE 10:e0140246

Liller CB, Walla A, Boer MP, Hedley P, Macaulay M et al (2017) Fine mapping of a major QTL for awn length in barley using a multiparent mapping population. Theor Appl Genet 130:269–281

Liu H, Bayer M, Druka A, Russell JR, Hackett CA, Poland J, Ramsay L, Hedley PE, Waugh R (2014) An evaluation of genotyping by sequencing (GBS) to map the Breviaristatum-e (ari-e) locus in cultivated barley. BMC Genomics 15:104

Lobell DB, Gourdji SM (2012) The influence of climate change on global crop productivity. Plant Physiol 160:1686–1697

Lombardo F, Yoshida H (2015) Interpreting lemma and palea homologies: a point of view from rice floral mutants. Front Plant Sci 6

Lord EM (1981) Cleistogamy: a tool for the study of floral morphogenesis, function and evolution. Bot Rev 47:421–449

Lundqvist U (2009) Eighty years of Scandinavian barley mutation genetics and breeding. In: Shu QY (ed) Induced plant mutations in the genomics era food and agricultural organization of the United Nations, Rome, pp 39–43

Lundqvist U, Franckowiak JD (2002) *Calcaroides-e*. Barley Genet Newsl 32:123

Lundqvist U, Franckowiak JD (2010a) *Compositum 1*. Barley Genet Newsl 40:118–119

Lundqvist U, Franckowiak JD (2010b) *Calcaroides-d*. Barley Genet Newsl 40:58–59

Lundqvist U, Franckowiak JD (2011) *Accordian rachis 4*. Barley Genet Newsl 41:201

Lundqvist U, Franckowiak JD (2014) *Calcaroides-b*. Barley Genet Newsl 44:197–198

Lundqvist U, Lundqvist A (1988) Induced intermedium mutants in barley: origin, morphology and inheritance. Hereditas 108:13–26

Lundqvist U, Franckowiak JD, Konishi T (1997) New and revised descriptions of barley genes. Barley Genet Newslett 26:22–516

Magallón S, Gómez-Acevedo S, Sánchez-Reyes LL, Hernández-Hernández T (2015) A metacalibrated time-tree documents the early rise of flowering plant phylogenetic diversity. New Phytol 207:437–453

Malcomber ST, Preston JC, Reinheimer R, Kossuth J, Kellogg EA (2006) Developmental gene evolution and the origin of grass inflorescence diversity. Adv Bot Res 44:425–481

Marcel TC, Varshney RK, Barbieri M, Jafary H, de Kock MJD, Graner A, Niks RE (2007) A high-density consensus map of barley to compare the distribution of QTLs for partial resistance to *Puccinia hordei* and of defence gene homologues. Theor Appl Genet 114:487–500

Mascher M, Muehlbauer GJ, Rokhsar DS, Chapman J, Schmutz J, Barry K, Muñoz-Amatriaín M, Close TJ, Wise RP, Schulman AH, Himmelbach A, Mayer KFX, Scholz U, Poland JA, Stein N, Waugh R (2013a) Anchoring and ordering NGS contig assemblies by population sequencing (POPSEQ). Plant J 76:718–727

Mascher M, Richmond TA, Gerhardt DJ, Himmelbach A, Clissold L, Sampath D, Ayling S, Steuernagel B, Pfeifer M, D'Ascenzo M, Akhunov ED, Hedley PE, Gonzales AM, Morrell PL, Kilian B, Blattner FR, Scholz U, Mayer KFX, Flavell AJ, Muehlbauer GJ, Waugh R, Jeddeloh JA, Stein N (2013b) Barley whole exome capture: a tool for genomic research in the genus *Hordeum* and beyond. Plant J 76:494–505

Mascher M, Jost M, Kuon J-E, Himmelbach A, Aßfalg A, Beier S, Scholz U, Graner A, Stein N (2014) Mapping-by-sequencing accelerates forward genetics in barley. Genome Biol 15:R78

Mascher M, Gundlach H, Himmelbach A, Beier S, Twardziok SO, Wicker T, Radchuk V, Dockter C, Hedley PE, Russell J, Bayer M, Ramsay L, Liu H, Haberer G, Zhang XQ, Zhang Q, Barrero RA, Li L, Taudien S, Groth M, Felder M, Hastie A, Šimková H, Staňková H, Vrána J, Chan S, Muñoz-Amatriaín M, Ounit R, Wanamaker S, Bolser D, Colmsee C, Schmutzer T, Aliyeva-Schnorr L, Grasso S, Tanskanen J, Chailyan A, Sampath D, Heavens D, Clissold L, Cao S, Chapman B, Dai F, Han Y, Li H, Li X, Lin C, McCooke JK, Tan C, Wang P, Wang S, Yin S, Zhou G, Poland JA, Bellgard MI, Borisjuk L, Houben A, Doležel J, Ayling S, Lonardi S, Kersey P, Langridge P, Muehlbauer GJ, Clark MD, Caccamo M, Schulman AH, Mayer KFX, Platzer M, Close TJ, Scholz U, Hansson M, Zhang G, Braumann I, Spannagl M, Li C, Waugh R, Stein N (2017) A chromosome conformation capture ordered sequence of the barley genome. Nature 544:427–433

Mason MG, Ross JJ, Babst BA, Wienclaw BN, Beveridge CA (2014) Sugar demand, not auxin, is the initial regulator of apical dominance. P Natl Acad Sci USA 111:6092–6097

Mayer K, Martis M, Hedley P, Simková H, Liu H, Morris J, Steuernagel B, Taudien S, Roessner S, Gundlach H, Kubaláková M, Suchánková P, Murat F, Felder M, Nussbaumer T, Graner A, Salse J, Endo T, Sakai H, Tanaka T, Itoh T, Sato K, Platzer M, Matsumoto T, Scholz U, Dolezel J, Waugh R, Stein N (2011) Unlocking the barley genome by chromosomal and comparative genomics. Plant cell 23:1249–1263

McKim SM, Stenvik G-E, Butenko MA, Kristiansen W, Cho SK et al (2008) The *BLADE-ON-PETIOLE* genes are essential for abscission zone formation in arabidopsis. Development 135:1537–1546

McSteen P, Laudencia-Chingcuanco D, Colasanti J (2000) A floret by any other name: control of meristem identity in maize. Trends Plant Sci 5:61–66

Mizuno N, Nitta M, Sato K, Nasuda S (2012) A wheat homologue of *PHYTOCLOCK 1* is a candidate gene conferring the early heading phenotype to einkorn wheat. Genes Genet Sys 87:357–367

Muller KJ, Romano N, Gerstner O, Garcia-Marotot F, Pozzi C et al (1995) The barley Hooded mutation caused by a duplication in a homeobox gene intron. Nature 374:727–730

Nair SK, Wang N, Turuspekov Y, Pourkheirandish M, Sinsuwongwat S, Chen G, Sameri M, Tagiri A, Honda I, Watanabe Y, Kanamori H, Wicker T, Stein N, Nagamura Y, Matsumoto T, Komatsuda T (2010) Cleistogamous flowering in barley arises from the suppression of microRNA-guided *HvAP2* mRNA cleavage. Proc Natl Acad Sci USA 107:490–495

Nečas J (1963) Inheritance of development of lateral florets in spikelets of barley spike. Biol Plant 5:89–99

Osnato M, Stile MR, Wang Y, Meynard D, Curiale S et al (2010) Cross talk between the KNOX and ethylene pathways is mediated by intron-binding transcription factors in barley. Plant Physiol 154:1616–1632

Pankin A, Campoli C, Dong X, Kilian B, Sharma R et al (2014) Mapping-by-sequencing identifies *HvPHYTOCHROME C* as a candidate gene for the *early maturity 5* locus modulating the circadian clock and photoperiodic flowering in barley. Genetics 198:383–396

Potokina E, Druka A, Luo ZW, Wise R, Waugh R, Kearsey M (2008) Gene expression quantitative trait locus analysis of 16,000 barley genes reveals a complex pattern of genome-wide transcriptional regulation. Plant J 53:90–101

Pourkheirandish M, Hensel G, Kilian B, Senthil N, Chen G, Sameri M, Azhaguvel P, Sakuma S, Dhanagond S, Sharma R, Mascher M, Himmelbach A, Gottwald S, Nair SK, Tagiri A, Yukuhiro F, Nagamura Y, Kanamori H, Matsumoto T, Willcox G, Middleton CP, Wicker T, Walther A, Waugh R, Fincher GB, Stein N, Kumlehn J, Sato K, Komatsuda T (2015) Evolution of the grain dispersal system in barley. Cell 162:527–539

Poursarebani N, Seidensticker T, Koppolu R, Trautewig C, Gawroński P et al (2015) The genetic basis of composite spike form in barley and 'miracle-wheat'. Genetics 201:155–165

Prusinkiewicz P, Erasmus Y, Lane B, Harder LD, Coen E (2007) Evolution and development of inflorescence architectures. Science 316:1452–1456

Pozzi C, Faccioli P, Terzi V, Stanca AM, Cerioli S et al (2000) Genetics of mutations affecting the development of a barley floral bract. Genetics 154:1335–1346

Prasad K, Sriram P, Kumar SC, Kushalappa K, Vijayraghavan U (2001) Ectopic expression of rice *OsMADS1* reveals a role in specifying the lemma and palea, grass floral organs analogous to sepals. Dev Genes Evol 211:281–290

Prasad K, Parameswaran S, Vijayraghavan U (2005) *OsMADS1*, a rice MADS-box factor, controls differentiation of specific cell types in the lemma and palea and is an early-acting regulator of inner floral organs. Plant J 43:915–928

Ramsay L, Comadran J, Druka A, Marshall DF, Thomas WTB, Macaulay M, MacKenzie K, Simpson C, Fuller J, Bonar N, Hayes PM, Lundqvist U, Franckowiak JD, Close TJ, Muehlbauer GJ, Waugh R (2011) *INTERMEDIUM-C*, a modifier of lateral spikelet fertility in barley, is an ortholog of the maize domestication gene *TEOSINTE BRANCHED 1*. Nat Genet 43:169–172

Rao NN, Prasad K, Kumar PR, Vijayraghavan U (2008) Distinct regulatory role for *RFL*, the rice *LFY* homolog, in determining flowering time and plant architecture. P Natl Acad Sci USA 105:3646–3651

Rebetzke GJ, Bonnett DG, Reynolds MP (2016) Awns reduce grain number to increase grain size and harvestable yield in irrigated and rainfed spring wheat. J Exp Bot 67:2573–2586

Remizowa MV, Rudall PJ, Choob VV, Sokoloff DD (2013) Racemose inflorescences of monocots: structural and morphogenetic interaction at the flower/inflorescence level. Ann Bot 112:1553–1566

Richardson A, Rebocho AB, Coen E (2016) Ectopic *KNOX* expression affects plant development by altering tissue cell polarity and identity. Plant Cell 28:2079–2096

Roig C, Pozzi C, Santi L, Müller J, Wang Y et al (2004) Genetics of barley hooded suppression. Genetics 167:439–448

Rossini L, Okagaki R, Druka A, Muehlbauer GJ (2014) Shoot and inflorescence architecture. In: Kumlehn J, Stein N (eds) Biotechnological approaches to barley improvement. Springer, Berlin, pp 55–80

Rossini L, Vecchietti A, Nicoloso L, Stein N, Franzago S et al (2006) Candidate genes for barley mutants involved in plant architecture: an *in silico* approach. Theor Appl Genet 112:1073–1085

Rostoks N, Mudie S, Cardle L, Russell J, Ramsay L, Booth A, Svensson JT, Wanamaker SI, Walia H, Rodriguez EM, Hedley PE, Liu H, Morris J, Close TJ, Marshall DF, Waugh R (2005) Genome-wide SNP discovery and linkage analysis in barley based on genes responsive to abiotic stress. Mol Genet Genomics 274:515–527

Russell J, Mascher M, Dawson IK, Kyriakidis S, Calixto C, Freund F, Bayer M, Milne I, Marshall-Griffiths

T, Heinen S, Hofstad A, Sharma R, Himmelbach A, Knauft M, van Zonneveld M, Brown JWS, Schmid K, Kilian B, Muehlbauer GJ, Stein N, Waugh R (2016) Exome sequencing of geographically diverse barley landraces and wild relatives gives insights into environmental adaptation. Nat Genet 48:1024–1030

Sakuma S, Salomon B, Komatsuda T (2011) The domestication syndrome genes responsible for the major changes in plant form in the Triticeae crops. Plant Cell Physiol 52:738–749

Sakuma S, Lundqvist U, Kakei Y, Thirulogachandar V, Suzuki T et al (2017) Extreme suppression of lateral floret development by a single amino acid change in the VRS1 transcription factor. Plant Physiol 175:1720–1731

Sánchez-Martín J, Steuernagel B, Ghosh S, Herren G, Hurni S, Adamski N, Vrána J, Kubaláková M, Krattinger SG, Wicker T, Doležel J, Keller B, Wulff BBH (2016) Rapid gene isolation in barley and wheat by mutant chromosome sequencing. Genome Biol 17:221

Sang T (2009) Genes and mutations underlying domestication transitions in grasses. Plant Physiol 149:63–70

Santi L, Wang Y, Stile MR, Berendzen K, Wanke D et al (2003) The GA octodinucleotide repeat binding factor BBR participates in the transcriptional regulation of the homeobox gene *Bkn3*. Plant J 34:813–826

Sawers RJH, Sheehan MJ, Brutnell TP (2005) Cereal phytochromes: targets of selection, targets for manipulation? Trends Plant Sci 10:138–143

Schaller CW, Qualset CO (1975) Isogenic analysis of productivity barley: interaction of genes affecting awn length and leaf-spotting. Crop Sci 15:378–382

Scholz F, Lehmann C (1961) Die gaterslebener mutanten der saatgerste in beziehung zur formenmannigfaltigkeit der art *Hordeum vulgare* L.s.l III. Die Kulturpflanze 9:230–272

Scotland RW, Wortley AH (2003) How many species of seed plants are there? Taxon 52:101–104

Smeekens S, Ma J, Hanson J, Rolland F (2010) Sugar signals and molecular networks controlling plant growth. Curr Opin Plant Biol 13:273–278

Sohlberg JJ, Myrenås M, Kuusk S, Lagercrantz U, Kowalczyk M et al (2006) *STY1* regulates auxin homeostasis and affects apical–basal patterning of the arabidopsis gynoecium. Plant J 47:112–123

Sreenivasulu N, Schnurbusch T (2012) A genetic playground for enhancing grain number in cereals. Trends Plant Sci 17:91–101

Stebbins GL, Yagil E (1966) The morphogenetic effects of the hooded gene in barley. I. The course of development in hooded and awned genotypes. Genetics 54:727–741

Stein N, Prasad M, Scholz U, Thiel T, Zhang H, Wolf M, Kota R, Varshney R, Perovic D, Grosse I, Graner A (2007) A 1,000-loci transcript map of the barley genome: new anchoring points for integrative grass genomics. Theor Appl Genet 114:823–839

Stenvik GE, Butenko MA, Urbanowicz BR, Rose JK, and Aalen RB (2006) Overexpression of INFLORESCENCE DEFICIENT IN ABSCISSION activates cell separation in vestigial abscission zones in Arabidopsis. Plant Cell 18:1467–1476

Sun Q, Zhou D-X (2008) Rice jmjC domain-containing gene *JMJ706* encodes H3K9 demethylase required for floral organ development. P Natl Acad Sci USA 105:13679–13684

Takahashi R (1955) The origin and evolution of cultivated barley. Academic 7

Takahashi R (1987) Genetic features of East Asian barleys. In: Yasuda S, Konishi T (eds) Barley genetics V. Proceedings of fifth international barley genetics symposium, Okayama, 1986. Sanyo Press Co., Okayama, pp 7–20

Takahashi R, Yasuda S (1971) Genetics of earliness and growth habit in barley. In: Nilan R (ed) Barley genetics II. Proceedings of the second international barley genetics symposium. Washington State University Press, Pullman, WA, pp 388–408

Takahashi R, Yamamoto J, Yasuda S, Itano Y (1953) Inheritance and linkage studies in barley. Ber Ohara Inst Landwirtschaftliche Forsch 10:29–53

Takeda K, Saito W (1988) Inheritance of the percentage of missing lateral florets in '*irregurale*' barley. Jap J Breed 38:72–80

Taketa S, Amano S, Tsujino Y, Sato T, Saisho D et al (2008) Barley grain with adhering hulls is controlled by an ERF family transcription factor gene regulating a lipid biosynthesis pathway. P Natl Acad Sci USA 105:4062–4067

Taketa S, Yuo T, Sakurai Y, Miyake S, Ichii M (2011) Molecular mapping of the *short awn 2* (*lks2*) and *dense spike 1* (*dsp1*) genes on barley chromosome 7H. Breed Sci 61:80–85

Talamè V, Bovina R, Sanguineti MC, Tuberosa R, Lundqvist U, Salvi S (2008) TILLMore, a resource for the discovery of chemically induced mutants in barley. Plant Biotechnol J 6:477–485

Tanto Hadado T, Rau D, Bitocchi E, Papa R (2010) Adaptation and diversity along an altitudinal gradient in Ethiopian barley (*Hordeum vulgare* L.) landraces revealed by molecular analysis. BMC Plant Biol 10:121

Teichmann T, Muhr M (2015) Shaping plant architecture. Front Plant Sci 6

Thirulogachandar V, Alqudah AM, Koppolu R, Rutten T, Graner A et al (2017) Leaf primordium size specifies leaf width and vein number among row-type classes in barley. Plant J 91:601–612

Tondelli A, Francia E, Visioni A, Comadran J, Mastrangelo AM et al (2014) QTLs for barley yield adaptation to Mediterranean environments in the 'Nure' × 'Tremois' biparental population. Euphytica 197:73–86

Trevaskis B, Tadege M, Hemming MN, Peacock WJ, Dennis ES et al (2007) *Short vegetative phase*-Like MADS-box genes inhibit floral meristem identity in barley. Plant Physiol 143:225–235

Turner A, Beales J, Faure S, Dunford RP, Laurie DA (2005) The Pseudo-response regulator *Ppd-H1* provides adaptation to photoperiod in barley. Science 310:1031–1034

van Esse GW, Walla A, Finke A, Koornneef M, Pecinka A et al (2017) *Six-rowed spike3* (*VRS3*) is a histone demethylase that controls lateral spikelet development in barley. Plant Physiol 174:2397–2408

Varshney R, Marcel T, Ramsay L, Russell J, Röder M, Stein N, Waugh R, Langridge P, Niks RE, Graner A (2007) A high density barley microsatellite consensus map with 775 SSR loci. Theor Appl Genet 114:1091–1103

Vegetti AA, Anton AM (1995) Some evolution trends in the inflorescence of Poaceae. Flora 190:225–228

von Bothmer R, Jacobsen N, Baden C, Jorgensen RB, Linde-Laursen I (1995) Systematic and ecogeographical studies on crop genepools; an ecogeographical study of the genus *Hordeum*. International plant genetic resources institute (IPGRI), Rome, Italy

von Korff M, Campoli C (2014) Genetic control of reproductive development in temperate cereals. In: Fornara F (ed) Advances in botanical research. Academic Press, London, UK, pp 131–152

von Korff M, Grando S, Del Greco A, This D, Baum M et al (2008) Quantitative trait loci associated with adaptation to Mediterranean dryland conditions in barley. Theor Appl Genet 117:653–669

von Ubisch G (1915) Analyse eines falles von bastardatavismus und faktoren-koppelung bei gerste. Zeitschrift für Induktive Abstammungs- und Vererbungslehre 14:226–237

Ward SP, Leyser O (2004) Shoot branching. Curr Opin Plant Biol 7:73–78

Wellmer F, Riechmann JL (2010) Gene networks controlling the initiation of flower development. Trends Genet 26:519–527

Wendt T, Holm PB, Starker CG, Christian M, Voytas DF, Brinch-Pedersen H, Holme IB (2013) TAL effector nucleases induce mutations at a pre-selected location in the genome of primary barley transformants. Plant Mol Biol 83:279–285

Wendt T, Holme I, Dockter C, Preuß A, Thomas W et al (2016) *HvDep1* is a positive regulator of culm elongation and grain size in barley and impacts yield in an environment-dependent manner. PLoS ONE 11: e0168924

Wenzl P, Li HB, Carling J, Zhou MX, Raman H, Paul E, Hearnden P, Maier C, Xia L, Caig V, Ovesna J, Cakir M, Poulsen D, Wang JP, Raman R, Smith KP, Muehlbauer GJ, Chalmers KJ, Kleinhofs A, Huttner E, Kilian A (2006) A high-density consensus map of barley linking DArT markers to SSR, RFLP and STS loci and agricultural traits. BMC Genomics 7:206

Whipple CJ (2017) Grass inflorescence architecture and evolution: the origin of novel signaling centers. New Phytol 216:367–372

Whipple CJ, Zanis MJ, Kellogg EA, Schmidt RJ (2007) Conservation of B class gene expression in the second whorl of a basal grass and outgroups links the origin of lodicules and petals. P Natl Acad Sci USA 104:1081–1086

Whipple CJ, Hall DH, DeBlasio S, Taguchi-Shiobara F, Schmidt RJ et al (2010) A conserved mechanism of bract suppression in the grass family. Plant Cell 22:565–578

Whipple CJ, Kebrom TH, Weber AL, Yang F, Hall D et al (2011) *grassy tillers1* promotes apical dominance in maize and responds to shade signals in the grasses. P Natl Acad Sci USA 108:E506–E512

Xu Y, Jia Q, Zhou G, Zhang X-Q, Angessa T, Broughton S, Yan G, Chang W, Li C (2017) Characterisation of the sdw1 semi-dwarf gene in barley. BMC Plant Biol 17:11

Yagil E, Stebbins GL (1969) The morphogenetic effects of the hooded gene in barley ii. Cytological and environmental factors affecting gene expression. Genetics 62:307–319

Yan L, Loukoianov A, Tranquilli G, Helguera M, Fahima T et al (2003) Positional cloning of the wheat vernalization gene *VRN1*. P Natl Acad Sci USA 100:6263–6268

Yan L, Loukoianov A, Blechl A, Tranquilli G, Ramakrishna W et al (2004) The Wheat *VRN2* gene is a flowering repressor down-regulated by vernalization. Science 303:1640–1644

Yan L, Fu D, Li C, Blechl A, Tranquilli G et al (2006) The wheat and barley vernalization gene *VRN3* is an orthologue of *FT*. P Natl Acad Sci USA 103:19581–19586

Yoshida H, Nagato Y (2011) Flower development in rice. J Exp Bot 62:4719–4730

Youssef HM, Koppolu R, Schnurbusch T (2012) Re-sequencing of *vrs1* and *int-c* loci shows that labile barleys (*Hordeum vulgare* convar. *labile*) have a six-rowed genetic background. Genet Resour Crop Ev 59:1319–1328

Youssef HM, Eggert K, Koppolu R, Alqudah AM, Poursarebani N, Fazeli A, Sakuma S, Tagiri A, Rutten T, Govind G, Lundqvist U, Graner A, Komatsuda T, Sreenivasulu N, Schnurbusch T (2017a) *VRS2* regulates hormone-mediated inflorescence patterning in barley. Nat Genet 49:157–161

Youssef HM, Mascher M, Ayoub MA, Stein N, Kilian B et al (2017b) Natural diversity of inflorescence architecture traces cryptic domestication genes in barley (*Hordeum vulgare* L.). Genet Resour Crop Ev 64:843–853

Yuo T, Yamashita Y, Kanamori H, Matsumoto T, Lundqvist U et al (2012) A SHORT INTERNODES (SHI) family transcription factor gene regulates awn elongation and pistil morphology in barley. J Exp Bot 63:5223–5232

Zakhrabekova S, Gough SP, Braumann I, Müller AH, Lundqvist J et al (2012) Induced mutations in circadian clock regulator *Mat-a* facilitated short-season adaptation and range extension in cultivated barley. P Natl Acad Sci USA 109:4326–4331

Zhang Y, Zhang F, Li XH, Baller JA, Qi YP, Starker CG, Bogdanove AJ, Voytas DF (2013) Transcription activator-like effector nucleases enable efficient plant genome engineering. Plant Physiol 161:20–27

Zhou Y, Lu D, Li C, Luo J, Zhu B-F, Zhu J, Shangguan Y, Wang Z, Sang T, Zhou B, Han B (2012) Genetic control of seed shattering in rice by the APETALA2 transcription factor SHATTERING ABORTION1. Plant Cell 24:1034–1048

Zhu Q-H, Hoque MS, Dennis ES, Upadhyaya NM (2003) *Ds* tagging of *BRANCHED FLORETLESS 1* (*BFL1*) that mediates the transition from spikelet to floret meristem in rice (*Oryza sativa* L.). BMC Plant Biol 3:6

Zohary D (1963) Proceedings of the first international barley genetics symposium, Wageningen: Barley genetics I. Pudoc Centre for Agricultural Publications and Documentations, Wageningen, The Netherlands, pp 27–31

Zohary D, Hopf M (2000) Domestication of plants in the old world: the origin and spread of cultivated plants in West Asia, Europe and the Nile Valley. Oxford University Press, Oxford

Genetics of Whole Plant Morphology and Architecture 13

Laura Rossini, Gary J. Muehlbauer, Ron Okagaki, Silvio Salvi and Maria von Korff

Abstract

Plant architectural features directly impact plant fitness and adaptation, and traits related to plant morphology and development represent important targets for crop breeding. Decades of mutagenesis research have provided a wealth of mutant resources, making barley (*Hordeum vulgare* L.) an interesting model for genetic dissection of grass morphology and architecture. Recent advances in genomics have propelled the identification of barley genes controlling different aspects of shoot and root development. In addition to gene discovery, it is important to understand the interplay between different developmental processes in order to support breeding of improved ideotypes for sustainable barley production under different climatic conditions. The purpose of the present chapter is to: (i) provide an overview of the morphology and development of shoot and root structures in barley; (ii) discuss novel insights into the genetic, molecular and hormonal mechanisms regulating root and shoot development and architecture; and (iii) highlight the genetic and physiological interactions among organs and traits with special focus on correlations between leaf and tiller development, flowering and tillering, as well as row-type and tillering.

L. Rossini (✉)
Università degli Studi di Milano, DiSAA, Via Celoria 2, 20133 Milan, Italy
e-mail: laura.rossini@unimi.it

G. J. Muehlbauer
Department of Plant and Microbial Biology, University of Minnesota, St. Paul, MN 55108, USA

G. J. Muehlbauer · R. Okagaki
Department of Agronomy and Plant Genetics, University of Minnesota, St. Paul, MN 55108, USA

S. Salvi
Department of Agricultural and Food Sciences, University of Bologna, Viale Fanin 44, 40127 Bologna, Italy

M. von Korff
Max Planck Institute for Plant Breeding Research, 50829 Cologne, Germany

M. von Korff
Institute for Plant Genetics, Heinrich-Heine-Universität Düsseldorf, 40225 Düsseldorf, Germany

M. von Korff
Cluster of Excellence in Plant Sciences, Heinrich-Heine-Universität Düsseldorf, 40255 Düsseldorf, Germany

N. Stein and G. J. Muehlbauer (eds.), *The Barley Genome*, Compendium of Plant Genomes,
https://doi.org/10.1007/978-3-319-92528-8_13

13.1 Barley Shoot and Root Morphology and Development

13.1.1 Shoot Development

13.1.1.1 Embryo and Shoot Apical Meristem Development

The barley embryo is composed of a scutellum, radicle, epicotyl (shoot apical meristem, leaf primordia, and coleoptile) and nodal region between the epicotyl and radicle (MacLeod and Palmer 1966). The shoot apical meristem (SAM) forms during embryogenesis and the activity of the SAM determine the overall shoot body plan. Barley meristems are composed of a clonally distinct outer tunica layer (L1) and inner corpus layer (L2), with the possibility of a three-layered SAM not excluded (Doring et al. 1999). The SAM consists of three zones including: (1) the peripheral zone which is the site of lateral organ initiation; (2) the central zone which contains pluripotent cells that maintain the meristem; and (3) the rib zone that produces the stem tissues. During embryogenesis, the coleoptile and three to four leaves develop that enclose the SAM, and an axillary bud forms in the coleoptile axil and the first leaf axil (Fig. 13.1).

13.1.1.2 Leaf, Stem and Tiller Development

Shoot development is governed by the SAM and axillary meristems (AXM). The SAM produces repeated segments known as phytomers, which are composed of an internode (stem segment), node, axillary bud, and a leaf (Weatherwax 1923; Sharman 1942; Bossinger et al. 1992; Forster et al. 2007).

Tiller development or lateral branching is characterized by three defined stages (Fig. 13.1; Schmitz and Theres 2005). The first stage involves localized cell divisions in leaf axils producing a small mound of meristematic stem cells referred to as the AXM. In the second stage, primordial leaves develop and an axillary bud is formed. At this stage, the AXM needs to balance stem cell maintenance and leaf initiation. Normally, one axillary bud develops in a leaf axil, but two may be occasionally observed (Fletcher and Dale 1974; Tavakol et al. 2015). At this stage axillary buds may become dormant, or

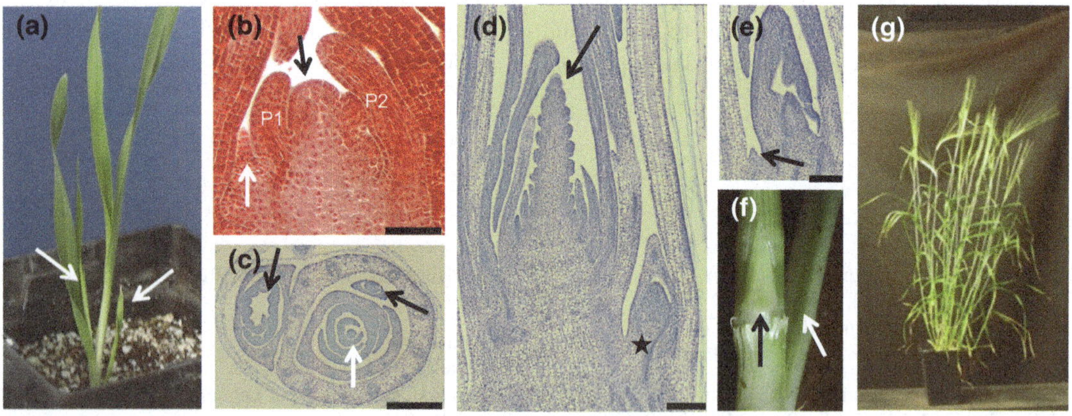

Fig. 13.1 Barley vegetative development. **a** Three-week-old seedling with two tillers. The arrows indicate tillers. **b** Longitudinal section through mature seed (photo courtesy of Allison Haaning) showing an apical meristem (black arrow), an axillary bud (white arrow) and two of the primordial leaves are designated P1 and P2. **c** Transverse section through a 7-day-old shoot, the black arrows indicate the two axillary buds and the white arrow indicates the shoot apex. **d** Longitudinal section of a 2-week-old shoot apex. The shoot apical meristem has transitioned to become an inflorescence meristem (black arrow). An axillary bud is indicated by the star. **e** Secondary axillary bud (arrow) developing in a tiller leaf axil. **f** Four-week-old seedling with leaves removed to show a tiller (white arrow) and an axillary bud (black arrow). **g** Adult wild-type (cv. Bowman) plant with many tillers. Scale bar = 200 μm

continue to the last stage and form a tiller. Usually, more axillary buds develop than tillers. The last stage is characterized by elongation of the internodes, leaf differentiation and tiller formation. Primary tillers are derived from AXMs forming in leaf axils from the first four or five leaves, and sometimes the coleoptile (Fletcher and Dale 1974; Kirby and Appleyard 1987). Internodes on the first several phytomers remain short throughout the life of the plant (basal internodes), and it is on these phytomers that axillary buds elongate and become tillers. Buds on above ground phytomers do not normally form branches. Primary tillers form in the leaf axil of the main culm, while secondary and tertiary tillers form in the leaf axils of primary and secondary tillers, respectively. This process occurs in a reiterative pattern to form the architecture of the plant. Tiller number is controlled by a combination of environmental (e.g., light, nitrogen, water, etc.) and genetic factors, and hormones. Each tiller and the main culm have the potential to form a seed-bearing inflorescence, demonstrating the importance of tiller number and vigor to yield.

Grass leaf development has been the focus of numerous studies (reviewed recently in Lewis and Hake 2016). Leaves initiate from a ring of periclinally dividing founder cells located on the flank (peripheral zone) of the meristem (Bossinger et al. 1992). The location of primordium initiation corresponds to the disk of insertion at the node (Sharman 1942). The barley leaf exhibits a strap-like appearance and is divided into the distal blade and proximal sheath. The sheath wraps around the culm and supports the plant while the blade is the major photosynthetic organ of the plant. Separating the blade and sheath is the ligular region, or collar, composed of the ligule and two auricles (Fig. 13.2). Growth and differentiation of the leaf proceeds via cell division and expansion patterns that occur in a basipetal (tip to base) wave such that when blade cells are fully differentiated the sheath cells are still dividing (Sylvester et al. 1990; Kołodziejek et al. 2006).

At the stem elongation phase, the apical meristem has acquired a reproductive identity (inflorescence meristem) and is pushed

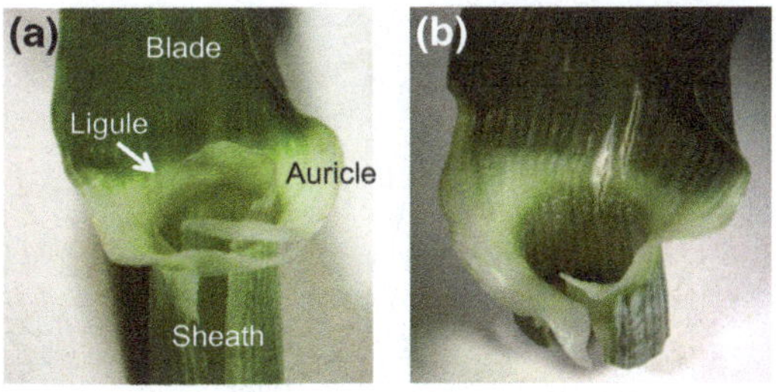

Fig. 13.2 Barley leaf morphology. **a** Wild-type leaf with blade, sheath, ligule and auricle. **b** *Cul4* mutant leaf exhibiting the lack of a ligule

acropetally by the growing aerial internodes. Basal internodes elongate first, followed by those located in more apical positions until elongation of the peduncle (the last internode below the spike) completes culm growth. A mature barley plant typically displays 5–8 elongated cylindrical and hollow internodes connected by solid nodes (Briggs 1978). When fully grown, each internode is generally longer than the internode immediately below it, except in some cases for the peduncle (Briggs 1978). Within the internode, the base is a meristematic zone (i.e., an intercalary meristem) characterized by delayed lignification and capacity for asymmetric growth to restore vertical stem orientation in case of lodging (Briggs 1978). Anatomically, the internode consists of different tissues: a silicified epidermis surrounds alternate vertical files of photosynthetic tissue and sclerenchyma fibers supporting vascular bundles (Briggs 1978). More internally a ring of sclerenchyma provides further support to the internode, encircling a parenchyma (containing additional vascular bundles), which is in direct contact with the pith cavity (Briggs 1978).

13.1.2 Root Development

13.1.2.1 Structures of Barley Root System

Similar to the other cereal species, barley is characterized by an embryonic and a post-embryonic root system (Fig. 13.3). The embryonic roots initiate directly in the embryo and emerge at germination or soon after and

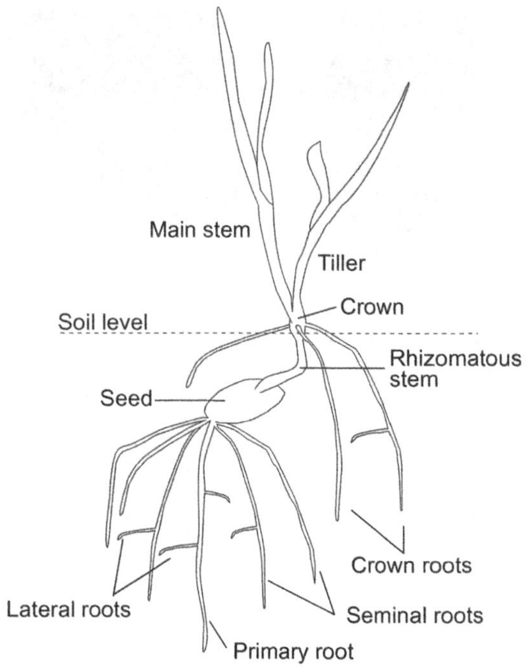

Fig. 13.3 Schematic diagram of a barley plant at tillering phase with emphasis on root system. Dimensions of different organs are not represented in scale

include the radicle (or primary root) and a variable number of seminal roots (ranging from 3 to 8, but mostly 5–6; Jackson 1922; Bergal and Clemencet 1962; Grando and Ceccarelli 1995; Robinson et al. 2016). All of these roots emerge from the basal pole of the embryo. The post-embryonic (or shoot-borne) system is composed of the crown (or nodal) roots arising from the lower stem nodes of the main shoot and tillers. Finally, axial roots (i.e., primary, seminal and nodal roots) of both embryonic and post-embryonic development give rise to lateral roots and this process can produce multiple branching levels. This multiple-axis root system typical of grasses is called homorhizic or fibrous.

Along each root, in longitudinal orientation, three distinct zones can be broadly recognized starting from the root tip: a meristematic zone proximal to the root cap where cells divide, an elongation zone, and a differentiation zone where root hairs and lateral roots are formed (Yu et al. 2016).

The transversal structure of a barley root as observable by optical microscopy on root cross sections (Fig. 13.4) shows three main portions: the epidermis, a multicellular cortex zone and the stele (vascular) zone. The cortex always ends internally with a single endodermis layer, followed by the pericycle, the first stele layer. In the stele, a variable number of central (mature) metaxylem and peripheral (early metaxylem) vessels can be observed, depending on the position and the root type. Primary, seminal and lateral roots have 4–5 cortical cell layers, and typically one large central and 6–9 smaller and circularly arranged peripheral metaxylem vessels (Luxová 1989; Knipfer and Fricke 2011). Crown roots have 7–8 cortical cell layers and 3–8 central and 12–16 peripheral metaxylem vessels when observed in sections at 20–300 mm from the root tip (Luxová 1989; Knipfer and Fricke 2011; Kotula et al. 2015). Phloem poles (or strands) alternate regularly with xylem vessels in the stele. Occasionally, and usually as a consequence of waterlogging and hypoxia events, aerenchyma develop following cortex cell breakdown (Pang et al. 2004; Kotula et al. 2015).

13.1.2.2 Cellular Basis of Root Development

Broadly speaking, cereal root architecture is shaped by four processes: initiation of axial roots (either in the embryo or at stem nodes), branching pattern, growth rate, and growth orientation in response to gravity and other stimuli. The modulation of these processes is mainly due to the action (and interaction) of two classes of phytohormones, auxin and cytokinins, in terms of synthesis, movement, perception and/or intracellular signal transduction (Lavenus et al. 2016).

The primary root primordium is initiated early in embryo development at the base of the scutellum node, while seminal root primordia (mostly five) appear later in embryo development, the first two near the primary root whereas the remaining form a triad more distantly in the scutellum at the level of the mesocotyl (Luxová 1986). The cellular origin of crown roots has not been investigated as thoroughly. In both maize (*Zea mays*) and rice (*Oryza sativa*), crown roots originate from cells next to the peripheral

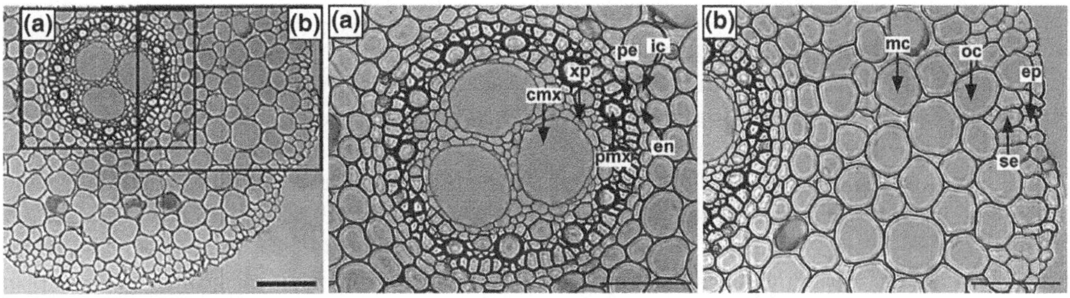

Fig. 13.4 Barley root anatomy. Cross-sections were taken at 50 mm from the root apex of approximate 85-mm-long young crown roots. **a** Magnification of the stele (or vascular) zone showing three central metaxylem (cmx) vessels. Twelve peripheral metaxylem (pmx) vessels are also visible. **b** Detail of the cortex zone with epidermis (ep) and sub-epidermis (se). oc, outer cortex; mc, middle cortex; ic, inner cortex; en, endodermis; pe, pericycle; xp, xylem parenchyma. Bars, 100 μm. Phloem elements are not highlighted in these cross sections. Reproduced with permission from Kotula et al. (2015). New Phytologist © 2015 New Phytologist Trust

cylinder of vascular bundles in the lower nodes of the stem (Atkinson et al. 2014), so it is expected that barley crown roots develop following the same pattern.

The root branching pattern is the result of cellular differentiation and growth processes, which give rise to lateral roots. In contrast to arabidopsis (*Arabidopsis thaliana*), where this process starts from the pericycle in correspondence with xylem poles, both pericycle and endodermis at phloem pole cells participate in forming the new lateral root in cereals (Smith and De Smet 2012; Yu et al. 2016). While arabidopsis has only two xylem poles, barley (and other cereals) can have >15 phloem poles, making the radial branching pattern much more complex and difficult to predict.

Root hairs develop as tubular outgrowths from a subset of epidermal cells called trichoblasts, count several billion per plant, and greatly extend the root surface area in contact with soil (Marzec et al. 2015). Root hair development is regarded as three-phase: the establishment of root epidermal patterning, the formation of the bulge on the root surface, and finally, root hair growth (Marzec et al. 2015). Approximately 50% of the epidermal cells develop root hairs in Poaceae so that root hair cells and hairless epidermal cells (atrichoblasts) alternate along each epidermal cell file of the mature root.

13.2 Genetic Control of Shoot Architecture and Development

Rich germplasm and mutant collections (e.g., Nordgen https://www.nordgen.org/bgs/index.php) represent valuable sources of genetic diversity that can be deployed in breeding and fundamental research into the molecular and physiological mechanisms of barley plant development. The following sections summarize current knowledge of loci and genes controlling various aspects of shoot and root morphology and architecture with special focus on correlations between different traits, as illustrated by specific case studies.

13.2.1 Genes Controlling Plant Height and Their Pleiotropic Effects

In cereals, plant height results from the number and length of elongated internodes and is conditioned by both endogenous and exogenous factors allowing the plant to adapt its growth to the environmental conditions. Genetic and physiological studies of stem elongation in different grass species highlighted the promoting role played by phytohormones such as gibberellins (GAs) and brassinosteroids (BRs) (Vriet et al. 2012; Hedden and Sponsel 2015) and the importance of auxin transport (Balzan et al. 2014). Manipulation of plant height has also been a strong focus for cereal breeders since the 1920s as shorter plants are more resistant to lodging, i.e. bending of the stem that causes loss of grain yield and quality (Salvi et al. 2013). The semi-dwarf varieties of the Green Revolution are the most emblematic example of ideotype breeding: mutations in different GA pathway genes were used to reduce plant height, enhance stem sturdiness and lodging resistance, improving fertilizer response and grain yield in rice and wheat (Hedden 2003). "Miracle rice" variety IR8 carries the recessive *semi-dwarf* (*sd-1*) trait, caused by a loss-of-function mutation of the GA20-oxidase2 gene (*OsGA20ox2*) controlling a key step in GA biosynthesis (Sasaki et al. 2002; Spielmeyer et al. 2002). In the case of wheat, semi-dwarf stature was attained through *Reduced height-1* (*Rht1*) alleles encoding gain-of-function forms of DELLA GA signaling repressors (Peng et al. 1999).

Reduction of plant height is a priority for barley yield improvement, especially in consideration of increased lodging risk as a consequence of climate change and extreme weather events (e.g., storms and strong winds). Hundreds of natural and induced barley mutants have been traditionally classified under different categories depending on alterations in plant height and other phenotypes: *breviaristatum (ari, lks), brachytic (brh), curly dwarf (cud), dense spike (dsp), vegetative dwarf (dwf), elongation (elo), erectoides (ert), gigas* (*gig*), *globosum* (*glo*), *GA-responsive dwarf (grd), GA sensitivity (gse), short culm (hcm), lazy dwarf (lzd), semi-minute dwarf (min), many noded dwarf (mnd), narrow leaf dwarf (nld), pyramidatum (pyr), semi-dwarf (sdw), single internode dwarf (sid), slender dwarf (sld), slender* (*sln*), *semi-brachytic (uzu), winding dwarf (wnd)*, and *Zeocriton (zeo)* (reviewed in Rossini et al. 2014; Dockter and Hansson 2015). The following two are examples of plant height mutants that have been characterized by phenotypes that are not related to plant height per se: the *ari* mutants which exhibit shorter awns compared to their wild-type counterparts and in some cases reduced plant height; and the *ert* mutants which exhibit compact, dense spikes and reduced plant stature due to shortening of rachis and culm internodes, respectively (Lundqvist 2014). However, mutants assigned to different categories often turn out to be allelic indicating this nomenclature may reflect pleiotropic biological functions (e.g., Dockter et al. 2014). For detailed information regarding barley plant height loci, the reader is referred to the recent excellent review by Dockter and Hansson (2015). The following discussion focuses on those loci for which the underlying genes have been identified, allowing us to link their molecular function to their phenotypic effects.

Advances in barley genomics are accelerating the cloning of barley plant height genes (Table 13.1), revealing in some cases their involvement in the GA and BR pathways. As these hormones regulate different aspects of plant development and physiology, genetic variants affecting their homeostasis, perception and signaling often lead to complex pleiotropic effects with important implications for breeding (Nadolska-Orczyk et al. 2017). A classic example is offered by the barley *semi-dwarf1/denso* (*sdw1*) locus, recently shown to encode *HvGA20ox2*, a homolog of the rice Green Revolution gene *Sd-1* (Xu et al. 2017). Analysis of diverse varieties identified four alleles with complete deletions or other gene mutations: the weak allele *sdw1.d* is widespread in European, US, Canadian and Australian malting barleys, but negative pleiotropic effects make the most severe alleles (e.g., *sdw1.a*) unsuitable for malting, limiting their use to feed barley (Xu et al. 2017). This is consistent with the major role of GAs during seed germination and reserve mobilization in the endosperm, processes that have a direct impact on malting and brewing (Hedden and Sponsel 2015). GA-related pleiotropic effects associated with the *sdw1* gene also include reduced thousand kernel weight, delayed flowering and maturity time, while results for other traits such as number of productive tillers and grain yield varied among different studies and may depend on genetic background or other factors (reviewed in Kuczyńska et al. 2013).

Table 13.1 Barley plant height mutant loci, whose genes have been identified

Locus	Gene	Function	References
*sdw1**	*HvGA20ox2*	GA biosynthesis	Xu et al. (2017)
grd5	*HvKAO*	GA biosynthesis cytochrome P450 ent-kaurenoic acid oxidase (CYP88A)	Helliwell et al. (2001)
gse1	*HvGID1*	GA receptor	Chandler et al. (2008)
sln1	*HvSLN1*	DELLA GA signaling repressor	Chandler et al. (2002)
spy1	*HvSPY1*	GA signaling repressor	Chandler and Harding (2013)
*uzu**	*HvBRI1*	BR receptor	Chono et al. (2003)
*ari-o/brh14**	*HvDIM*	BR pathway Δ5-sterol-Δ24-reductase DIMINUTO	Dockter et al. (2014)
ari-u/ert-t/brh3	*HvBRD*	BR pathway brassinosteroid-6-oxidase	Dockter et al. (2014)
brh13/brh18	*HvCPD*	BR pathway C-23α-hydroxylase cytochrome P450 90A1 (CYP90A1)	Dockter et al. (2014)
*ari-e**	*HvDEP1*	Heterotrimeric G protein AGG3-type subunit	Wendt et al. (2016)
brh1, ari-m		Heterotrimeric G protein α-subunit	Braumann et al. (2018), Ito et al. (2017)
Zeo	*HvAP2*	AP2/ERF transcription factor	Houston et al. (2013)
nld1	*HvNS*	WUSCHEL-RELATED HOMEOBOX transcription factor	Yoshikawa et al. (2016)
hvd14.d	*HvD14*	SL receptor α/β hydrolase	Marzec et al. (2016)

*Used in barley breeding (Kuczyńska et al. 2013). *BR* brassinosteroid, *GA* gibberellin, *SL* strigolactone

Further insight into the barley GA pathway came from the characterization of *GA-responsive dwarf* (*grd*), *GA sensitivity* (*gse*), *elongation* (*elo*), and *slender* (*sln*) mutants (Chandler and Robertson 1999). The *Grd5* locus encodes HvKAO, the cytochrome P450 ent-kaurenoic acid oxidase (CYP88A), which catalyzes a key step in GA biosynthesis (Helliwell et al. 2001). *Gse1* was shown to correspond to the *HvGID1* gene, the barley orthologue of the rice GA receptor gene (Chandler et al. 2008). In contrast to dwarf *grd5* and *gse1* mutants, loss-of-function *sln* mutant plants exhibit excessive stem growth due to a constitutive GA response; conversely, dominant mutations of the same locus result in a dwarf phenotype similar to wheat *Rht1*, arabidopsis *GAI/RGA* and maize *D8* (Peng et al. 1999; Chandler et al. 2002). Despite overall conservation of this GA signaling mechanism, some differences emerged in the barley GA response and other species and in contrast to wheat the *Sln1* locus has not been used in barley breeding (Kuczyńska et al. 2013). A genetic screen for suppressors of GA-defective dwarf mutants identified a single aminoacid substitution in the *Spindly1* (*Spy1*) barley gene, encoding a negative regulator of GA signaling: in the *spy1a* mutant enhanced GA signaling leads to a higher leaf elongation rate, but effects on plant height were not described in detail (Chandler and Harding 2013).

Early formulations of ideotype breeding focused on selecting for morphological characteristics to improve yield (Donald 1968). Subsequently, this definition has broadened to include a range of physiological, phenological, and architectural features that are proposed to improve crop productivity (e.g., see the following reviews on rice breeding programs: Peng et al. 2008; Jeon et al. 2011; Khush 2013; Sharma et al. 2013). An interesting development is the "smart canopy" model, which aims at optimizing the performance of the whole crop community by maximizing photosynthetic efficiency through a combination of biochemical and morphological traits (Ort et al. 2015). Among them, erect leaf angle (especially of apical leaves) was proposed to allow better penetration of solar radiation through the canopy enhancing photosynthesis in lower leaves: this trait was demonstrated to increase biomass production and grain yield in dense planting conditions (Sinclair and Sheehy 1999; Sakamoto et al. 2006). The BR pathway is a promising target for the development of crop plants that combine reduced plant stature and narrow leaf angle (Dockter and Hansson 2015). The role of BRs in culm elongation is well established and genes involved in BR biosynthesis and signaling have been identified from dwarf mutants in rice (Zhang et al. 2014). BRs also condition leaf inclination: in rice, cells on the adaxial side of the lamina-joint expand in response to BR, making the blade bend away from the stem axis and mutants impaired in BR biosynthesis or signaling have more erect leaves (Cao and Chen 1995; Yamamuro et al. 2000; Sakamoto et al. 2006). The first BR gene identified in barley was *semi-brachytic* (*uzu*), encoding the ortholog of the BRASSINOSTEROID1 (BRI1) hormone receptor (Chono et al. 2003). The recessive *uzu1.a* allele has long been used to confer reduced plant height and improved lodging resistance in winter barley cultivars in Japan, the Korean peninsula and China (Saisho et al. 2004). In addition, leaf erectness and decreased tillering make *uzu* barleys suitable for dense planting conditions. While the *uzu1.a* mutation has temperature-conditional negative phenotypic effects, future breeding may take advantage of a newly identified allele less sensitive to elevated temperatures (Dockter et al. 2014). A recent screen for BR-related mutants in barley identified a number of lines exhibiting reduced plant stature (especially reduced length of apical culm internodes) and acute leaf angle, as well as short rachis internodes and short awns, associating some of them to different mutations in orthologs of BR-related genes (Dockter et al. 2014; Table 13.1). Decreased plant height, strong culms, erect growth habit, and narrow leaf inclination angle are also characteristics of recessive mutants of the *ari-e* locus, which was shown to encode *HvDep1*, a heterotrimeric G-protein AGG3-type (Gγ) subunit gene related to rice *DENSE AND ERECT PANICLE* (Wendt et al. 2016). These ideotype traits contributed to the success of the malting barley cultivar Golden Promise, which harbors a frameshift mutation in the gene resulting in a truncated protein (allele *ari-e.GP*, Wendt et al. 2016). Heterotrimeric G protein complexes consist of Gα, Gβ and Gγ subunits and mediate signal transduction from G-protein-coupled receptors to downstream effectors. Interestingly, another dwarf barley mutant locus, *brachytic 1* (*brh1*) was recently associated with mutations in a gene encoding the α subunit of heterotrimeric G protein (Gα), reinforcing the role of this signaling complex in controlling barley plant stature (Ito et al. 2017; Braumann et al. 2018).

Together with BR-related mutants, further evidence for partially independent regulation of internode elongation is offered by the *narrow leafed dwarf1 (nld1)* mutant, characterized by reduced length of the apical internodes of the culm (Yoshikawa et al. 2016). *Nld1* acts in control of leaf margin development and encodes HvNS, a WUSCHEL-RELATED HOMEBOX transcription factor homologous to the redundant maize NARROW SHEATH1 (NS1) and NS2 factors (Nardmann et al. 2004; Yoshikawa et al. 2016). NS-defective maize mutants also exhibit defects in internode elongation and leaf margin formation suggesting that NS control phytomer development coordinating medio-lateral leaf development and internode growth (Scanlon et al. 1996). However, in maize *ns* mutants, basal internodes are affected, contrary to barley *nld1* mutants (Scanlon et al. 1996).

Another transcription factor implicated in control of internode elongation is HvAP2. The corresponding transcript is a target of miR172: mutations in the miRNA binding site (*Zeocriton, Zeo1, Zeo2, Zeo3*) result in semidominant phenotypes including dwarfism and spike compactness, a trait favored by breeders who wish to reduce the amount of moisture retained by the spike to prevent pre-harvest sprouting and fungal infections (Houston et al. 2013; Dockter and Hansson 2015).

Recently, through a reverse genetics TILLING approach a new barley dwarf mutant was identified and associated to a mutation in the *HvDWARF14* gene (*HvD14*): plants homozygous for the *hvd14.d* allele showed significantly reduced plant height, increased tiller number, lower seminal root extension and higher lateral root density compared to parent cultivar Sebastian (Marzec et al. 2016). *HvD14* encodes an α/β hydrolase highly related to rice D14, the receptor of strigolactones (SL), a class of plant hormones that control different shoot branching and architectural traits (Waters et al. 2017). *HvD14* represents the first functionally characterized SL-pathway gene in barley and like its rice ortholog is proposed to negatively control tiller number (Marzec et al. 2016). Future research on the characterization of other SL-pathway genes might thus provide new allelic diversity to develop barley cultivars combining reduced stature and increased tillering to augment barley biomass production.

13.2.2 Genes Controlling Tillering

Barley researchers have identified and characterized a large number of mutants with either an increased or decreased number of tillers (Table 13.2; Fig. 13.5). Many of these mutants have been backcrossed into the cultivar Bowman to provide a common genetic background for comparing phenotypes. In many cases the near-isogenic lines have been used to locate the mutant allele to a small genetic interval (Druka et al. 2011). There are four general classes of tillering mutants. One class that does not develop tillers but only develops a single main culm is characterized by mutations in *Uniculm2* (*Cul2*; Babb and Muehlbauer 2003). Another class exhibits a small number of tillers and is characterized by mutations in *Low number of tillers1* (*Lnt1*; Dabbert et al. 2010), *Uniculme4* (*Cul4*; Tavakol et al. 2015), and *Absent lower laterals1* (*Als1*; Dabbert et al. 2009). A third class of mutants exhibit a moderate reduction in tiller number and are defined by mutations in *Intermedium-b* (*Int-b*; Babb and Muehlbauer 2003), *Opposite spike1* (*Ops1*; Okagaki and Muehlbauer unpublished results) and *Semi-brachytic* (*Uzu*; Babb and Muehlbauer 2003; Dockter et al. 2014). The fourth class of mutants exhibits an increase in tiller number and include mutations in *Granum-a* (*Gra-a*; Babb and Muehlbauer 2003), *Grassy tillers* (*Grassy*; Rossini et al. 2014), *Narrow leafed dwarf1* (*Nld1*; Yoshikawa et al. 2016), *HvD14* (Marzec et al. 2016), *Many noded dwarf1/5* (*Mnd1/5*), *Many noded dwarf3* (*Mnd3*; Franckowiak and Lundqvist 2002) and *Many noded dwarf 4/6* (Mascher et al. 2014).

Several studies have described the genetic and morphological characterization of some of the tillering mutants. Mutations in *Cul2* result in the lack of tiller development due to the inability to develop an axillary bud, indicating that *Cul2* regulates axillary bud formation (Babb and Muehlbauer 2003). Mutations in *Lnt1*, *Cul4* and *Als1* result in one to a few tillers due to the inability to form axillary meristems on primary tillers (Dabbert et al. 2009, 2010; Tavakol et al. 2015). Low-tillering *cul2*, *lnt1*, *cul4,* and *als1* mutants combined with the high-tillering mutants *gra-a*, *mnd1* and *mnd6* in double mutant combinations result in the respective low-tillering phenotype (Babb and Muehlbauer 2003; Dabbert et al. 2009, 2010). Interestingly, mutations in *Eligulum-a* that exhibit dwarf, liguleless and low-tillering phenotypes suppress the *cul2* mutant phenotype, resulting in tiller development on *cul2* mutants (Okagaki et al. 2018).

Surprisingly, few of the barley tillering genes identified to date correspond to expected genes or pathways identified in other species. *HvD14* is a SL signaling pathway gene, and mutant alleles confer a high-tillering dwarf phenotype similar to rice SL mutants (Marzec et al. 2016). *Mnd4/6* encodes a cytochrome P450 protein related to the rice heterochronic mutant *plastochron1* (Mascher et al. 2014). Tillering defects are associated also with mutations in the *Uzu* gene, encoding the ortholog of the arabidopsis *BRI1* and rice *D61* genes (Babb and Muehlbauer 2003; Chono et al. 2003; Dockter et al. 2014).

Table 13.2 Barley tillering mutants

Gene name	Gene annotation	Vegetative phenotypes	References
Uniculm2 (Cul2)	No candidate	No tillers, thick culm, wide leaves	Babb and Muehlbauer (2003)
Eligulum-a (Eli-a)	Unknown RNaseH-like domain protein	Suppresses *cul2*, thick culms, dwarf, wide leaves, liguleless	Okagaki et al. (2018), Lundqvist and Franckowiak (2013)
Absent lower laterals 1 (Als1)	No candidate gene	Low tillering, thick culms, wide leaves	Babb and Muehlbauer (2003), Dabbert et al. (2009)
Uniculme4 (Cu4)	*HvBlade-on-petiole*	Low tillering, thick culms, wide leaves liguleless	Babb and Muehlbauer (2003), Tavakol et al. (2015)
Low number of tillers1 (LNT1)	*JuBel2*	Low tillering, thick culms, wide leaves	Babb and Muehlbauer (2003), Dabbert et al. (2010)
Intermedium-b (Int-b)	No candidate gene	Reduced tiller number	Babb and Muehlbauer (2003), Lundqvist and Franckowiak (2014)
Opposite spikelets1 (Ops1)	Candidate—*HvBarren stalk fastigiate1* (Hv*Baf1*)	Reduced tiller number	Lundqvist and Franckowiak (2013), Okagaki and Muehlbauer unpublished results
Many noded dwarf1/5 (Mnd1/5)	No candidate gene	Many tillers, dwarf, narrow leaves	Babb and Muehlbauer (2003), Lundqvist and Franckowiak (2013)
Many noded dwarf 4/6 (Mnd4/6)	Cytochrome P450	Many tillers, semi-dwarf, narrow leaf	Mascher et al. (2014), Franckowiak and Lundqvist (2002)
Many noded dwarf 3 (Mnd3)	No candidate gene	Many tillers, dwarf, narrow leaf	Franckowiak and Lundqvist (2002)
Granum-a (Gra-a)	No candidate gene	Many tillers, dwarf, narrow leaf	Babb and Muehlbauer (2003)
Grassy tillers	No candidate gene	Many tillers, narrow leaf	Rossini et al. (2014)
HvD14	*D14* α/β-hydrolase	Many tillers, dwarf	Marzec et al. (2016)
Intermedium spike-m.85 (Int-m)	No candidate gene	Many tillers, semi-dwarf	Babb and Muehlbauer (2003), Dahleen et al. (2007)
Intermedium spike-c (Int-c)	*HvTEOSINTE BRANCHED1 (HvTB1)*	Increased tillering in seedlings	Ramsay et al. (2011)

The majority of the classic barley tillering mutants that have been cloned correspond to genes linked to other functions or had previously unknown functions. For example, *Cul4* encodes a barley homolog of the Arabidopsis boundary genes *BLADE-ON-PETIOLE* (*BOP1* and *BOP2*), and *Lnt1* is believed to encode the JuBEL2 protein, a homolog of the arabidopsis *PENNYWISE* protein (Dabbert et al. 2010; Tavakol et al. 2015). Mutations in *BOP* or *PNY* do not confer a strong axillary meristem or branching phenotype in Arabidopsis (Smith and Hake 2003; Norberg et al. 2005). In addition, barley *Eli-a* encodes an unknown protein that is highly conserved in

Fig. 13.5 Wildtype and mutant tillering phenotypes. **a** Wild-type Bowman cultivar with tillers. **b** *Cul2* mutant with no tillers. **c** *Lnt1* mutant with few tillers. **d** *Cul4* mutant with few tillers. **e** *Nld1* mutant with many tillers

plants but a mutant phenotype has not been observed in other plants (Okagaki et al. 2018). Also, map positions of known barley tillering mutants generally do not coincide with obvious candidate genes (Hussien et al. 2014). Another complication is raised by the *Nld1* gene (Yoshikawa et al. 2016). Barley *nld1* mutants increase tillering (Okagaki and Muehlbauer unpublished data), which has not been observed in the maize *NS* mutant (Scanlon et al. 1996). However, common maize inbred lines do not tiller and an effect on vegetative axillary meristems in maize *NS* mutants, and potentially other mutants, is unlikely to be detected. The maize *TEOSINTE BRANCHED* (*TB1*) gene is a major determinant of tiller number that regulates axillary bud outgrowth in maize; homologs in sorghum (*Sorghum bicolor*) and arabidopsis play similar roles (Doebley et al. 1997; Aguilar-Martínez et al. 2007; Kebrom et al. 2010). The barley *TB1* homolog, *Intermedium-c* (*Int-c*), was shown to control spike architecture; however, *int-c* mutants do not increase adult tiller number although tillers do develop faster (Ramsay et al. 2011). Most of the genes correspond to previously isolated genes in other species, however the barley mutant phenotypes reveal additional roles for these genes that were not observed in other plants.

13.2.3 Genetic Relationships Between Leaf and Axillary Meristem Development

A series of mutants that disrupt leaf patterning have been identified and characterized in barley. These include mutations that result in disruption of the blade-sheath boundary and exhibit a liguleless phenotype [e.g., *cul4*, *liguleless* (*lig*), and *eli-a*]. Other mutations alter the shape of the leaf and include *nld1* and *broad leaf1 (blf1)* that exhibit narrow and broad leaves, respectively. *Blf1* encodes an INDETERMINATE DOMAIN protein and reduces leaf width by restricting cell division (Jost et al. 2016).

Recent evidence in barley and other plant species are beginning to reveal a genetic connection between lateral organ initiation and leaf patterning (Johnston et al. 2014; Tavakol et al. 2015). Plants carrying mutations in *Cul4* are low tillering and liguleless (Fig. 13.2; Franckowiak and Lundqvist 2002). RNA in situ hybridizations showed that *Cul4* is expressed in the leaf axil and AXM, and during ligule development, indicating that *Cul4* is a boundary gene that establishes the position of the axillary meristem and ligule development. In addition, *eli-a*, originally described as a liguleless leaf mutant

(Franckowiak and Lundqvist 2002) is expressed in the preligular band and the developing ligule, the leaf margin, and surrounding vascular bundles, and was determined to promote tillering in the *cul2* mutant background (Okagaki et al. 2018). Characterization of these barley mutants complemented maize transcriptome studies describing a set of genes expressed at both the ligular region and lateral organ boundaries (Johnston et al. 2014) and work in eudicots proposing a conserved genetic control of leaf and axillary meristem boundary development (Busch et al. 2011). Another leaf trait linked with tillering is leaf width. For example, barley high-tillering mutants, *gra-a*, *mnd1*, *mnd6*, and *nld1*, have narrow leaves (Okagaki and Muehlbauer unpublished results), whereas the low-tillering mutants *cul2*, *cul4*, *als1*, and *lnt1* all exhibit broad leaves (Okagaki and Muehlbauer unpublished results). Taken together, analyses of barley mutants are revealing how tiller development and leaf patterning are regulated by a shared set of genes acting at multiple organ boundaries throughout the shoot.

13.2.4 Correlation Between Flowering Time, Leaf and Tiller Development

Variation in tillering is also affected by genetic variation in reproductive development. Studies in biparental and diverse populations have identified QTLs or marker-trait associations for tillering close to flowering time genes (Karsai et al. 1999; Borràs et al. 2009; Alqudah et al. 2016; Ogrodowicz et al. 2017). A strong vernalization requirement and reduced photoperiod sensitivity are commonly correlated with an increase in tillering in barley (Karsai et al. 1999; Wang et al. 2010). Accordingly, QTLs for tillering co-locate with the major vernalization genes *Vrn-H1* and *Vrn-H2* and the photoperiod response gene *Ppd-H1* (Karsai et al. 1999; von Korff et al. 2006; Wang et al. 2010). *Vrn-H2* is a strong inhibitor of *Vrn-H3* (a homolog of arabidopsis *FLOWERING LOCUS T*, *FT*) and thus flowering time under long photoperiods, whereas *Vrn-H1*A upregulates *Vrn-H3* and triggers flowering. *Vrn-H1* is induced during vernalization and represses *Vrn-H2* to allow for the expression of *Vrn-H3/FT1* (Yan et al. 2003, 2004). A deletion of the *Vrn-H2* locus and deletions in the first regulatory intron of *Vrn-H1* are prevalent in spring barley, which flowers without vernalization (Hemming et al. 2009; Rollins et al. 2013). In barley or wheat populations segregating for natural variation at *VRN1* or *VRN2,* the duration of vernalization has a significant influence on a number of developmental traits including tiller number (Kato et al. 2000; Karsai et al. 2005, 2006). *Ppd-H1* encodes a PSEUDO-RESPONSE-REGULATOR (PRR) protein, which is homologous to the arabidopsis PRR3/PRR7 of the circadian clock, and characterized by a pseudoreceiver and a CCT (CONSTANS, CONSTANS-like, and TOC1) domain (Turner et al. 2005). The ancestral, dominant form of *Ppd-H1* confers an acceleration of flowering under increasing day length and is commonly found in Mediterranean barley genotypes. A recessive mutation in the CCT domain of *ppd-H1* has been selected in spring cultivars grown in temperate agricultural areas. This variant leads to a minor delay in the vegetative to reproductive phase transition, but a strong delay of the inflorescence development in spring barley (Alqudah et al. 2014; Digel et al. 2015). Under long photoperiods, *Ppd-H1* upregulates *Vrn-H3.* Proteins homologous to VRN-H3 such as arabidopsis FT and rice Hd3a move from the leaves through the phloem to the shoot apical meristem, where they induce the transition from vegetative to reproductive growth (Corbesier et al. 2007; Tamaki et al. 2007). *Ppd-H1*, *Vrn-H1* and *Vrn-H2* likely affect tillering by controlling the expression levels of *Vrn-H3/FT1* and consequently the flowering signal in the shoot apical meristem. A strong flowering signal increases apical dominance, a term used to describe the inhibition of axillary bud outgrowth by the main stem (e.g., Cline 1997). The apical meristem produces the plant hormone auxin (IAA) that inhibits axillary bud growth by preventing the synthesis or utilization of cytokinin within axillary buds (Müller and Leyser 2011). In this context, it is interesting to note that in arabidopsis, *FT* interacts with *BRANCHED 1* (*BRC1*, a homolog of maize TB1) in axillary buds to inhibit early floral transition (Aguilar-Martínez et al. 2007; Hiraoka et al. 2013; Niwa et al. 2013). The function of *BRC1*/*TB1* is conserved among distantly related angiosperm species, where the gene integrates signals controlling auxin induced apical dominance and thus branching (Doebley et al. 1997). However, future research is needed to further elucidate the molecular networks integrating reproductive development and tiller number in grasses.

It has been shown in many cereal crops that early flowering does not only correlate with low tillering but also a reduction in leaf size. A recent genome wide association study in winter barley has demonstrated that the largest effect on leaf length and width colocated with *Ppd-H1* (Digel et al. 2016). The effect of *Ppd-H1* on leaf size was verified in near-isogenic lines that were characterized by a dominant *Ppd-H1* allele introgressed into spring barley cultivars. The effect of *Ppd-H1* on leaf growth was only observed under long days and correlated with its effect on reproductive development. Differences in the rate of leaf emergence, in the duration of leaf growth and in the final leaf cell number suggested that *Ppd-H1* affected leaf size by influencing the rate of age-dependent progression of leaf development (Digel et al. 2016). Interestingly, a recent study also showed that the row-type gene *Vrs1* influences leaf size by controlling cell proliferation of leaf primordia cells (Thirulogachandar et al. 2017). The authors demonstrated that *Vrs1* negatively affects the development of leaf primordia in two-rowed cultivars, probably by suppressing proliferation of leaf founder cells that give rise to the leaf primordia. This work provides another example for the coordination of spike and leaf development in barley.

Detailed physiological studies on the function of genes controlling spike development will shed more light on their role in shaping different plant organs. Pleiotropic effects of these genes on tiller number and leaf size might be important to optimize source-sink relationships and maximize yield.

13.2.5 Genetic Correlations Between Row-Type and Tiller Number

Shoot and spike branching determine aboveground plant architecture and the major yield components: the number of spikes per plant, the number of seeds per spike and seed weight. It is well known that these yield components are often negatively correlated, i.e., increased spike number negatively correlates with seed weight (Fonesca and Patterson 1968; Gebeyehou et al. 1982). These negative correlations between yield component traits have been attributed to competition of different plant organs for limited resources (Adams 1967). However, the identification and characterization of genes affecting either shoot or spike branching have challenged this assumption and suggested that both traits might be genetically linked. For instance, the *TB1* gene controls tiller and spike morphology (Studer et al. 2011). In cultivated maize, a natural overexpressing allele was introgressed into cultivated germplasm to decrease tillering and improve yield (Studer et al. 2011). The barley homolog of *TB1*, *Int-c* has been identified as a major gene controlling lateral spikelet development that also represses the number of tillers in barley early in development (Ramsay et al. 2011). A recent study has examined the pleiotropic effects of a number of row-type genes on tiller number in barley to better understand the genetic correlations between individual yield components (Liller et al. 2015). In two-rowed barley only the central spikelet is fertile while in six-rowed barley all three spikelets per rachis node develop and set seeds. The genes underlying the six-rowed spike (*vrs*) loci *Vrs1* and *Vrs4* encode transcription factors of the HD-ZIP (Hv. Hox1) and LATERAL ORGAN BOUNDARY families, respectively, and repress the development of lateral spikelets (Komatsuda et al. 2007; Koppolu et al. 2013). *Intermedium* (*int*) mutants display enlarged lateral spikelets, which sometimes develop into seeds depending on the position on the spike and the environment (Lundqvist and Lundqvist 1987, 1988). The *int-c* and *int-a* (*vrs3*) genotypes with mutations in *TB1* and a *histone demethylase*, respectively, are characterized by a reduced upregulation of *Vrs1* and a correlated development of lateral spikelets (Bull et al. 2017; van Esse et al. 2017). Based on differences in the correlation between the number of seeds and tillers, Liller et al. (2015) identified two mutant groups. The first group comprised mutants with a higher number of seeds and reduced tiller number expressed either early in development or at full maturity. Mutants in the second group were distinguished by a reduction in seeds per spike and in tiller number and consequently positive correlations between both traits.

Negative associations between seed and tiller number were also found in the wheat *tiller inhibition* (*tin*) mutant, which is characterized by larger spikes and a higher grain number per spikelet and a low tiller number (Kebrom et al. 2012). The spike phenotype can be induced by removal of tillers after their initiation in the wild-type plant (Kebrom et al. 2012). This demonstrates that different plant organs can compete for the same resources, resulting in a decrease of tillers and an increase in seed number. Similarly, Alqudah et al. (2016) found that the six-rowed *vrs1* mutant with developed lateral spikelets had significantly less tillers at the flag leaf and heading stages compared to their wild-type (*Vrs1*) parents. Liller et al. (2015), however, found that *vrs1* mutants displayed a reduced tiller number at maturity, but not at earlier stages of development. The authors, therefore, concluded that *vrs1* does not directly control the outgrowth of axillary buds and the reduction in tiller number was likely caused by resource competition between tillers and seeds. The additional row-type mutants, however, either displayed altered tillering already at early developmental stages or positive correlations between seed number per spike and tiller number. For example, *int-c* mutants produced more tillers at early developmental stages (Ramsay et al. 2011; Liller et al. 2015), suggesting a direct effect of this gene on tiller number. Similarly, loss of function mutations in the homologous maize *TB1* or rice *FINE CULM 1* (*FC1*) genes increase the number of tillers (Doebley et al. 1997; Goto et al. 2005). While the barley *int-c* mutant was high tillering at early developmental stages, it ceased to develop new tillers after reaching the flag leaf stage when tiller numbers were decreased compared to the wild type (Liller et al. 2015). This suggested that either *Int-c* exerted different effects on axillary buds of different order (primary, secondary, tertiary), or its effect changed over development.

Alternatively, reduced tillering mutants such as *als*, *lnt1*, and *int-b* exhibited a reduction in seed number. This suggested that the causative mutations may directly influence the initiation or outgrowth of different meristematic tissues in barley, rather than indirectly through resource partitioning. Similarly, in maize it was shown that the basic helix-loop-helix protein *BARREN STALK 1* (*BA1*) regulates vegetative and reproductive lateral meristem development (Ritter et al. 2002; Gallavotti et al. 2004). Furthermore, the rice gene *MONOCULM 1* (*MOC1*), a homolog of the arabidopsis boundary gene *LATERAL SUPPRESSOR* (*LAS*), controls shoot as well as inflorescence architecture (Li et al. 2003). Consequently, these genes may have large effects on seed number as they control the number of seeds per spike and the number of spike-bearing tillers. Taken together, there is increasing evidence that the development of different plant organs is governed by the same genes. Knowledge of these genes and their functions in different plant organs is important to break the genetic correlations of shoot and spike phenotypes to improve yield.

13.3 Genetic Control of Root Architecture and Anatomy

The genetic analysis of root development in barley lags behind those of other systems given that just two mutants affecting roots (*curly2* = *cu2*, and *few roots1* = *fer1*, Lundqvist and

Franckowiak 2014) are included out of hundreds described for shoot characteristics. However, at least two mutagenized populations were screened and dozens of root mutants were identified, although those mutants are mostly not yet genetically analyzed. For example, in the "Optic"-based EMS-mutagenized population (Caldwell et al 2004), White et al. (2009) reported the identification of 30 lines showing a root phenotype, with agravitropic/curly roots and fewer or shorter root hairs. Similarly, in a NaN_3-treated population in the Morex genetic background (Talamè et al. 2008), 70 lines showing aberrant root morphology were identified where the prevalent phenotypes were short and/or curly seminal roots (Bovina et al. 2011).

13.3.1 Seminal Roots

The genetics of seminal root development has been well investigated in maize (Salvi 2017), where the mutants *rtcs* and *rum1*, which genes code for a LOB-domain and an Aux/IAA protein, respectively, abolish the initiation of seminal roots (Taramino et al. 2007; von Behrens et al. 2011). Plants carrying the *rtcs* mutant also lack crown and brace roots, causing severe plant lodging. Interestingly, the yet unmapped barley mutant *Fer1* (Linde-Laursen 1977) was described as having 1–3 (instead of 5–6) seminal roots and reduced or no nodal roots, associated with plant lodging, suggesting a similarity with maize *rtcs*. Five QTLs for number of seminal roots were mapped in a "ND24260" × "Flagship" double haploid (DH) barley population, with the major QTL mapping on chromosome 5HL. This QTL co-located with QTLs for root angle and drought tolerance in barley, and for number of seminal roots in wheat; an expansin gene was suggested as a candidate gene (Robinson et al. 2016). In two additional studies, a barley mutant displaying long roots in hydroponics was speculatively linked with disturbed hydrotropism (Martinez et al. 2004), and a cell wall-associated receptor-like kinase gene was associated with root length in control and high-salt stress conditions (Kaur et al. 2013).

13.3.2 Nodal and Lateral Roots

No specific mutants or genes controlling nodal or lateral root development have been analyzed in barley, whereas mutants lacking crown roots have been isolated in rice and maize with auxin-related functions showing an important regulatory role (Orman-Ligeza et al. 2013; Yu et al. 2016). In barley, several QTLs for nodal root system size were identified with two major loci co-mapping with the semi dwarfing genes *sdw1* and *ari-e.GP*, highlighting the potential pleiotropic effects of shoot-oriented breeding activities on root architecture in modern cultivars (Chloupek et al. 2006). Specifically, in the *sdw1* mutant reduced plant height was clearly associated with larger root size (Chloupek et al. 2006).

A source of allelic variation for nodal (as well as seminal) root traits is wild barley (*Hordeum vulgare* ssp. *spontaneum*). Variation among wild barley accessions and between cultivated and wild barley has been described in terms of number of nodal roots, lateral roots, and total root biomass (Grando and Ceccarelli 1995, Zhao et al. 2010). QTLs associated with root phenotypes have been mapped in introgression libraries derived from cultivated x wild barley accessions. Naz et al. (2012, 2014) characterized introgression lines of the wild barley accession ISR42-8 (a wild barley accession from semi-desert areas of Israel) in the Scarlett cultivar background. Noteworthy, a major QTL for root system size on the distal part of chromosome 5 was found to overlap with a major QTL for root length identified in a wild × wild barley mapping population (Chen et al. 2009).

13.3.3 Growth Angle

The genetic control of root growth angle is an active area of research given that a number of modeling and physiological studies identified a steep root angle as associated with higher resource use efficiency and productivity (Lynch 2013). One gravitropic mutant in the Morex genetic background (Talamé et al. 2008), *enhanced gravitropic1* (*eg1*), was mapped on chromosome 6H

(Salvi, unpublished results). *egl* showed reinforced response to gravitropism both in seminal and lateral roots, globally resulting in a narrower and steeper root apparatus. QTLs for root angle were identified in an experimental cross involving a wild-derived introgression line (Sayed et al. 2017). The strongest QTL was detected at marker locus *bPb-8558* (7.52 cM) on chromosome 7H where the introgression of an exotic allele resulted in a 31.6% increase in root angle and around 47% of the roots exhibited an angle greater than 45 degrees. Recently natural allelic variation at the key flowering time regulatory gene *VRN-H1* was causally linked to root architectural changes (Voss-Fels et al. 2018). Interestingly, an early flowering *VRN-H1* allele was associated with narrow root growth at the seedling stage and prolonged root growth at the deepest soil level during grain filling. The authors suggested that this allele likely contributed to the adaptation of some barley cultivars to dry environments by providing flowering-mediated drought escape and improved root water acquisition (Voss-Fels et al. 2018).

13.3.4 Root Hairs

In an extensive effort, 19 root hair mutants were subjected to detailed morphological and genetic analyses and shown to correspond to nine independent loci (Kwasniewski et al. 2013; Chmielewska et al. 2014). Phenotypically, these mutants were classified into the following categories: mutants with no root hairs (*rhl*, *root hairless*), mutants with root hairs arrested at the initial stage of development (*root hair primordial*; *rhp*), mutants with short root hairs (*rhs*, *root hair short*) and mutants with sparsely located root hairs of different lengths (*rhi*, *root hair irregular*). Differential expression of a root expansin (*HvEXPB1*) gene was shown to cosegregate with the root hair phenotype associated with *rhl1.a* (Kwasniewski and Szarejko 2006). EXPA and EXPB are expansin proteins potentially involved in cell wall-loosening and cell enlargement (Cosgrove 2015).

The root hairless mutant *bald root barley* (*brb* or *rhl1.d*) was utilized in comparative analyses at various P availability conditions. The results suggested that root hairs are important in P acquisition and plant survival in low-P environments, but may be dispensable under high-P conditions (Gahoonia and Nielsen 2003). Using the same mutant, Zheng et al. (2011) demonstrated a role for root hairs in heavy metal acquisition. Unexpectedly, the same mutant was tested for response to soil water deficit and showed essentially no penalty in shoot growth when compared to wild type, leading to the implication that the absence of root hairs does not affect the plant response to reduced soil moisture (Dodd and Diatloff 2016). Recent experiments based on root hair mutants confirmed the strong interconnection between root hairs and root-associated microbes in barley (Robertson-Albertyn et al. 2017).

13.3.5 Genetics of Rhizosheath

An interesting root feature for which natural variation was observed in barley is the rhizosheath, defined as the weight of soil that adheres strongly to roots on uprooting. This trait was first noted in desert grasses by A. G. Tansley more than 100 years ago (George et al. 2014), and has since been associated with stress tolerance and nutrient uptake. The rhizosheath results from many factors including root hairs, production of root-microbe mucilage, and symbiotic or otherwise root-associated bacteria and fungi. In an association mapping population of 144 European two-row spring barley genotypes, variation for rhizosheath weight was found to be under genetic control. A major rhizosheath QTL apparently not related with roothairs was mapped on chromosome 2H at 112 cM (George et al. 2014). However, in a subsequent study using a different population, rhizosheath traits were not clearly associated with phosphorus use efficiency (Gong and McDonald 2017).

13.3.6 Aerenchyma

A large range of phenotypic variation for root aerenchyma across the genus *Hordeum* has been documented by Garthwaite et al. (2003), who highlighted *Hordeum marinum* as particularly rich in aerenchyma and resistant to waterlogging. Broughton et al. (2015) reported a major QTL for root porosity (the percentage of gas volume per root volume, widely used as an indicator of aerenchyma formation) and waterlogging tolerance on chromosome 4H. A major QTL for aerenchyma and porosity was mapped in the same map position in two unrelated crosses (Zhang et al. 2016, 2017), the latter involving a *H. spontaneum* accession. Intriguingly, Broughton et al. (2015) observed that the porosity QTL on 4H was syntenic with both the rice *Sub1A-1* gene and maize QTL, *Qaer1.02-3*. Both loci affect ethylene biosynthesis and responsiveness, although in seemingly different ways. Additional investigation is required in order to clarify the actual mechanisms involved.

13.4 Concluding Remarks and Future Prospects

Plant ideotypes have been defined as "a combination of morphological and/or physiological traits, or their genetic bases, optimizing crop performance to a particular biophysical environment, crop management, and end-use" (Martre et al. 2015). In ideotype breeding, special attention should be dedicated to compensatory mechanisms in the development of different plant organs and interrelationships among traits (Peng et al. 2008). The application of novel genomic approaches to the analysis of mutants and natural genetic diversity is accelerating the discovery of genes controlling barley plant architecture, beginning to clarify the molecular and physiological mechanisms that subtend development and morphogenesis and the correlations among different traits. Future research in this area will be essential to provide the knowledge and allelic diversity needed to design crop ideotypes that combine favorable features for different end-uses.

Acknowledgements LR wishes to acknowledge FACCE ERA-NET funding under projects BarPLUS (ERA-NET FACCE SURPLUS grant no. 93) and ClimBar (ERA-NET FACCE on Climate Smart Agriculture) for supporting research on genetics of barley plant architecture in her laboratory. MK acknowledges funding by the German Cluster of Excellence on Plant Sciences (CEPLAS) EXC1028, the Priority Programme (SPP1530 Flowering time control—from natural variation to crop improvement) and the Max Planck Society.

Authors' Contributions LR, GJM, SS and MK conceived the layout of the chapter. GJM and RO wrote Sects. 13.1.1, 13.2.2 and 13.2.3, prepared Figs. 13.1, 13.2, 13.5 and Table 13.2. SS wrote Sects. 13.1.2 and 13.3, prepared Figs. 13.3 and 13.4. MK wrote Sects. 13.2.4 and 13.2.5. LR wrote Sects. 13.2.1 and 13.4, prepared Table 13.1, and integrated contributions from other authors. All authors reviewed and approved the final version of the chapter.

References

Adams MW (1967) Basis of yield components compensation in crop plants with special reference to field bean (*Phaseolus vulgaris* L.). Crop Sci 7(5):505–510

Aguilar-Martínez JA, Poza-Carrión C, Cubas P (2007) *Arabidopsis BRANCHED1* acts as an integrator of branching signals within axillary buds. Plant Cell 19 (2):458–472

Alqudah AM, Koppulu R, Wolde GM, Graner A, Schnurbusch T (2016) The genetic architecture of barley plant stature. Front Genet 7:117

Atkinson JA, Rasmussen A, Traini R, Voß U, Sturrock C, Mooney SJ, Well DM, Bennett MJ (2014) Branching out in roots: uncovering form, function, and regulation. Plant Physiol 166(2):538–550

Alqudah AM, Sharma R, Pasam RK, Graner A, Kilian B, Schnurbusch T (2014) Genetic dissection of photoperiod response based on GWAS of pre-anthesis phase duration in spring barley. PLoS One 9(11):e113120

Babb S, Muehlbauer GJ (2003) Genetic and morphological characterization of the barley *uniculm2* (*cul2*) mutant. Theor Appl Genet 106(5):846–857

Balzan S, Johal GS, Carraro N (2014) The role of auxin transporters in monocots development. Front Plant Sci 5:393

Bergal P, Clemencet M (1962) The botany of the barley plant. In: Cook AH (ed) Barley and malt: biology, biochemistry and technology. Academic Press, New York

Borràs G, Romagosa I, van Eeuwijk F, Slafer G (2009) Genetic variability in the duration of pre-heading phases and relationships with leaf appearance and tillering dynamics in a barley population. Field Crop Res 113(2):95–104

Bossinger G, Lundqvist U, Rohde W, Salamini F (1992) Genetics of plant development in barley. In: Munck L (ed) Barley genetics VI, vol II. Munksgaard International Publishers, Copenhagen, pp 989–1017

Bovina R, Talamè V, Ferri M, Tuberosa R, Chmielewska B, Szarejko I, Sanguineti MC (2011) Identification of root morphology mutants in barley. Plant Genet Res 9(2):357–360

Braumann I, Dockter C, Beier S, Himmelbach A, Lok F, Lundqvist U, Skadhauge B, Stein N, Zakhrabekova S, Shou R, Hansson M (2018) Mutations in the gene of the Gα subunit of the heterotrimeric G protein are the cause for the *brachytic1* semi-dwarf phenotype in barley and applicable for practical breeding. Hereditas 155:10

Briggs DE (1978) The morphology of barley; the vegetative phase. In: "Barley" Briggs DE, Chapman and Hall, London, pp 1–38. ISBN-13: 978-94-009-5717-6

Broughton S, Zhou G, Teakle N, Matsuda R, Zhou M, O'Leary R, Colmer T, Li C (2015) Waterlogging tolerance is associated with root porosity in barley (*Hordeum vulgare* L.). Mol Breed 35(1):27

Bull H, Casao MC, Zwirek M, Flavell AJ, Thomas WT, Guo W et al (2017) Barley *SIX-ROWED SPIKE3* encodes a putative Jumonji C-type H3K9me2/me3 demethylase that represses lateral spikelet fertility. Nature Commun 8(1):936

Busch BL, Schmitz G, Rossmann S, Piron F, Ding J et al (2011) Shoot branching and leaf dissection in tomato are regulated by homologous gene modules. Plant Cell 23(10):3595–3609

Caldwell DG, McCallum N, Shaw P, Muehlbauer GJ, Marshall DF, Waugh R (2004) A structured mutant population for forward and reverse genetics in barley (*Hordeum vulgare* L.). Plant J 40(1):143–150

Cao H, Chen S (1995) Brassinosteroid-induced rice lamina joint inclination and its relation to indole-3-acetic acid and ethylene. Plant Growth Regul 16(2):189–196

Chandler PM, Harding CA (2013) 'Overgrowth' mutants in barley and wheat: new alleles and phenotypes of the 'Green Revolution' DELLA gene. J Exp Bot 64 (6):1603–1613

Chandler PM, Robertson M (1999) Gibberellin dose-response curves and the characterization of dwarf mutants of barley. Plant Physiol 120(2):623–632

Chandler PM, Marion-Poll A, Ellis M, Gubler F (2002) Mutants at the *Slender1* locus of barley cv. Himalaya. Molecular and physiological characterization. Plant Physiol 129(1):181–190

Chandler PM, Harding CA, Ashton AR, Mulcair MD, Dixon NE, Mander LN (2008) Characterization of gibberellin receptor mutants of barley (*Hordeum vulgare* L.). Mol Plant 1(2):285–294

Chen G, Krugman T, Fahima T, Chen K, Hu Y, Roder M, Nevo E, Korol A (2009) Chromosomal regions controlling seedling drought resistance in Israeli wild barley, *Hordeum spontaneum* C. Koch. Genet Resour Crop Evol 57(1):85–99

Chloupek O, Forster BP, Thomas WT (2006) The effect of semi-dwarf genes on root system size in field-grown barley. Theor Appl Genet 112(5):779–786

Chmielewska B, Janiak A, Karcz J, Guzy-Wrobelska J, Forster BP, Nawrot M, Rusek A, Smyda P, Kedziorski P, Maluszynski M, Szarejko I (2014) Morphological, genetic and molecular characteristics of barley root hair mutants. J Appl Genet 55 (4):433–447

Chono M, Honda I, Zeniya H, Yoneyama K, Saisho D, Takeda K et al (2003) A semidwarf phenotype of barley *uzu* results from a nucleotide substitution in the gene encoding a putative brassinosteroid receptor. Plant Physiol 133(3):1209–1219

Cline MG (1997) Concepts and terminology of apical dominance. Am J Bot 84(8):1064–1069

Corbesier L, Vincent C, Jang S, Fornara F, Fan Q, Searle I, Giakountis A, Farrona S, Gissot L, Turnbull C, Coupland G (2007) FT protein movement contributes to long-distance signaling in floral induction of *Arabidopsis*. Science 316(5827):1030–1033

Cosgrove DJ (2015) Plant expansins: diversity and interactions with plant cell wall. Curr Opin Plant Biol 25:162–172

Dabbert T, Okagaki R, Cho S, Boddu J, Muehlbauer GJ (2009) The genetics of barley low-tillering mutants: *absent lower laterals* (*als*). Theor Appl Genet 118 (7):1351–1360

Dabbert T, Okagaki RJ, Cho S, Heinen S, Boddu J, Muehlbauer GJ (2010) The genetics of barley low-tillering mutants: *low number of tillers-1* (*lnt1*). Theor Appl Genet 121(4):705–717

Dahleen L, Franckowiak JD, Lundqvist U (2007) Barley Genet Newslett 37. URL: https://wheat.pw.usda.gov/ggpages/bgn/37/index.html

Digel B, Pankin A, von Korff M (2015) Global transcriptome profiling of developing leaf and shoot apices reveals distinct genetic and environmental control of floral transition and inflorescence development in barley. Plant Cell 27(9):2318–2334

Digel B, Tavakol E, Verderio G, Tondelli A, Xu X, Cattivelli L, Rossini L, von Korff M (2016) *Photoperiod-H1* (*Ppd-H1*) controls leaf size. Plant Physiol 172(1):405–415

Dockter C, Hansson M (2015) Improving barley culm robustness for secured crop yield in a changing climate. J Exp Bot 66(12):3499–3509

Dockter C, Gruszka D, Braumann I, Druka A, Druka I, Franckowiak J, Gough SP, Janeczko A, Kurowska M, Lundqvist J, Lundqvist U, Marzec M, Matyszczak I, Müller AH, Oklestkova J, Schulz B, Zakhrabekova S, Hansson M (2014) Induced variations in brassinosteroid genes define barley height and sturdiness, and expand the green revolution genetic toolkit. Plant Physiol 166(4):1912–1927

Dodd IC, Diatloff E (2016) Enhanced root growth of the *brb* (*bald root barley*) mutant in drying soil allows similar shoot physiological responses to soil water deficit as wild-type plants. Funct Plant Biol 43 (2):199–206

Doebley J, Stec A, Hubbard L (1997) The evolution of apical dominance in maize. Nature 386(6624):485–488

Donald CM (1968) The breeding of crop ideotypes. Euphytica 17(3):385–403

Doring HP, Lin J, Uhrig H, Salamini F (1999) Clonal analysis of the development of the barley (*Hordeum vulgare* L.) leaf using periclinal chlorophyll chimeras. Planta 207(3):335–342

Druka A, Franckowiak J, Lundqvist U, Bonar N, Alexander J et al (2011) Genetic dissection of barley morphology and development. Plant Physiol 155 (2):617–627

Fletcher GM, Dale JE (1974) Growth of tiller buds in barley: effects of shade treatment and mineral nutrition. Ann Bot 38(1):63–76

Fonesca S, Patterson L (1968) Hybrid vigour in seven parental diallel crosses in common winter wheat. Crop Sci 8:85–88

Forster BP, Franckowiak JD, Lundqvist U, Lyon J, Pitkethly I, Thomas WTB (2007) The barley phytomer. Ann Bot 100(4):725–733

Franckowiak JD, Lundqvist U (2002) Barley Genet Newslett 32. URL: https://wheat.pw.usda.gov/ggpages/bgn/32/index.html

Gahoonia TS, Nielsen NE (2003) Phosphorus (P) uptake and growth of a root hairless barley mutant (bald root barley, brb) and wild type in low-and high-P soils. Plant, Cell Environ 26(10):1759–1766

Gallavotti A, Zhao Q, Kyozuka J, Meeley RB, Ritter MK, Doebley JF, Pè ME, Schmidt RJ (2004) The role of *barren stalk1* in the architecture of maize. Nature 432 (7017):630–635

Garthwaite AJ, von Bothmer R, Colmer TD (2003) Diversity in root aeration traits associated with waterlogging tolerance in the genus *Hordeum*. Funct Plant Biol 30(8):875–889

Gebeyehou G, Knott DR, Baker RJ (1982) Relationships among duration of vegetative and grain filling phases, yield components and grain yield in durum wheat cultivars. Crop Sci 22(2):287–290

George TS, Brown LK, Ramsay L, White PJ, Newton AC, Bengough AG, Russell J, Thomas WTB (2014) Understanding the genetic control and physiological traits associated with rhizosheath production by barley (*Hordeum vulgare*). New Phytol 203(1):195–205

Gong X, McDonald G (2017) QTL mapping of root traits in phosphorus-deficient soils reveals important genomic regions for improving NDVI and grain yield in barley. Theor Appl Genet 130(9):1885–1902

Goto Y, Tanabe M, Ishibashi T, Tsutsumi N, Yoshimura A, Nemoto K (2005) Tillering behavior of the rice *fine culm 1* mutant. Plant Prod Sci 8 (1):68–70

Grando S, Ceccarelli S (1995) Seminal root morphology and coleoptile length in wild (*Hordeum vulgare* ssp. *spontaneum*) and cultivated (*Hordeum vulgare* ssp. *vulgare*) barley. Euphytica 86(1):73–80

Hedden P (2003) The genes of the Green Revolution. Trends Genet 19(1):5–9

Hedden P, Sponsel V (2015) A century of Gibberellin research. J Plant Growth Regul 34(4):740–760

Helliwell CA, Chandler PM, Poole A, Dennis ES, Peacock WJ (2001) The CYP88A cytochrome P450, ent-kaurenoic acid oxidase, catalyzes three steps of the gibberellin biosynthesis pathway. Proc Natl Acad Sci USA 98(4):2065–2070

Hemming M, Fieg S, Peacock WJ, Dennis ES, Trevaskis B (2009) Regions associated with repression of the barley (*Hordeum vulgare*) *VERNALIZATION1* gene are not required for cold induction. Mol Genet Genomics 282(2):107–117

Hiraoka K, Yamaguchi A, Abe M, Araki T (2013) The florigen genes *FT* and *TSF* modulate lateral shoot outgrowth in *Arabidopsis thaliana*. Plant Cell Physiol 54(3):352–368

Houston K, McKim SM, Comadran J, Bonar N, Druka I, Uzrek N et al (2013) Variation in the interaction between alleles of *HvAPETALA2* and microRNA172 determines the density of grains on the barley inflorescence. Proc Natl Acad Sci USA 110 (41):16675–16680

Hussien A, Tavakol E, Horner DS, Munoz-Amatriain M, Muehlbauer GJ, Rossini L (2014) Genetics of tillering in rice and barley. Plant Genome 7(1):1–20

Ito A, Yasuda A, Yamaoka K, Ueda M, Nakayama A, Takasuto S, Honda I (2017) *Brachytic1* of barley (*Hordeum vulgare* L.) encodes the α subunit of heterotrimeric G protein. J Plant Physiol 213:209–215

Jackson VG (1922) Anatomical structure of the roots of barley. Ann Bot 36(141):21–39

Jeon JS, Jung KH, Kim HB, Suh JP, Khush GS (2011) Genetic and molecular insights into the enhancement of rice yield potential. J Plant Biol 54(1):1–9

Johnston R, Wang M, Sun Q, Sylvester AW, Hake S, Scanlon MJ (2014) Transcriptomic analyses indicate that maize ligule development recapitulates gene expression patterns that occur during lateral organ initiation. Plant Cell 26(12):4718–4732

Jost M, Hensel G, Kappel C, Druka A, Sicard A, Hohmann U, Beier S, Himmelbach A, Waugh R, Kumlehn J, Stein N, Lenhard M (2016) The INTERMEDIATE DOMAIN protein BROAD LEAF1 limits barley leaf width by restricting lateral proliferation. Curr Biol 26(7):903–909

Karsai I, Eszaros KM, Szucs P, Hayes PM, Lang L, Bedo Z (1999) Effects of loci determining photoperiod sensitivity (*Ppd-H1*) and vernalization response (*Sh2*) on agronomic traits in the 'Dicktoo' × 'Morex' barley mapping population. Plant Breed 118(5):399–403

Karsai I, Szücs P, Meszaros K, Filichkina T, Hayes PM, Skinner JS, Lang L, Bedo Z (2005) The *Vrn-H2* locus is a major determinant of flowering time in a facultative × winter growth habit barley (*Hordeum vulgare* L.) mapping population. Theor Appl Genet 110(8):1458–1466

Karsai I, Meszaros K, Szücs P, Hayes PM, Lang L, Bedo Z (2006) The influence of photoperiod on the

Vrn-H2 locus (4H) which is a major determinant of plant development and reproductive fitness traits in a facultative × winter barley (*Hordeum vulgare* L.) mapping population. Plant Breed 125(5):468–472

Kato K, Miura H, Sawada S (2000) Mapping QTLs controlling grain yield and its components on chromosome 5A of wheat. Theor Appl Genet 101 (7):1114–1121

Kaur R, Singh K, Singh J (2013) A root-specific wall-associated kinase gene, *HvWAK1*, regulates root growth and is highly divergent in barley and other cereals. Funct Integr Genomics 13(2):167–177

Kebrom TH, Brutnell TP, Finlayson SA (2010) Suppression of sorghum axillary bud outgrowth by shade, phyB and defoliation signalling pathways. Plant Cell Environ 33(1):45–58

Kebrom TH, Chandler PM, Swain SM, King RW, Richards RA, Spielmeyer W (2012) Inhibition of tiller bud outgrowth in the tin mutant of wheat is associated with precocious internode development. Plant Physiol 160(1):308–318

Khush GS (2013) Strategies for increasing the yield potential of cereals: case of rice as an example. Plant Breed 132(5):433–436

Kirby E, Appleyard M (1987) Cereal development guide, 2nd edn. Arable Unit, Stoneleigh, Warwickshire

Knipfer T, Fricke W (2011) Water uptake by seminal and adventitious roots in relation to whole-plant water flow in barley (*Hordeum vulgare* L.). J Exp Bot 62 (2):717–733

Kołodziejek I, Wałęza M, Mostowska A (2006) Morphological, histochemical and ultrastructural indicators of maize and barley leaf senescence. Biol Plant 50 (4):565–573

Komatsuda T, Pourkheirandish M, He CF, Azhaguvel P, Kanamori H, Perovic D et al (2007) Six-rowed barley originated from a mutation in a homeodomain-leucine zipper I-class homeobox gene. Proc Natl Acad Sci USA 104(4):1424–1429

Koppolu R, Anwar N, Sakuma S, Tagiri A, Lundqvist U, Pourkheirandish M et al (2013) *Six-rowed spike4* (*Vrs4*) controls spikelet determinacy and row-type in barley. Proc Natl Acad Sci USA 110(32):13198–13203

Kotula L, Clode PL, Striker GG, Pedersen O, Läuchli A, Shabala S, Colmer TD (2015) Oxygen deficiency and salinity affect cell specific ion concentrations in adventitious roots of barley (*Hordeum vulgare*). New Phytol 208(4):1114–1125

Kuczyńska A, Surma M, Adamski T, Mikołajczak K, Krystkowiak K, Ogrodowicz P (2013) Effects of the semi-dwarfing *sdw1/denso* gene in barley. J Appl Genet 54(4):381–390

Kwasniewski M, Szarejko I (2006) Molecular cloning and characterization of beta-expansion gene related to root hair formation in barley. Plant Physiol 141(3):1149–1158

Kwasniewski M, Nowakowska U, Szumera J, Chwialkowska K, Szarejko I (2013) iRootHair: a comprehensive root hair genomics database. Plant Physiol 161:28–35. http://www.iroothair.org/. Accessed 6 Mar 2018

Lavenus J, Guyomarc'h S, Laplaze L (2016) *PIN* transcriptional regulation shapes root system architecture. Trends Plant Sci 21(3):175–177

Lewis MW, Hake S (2016) Keep on growing: building and patterning leaves in the grasses. Curr Opin Plant Biol 29:80–86

Li XY, Qian Q, Fu ZM, Wang YH, Xiong GS, Zeng D et al (2003) Control of tillering in rice. Nature 422 (6932):618–621

Liller CB, Neuhaus R, von Korff M, Koornneef M, van Esse W (2015) Mutations in barley row type genes have pleiotropic effects on shoot branching. PLoS ONE 10(10):e0140246

Linde-Laursen I (1977) Barley mutant with few roots. Barley Genet Newslett 7:43–45

Lundqvist U (2014) Scandinavian mutation research in barley—a historical review. Hereditas 151(6):123–131

Lundqvist U, Franckowiak JD (2013) Barley Genet Newslett 43. URL: https://wheat.pw.usda.gov/ggpages/bgn/43/index.html

Lundqvist U, Franckowiak JD (2014) Barley Genet Newslett 44. URL: https://wheat.pw.usda.gov/ggpages/bgn/44/index.html

Lundqvist U, Lundqvist A (1987) An intermedium gene present in a commercial 6-row variety of barley. Hereditas 107(2):131–135

Lundqvist U, Lundqvist A (1988) Induced intermedium mutants in barley—origin. Morphology and inheritance. Hereditas 108(1):13–26

Luxová M (1986) The seminal root primordia in barley and the participation of their non-meristematic cells in root construction. Biol Plantarum 28(3):161–167

Luxová M (1989) The vascular system in the roots of barley and its hydraulic aspects. In: Loughman BC, Gasparikova O, Kolek J (eds) Structural and functional aspects of transport in roots. Kluwer Academic Publisher, Dordrecht, pp 15–20

Lynch JP (2013) Steep, cheap and deep: an ideotype to optimize water and N acquisition by maize root systems. Ann Bot 112(2):347–357

MacLeod AM, Palmer GH (1966) The embryo of barley in relation to modification of the endosperm. J Inst Brew 72(6):580–589

Martınez AE, Franzone PM, Aguinaga A, Polenta G, Murray R, Prina AR (2004) A nuclear gene controlling seminal root growth response to hydroponic cultivation in barley. Environ Exp Bot 51 (2):133–144

Martre P, Quilot-Turion B, Luquet D, Ould-Sidi Memmah MM, Chenu K, Debaeke P (2015) Model-assisted phenotyping and ideotype design. In: Sadras V, Calderini D (eds) Crop physiology. Applications for genetic improvement and agronomy, 2nd edn. Academic Press, London

Marzec M, Melzer M, Szarejko I (2015) Root hair development in the grasses: what we already know and what we still need to know. Plant Physiol 168 (2):407–414

Marzec M, Gruszka D, Tylec P, Szarejko I (2016) Identification and functional analysis of the *HvD14* gene involved in strigolactone signaling in *Hordeum vulgare*. Physiol Plant 158(3):341–355

Mascher M, Jost M, Kuon J-E, Himmelbach A, Aßfalg A, Beier S, Scholz U, Graner A, Stein N (2014) Mapping-by-sequencing accelerates forward genetics in barley. Genome Biol 15:R78

Müller D, Leyser O (2011) Auxin, cytokinin and the control of shoot branching. Ann Bot 107(7):1203–1212

Nadolska-Orczyk A, Rajchel IK, Orczyk W, Gasparis S (2017) Major genes determining yield-related traits in wheat and barley. Theor Appl Genet 130(6):1081

Nardmann J, Ji J, Werr W, Scanlon MJ (2004) The maize duplicate genes *narrow sheath1* and *narrow sheath2* encode a conserved homeobox gene function in a lateral domain of shoot apical meristems. Development 131(12):2827–2839

Naz AA, Ehl A, Pillen K, Léon J (2012) Validation for root-related quantitative trait locus effects of wild origin in the cultivated background of barley (*Hordeum vulgare* L.). Plant Breed 131(3):392–398

Naz AA, Arifuzzaman M, Muzammil S, Pillen K, Léon J (2014) Wild barley introgression lines revealed novel QTL alleles for root and related shoot traits in the cultivated barley (*Hordeum vulgare* L.). BMC Genet 15(1):107

Niwa M, Daimon Y, Kurotani K, Higo A, Pruneda-Paz JL, Breton G, Mitsuda N, Kay SA, Ohme-Takagi M, Endo M, Araki T (2013) BRANCHED1 Interacts with FLOWERING LOCUS T to repress the floral transition of the axillary meristems in *Arabidopsis*. Plant Cell 25(4):1228–1242

Norberg M, Holmlund M, Nilsson O (2005) The *BLADE ON PETIOLE* genes act redundantly to control the growth and development of lateral organs. Development 132(9):2203–2213

Ogrodowicz P, Adamski T, Mikołajczak K, Kuczyńska A, Surma M, Krajewski P, Sawikowska A, Górny AG, Gudyś K, Szarejko I, Guzy-Wróbelska J, Krystkowiak K (2017) QTLs for earliness and yield-forming traits in the Lubuski × CamB barley RIL population under various water regimes. J Appl Genet 58(1):49–65

Okagaki RJ, Haaning A, Bilgic H, Heinen S, Druka A, Bayer M, Waugh R, Muehlbauer GJ (2018) *ELIGULUM-A* regulates lateral branch and leaf development in barley. Plant Physiol 176:2750–2760

Orman-Ligeza B, Parizot B, Gantet PP, Beeckman T, Bennett MJ, Draye X (2013) Post-embryonic root organogenesis in cereals: branching out from model plants. Trends Plant Sci 18(8):459–467

Ort DR, Merchant SS, Alric J, Barkan A, Blankenship RE, Bock R et al (2015) Redesigning photosynthesis to sustainably meet global food and bioenergy demand. Proc Natl Acad Sci USA 112(28):8529–8536

Pang J, Zhou M, Mendham N, Shabala S (2004) Growth and physiological responses of six barley genotypes to waterlogging and subsequent recovery. Aust J Agric Res 55(8):895–906

Peng J, Richards DE, Hartley NM, Murphy GP, Devos KM, Flintham JE, Beales J, Fish LJ, Worland AJ, Pelica F, Sudhakar D (1999) 'Green revolution' genes encode mutant gibberellin response modulators. Nature 400(6741):256–261

Peng S, Khush GS, Virk P, Tang Q, Zou Y (2008) Progress in ideotype breeding to increase rice yield potential. Field Crop Res 108(1):32–38

Ramsay L, Comadran J, Druka A, Marshall DF, Thomas WTB, Macaulay M et al (2011) *INTERMEDIUM-C*, a modifier of lateral spikelet fertility in barley, is an ortholog of the maize domestication gene *TEOSINTE BRANCHED 1*. Nat Genet 43(2):169–172

Ritter MK, Padilla CM, Schmidt RJ (2002) The maize mutant *barren stalk1* is defective in axillary meristem development. Am J Bot 89(2):203–210

Robertson-Albertyn S, Alegria Terrazas R, Balbirnie K, Blank M, Janiak A, Szarejko I, Chmielewska B, Karcz J, Morris J, Hedley PE, George TS, Bulgarelli D (2017) Root hair mutations displace the barley rhizosphere microbiota. Front Plant Sci 8:1094

Robinson H, Hickey L, Richard C, Mace E, Kelly A, Borrell A, Franckowiak J, Fox G (2016) Genomic regions influencing seminal root traits in barley. Plant Genome 9(1)

Rollins JA, Drosse B, Mulki MA, Grando S, Baum M, Singh M, Ceccarelli S, von Korff M (2013) Variation at the vernalisation genes *Vrn-H1* and *Vrn-H2* determines growth and yield stability in barley (*Hordeum vulgare*) grown under dryland conditions in Syria. Theor Appl Genet 126(11):2803–2824

Rossini L, Okagaki R, Druka A, Muehlbauer GJ (2014) Shoot and inflorescence architecture. In: Kumlehn J, Stein N (eds) Biotechnological approaches to barley improvement. Springer, Berlin, pp 55–80. ISBN 978-3-662-44405-4

Saisho D, Tanno K, Chono M, Honda I, Kitano H, Takeda K (2004) Spontaneous brassinolide-insensitive barley mutants "*uzu*" adapted to East Asia. Breed Sci 54(4):409–416

Sakamoto T, Morinaka Y, Ohnishi T, Sunohara H, Fujioka S, Ueguchi-Tanaka M et al (2006) Erect leaves caused by brassinosteroid deficiency increase biomass production and grain yield in rice. Nat Biotech 24(1):105–109

Salvi S (2017) An evo-devo perspective on root genetic variation in cereals. J Exp Bot 68(3):351–354

Salvi S, Porfiri O, Ceccarelli S (2013) Nazareno Strampelli, the 'Prophet' of the green revolution. J Agric Sci 151(1):1–5

Sasaki A, Ashikari M, Ueguchi-Tanaka M, Itoh H, Nishimura A, Swapan D, Ishiyama K, Saito T, Kobayashi M, Khush GS, Kitano H (2002) Green revolution: a mutant gibberellin-synthesis gene in rice. Nature 416(6882):701–702

Sayed MA, Hamada A, Lèon J, Naz AA (2017) Genetic mapping reveals novel exotic QTL alleles for seminal root architecture in barley advanced backcross double haploid population. Euphytica 213(1):2

Scanlon MJ, Schneeberger RG, Freeling M (1996) The maize mutant *narrow sheath* fails to establish leaf margin identity in a meristematic domain. Development 122(6):1683–1691

Schmitz G, Theres K (2005) Shoot and inflorescence branching. Curr Opin Plant Biol 8(5):506–511

Sharma D, Sanghera GS, Sahu P, Sahu P, Parikh M, Sharma B et al (2013) Tailoring rice plants for sustainable yield through ideotype breeding and physiological interventions. Afr J Agric Res 8 (40):5004–5019

Sharman BC (1942) Developmental anatomy of the shoot *Zea mays* L. Ann Bot 6(2):245–282

Sinclair TR, Sheehy JE (1999) Erect leaves and photosynthesis in rice. Science 283(5407):1456–1456

Smith S, De Smet Ive (2012) Root system architecture: insights from Arabidopsis and cereal crops. Philos Trans R Soc Lond B Biol Sci 367(1595):1441–1452

Smith HMS, Hake S (2003) The interaction of two homeobox genes, *BREVIPEDICELLUS* and *PENNYWISE*, regulates internode patterning in the Arabidopsis inforescence. Plant Cell 15(8):1717–1727

Spielmeyer W, Ellis MH, Chandler PM (2002) Semidwarf (sd-1), "green revolution" rice, contains a defective gibberellin 20-oxidase gene. Proc Natl Acad Sci USA 99(13):9043–9048

Studer A, Zhao Q, Ross-Ibarra J, Doebley J (2011) Identification of a functional transposon insertion in the maize domestication gene *tb1*. Nat Genet 43 (11):1160–1164

Sylvester AW, Cande WZ, Freeling M (1990) Division and differentiation during normal and *liguleless-1* maize leaf development. Development 110(3):985–1000

Talamè V, Bovina R, Sanguineti MC, Tuberosa R, Lundqvist U, Salvi S (2008) TILLMore, a resource for the discovery of chemically induced mutants in barley. Plant Biotechnol J 6(5):477–485

Tamaki S, Matsuo S, Wong HL, Yokoi S, Shimamoto K (2007) Hd3a protein is a mobile flowering signal in rice. Science 316(5827):1033–1036

Taramino G, Sauer M, Stauffer JL, Multani D, Niu XM, Sakai H, Hochholdinger F (2007) The maize (*Zea mays* L.) *RTCS* gene encodes a LOB domain protein that is a key regulator of embryonic seminal and postembryonic shoot-borne root initiation. Plant J 50 (4):649–659

Tavakol E, Okagaki R, Verderio G, Shariati JV, Hussien A et al (2015) The barley *Uniculme4* gene encodes a BLADE-ON-PETIOLE-Like protein that controls tillering and leaf patterning. Plant Physiol 168 (1):164–174

Thirulogachandar V, Alqudah AM, Koppolu R, Rutten T, Graner A, Hensel G, Kumlehn J, Bräutigam A, Sreenivasulu N, Schnurbusch T, Kuhlmann M (2017) Leaf primordium size specifies leaf width and vein number among row-type classes in barley. Plant J 91(4):601–612

Turner A, Beales J, Faure S, Dunford RP, Laurie DA (2005) The pseudo-response regulator *Ppd-H1* provides adaptation to photoperiod in barley. Science 310 (5750):1031–1034

van Esse W, Walla A, Finke A, Koornneef M, Pecinka A, von Korff M (2017) Six-rowed spike 3 (VRS3) is a histone demethylase that controls lateral spikelet development in barley. Plant Physiol 174(4):2397–2408

von Behrens I, Komatsu M, Zhang Y, Berendzen KW, Niu X, Sakai H, Taramino G, Hochholdinger F (2011) Rootless with undetectable meristem 1 encodes a monocot-specific AUX/IAA protein that controls embryonic seminal and post-embryonic lateral root initiation in maize. Plant J 66(2):341–353

von Korff M, Wang H, Léon J, Pillen K (2006) AB-QTL analysis in spring barley: II. Detection of favourable exotic alleles for agronomic traits introgressed from wild barley (*H. vulgare* ssp. *spontaneum*). Theor Appl Genet 112(7):1221–1231

Voss-Fels KP, Robinson H, Mudge SR, Richard C, Newman S, Wittkop B, Stahl A, Friedt W, Frisch M, Gabur I, Miller-Cooper A, Campbell BC, Kelly A, Fox G, Christopher J, Christopher M, Chenu K, Franckowiak J, Mace ES, Borrell AK, Eagles H, Jordan DR, Botella JR, Hammer G, Godwin ID, Trevaskis B, Snowdon RJ, Hickey LT (2018) *VERNALIZATION1* modulates root system architecture in wheat and barley. Mol Plant 11(1):226–229

Vriet C, Russinova E, Reuzeaua C (2012) Boosting crop yields with plant steroids. Plant Cell 24(3):842–857

Wang G, Schmalenbach I, von Korff M, Léon J, Kilian B, Rode J, Pillen K (2010) Association of barley photoperiod and vernalization genes with QTLs for flowering time and agronomic traits in a BC2DH population and a set of wild barley introgression lines. Theor Appl Genet 120(8):1559–1574

Waters MT, Gutjahr C, Bennett T, Nelson DC (2017) Strigolactone signaling and evolution. Annu Rev Plant Biol 68:291–322

Weatherwax P (1923) The story of the maize plant. University of Chicago Press, Chicago

Wendt T, Holme I, Dockter C, Preuû A, Thomas W, Druka A et al (2016) *HvDep1* Is a positive regulator of culm elongation and grain size in barley and impacts yield in an environment-dependent manner. PLoS ONE 11(12):e0168924

White PJ, Bengough AG, Bingham IJ, George TS, Karley AJ, Valentine TA (2009) Induced mutations affecting root architecture and mineral acquisition in barley. In: Shu QY (ed) Induced plant mutations in the genomics era. Joint FAO/IAEA division of nuclear techniques in food and agriculture. IAEA, Vienna, pp 338–340

Xu Y, Jia Q, Zhou G, Zhang XQ, Angessa T, Broughton S, Yan G, Zhang W, Li C (2017) Characterization of the *sdw1* semi-dwarf gene in barley. BMC Plant Biol 17(1):11

Yamamuro C, Ihara Y, Wu X, Noguchi T, Fujioka S, Takatsuto S et al (2000) Loss of function of a rice *brassinosteroid insensitive1* homolog prevents internode elongation and bending of the lamina joint. Plant Cell 12(9):1591–1605

Yan L, Loukoianov A, Tranquilli G, Helguera M, Fahima T et al (2003) Positional cloning of the wheat vernalization gene *VRN1*. Proc Natl Acad Sci USA 100(10):6263–6268

Yan L, Loukoianov A, Blechl A, Tranquilli G, Ramakrishna W et al (2004) The wheat *VRN2* gene is a flowering repressor down-regulated by vernalization. Science 303(5664):1640–1644

Yoshikawa T, Tanaka SY, Masumoto Y, Nobori N, Ishii H, Hibara KI et al (2016) Barley *NARROW LEAFED DWARF1* encoding a WUSCHEL-RELATED HOMEOBOX 3 (WOX3) regulates the marginal development of lateral organs. Breed Sci 66(3):416–424

Yu P, Gutjahr C, Li C, Hochholdinger F (2016) Genetic control of lateral root formation in cereals. Trends Plant Sci 21(11):951–961

Zhang C, Bai MY, Chong K (2014) Brassinosteroid-mediated regulation of agronomic traits in rice. Plant Cell Rep 33(5):683–696

Zhang X, Zhou G, Shabala S, Koutoulis A, Shabala L, Johnson P, Li C, Zhou M (2016) Identification of aerenchyma formation related QTL in barley that can be effective in breeding for waterlogging tolerance. Theor Appl Genet 129(6):1167–1177

Zhang X, Fan Y, Shabala S, Koutoulis A, Shabala L, Johnson P, Hu H, Zhou M (2017) A new major-effect QTL for waterlogging tolerance in wild barley (*H. spontaneum*). Theor Appl Genet 130(8):1559–1568

Zhao J, Sun H, Dai H, Zhang G, Wu F (2010) Differences in response to drought stress among Tibet wild barley genotypes. Euphytica 172(3):395–403

Zheng R, Li H, Jiang R, Römheld V, Zhang F, Zhao FJ (2011) The role of root hairs in cadmium acquisition by barley. Environ Pollut 159(2):408–415

A Genomic View of Biotic Stress Resistance

14

J. Mitch Elmore, Dragan Perovic, Frank Ordon, Patrick Schweizer and Roger P. Wise

Abstract

Over the last 100 years, research in barley disease resistance has progressed from classical genetic approaches up to molecular genetics and genome-wide functional profiling. Along the way, a multitude of well-characterized genetic resources have paved the way for mechanistic investigations into barley immunity. Seminal discoveries in barley have been translated to other crops, thus proving it to be an excellent system for plant disease research. Recent access to high-quality barley and pathogen genomes has empowered large-scale functional studies and facilitated the development of new technologies to associate traits with genes. Continued integration of existing germplasm with anchored sequence data promises to expedite mapping and functional characterization of important disease traits. In this chapter, we offer a genomic view of barley biotic stress with a focus on different types of resistance, current genomic approaches to dissect immune responses, and future prospects for the field.

14.1 Biotic Threats and Disease Resistance

Across its wide agricultural range, a variety of invertebrate pests and microbial pathogens can impact barley production resulting in significant declines in grain yield and quality (Paulitz and Steffenson 2010; Schweizer 2014). Microbial pathogens are generally divided into categories based on their lifestyle: biotrophs require living host tissue for growth and reproduction, necrotrophs feed on dead tissue, and hemibiotrophs exhibit both biotrophic and necrotrophic habits. Fungal and viral diseases are major threats to barley production and will be the focus of this chapter. The relative importance of specific diseases may vary substantially over time and geographic region due to differences in environmental conditions, cultivars, and pathogen virulence (Paulitz and Steffenson 2010). Cereal rust caused

J. M. Elmore · R. P. Wise (✉)
Corn Insects and Crop Genetics Research, USDA—Agriculture Research Service, Iowa State University, Ames, IA 50011, USA
e-mail: roger.wise@ars.usda.gov

J. M. Elmore · R. P. Wise
Department of Plant Pathology and Microbiology, Iowa State University, Ames, IA 50011, USA

D. Perovic · F. Ordon
Julius Kuehn-Institute (JKI), Federal Research Centre for Cultivated Plants, Institute for Resistance Research and Stress Tolerance, Erwin-Baur-Straße 27, 06484 Quedlinburg, Germany

P. Schweizer
Pathogen-Stress Genomics Lab, Leibniz Institute of Plant Genetics and Crop Plant Research (IPK), OT Gatersleben, Corrensstrasse 3, Stadt, 06466 Seeland, Germany

N. Stein and G. J. Muehlbauer (eds.), *The Barley Genome*, Compendium of Plant Genomes,
https://doi.org/10.1007/978-3-319-92528-8_14

by *Puccinia* spp. and powdery mildew caused by *Blumeria graminis* f. sp. *hordei* are persistent problems in most barley growing regions. *Bipolaris sorokiniana* spot blotch, *Fusarium* spp. head blight, *Pyrenophora graminea* leaf stripe, *Pyrenphora teres* net blotch, *Ramularia collo-cygni* leaf spot, and *Rynchosporium commune* leaf scald are other economically important diseases caused by fungi (Paulitz and Steffenson 2010). The aphid-transmitted *Barley yellow dwarf virus* (BYDV) and *Cereal yellow dwarf virus* (CYDV), and the soil-borne *Barley yellow mosaic virus* (BaYMV) and *Barley mild mosaic virus* (BaMMV) can also cause major yield losses (Ordon and Kühne 2014).

14.1.1 Genome-Encoded Immune Receptors

Plants produce pre-formed physical and chemical barriers to pathogen ingress. In addition, plant genomes encode an innate immune system composed of different types of receptors that actively recognize and respond to pathogens (Jones and Dangl 2006; Dodds and Rathjen 2010). Extracellular pattern recognition receptors (PRRs) at the cell surface monitor the apoplast for "non-self" pathogen-associated molecular patterns (PAMPs) such as fungal chitin. PRRs can also recognize "modified-self" damage-associated molecular patterns (DAMPs) such as degraded fragments of the plant cell wall or endogenous peptides that are expressed during pathogen attack (Zipfel 2014). Plant PRRs are plasma membrane-localized receptor-like kinases or receptor-like proteins containing extracellular domains that bind specific ligands. In response to PAMPs, barley activates signaling networks associated with MAP kinase cascades, production of reactive oxygen species, vesicle trafficking and secretion, cell-wall reinforcement, and transcriptional reprogramming that culminate in pattern-triggered immunity (PTI) (Hückelhoven 2007; Hückelhoven and Seidl 2016; Scheler et al. 2016). PTI is considered one of the first lines of inducible pathogen defense and likely contributes to a basal level of resistance to most pathogens (Zipfel 2014). Currently little is known about PRRs in barley (Tanaka et al. 2010; Hückelhoven and Seidl 2016; Rajaraman et al. 2016), but downstream factors affecting pathogen invasion or spread have been identified via forward genetics and functional genomics approaches (Collins et al. 2003; Hu et al. 2009; Douchkov et al. 2014, 2016).

Intracellular receptor proteins comprised of amino-terminal coiled (CC) or Toll/interleukin-1 receptor homology (TIR) domains, a central nucleotide binding NB-ARC domain, and a carboxy-terminal leucine-rich repeat domain, termed NLRs, control another aspect of pathogen recognition in plants (Jacob et al. 2013; Cui et al. 2015). Interestingly, TIR-NLRs are absent in the monocot lineage including barley (Jacob et al. 2013). Recent molecular characterization and genome-wide analyses across flowering plants have revealed the presence of many NLRs with additional, "integrated" domains hypothesized to facilitate pathogen recognition and represent about 5–10% of NLRs encoded in plant genomes (Cesari et al. 2014; Wu et al. 2015; Sarris et al. 2016).

NLR receptors detect the presence of pathogen virulence proteins, termed effectors, inside host cells. Effectors function to manipulate host cell physiology in order to suppress defense, and accommodate growth and reproduction of the pathogen. Effectors produced by biotrophs often target PTI signaling proteins to suppress host defense, while necrotroph effectors can induce plant cell death (Lo Presti et al. 2015). NLRs recognize effectors either through direct interaction or indirectly by monitoring the status of additional host proteins that are targeted by effectors (Cui et al. 2015). NLR activation results in effector-triggered immunity (ETI). ETI is associated with many of the same cellular responses as PTI but is executed more intensely and is often accompanied by programmed cell death or the hypersensitive reaction (HR) at sites of infection, resulting in near-complete disease resistance (Tsuda and Katagiri 2010). Research in barley has been critical for dissecting plant NLR biology and mechanisms underlying ETI (discussed in Sect. 14.4).

While the PTI/ETI conceptualization has been a useful model for analyzing plant immunity, it is increasingly clear that this classification fails to capture the full scope of plant-microbe interactions (Jones and Dangl 2006; Thomma et al. 2011; Pritchard and Birch 2014). For instance, apoplastic effectors can be recognized by PRR-type receptors and PAMPs can activate ETI-like responses including HR (Cook et al. 2015; Thomma et al. 2011). There can be a great deal of overlap in the magnitude of protection mediated by different PRR-type and NLR-type immune receptors with both types activating "weak" or "strong" responses (Boyd et al. 1995; Thomma et al. 2011). Also, it has become clear that significant overlap exists in the defense outputs activated by different classes of immune receptors (Caldo et al. 2004; Thomma et al. 2011). Therefore, the plant immune system can be considered a "continuous" surveillance system for microbial invasion triggered by a range of pathogen molecules with varying defense outputs that do not depend strictly on the recognized molecule or type of immune receptor (Cook et al. 2015).

14.1.2 Distinguishing Different Types of Disease Resistance

Disease resistance can be classified broadly based on different phenotypic outcomes and the underlying genetics of both host and pathogen (Zhang et al. 2013; Roux et al. 2014). The relative impact of structural and chemical (in)compatibilities, PRR- and NLR-type immune receptors, and various downstream cellular defenses may vary depending on the specific plant–microbe interaction (Schulze-Lefert and Panstruga 2011). Thus, the general definitions described below fall on a spectrum of disease resistance and could be achieved via overlapping mechanisms.

Qualitative disease resistance is nearly complete and renders the plant fully immune to infection by particular pathogen isolates (Zhang et al. 2013). This type of immunity is typically genetically dominant and conferred by a single resistance (*R*) gene with a major effect on the disease outcome. *R*-genes often encode NLR proteins that activate ETI signaling upon recognition of specific effectors expressed by a pathogen isolate. However, some PRR-type receptors (Song et al. 1995) and other non-receptor genes (Krattinger et al. 2009b) have been designated as *R*-genes due to their measurable effect on reducing disease incidence. Qualitative disease resistance is often *race-specific*: pathogen isolates with variable effector repertoires can escape detection and cause disease on plant varieties with *R*-genes effective against other races of the same pathogen. Indeed, many plant pathogens are highly adaptive and numerous examples exist where major *R*-gene resistance has broken down in the field, possibly via loss or modification of a corresponding effector in the pathogen population (McDonald and Linde 2002).

Quantitative disease resistance (QDR) is often observed when sources of qualitative resistance are not present in the host population (Roux et al. 2014). QDR is incomplete and results in a reduction but not elimination of disease symptoms. QDR can be controlled by multiple genes, each with a partial effect on the resistance phenotype (Poland et al. 2009; Niks et al. 2015). QDR is often *race nonspecific* and therefore active against all races of a pathogen. In a segregating population, QDR results in a continuous or near-continuous distribution of phenotypes ranging from "more resistant" to "more susceptible" (St. Clair 2010). Quantitative trait loci (QTL) mapping is used to localize QDR genes on the genome but associating specific genes with QTL remains a challenge (Niks et al. 2015). At a mechanistic level many forms of QDR appear to be a manifestation of PTI, reflecting specific cellular events that reduce pathogen infection. Other forms of QDR appear to be specific to particular pathogen isolates and might be controlled by weaker alleles of *R*-genes (Marcel et al. 2008; French et al. 2016). It is generally thought that QDR can be more durable than *R*-gene resistance, in part, due to its polygenic nature where each gene may control a different aspect of the defense response (Poland et al. 2009; French et al. 2016).

Nonhost resistance (NHR) represents a special case of immunity and is the most common in nature. Most plants are resistant to most pathogens, so disease is relatively rare when considering all potential interactions amongst plants and microbes. NHR is defined as full immunity of an entire plant species to all genotypes of a pathogen species (Niks and Marcel 2009). There has been considerable interest in defining mechanisms of NHR with hopes of engineering plants with this seemingly robust form of immunity. Barley and other cereals represent good model systems for defining the genetic basis of NHR due to closely related pathogen (sub)species being reciprocal non-pathogens on closely related crops (Jafary et al. 2008; Aghnoum and Niks 2010; Douchkov et al. 2014). A combination of genes and mechanisms likely operate during NHR including structural barriers, inappropriate chemical cues for pathogen development, and activation of robust PTI and/or ETI responses (Senthil-Kumar and Mysore 2013; Schulze-Lefert and Panstruga 2011).

With respect to barley viruses, tolerance has been observed in some cases, for example against BYDV. Tolerance is the ability of the host to limit the symptoms associated with disease without limiting viral accumulation. Selecting for tolerance reduces outward symptoms and enhances agronomic traits such as yield under viral load (Ordon and Perovic 2013). Pyramiding of loci conferring tolerance has resulted in quantitative resistance (a reduction in viral titer) against BYDV (Riedel et al. 2011; Ordon and Perovic 2013). Further studies have to be conducted to investigate the mechanisms responsible for this observation.

14.2 Downstream Applications of the Barley Reference Genome in Gene Isolation and Detection of Natural Variation

Development of a physical map and a reference sequence by the International Barley Genome Sequencing Consortium (IBSC 2012; Mascher et al. 2013a, 2017) has facilitated many downstream applications. These include the development of high-throughput genotyping platforms, e.g., Illumina Golden Gate Oligo Pooled Assay (Close et al. 2009), Illumina 9K iSelect array (Comadran et al. 2012) and the Illumina 50K iSelect chip (R. Waugh, personal communication), the barley Genome Zipper (Mayer et al. 2011), and exome-capture-based sequencing methods (Mascher et al. 2013b). Using these technologies in conjunction with the linear-ordered genome sequence, researchers and breeders can now associate traits of interest with a precise genomic location and identify candidate genes within the region almost instantaneously.

14.2.1 Fast Mapping, Development of "Gene-Based" Markers and Candidate Gene Identification

Barley genetics has a long history of using molecular markers and functional analysis to identify genes underlying important biological and agronomic traits. The reference genome will facilitate sequencing of phenotyped resistant and susceptible sub-pools from segregating populations followed by subsequent mapping using methods such as next-generation mapping (NGM) (Austin et al. 2011) and Population Sequencing (POPSEQ) (Mascher et al. 2013a). At the same time, high-resolution mapping of resistance and tolerance genes using high-density Single nucleotide Polymorphism (SNP) platforms, e.g., Illumina iSelect 9K and 50K arrays or GBS (Poland et al. 2012) in combination with bulk segregant analysis (BSA) (Michelmore et al. 1991) is nowadays a matter of months. Mapped genes can be immediately positioned at the barley reference genome. For precisely mapped resistance genes, knowledge of the barley sequence allows a rapid identification of candidate genes, followed by their validation by allele-specific sequencing in large populations, by mutant analysis, and by transient or stable transgenic complementation approaches.

Targeted sequence-based capture methods have been used in conjunction with next-generation sequencing to rapidly characterize disease-associated loci. *R*-gene enrichment and sequencing (RenSeq) (Jupe et al. 2013; Andolfo et al. 2014) offers the unique opportunity to generate the required high-resolution information by capturing NLR sequences to identify candidate resistance genes. MutRenSeq combines mutagenesis with exome-capture and sequencing and has rapidly identified *R*-genes in wheat (Steuernagel et al. 2016). Exome QTL-seq has been employed recently to map QTL for net blotch resistance (Hisano et al. 2017). In addition, non-exome based complexity reduction strategies using chromosome sorting have been used to identify *R*-gene mutations in wheat and barley (Sánchez-Martín et al. 2016).

In addition to fast chromosomal assignment of novel genes conferring resistance or tolerance to pathogens, the barley genome sequence will allow a precise localisation of already mapped genes. Different types of mapping populations, various marker technologies used during the last 25 years, different modes of inheritance of various marker types, rendered the comparison of the exact positions of respective genes difficult. Today, if respective marker sequences are known, the gene-harboring interval can be assigned precisely to the reference genome, giving information whether respective genes may be closely linked or be alleles of the same gene.

14.2.2 Using Natural Variation to Discover Novel Resistance and Tolerance Alleles

One of the most important applications of the reference barley genome is deciphering the natural variation in barley and related species. The availability of gene-space- or genome sequences for different barley cultivars, landraces, and wild relatives will facilitate the rapid mapping and isolation of genes conferring valuable agronomic traits. These sequences will provide new markers, including SNPs, insertions/deletions (indels) and copy number variants (CNVs), which may be used in genome-wide association studies (GWAS) for the identification and fine mapping of genes and QTLs for disease resistance/tolerance. Millions of informative markers (SNPs, indels, and CNVs), and structural variations, such as duplications, inversions, transpositions, etc. identified through genome comparisons will promote investigations into the genetic and molecular bases of resistance and tolerance mechanisms. Additionally, genome sequences of *Hordeum*-related species such as *Brachypodium distachyon*, *Triticum aestivum* L. *subsp. aestivum* and wheat relatives, i.e., *T. urartu*, *T. monoccocum*, *T. turgidum* L. subsp. *durum* and *T. turgidum* ssp. *dicoccoides* will promote comparative genomics and provide insights into the evolution of gene families that are responsible for disease resistance and tolerance. Thus, soon it will be possible to thoroughly exploit the rich barley and related cereal germplasm resources by identifying gene variants involved in disease resistance and introducing them into elite cultivars using targeted breeding or precision genome editing.

The use of Genotyping-by-Sequencing (GBS) (Poland et al. 2012) in characterizing *Hordeum bulbosum* introgression lines carrying resistance to soil-borne viruses and BYDV has been successfully employed to allocate the introgressed genomic segments (Wendler et al. 2015) and to genotype the $Rym16^{Hb}$ mapping population (Wendler et al. 2014). Furthermore, historical multi-pathogen phenotypic data can be combined with GBS to more precisely localize QTL for resistance in existing introgression populations (von Korff et al. 2005; Yun et al. 2005; Shtaya et al. 2007).

The most powerful method for the detection of novel natural variants is whole genome shotgun sequencing (WGS) (Beier et al. 2016) of resistant and tolerant genotypes. However, WGS an expensive and bioinformatically demanding endeavor due to the large size of the barley genome and the abundance of repetitive elements. In order to reduce costs associated with WGS and to enrich datasets with the most useful sequence information, exome-capture methods have been developed (Mascher et al. 2013b; Russell et al. 2016). Compared with WGS, exome sequencing is relatively cheaper and less data-intensive, but because only protein-coding genes are sequenced, information that could affect phenotypes (e.g., regulatory regions) is excluded. Nonetheless, exome sequencing of resistant and susceptible subpopulations can still narrow down the genomic location of causal allelic variants. Sequencing of the exomes in a collection of 267 geo-referenced landraces and wild barley accessions (Russell et al. 2016) already has demonstrated the power of this method in the discovery of hidden natural variation. The ongoing European project WHEALBI (N. Stein, personal communication), in which an additional 500 barley cultivars and landraces have been subjected to exome sequencing, will enhance our ability to associate traits with genes. Furthermore, data from a current 50 barley variety "pan-genome" project using WGS (N. Stein, personal communication) can be used to mine for novel resistance alleles. These collections are ideal sources for a high-resolution GWAS analysis and will fully pay off upon completion of detailed disease phenotyping with multiple pathogens. Polymorphisms identified using the above-mentioned methods in combination with the barley reference sequence can be used for development of high-throughput PCR based assays such as pyrosequencing (Silvar et al. 2011) or competitive allele-specific polymerase chain reaction (KASP) (Allen et al. 2011) which are the basis for marker-assisted selection (MAS) procedures for disease resistance.

14.2.3 Leveraging the Transcriptome to Target Regulators of Disease Resistance

One of the major tools for functional genomics of disease resistance is RNA-sequence—based transcriptome profiling (RNA-seq). For example, comparison of transcriptome profiles from barley lines after virus inoculation and infestation by non-viruliferous vectors can provide insights into the gene expression differences associated with virus replication and movement through infected plants. RNA-seq transcriptome profiling of resistant and susceptible barley lines enables simultaneous investigation into host and pathogen gene expression during infection, facilitating the identification of genes associated with both host immunity and pathogen virulence. Genes differentially expressed during infection can then be evaluated for their roles in resistance or susceptibility via functional assays (more in Sect. 14.5.2).

Genetical genomics, or expression quantitative trait loci (eQTLs) mapping, enables one to query thousands of expression level polymorphisms, i.e., the molecular phenotypes of genes (Jansen and Nap 2001; Rockman and Kruglyak 2006; Chen and Kendziorski 2007). By profiling the transcriptomes of allelic variants in a segregating population, linkage-based eQTL analysis can identify the chromosomal positions of *key regulators* for a particular condition (Jansen and Nap 2001; Rockman and Kruglyak 2006; Williams et al. 2007; Smith and Kruglyak 2008; Kliebenstein 2009). In barley, eQTL approaches have identified candidate genes for quantitative resistance against leaf and stem rusts (Chen et al. 2010; Druka et al. 2008b; Moscou et al. 2011b) and uncovered a defense gene regulon activated by diverse powdery mildew resistance loci (Surana et al. 2017).

One eQTL study dissected the genetic control of immunity-associated gene expression in the Q21861 × SM89010 doubled haploid population (Steffenson et al. 1995) challenged with *Puccinia graminis* f. sp. *tritici* TTKSK, commonly referred to as Ug99 stem rust (Moscou et al. 2011b). Upon challenge with Ug99, over 360 defense-associated genes transferred regulatory control from several unlinked loci to a single chromosomal region on barley chromosome 2HS. This *trans* eQTL hotspot also co-localized with an enhancer of adult plant resistance to Ug99.

Likewise, when the Q21861 × SM89010 population was infected with *B. graminis* f. sp. *hordei* (*Bgh*), two highly significant clusters of *trans* eQTL were identified near the telomeric ends of chromosome 2HL (unlinked to the Ug99 hotspot above) and chromosome 1HS. Within these clusters reside diverse resistance loci derived originally from the barley landrace, *Hordeum laevigatum* (*MlLa*), and *H. vulgare* cv. Algerian (*Mla1*); and associate with the altered expression of 961 and 3296 genes during penetration of host epidermal cells or development of haustorial feeding structures, respectively. Intriguingly, regulatory control of transcript levels for 299 of the 961 genes is shifted from *MlLa* on 2HL to *Mla1* on 1HS as infection progressed (Surana et al. 2017). Genes regulated by alternate chromosomal positions appear to be repurposed as part of a conserved immune regulon to respond to different stages of pathogen attack in a temporal manner.

As described above, expression profiling of populations segregating for disease traits enables the prediction of defense-associated genes and their *trans* regulatory regions (Smith and Kruglyak 2008; Moscou et al. 2011b; Chen et al. 2010; Druka et al. 2008b; Surana et al. 2017). Thus, eQTL analyses can provide a set of testable hypotheses for investigating the regulation of disease resistance. The key is to physically link pathogen-induced *trans* gene regulation to the immune response, for instance, by integrating expression and phenotypic QTL data (Druka et al. 2008a, b; Chen et al. 2010; Moscou et al. 2011b). This approach in conjunction with the ordered genome sequence can result in the identification of candidate *R*-genes, regulatory transcription factors that control defense gene modules, immune-active miRNA/siRNAs and their targets, or variations of chromatin-based regulation. Integration of different data layers will ultimately result in a holistic view of barley interactions with pathogens.

14.3 Genomic Tools in Breeding for Virus Resistance

About 40 different viruses are known to infect barley naturally (Lapierre and Signoret 2004). Of these, soil-borne *Barley yellow mosaic virus* (BaYMV) and *Barley mild mosaic virus* (BaMMV) as well as the aphid-transmitted *Barley yellow dwarf virus* (BYDV), and *Cereal yellow dwarf virus* (CYDV) cause high yield losses in susceptible barley crops (Ordon and Perovic 2013; Ordon and Kühne 2014). Currently, phytopathogenic viruses cannot be controlled directly by chemicals and the control of insect vectors has become more difficult due to increasing insecticide resistance. Therefore, genetic resistance is essential to reduce disease incidence to these viruses as well as to soil-borne viruses whose plasmodiophorid vectors also currently cannot be combated by chemicals. This holds especially true in the background of climate change creating favorable conditions for virus vectoring insects, in Northern Europe, for example.

Domestication and breeding activities have shaped the genome of barley and reduced the level of genetic diversity available in the cultivated barley germplasm, which is presenting a major limiting factor in developing new cultivars resistant to biotic and abiotic stress (Tanksley and McCouch 1997). Regarding virus resistance genes, a major consequence is the depletion of resistance genes in modern elite lines that can be compensated by resistance re-introduction from the primary gene pool (Ordon and Perovic 2013) or by introgression from the secondary gene pool represented by *H. bulbosum* (Perovic et al. 2013; Wendler et al. 2014, 2015).

Barley geneticists, virologists and breeders are in a position in which users of the first sequenced model plants, i.e., Arabidopsis and rice, have been for more than 15 years. Knowledge of the whole barley genome sequence will not only promote breeding but also provide an extraordinary opportunity for basic biological research, i.e., accelerated isolation of resistance genes (Ordon and Perovic 2013) and rapid detection of natural variation by allele-mining (Kaur et al. 2008; Hofinger et al. 2011; Yang et al. 2017). Since elite barley cultivars show limited genetic diversity regarding virus resistance and tolerance genes, exotic accessions and wild barley relatives, e.g., *H. vulgare ssp.* spontaneum and

H. bulbosum, stored in gene banks, provide a rich source of useful genetic/allelic variation.

14.3.1 Mapped Resistance Genes to Viral Diseases of Barley

Molecular markers closely linked to loci conferring resistance to BaMMV/BaYMV or tolerance/resistance to BYDV are available today and are widely applied in barley breeding (Ruge et al. 2003; Ruge-Wehling et al. 2006; Scholz et al. 2009; Ordon and Perovic 2013; Perovic et al. 2014; Yang et al. 2013, 2017). Up to now, 18 major resistance genes corresponding to at least 9 different genetic loci have been identified that confer resistance to either BaYMV or BaMMV, or to several strains of both bymovirus species (Kai et al. 2012; Perovic et al. 2014). Out of these, 16 genes originate from the primary gene pool and are recessively inherited, except *Rym17* which is dominant (Kai et al. 2012). Two additional dominant resistance genes, *Rym14* and *Rym16,* were introgressed from *H. bulbosum* (Ruge et al. 2003; Ruge-Wehling et al. 2006). With respect to BYDV, five genes conferring resistance/tolerance are known: *ryd1*, *Ryd2* and *Ryd3* originating from the primary gene pool (Ordon and Kühne 2014), as well as *Ryd4* (Scholz et al. 2009) and an additional QTL (Perovic et al. 2013) derived from *H. bulbosum.*

To date, two genes effective against BaMMV/BaYMV have been isolated using a map-based cloning approach. These recessive resistance genes correspond to two different host factors needed for virus accumulation. The *rym11/11* locus encodes Protein Disulfide Isomerase Like 5-1 (Hv-PDIL5-1) which is speculated to function as a chaperone in correct folding of virus proteins (Verchot 2012; Yang et al. 2014a, b). The *rym4/5* gene encodes Eukaryotic Translation Initiation Factor 4E (Hv-eIF4E), which putatively functions in assisting the translation initiation of a bymovirus precursor protein (Kanyuka et al. 2005; Stein et al. 2005; Moury et al. 2014; Sanfacon 2015). Genes specifying resistance or tolerance to BYDV/CYDV have been mapped but not yet been isolated.

14.4 Archetype Genes and Pathways for Qualitative Disease Resistance

While many traits for qualitative or race-specific resistance have been identified and mapped in barley, relatively few of the causal genes have been cloned (Chełkowski et al. 2003; Friedt and Ordon 2007). Several features of barley have hindered physical mapping projects thus far: (1) the 5.1Gb genome requires very large libraries to achieve appropriate coverage; (2) repetitive sequences account for up to 84% of the genome, making marker development a challenge; and (3) suppressed recombination is common, especially within centromeric and pericentromeric regions which comprise nearly 50% of the genome (Künzel et al. 2000; Krattinger et al. 2009a; Graner et al. 2010; IBSC 2012; Mascher et al. 2017). The availability of genome-enabled tools (Sect. 14.2) has alleviated some of these problems and promises to further accelerate map-based cloning projects.

Initial annotation of the barley cv. Morex genome identified 191 NLR genes presumed to control pathogen recognition (IBSC 2012). Recent re-analyses of the barley *R*-gene complement has detected 336 NB-ARC encoding genes (Sarris et al. 2016) and targeted RenSeq efforts have annotated more than 400 NB-ARC genes in Morex (J. Bettgenhaeuser, I. Hernanadez-Pinzon, M. J. Moscou, personal communication). NLR genes tend to cluster at the distal ends of barley chromosomes especially on the arms of 1HS, 2HS, 6HL, and 7HS (IBSC 2012). The enrichment of NLRs at specific genomic locations can favor recombination and provides one potential mechanism for sequence and functional diversification (Hulbert et al. 2001; IBSC 2012; Jacob et al. 2013).

14.4.1 Mla—Exemplary Diversity at an R-gene Locus

Some of the best-characterized plant NLRs belong to the barley *Mla* gene family conferring resistance against different isolates of powdery mildew, *Bgh* (Halterman et al. 2001; Zhou et al. 2001; Halterman and Wise 2004). Around 30 unique *Mla* specificities against *Bgh* have been described thus far which map to the short arm of chromosome 1H (Jørgensen and Wolfe 1994; Seeholzer et al. 2010). Remarkably, *Mla* orthologs at syntenic regions of the homeologous chromosome 1 of wheat and rye confer resistance to not only wheat powdery mildew (*B. graminis* f. sp. *tritici*) (Jordan et al. 2011) but also wheat stem rust (*P. graminis* f. sp. *tritici*) (Periyannan et al. 2013; Mago et al. 2015; Cesari et al. 2016). These discoveries suggest that the cereal *Mla* locus is a powerful incubator for the evolution of disease resistance specificities (Mago et al. 2015; Krattinger and Keller 2016). The ability of *Mla* homologs to recognize different races of the same species and distantly related fungal pathogens makes it a valuable model to study *R*-gene evolution. This diversity also indicates a potential to engineer new recognition specificities as more is known about the structural basis for effector recognition.

Functional *Mla* alleles exist in repulsion from each other and are rarely detected in the same barley genotype (Jørgensen and Wolfe 1994; Seeholzer et al. 2010). Sequencing of a 261-kb BAC contig from barley cultivar Morex revealed that the *Mla* locus is complex and comprised of three distinct families of NLRs nested among multiple classes of transposable elements and repetitive DNA (Wei et al. 2002). Most barley and wheat accessions contain 1–8 paralogous genes belonging to the *Mla* family although not all genes are expressed (Wei et al. 1999, 2002; Seeholzer et al. 2010; Jordan et al. 2011). The family has expanded to around 20 members in rye (Mago et al. 2004, 2015). Sequence analysis indicates that the resistance specificities of barley, wheat and rye *Mla* orthologs evolved after the divergence of these species ca. 9 million years ago (Mago et al. 2015; Krattinger and Keller 2016). Examination of the *Mla* region in Morex has supported a model where multiple rounds of duplication, inversion, transposon invasion, and repeat propagation have influenced its current state (Wei et al. 2002). Together, these observations suggest that the genomic region where *Mla* resides is highly active and permits the diversification of resistance specificities in response to pathogen pressures. As more Triticeae genomes are sequenced, it will be interesting to reconstruct the evolutionary trajectories of this dynamic locus.

Mla alleles exhibit high levels of sequence identity, especially in the region encoding the CC and NB-ARC domains (Halterman et al. 2001; Halterman and Wise 2004; Zhou et al. 2001; Wei et al. 2002; Shen et al. 2003; Halterman et al. 2003; Seeholzer et al. 2010). By contrast, the region encoding the LRR domain is much more variable and sites of positive selection in *Mla* are located nearly exclusively in the LRR region. These findings, along with domain swap experiments, indicate that the LRR is under strong diversifying selection and determines the recognition specificity of MLA proteins, possibly through direct binding with the cognate effectors (Seeholzer et al. 2010; Shen et al. 2003). Effectors from *Bgh* have been cloned using a map-based approach (AVR_{a10} corresponding to MLA10) (Ridout et al. 2006) and a transcriptome-wide association approach (AVR_{a1} and AVR_{a13} recognized by MLA1 and MLA13, respectively) (Lu et al. 2016). The availability of both host and pathogen reference genomes facilitates such genome- and/or transcriptome-wide investigations into the natural variation of diverse barley accessions and *Bgh* isolates. This promises to accelerate matching of additional *Mla* receptors with their recognized effectors which should illuminate co-evolutionary processes governing host-pathogen interactions.

Much is known about how *Mla* activates disease resistance signaling. Many, but not all, *Mla* alleles require *Rar1*, first identified in barley and later found to be generally required for many plant NLRs as part of a conserved chaperone complex that regulates R-protein stability and signaling (Shirasu et al. 1999; Zhou et al. 2001;

Shen et al. 2003; Bieri et al. 2004; Halterman and Wise 2004; Shirasu 2009). In the cell, MLA10 localizes to the cytoplasm and nucleus where it controls distinct signaling events in each compartment (Shen et al. 2007; Bai et al. 2012). Cytoplasmic MLA10 appears to activate cell death while the nuclear pool of MLA10 interacts with MYB and WRKY transcription factors to regulate defense gene expression sufficient for disease resistance (Shen et al. 2007; Bai et al. 2012; Chang et al. 2013).

Development of the Barley1 GeneChip, the first commercial microarray platform for a crop plant (Close et al. 2004), enabled genome-wide surveys into the transcriptional reprogramming that occurs during powdery mildew infection of *Mla* resistant and corresponding susceptible mutants (Caldo et al. 2004, 2006; Moscou et al. 2011a). These studies provided some of the first transcriptional evidence for a link between basal (PTI) and isolate-specific (ETI) resistance responses. It was shown that virulent pathogens suppress basal defense gene expression, whereas MLA activation boosts expression of these defense genes (Caldo et al. 2004, 2006). The core *Mla* transcriptional regulon was also predicted (Moscou et al. 2011a). One limitation of these microarray analyses was that only changes in host gene expression could be analyzed, but the use of RNA-seq now allows the simultaneous interrogation of both host and pathogen transcriptomes during infection.

14.4.2 Rdg2a—Seed-Based Resistance in the Absence of HR

The hemibiotrophic fungus *P. graminea* causes leaf-stripe disease on barley and is transmitted by seed. The race-specific resistance gene *Rdg2a* was cloned from chromosome 7HS using a map-based approach and validated using stable transformation of a susceptible variety (Bulgarelli et al. 2004, 2010). *Rdg2a* encodes a NLR protein that resides in a gene cluster with two additional NLR genes with high sequence identity in close proximity (<50 kb) indicating recent duplication events. Interestingly, *Rdg2a*-based resistance operates in the embryo of infected seed and is associated with the accumulation of phenolic compounds in the plant cell wall coinciding with termination of *P. graminea* hyphal growth (Haegi et al. 2008). However, resistance is not associated with hypersensitive cell death in the embryo, indicating that other defense mechanisms are at work (Bulgarelli et al. 2010).

14.4.3 Rpg1—Durable, Qualitative Resistance to Stem Rust

The *Rpg1* gene confers qualitative resistance to most races of the wheat stem rust pathogen *P. graminis* f. sp. *tritici* (*Pgt*). Unlike many major resistance genes, *Rpg1* has been quite durable in the field since the early 1940s but emergence of *Rpg1*-virulent *Pgt* races QCCJ and TTKSK threaten barley production throughout the world (Steffenson 1992; Steffenson and Jin 2006). *Rpg1* resides on chromosome 7HS and was cloned using a map-based approach (Brueggeman et al. 2002; Horvath et al. 2003). Interestingly, the RPG1 protein is not a NLR but exhibits a tandem protein kinase architecture (Brueggeman et al. 2002). Molecular characterization of RPG1-dependent signaling and identification of the corresponding avirulence proteins from *Pgt* indicate that RPG1 acts as a receptor and/or early transducer of defense signaling upon recognition of *Pgt* spores (Nirmala et al. 2006, 2007, 2010, 2011).

Mutagenesis of *Rpg1*-containing barley led to the identification of a series of mutants, *rpr1–rpr7,* that are required for *P. graminis* resistance (Zhang et al. 2006; Gill et al. 2016). Using the Barley1 GeneChip platform (Close et al. 2004), a transcript-based cloning approach was employed to identify *rpr1* gene candidates in this fast-neutron induced mutant (Zhang et al. 2006). Microarray analysis comparing wild type and *rpr1* identified eight genes that were highly down-regulated in the mutant. Three of these genes, including a putative receptor-like protein kinase, were found to be deleted in *rpr1* and all deletions co-segregated with the *rpr1* phenotype on chromosome 4H, suggesting one or more are required for *Rpg1* resistance (Zhang et al. 2006). The use of genome-enabled tools (discussed in Sect. 14.2) will likely result in the identification and characterization of *Rpr* genes in the near future.

14.4.4 rpg4/Rpg5—A Complex Resistance System Against Stem Rusts

Genetic and molecular dissection of the *rpg4/Rpg5* resistance locus has revealed complex interactions among multiple genes that control the barley response to stem rust. The *rpg4* locus on chromosome 5HL confers recessive resistance to many wheat stem rust races, including the *Rpg1*-virulent *Pgt* races QCCJ and TTKSK (Jin et al. 1994; Steffenson et al. 2009). Co-segregating with *rpg4* is *Rpg5*, which controls dominant resistance to the rye stem rust pathogen *P. graminis* f. sp. *secalis (Pgs)* isolate 92-MN-90 (Brueggeman et al. 2008). *Rpg5* was isolated via map-based cloning and found to encode a NLR protein with an integrated carboxy-terminal protein kinase domain (Brueggeman et al. 2008). Recent work has also implicated *Rpg5* in resistance of barley as a near nonhost to the heterologous oat stem rust pathogen *P. graminis* f. sp. *avenae* (Dracatos et al. 2015).

The exact gene(s) controlling *rpg4*-mediated resistance had remained elusive until recently. High-resolution mapping of three recombinant populations revealed that *rpg4* resistance is controlled by two distinct, but tightly linked, genetic intervals (*rpg4*-mediated resistance locus RMRL1 and RMRL2) (Wang et al. 2013). Located within RMRL1 is *Rpg5*, an actin depolymerizing factor-like *Adf3* gene, and a typical NLR gene designated *Rga1*. Surprisingly, virus-induced gene silencing of individual genes at RMRL1 resulted in conversion of *rpg4* plants from resistant to susceptible, indicating that all three genes are required for resistance (Wang et al. 2013). Additionally, an unknown gene(s) designated *rpg4* modifier element 1 (*Rme1*) at RMRL2 is required for *rpg4* defense (Wang et al. 2013). Thus, *rpg4*-mediated defense is regulated by at least four genes within a small region of chromosome 5HL.

Diversity analysis of RMRL1 *Rpg5, Adf3,* and *Rga1* alleles in a panel of wild and domesticated barley indicated that *Rpg5* polymorphisms alone likely control *rpg4* resistance, thus implicating it as the *bona fide R*-gene (Arora et al. 2013). Intriguingly, the "recessive" nature of *rpg4* could be due to an insertion/deletion event that replaced the kinase domain sequence of *Rpg5* with an apparently functional protein phosphatase 2C (*PP2C*) gene in many susceptible genotypes including the original susceptible parents of mapping populations (Wang et al. 2013; Arora et al. 2013). Genetic data indicate that this *PP2C* functions as a dominant susceptibility factor and thus might suppress RPG5 activity in *Rpg5/PP2C* heterozygous plants (Wang et al. 2013).

The integrated kinase domain of the RPG5 NLR and requirement of *Rga1* at the same locus suggest that the two NLRs interact biochemically to activate resistance upon recognition of an effector from *Pgt* stem rust (Wang et al. 2013; Cesari et al. 2014; Wu et al. 2015). Because *Rpg5* is implicated in recognition of at least three formae speciales of stem rust whose primary hosts are wheat, rye, and oat, it is tempting to speculate that a conserved effector from these pathogens is recognized directly by RPG5, possibly via its integrated kinase domain. It is also possible that unrelated effectors are recognized directly by RPG5 or converge on a common host target that is guarded by RPG5. The identification of the effector(s) recognized by *Rpg5* will further our understanding of this important locus and will inform efforts to transfer *Rpg5* immunity to other cereals.

14.5 Quantitative Disease Resistance and Nonhost Resistance

PAMP—or pattern-triggered immunity (PTI) of barley is suppressed by secreted effector molecules from host-adapted pathogens (Deslandes and Rivas 2012; Doehlemann et al. 2014).

The residual level of PTI reflects what breeders usually call quantitative disease resistance (QDR). QDR has been described in a multitude of, mostly fungal, barley-pathogen interactions (Schweizer and Stein 2011). There are two obvious ways of improving QDR to pathogenic fungi: by either enhancing the PTI signaling and/or outputs in order to override effector-mediated defense suppression, or by rendering plants less responsive to co-opting effectors and thus, less vulnerable to attack. The former can for example be achieved by introgressing a novel PAMP receptor from plant genetic resources such as wild barley *H. vulgare* ssp. *spontaneum* or *H. bulbosum*, or by improving on the plant's performance to rapidly build up local cell-wall appositions (papillae) as a universal penetration barrier (Chowdhury et al. 2014). The latter can be achieved by knocking out, mutagenizing or introgressing more resilient alleles of so-called "host susceptibility factors" such as *Mlo,* which encodes a seven transmembrane-domain protein. Strongly compromised or null-mutant alleles of *Mlo* (e.g. *mlo-11*) have been used in spring barley breeding programs and are still effective after more than 30 years of use (Acevedo-Garcia et al. 2014).

Nonhost resistance (NHR) of barley to a number of important fungal pathogens has been described including the wheat powdery mildew fungus *B. graminis* f. sp. *tritici*, different rust fungi (*Puccinia* sp.), *Magnaporthe grisea* (*Setaria* isolate, related to the rice blast fungus *Magnaporthe oryzae*), and *Phakopsora pachyrhizi* (soybean asian rust). There currently exist two concepts for explaining the fascinating phenomenon of NHR, which renders a plant species immune or at least highly resistant to all known races or isolates of non-adapted pathogens invading other plant species. First, a nonhost plant can possess one specific or a stack of NLR-type resistance protein(s) (see previous section) recognizing either an essential, conserved effector molecule that is critical for pathogenesis, or perceiving a sufficiently complex array of effectors, which would leave sporadic mutations in one or the other without co-evolutionary selection gain due to the redundancy of pathogen perception by the nonhost. Second, the nonhost can have escaped effector-mediated defense suppression or co-option after speciation and during the subsequent, separated co-evolution with the new pathogen forms or species that were "pulled along" during plant speciation. This model predicts that effector molecules from non-adapted pathogens are ineffective or at least strongly reduced in efficacy and therefore, allow the plant to fully express a robust PTI response. The two hypotheses of nonhost resistance were put into a co-evolutionary concept suggesting that *R*-gene stacks are more important for close nonhost species, whereas release of PTI suppression is predominant in more ancient nonhost interactions (Schulze-Lefert and Panstruga 2011; Stam et al. 2014).

Evidence has been obtained in barley for both quantitative species-nonspecific and -specific factors in mediating NHR to *Puccinia* spp. Barley is considered an intermediate host or near nonhost to many rust species whose primary host is a closely related cereal like wheat. Atienza and colleagues (2004) accumulated genes for susceptibility to many heterologous rusts into one experimental line called SusPtrit. SusPtrit was then crossed with nonhost immune genotypes to isolate sources of nonhost resistance (Jafary et al. 2006, 2008; Niks and Marcel 2009). These studies found that nonhost resistance to *Puccinia* is controlled by QTL with different, but overlapping, specificities for the rust species studied. Furthermore, different immune cultivars carried distinct QTL for resistance against the same rust species, indicating a variety of genes can control nonhost resistance even to the same pathogen (Jafary et al. 2008). Evidence for major *R*-gene qualitative resistance associated with HR (Jafary et al. 2008) as well as isolate-specific quantitative resistance without HR was also uncovered (Dracatos et al. 2016), suggesting that gene-for-gene interactions might also contribute to nonhost resistance in some cases. Cloning of the genes underlying the identified QTL will significantly increase our understanding of molecular mechanisms controlling NHR and QDR in barley.

14.5.1 Genome-Wide Analyses of QDR and NHR

There are several options to address QDR and NHR in barley at a genome-wide scale. One possibility is to perform *meta*-QTL studies by referring to genetic mapping data from the past 25 years (Fig. 14.1) (Schweizer and Stein 2011). The historical RFLP- and SSR markers can now be precisely placed onto the barley reference sequence (IBSC 2012; Mascher et al. 2017) together with more recent SNP markers. This will allow searching for the most robust and potentially multi-disease resistance QTL as an entry point for high-resolution mapping and eventual gene cloning. To our knowledge, a causative gene for any resistance QTL has yet to be molecularly isolated in barley.

Quantitative resistance is a trait of moderate genetic complexity, usually being inherited by less than ten QTL in plant genotypes carrying agronomically interesting levels of resistance. This allows for stacked introgression of QTL from plant genetic resources by marker-assisted selection. At the same time, genome-wide background selection can be performed in order to reduce the load of non-adapted genome in the backcrossing approach, which will accelerate the recovery of elite cultivar performance (Francia et al. 2005). It will have to be carefully tested, however, to what extent QDR enhancement might be associated with physiological costs or trade-offs, which are directly linked to the function of the target genes and not to linkage drag. Examples of resistance genes impacting cultivar performance include *mlo* in barley (McGrann et al. 2014), and *Sr2* as well as *Lr34* in wheat (Juliana et al. 2015). One of the difficulties, though, might be related to the fact that resistance QTL are not preferentially located in distal chromosome regions, in contrast to many NLR genes and race-specific resistance loci (IBSC 2012). Due to the suppressed recombination frequency around barley centromeres, some QTL might therefore disqualify themselves for non-transgenic introgression approaches.

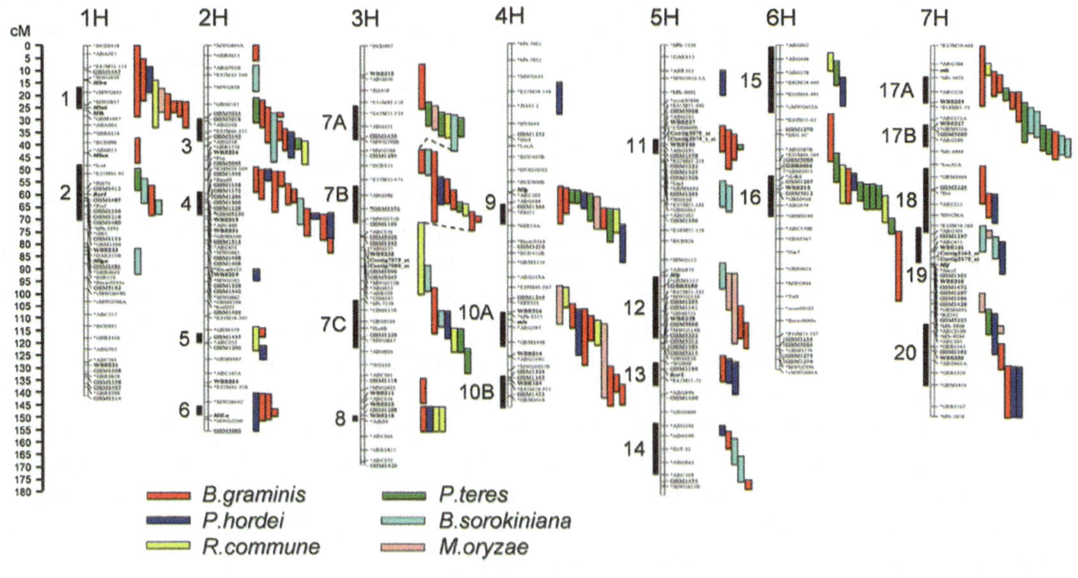

Fig. 14.1 Meta-quantitative trait loci (QTL) for resistance of barley to major fungal pathogens. QTL positions projected onto the "Barley, integrated, Marcel 2009" are indicated by bars to the right of the respective chromosomes. They either span the distance between nearby QTL-flanking markers or indicate peak-marker positions ±5 cM. QTL extending over more than 30 cM were excluded from the analysis. Meta-QTL calculated from midpoint positions of overlapping QTL are indicated by black bars to the left of chromosomes. Meta-QTL numbers followed by letters indicate the presence of several neighboring meta-QTL located within more complex QTL regions. Figure adapted from Schweizer and Stein (2011)

14.5.2 Functional Genomics Identifies QDR- and NHR-Associated Genes

A more direct way to identifying genes that are relevant for the interaction of barley with its major fungal pathogens may be to use functional genomics approaches. One possible entry point into the selection of candidate genes out of the approximately 39,000 proposed gene models in barley is transcript profiling, which was extensively used and led to the identification of a subset of pathogen- and *R* gene-regulated transcripts, mostly from different interactions with powdery mildew fungi (Caldo et al. 2004, 2006; Zierold et al. 2005; Zellerhoff et al. 2010). In these studies, up to 10% of all barley genes were found to be regulated by any given pathogen. This is still a large number and clearly too much for labor-intensive, single-gene approaches such as stable transgenic plants over-expressing or silencing the respective candidate gene, TILLING mutants, or CRISPR/Cas9 mutagenesis.

To efficiently screen hundreds of candidate genes for an effect in barley-powdery mildew interactions, Douchkov and associates (2005) developed a high-throughput transient-induced gene silencing (TIGS) system. TIGS of approximately 1200 pre-selected, differentially expressed barley genes resulted in the identification of 96 genes affecting the interaction in a statistically significant manner (Douchkov et al. 2014). Intriguingly, approximately two thirds of the genes behaved as susceptibility-related factors because TIGS enhanced resistance to the adapted barley-powdery mildew fungus.

The identification of such susceptibility-related genes opens up a fascinating option for gaining durable resistance by site-directed mutagenesis of such factors. Recent studies have employed transgenic approaches using TALEN- and CRISPR/Cas9 mediated gene editing (Wang et al. 2014) and non-transgenic TILLING-based methods to create *mlo* mutants at all alleles in hexaploid wheat (Acevedo-Garcia et al. 2016). Proof of concept for precision gene editing was recently obtained in barley (Budhagatapalli et al. 2015; Lawrenson et al. 2015). As known in the case of *mlo* mutagenesis, generating null (knock-out) variants of host-susceptibility factors may compromise overall plant performance and, as consequence, yield. However, breeders could reduce yield penalty of *mlo* by careful background selection resulting in cultivars that are most resilient to the undesired side effects (enhanced susceptibility to necrotrophic-pathogen attack and to abiotic stress). Recently, a tempered natural *mlo*-allele was described, which proved valuable for resisting extremely virulent, novel races of the barley-powdery mildew fungus in Western Australia yet allowing good abiotic stress resistance (Spies et al. 2012; Ge et al. 2016). This example also demonstrates the value of mining in barley genetic resources for advantageous alleles.

A genome-wide TIGS screen was also performed for NHR to the wheat powdery mildew fungus. This yielded ten *Rnr* (for: *Required for nonhost resistance*) candidate genes causing partial break down of NHR (Douchkov et al. 2014). Three *Rnr* genes (*Rnr5*, *Rnr6* and *Rnr8*) are currently under deeper investigation, using stable transgenic plants and CRISPR/Cas9-induced mutations. The three genes encode a *Triticeae*-specific, partial duplicate of the E3 ligase *HvPUB15* for proteosomal protein degradation (Rajaraman 2016), a cellulose synthase-like protein (Douchkov et al. 2016), and a putative PAMP receptor of the LRR-receptor kinase type of proteins (Rajaraman et al. 2016). Besides the *Rnr* genes, *mlo*, *Ror2* and *Bax-inhibitor 1* were described as factors of NHR against wheat powdery mildew (Elliott et al. 2002; Huckelhoven et al. 2003; Trujillo et al. 2004). Several of these gene candidates did not only affect NHR- but also QDR suggesting a functional similarity of the two forms of resistance and strengthening the model of NHR as a strong manifestation of PTI in the absence of effector-mediated defense suppression.

14.5.3 Exploiting NHR in Barley

How can barley translational research benefit from the above-mentioned discoveries in NHR?

Clearly, if we will be able to transfer nonhost resistance-like traits we might achieve longer durability of resistance. The problem of the above-mentioned results on *Rnr*- and related genes is that they tell us something about mechanisms of NHR in barley—a "solved problem"—but are not directly applicable for the improvement of host resistance. What would be required is broadening this approach to close relatives of barley from the *Triticeae* tribe of grasses searching for functional homologs of *Rnr* genes against barley pathogens and then transfer them into barley. This would have to be a transgenic approach and therefore, not economically viable in a mid- to long-term perspective due to consumer opposition in many countries to marketing of transgenic plant products. On top of this difficulty comes a scientific argument against the direct transfer of NHR genes from closely related plant species: A transcriptome comparison of barley and wheat responding either as hosts or as nonhosts to powdery mildew attack revealed a clear decline in the overlap of those orthologous genes that were differentially regulated between hosts and nonhosts, compared to the more generally pathogen-regulated set of genes (Rajaraman 2016). Thus, how a close relative gained NHR after speciation may not be predictable from studying just one species. There rather appear to exist many ways to NHR, even against different *formae speciales* or pathovars of the same pathogen species.

Maybe a more viable, non-transgenic approach to using NHR genes might be by genome introgressions from the wild relative *H. bulbosum* belonging to the secondary gene pool of cultivated barley. Indeed, a number of interesting resistance loci against important barley diseases were discovered in *H. bulbosum* introgression lines (Shtaya et al. 2007). The map-based identification of the causative genes, however, represents a big challenge due to severe suppression of recombination in most "wild" loci introgressed by wide crosses. At present, the use of *bona fide* nonhost-resistance genes appears out of reach for achieving durable resistance in barley. More realistic might be to identify those NHR genes, which are also relevant to host quantitative resistance and which might, therefore, confer more durable resistance. It will be fascinating to find out if such bi-functional genes might include host-susceptibility factors that escaped effector manipulation after speciation into host/nonhost. Such genes may indeed exist; a receptor-like kinase contributing to NHR against the wheat powdery mildew while acting as host-susceptibility factor for barley-powdery mildew fungus was recently identified (A. Himmelbach, D. Douchkov and P. Schweizer, unpublished). The identification of causative SNP for such putative escapes followed by gene editing bears the promise of highly targeted introgression of a nonhost-like status into barley. Another attractive option from genome-wide approaches to QDR will be the introgression of potentially valuable, resistance-associated and ideally functionally validated, alleles of relevant host genes from barley genetic resources of the primary gene pool by backcrossing (Fig. 14.2). Such backcrossing approaches will benefit from genome-wide background selection to regain elite cultivar performance more rapidly than was possible in the "*pre*-genome era" of barley research and breeding.

14.6 Conclusions and Future Prospects

The availability of a high-quality barley genome reference sequence is revolutionizing barley genetics by accelerating the identification and use of rare allelic variants in classical breeding schemes, and facilitating efficient marker-assisted backcrossing (MABC), marker-assisted selection (MAS) and pyramiding of genetic factors responsible for resistance and tolerance to important barley diseases. The barley genome sequence will accelerate the discovery of new genes, facilitate allele-mining and the development of large-scale SNP genotyping platforms. More efficient use of genetics and breeding methods and large-scale structural- and functional genomic studies in combination with

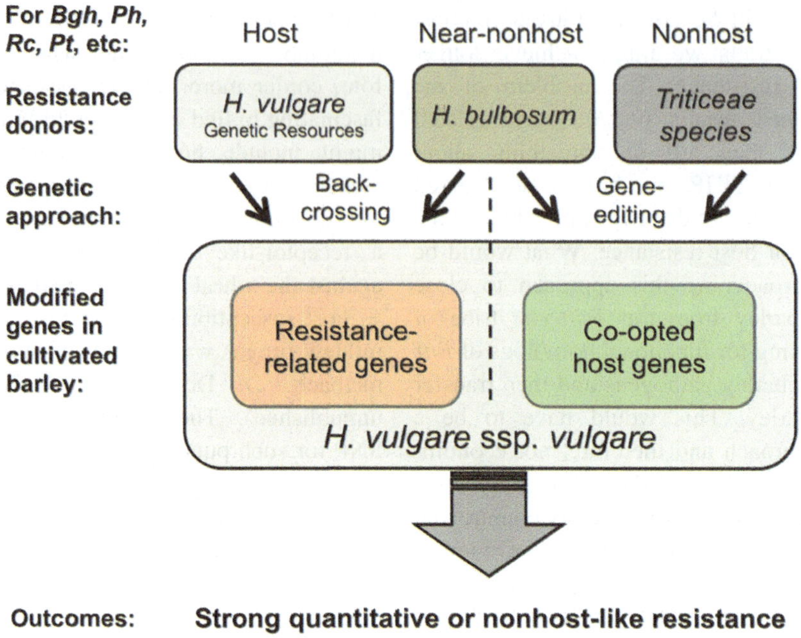

Fig. 14.2 Approaches to introgress or engineer quantitative host- or nonhost-like resistance into cultivated barley. Genetic resources of *H. vulgare*, including the wild relative *H. vulgare* ssp. *spontaneum*, are valuable donors of quantitative host resistance. Alleles of validated, resistance-related genes can be introgressed by marker-assisted backcrossing. From *Triticeae* species that cannot be crossed with barley, susceptibility-related and possibly co-opted genes can be edited by RNA-guided Cas9 to gain nonhost-like resilience to effector-mediated manipulation. Both backcrossing and gene editing are feasible using the secondary gene pool from *H. bulbosum*, which exhibits strong resistance to many pathogens of cultivated barley and which might be considered as near-nonhost. The described approaches are limited to barley pathogens such as *B. graminis* f. sp. *hordei* (*Bgh*), *Rhynchosporium commune* (*Rc*), *P. hordei* (*Ph*), or *Pyrenophora teres* (*Pt*) that cannot infect other *Triticeae* species

system-biology approaches will promote the rapid improvement of cultivars and the functional analysis of genes controlling disease resistance.

Accordingly, gene isolation will be accelerated by reducing the time from gene mapping to the identification and functional validation of candidate genes as well as reducing the costs for accomplishing the complete procedure. In this respect, isolation of the first virus resistance gene, *rym4/5* (Stein et al. 2005) has taken almost 15 years, while using the first draft of the barley genome helped to isolate the *rym1/11* locus (Yang et al. 2014a, b) in about 5 years. It is expected that newly targeted resistance genes will be cloned in a time frame from three to 5 years by applying map-based cloning procedures.

Functional genomics of barley and its pathogens will expedite the characterization of resistance traits and pathogen virulence strategies. Successful cloning of genes conferring resistance to stem rust caused by *P. graminis* spp., leaf-stripe disease caused by *P. graminea*, and the powdery mildew pathogen *B. graminis* f. sp. *hordei* is leading to a greater understanding of *R*-gene evolution and diverse mechanisms of cereal immunity. Barley and related cereals are proving to be good model systems for investigations into QDR and NHR, and the cloning of causal genes is imminent. Barley loci associated with resistance can be complex, often controlled by many genes that work together for pathogen recognition, downstream signaling, and ensuing defense responses. As more resistance and defense-associated genes are cloned, it grows our genetic toolbox for the rational design of durable disease resistance.

Acknowledgements Research supported in part by the National Science Foundation—Plant Genome Research Program grants 13-39348 and USDA-Agricultural Research Service project 3625-21000-060-00D to RPW, by the German Ministry of Education and Research (grants GABI-nonhost; GABI-phenome) and by German DFG (grants ERA-PG TritNONHOST; ERA-CAPS DURESTrit) to PS, and by the Federal Ministry of Education and Research (0314000D, 0315708B, 031A323B) to FO. JME is supported by a USDA-ARS Postdoctoral Research Associateship and NIFA-AFRI Postdoctoral Fellowship 2017-67012-26086. DP is supported by the Federal Ministry of Food and Agriculture (FKZ 2814601913). The funders had no role in study design, data collection and analysis, decision to publish, or preparation of the manuscript.Mention of trade names or commercial products in this publication is solely for the purpose of providing scientific information and does not imply recommendation or endorsement by the U.S. Department of Agriculture or the National Science Foundation. USDA is an equal opportunity provider and employer.

References

Acevedo-Garcia J, Kusch S, Panstruga R (2014) Magical mystery tour: MLO proteins in plant immunity and beyond. New Phytol 204(2):273–281. https://doi.org/10.1111/nph.12889

Acevedo-Garcia J, Spencer D, Thieron H, Reinstädler A, Hammond-Kosack K, Phillips AL, Panstruga R (2016) *mlo*-based powdery mildew resistance in hexaploid bread wheat generated by a non-transgenic TILLING approach. Plant Biotechnol J. https://doi.org/10.1111/pbi.12631

Aghnoum R, Niks RE (2010) Specificity and levels of nonhost resistance to nonadapted *Blumeria graminis* forms in barley. New Phytol 185(1):275–284. https://doi.org/10.1111/j.1469-8137.2009.03039.x

Allen AM, Barker GLA, Berry ST, Coghill JA, Gwilliam R, Kirby S, Robinson P, Brenchley RC, D'Amore R, McKenzie N, Waite D, Hall A, Bevan M, Hall N, Edwards KJ (2011) Transcript-specific, single-nucleotide polymorphism discovery and linkage analysis in hexaploid bread wheat (*Triticum aestivum* L.). Plant Biotechnol J 9(9):1086–1099. https://doi.org/10.1111/j.1467-7652.2011.00628.x

Andolfo G, Jupe F, Witek K, Etherington GJ, Ercolano MR, Jones JDG (2014) Defining the full tomato NB-LRR resistance gene repertoire using genomic and cDNA RenSeq. BMC Plant Biol 14. https://doi.org/10.1186/1471-2229-14-120

Arora D, Gross T, Brueggeman R (2013) Allele characterization of genes required for *rpg4*-mediated wheat stem rust resistance identifies *Rpg5* as the *R* Gene. Phytopathology 103(11):1153–1161. https://doi.org/10.1094/PHYTO-01-13-0030-R

Atienza SG, Jafary H, Niks RE (2004) Accumulation of genes for susceptibility to rust fungi for which barley is nearly a nonhost results in two barley lines with extreme multiple susceptibility. Planta 220(1):71–79. https://doi.org/10.1007/s00425-004-1319-1

Austin RS, Vidaurre D, Stamatiou G, Breit R, Provart NJ, Bonetta D, Zhang JF, Fung P, Gong YC, Wang PW, McCourt P, Guttman DS (2011) Next-generation mapping of *Arabidopsis* genes. Plant J 67(4):715–725. https://doi.org/10.1111/j.1365-313X.2011.04619.x

Bai S, Liu J, Chang C, Zhang L, Maekawa T, Wang Q, Xiao W, Liu Y, Chai J, Takken FLW, Schulze-Lefert P, Shen Q-H (2012) Structure-function analysis of barley NLR immune receptor MLA10 reveals Its cell compartment specific activity in cell death and disease resistance. PLoS Pathog 8(6):e1002752. https://doi.org/10.1371/journal.ppat.1002752

Beier S, Himmelbach A, Schmutzer T, Felder M, Taudien S, Mayer KFX, Platzer M, Stein N, Scholz U, Mascher M (2016) Multiplex sequencing of bacterial artificial chromosomes for assembling complex plant genomes. Plant Biotechnol J 14(7):1511–1522. https://doi.org/10.1111/pbi.12511

Bieri S, Mauch S, Shen Q-H, Peart J, Devoto A, Casais C, Ceron F, Schulze S, Steinbiß H-H, Shirasu K, Schulze-Lefert P (2004) RAR1 positively controls steady state levels of barley MLA resistance proteins and enables sufficient MLA6 accumulation for effective resistance. Plant Cell 16(12):3480–3495

Boyd LA, Smith PH, Foster EM, Brown JKM (1995) The effects of allelic variation at the *Mla* resistance locus in barley on the early development of *Erysiphe graminis* f.sp. *hordei* and host responses. Plant J 7(6):959–968. https://doi.org/10.1046/j.1365-313X.1995.07060959.x

Brueggeman R, Rostoks N, Kudrna D, Kilian A, Han F, Chen J, Druka A, Steffenson B, Kleinhofs A (2002) The barley stem rust-resistance gene *Rpg1* is a novel disease-resistance gene with homology to receptor kinases. Proc Natl Acad Sci 99(14):9328–9333

Brueggeman R, Druka A, Nirmala J, Cavileer T, Drader T, Rostoks N, Mirlohi A, Bennypaul H, Gill U, Kudrna D, Whitelaw C, Kilian A, Han F, Sun Y, Gill K, Steffenson B, Kleinhofs A (2008) The stem rust resistance gene *Rpg5* encodes a protein with nucleotide-binding-site, leucine-rich, and protein kinase domains. Proc Natl Acad Sci 105(39):14970–14975

Budhagatapalli N, Rutten T, Gurushidze M, Kumlehn J, Hensel G (2015) Targeted modification of gene function exploiting homology-directed repair of TALEN-mediated double-strand breaks in barley. G3: Genes|Genomes|Genet 5(9):1857

Bulgarelli D, Collins NC, Tacconi G, Dellaglio E, Brueggeman R, Kleinhofs A, Stanca AM, Valè G (2004) High-resolution genetic mapping of the leaf stripe resistance gene *Rdg2a* in barley. Theor Appl Genet 108(7):1401–1408. https://doi.org/10.1007/s00122-003-1557-9

Bulgarelli D, Biselli C, Collins NC, Consonni G, Stanca AM, Schulze-Lefert P, Valè G (2010)

The CC-NB-LRR-type *Rdg2a* resistance gene confers immunity to the seed-borne barley leaf stripe pathogen in the absence of hypersensitive cell death. PLoS ONE 5(9):e12599. https://doi.org/10.1371/journal.pone.0012599

Caldo RA, Nettleton D, Wise RP (2004) Interaction-dependent gene expression in *Mla*-specified response to barley powdery mildew. Plant Cell 16(9):2514–2528

Caldo RA, Nettleton D, Peng J, Wise RP (2006) Stage-specific suppression of basal defense discriminates barley plants containing fast- and delayed-acting *Mla* powdery mildew resistance alleles. Mol Plant Microbe Interact 19(9):939–947. https://doi.org/10.1094/MPMI-19-0939

Cesari S, Bernoux M, Moncuquet P, Kroj T, Dodds P (2014) A novel conserved mechanism for plant NLR protein pairs: the 'integrated decoy' hypothesis. Front Plant Sci 5:606

Cesari S, Moore J, Chen C, Webb D, Periyannan S, Mago R, Bernoux M, Lagudah ES, Dodds PN (2016) Cytosolic activation of cell death and stem rust resistance by cereal MLA-family CC–NLR proteins. Proc Natl Acad Sci 113(36):10204–10209

Chang C, Yu D, Jiao J, Jing S, Schulze-Lefert P, Shen Q-H (2013) Barley MLA immune receptors directly interfere with antagonistically acting transcription factors to initiate disease resistance signaling. Plant Cell 25(3):1158–1173

Chełkowski J, Tyrka M, Sobkiewicz A (2003) Resistance genes in barley (*Hordeum vulgare* L.) and their identification with molecular markers. J Appl Genet 44(3):291–309

Chen M, Kendziorski C (2007) A statistical framework for expression quantitative trait loci mapping. Genetics 177(2):761

Chen X, Hackett CA, Niks RE, Hedley PE, Booth C, Druka A, Marcel TC, Vels A, Bayer M, Milne I, Morris J, Ramsay L, Marshall D, Cardle L, Waugh R (2010) An eQTL analysis of partial resistance to *Puccinia hordei* in barley. PLoS ONE 5(1):e8598. https://doi.org/10.1371/journal.pone.0008598

Chowdhury J, Henderson M, Schweizer P, Burton RA, Fincher GB, Little A (2014) Differential accumulation of callose, arabinoxylan and cellulose in nonpenetrated versus penetrated papillae on leaves of barley infected with *Blumeria graminis* f. sp *hordei*. New Phytol 204(3):650–660. https://doi.org/10.1111/nph.12974

Close TJ, Wanamaker SI, Caldo RA, Turner SM, Ashlock DA, Dickerson JA, Wing RA, Muehlbauer GJ, Kleinhofs A, Wise RP (2004) A new resource for cereal genomics: 22K barley GeneChip comes of age. Plant Physiol 134(3):960–968

Close TJ, Bhat PR, Lonardi S, Wu Y, Rostoks N, Ramsay L, Druka A, Stein N, Svensson JT, Wanamaker S, Bozdag S, Roose ML, Moscou MJ, Chao S, Varshney RK, Szűcs P, Sato K, Hayes PM, Matthews DE, Kleinhofs A, Muehlbauer GJ, DeYoung J, Marshall DF, Madishetty K, Fenton RD, Condamine P, Graner A, Waugh R (2009) Development and implementation of high-throughput SNP genotyping in barley. BMC Genom 10(1):582. https://doi.org/10.1186/1471-2164-10-582

Collins NC, Thordal-Christensen H, Lipka V, Bau S, Kombrink E, Qiu J-L, Huckelhoven R, Stein M, Freialdenhoven A, Somerville SC, Schulze-Lefert P (2003) SNARE-protein-mediated disease resistance at the plant cell wall. Nature 425(6961):973–977. http://www.nature.com/nature/journal/v425/n6961/suppinfo/nature02076_S1.html

Comadran J, Kilian B, Russell J, Ramsay L, Stein N, Ganal M, Shaw P, Bayer M, Thomas W, Marshall D, Hedley P, Tondelli A, Pecchioni N, Francia E, Korzun V, Walther A, Waugh R (2012) Natural variation in a homolog of *Antirrhinum CENTRORADIALIS* contributed to spring growth habit and environmental adaptation in cultivated barley. Nat Genet 44(12):1388–1392. https://doi.org/10.1038/ng.2447

Cook DE, Mesarich CH, Thomma BPHJ (2015) Understanding plant immunity as a surveillance system to detect invasion. Annu Rev Phytopathol 53(1):541–563. https://doi.org/10.1146/annurev-phyto-080614-120114

Cui H, Tsuda K, Parker JE (2015) Effector-triggered immunity: from pathogen perception to robust defense. Annu Rev Plant Biol 66(1):487–511. https://doi.org/10.1146/annurev-arplant-050213-040012

Deslandes L, Rivas S (2012) Catch me if you can: bacterial effectors and plant targets. Trends Plant Sci 17(11):644–655. https://doi.org/10.1016/j.tplants.2012.06.011

Dodds PN, Rathjen JP (2010) Plant immunity: towards an integrated view of plant–pathogen interactions. Nat Rev Genet 11(8):539–548

Doehlemann G, Requena N, Schaefer P, Brunner F, O'Connell R, Parker JE (2014) Reprogramming of plant cells by filamentous plant-colonizing microbes. New Phytol 204(4):803–814. https://doi.org/10.1111/nph.12938

Douchkov D, Nowara D, Zierold U, Schweizer P (2005) A high-throughput gene-silencing system for the functional assessment of defense-related genes in barley epidermal cells. Mol Plant Microbe Interact 18(8):755–761

Douchkov D, Lück S, Johrde A, Nowara D, Himmelbach A, Rajaraman J, Stein N, Sharma R, Kilian B, Schweizer P (2014) Discovery of genes affecting resistance of barley to adapted and non-adapted powdery mildew fungi. Genome Biol 15(12):518. https://doi.org/10.1186/s13059-014-0518-8

Douchkov D, Lueck S, Hensel G, Kumlehn J, Rajaraman J, Johrde A, Doblin MS, Beahan CT, Kopischke M, Fuchs R, Lipka V, Niks RE, Bulone V, Chowdhury J, Little A, Burton RA, Bacic A, Fincher GB, Schweizer P (2016) The barley (*Hordeum vulgare*) cellulose synthase-like D2 gene (*HvCslD2*) mediates penetration resistance to host-adapted and nonhost isolates of the powdery mildew fungus. New Phytol 212(2):421–433. https://doi.org/10.1111/nph.14065

Dracatos P, Singh D, Fetch T, Park R (2015) Resistance to *Puccinia graminis* f. sp. *avenae* in barley is associated with the *Rpg5* locus. Phytopathology 105 (4):490–494. https://doi.org/10.1094/PHYTO-08-14-0224-R

Dracatos PM, Nansamba M, Berlin A, Park RF, Niks RE (2016) Isolate specificity and polygenic inheritance of resistance in barley to the heterologous rust pathogen *Puccinia graminis* f. sp. *avenae*. Phytopathology 106 (9):1029–1037. https://doi.org/10.1094/PHYTO-10-15-0264-R

Druka A, Druka I, Centeno AG, Li H, Sun Z, Thomas WTB, Bonar N, Steffenson BJ, Ullrich SE, Kleinhofs A, Wise RP, Close TJ, Potokina E, Luo Z, Wagner C, Schweizer GF, Marshall DF, Kearsey MJ, Williams RW, Waugh R (2008a) Towards systems genetic analyses in barley: integration of phenotypic, expression and genotype data into GeneNetwork. BMC Genet 9(1):73. https://doi.org/10.1186/1471-2156-9-73

Druka A, Potokina E, Luo Z, Bonar N, Druka I, Zhang L, Marshall DF, Steffenson BJ, Close TJ, Wise RP, Kleinhofs A, Williams RW, Kearsey MJ, Waugh R (2008b) Exploiting regulatory variation to identify genes underlying quantitative resistance to the wheat stem rust pathogen *Puccinia graminis* f. sp. *tritici* in barley. Theor Appl Genet 117(2):261–272. https://doi.org/10.1007/s00122-008-0771-x

Elliott C, Zhou FS, Spielmeyer W, Panstruga R, Schulze-Lefert P (2002) Functional conservation of wheat and rice *Mlo* orthologs in defense modulation to the powdery mildew fungus. Mol Plant Microbe Interact 15(10):1069–1077

Francia E, Tacconi G, Crosatti C, Barabaschi D, Bulgarelli D, Dall'Aglio E, Vale G (2005) Marker assisted selection in crop plants. Plant Cell Tissue Organ Cult 82(3):317–342. https://doi.org/10.1007/s11240-005-2387-z

French E, Kim B-S, Iyer-Pascuzzi AS (2016) Mechanisms of quantitative disease resistance in plants. Semin Cell Dev Biol 56:201–208. https://doi.org/10.1016/j.semcdb.2016.05.015

Friedt W, Ordon F (2007) Molecular markers for gene pyramiding and disease resistance breeding in barley. In: Varshney RK, Tuberosa R (eds) Genomics-assisted crop improvement, Genomics applications in crops, vol 2. Springer, Dordrecht, pp 81–101. https://doi.org/10.1007/978-1-4020-6297-1_4

Ge XT, Deng WW, Lee ZZ, Lopez-Ruiz FJ, Schweizer P, Ellwood SR (2016) Tempered *mlo* broad-spectrum resistance to barley powdery mildew in an Ethiopian landrace. Sci Rep 6:10. https://doi.org/10.1038/srep29558

Gill U, Brueggeman R, Nirmala J, Chai Y, Steffenson B, Kleinhofs A (2016) Molecular and genetic characterization of barley mutants and genetic mapping of mutant *rpr2* required for *Rpg1*-mediated resistance against stem rust. Theor Appl Genet 129(8):1519–1529. https://doi.org/10.1007/s00122-016-2721-3

Graner A, Kilian A, Kleinhofs A (2010) Barley genome organization, mapping, and synteny. In: Barley. Wiley-Blackwell, pp 63–84. https://doi.org/10.1002/9780470958636.ch3

Haegi A, Bonardi V, Dall'Aglio E, Glissant D, Tumino G, Collins NC, Bulgarelli D, Infantino A, Stanca AM, Delledonne M, ValÈ G (2008) Histological and molecular analysis of *Rdg2a* barley resistance to leaf stripe. Mol Plant Pathol 9(4):463–478. https://doi.org/10.1111/j.1364-3703.2008.00479.x

Halterman DA, Wise RP (2004) A single-amino acid substitution in the sixth leucine-rich repeat of barley MLA6 and MLA13 alleviates dependence on RAR1 for disease resistance signaling. Plant J 38(2): 215–226. https://doi.org/10.1111/j.1365-313X.2004.02032.x

Halterman D, Zhou F, Wei F, Wise RP, Schulze-Lefert P (2001) The MLA6 coiled-coil, NBS-LRR protein confers *AvrMla6*-dependent resistance specificity to *Blumeria graminis* f. sp. *hordei* in barley and wheat. Plant J 25(3):335–348. https://doi.org/10.1046/j.1365-313x.2001.00982.x

Halterman DA, Wei F, Wise RP (2003) Powdery mildew-induced *Mla* mRNAs are alternatively spliced and contain multiple upstream open reading frames. Plant Physiol 131(2):558–567

Hisano H, Sakamoto K, Takagi H, Terauchi R, Sato K (2017) Exome QTL-seq maps monogenic locus and QTLs in barley. BMC Genom 18(1):125. https://doi.org/10.1186/s12864-017-3511-2

Hofinger BJ, Russell JR, Bass CG, Baldwin T, Dos Reis M, Hedley PE, Li Y, Macaulay M, Waugh R, Hammond-Kosack KE, Kanyuka K (2011) An exceptionally high nucleotide and haplotype diversity and a signature of positive selection for the *eIF4E* resistance gene in barley are revealed by allele mining and phylogenetic analyses of natural populations. Mol Ecol 20(17):3653–3668. https://doi.org/10.1111/j.1365-294X.2011.05201.x

Horvath H, Rostoks N, Brueggeman R, Steffenson B, von Wettstein D, Kleinhofs A (2003) Genetically engineered stem rust resistance in barley using the *Rpg1* gene. Proc Natl Acad Sci 100(1):364–369

Hu P, Meng Y, Wise RP (2009) Functional contribution of chorismate synthase, anthranilate synthase, and chorismate mutase to penetration resistance in barley–powdery mildew interactions. Mol Plant Microbe Interact 22(3):311–320. https://doi.org/10.1094/MPMI-22-3-0311

Hückelhoven R (2007) Cell wall–associated mechanisms of disease resistance and susceptibility. Annu Rev Phytopathol 45(1):101–127. https://doi.org/10.1146/annurev.phyto.45.062806.094325

Hückelhoven R, Seidl A (2016) PAMP-triggered immune responses in barley and susceptibility to powdery mildew. Plant Signaling Behav 11(7):e1197465. https://doi.org/10.1080/15592324.2016.1197465

Huckelhoven R, Dechert C, Kogel KH (2003) Overexpression of barley BAX inhibitor 1 induces breakdown of *mlo*-mediated penetration resistance to *Blumeria graminis*. Proc Natl Acad Sci 100(9):5555–5560

Hulbert SH, Webb CA, Smith SM, Sun Q (2001) Resistance gene complexes: evolution and utilization. Annu Rev Phytopathol 39(1):285–312

IBSC, International Barley Sequencing Consortium (2012) A physical, genetic and functional sequence assembly of the barley genome. Nature 491(7426): 711–716. http://www.nature.com/nature/journal/v491/n7426/abs/nature11543.html#supplementary-information

Jacob F, Vernaldi S, Maekawa T (2013) Evolution and conservation of plant NLR functions. Front Immunol 4:297

Jafary H, Szabo LJ, Niks RE (2006) Innate nonhost immunity in barley to different heterologous rust fungi is controlled by sets of resistance genes with different and overlapping specificities. Mol Plant Microbe Interact 19(11):1270–1279. https://doi.org/10.1094/MPMI-19-1270

Jafary H, Albertazzi G, Marcel TC, Niks RE (2008) High diversity of genes for nonhost resistance of barley to heterologous rust fungi. Genetics 178(4):2327

Jansen RC, Nap J-P (2001) Genetical genomics: the added value from segregation. Trends Genet 17(7): 388–391. https://doi.org/10.1016/S0168-9525(01)02310-1

Jin Y, Steffenson BJ, Miller JD (1994) Inheritance of resistance to pathotypes QCC and MCC of *Puccinia graminis* f. sp. *tritici* in barley line Q21861 and temperature effects on the expression of resistance. Phytopathology 84(5):452–455

Jones JDG, Dangl JL (2006) The plant immune system. Nature 444(7117):323–329

Jordan T, Seeholzer S, Schwizer S, Töller A, Somssich IE, Keller B (2011) The wheat *Mla* homologue *TmMla1* exhibits an evolutionarily conserved function against powdery mildew in both wheat and barley. Plant J 65(4):610–621. https://doi.org/10.1111/j.1365-313X.2010.04445.x

Jørgensen JH, Wolfe M (1994) Genetics of powdery mildew resistance in barley. Crit Rev Plant Sci 13(1): 97–119. https://doi.org/10.1080/07352689409701910

Juliana P, Rutkoski JE, Poland JA, Singh RP, Murugasamy S, Natesan S, Barbier H, Sorrells ME (2015) Genome-wide association mapping for leaf tip necrosis and pseudo-black chaff in relation to durable rust resistance in wheat. Plant Genome 8(2). https://doi.org/10.3835/plantgenome2015.01.0002

Jupe F, Witek K, Verweij W, Śliwka J, Pritchard L, Etherington GJ, Maclean D, Cock PJ, Leggett RM, Bryan GJ, Cardle L, Hein I, Jones JDG (2013) Resistance gene enrichment sequencing (RenSeq) enables reannotation of the NB-LRR gene family from sequenced plant genomes and rapid mapping of resistance loci in segregating populations. Plant J 76 (3):530–544. https://doi.org/10.1111/tpj.12307

Kai H, Takata K, Tsukazaki M, Furusho M, Baba T (2012) Molecular mapping of *Rym17*, a dominant and *rym18* a recessive barley yellow mosaic virus (BaYMV) resistance genes derived from *Hordeum vulgare* L. Theor Appl Genet 124(3):577–583. https://doi.org/10.1007/s00122-011-1730-5

Kanyuka K, Druka A, Caldwell DG, Tymon A, McCallum N, Waugh R, Adams MJ (2005) Evidence that the recessive bymovirus resistance locus *rym4* in barley corresponds to the eukaryotic translation initiation factor 4E gene. Mol Plant Pathol 6(4):449–458. https://doi.org/10.1111/j.1364-3703.2005.00294.x

Kaur N, Street K, Mackay M, Yahiaoui N, Keller B (2008) Molecular approaches for characterization and use of natural disease resistance in wheat. Eur J Plant Pathol 121(3):387–397. https://doi.org/10.1007/s10658-007-9252-3

Kliebenstein D (2009) Quantitative genomics: analyzing intraspecific variation using global gene expression polymorphisms or eQTLs. Annu Rev Plant Biol 60:93–114

Krattinger SG, Keller B (2016) Molecular genetics and evolution of disease resistance in cereals. New Phytol 212(2):320–332. https://doi.org/10.1111/nph.14097

Krattinger S, Wicker T, Keller B (2009a) Map-based cloning of genes in *Triticeae* (Wheat and Barley). In: Muehlbauer GJ, Feuillet C (eds) Genetics and genomics of the triticeae. Springer, New York, pp 337–357. https://doi.org/10.1007/978-0-387-77489-3_12

Krattinger SG, Lagudah ES, Spielmeyer W, Singh RP, Huerta-Espino J, McFadden H, Bossolini E, Selter LL, Keller B (2009b) A putative ABC transporter confers durable resistance to multiple fungal pathogens in wheat. Science 323(5919):1360

Künzel G, Korzun L, Meister A (2000) Cytologically integrated physical restriction fragment length polymorphism maps for the barley genome based on translocation breakpoints. Genetics 154(1):397

Lapierre H, Signoret PA (2004) Viruses and virus diseases of *Poaceae* (*Gramineae*). Editions Quae, France

Lawrenson T, Shorinola O, Stacey N, Li C, Østergaard L, Patron N, Uauy C, Harwood W (2015) Induction of targeted, heritable mutations in barley and *Brassica oleracea* using RNA-guided Cas9 nuclease. Genome Biol 16(1):258. https://doi.org/10.1186/s13059-015-0826-7

Lo Presti L, Lanver D, Schweizer G, Tanaka S, Liang L, Tollot M, Zuccaro A, Reissmann S, Kahmann R (2015) Fungal effectors and plant susceptibility. Annu Rev Plant Biol 66(1):513–545. https://doi.org/10.1146/annurev-arplant-043014-114623

Lu X, Kracher B, Saur IML, Bauer S, Ellwood SR, Wise R, Yaeno T, Maekawa T, Schulze-Lefert P (2016) Allelic barley MLA immune receptors recognize sequence-unrelated avirulence effectors of the powdery mildew pathogen. Proc Natl Acad Sci 113 (42):E6486–E6495

Mago R, Spielmeyer W, Lawrence GJ, Ellis JG, Pryor AJ (2004) Resistance genes for rye stem rust (*SrR*) and barley powdery mildew (*Mla*) are located in syntenic regions on short arm of chromosome. Genome 47 (1):112–121. https://doi.org/10.1139/g03-096

Mago R, Zhang P, Vautrin S, Šimková H, Bansal U, Luo M-C, Rouse M, Karaoglu H, Periyannan S, Kolmer J, Jin Y, Ayliffe MA, Bariana H, Park RF, McIntosh R, Doležel J, Bergès H, Spielmeyer W, Lagudah ES, Ellis JG, Dodds PN (2015) The wheat *Sr50* gene reveals rich diversity at a cereal disease resistance locus. Nat Plants 1:15186. https://doi.org/10.1038/nplants.2015.186

Marcel TC, Gorguet B, Ta MT, Kohutova Z, Vels A, Niks RE (2008) Isolate specificity of quantitative trait loci for partial resistance of barley to *Puccinia hordei* confirmed in mapping populations and near-isogenic lines. New Phytol 177(3):743–755. https://doi.org/10.1111/j.1469-8137.2007.02298.x

Mascher M, Muehlbauer GJ, Rokhsar DS, Chapman J, Schmutz J, Barry K, Munoz-Amatriain M, Close TJ, Wise RP, Schulman AH, Himmelbach A, Mayer KFX, Scholz U, Poland JA, Stein N, Waugh R (2013a) Anchoring and ordering NGS contig assemblies by population sequencing (POPSEQ). Plant J 76(4):718–727. https://doi.org/10.1111/tpj.12319

Mascher M, Richmond TA, Gerhardt DJ, Himmelbach A, Clissold L, Sampath D, Ayling S, Steuernagel B, Pfeifer M, D'Ascenzo M, Akhunov ED, Hedley PE, Gonzales AM, Morrell PL, Kilian B, Blattner FR, Scholz U, Mayer KFX, Flavell AJ, Muehlbauer GJ, Waugh R, Jeddeloh JA, Stein N (2013b) Barley whole exome capture: a tool for genomic research in the genus *Hordeum* and beyond. Plant J 76(3):494–505. https://doi.org/10.1111/tpj.12294

Mascher M, Gundlach H, Himmelbach A, Beier S, Twardziok SO, Wicker T, Radchuk V, Dockter C, Hedley PE, Russell J, Bayer M, Ramsay L, Liu H, Haberer G, Zhang X-Q, Zhang Q, Barrero RA, Li L, Taudien S, Groth M, Felder M, Hastie A, Šimková H, Staňková H, Vrána J, Chan S, Muñoz-Amatriaín M, Ounit R, Wanamaker S, Bolser D, Colmsee C, Schmutzer T, Aliyeva-Schnorr L, Grasso S, Tanskanen J, Chailyan A, Sampath D, Heavens D, Clissold L, Cao S, Chapman B, Dai F, Han Y, Li H, Li X, Lin C, McCooke JK, Tan C, Wang P, Wang S, Yin S, Zhou G, Poland JA, Bellgard MI, Borisjuk L, Houben A, Doležel J, Ayling S, Lonardi S, Kersey P, Langridge P, Muehlbauer GJ, Clark MD, Caccamo M, Schulman AH, Mayer KFX, Platzer M, Close TJ, Scholz U, Hansson M, Zhang G, Braumann I, Spannagl M, Li C, Waugh R, Stein N (2017) A chromosome conformation capture ordered sequence of the barley genome. Nature 544 (7651):427–433. https://doi.org/10.1038/nature22043

Mayer KFX, Martis M, Hedley PE, Simkova H, Liu H, Morris JA, Steuernagel B, Taudien S, Roessner S, Gundlach H, Kubalakova M, Suchankova P, Murat F, Felder M, Nussbaumer T, Graner A, Salse J, Endo T, Sakai H, Tanaka T, Itoh T, Sato K, Platzer M, Matsumoto T, Scholz U, Dolezel J, Waugh R, Stein N (2011) Unlocking the barley genome by chromosomal and comparative genomics. Plant Cell 23(4):1249–1263. https://doi.org/10.1105/tpc.110.082537

McDonald BA, Linde C (2002) Pathogen population genetics, evolutionary potential, and durable resistance. Annu Rev Phytopathol 40(1):349–379

McGrann GRD, Stavrinides A, Russell J, Corbitt MM, Booth A, Chartrain L, Thomas WTB, Brown JKM (2014) A trade off between *mlo* resistance to powdery mildew and increased susceptibility of barley to a newly important disease, *Ramularia* leaf spot. J Exp Bot 65(4):1025–1037

Michelmore RW, Paran I, Kesseli RV (1991) Identification of markers linked to disease-resistance genes by bulked segregant analysis: a rapid method to detect markers in specific genomic regions by using segregating populations. Proc Natl Acad Sci 88(21):9828–9832. https://doi.org/10.1073/pnas.88.21.9828

Moscou MJ, Lauter N, Caldo RA, Nettleton D, Wise RP (2011a) Quantitative and temporal definition of the *Mla transcriptional* regulon during barley–powdery mildew interactions. Mol Plant Microbe Interact 24 (6):694–705. https://doi.org/10.1094/MPMI-09-10-0211

Moscou MJ, Lauter N, Steffenson B, Wise RP (2011b) Quantitative and qualitative stem rust resistance factors in barley are associated with transcriptional suppression of defense regulons. PLoS Genet 7(7):e1002208. https://doi.org/10.1371/journal.pgen.1002208

Moury B, Charron C, Janzac B, Simon V, Gallois JL, Palloix A, Caranta C (2014) Evolution of plant eukaryotic initiation factor 4E (eIF4E) and potyvirus genome-linked protein (VPg): A game of mirrors impacting resistance spectrum and durability. Infect Genet Evol 27:472–480. https://doi.org/10.1016/j.meegid.2013.11.024

Niks RE, Marcel TC (2009) Nonhost and basal resistance: how to explain specificity? New Phytol 182(4): 817–828. https://doi.org/10.1111/j.1469-8137.2009.02849.x

Niks RE, Qi X, Marcel TC (2015) Quantitative resistance to biotrophic filamentous plant pathogens: concepts, misconceptions, and mechanisms. Annu Rev Phytopathol 53(1):445–470. https://doi.org/10.1146/annurev-phyto-080614-115928

Nirmala J, Brueggeman R, Maier C, Clay C, Rostoks N, Kannangara CG, von Wettstein D, Steffenson BJ, Kleinhofs A (2006) Subcellular localization and functions of the barley stem rust resistance receptor-like serine/threonine-specific protein kinase RPG1. Proc Natl Acad Sci 103(19):7518–7523

Nirmala J, Dahl S, Steffenson BJ, Kannangara CG, von Wettstein D, Chen X, Kleinhofs A (2007) Proteolysis of the barley receptor-like protein kinase RPG1 by a proteasome pathway is correlated with *Rpg1*-mediated stem rust resistance. Proc Natl Acad Sci 104 (24):10276–10281

Nirmala J, Drader T, Chen X, Steffenson B, Kleinhofs A (2010) Stem rust spores elicit rapid RPG1 phosphorylation. Mol Plant Microbe Interact 23(12):1635–1642. https://doi.org/10.1094/MPMI-06-10-0136

Nirmala J, Drader T, Lawrence PK, Yin C, Hulbert S, Steber CM, Steffenson BJ, Szabo LJ, von Wettstein D, Kleinhofs A (2011) Concerted action of two avirulent spore effectors activates *Reaction to Puccinia graminis 1* (*Rpg1*)-mediated cereal stem rust resistance. Proc Natl Acad Sci 108(35):14676–14681

Ordon F, Kühne T (2014) Response to viral pathogens. In: Kumlehn J, Stein N (eds) Biotechnological approaches to barley improvement. Springer, Heidelberg, pp 181–196. https://doi.org/10.1007/978-3-662-44406-1_10

Ordon F, Perovic D (2013) Virus resistance in barley. In: Translational genomics for crop breeding. Wiley Ltd, pp 63–75. https://doi.org/10.1002/9781118728475.ch5

Paulitz TC, Steffenson BJ (2010) Biotic stress in barley: disease problems and solutions. In: Barley. Wiley-Blackwell, pp 307–354. https://doi.org/10.1002/9780470958636.ch11

Periyannan S, Moore J, Ayliffe M, Bansal U, Wang X, Huang L, Deal K, Luo M, Kong X, Bariana H, Mago R, McIntosh R, Dodds P, Dvorak J, Lagudah E (2013) The gene *Sr33*, an ortholog of barley *Mla* genes, encodes resistance to wheat stem rust race Ug99. Science 341(6147):786

Perovic D, Kopahnke D, Steffenson BJ, Förster J, König J, Kilian B, Plieske J, Durstewitz G, Korzun V, Kraemer I, Habekuss A, Johnston P, Pickering R, Ordon F (2013) Genetic fine mapping of a novel leaf rust resistance gene and a Barley Yellow Dwarf Virus tolerance (BYDV) introgressed from *Hordeum bulbosum* by the use of the 9K iSelect chip. In: Zhang G, Li C, Liu X (eds) Advance in barley sciences: Proceedings of 11th international barley genetics symposium. Springer, Dordrecht, pp 269–284. https://doi.org/10.1007/978-94-007-4682-4_23

Perovic D, Kramer I, Habekuss A, Perner K, Pickering R, Proeseler G, Kanyuka K, Ordon F (2014) Genetic analyses of BaMMV/BaYMV resistance in barley accession HOR4224 result in the identification of an allele of the translation initiation factor 4e (Hv-eIF4E) exclusively effective against *Barley mild mosaic virus* (BaMMV). Theor Appl Genet 127(5):1061–1071. https://doi.org/10.1007/s00122-014-2279-x

Poland JA, Balint-Kurti PJ, Wisser RJ, Pratt RC, Nelson RJ (2009) Shades of gray: the world of quantitative disease resistance. Trends Plant Sci 14(1):21–29. https://doi.org/10.1016/j.tplants.2008.10.006

Poland JA, Brown PJ, Sorrells ME, Jannink JL (2012) Development of high-density genetic maps for barley and wheat using a novel two-enzyme genotyping-by-sequencing approach. PLOS ONE 7(2). https://doi.org/10.1371/journal.pone.0032253

Pritchard L, Birch PRJ (2014) The zigzag model of plant–microbe interactions: is it time to move on? Mol Plant Pathol 15(9):865–870. https://doi.org/10.1111/mpp.12210

Rajaraman J (2016) Discovery and validation of genes for quantitative host- and nonhost-resistance in barley and wheat to powdery mildew attack. Dissertation, Martin-Luther University Halle-Wittenberg urn:nbn:de:gbv:3:4-18776:1-341

Rajaraman J, Douchkov D, Hensel G, Stefanato FL, Gordon A, Ereful N, Caldararu OF, Petrescu A-J, Kumlehn J, Boyd LA, Schweizer P (2016) An LRR/Malectin receptor-like kinase mediates resistance to non-adapted and adapted powdery mildew fungi in barley and wheat. Front Plant Sci 7(1836). https://doi.org/10.3389/fpls.2016.01836

Ridout CJ, Skamnioti P, Porritt O, Sacristan S, Jones JDG, Brown JKM (2006) Multiple avirulence paralogues in cereal powdery mildew fungi may contribute to parasite fitness and defeat of plant resistance. Plant Cell 18(9):2402–2414

Riedel C, Habekuß A, Schliephake E, Niks R, Broer I, Ordon F (2011) Pyramiding of *Ryd2* and *Ryd3* conferring tolerance to a German isolate of Barley yellow dwarf virus-PAV (BYDV-PAV-ASL-1) leads to quantitative resistance against this isolate. Theor Appl Genet 123(1):69. https://doi.org/10.1007/s00122-011-1567-y

Rockman MV, Kruglyak L (2006) Genetics of global gene expression. Nat Rev Genet 7(11):862–872

Roux F, Voisin D, Badet T, Balagué C, Barlet X, Huard-Chauveau C, Roby D, Raffaele S (2014) Resistance to phytopathogens *e tutti quanti*: placing plant quantitative disease resistance on the map. Mol Plant Pathol 15(5):427–432. https://doi.org/10.1111/mpp.12138

Ruge B, Linz A, Pickering R, Proeseler G, Greif P, Wehling P (2003) Mapping of *Rym14(Hb)*, a gene introgressed from *Hordeum bulbosum* and conferring resistance to BaMMV and BaYMV in barley. Theor Appl Genet 107(6):965–971. https://doi.org/10.1007/s00122-003-1339-4

Ruge-Wehling B, Linz A, Habekuss A, Wehling P (2006) Mapping of *Rym16(Hb)*, the second soil-borne virus-resistance gene introgressed from *Hordeum bulbosum*. Theor Appl Genet 113(5):867–873. https://doi.org/10.1007/s00122-006-0345-8

Russell J, Mascher M, Dawson IK, Kyriakidis S, Calixto C, Freund F, Bayer M, Milne I, Marshall-Griffiths T, Heinen S, Hofstad A, Sharma R, Himmelbach A, Knauft M, van Zonneveld M, Brown JWS, Schmid K, Kilian B, Muehlbauer GJ, Stein N, Waugh R (2016) Exome sequencing of geographically diverse barley landraces and wild relatives gives insights into environmental adaptation. Nat Genet 48(9):1024± . https://doi.org/10.1038/ng.3612

Sánchez-Martín J, Steuernagel B, Ghosh S, Herren G, Hurni S, Adamski N, Vrána J, Kubaláková M, Krattinger SG, Wicker T, Doležel J, Keller B, Wulff BBH (2016) Rapid gene isolation in barley and wheat by mutant chromosome sequencing. Genome Biol 17(1):221. https://doi.org/10.1186/s13059-016-1082-1

Sanfacon H (2015) Plant translation factors and virus resistance. Viruses 7(7):3392–3419. https://doi.org/10.3390/v7072778

Sarris PF, Cevik V, Dagdas G, Jones JDG, Krasileva KV (2016) Comparative analysis of plant immune receptor architectures uncovers host proteins likely targeted by pathogens. BMC Biol 14(1):8. https://doi.org/10.1186/s12915-016-0228-7

Scheler B, Schnepf V, Galgenmüller C, Ranf S, Hückelhoven R (2016) Barley disease susceptibility factor RACB acts in epidermal cell polarity and positioning of the nucleus. J Exp Bot 67(11):3263–3275

Scholz M, Ruge-Wehling B, Habekuss A, Schrader O, Pendinen G, Fischer K, Wehling P (2009) *Ryd4 (Hb)*: a novel resistance gene introgressed from *Hordeum bulbosum* into barley and conferring complete and dominant resistance to the barley yellow dwarf virus. Theor Appl Genet 119(5):837–849. https://doi.org/10.1007/s00122-009-1093-3

Schulze-Lefert P, Panstruga R (2011) A molecular evolutionary concept connecting nonhost resistance, pathogen host range, and pathogen speciation. Trends Plant Sci 16(3):117–125. https://doi.org/10.1016/j.tplants.2011.01.001

Schweizer P (2014) Host and nonhost response to attack by fungal pathogens. In: Kumlehn J, Stein N (eds) Biotechnological approaches to barley improvement. Springer, Heidelberg, pp 197–235. https://doi.org/10.1007/978-3-662-44406-1_11

Schweizer P, Stein N (2011) Large-scale data integration reveals colocalization of gene functional groups with meta-QTL for multiple disease resistance in barley. Mol Plant Microbe Interact 24(12):1492–1501. https://doi.org/10.1094/mpmi-05-11-0107

Seeholzer S, Tsuchimatsu T, Jordan T, Bieri S, Pajonk S, Yang W, Jahoor A, Shimizu KK, Keller B, Schulze-Lefert P (2010) Diversity at the *Mla* powdery mildew resistance locus from cultivated barley reveals sites of positive selection. Mol Plant Microbe Interact 23(4):497–509. https://doi.org/10.1094/MPMI-23-4-0497

Senthil-Kumar M, Mysore KS (2013) Nonhost resistance against bacterial pathogens: retrospectives and prospects. Annu Rev Phytopathol 51(1):407–427. https://doi.org/10.1146/annurev-phyto-082712-102319

Shen Q-H, Zhou F, Bieri S, Haizel T, Shirasu K, Schulze-Lefert P (2003) Recognition specificity and RAR1/SGT1 dependence in barley *Mla* disease resistance genes to the powdery mildew fungus. Plant Cell 15(3):732–744

Shen Q-H, Saijo Y, Mauch S, Biskup C, Bieri S, Keller B, Seki H, Ülker B, Somssich IE, Schulze-Lefert P (2007) Nuclear activity of MLA immune receptors links isolate-specific and basal disease-resistance responses. Science 315(5815):1098

Shirasu K (2009) The HSP90-SGT1 chaperone complex for NLR immune sensors. Annu Rev Plant Biol 60:139–164

Shirasu K, Lahaye T, Tan M-W, Zhou F, Azevedo C, Schulze-Lefert P (1999) A novel class of eukaryotic zinc-binding proteins is required for disease resistance signaling in barley and development in *C. elegans*. Cell 99(4):355–366. https://doi.org/10.1016/S0092-8674(00)81522-6

Shtaya MJY, Sillero JC, Flath K, Pickering R, Rubiales D (2007) The resistance to leaf rust and powdery mildew of recombinant lines of barley (*Hordeum vulgare* L.) derived from *H. vulgare* x *H. bulbosum* crosses. Plant Breed 126(3):259–267

Silvar C, Perovic D, Casas AM, Igartua E, Ordon F (2011) Development of a cost-effective pyrosequencing approach for SNP genotyping in barley. Plant Breed 130(3):394–397. https://doi.org/10.1111/j.1439-0523.2011.01853.x

Smith EN, Kruglyak L (2008) Gene–environment interaction in yeast gene expression. PLoS Biol 6(4):e83. https://doi.org/10.1371/journal.pbio.0060083

Song W-Y, Wang G-L, Chen L-L, Kim H-S, Pi L-Y, Holsten T, Gardner J, Wang B, Zhai W-X, Zhu L-H, Fauquet C, Ronald P (1995) A receptor kinase-like protein encoded by the rice disease resistance gene, *Xa21*. Science 270(5243):1804

Spies A, Korzun L, Bayles R, Rajaraman J, Himmelbach A, Hedley PE, Schweizer P (2012) Allele mining in barley genetic resources reveals genes of race-nonspecific powdery mildew resistance. Front Plant Sci 2(Article 113)

St.Clair DA (2010) Quantitative disease resistance and quantitative resistance loci in breeding. Annu Rev Phytopathol 48(1):247–268. https://doi.org/10.1146/annurev-phyto-080508-081904

Stam R, Mantelin S, McLellan H, Thilliez G (2014) The role of effectors in non host resistance to filamentous plant pathogens. Front Plant Sci 5. https://doi.org/10.3389/fpls.2014.00582

Steffenson BJ (1992) Analysis of durable resistance to stem rust in barley. Euphytica 63(1):153–167. https://doi.org/10.1007/BF00023920

Steffenson BJ, Jin Y (2006) Resistance to race TTKS of *Puccinia graminis* f. sp. *tritici* in barley (Abstr.). Phytopathology 96(S110)

Steffenson BJ, Jin Y, Rossnagel BG, Rasmussen JB, Kao K (1995) Genetics of multiple disease resistance in a doubled-haploid population of barley. Plant Breed 114(1):50–54

Steffenson BJ, Jin Y, Brueggeman RS, Kleinhofs A, Sun Y (2009) Resistance to stem rust race TTKSK maps to the *rpg4/Rpg5* complex of chromosome 5H of barley. Phytopathology 99(10):1135–1141. https://doi.org/10.1094/PHYTO-99-10-1135

Stein N, Perovic D, Kumlehn J, Pellio B, Stracke S, Streng S, Ordon F, Graner A (2005) The eukaryotic translation initiation factor 4E confers multiallelic recessive *Bymovirus* resistance in *Hordeum vulgare* (L.). Plant J 42(6):912–922. https://doi.org/10.1111/j.1365-313X.2005.02424.x

Steuernagel B, Periyannan SK, Hernandez-Pinzon I, Witek K, Rouse MN, Yu G, Hatta A, Ayliffe M, Bariana H, Jones JDG, Lagudah ES, Wulff BBH (2016) Rapid cloning of disease-resistance genes in plants using mutagenesis and sequence capture. Nat

Biotechnol 34(6):652–655. https://doi.org/10.1038/nbt.3543

Surana P, Xu R, Fuerst G, Chapman AVE, Nettleton D, Wise RP (2017) Inter-chromosomal transfer of immune regulation during infection of barley with the powdery mildew pathogen. G3: Genes|Genomes|Genet Early online. https://doi.org/10.1534/g3.117.300125

Tanaka S, Ichikawa A, Yamada K, Tsuji G, Nishiuchi T, Mori M, Koga H, Nishizawa Y, O'Connell R, Kubo Y (2010) *HvCEBiP*, a gene homologous to rice chitin receptor *CEBiP*, contributes to basal resistance of barley to *Magnaporthe oryzae*. BMC Plant Biol 10(1):288. https://doi.org/10.1186/1471-2229-10-288

Tanksley SD, McCouch SR (1997) Seed banks and molecular maps: unlocking genetic potential from the wild. Science 277(5329):1063–1066. https://doi.org/10.1126/science.277.5329.1063

Thomma BPHJ, Nürnberger T, Joosten MHAJ (2011) Of PAMPs and effectors: the blurred PTI-ETI dichotomy. Plant Cell 23(1):4–15

Trujillo M, Troeger M, Niks RE, Kogel KH, Huckelhoven R (2004) Mechanistic and genetic overlap of barley host and non-host resistance to *Blumeria graminis*. Mol Plant Pathol 5(5):389–396

Tsuda K, Katagiri F (2010) Comparing signaling mechanisms engaged in pattern-triggered and effector-triggered immunity. Curr Opin Plant Biol 13(4):459–465. https://doi.org/10.1016/j.pbi.2010.04.006

Verchot J (2012) Cellular chaperones and folding enzymes are vital contributors to membrane bound replication and movement complexes during plant RNA virus infection. Front Plant Sci 3. https://doi.org/10.3389/fpls.2012.00275

von Korff M, Wang H, Léon J, Pillen K (2005) AB-QTL analysis in spring barley. I. Detection of resistance genes against powdery mildew, leaf rust and scald introgressed from wild barley. Theor Appl Genet 111(3):583–590. https://doi.org/10.1007/s00122-005-2049-x

Wang X, Richards J, Gross T, Druka A, Kleinhofs A, Steffenson B, Acevedo M, Brueggeman R (2013) The *rpg4*-mediated resistance to wheat stem rust (*Puccinia graminis*) in barley (*Hordeum vulgare*) requires *Rpg5*, a second NBS-LRR gene, and an actin depolymerization factor. Mol Plant Microbe Interact 26(4):407–418. https://doi.org/10.1094/MPMI-06-12-0146-R

Wang Y, Cheng X, Shan Q, Zhang Y, Liu J, Gao C, Qiu J-L (2014) Simultaneous editing of three homoeoalleles in hexaploid bread wheat confers heritable resistance to powdery mildew. Nat Biotechnol 32(9):947–951. https://doi.org/10.1038/nbt.2969

Wei F, Gobelman-Werner K, Morroll SM, Kurth J, Mao L, Wing R, Leister D, Schulze-Lefert P, Wise RP (1999) The *Mla* (Powdery Mildew) resistance cluster Is associated with three NBS-LRR gene families and suppressed recombination within a 240-kb DNA interval on chromosome 5S (1HS) of barley. Genetics 153(4):1929

Wei F, Wing RA, Wise RP (2002) Genome dynamics and evolution of the *Mla* (powdery mildew) resistance locus in barley. Plant Cell 14(8):1903–1917

Wendler N, Mascher M, Noh C, Himmelbach A, Scholz U, Ruge-Wehling B, Stein N (2014) Unlocking the secondary gene-pool of barley with next-generation sequencing. Plant Biotechnol J 12(8):1122–1131. https://doi.org/10.1111/pbi.12219

Wendler N, Mascher M, Himmelbach A, Johnston P, Pickering R, Stein N (2015) *Bulbosum* to go: a toolbox to utilize *Hordeum vulgare/bulbosum* introgressions for breeding and beyond. Mol Plant 8(10):1507–1519. https://doi.org/10.1016/j.molp.2015.05.004

Williams RBH, Chan EKF, Cowley MJ, Little PFR (2007) The influence of genetic variation on gene expression. Genome Res 17(12):1707–1716

Wu C-H, Krasileva K, Banfield M, Terauchi R, Kamoun S (2015) The "sensor domains" of plant NLR proteins: more than decoys? Front Plant Sci 6:134

Yang P, Perovic D, Habekuss A, Zhou RN, Graner A, Ordon F, Stein N (2013) Gene-based high-density mapping of the gene *rym7* conferring resistance to *Barley mild mosaic virus* (BaMMV). Mol Breed 32(1):27–37. https://doi.org/10.1007/s11032-013-9842-z

Yang P, Habekuss A, Ordon F, Stein N (2014a) Analysis of bymovirus resistance genes on proximal barley chromosome 4HL provides the basis for precision breeding for BaMMV/BaYMV resistance. Theor Appl Genet 127(7):1625–1634. https://doi.org/10.1007/s00122-014-2324-9

Yang P, Lupken T, Habekuss A, Hensel G, Steuernagel B, Kilian B, Ariyadasa R, Himmelbach A, Kumlehn J, Scholz U, Ordon F, Stein N (2014b) *PROTEIN DISULFIDE ISOMERASE LIKE 5-1* is a susceptibility factor to plant viruses. Proc Natl Acad Sci 111(6):2104–2109. https://doi.org/10.1073/pnas.1320362111

Yang P, Hofinger B, Kanyuka K, Kilian B, Graner A, Ordon F, Stein N (2017) Sequence diversification in recessive alleles of two host factor genes suggests adaptive selection for bymovirus resistance in cultivated barley from East Asia. Theor Appl Genet 130:331–344

Yun SJ, Gyenis L, Hayes PM, Matus I, Smith KP, Steffenson BJ, Muehlbauer GJ (2005) Quantitative trait loci for multiple disease resistance in wild barley. Crop Sci 45(6):2563–2572. https://doi.org/10.2135/cropsci2005.0236

Zellerhoff N, Himmelbach A, Dong W, Bieri S, Schaffrath U, Schweizer P (2010) Nonhost resistance of barley to different fungal pathogens is associated with largely distinct, quantitative transcriptional responses. Plant Physiol 152(4):2053–2066

Zhang L, Fetch T, Nirmala J, Schmierer D, Brueggeman R, Steffenson B, Kleinhofs A (2006) *Rpr1*, a gene required for *Rpg1*-dependent resistance to stem rust in barley. Theor Appl Genet 113(5):847–855. https://doi.org/10.1007/s00122-006-0342-y

Zhang Y, Lubberstedt T, Xu M (2013) The genetic and molecular basis of plant resistance to pathogens. J Genet Genomics 40(1):23–35. https://doi.org/10.1016/j.jgg.2012.11.003

Zhou F, Kurth J, Wei F, Elliott C, Valè G, Yahiaoui N, Keller B, Somerville S, Wise R, Schulze-Lefert P (2001) Cell-autonomous expression of barley *Mla1* confers race-specific resistance to the powdery mildew fungus via a *Rar1*-independent signaling pathway. Plant Cell 13(2):337–350

Zierold UWE, Scholz UWE, Schweizer P (2005) Transcriptome analysis of mlo-mediated resistance in the epidermis of barley. Mol Plant Pathol 6(2):139–151. https://doi.org/10.1111/j.1364-3703.2005.00271.x

Zipfel C (2014) Plant pattern-recognition receptors. Trends Immunol 35(7):345–351. https://doi.org/10.1016/j.it.2014.05.004

15 Genomic and Genetic Studies of Abiotic Stress Tolerance in Barley

Stephanie Saade, Sónia Negrão, Darren Plett, Trevor Garnett and Mark Tester

Abstract

Barley is a resilient crop plant with higher tolerance than other cereal plants for several types of abiotic stress. In this chapter, we describe the genetic components underlying barley's response to abiotic stresses, including soil acidity, boron toxicity, soil salinity, drought, temperature, and nutrient deficiency. We describe typical symptoms observed in barley in response to these stresses. We enumerate the major qualitative trait loci (QTLs) identified so far, such as *FR-H1* and *FR-H2* for low-temperature tolerance. We also discuss candidate genes that are the basis for stress tolerance, such as *HVP10*, which underlies the *HvNax3* locus for salinity tolerance. Although knowledge about barley's responses to abiotic stresses is far from complete, the genetic diversity in cultivated barley and its close wild relatives could be further exploited to improve stress tolerance. To this end, the release of the barley high-quality reference genome provides a powerful tool to facilitate identification of new genes underlying barley's relatively high tolerance to several abiotic stresses.

15.1 Introduction

Barley (*Hordeum vulgare* ssp. *vulgare*) is cultivated in both high-yielding, high-input agricultural systems, and in marginal, low-input agricultural environments. Barley was domesticated 10,000 years ago in the Fertile Crescent (Badr et al. 2000); however, its domestication was not confined to a single center of origin (Allaby 2015; Zohary 1999). In fact, Poets et al. (2015) recently showed that cultivated barley results from multiple sources of wild populations. Thus, the different geographic environments to which barley and its wild relatives have been exposed have resulted in a mosaic of adaptive variants containing tolerance to abiotic stresses (Allaby 2015). Barley is thus considered to be more resilient than other crops [for further

S. Saade · S. Negrão · M. Tester (✉)
Division of Biological and Environmental Sciences and Engineering (BESE), King Abdullah University of Science and Technology (KAUST), Thuwal 23955-6900, Saudi Arabia
e-mail: mark.tester@kaust.edu.sa

D. Plett · T. Garnett
School of Agriculture, Food and Wine, Waite Research Institute, University of Adelaide, PMB1, Glen Osmond, SA 5064, Australia

D. Plett
Australian Centre for Plant Functional Genomics, Waite Research Institute, University of Adelaide, PMB1, Glen Osmond, SA 5064, Australia

T. Garnett
The Plant Accelerator, Australian Plant Phenomics Facility, Waite Research Institute, The University of Adelaide, PMB 1, Glen Osmond, SA 5064, Australia

N. Stein and G. J. Muehlbauer (eds.), *The Barley Genome*, Compendium of Plant Genomes,
https://doi.org/10.1007/978-3-319-92528-8_15

details, see Newton et al. (2011)]. Despite this resilience, barley is affected by several abiotic stresses that decrease both yield and grain quality, which in turn result in considerable economic losses (e.g., the malting industry is directly affected by the yield and quality of barley). To breed for increased abiotic stress tolerance, we must harness the wealth of genetic variation provided by cultivated barley and its wild relatives. For instance, a wild allele from one *H. vulgare* ssp. *spontaneum* donor contributed to a 30% increase in barley yield under high soil salinity conditions in the field (Saade et al. 2016).

In this chapter, we describe the geographic extent of each abiotic stress that affects barley, the general symptoms barley exhibits in response to these stresses, mechanisms of stress tolerance and their molecular bases, and future research prospects and opportunities offered by the release of the high-quality genome sequence of barley (Mascher et al. 2017). Here, we review studies of barley's responses to single stresses although we know that, under field conditions, barley plants are often exposed to more than one stress. For example, heat stress and drought can co-occur in arid regions (Savin et al. 1996) and boron toxicity is often observed at the same time as soil salinity and alkalinity (Goldberg 1997). While work on combined stresses is beyond the scope of this review, we aim to provide a comprehensive review of the major advances in research on the genetic mechanisms affecting barley's tolerance to abiotic stresses.

15.2 Soil Acidity

Acidity in soil affects all the world's regions to some extent; it is a major problem in developing countries where management practices are often unable to ameliorate pH decreases (FAO 2015). It was estimated that soil acidity affects over 40% of the world's arable land (Kochian 1995). Aluminum (Al) is the main constraint in acidic soils (Foy et al. 1978). Under acidic conditions, Al is solubilized to the toxic Al^{3+}, which inhibits root elongation and affects absorption of several plant nutrients such as Ca, Mg, P, and K (see Zhang and Li (2016) and Foy et al. (1978) for details). Symptoms of Al toxicity in plants include root thickening and darkening; phosphorus deficiency-like symptoms in the shoots including overall stunting with small, dark green leaves and late maturity; purpling of stems, leaves and leaf veins; and yellowing and death of leaf tips.

Barley is among the most Al-sensitive of the cereal crops, although genetic variation in Al tolerance exists in barley's germplasm (Reid et al. 1969). The differences in Al tolerance in barley genotypes have been attributed to differences in the release of organic acid anions, such as malate and citrate, from root apices (Ma et al. 2001; Ryan et al. 2001). Once these organic acid anions are released from root cells, they can bind Al^{3+}, reducing its availability and thus toxicity (Zhou et al. 2013). The genetic mechanisms controlling the release of organic acid anions are likely associated with allelic variation at several loci.

Early research on the genetic control of Al tolerance in barley using the Dayton (Al-tolerant) and Smooth Awn 86 (Al-sensitive) genotypes detected a major locus, *Alp*, on the long arm of chromosome 4H (4HL) (Minella and Sorrells 1992; Reid 1971). Another early study found that a single gene, named *Pht*, on chromosome 4H was responsible for tolerance to high acidity in the soil (Stolen and Andersen 1978). The same chromosomal location on chromosome 4HL also included the Al tolerance locus, *Alt*, which was mapped using amplified fragment length polymorphism (AFLP) in an F_2 progeny derived from a cross between the Yambla (moderately Al-tolerant) and WB229 (Al-tolerant) genotypes (Raman et al. 2002). Later, Minella and Sorrells (1997) showed that tolerance to low pH (*Pht*) and tolerance to aluminum (*Alp*) are controlled by the same locus. Ma et al. (2004) used F_2 seedlings from a cross between the Murasakimochi (Al-tolerant) and Morex (Al-sensitive) genotypes to identify microsatellite markers associated with Al tolerance and citrate secretion on chromosome 4H. High-resolution mapping of *Alp* suggested that *HvMATE*, which encodes a multidrug and toxic compound extrusion (MATE) protein, may underlie Al tolerance (Wang et al. 2007).

In 2007, Furukawa et al. (2007) identified *HvAACT1* as the gene underlying the Al tolerance locus. This gene was found to encode a citrate transport protein located on the plasma membrane and was identified by fine mapping using the F_4 recombinants between the Murasakimochi and Morex genotypes. *HvAACT1* is a member of the MATE family; it is constitutively expressed in the roots of the Al-tolerant barley cultivar Murasakimochi (Furukawa et al. 2007). Interestingly, although expression of *HvAACT1* was not induced by Al, a good correlation was found between Al tolerance and citrate release from barley root apices; the *HvAACT1* expression level was determined to be higher in Al-tolerant genotypes (Furukawa et al. 2007). Constitutive overexpression of *HvAACT1* in transgenic barley and wheat plants demonstrated the ability of HvAACT1 to increase citrate efflux and Al tolerance in acidic soils (Zhou et al. 2013). Also, barley plants transformed with sorghum *SbMATE* showed an increased Al-dependent citrate efflux from root apices and Al tolerance when compared with null segregant plants (Zhou et al. 2014). The presence of a 1-kb insertion in the 5′ untranslated region (UTR) was later linked to the higher expression of *HvAACT1* in Al-tolerant genotypes (Fujii et al. 2012). This mutation altered protein expression from root pericycle cells in sensitive genotypes to the root apices of Al-tolerant genotypes. This finding affects our understanding of the original role of *HvAACT1* in facilitating the translocation of iron from roots to shoots (Fujii et al. 2012). More recently, another allele was found in *HvAACT1* of the Al-tolerant Chinese genotype CXHKSL, but the mechanisms leading to this tolerance are yet to be confirmed (Ma et al. 2016).

In wheat, Al tolerance, controlled by a gene located in chromosome 4DL and expressed constitutively in root apices, depends on malate efflux from root apices (Raman et al. 2005; Sasaki et al. 2004). Indeed, the transgenic Golden Promise (Al-sensitive) barley genotype that expresses *TaALMT1* under ubiquitin promoter exhibits Al-activated malate efflux and enhanced Al tolerance (Delhaize et al. 2004). Zhou et al. (2014) showed that barley lines transformed with wheat *TaALMT1* were more Al-tolerant than barley lines overexpressing *HvAAC1* and lines transformed with *SbMATE.* The closest homolog of *TaALMT1* in barley is *HvALMT1*, which is located on chromosome 2H and is mainly expressed in stomatal guard cells (Gruber et al. 2010). In *Xenopus* oocytes, HvALMT1 is weakly activated by Al^{3+}, facilitating the movement of malate and other ions across oocyte membranes. Thus, *HvALMT1* has been proposed to facilitate organic anion transport to regulate turgor in stomatal guard cells and root elongation zones (Gruber et al. 2010).

Using a genome-wide association study, Zhou et al. (2016) found 22 QTLs associated with Al tolerance, including *HvAAC1*, but also presented unknown QTLs underlying acid soil resistance. Although the genetic mechanisms underlying Al tolerance in barley are quite well known, it is likely that publication of the barley high-quality genome sequence (Mascher et al. 2017) will facilitate the discovery of other genes related to the release of organic acid anions from root apices and new alleles of *HvAACT1*. In the field, barley farmers should follow effective soil management practices and use Al-tolerant germplasm to overcome the negative effects of acidic soils on yield.

15.3 Boron Toxicity

The range in which soil is considered neither deficient of boron (B) nor toxic from B, an essential plant nutrient, is quite narrow (Goldberg 1997). In arid and semiarid regions with low rainfall, soluble B is only partially leached from the subsoil, and the problem of high B concentration in the soil (B toxicity) becomes evident (Reid 2010; Ryan et al. 1998). In particular, plants in Australia (Cartwright et al. 1986) and some areas of the Middle East, such as Syria (Ryan et al. 1998), suffer from B toxicity. Symptoms of B toxicity include chlorosis and necrosis starting at the leaf tips. Brown lesions also appear in the margins of the oldest leaves and progressively affect the whole shoot

(Jefferies et al. 1999). Also, the growth of shoots and roots is retarded, which results in reduced yield (Holloway and Alston 1992; McDonald et al. 2010). B toxicity in barley is managed by the use of B-tolerant genotypes that can maintain low B concentrations in the shoots (Yau and Ryan 2008). For example, Sahara 3771 (a B-tolerant Algerian landrace) is a promising source of B tolerance alleles because it can maintain growth under high B conditions. Under the same conditions, Schooner (a B-sensitive genotype) showed shoot and root biomass reduction by 79 and 51%, respectively, compared with plants grown under control conditions. Furthermore, the roots of Sahara 3771 plants had lower B concentrations than in the external solution, whereas the roots of Schooner plants had equal B concentrations to the growth solution. Hayes and Reid (2004) concluded that the mechanism responsible for B tolerance in Sahara 3771 was related to the B efflux from the roots.

Jefferies et al. (1999) used restriction fragment length polymorphism (RFLP) markers and doubled-haploid lines resulting from the cross between the Sahara (B-tolerant) and Clipper (B-sensitive) genotypes to map regions of the barley genome associated with B tolerance. They found four regions (on chromosomes 2H, 3H, 4H, and 6H) associated with different traits involved in B tolerance. Regions on chromosomes 4HL and 3HS associated with relative root length (the root length at a B concentration of 0 mg/L relative to the root length at 100 mg/L of B) accounted for 39% of the trait variation. Other regions on chromosomes 2H and 4H associated with the leaf symptom score accounted for 38% of the observed variation. The region on chromosome 4H was also associated with shoot dry mass, and together with a region on chromosome 6H, accounted for 53% of the variation observed in the B concentration in shoots. The alleles coming from Sahara conferred higher relative root length, lower leaf symptoms, higher dry mass, and lower B concentration in the shoots at the respective loci (Jefferies et al. 1999). Sutton et al. (2007) identified *HvBot1* as the gene underlying the 4H QTL responsible for B tolerance in Sahara, with higher transcript levels of *HvBot1* related to B tolerance (Sutton et al. 2007). Hayes et al. (2015) found that the nine Saharan accessions (Sahara 3763–Sahara 3771) from the Australian barley germplasm collection had tandemly duplicated Sahara alleles that conferred B tolerance. Genotypes carrying the duplicated (estimated four copies) Sahara allele of *HvBot1* had higher gene expression in the roots (more than a 200-fold increase) (Hayes et al. 2015), where it functions as an efflux-type borate anion transporter (Sutton et al. 2007). However, more investigation of *HvBot1* is needed to clarify the mechanism of B efflux (Hayes and Reid 2004).

More recently, Hayes et al. (2015) found eight synonymous single nucleotide polymorphisms (SNPs) in the *HvBot1* coding regions of the Clipper and Sahara alleles, although all were silent mutations. They therefore attributed the difference between alleles solely to expression levels. Emebiri et al. (2009) used marker-assisted selection (MAS) to introgress the favorable Sahara alleles at the 4H locus to the breeding line VB9104, which is adapted to conditions in southern Australia. However, results from field experiments conducted at four locations on the generated nearly isogenic lines were inconsistent because, in some locations, genotypes with the B tolerance allele had higher yields, whereas, in other locations, these genotypes had lower yields, suggesting a significant genotype-by-environment interaction. However, malting quality was not affected, suggesting that the introgression of the B tolerance allele had no penalty on quality (Emebiri et al. 2009). Similarly, McDonald et al. (2010) found that introgressing B tolerance alleles from Sahara at the 2H and 4H loci had little or no effect on yield gain in the field and that yield gain was largely affected by site location. This recombinant inbred line (RIL) population had yields similar to or lower than their progenitors although they exhibited reduced B toxicity symptoms and reduced B concentrations in their shoots. This result could be because B-tolerant plants grow vigorously during early stages, thus using up the limited water available in arid regions and therefore affecting plants at the reproductive stage. Another reason could be that B toxicity is often accompanied by other constraints, such as high salinity. In fact, McDonald (2006) showed earlier that variation in yield in soils in which both stresses occurred simultaneously is due to salinity rather than to B toxicity. Tissue tolerance to B and the ability to overcome other soil constraints are both necessary for increasing yield under high B concentrations (McDonald et al. 2010).

The B tolerance gene underlying the 6H QTL, identified by Jefferies et al. (1999), was cloned by Schnurbusch et al. (2010). This gene, *HvNIP2;1*, encodes an aquaporin that belongs to the subfamily of nodulin-26-like intrinsic proteins (NIPs). NIPs are channel proteins that allow transport of water and other small solutes such as boric acid (Takano et al. 2006). The expression of *HvNIP2;1* was found to be specific to the roots and to increase the permeability of the plasma membrane to B in heterologous systems (i.e., *Xenopus oocytes* and yeast cells). In addition, the roots of Clipper (B-sensitive genotype) plants had 15-fold more *HvNIP2;1* transcripts compared with those of B-tolerant Sahara. Differences in the regulatory sequence regions, upstream of the ATG codon, may explain the differences in transcript levels of *HvNIP2;1* between the two genotypes. One mechanism of barley tolerance under high B concentrations in the soil could be the reduced expression of *HvNIP2;1* to limit the passive influx of B from the roots (Schnurbusch et al. 2010). This mechanism, combined with high expression of *HvBot1*, may increase B tolerance. In B-limited environments, higher *HvNIP2;1* expression would be beneficial for plants to avoid deficiency in this nutrient (Schnurbusch et al. 2010; Takano et al. 2006). Hayes et al. (2015) screened 65 barley genotypes for differences in the coding sequence and transcription level of *HvNIP2;1* and found that the open reading frame of *HvNIP2;1* was highly conserved. However, Sahara genotypes presented a unique SNP in the 5′ UTR region of *HvNIP2;1*. This SNP, 44 bp upstream of the start codon, resulted in a shorter upstream open reading frame that translated into a short polypeptide, resulting in lower expression in Sahara (Hayes et al. (2015). Limited knowledge is available about the genes underlying the 2H and 3H QTL. *HvBot2*, a B-transporter gene, has been suggested as a candidate gene underlying the 3H QTL because Sahara 3771 has a deletion in the 3′ end of the coding sequence, which may disrupt the protein function. The publication of the barley high-quality genome sequence (Mascher et al. 2017) will allow advancement of the research needed to clarify the B tolerance mechanisms underlying the 2H and 3H mechanisms.

15.4 Soil Salinity

Soil salinity is a major constraint on agriculture and impacts food security and political stability. Currently, 1128 Mha, including 20% of irrigated lands (http://www.fao.org/water/en/), are estimated to be affected by soil salinity. The largest such area, of 189 Mha, is located in the Middle East (Wicke et al. 2011). These numbers are expected to increase due to climate change and poor irrigation practices (Hayes et al. 2015). Qadir et al. (2014) estimated that an annual economic loss of US$27.3 billion is due to soil salinity. Salinity tolerance is a complex polygenic mechanism (Roy et al. 2014) that enables plants to maintain growth and to produce photoassimilates under saline conditions (Munns 2002). The response of plants to salinity includes several tolerance mechanisms, namely shoot ion-independent tolerance (or osmotic tolerance), ionic tolerance and tissue tolerance (Munns and Tester 2008; Roy et al. 2014). Numerous genes have been proposed to contribute to salinity tolerance traits (Roy et al. 2014).

Barley is the most salt-tolerant cereal crop (Munns and Tester 2008). This distinction for barley could partially be due to salinity tolerance in its tissues. For example, evidence exists that is consistent with barley having better ability than durum wheat to compartmentalize Na^+ in vacuoles (James et al. 2006). A high cytoplasmic K^+/Na^+ ratio in mesophyll cells would allow barley to maintain its photosynthetic capacity under saline conditions (James et al. 2006;

Wu et al. 2013). Besides the ability of barley to accumulate and tolerate high levels of Na^+, Na^+ exclusion is still an important mechanism in barley salinity tolerance. Halophytic relatives of barley have been used in several studies to understand the differences in the mechanisms of salinity tolerance between species (Islam et al. 2007; Maršálová et al. 2016). For example, Garthwaite et al. (2005) reported that seven wild *Hordeum* species were better able to exclude Na^+ from their shoots compared with cultivated barley, when grown on nutrient solutions with different NaCl concentrations. Hence, improving cultivated barley salinity tolerance could be done by making use of exotic germplasm. In fact, *Nax3* and *Nax4* loci, described below, were identified using a landrace (Sahara 3771) and a wild barley (*Hordeum spontaneum*) accessions.

Barley wild relatives are an important source of genes to increase salinity tolerance. For example, Saade et al. (2016) identified a wild allele on chromosome 2H that resulted in a 30% yield increase under saline conditions. Also, the overexpression of *HsCBL8* gene from a *H. spontaneum* line native to the Qinghai–Tibet Plateau resulted in an increased salinity tolerance in rice (Guo et al. 2016).

The ability of barley to retain K^+ in its cells under saline conditions is possible through high-affinity K^+ transporters, such as HvHAK1 (Santa-Maria et al. 1997). In fact, *HvHAK1* expression was higher in the shoots and roots of K305 (a salt-tolerant genotype) compared with I743 (a salt-sensitive genotype) (Ligaba and Katsuhara 2010). Na^+ is sequestered in the vacuoles away from the cytoplasmic machinery via Na^+/H^+ antiporters localized on the tonoplasts (vacuolar membranes). The electrochemical gradient generated by the vacuolar H^+-inorganic pyrophosphatase (V-PPase) and the vacuolar H^+-ATPase (V-ATPase) provides the driving force for this movement (Rea and Poole 1993; Sze et al. 1992). Fukuda et al. (2004) studied the effects of 400 mM of mannitol (osmotic effect) and 200 mM of NaCl (ionic effect) on the expression of two V-PPases (*HVP1* and *HVP10*), V-ATPase subunit A (*HvVHA-A*) and Na^+/H^+ antiporter (*HvNHX1*). One copy of each gene was detected in the barley genome. The transcript level of *HvNHX1* increased fivefold in the roots of barley plants treated with 200 mM of NaCl compared with the roots of control plants, suggesting that *HvNHX1* plays a role in salinity tolerance. Treatment with 400 mM mannitol increased the transcript level of *HvNHX1* fourfold compared with the transcript level in control plants. HvNHX1 protein could thus be transporting ions other than Na^+ into the vacuoles. In fact, AtNHX1 has a K^+/H^+ exchange function (Bassil et al. 2011; Venema et al. 2002) and plays an important role in turgor regulation and stomatal function through K^+ uptake at the tonoplast (Barragan et al. 2012). V-PPase *HVP10* exhibited a higher transcript level in roots than in shoots, whereas *HVP1* exhibited higher shoot expression. *HVP10* and *HvVHA-A* transcript levels in the roots increased in response to the addition of NaCl and decreased in response to the addition of mannitol, while the transcript level of *HVP1* increased in response to the addition of both. *HVP10*, *HvVHA-A*, and *HVP1* transcript levels did not change in the shoots (Fukuda et al. 2004).

HVP10 was proposed as a strong candidate underlying *HvNax3*, a locus on the short arm of chromosome 7H identified in a biparental population resulting from a cross between the wild barley CPI-71284-48 (*H. vulgare* ssp. spontaneum) and the cultivar Barque-73 (Shavrukov et al. 2010, 2013). *HvNax3* was found to be associated with Na^+ exclusion, explaining 10–25% difference in leaf Na^+ of plants grown hydroponically under salinity (Shavrukov et al. 2013). The importance of vacuolar pyrophosphatase in improving salinity tolerance was best illustrated by Schilling et al. (2014) who found that transgenic barley genotypes expressing the arabidopsis vacuolar pyrophosphatase gene (*AVP1*) had a larger shoot biomass (30–42%) compared with wild-type genotypes under salinity conditions in the field. Furthermore, transgenic barley *AVP1* genotypes had a higher yield as indicated by a larger number of heads per plant (16–58%), grains per plant (76–85%), average grain weight (29–43%), and grain yield per plant (79–87%) (Schilling et al. 2014).

A third vacuolar Na^+/H^+ antiporter, *HvNHX2*, was isolated in barley (Vasekina et al. 2005) and increased salinity tolerance was observed when *HvNHX2* was overexpressed in arabidopsis (Bayat et al. 2011) and potato plants (Bayat et al. 2010).

The role played by the so-called high-affinity potassium transporters (HKT) in the ionic component of salinity tolerance has been clearly established. In fact, *TmHKT1;4-A2* and *TmHKT1;5-A* were suggested as candidate genes underlying two sodium exclusion loci, *Nax1* and *Nax2*, respectively. When grown in saline fields, durum wheat genotypes carrying the *TmHKT1;5-A* locus had lower concentrations of leaf Na^+ and 25% higher grain yield compared with near-isogenic lines lacking *TmHKT1;5-A*. More importantly, *TmHKT1;5-A* had no effect on yield in low-salinity soils (Munns et al. 2012). HKTs can be classified into two groups based on their amino acid sequences (Platten et al. 2006). Class I HKTs are more likely to be Na^+-transporters whereas class II HKTs (high-affinity Na^+) function as Na^+/K^+ symporters or Na^+- or K^+-uniporters (Hauser and Horie 2010; Maser et al. 2002). Furthermore, some class I HKT transporters are localized in the plasma membrane of xylem parenchyma cells in the root stele, allowing them to retrieve Na^+ from the xylem and to prevent Na^+ from reaching the shoot (Byrt et al. 2007). Class II HKT proteins, which are localized at the root epidermis/cortex, play an important role in plant growth under K^+-limited conditions, and they may be downregulated under salinity conditions (Huang et al. 2008).

Identifying the specific functions of HKT has been challenging and controversial, especially when heterologous systems are used (Haro et al. 2005). In barley, two HKT proteins have been characterized: HvHKT1;5 (class I) and HvHKT2;1 (class II) (Hauser and Horie 2010). Mian et al. (2011) reported that *HvHKT2;1* was predominantly expressed in the root cortex and cotransported Na^+ and K^+ when expressed in *Xenopus* oocytes. Also, transgenic barley lines overexpressing *HvHKT2;1* had a higher relative growth rate compared with wild plants under saline conditions (Mian et al. 2011). Furthermore, *HKT2;1/2-like*, *HKT2;3/4-like*, *HKT1;1/2-like*, *HKT1;3-like*, and *HKT1;4-like* genes have been mapped on respective regions of barley chromosomes 7HL, 7HL, 2HL, 6HS, and 2HL. These regions are in synteny with regions in rice that contain HKT genes (*OsHKT2;1*, *OsHKT2;4*, *OsHKT1;1* & *OsHKT1;2*, *OsHKT1;3*, and *OsHKT1;4*, respectively); they could thus harbor rice HKT orthologs (Huang et al. 2008).

Little is known about the shoot ion-independent tolerance mechanism in plants (Roy et al. 2014). Rapid and long-distance sensing seems to require reactive oxygen species and calcium waves (Choi et al. 2014; Gilroy et al. 2014). The salt-overly sensitive (SOS) pathway is a putative signaling mechanism in salinity response (Munns and Tester 2008). In barley, the homolog of arabidopsis SOS3, *HvCBL4*, has been identified as a candidate gene underlying the *HvNax4* locus (Rivandi et al. 2011). *HvNax4* was initially identified using a mapping population of doubled-haploid lines derived from a cross between Clipper and Sahara 3771. It is located on chromosome 1HL (Lonergan et al. 2009). Rivandi et al. (2011) characterized *HvNax4* using the same population. They found that *HvNax4* governed the Na^+-concentration in the shoots of plants grown in pots, but it had no significant effect on plants grown under hydroponic conditions. Moreover, the effect of *HvNax4* on the Na^+ content in shoots varied with soil type, suggesting that *HvNax4* expression was strongly influenced by the environment. Furthermore, the mRNA expression of *HvCBL4* did not vary between Clipper and Sahara 3771, but an amino acid residue change was proposed to alter the protein function between the two genotypes. Because *HvNax4* expression was found to be highly dependent on the environment and because it did not significantly affect plant biomass, the agronomical importance of this locus has not yet been determined. Further progress in research on the shoot ion-independent tolerance mechanism in barley is expected due to the advances in high-throughput phenotyping methodologies (Saade et al., unpublished data). Such research employing a wide range of barley genotypes may clarify which genomic regions (and/or genes) are responsible for the shoot ion-independent component of salinity tolerance.

One should note that besides the QTLs presented here, a large body of genetic studies has identified QTLs associated with traits contributing to salinity tolerance in barley. For instance, Long et al. (2013) identified two QTLs on chromosome 3H (126.3 cM) and 6H (60.2 cM) that are associated with salinity tolerance, defined as the biomass production under saline relative to nonsaline conditions. Other study focused on plant survival and leaf chlorosis, and identified a QTL on chromosome 4H (145 cM) associated with these traits (Fan et al. 2016b). Liu et al. (2017) related stomatal traits and yield to salinity tolerance, and found QTLs on chromosomes 1H (11 cM) and 3H (58.9 cM) associated with grain yield, and QTLs on chromosome 1H (121.4 cM) and 2H (10.3 cM) associated with stomatal pore area and stomatal conductance, respectively. Moreover, QTLs associated with salinity tolerance were identified at different developmental stages such as salinity tolerance at germination (Mano and Takeda 1997), seedling (Ellis et al. 2002; Mano and Takeda 1997), and reproductive stage (Liu et al. 2017; Xue et al. 2009). Barati et al. (2017) identified several QTLs contributing to agronomic performance in the field under nonsaline and saline conditions using a doubled-haploid population derived from Clipper and Sahara. The barley high-quality reference genome (Mascher et al. 2017) will allow the identification of the candidate genes underlying many of these described QTLs. In addition, the publication of the high-quality barley genome sequence (Mascher et al. 2017) allows the identification of homologs of important players in salinity tolerance, such as the Na^+/Ca^{2+} exchanger-like protein AtNCL (Wang et al. 2012a). Also, salt tolerance genes that are absent from the barley genome may be identified. For instance, AtCIPK16, a calcineurin B-like protein interacting protein kinase that reduces shoot Na^+ in arabidopsis (Roy et al. 2013), does not have a homolog in barley (Amarasinghe et al. 2016). Moreover, the genetic position of some barley genes involved in salinity tolerance, such as *HVP1* and other candidate genes underlying QTL regions associated with salinity tolerance (Long et al. 2013; Ma et al. 2015), can be further pinpointed using the barley high-quality reference genome (Mascher et al. 2017).

15.5 Drought

In a general sense, drought can be defined as a lack of precipitation over a long period of time affecting certain activities and sectors (National Drought Mitigation Center (NDMC), http://drought.unl.edu/DroughtBasics/WhatisDrought.aspx). However, drought is a complex phenomenon that can be divided into four categories: meteorological, hydrological, agricultural, and socioeconomic. In this section, we discuss agricultural drought (hereafter drought), which refers to a condition in which soil moisture does not meet the needs of a particular crop at a particular growth stage. The occurrence of drought has become more frequent due to climate change; this increasing frequency has severe consequences for agriculture and, consequently, for economic stability and food security. The impact of drought on crop yield is enormous. For example, severe drought in Kenya caused wheat yield to decrease by 45% in 2009 compared with the yield in 2010. The 2011 drought that affected the grain belt and large parts of the United States increased food prices across the globe (http://www.fao.org/docrep/017/aq191e/aq191e.pdf).

Symptoms of drought stress include slower plant growth and leaf expansion, deeper root systems, stomatal closure to reduce transpiration (which in turn increases leaf temperature and affects photosynthesis) and the accumulation of compatible solutes such as proline, sorbitol, and glycine betaine to adjust the cytoplasmic osmotic potential (Baerenfaller et al. 2012; Daszkowska-Golec and Szarejko 2013; Farooq et al. 2009; Singh et al. 2015; Tombesi et al. 2015; Uga et al. 2013). A pot experiment, using barley landraces from Iran, showed that drought can decrease the relative water content in leaves, the midday leaf water potential, and stomatal conductance (by 90%) compared with plants in well-watered pots, thus reducing photosynthetic activity (Robredo et al. 2007, 2010)

Another pot experiment, using 11-day-old barley seedlings showed that root mass aggregated towards the bottom of the pot and adventitious root formation was halted under drought conditions. In addition, a 10-fold increase in proline, involved in osmotic adjustment, was observed in barley roots in response to drought, as well as a severe reduction in the shoot-to-root ratio (Sicher et al. 2012).

Because of the importance of drought stress, several studies have focused on genetic mapping of drought-associated traits. Fan et al. (2015) examined a mapping population derived from a cross between TX9425 (a drought-tolerant Chinese landrace) and Franklin (a drought-sensitive genotype). They used wilting as the trait to evaluate drought response. This work led to the identification of a QTL on chromosome 5H that was independent of two other developmental traits—awn length and heading date. The gene underlying this QTL was suggested to be 9-cis-epoxycarotenoid dioxygenase 2 (*HvNCED2*), which is involved in the synthesis of abscisic acid (Fan et al. 2015). Mikolajczak et al. (2017) identified 103 QTLs associated with yield-related traits when 100 RILs from a cross between European and Syrian cultivars were grown under different water regimes.

Roots play a crucial role in acquiring water under drought stress. In fact, the smaller size of the root system in spring barley was associated with lower grain yield in dry environments (Chloupek et al. 2010). Reinert et al. (2016) used a barley diversity panel composed of modern cultivars, landraces, and wild accessions to identify a QTL on chromosome 5H that was induced by drought and associated with the dry weight of roots. *HvCBF10B* and *HvCBF10A* were suggested as candidate genes underlying this QTL because their conserved domains harbored a large deletion of 37 amino acids in wild barley compared with cultivated genotypes.

RILs from the mapping population between Tadmor (a drought-tolerant Syrian genotype) and Er/Apm (a moderately drought-tolerant genotype) were grown in pots in growth chambers under control and drought conditions. A QTL on chromosome 7H was associated with the relative water content and osmotic potential of leaves (Teulat et al. 1997, 1998, 2001a). It was cosegregated with the *Acl3* locus (Hansen and von Wettstein-Knowles 1991). *Acl3* encodes barley acyl carrier protein III, which is involved in the synthesis of the fatty acyl chain (Hansen and von Wettstein-Knowles 1991). *Acl3* may play a role in membrane protection or fluidity under stress conditions and thus improve drought tolerance (Teulat et al. 1998). Another study associated this QTL with carbon isotope discrimination in the field, which is a measure of transpiration efficiency (Teulat et al. 2002). Under field conditions, Teulat et al. (2001b) and von Korff et al. (2008) evaluated the same RIL population for agronomic traits across different Mediterranean environments with varying drought severity. The genetic effects on flowering time, kernel weight, and plant height were stable across these environments, and QTLs associated with these traits were detected in both studies with comparable effects. However, yield seemed to be more influenced by the environment than other traits. In particular, the Tadmor allele at the pHva1 marker had a favorable effect on grain yield in the locations with more severe drought (von Korff et al. 2008). This marker colocates with *HVA1* gene on chromosome 1H, which also plays a role in low-temperature tolerance (Tondelli et al. 2006; von Korff et al. 2008). The same locus also had a significant effect on the wilting score in a mapping population derived from a cross between Scarlett (a drought-sensitive German genotype) and ISR42-8 (a wild accession of *H. spontaneum*). The *HVA1* QTL explained 12% of the phenotypic variation, with the wild allele contributing to a 17% decrease in the wilting score relative to the domesticated allele (Sayed et al. 2012). Transgenic studies showed that wheat, rice, and oat transformed with barley *HVA1* had higher drought and salinity tolerance than control plants had (Babu et al. 2004; Bahieldin et al. 2005; Oraby et al. 2005; Xu et al. 1996).

Other studies, such as Suprunova et al. (2007), identified *Hsdr4* (*H. spontaneum dehydration-responsive 4*), which codes for a Rho-GTPase-activating-like protein and is involved in drought tolerance in wild barley. The arabidopsis Rho GTPase homolog (*AtRac1*) is known to play a central role in abscisic-acid-mediated stomatal closure (Lemichez et al. 2001). Besides stomatal closure, the cuticle also plays a role in plants' adaptation to drought because it limits water loss. In barley, several *eceriferum* (*cer*) genes involved in the deposition of epicuticular waxes were detected in a mutagenesis study (Lundqvist and Lundqvist 1988). Recently, *Cer-cqu* (*Cer-c*, *Cer-q*, and *Cer-u*) has been identified on chromosome 2HS as a cluster of three separate genes encoding: (1) a chalcone synthase-like polyketide synthase, (2) a lipase/carboxyl transferase, and (3) a P450 enzyme (Hen-Avivi et al. 2016; Schneider et al. 2016). Chen et al. (2004) identified a spontaneous wild barley mutant genotype, *eibi1*, which was found to be hypersensitive to drought with a low capacity to retain leaf water. The hypersensitivity of this genotype was due to a thin cuticle that resulted from low cutin deposition. The *eibi1* gene was mapped to the pericentromeric region of chromosome 3H (Chen et al. 2009), and *HvABCG31*, an ATP-binding cassette subfamily G (ABCG) transporter, was identified as the gene underlying *eibi1* (Chen et al. 2011).

Drought-induced root and leaf proteomic and metabolomic changes revealed that several proteins and metabolites, namely HSP70, proline, carbohydrates and ascorbic acid, accumulated in response to drought in drought-sensitive genotypes, but these proteins and metabolites were constitutively elevated in drought-tolerant genotypes. Hence, biochemical predisposition could confer increased tolerance to drought (Chmielewska et al. 2016). In fact, several barley genes have been successfully used to improve drought tolerance in barley and in other crops, e.g., overexpression of the isoform *HvSNAC1* in barley (Al Abdallat et al. 2014) and overexpression of *HvCBF4* in rice (Oh et al. 2007).

Many traits, such as relative water content, have been shown to work as selection parameters for drought in barley (Matin et al. 1989). However, other traits have the potential to be used to map drought tolerance loci. F_v/F_m, the ratio of variable to maximal fluorescence in a dark-adapted state, evaluates the effect of stress on the reaction centers of Photosystem II and works as a selection criterion for drought tolerance (Guo et al. 2008). This trait has been used in a mapping population under greenhouse conditions (Guo et al. 2008). Two QTLs (116 and 135.7 cM) on the long arm of chromosome 2H respectively explained 15 and 9% of F_v/F_m phenotypic variation under drought conditions during the post-flowering stage. However, the use of F_v/F_m as a possible selection criterion should be validated in other populations.

Several studies have detected QTLs involved in drought tolerance under controlled and field conditions using different genetic material (Baum et al. 2003; Diab et al. 2004; Guo et al. 2008; Honsdorf et al. 2014; Lakew et al. 2013; Peighambari et al. 2005; Rollins et al. 2013; Talame et al. 2004; Teulat et al. 1997, 1998, 2001a, b, 2002, 2003; von Korff et al. 2008). However, these studies identify only the QTLs underlying the traits of interest; they do not identify genes underlying drought tolerance in barley. It is possible that many of these studies could be reanalyzed using new markers to validate discovered QTLs or to identify new QTLs using the improved barley genome sequence (Mascher et al. 2017). As such, the use of more markers and better physical maps of barley provided by the high-quality reference genome (Mascher et al. 2017) should lead to more accurate identification of QTLs and their underlying genes involved in drought tolerance. The publication of the high-quality barley genome sequence (Mascher et al. 2017) allows the identification of homologs of important players in drought tolerance. For example, the in silico and expression analyses of barley calcium-dependent protein kinases (CDPKs) demonstrated the involvement of

CDPKs in barley signaling pathways in response to drought (Fedorowicz-Strońska et al. 2017).

15.6 Temperature Stress

Temperature stress has a strong effect on the growth and development of barley. Temperature can be divided into three ranges: (i) low-temperature range, which includes cold and freezing; (ii) optimum temperature range at which the plants grow and develop normally; and (iii) high-temperature range, which is associated with heat stress.

15.6.1 Low-Temperature Stress

Low-temperature stress, which includes cold (also known as chilling, above freezing point temperature) and freezing (below freezing point temperature), has a severe effect on small-grain crop plants in temperate climates (Kosova et al. 2011). In regions such as southern Australia, where winter temperatures are not low enough to cause freezing damage at the vegetative stage, radiation frost constitutes a problem at the reproductive stage. Radiation frost occurs during clear nights when the plant canopy is receiving less heat than what is radiated away, causing temperatures to drop below zero. Yield is severely affected by radiation frost as a result of floret and spike abortion as well as damage to developing grains (Reinheimer et al. 2004; Zheng et al. 2015). According to the 2006 and 2012 reports from the Grains Research & Development Corporation (https://grdc.com.au/uploads/documents/frost.pdf, https://grdc.com.au/uploads/documents/GRDC-FS-CrackingWheatsToughestNuts.pdf), frost causes yearly losses of more than $33 million in crop production in southern Australia and Victoria and about $100 million in barley and wheat production in northern New South Wales and Queensland.

Barley, like other temperate cereals, has an increase in freezing tolerance when exposed to cold at vegetative stages, a phenomenon known as cold acclimation or hardening. Under cold stress, plants have a short-term response including cold acclimation, and a long-term response, which involves developmental responses such as vernalization (Kosova et al. 2011; Pecchioni et al. 2014). Barley can be classified into spring, facultative, and winter types depending on its growth habitat, tolerance to low temperature, requirements for vernalization, and sensitivity to photoperiod (Tondelli et al. 2014). Photoperiod sensitivity and vernalization play roles in low-temperature tolerance by delaying the transition from the vegetative to the reproductive stage, thus delaying flowering, which allows plants to survive low-temperature stress (Fowler et al. 1996a, b, 2001; Mahfoozi et al. 2001a, b). In fact, genes that confer low-temperature-tolerance are expressed during the vegetative stage (Fowler et al. 1996b; Mahfoozi et al. 2001b), and processes that prolong this stage increase low-temperature tolerance. Some mechanisms that may contribute to freezing tolerance include the ability to prevent or reverse proteins' denaturation and to reduce the physical damage of dehydration resulting from freezing-induced intercellular ice formation (Snyder and Melo-Abreu 2005; Thomashow 1998; Yadav 2010). Another important mechanism in freezing tolerance is the stabilization of membranes. In fact, the lipid composition of membranes and the accumulation of sucrose and other simple sugars seem to change in response to cold acclimation (Thomashow 1998). For instance, *blt4*, a barley gene involved in freezing tolerance, encodes a putative lipid transfer protein and plays a role in modifying the lipid composition of membranes (Thomashow 1998). Because *blt4* expression is also induced under drought stress, this gene could be responsive to dehydration (Dunn et al. 1991). Hydrophilic polypeptides, encoded by cold-responsive (COR) genes, seem to contribute to membrane stabilization and increase stress tolerance during freezing (Thomashow 1998).

Genetic mapping of barley has identified loci associated with freezing tolerance on chromosomes 1H, 2H, 5H, and 6H. A doubled-haploid population resulting from a cross between a winter barley (Nure) and spring barley (Tremois) was assessed for low-temperature tolerance using three traits: winter survival in the field, frost resistance (controlled freeze test), and F_v/F_m. Two major QTLs were associated with the three traits and mapped on the long arm of chromosome 5H (Francia et al. 2004). The first QTL (*FR-H1*) was previously mapped by Hayes et al. (1993) in a barley population of Diktoo (winter) crossed with Morex (spring), which was assessed for field survival and the lethal temperature at which 50% of the plants die (LT_{50}). Interestingly, *FR-H1* was found to overlap with the *VRN-H1* vernalization locus. Whether this association of both vernalization requirement and freezing tolerance is a result of having linked genes (Francia et al. 2004; Hayes et al. 1993) or having one gene with pleiotropic effects (Tondelli et al. 2014; von Zitzewitz et al. 2011) is still a debatable question (Fisk et al. 2013). If one considers the pleiotropy effect, then *HvBM5A* would be the candidate gene underlying *VRN-H1* and *FR-H1* (von Zitzewitz et al. 2005). *HvBM5A*, the barley ortholog of *TmAP1*, is a MADS-box transcription factor (von Zitzewitz et al. 2005; Yan et al. 2003), and spring accessions harbor a deletion in their first intron (Fu et al. 2005). Besides freezing tolerance at the vegetative stage (Francia et al. 2004; Hayes et al. 1993), the *FR-H1/VRN-H1* locus is also associated with frost tolerance at the reproductive stage. In fact, *FR-H1* has been associated with frost-induced floret sterility and grain damage in three different mapping populations (Reinheimer et al. 2004).

Another QTL specifically associated with frost-induced floret sterility was mapped on the long arm of chromosome 2H (Reinheimer et al. 2004). Approximately 30 cM away from *FR-H1*, the second QTL (*FR-H2*) detected by Francia et al. (2004) mapped to the same region where QTLs for the accumulation of two cold-responsive proteins, HvCOR14b (Crosatti et al. 1999) and TMC-Ap3 (Baldi et al. 1999; Mastrangelo et al. 2000), mapped. An *HvCBF* transcription factor was suggested as a candidate underlying *FR-H2* because *HvCBF4* was the peak marker and at least 12 *HvCBF* of the 20 barley CBFs mapped to *FR-H2* (Francia et al. 2004, 2007; Pasquariello et al. 2014). Tondelli et al. (2006) suggested that *HvCBF* genes rather than their regulators, i.e., the inducer of CBF expression (*ICE1*) and fiery-1 (*FRY1*), were candidates for the *FR-H2* locus. The higher copy number of CBF genes in winter genotypes compared with those in spring suggests the involvement of copy number/gene duplication in the degree of low-temperature tolerance in temperate climate cereals (Knox et al. 2010; Tondelli et al. 2011). In fact, transgenic barley overexpressing *TaDREB2* and *TaDREB3* exhibited increased drought and frost tolerance, as well as higher expression of stress-responsive genes involved in protecting cells from damage and desiccation (Morran et al. 2011).

The genetic mechanism of freezing tolerance exhibited dynamic control between *FR-H1* and *FR-H2*. *CBF* genes at *FR-H2* were more highly expressed in barley genotypes carrying the winter allele (*vrn-h1*) at the *VRN-H1* locus compared with genotypes carrying the spring allele. In addition, the expression of CBF was reduced after vernalization (in vernalization-requiring plants), indicating that *VRN-H1* attenuated the expression of *FR-H2* (Stockinger et al. 2007). The involvement of *FR-H1* and *FR-H2* in cold acclimation was detected using a genome-wide association mapping approach, where the presence of specific alleles (winter alleles) at these loci conferred maximum low-temperature tolerance (von Zitzewitz et al. 2011). The allelic variation at *FR-H1* and *FR-H2* was also the main determinant of frost tolerance in a panel of a European two-row spring barley population (Tondelli et al. 2014). Besides *FR-H1* and *FR-H2*, Fisk et al. (2013) mapped another QTL, *FR-H3*, which is associated with freezing tolerance on the short arm of chromosome 1H.

Research on low-temperature tolerance in barley is far from complete. Extensive high-quality research, including proteomics and metabolomics studies, is warranted to identify the genes involved in each of the two components of low-temperature tolerance, cold tolerance and freezing/frost tolerance. The barley high-quality reference genome (Mascher et al. 2017) can significantly contribute to a better understanding of the genetic mechanisms underlying low-temperature stress. For instance, the genes underlying the *Fr-H3* QTL still need to be identified (Fisk et al. 2013). Also, whether or not freezing tolerance and vernalization requirements are due to a pleiotropic effect or linkage still needs to be clarified. Interestingly, copy number variation also plays a role in frost tolerance. Francia et al. (2016) found that within a collection of 41 barley genotypes, the genotypes with higher copy numbers at *HvCBF2* and *HvCBF4* showed increased frost tolerance (Francia et al. 2016). Therefore, sequencing different barley varieties could help in identifying variations in copy number and their role in enhancing stress tolerance.

15.6.2 High-Temperature Stress

The negative impact of climate change on crop yield has been established, with a projected increase in global temperature on the order of ~4 °C by the late twenty-first century (IPCC 2014). High temperature (hereafter heat stress) affects the duration of the growing season, geographic distribution (Dawson et al. 2015), and malting quality of barley (Savin et al. 1996). The effects of heat stress depend on several factors, such as stress intensity (temperature in degrees), duration and rate of temperature increase (Wahid et al. 2007). Symptoms of heat stress in barley include the reduction of yield and yield components, such as number of tillers per plant, spike length and thousand grain weight, and a lower final starch concentration (possibly caused by irreversible inactivation of sucrose synthase) (Abou-Elwafa and Amein 2016; Högy et al. 2013; MacLeod and Duffus 1988; Wallwork et al. 1998). Nevertheless, heat stress in barley is still an understudied topic, and further research is expected to take place due to climate change.

Barley, like other cereal crops, is particularly sensitive to heat stress during panicle development and meiosis, with high temperatures (over several days) causing abnormal pollen development and complete sterility (Sakata et al. 2000). Oshino et al. (2007) studied the effect of high-temperature injury on anther development in barley and observed a premature progression of early developmental stages and fate (e.g., progression to meiosis of pollen-mother cells). Transcriptional analysis suggested that the genetic control involved in the cessation of anther cell proliferation was related to the repression of cell proliferation factors, such as histones, replication licensing factors, and DNA polymerases (Oshino et al. 2007). Gene expression and differentiation of anthers were studied under normal (20 °C day/15 °C night) and heat stress (30 °C day/25 °C night) conditions, revealing that genes that are active in the anthers under normal temperature conditions, such as H3, H4, and glycine-rich RNA-binding protein genes, were transcriptionally inhibited under heat stress (Abiko et al. 2005). After anthesis, a field trial in Iran showed a 17% reduction in grain yield in barley under heat stress and significantly affected parameters, including the translocation of photosynthates to the grain, starch synthesis, and deposition in the developing grain (Modhej et al. 2015). Field trials have shown grain yield reduction caused by heat stress causing a higher penalty when temperature increased during stem elongation (Ugarte et al. 2007). In another field experiment using a panel of 138 spring barley genotypes, Ingvordsen et al. (2015b) corroborated the previously found decrease in grain yield (55.8%) caused by heat stress (Högy et al. 2013).

One field study in Egypt comprising 320 wild barley accessions and six local genotypes showed that several yield components, such as days to flowering, plant height and thousand kernel weight (among others), were affected by heat stress (Abou-Elwafa and Amein 2016). The screening of such a large number of individuals in the primary gene pool of barley quantified the effects of heat stress and grouped genotypes into two distinct groups based on heat stress tolerance. Cluster analysis revealed that group 1 consisted of 112 barley accessions mainly originating from cold regions while group 2 consisted of 224 barley genotypes with variable degrees of tolerance to heat stress (Abou-Elwafa and Amein 2016). These valuable data may provide the basis for future mapping populations and/or genome-wide association studies that could reveal the genetic mechanisms underlying heat stress in barley.

Also, Phillips et al. (2015) examined barley's recombination landscape and observed that high temperature significantly changed the patterns of recombination only in male meiosis. Transcriptome changes in developing barley seeds in response to heat stress revealed the downregulation of genes related to storage compound biosynthesis and cell growth (Mangelsen et al. 2011). Metadata analysis also showed that seed embryo and endosperm were the primary locations of heat stress response genes and an overlap of heat- and drought-responsive genes in barley caryopses (Mangelsen et al. 2011).

The relative importance of day and night heat stress is still under debate. Garcia et al. (2015) observed approximately 7% reduction in grain yield per degree of night temperature increase during the critical period of grain filling. This yield reduction was caused by an accelerated development, which caused increased tillering, death and impacted biomass and grain production.

Despite the undeniable importance of heat stress, limited knowledge is available on the genetic mechanisms controlling barley's response to high temperature. Heat shock proteins (HSP) are the obvious candidates underlying the plant's heat stress response. Xia et al. (2013) investigated the allelic variation of *HSP17.8*, which encodes an HSP ubiquitously produced in response to heat stress, and its association with agronomic traits in barley. Nine haplotypes for *HSP17.8* were identified, with wild accessions exhibiting greater allelic diversity than found in cultivated barley. These results suggest the role of HSP17.8 in response to heat stress as well as to drought. In addition to HSP, miRNAs are heat inducible and may play a role in barley thermotolerance through downregulation of their target genes (Kruszka et al. 2014).

To date, numerous studies have identified QTLs associated with general agronomic traits (e.g., yield and plant height) although little work has been done on traits specifically associated with heat stress. Gous et al. (2016) mapped a doubled-haploid population derived from a cross between ND24260 (stay-green genotype) and Flagship (high-quality malting genotype) to identify QTLs associated with the stay-green trait under abiotic stress conditions. Ten QTLs were identified, six of which were associated with heat stress and four with drought (Gous et al. 2016). In addition, Ingvordsen et al. (2015a) performed a genome-wide association study on spring barley genotypes under heat stress and elevated CO_2 and identified a QTL associated with grain yield under heat stress on chromosome 2H. However, these loci should be further validated in field trials and the genes underlying these regions should be identified. Most strategies to avoid heat stress are related to early sowing and shorter season cultivars. This may be due to the complexity of heat stress and its strong interaction with the environment and to limited knowledge about heat-tolerant genotypes and the genetic regions underlying heat stress. The opportunity provided by the barley high-quality reference genome (Mascher et al. 2017) to identify new markers is expected to provide considerable advances in genetic mapping for heat stress. Studies with a design similar to the one adopted by Abou-Elwafa and Amein (2016) and Ingvordsen et al. (2015b) can be extremely valuable to future genetic mapping studies specifically targeting heat stress.

15.7 Nutrient Deficiency (Nitrogen and Phosphorus) Stress

Crop nutrition is a key determinant of yield potential. An adequate supply of essential macro- and micronutrients is required for crops to achieve sufficient vegetative growth to ensure the development of nutritious grain. Fertilizer is a major input cost for farmers but is nevertheless often applied in excess of a crop's requirements. For example, estimates are that cereal crops take up only 40–60% of the applied nitrogen (N) fertilizer and that unused N may be leached from the soil or lost to volatilization (Raun and Johnson 1999; Sylvester-Bradley and Kindred 2009). The impact of this "lost" N is of critical concern as a source for major economic loss as an environmental pollutant that leads to release of potent greenhouse gases and eutrophication of aquatic environments (Zhang et al. 2015).

Insufficient availability of N and phosphorus (P) has the greatest impact on yield potential of crop plants (Hawkesford et al. 2012). Such deficiency directly limits yield potential as affected plants have a lower capacity for radiation capture and hence grain production, regardless of the application of fertilizer during the reproductive growth stage. The nutrient required in greatest supply is N because it is an integral component of major plant macromolecules. N-limited crops may yellow, particularly older leaves, due to loss of chlorophyll. P-deficient crops may, on the contrary, have older dark green leaves due to the accumulation of sugars and other compounds, and their root growth can be limited, reducing uptake capacity for other nutrients and water.

Nutrient use efficiency (NUE) is most commonly measured by relating crop yield to the amount, both applied and residual in the soil, of a particular nutrient available to the crop (Good et al. 2004). NUE may be improved through changing agronomic practices, such as the timing of fertilizer application and crop rotations (Fageria and Baligar 2005). The limited capacity for nutrient uptake and utilization in plants presents a real opportunity to improve NUE. In high input, high rainfall areas, what determines NUE is very different than in low input, dry areas (Cossani et al. 2010). In a similar way, increasing P use efficiency in P-fixing soils is very different than in P-available soils. Within these scenarios, the weather at a site influences NUE from season to season. However, there are a few underlying mechanisms which, if improved, would work across environments and management scenarios. First, in some environments, uptake efficiency may be improved by altering the architectural traits of roots such that the crop may access fertilizer deep in the soil profile or at shallow depths (Garnett et al. 2009). The families of transporters mediating both low- and high-affinity uptake of N and P have been described (Smith et al. 2003; Wang et al. 2012b), with the barley NRT2 and Pht1 transporters among the first N and P transporters identified and characterized in cereals (Rae et al. 2003; Vidmar et al. 2000a, b). Improving the efficiency of these transporters through the selection of superior alleles has improved uptake of target nutrients in rice (Fan et al. 2016a; Hu et al. 2015). Another important aspect for P uptake is the plant's interaction with mycorrhizae. Selecting varieties of barley and mycorrhizae that maximize P uptake in an agricultural environment is a possible strategy to improve NUE (Grace et al. 2009).

Similarly, there is scope to improve the utilization efficiency of nutrients once they have been taken up from the soil. Much is known about the biochemical pathways in plants that are responsible for the conversion of inorganic nutrients into the organic building blocks for protein synthesis (Xu et al. 2012). The role of glutamine synthetase (GS) in N assimilation in barley has been well documented (Avila-Ospina et al. 2015; Goodall et al. 2013) and *GS* is regarded as a candidate gene in both reverse and forward genetic approaches to improve NUE (Hirel et al. 2007; Thomsen et al. 2014). However, the gene encoding alanine aminotransferase (AlaAT) in barley has exhibited the most promise to increase NUE via overexpression because it has improved NUE in canola and rice (Good et al. 2007; Shrawat et al. 2008). Importantly, uptake and utilization systems are often intricately intertwined such that the status of one system provides important feedback signaling, which alters the capacity of the other. In general, the remobilization of nutrients, particularly N, stored in vacuoles, proteins and other molecules in older leaves is another determinant of NUE for which underlying genetic loci have been identified in barley (Havé et al. 2016; Hollmann et al. 2014; Mickelson et al. 2003; Yang et al. 2004). However, for malting barley production, it is important the grain has 9–12% protein content, as excess protein interferes with the malting process (Bertholdsson 1999). Grain N has been the focus of most efforts to identify genetic regions regulating NUE in barley and a locus has been identified on chromosome 6 that regulates leaf senescence and N reallocation (Bezant et al.

1997; Cai et al. 2013; Heidlebaugh et al. 2008; Jukanti and Fischer 2008; Jukanti et al. 2008). The newly published high-quality barley genomic sequence (Mascher et al. 2017) will aid in efforts to identify the causal gene(s).

Genetic improvements in NUE have been made "inadvertently" during yield-based selection in barley breeding over the last 75 years, even though breeding trials were conducted with extensive application of N fertilizer (Bingham et al. 2012). However, these gains are not sufficient to alleviate concerns over rising fertilizer costs and increasing environmental regulations of fertilizer application. Variation in NUE has been identified in adapted barley varieties (Anbessa et al. 2009, 2010; Beatty et al. 2010); however, variation in the transcriptomic responses of contrasting Tibetan wild barley lines to low N or K barley lines suggests that this collection may be a source of greater variation in NUE and component traits (Quan et al. 2016; Wei et al. 2016; Zeng et al. 2014). To assess the potential of exotic germplasm to be a source of superior NUE more fully will require its incorporation into more advanced populations such as the nested association mapping (NAM) or multi-parent advanced generation intercross (MAGIC) populations (Nice et al. 2016; Sannemann et al. 2015) to reduce the "wildness" of the material, which can create artifacts in data due to issues such as phenology (Karley et al. 2011). A field-based analysis of a barley association mapping panel revealed how large the environmental influence is when measuring traits like P uptake and assimilation efficiency, requiring multiple site-years of data to produce meaningful genetic associations (George et al. 2011; George et al. 2014). The studies also showed the potential for identifying surrogate traits for NUE, such as rhizosheath production, as a way to screen more quickly for NUE potential.

There has been very little success in improving barley NUE through genetic modification approaches (McAllister et al. 2012). It is clear that improved genetic modification tools will be necessary and the identification of N- and P-responsive promoter elements will play an important role in fine-tuning transgene expression to change nutrient supply in the field (Schünmann et al. 2004a, b). The arrival of the barley high-quality reference genome (Mascher et al. 2017) will help identify further regulatory motifs and targets for application of genome editing technologies, such as CRISPR/CAS (Teotia et al. 2016).

15.8 Other Stresses

Other stresses, mainly waterlogging and manganese (Mn) and cadmium (Cd) toxicity, affect barley to a lesser extent.

Cultivated barley shows a genetic diversity in waterlogging tolerance, and this tolerance varies according to life stage, with barley being more susceptible to waterlogging at preemergence, seedling growth, and reproductive stages (Setter and Waters 2003). Breeding for waterlogging tolerance in barley is still in its early stages with few studies examining waterlogging tolerance in barley. Li et al. (2008) mapped waterlogging tolerance using two doubled-haploid populations and targeting traits such as leaf chlorosis, plant survival, and plant biomass reduction. Some QTLs (on chromosomes 1H, 3H, and 7H) were significant in both populations and some QTLs associated with more than one trait, such as the 4H QTL, which associated with leaf chlorosis and plant biomass. In addition to these mapping studies, Mendiondo et al. (2016) showed that HvPRT6, an N-end rule pathway N-recognin E3 ligase PROTEOLYSIS6, played a role in waterlogging tolerance in barley. In fact, the role of the N-end rule pathway in controlling plant response to hypoxia, which results from waterlogging, has already been established in arabidopsis. Transgenic barley plants with decreased *HvPRT6* expression exhibited increased waterlogging tolerance as manifested by biomass maintenance, increased yield, and decreased chlorophyll degradation under stress (Mendiondo et al. 2016).

Although barley is sensitive to Mn toxicity, a large genetic variation in Mn tolerance exists among cultivated genotypes (Hebbern et al. 2005). This variation could be harnessed to identify chromosomal regions not yet identified that are involved in Mn tolerance. Huang et al. (2015) proposed that breeding for Mn tolerance could improve waterlogging tolerance in barley, which could be due to the decrease in soil redox potential under waterlogging, which in turn increases the soil Mn^{2+} concentration and causes Mn toxicity effects in plants. On the other hand, Leplat et al. (2016) examined 248 barley varieties that were cultivated in six distinct environments prone to Mn deficiency, and identified several putative QTLs. Then, the authors were able to explore putative candidate genes in the flanking region of the SNP associations, using the barley genome sequence.

Cadmium is a heavy metal usually associated with polluted soils. It can affect human health, and barley grains have the lowest threshold value for Cd concentration for food safety among cereal crops (Wu et al. 2015). Barley genotypes have differences in tolerance to Cd toxicity (Chen et al. 2008; Persson et al. 2006). A genome-wide association study demonstrated a large genotypic variation in the Cd concentration in different organs and detected several QTLs associated with Cd accumulation, some of which were

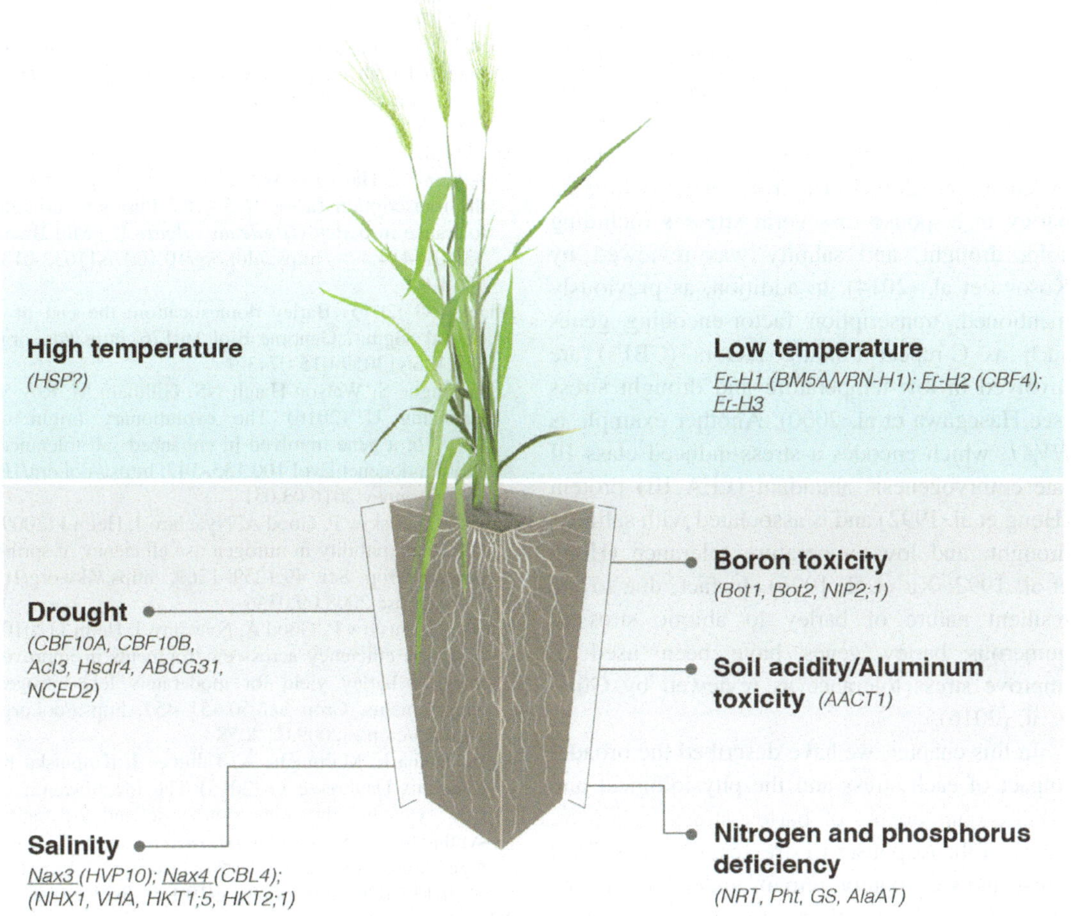

Fig. 15.1 The genetic components of barley tolerance to abiotic stresses. The major abiotic stresses reviewed in this chapter are shown in bold; examples of QTLs associated with each stress are italicized; and the candidate genes underlying the tolerance mechanisms are italicized and in parentheses

organ-specific (Wu et al. 2015). Furthermore, Cao et al. (2014) used microarray expression profiling on two genotypes contrasting in their Cd tolerance to identify novel genes that might play a role in various mechanisms of Cd tolerance.

15.9 Conclusions

The responses of plants to abiotic stresses are complex. While certain tolerance mechanisms are specific to a particular stress (e.g., Na^+ exclusion from the leaf blades under salinity), plants have to cope with the co-occurrence of multiple stresses under field conditions. Co-occurrence of stresses can trigger common responses, with some molecular pathways likely to play roles in more than one stress, e.g., pathways involving reactive oxygen species. In barley, three out of the 13 dehydrin genes were induced by low temperature and drought (Tommasini et al. 2008). The role of dehydrins in barley in response to several stresses including cold, drought, and salinity was reviewed by Kosova et al. (2014). In addition, as previously mentioned, transcription factor-encoding genes such as C-repeat binding factors (CBFs) are involved in low-temperature and drought stress (see Hasegawa et al. 2000). Another example is *HVA1*, which encodes a stress-induced class III late embryogenesis abundant (LEA III) protein (Hong et al. 1992) and is associated with salinity, drought, and low-temperature tolerance (Hong et al. 1992; Xu et al. 1996). In fact, due to the resilient nature of barley to abiotic stresses, numerous barley genes have been used to improve stress tolerance as reviewed by Gürel et al. (2016).

In this chapter, we have described the broader impact of each stress and the physiological and genetic components of barley stress responses. Many of the responses to stresses, such as to heat stress and Cd toxicity, remain understudied. The previously released barley genome sequence (Mayer et al. 2012) enabled the identification of candidate genes involved in stress tolerance. The completion of the barley high-quality reference genome (Mascher et al. 2017) will undoubtedly contribute to an improved and more comprehensive picture of barley genetics and genomics in response to abiotic stress (Fig. 15.1).

Acknowledgements We thank Ivan Gromicho, scientific illustrator from King Abdullah University of Science and Technology, for the scientific illustration in this book chapter. Financial support from King Abdullah University of Science and Technology (KAUST) is gratefully acknowledged.

References

Abiko M et al (2005) High-temperature induction of male sterility during barley (*Hordeum vulgare* L.) anther development is mediated by transcriptional inhibition. Sex Plant Reprod 18:91–100. https://doi.org/10.1007/s00497-005-0004-2

Abou-Elwafa SF, Amein KA (2016) Genetic diversity and potential high temperature tolerance in barley (*Hordeum vulgare*). WJAS 4:1–8. Doi:https://doi.org/10.12691/wjar-4-1-1

Al Abdallat AM, Ayad JY, Abu Elenein JM, Al Ajlouni Z, Harwood WA (2014) Overexpression of the transcription factor *HvSNAC1* improves drought tolerance in barley (*Hordeum vulgare* L.) Mol Breed 33:401–414. https://doi.org/10.1007/s11032-013-9958-1

Allaby RG (2015) Barley domestication: the end of a central dogma? Genome Biol 16:176. https://doi.org/10.1186/s13059-015-0743-9

Amarasinghe S, Watson-Haigh NS, Gilliham M, Roy S, Baumann U (2016) The evolutionary origin of *CIPK16*: a gene involved in enhanced salt tolerance. Mol Phylogenet Evol 100:135–147. https://doi.org/10.1016/j.ympev.2016.03.031

Anbessa Y, Juskiw P, Good A, Nyachiro J, Helm J (2009) Genetic variability in nitrogen use efficiency of spring barley. Crop Sci 49:1259–1269. https://doi.org/10.2135/cropsci2008.09.0566

Anbessa Y, Juskiw P, Good A, Nyachiro J, Helm J (2010) Selection efficiency across environments in improvement of barley yield for moderately low nitrogen environments. Crop Sci 50:451–457. https://doi.org/10.2135/cropsci2009.02.0058

Avila-Ospina L, Marmagne A, Talbotec J, Krupinska K, Masclaux-Daubresse C (2015) The identification of new cytosolic glutamine synthetase and asparagine synthetase genes in barley (*Hordeum vulgare* L.), and their expression during leaf senescence. J Exp Bot 66:2013–2026. https://doi.org/10.1093/jxb/erv003

Babu RC, Zhang JX, Blum A, Ho THD, Wu R, Nguyen HT (2004) *HVA1*, a LEA gene from barley confers dehydration tolerance in transgenic rice (*Oryza sativa* L.) via cell membrane protection. Plant Sci 166:855–862. https://doi.org/10.1016/j.plantsci.2003.11.023

Badr A et al (2000) On the origin and domestication history of barley (*Hordeum vulgare*). Mol Biol Evol 17:499–510. https://doi.org/10.1093/oxfordjournals.molbev.a026330

Baerenfaller K et al (2012) Systems-based analysis of Arabidopsis leaf growth reveals adaptation to water deficit. Mol Syst Biol 8:606. https://doi.org/10.1038/msb.2012.39

Bahieldin A et al (2005) Field evaluation of transgenic wheat plants stably expressing the *HVA1* gene for drought tolerance. Physiol Plant 123:421–427. https://doi.org/10.1111/j.1399-3054.2005.00470.x

Baldi P, Grossi M, Pecchioni N, Vale G, Cattivelli L (1999) High expression level of a gene coding for a chloroplastic amino acid selective channel protein is correlated to cold acclimation in cereals. Plant Mol Biol 41:233–243. https://doi.org/10.1023/A:1006375332677

Barati A, Moghaddam M, Mohammadi SA, Ghazvini HA, Sadeghzadeh B (2017) Identification of QTLs associated with agronomic and physiological traits under salinity stress in barley. J Agric Sci Technol-Iran 19:185–200

Barragan V et al (2012) Ion exchangers NHX1 and NHX2 mediate active potassium uptake into vacuoles to regulate cell turgor and stomatal function in *Arabidopsis*. Plant Cell 24:1127–1142. https://doi.org/10.1105/tpc.111.095273

Bassil E et al (2011) The *Arabidopsis* Na^+/H^+ antiporters NHX1 and NHX2 control vacuolar pH and K^+ homeostasis to regulate growth, flower development, and reproduction. Plant Cell 23:3482–3497. https://doi.org/10.1105/tpc.111.089581

Baum M, Grando S, Backes G, Jahoor A, Sabbagh A, Ceccarelli S (2003) QTLs for agronomic traits in the Mediterranean environment identified in recombinant inbred lines of the cross 'Arta' × *H. spontaneum* 41-1. Theor Appl Genet 107:1215–1225. https://doi.org/10.1007/s00122-003-1357-2

Bayat F et al (2010) Potato plants bearing a vacuolar Na^+/H^+ antiporter HvNHX2 from barley are characterized by improved salt tolerance. Russ J Plant Physiol 57:696–706. https://doi.org/10.1134/s1021443710050134

Bayat F, Shiran B, Belyaev DV (2011) Overexpression of *HvNHX2*, a vacuolar Na^+/H^+ antiporter gene from barley, improves salt tolerance in *Arabidopsis thaliana*. Aust J Crop Sci 5:428–432

Beatty PH, Anbessa Y, Juskiw P, Carroll RT, Wang J, Good AG (2010) Nitrogen use efficiencies of spring barley grown under varying nitrogen conditions in the field and growth chamber. Ann Bot 105:1171–1182. https://doi.org/10.1093/aob/mcq025

Bertholdsson NO (1999) Characterization of malting barley cultivars with more or less stable grain protein content under varying environmental conditions. Eur J Agron 10:1–8. https://doi.org/10.1016/S1161-0301(98)00043-4

Bezant H, Laurie DA, Pratchett N, Chojecki J, Kearsey MJ (1997) Mapping of QTL controlling NIR predicted hot water extract and grain nitrogen content in a spring barley cross using marker-regression. Plant Breed 116:141–145. https://doi.org/10.1111/j.1439-0523.1997.tb02168.x

Bingham IJ, Karley AJ, White PJ, Thomas WTB, Russell JR (2012) Analysis of improvements in nitrogen use efficiency associated with 75 years of spring barley breeding. Eur J Agron 42:49–58. https://doi.org/10.1016/j.eja.2011.10.003

Byrt CS et al (2007) HKT1;5-like cation transporters linked to Na^+ exclusion loci in wheat, *Nax2* and *Kna1*. Plant Physiol 143:1918–1928. https://doi.org/10.1104/pp.106.093476

Cai S, Yu G, Chen X, Huang Y, Jiang X, Zhang G, Jin X (2013) Grain protein content variation and its association analysis in barley. BMC Plant Biol 13:1–11. https://doi.org/10.1186/1471-2229-13-35

Cao FB, Chen F, Sun HY, Zhang GP, Chen ZH, Wu FB (2014) Genome-wide transcriptome and functional analysis of two contrasting genotypes reveals key genes for cadmium tolerance in barley. BMC Genomics 15. https://doi.org/10.1186/1471-2164-15-611

Cartwright B, Zarcinas BA, Spouncer LR (1986) Boron toxicity in South Australian barley crops. Aust J Agric Res 37:351–359. https://doi.org/10.1071/Ar9860351

Chen GX et al (2009) Mapping of the *eibi1* gene responsible for the drought hypersensitive cuticle in wild barley (*Hordeum spontaneum*). Breed Sci 59:21–26. https://doi.org/10.1270/jsbbs.59.21

Chen G, Sagi M, Weining S, Krugman T, Fahima T, Korol AB, Nevo E (2004) Wild barley *eibi1* mutation identifies a gene essential for leaf water conservation. Planta 219:684–693. https://doi.org/10.1007/s00425-004-1277-7

Chen F, Wang F, Zhang G, Wu F (2008) Identification of barley varieties tolerant to cadmium toxicity. Biol Trace Elem Res 121:171–179. https://doi.org/10.1007/s12011-007-8042-2

Chen G et al (2011) An ATP-binding cassette subfamily G full transporter is essential for the retention of leaf water in both wild barley and rice. Proc Natl Acad Sci USA 108:12354–12359. https://doi.org/10.1073/pnas.1108444108

Chloupek O, Dostal V, Streda T, Psota V, Dvorackova O (2010) Drought tolerance of barley varieties in relation to their root system size. Plant Breed 129:630–636. https://doi.org/10.1111/j.1439-0523.2010.01801.x

Chmielewska K et al (2016) Analysis of drought-induced proteomic and metabolomic changes in barley (*Hordeum vulgare* L.) leaves and roots unravels some aspects of biochemical mechanisms involved in drought tolerance. Front Plant Sci 7. https://doi.org/10.3389/fpls.2016.01108

Choi WG, Toyota M, Kim SH, Hilleary R, Gilroy S (2014) Salt stress-induced Ca^{2+} waves are associated with rapid, long-distance root-to-shoot signaling in plants. Proc Natl Acad Sci USA 111:6497–6502. https://doi.org/10.1073/pnas.1319955111

Cossani CM, Slafer GA, Savin R (2010) Co-limitation of nitrogen and water, and yield and resource-use efficiencies of wheat and barley. Crop Pasture Sci 61:844–851. https://doi.org/10.1071/cp10018

Crosatti C, de Laureto PP, Bassi R, Cattivelli L (1999) The interaction between cold and light controls the expression of the cold-regulated barley gene *cor14b* and the accumulation of the corresponding protein. Plant Physiol 119:671–680. https://doi.org/10.1104/pp.119.2.671

Daszkowska-Golec A, Szarejko I (2013) Open or close the gate—stomata action under the control of phytohormones in drought stress conditions. Front Plant Sci 4:138. https://doi.org/10.3389/fpls.2013.00138

Dawson IK, Russell J, Powell W, Steffenson B, Thomas WTB, Waugh R (2015) Barley: a translational model for adaptation to climate change. New Phytol 206:913–931. https://doi.org/10.1111/nph.13266

Delhaize E, Ryan PR, Hebb DM, Yamamoto Y, Sasaki T, Matsumoto H (2004) Engineering high-level aluminum tolerance in barley with the *ALMT1* gene. Proc Natl Acad Sci USA 101:15249–15254. https://doi.org/10.1073/pnas.0406258101

Diab AA, Teulat-Merah B, This D, Ozturk NZ, Benscher D, Sorrells ME (2004) Identification of drought-inducible genes and differentially expressed sequence tags in barley. Theor Appl Genet 109:1417–1425. https://doi.org/10.1007/s00122-004-1755-0

Dunn MA, Hughes MA, Zhang L, Pearce RS, Quigley AS, Jack PL (1991) Nucleotide sequence and molecular analysis of the low temperature induced cereal gene, *BLT4*. Mol Gen Genet 229:389–394

Ellis RP et al (2002) Phenotype/genotype associations for yield and salt tolerance in a barley mapping population segregating for two dwarfing genes. J Exp Bot 53:1163–1176. https://doi.org/10.1093/jexbot/53.371.1163

Emebiri L, Michael P, Moody D (2009) Enhanced tolerance to boron toxicity in two-rowed barley by marker-assisted introgression of favourable alleles derived from Sahara 3771. Plant Soil 314:77–85. https://doi.org/10.1007/s11104-008-9707-0

Fageria NK, Baligar VC (2005) Enhancing nitrogen use efficiency in crop plants. In: Advances in agronomy, vol 88. Elsevier, pp 97–185. https://doi.org/10.1016/s0065-2113(05)88004-6

Fan Y, Shabala S, Ma YL, Xu RG, Zhou MX (2015) Using QTL mapping to investigate the relationships between abiotic stress tolerance (drought and salinity) and agronomic and physiological traits. BMC Genomics 16. https://doi.org/10.1186/s12864-015-1243-8

Fan X et al (2016a) Overexpression of a pH-sensitive nitrate transporter in rice increases crop yields. Proc Natl Acad Sci USA 113:7118–7123. https://doi.org/10.1073/pnas.1525184113

Fan Y, Zhou G, Shabala S, Chen ZH, Cai S, Li C, Zhou M (2016b) Genome-wide association study reveals a new QTL for salinity tolerance in barley (*Hordeum vulgare* L.). Front Plant Sci 7:946. https://doi.org/10.3389/fpls.2016.00946

FAO (2015) Global soil status, processes and trends. In: Status of the world's soil resources (SWSR)—main report

Farooq M, Wahid A, Lee DJ, Ito O, Siddique KHM (2009) Advances in drought resistance of rice. Crit Rev Plant Sci 28:199–217. https://doi.org/10.1080/07352680902952173

Fedorowicz-Strońska O, Koczyk G, Kaczmarek M, Krajewski P, Sadowski J (2017) Genome-wide identification, characterisation and expression profiles of calcium-dependent protein kinase genes in barley (*Hordeum vulgare* L.). J Appl Genet 58:11–22. https://doi.org/10.1007/s13353-016-0357-2

Fisk SP et al (2013) *FR-H3*: a new QTL to assist in the development of fall-sown barley with superior low temperature tolerance. Theor Appl Genet 126:335–347. https://doi.org/10.1007/s00122-012-1982-8

Fowler DB, Chauvin LP, Limin AE, Sarhan F (1996a) The regulatory role of vernalization in the expression of low-temperature-induced genes in wheat and rye. Theor Appl Genet 93:554–559. https://doi.org/10.1007/Bf00417947

Fowler DB, Limin AE, Wang SY, Ward RW (1996b) Relationship between low-temperature tolerance and vernalization response in wheat and rye. Can J Plant Sci 76:37–42. https://doi.org/10.4141/cjps96-007

Fowler DB, Breton G, Limin AE, Mahfoozi S, Sarhan F (2001) Photoperiod and temperature interactions regulate low-temperature-induced gene expression in barley. Plant Physiol 127:1676–1681. https://doi.org/10.1104/pp.010483

Foy CD, Chaney RL, White MC (1978) The physiology of metal toxicity in plants. Annu Rev Plant Physiol 29:511–566. https://doi.org/10.1146/annurev.pp.29.060178.002455

Francia E et al (2004) Two loci on chromosome 5H determine low-temperature tolerance in a 'Nure' (winter) × 'Tremois' (spring) barley map. Theor Appl Genet 108:670–680. https://doi.org/10.1007/s00122-003-1468-9

Francia E et al (2007) Fine mapping of a *HvCBF* gene cluster at the frost resistance locus *Fr-H2* in barley. Theor Appl Genet 115:1083–1091. https://doi.org/10.1007/s00122-007-0634-x

Francia E et al (2016) Copy number variation at the *HvCBF4-HvCBF2* genomic segment is a major component of frost resistance in barley. Plant Mol Biol 92:161–175. https://doi.org/10.1007/s11103-016-0505-4

Fu DL et al (2005) Large deletions within the first intron in *VRN-1* are associated with spring growth habit in barley and wheat. Mol Genet Genomics 273:54–65. https://doi.org/10.1007/s00438-004-1095-4

Fujii M et al (2012) Acquisition of aluminium tolerance by modification of a single gene in barley. Nat Commun 3:713. https://doi.org/10.1038/ncomms1726

Fukuda A, Chiba K, Maeda M, Nakamura A, Maeshima M, Tanaka Y (2004) Effect of salt and osmotic stresses on the expression of genes for the vacuolar H^+-pyrophosphatase, H^+-ATPase subunit A, and Na^+/H^+ antiporter from barley. J Exp Bot 55:585–594. https://doi.org/10.1093/jxb/erh070

Furukawa J et al (2007) An aluminum-activated citrate transporter in barley. Plant Cell Physiol 48:1081–1091. https://doi.org/10.1093/pcp/pcm091

Garcia GA, Dreccer MF, Miralles DJ, Serrago RA (2015) High night temperatures during grain number determination reduce wheat and barley grain yield: a field study. Glob Change Biol 21:4153–4164. https://doi.org/10.1111/gcb.13009

Garnett T, Conn V, Kaiser BN (2009) Root based approaches to improving nitrogen use efficiency in plants. Plant, Cell Environ 32:1272–1283. https://doi.org/10.1111/j.1365-3040.2009.02011.x

Garthwaite AJ, von Bothmer R, Colmer TD (2005) Salt tolerance in wild *Hordeum* species is associated with restricted entry of Na^+ and Cl^- into the shoots. J Exp Bot 56:2365–2378. https://doi.org/10.1093/jxb/eri229

George TS, Brown LK, Newton AC, Hallett PD, Sun BH, Thomas WTB, White PJ (2011) Impact of soil tillage on the robustness of the genetic component of variation in phosphorus (P) use efficiency in barley (*Hordeum vulgare* L.). Plant Soil 339:113–123. https://doi.org/10.1007/s11104-009-0209-5

George TS et al (2014) Understanding the genetic control and physiological traits associated with rhizosheath production by barley (*Hordeum vulgare*). New Phytol 203:195–205. https://doi.org/10.1111/nph.12786

Gilroy S, Suzuki N, Miller G, Choi WG, Toyota M, Devireddy AR, Mittler R (2014) A tidal wave of signals: calcium and ROS at the forefront of rapid systemic signaling. Trends Plant Sci 19:623–630. https://doi.org/10.1016/j.tplants.2014.06.013

Goldberg S (1997) Reactions of boron with soils. Plant Soil 193:35–48. https://doi.org/10.1023/A:1004203723343

Good AG, Shrawat AK, Muench DG (2004) Can less yield more? Is reducing nutrient input into the environment compatible with maintaining crop production? Trends Plant Sci 9:597–605. https://doi.org/10.1016/j.tplants.2004.10.008

Good AG et al (2007) Engineering nitrogen use efficiency with alanine aminotransferase. Can J Bot 85:252–262. https://doi.org/10.1139/B07-019

Goodall AJ, Kumar P, Tobin AK (2013) Identification and expression analyses of cytosolic glutamine synthetase genes in barley (*Hordeum vulgare* L.). Plant Cell Physiol 54:492–505. https://doi.org/10.1093/pcp/pct006

Gous PW, Hickey L, Christopher JT, Franckowiak J, Fox GP (2016) Discovery of QTL for stay-green and heat-stress in barley (*Hordeum vulgare*) grown under simulated abiotic stress conditions. Euphytica 207:305–317. https://doi.org/10.1007/s10681-015-1542-9

Grace EJ, Cotsaftis O, Tester M, Smith FA, Smith SE (2009) Arbuscular mycorrhizal inhibition of growth in barley cannot be attributed to extent of colonization, fungal phosphorus uptake or effects on expression of plant phosphate transporter genes. New Phytol 181:938–949. https://doi.org/10.1111/j.1469-8137.2008.02720.x

Gruber BD et al (2010) *HvALMT1* from barley is involved in the transport of organic anions. J Exp Bot 61:1455–1467. https://doi.org/10.1093/jxb/erq023

Guo PG, Baum M, Varshney RK, Graner A, Grando S, Ceccarelli S (2008) QTLs for chlorophyll and chlorophyll fluorescence parameters in barley under post-flowering drought. Euphytica 163:203–214. https://doi.org/10.1007/s10681-007-9629-6

Guo W, Chen T, Hussain N, Zhang G, Jiang L (2016) Characterization of salinity tolerance of transgenic rice lines harboring *HsCBL8* of wild barley (*Hordeum spontaneum*) line from Qinghai-Tibet Plateau. Front Plant Sci 7:1678. https://doi.org/10.3389/fpls.2016.01678

Gürel F, Öztürk ZN, Uçarlı C, Rosellini D (2016) Barley genes as tools to confer abiotic stress tolerance in crops. Front Plant Sci 7. https://doi.org/10.3389/fpls.2016.01137

Hansen L, von Wettstein-Knowles P (1991) The barley genes *Acl1* and *Acl3* encoding acyl carrier proteins I and III are located on different chromosomes. Mol Gen Genet 229:467–478

Haro R, Banuelos MA, Senn ME, Barrero-Gil J, Rodriguez-Navarro A (2005) HKT1 mediates sodium uniport in roots. Pitfalls in the expression of *HKT1* in yeast. Plant Physiol 139:1495–1506. https://doi.org/10.1104/pp.105.067553

Hasegawa PM, Bressan RA, Zhu JK, Bohnert HJ (2000) Plant cellular and molecular responses to high salinity. Annu Rev Plant Physiol Plant Mol Biol 51:463–499. https://doi.org/10.1146/annurev.arplant.51.1.463

Hauser F, Horie T (2010) A conserved primary salt tolerance mechanism mediated by HKT transporters: a mechanism for sodium exclusion and maintenance of high K^+/Na^+ ratio in leaves during salinity stress. Plant, Cell Environ 33:552–565. https://doi.org/10.1111/j.1365-3040.2009.02056.x

Havé M, Marmagne A, Chardon F, Masclaux-Daubresse C (2016) Nitrogen remobilisation during leaf senescence: lessons from Arabidopsis to crops. J Exp Bot. https://doi.org/10.1093/jxb/erw365

Hawkesford M, Horst W, Kichey T, Lambers H, Schjoerring J, Møller IS, White P (2012) Functions of macronutrients. In: Marschner's mineral nutrition of higher plants, 3 edn. Elsevier, San Diego, pp 135–189. https://doi.org/10.1016/b978-0-12-384905-2.00006-6

Hayes JE, Reid RJ (2004) Boron tolerance in barley is mediated by efflux of boron from the roots. Plant Physiol 136:3376–3382. https://doi.org/10.1104/pp.103.037028

Hayes PM, Blake T, Chen TH, Tragoonrung S, Chen F, Pan A, Liu B (1993) Quantitative trait loci on barley (*Hordeum vulgare* L.) chromosome 7 associated with components of winterhardiness. Genome 36:66–71. https://doi.org/10.1139/g93-009

Hayes JE, Pallotta M, Garcia M, Oz MT, Rongala J, Sutton T (2015) Diversity in boron toxicity tolerance of Australian barley (*Hordeum vulgare* L.) genotypes. BMC Plant Biol 15. https://doi.org/10.1186/s12870-015-0607-1

Hebbern CA, Pedas P, Schjoerring JK, Knudsen L, Husted S (2005) Genotypic differences in manganese efficiency: field experiments with winter barley (*Hordeum vulgare* L.). Plant Soil 272:233–244. https://doi.org/10.1007/s11104-004-5048-9

Heidlebaugh NM, Trethewey BR, Jukanti AK, Parrott DL, Martin JM, Fischer AM (2008) Effects of a barley (*Hordeum vulgare*) chromosome 6 grain protein content locus on whole-plant nitrogen reallocation under two different fertilisation regimes. Funct Plant Biol 35:619–632. https://doi.org/10.1071/FP07247

Hen-Avivi S et al (2016) A metabolic gene cluster in the wheat *W1* and the barley *Cer-cqu* loci determines β-diketone biosynthesis and glaucousness. Plant Cell 28:1440–1460. https://doi.org/10.1105/tpc.16.00197

Hirel B, Le Gouis J, Ney B, Gallais A (2007) The challenge of improving nitrogen use efficiency in crop plants: towards a more central role for genetic variability and quantitative genetics within integrated approaches. J Exp Bot 58:2369–2387. https://doi.org/10.1093/jxb/erm097

Högy P, Poll C, Marhan S, Kandeler E, Fangmeier A (2013) Impacts of temperature increase and change in precipitation pattern on crop yield and yield quality of barley. Food Chem 136:1470–1477. https://doi.org/10.1016/j.foodchem.2012.09.056

Hollmann J, Gregersen PL, Krupinska K (2014) Identification of predominant genes involved in regulation and execution of senescence-associated nitrogen remobilization in flag leaves of field grown barley. J Exp Bot. https://doi.org/10.1093/jxb/eru094

Holloway RE, Alston AM (1992) The effects of salt and boron on growth of wheat. Aust J Agric Res 43:987–1001. https://doi.org/10.1071/Ar9920987

Hong B, Barg R, Ho TH (1992) Developmental and organ-specific expression of an ABA- and stress-induced protein in barley. Plant Mol Biol 18:663–674. https://doi.org/10.1007/BF00020009

Honsdorf N, March TJ, Berger B, Tester M, Pillen K (2014) High-throughput phenotyping to detect drought tolerance QTL in wild barley introgression lines. PLOS ONE 9:e97047. https://doi.org/10.1371/journal.pone.0097047

Hu B et al (2015) Variation in *NRT1.1B* contributes to nitrate-use divergence between rice subspecies. Nat Genet 47:834–838. https://doi.org/10.1038/ng.3337

Huang S, Spielmeyer W, Lagudah ES, Munns R (2008) Comparative mapping of *HKT* genes in wheat, barley, and rice, key determinants of Na^+ transport, and salt tolerance. J Exp Bot 59:927–937. https://doi.org/10.1093/jxb/ern033

Huang X, Shabala S, Shabala L, Rengel Z, Wu X, Zhang G, Zhou M (2015) Linking waterlogging tolerance with Mn^{2+} toxicity: a case study for barley. Plant Biol 17:26–33. https://doi.org/10.1111/plb.12188

Ingvordsen CH, Backes G, Lyngkjaer MF, Peltonen-Sainio P, Jahoor A, Mikkelsen TN, Jorgensen RB (2015a) Genome-wide association study of production and stability traits in barley cultivated under future climate scenarios. Mol Breed 35. https://doi.org/10.1007/s11032-015-0283-8

Ingvordsen CH et al (2015b) Significant decrease in yield under future climate conditions: stability and production of 138 spring barley accessions. Eur J Agron 63:105–113. https://doi.org/10.1016/j.eja.2014.12.003

IPCC (2014) Climate change 2014: synthesis report. IPCC, Geneva, Switzerland

Islam S, Malik AI, Islam AK, Colmer TD (2007) Salt tolerance in a *Hordeum marinum-Triticum aestivum* amphiploid, and its parents. J Exp Bot 58:1219–1229. https://doi.org/10.1093/jxb/erl293

James RA, Munns R, Von Caemmerer S, Trejo C, Miller C, Condon T (2006) Photosynthetic capacity is related to the cellular and subcellular partitioning of Na^+, K^+ and Cl^- in salt-affected barley and durum wheat. Plant, Cell Environ 29:2185–2197. https://doi.org/10.1111/j.1365-3040.2006.01592.x

Jefferies SP et al (1999) Mapping of chromosome regions conferring boron toxicity tolerance in barley (*Hordeum vulgare* L.). Theor Appl Genet 98:1293–1303. https://doi.org/10.1007/s001220051195

Jukanti AK, Fischer AM (2008) A high-grain protein content locus on barley (*Hordeum vulgare*) chromosome 6 is associated with increased flag leaf proteolysis and nitrogen remobilization. Physiol Plant 132:426–439. https://doi.org/10.1111/j.1399-3054.2007.01044.x

Jukanti AK, Heidlebaugh NM, Parrott DL, Fischer IA, McInnerney K, Fischer AM (2008) Comparative transcriptome profiling of near-isogenic barley (*Hordeum vulgare*) lines differing in the allelic state of a major grain protein content locus identifies genes with possible roles in leaf senescence and nitrogen reallocation. New Phytol 177:333–349. https://doi.org/10.1111/j.1469-8137.2007.02270.x

Karley AJ, Valentine TA, Squire GR (2011) Dwarf alleles differentially affect barley root traits influencing nitrogen acquisition under low nutrient supply. J Exp Bot 62:3917–3927. https://doi.org/10.1093/jxb/err089

Knox AK, Dhillon T, Cheng HM, Tondelli A, Pecchioni N, Stockinger EJ (2010) *CBF* gene copy number variation at *Frost Resistance-2* is associated with levels of freezing tolerance in temperate-climate cereals. Theor Appl Genet 121:21–35. https://doi.org/10.1007/s00122-010-1288-7

Kochian LV (1995) Cellular mechanisms of aluminum toxicity and resistance in plants. Annu Rev Plant Physiol Plant Mol Biol 46:237–260. https://doi.org/10.1146/annurev.arplant.46.1.237

Kosova K, Vitamvas P, Prasil IT (2011) Expression of dehydrins in wheat and barley under different temperatures. Plant Sci 180:46–52. https://doi.org/10.1016/j.plantsci.2010.07.003

Kosova K, Vitamvas P, Prasil IT (2014) Wheat and barley dehydrins under cold, drought, and salinity—what can LEA-II proteins tell us about plant stress response? Front Plant Sci 5:343. https://doi.org/10.3389/fpls.2014.00343

Kruszka K et al (2014) Transcriptionally and post-transcriptionally regulated microRNAs in heat stress response in barley. J Exp Bot 65:6123–6135. https://doi.org/10.1093/jxb/eru353

Lakew B, Henry RJ, Ceccarelli S, Grando S, Eglinton J, Baum M (2013) Genetic analysis and phenotypic associations for drought tolerance in *Hordeum spontaneum* introgression lines using SSR and SNP markers. Euphytica 189:9–29. https://doi.org/10.1007/s10681-012-0674-4

Lemichez E, Wu Y, Sanchez JP, Mettouchi A, Mathur J, Chua NH (2001) Inactivation of AtRac1 by abscisic acid is essential for stomatal closure. Gene Dev 15:1808–1816. https://doi.org/10.1101/gad.900401

Leplat F, Pedas PR, Rasmussen SK, Husted S (2016) Identification of manganese efficiency candidate genes in winter barley (*Hordeum vulgare*) using genome wide association mapping. BMC Genom 17:775. https://doi.org/10.1186/s12864-016-3129-9

Li HB, Vaillancourt R, Mendham N, Zhou MX (2008) Comparative mapping of quantitative trait loci associated with waterlogging tolerance in barley (*Hordeum vulgare* L.). BMC Genomics 9. https://doi.org/10.1186/1471-2164-9-401

Ligaba A, Katsuhara M (2010) Insights into the salt tolerance mechanism in barley (*Hordeum vulgare*) from comparisons of cultivars that differ in salt sensitivity. J Plant Res 123:105–118. https://doi.org/10.1007/s10265-009-0272-2

Liu X et al (2017) QTLs for stomatal and photosynthetic traits related to salinity tolerance in barley. BMC Genomics 18:9. https://doi.org/10.1186/s12864-016-3380-0

Lonergan PF, Pallotta MA, Lorimer M, Paull JG, Barker SJ, Graham RD (2009) Multiple genetic loci for zinc uptake and distribution in barley (*Hordeum vulgare*). New Phytol 184:168–179. https://doi.org/10.1111/j.1469-8137.2009.02956.x

Long NV, Dolstra O, Malosetti M, Kilian B, Graner A, Visser RGF, van der Linden CG (2013) Association mapping of salt tolerance in barley (*Hordeum vulgare* L.). Theor Appl Genet 126:2335–2351. https://doi.org/10.1007/s00122-013-2139-0

Lundqvist U, Lundqvist A (1988) Mutagen specificity in barley for 1580 *eceriferum* mutants localized to 79 loci. Hereditas 108:1–12. https://doi.org/10.1111/j.1601-5223.1988.tb00676.x

Ma JF, Ryan PR, Delhaize E (2001) Aluminium tolerance in plants and the complexing role of organic acids. Trends Plant Sci 6:273–278. https://doi.org/10.1016/S1360-1385(01)01961-6

Ma JF, Nagao S, Sato K, Ito H, Furukawa J, Takeda K (2004) Molecular mapping of a gene responsible for Al-activated secretion of citrate in barley. J Exp Bot 55:1335–1341. https://doi.org/10.1093/jxb/erh152

Ma Y, Shabala S, Li C, Liu C, Zhang W, Zhou M (2015) Quantitative trait loci for salinity tolerance identified under drained and waterlogged conditions and their association with flowering time in barley (*Hordeum vulgare*. L). PLoS ONE 10:e0134822. https://doi.org/10.1371/journal.pone.0134822

Ma Y et al (2016) A new allele for aluminium tolerance gene in barley (*Hordeum vulgare* L.). BMC Genom 17:186. https://doi.org/10.1186/s12864-016-2551-3

MacLeod L, Duffus C (1988) Reduced starch content and sucrose synthase activity in developing endosperm of barley plants grown at elevated temperatures. Funct Plant Biol 15:367–375

Mahfoozi S, Limin AE, Fowler DB (2001a) Developmental regulation of low-temperature tolerance in winter wheat. Ann Bot 87:751–757. https://doi.org/10.1006/anbo.2001.1403

Mahfoozi S, Limin AE, Fowler DB (2001b) Influence of vernalization and photoperiod responses on cold hardiness in winter cereals. Crop Sci 41:1006–1011. https://doi.org/10.2135/cropsci2001.4141006x

Mangelsen E, Kilian J, Harter K, Jansson C, Wanke D, Sundberg E (2011) Transcriptome analysis of high-temperature stress in developing barley caryopses: early stress responses and effects on storage compound biosynthesis. Mol Plant 4:97–115. https://doi.org/10.1093/mp/ssq058

Mano Y, Takeda K (1997) Mapping quantitative trait loci for salt tolerance at germination and the seedling stage in barley (*Hordeum vulgare* L). Euphytica 94:263–272. https://doi.org/10.1023/A:1002968207362

Maršálová L, Vítámvás P, Hynek R, Prášil IT, Kosová K (2016) Proteomic response of *Hordeum vulgare* cv. Tadmor and Hordeum marinum to salinity stress: similarities and differences between a glycophyte and a halophyte. Front Plant Sci 7:1154. https://doi.org/10.3389/fpls.2016.01154

Mascher M et al (2017) A chromosome conformation capture ordered sequence of the barley genome. Nature 544:427–433. https://doi.org/10.1038/nature22043

Maser P et al (2002) Glycine residues in potassium channel-like selectivity filters determine potassium selectivity in four-loop-per-subunit HKT transporters from plants. Proc Natl Acad Sci USA 99:6428–6433. https://doi.org/10.1073/pnas.082123799

Mastrangelo AM, Baldi P, Mare C, Terzi V, Galiba G, Cattivelli L, Di Fonzo N (2000) The cold dependent accumulation of COR TMC-AP3 in cereals with contrasting, frost tolerance is regulated by different mRNA expression and protein turnover. Plant Sci 156:47–54. https://doi.org/10.1016/S0168-9452(00)00228-4

Matin MA, Brown JH, Ferguson H (1989) Leaf water potential, relative water content, and diffusive resistance as screening techniques for drought resistance in barley. Agron J 81:100–105. https://doi.org/10.2134/agronj1989.00021962008100010018x

Mayer KFX et al (2012) A physical, genetic and functional sequence assembly of the barley genome. Nature 491:711–716. https://doi.org/10.1038/nature11543

McAllister CH, Beatty PH, Good AG (2012) Engineering nitrogen use efficient crop plants: the current status.

Plant Biotechnol J 10:1011–1025. https://doi.org/10.1111/j.1467-7652.2012.00700.x
McDonald GK (2006) Effects of soil properties on variation in growth, grain yield and nutrient concentration of wheat and barley. Aust J Exp Agr 46:93–105. https://doi.org/10.1071/ea04015
McDonald GK, Eglinton JK, Barr AR (2010) Assessment of the agronomic value of QTL on chromosomes 2H and 4H linked to tolerance to boron toxicity in barley (*Hordeum vulgare* L.). Plant Soil 326:275–290. https://doi.org/10.1007/s11104-009-0006-1
Mendiondo GM et al (2016) Enhanced waterlogging tolerance in barley by manipulation of expression of the N-end rule pathway E3 ligase PROTEOLYSIS6. Plant Biotechnol J 14:40–50. https://doi.org/10.1111/pbi.12334
Mian A, Oomen RJ, Isayenkov S, Sentenac H, Maathuis FJ, Véry AA (2011) Over-expression of an Na^+-and K^+-permeable HKT transporter in barley improves salt tolerance. Plant J 68:468–479. https://doi.org/10.1111/j.1365-313X.2011.04701.x
Mickelson S, See D, Meyer FD, Garner JP, Foster CR, Blake TK, Fischer AM (2003) Mapping of QTL associated with nitrogen storage and remobilization in barley (*Hordeum vulgare* L.) leaves. J Exp Bot 54:801–812. https://doi.org/10.1093/jxb/erg084
Mikolajczak K et al (2017) Quantitative trait loci for plant height in Maresi × CamB barley population and their associations with yield-related traits under different water regimes. J Appl Genet 58:23–35. https://doi.org/10.1007/s13353-016-0358-1
Minella E, Sorrells ME (1992) Aluminum tolerance in barley: Genetic relationships among genotypes of diverse origin. Crop Sci 32:593–598
Minella E, Sorrells ME (1997) Inheritance and chromosome location of *Alp*, a gene controlling aluminum tolerance in 'Dayton' barley. Plant Breed 116:465–469. https://doi.org/10.1111/j.1439-0523.1997.tb01032.x
Modhej A, Farhoudi R, Afrous A (2015) Effect of post-anthesis heat stress on grain yield of barley, durum and bread wheat genotypes. JSRAD 2:127–131
Morran S et al (2011) Improvement of stress tolerance of wheat and barley by modulation of expression of DREB/CBF factors. Plant Biotechnol J 9:230–249. https://doi.org/10.1111/j.1467-7652.2010.00547.x
Munns R (2002) Comparative physiology of salt and water stress. Plant, Cell Environ 25:239–250. https://doi.org/10.1046/j.0016-8025.2001.00808.x
Munns R, Tester M (2008) Mechanisms of salinity tolerance. Annu Rev Plant Biol 59:651–681. https://doi.org/10.1146/annurev.arplant.59.032607.092911
Munns R et al (2012) Wheat grain yield on saline soils is improved by an ancestral Na^+ transporter gene. Nat Biotechnol 30:360–364. https://doi.org/10.1038/nbt.2120
Newton AC et al (2011) Crops that feed the world 4. Barley: a resilient crop? Strengths and weaknesses in the context of food security. Food Secur 3:141–178. https://doi.org/10.1007/s12571-011-0126-3
Nice LM et al (2016) Development and genetic characterization of an advanced backcross-nested association mapping (AB-NAM) population of wild × cultivated barley. Genetics 203:1453–1467. https://doi.org/10.1534/genetics.116.190736
Oh SJ, Kwon CW, Choi DW, Song SI, Kim JK (2007) Expression of barley *HvCBF4* enhances tolerance to abiotic stress in transgenic rice. Plant Biotechnol J 5:646–656. https://doi.org/10.1111/j.1467-7652.2007.00272.x
Oraby HF, Ransom CB, Kravchenko AN, Sticklen MB (2005) Barley *HVA1* gene confers salt tolerance in R3 transgenic oat. Crop Sci 45:2218–2227. https://doi.org/10.2135/cropsci2004-0605
Oshino T, Abiko M, Saito R, Ichiishi E, Endo M, Kawagishi-Kobayashi M, Higashitani A (2007) Premature progression of anther early developmental programs accompanied by comprehensive alterations in transcription during high-temperature injury in barley plants. Mol Genet Genomics 278:31–42. https://doi.org/10.1007/s00438-007-0229-x
Pasquariello M et al (2014) The barley frost resistance-H2 locus. Funct Integr Genomics 14:85–100. https://doi.org/10.1007/s10142-014-0360-9
Pecchioni N et al (2014) Genomics of low-temperature tolerance for an increased sustainability of wheat and barley production. In: Genomics of plant genetic resources. Springer, Netherlands, pp 149–183. https://doi.org/10.1007/978-94-007-7575-6_6
Peighambari SA, Samadi BY, Nabipour A, Charmet G, Sarrafi A (2005) QTL analysis for agronomic traits in a barley doubled haploids population grown in Iran. Plant Sci 169:1008–1013. https://doi.org/10.1016/j.plantsci.2005.05.018
Persson DP et al (2006) Multi-elemental speciation analysis of barley genotypes differing in tolerance to cadmium toxicity using SEC-ICP-MS and ESI-TOF-MS. J Anal At Spectrom 21:996–1005. https://doi.org/10.1039/b608701a
Phillips D et al (2015) The effect of temperature on the male and female recombination landscape of barley. New Phytol 208:421–429. https://doi.org/10.1111/nph.13548
Platten JD et al (2006) Nomenclature for HKT transporters, key determinants of plant salinity tolerance. Trends Plant Sci 11:372–374. https://doi.org/10.1016/j.tplants.2006.06.001
Poets AM, Fang Z, Clegg MT, Morrell PL (2015) Barley landraces are characterized by geographically heterogeneous genomic origins. Genome Biol 16:173. https://doi.org/10.1186/s13059-015-0712-3
Qadir M et al (2014) Economics of salt-induced land degradation and restoration. Nat Resour Forum 38:282–295. https://doi.org/10.1111/1477-8947.12054
Quan X, Zeng J, Ye L, Chen G, Han Z, Shah JM, Zhang G (2016) Transcriptome profiling analysis for two Tibetan wild barley genotypes in responses to low nitrogen. BMC Plant Biol 16:1–16. https://doi.org/10.1186/s12870-016-0721-8

Rae AL, Cybinski DH, Jarmey JM, Smith FW (2003) Characterization of two phosphate transporters from barley; evidence for diverse function and kinetic properties among members of the Pht1 family. Plant Mol Biol 53:27–36. https://doi.org/10.1023/b:plan.0000009259.75314.15

Raman H, Moroni S, Sato K, Read J, Scott J (2002) Identification of AFLP and microsatellite markers linked with an aluminium tolerance gene in barley (*Hordeum vulgare* L.). Theor Appl Genet 105:458–464. https://doi.org/10.1007/s00122-002-0934-0

Raman H et al (2005) Molecular characterization and mapping of *ALMT1*, the aluminium-tolerance gene of bread wheat (*Triticum aestivum* L.). Genome 48:781–791. https://doi.org/10.1139/g05-054

Raun WR, Johnson GV (1999) Improving nitrogen use efficiency for cereal production. Agron J 91:357–363. https://doi.org/10.2134/agronj1999.00021962009100030001x

Rea PA, Poole RJ (1993) Vacuolar H^+-translocating pyrophosphatase. Annu Rev Plant Physiol Plant Mol Biol 44:157–180. https://doi.org/10.1146/annurev.arplant.44.1.157

Reid DA (1971) Genetic control of reaction to aluminum in winter barley. In: Second international barley genetics symposium, Washington State University, pp 409–413

Reid R (2010) Can we really increase yields by making crop plants tolerant to boron toxicity? Plant Sci 178:9–11. https://doi.org/10.1016/j.plantsci.2009.10.006

Reid DA, Jones GD, Armiger WH, Foy CD, Koch EJ, Starling TM (1969) Differential aluminum tolerance of winter barley varieties and selections in associated greenhouse and field experiments. Agron J 61:218–222. https://doi.org/10.2134/agronj1969.00021962006100020014x

Reinert S, Kortz A, Leon J, Naz AA (2016) Genome-wide association mapping in the global diversity set reveals new QTL controlling root system and related shoot variation in barley. Front Plant Sci 7:1061. https://doi.org/10.3389/fpls.2016.01061

Reinheimer JL, Barr AR, Eglinton JK (2004) QTL mapping of chromosomal regions conferring reproductive frost tolerance in barley (*Hordeum vulgare* L.). Theor Appl Genet 109:1267–1274. https://doi.org/10.1007/s00122-004-1736-3

Rivandi J, Miyazaki J, Hrmova M, Pallotta M, Tester M, Collins NC (2011) A *SOS3* homologue maps to *HvNax4*, a barley locus controlling an environmentally sensitive Na^+ exclusion trait. J Exp Bot 62:1201–1216. https://doi.org/10.1093/jxb/erq346

Robredo A, Perez-Lopez U, de la Maza HS, Gonzalez-Moro B, Lacuesta M, Mena-Petite A, Munoz-Rueda A (2007) Elevated CO_2 alleviates the impact of drought on barley improving water status by lowering stomatal conductance and delaying its effects on photosynthesis. Environ Exp Bot 59:252–263. https://doi.org/10.1016/j.envexpbot.2006.01.001

Robredo A, Perez-Lopez U, Lacuesta M, Mena-Petite A, Munoz-Rueda A (2010) Influence of water stress on photosynthetic characteristics in barley plants under ambient and elevated CO_2 concentrations. Biol Plant 54:285–292. https://doi.org/10.1007/s10535-010-0050-y

Rollins JA et al (2013) Variation at the vernalisation genes *Vrn-H1* and *Vrn-H2* determines growth and yield stability in barley (*Hordeum vulgare*) grown under dryland conditions in Syria. Theor Appl Genet 126:2803–2824. https://doi.org/10.1007/s00122-013-2173-y

Roy SJ, Huang W, Wang XJ, Evrard A, Schmockel SM, Zafar ZU, Tester M (2013) A novel protein kinase involved in Na^+ exclusion revealed from positional cloning. Plant, Cell Environ 36:553–568. https://doi.org/10.1111/j.1365-3040.2012.02595.x

Roy SJ, Negrão S, Tester M (2014) Salt resistant crop plants. Curr Opin Biotechnol 26:115–124. https://doi.org/10.1016/j.copbio.2013.12.004

Ryan J, Singh M, Yau SK (1998) Spatial variability of soluble boron in Syrian soils. Soil Tillage Res 45:407–417. https://doi.org/10.1016/S0933-3630(97)00037-8

Ryan P, Delhaize E, Jones D (2001) Function and mechanism of organic anion exudation from plant roots. Annu Rev Plant Physiol Plant Mol Biol 52:527–560. https://doi.org/10.1146/annurev.arplant.52.1.527

Saade S et al. (2016) Yield-related salinity tolerance traits identified in a nested association mapping (NAM) population of wild barley. Sci Rep 6:32586. https://doi.org/10.1038/srep32586

Sakata T, Takahashi H, Nishiyama I, Higashitani A (2000) Effects of high temperature on the development of pollen mother cells and microspores in barley *Hordeum vulgare* L. J Plant Res 113:395–402. https://doi.org/10.1007/Pl00013947

Sannemann W, Huang BE, Mathew B, Léon J (2015) Multi-parent advanced generation inter-cross in barley: high-resolution quantitative trait locus mapping for flowering time as a proof of concept. Mol Breed 35:86. https://doi.org/10.1007/s11032-015-0284-7

Santa-Maria GE, Rubio F, Dubcovsky J, Rodriguez-Navarro A (1997) The *HAK1* gene of barley is a member of a large gene family and encodes a high-affinity potassium transporter. Plant Cell 9:2281–2289. https://doi.org/10.1105/tpc.9.12.2281

Sasaki T et al (2004) A wheat gene encoding an aluminum-activated malate transporter. Plant J 37: 645–653. https://doi.org/10.1111/j.1365-313X.2003.01991.x

Savin R, Stone P, Nicolas M (1996) Responses of grain growth and malting quality of barley to short periods of high temperature in field studies using portable chambers. Aust J Agric Res 47:465–477. https://doi.org/10.1071/AR9960465

Sayed MA, Schumann H, Pillen K, Naz AA, Leon J (2012) AB-QTL analysis reveals new alleles associated to proline accumulation and leaf wilting under drought stress conditions in barley (*Hordeum vulgare* L.). BMC Genet 13. https://doi.org/10.1186/1471-2156-13-61

Schilling RK, Marschner P, Shavrukov Y, Berger B, Tester M, Roy SJ, Plett DC (2014) Expression of the

Arabidopsis vacuolar H^+-pyrophosphatase gene (*AVP1*) improves the shoot biomass of transgenic barley and increases grain yield in a saline field. Plant Biotechnol J 12:378–386. https://doi.org/10.1111/pbi.12145

Schneider LM et al (2016) The *Cer-cqu* gene cluster determines three key players in a β-diketone synthase polyketide pathway synthesizing aliphatics in epicuticular waxes. J Exp Bot 67:2715–2730. https://doi.org/10.1093/jxb/erw105

Schnurbusch T et al (2010) Boron toxicity tolerance in barley through reduced expression of the multifunctional aquaporin HvNIP2;1. Plant Physiol 153:1706–1715. https://doi.org/10.1104/pp.110.158832

Schünmann PHD, Richardson AE, Smith FW, Delhaize E (2004a) Characterization of promoter expression patterns derived from the Pht1 phosphate transporter genes of barley (*Hordeum vulgare* L.). J Exp Bot 55:855–865. https://doi.org/10.1093/jxb/erh103

Schünmann PHD, Richardson AE, Vickers CE, Delhaize E (2004b) Promoter analysis of the barley *Pht1;1* phosphate transporter gene identifies regions controlling root expression and responsiveness to phosphate deprivation. Plant Physiol 136:4205–4214. https://doi.org/10.1104/pp.104.045823

Setter TL, Waters I (2003) Review of prospects for germplasm improvement for waterlogging tolerance in wheat, barley and oats. Plant Soil 253:1–34. https://doi.org/10.1023/A:1024573305997

Shavrukov Y et al (2010) *HvNax3*—a locus controlling shoot sodium exclusion derived from wild barley (*Hordeum vulgare* ssp. *spontaneum*). Funct Integr Genomics 10:277–291. https://doi.org/10.1007/s10142-009-0153-8

Shavrukov Y, Bovill J, Afzal I, Hayes JE, Roy SJ, Tester M, Collins NC (2013) *HVP10* encoding V-PPase is a prime candidate for the barley *HvNax3* sodium exclusion gene: evidence from fine mapping and expression analysis. Planta 237:1111–1122. https://doi.org/10.1007/s00425-012-1827-3

Shrawat AK, Carroll RT, DePauw M, Taylor GJ, Good AG (2008) Genetic engineering of improved nitrogen use efficiency in rice by the tissue-specific expression of *alanine aminotransferase*. Plant Biotechnol J 6:722–732. https://doi.org/10.1111/j.1467-7652.2008.00351.x

Sicher RC, Timlin D, Bailey B (2012) Responses of growth and primary metabolism of water-stressed barley roots to rehydration. J Plant Physiol 169:686–695. https://doi.org/10.1016/j.jplph.2012.01.002

Singh M, Kumar J, Singh S, Singh VP, Prasad SM (2015) Roles of osmoprotectants in improving salinity and drought tolerance in plants: a review. Rev Environ Sci Biotechnol 14:407–426. https://doi.org/10.1007/s11157-015-9372-8

Smith FW, Mudge SR, Rae AL, Glassop D (2003) Phosphate transport in plants. Plant Soil 248:71–83. https://doi.org/10.1023/a:1022376332180

Snyder RL, Melo-Abreu JPd (2005) Frost protection: fundamentals, practice and economics. Environment and natural resources series. Food and Agriculture Organization of the United Nations, Rome

Stockinger EJ, Skinner JS, Gardner KG, Francia E, Pecchioni N (2007) Expression levels of barley *Cbf* genes at the *Frost resistance-H2* locus are dependent upon alleles at *Fr-H1* and *Fr-H2*. Plant J 51:308–321. https://doi.org/10.1111/j.1365-313X.2007.03141.x

Stolen O, Andersen S (1978) Inheritance of tolerance to low soil pH in barley. Hereditas 88:101–105. https://doi.org/10.1111/j.1601-5223.1978.tb01608.x

Suprunova T, Krugman T, Distelfeld A, Fahima T, Nevo E, Korol A (2007) Identification of a novel gene (*Hsdr4*) involved in water-stress tolerance in wild barley. Plant Mol Biol 64:17–34. https://doi.org/10.1007/s11103-006-9131-x

Sutton T et al (2007) Boron-toxicity tolerance in barley arising from efflux transporter amplification. Science 318:1446–1449. https://doi.org/10.1126/science.1146853

Sylvester-Bradley R, Kindred DR (2009) Analysing nitrogen responses of cereals to prioritize routes to the improvement of nitrogen use efficiency. J Exp Bot 60:1939–1951. https://doi.org/10.1093/jxb/erp116

Sze H, Ward JM, Lai SP (1992) Vacuolar H^+-translocating ATPases from plants: Structure, function, and isoforms. J Bioenerg Biomembr 24:371–381. https://doi.org/10.1007/Bf00762530

Takano J, Wada M, Ludewig U, Schaaf G, von Wiren N, Fujiwara T (2006) The *Arabidopsis* major intrinsic protein NIP5;1 is essential for efficient boron uptake and plant development under boron limitation. Plant Cell 18:1498–1509. https://doi.org/10.1105/tpc.106.041640

Talame V et al (2004) Identification of *Hordeum spontaneum* QTL alleles improving field performance of barley grown under rainfed conditions. Ann Appl Biol 144:309–319. https://doi.org/10.1111/j.1744-7348.2004.tb00346.x

Teotia S, Singh D, Tang X, Tang G (2016) Essential RNA-based technologies and their applications in plant functional genomics. Trends Biotechnol 34:106–123. https://doi.org/10.1016/j.tibtech.2015.12.001

Teulat B, Monneveux P, Wery J, Borries C, Souyris I, Charrier A, This D (1997) Relationships between relative water content and growth parameters under water stress in barley: a QTL study. New Phytol 137:99–107. https://doi.org/10.1046/j.1469-8137.1997.00815.x

Teulat B et al (1998) Several QTLs involved in osmotic adjustment trait variation in barley (*Hordeum vulgare* L.). Theor Appl Genet 96:688–698. https://doi.org/10.1007/s001220050790

Teulat B, Borries C, This D (2001a) New QTLs identified for plant water status, water-soluble carbohydrate and osmotic adjustment in a barley population grown in a growth-chamber under two water regimes. Theor Appl Genet 103:161–170. https://doi.org/10.1007/s001220000503

Teulat B, Merah O, Souyris I, This D (2001b) QTLs for agronomic traits from a Mediterranean barley progeny grown in several environments. Theor Appl Genet 103:774–787. https://doi.org/10.1007/s001220100619

Teulat B, Merah O, Sirault X, Borries C, Waugh R, This D (2002) QTLs for grain carbon isotope discrimination in field-grown barley. Theor Appl Genet 106:118–126. https://doi.org/10.1007/s00122-002-1028-8

Teulat B, Zoumarou-Wallis N, Rotter B, Ben Salem M, Bahri H, This D (2003) QTL for relative water content in field-grown barley and their stability across Mediterranean environments. Theor Appl Genet 108:181–188. https://doi.org/10.1007/s00122-003-1417-7

Thomashow MF (1998) Role of cold-responsive genes in plant freezing tolerance. Plant Physiol 118:1–7. https://doi.org/10.1104/pp.118.1.1

Thomsen HC, Eriksson D, Møller IS, Schjoerring JK (2014) Cytosolic glutamine synthetase: a target for improvement of crop nitrogen use efficiency? Trends Plant Sci 19:656–663. https://doi.org/10.1016/j.tplants.2014.06.002

Tombesi S et al (2015) Stomatal closure is induced by hydraulic signals and maintained by ABA in drought-stressed grapevine. Sci Rep 5. https://doi.org/10.1038/srep12449

Tommasini L et al (2008) Dehydrin gene expression provides an indicator of low temperature and drought stress: transcriptome-based analysis of Barley (*Hordeum vulgare* L.). Funct Integr Genomics 8:387–405. https://doi.org/10.1007/s10142-008-0081-z

Tondelli A et al (2006) Mapping regulatory genes as candidates for cold and drought stress tolerance in barley. Theor Appl Genet 112:445–454. https://doi.org/10.1007/s00122-005-0144-7

Tondelli A, Francia E, Barabaschi D, Pasquariello M, Pecchioni N (2011) Inside the *CBF* locus in *Poaceae*. Plant Sci 180:39–45. https://doi.org/10.1016/j.plantsci.2010.08.012

Tondelli A et al (2014) Allelic variation at *Fr-H1/Vrn-H1* and *Fr-H2* loci is the main determinant of frost tolerance in spring barley. Environ Exp Bot 106:148–155. https://doi.org/10.1016/j.envexpbot.2014.02.014

Uga Y et al (2013) Control of root system architecture by *DEEPER ROOTING 1* increases rice yield under drought conditions. Nat Genet 45:1097–1102. https://doi.org/10.1038/ng.2725

Ugarte C, Calderini DF, Slafer GA (2007) Grain weight and grain number responsiveness to pre-anthesis temperature in wheat, barley and triticale. Field Crops Res 100:240–248. https://doi.org/10.1016/j.fcr.2006.07.010

Vasekina AV, Yershov PV, Reshetova OS, Tikhonova TV, Lunin VG, Trofimova MS, Babakov AV (2005) Vacuolar Na^+/H^+ antiporter from barley: identification and response to salt stress. Biokhim Biochem (Mosc) 70:100–107. https://doi.org/10.1007/pl00021768

Venema K, Quintero FJ, Pardo JM, Donaire JP (2002) The *Arabidopsis* Na^+/H^+exchanger AtNHX1 catalyzes low affinity Na^+ and K^+ transport in reconstituted liposomes. J Biol Chem 277:2413–2418. https://doi.org/10.1074/jbc.M105043200

Vidmar JJ, Zhuo D, Siddiqi MY, Glass ADM (2000a) Isolation and characterization of *HvNRT2.3* and *HvNRT2.4*, cDNAs encoding high-affinity nitrate transporters from roots of barley. Plant Physiol 122:783–792. https://doi.org/10.1104/pp.122.3.783

Vidmar JJ, Zhuo D, Siddiqi MY, Schjoerring JK, Touraine B, Glass ADM (2000b) Regulation of high-affinity nitrate transporter genes and high-affinity nitrate influx by nitrogen pools in roots of barley. Plant Physiol 123:307–318. https://doi.org/10.1104/pp.123.1.307

von Korff M, Grando S, Del Greco A, This D, Baum M, Ceccarelli S (2008) Quantitative trait loci associated with adaptation to Mediterranean dryland conditions in barley. Theor Appl Genet 117:653–669. https://doi.org/10.1007/s00122-008-0787-2

von Zitzewitz J et al (2005) Molecular and structural characterization of barley vernalization genes. Plant Mol Biol 59:449–467. https://doi.org/10.1007/s11103-005-0351-2

von Zitzewitz J et al (2011) The genetics of winterhardiness in barley: perspectives from genome-wide association mapping. Plant Genome 4:76–91. https://doi.org/10.3835/plantgenome2010.12.0030

Wahid A, Gelani S, Ashraf M, Foolad MR (2007) Heat tolerance in plants: an overview. Environ Exp Bot 61:199–223. https://doi.org/10.1016/j.envexpbot.2007.05.011

Wallwork M, Logue S, MacLeod L, Jenner C (1998) Effects of a period of high temperature during grain filling on the grain growth characteristics and malting quality of three Australian malting barleys. Aust J Agric Res 49:1287–1296. https://doi.org/10.1071/A98004

Wang JP et al (2007) High-resolution mapping of the *Alp* locus and identification of a candidate gene *HvMATE* controlling aluminium tolerance in barley (*Hordeum vulgare* L.). Theor Appl Genet 115:265–276. https://doi.org/10.1007/s00122-007-0562-9

Wang P, Li ZW, Wei JS, Zhao ZL, Sun DY, Cui SJ (2012a) A Na^+/Ca^{2+} exchanger-like protein (AtNCL) involved in salt stress in *Arabidopsis*. J Biol Chem 287:44062–44070. https://doi.org/10.1074/jbc.M112.351643

Wang Y-Y, Hsu P-K, Tsay Y-F (2012b) Uptake, allocation and signaling of nitrate. Trends Plant Sci 17:458–467. https://doi.org/10.1016/j.tplants.2012.04.006

Wei Z et al (2016) Comparative transcriptome analysis revealed genes commonly responsive to varied nitrate stress in leaves of Tibetan hulless barley. Front Plant Sci 7:1067. https://doi.org/10.3389/fpls.2016.01067

Wicke B, Smeets E, Dornburg V, Vashev B, Gaiser T, Turkenburg W, Faaij A (2011) The global technical and economic potential of bioenergy from salt-affected soils. Energ Environ Sci 4:2669–2681. https://doi.org/10.1039/c1ee01029h

Wu H, Shabala L, Barry K, Zhou M, Shabala S (2013) Ability of leaf mesophyll to retain potassium

correlates with salinity tolerance in wheat and barley. Physiol Plant. https://doi.org/10.1111/ppl.12056
Wu DZ, Sato K, Ma JF (2015) Genome-wide association mapping of cadmium accumulation in different organs of barley. New Phytol 208:817–829. https://doi.org/10.1111/nph.13512
Xia YS et al (2013) Single nucleotide polymorphisms in *HSP17.8* and their association with agronomic traits in barley. PLOS ONE 8. https://doi.org/10.1371/journal.pone.0056816
Xu D, Duan X, Wang B, Hong B, Ho T, Wu R (1996) Expression of a late embryogenesis abundant protein gene, *HVA1*, from barley confers tolerance to water deficit and salt stress in transgenic rice. Plant Physiol 110:249–257. https://doi.org/10.1104/pp.110.1.249
Xu G, Fan X, Miller AJ (2012) Plant nitrogen assimilation and use efficiency. Annu Rev Plant Biol 63:153–182. https://doi.org/10.1146/annurev-arplant-042811-105532
Xue DW et al (2009) Identification of QTLs associated with salinity tolerance at late growth stage in barley. Euphytica 169:187–196. https://doi.org/10.1007/s10681-009-9919-2
Yadav SK (2010) Cold stress tolerance mechanisms in plants. Agron Sustain Dev 30(3):515–527. https://doi.org/10.1051/agro/2009050
Yan L, Loukoianov A, Tranquilli G, Helguera M, Fahima T, Dubcovsky J (2003) Positional cloning of the wheat vernalization gene *VRN1*. Proc Natl Acad Sci USA 100:6263–6268. https://doi.org/10.1073/pnas.0937399100
Yang L, Mickelson S, See D, Blake TK, Fischer AM (2004) Genetic analysis of the function of major leaf proteases in barley (*Hordeum vulgare* L.) nitrogen remobilization. J Exp Bot 55:2607–2616. https://doi.org/10.1093/jxb/erh267
Yau SK, Ryan J (2008) Boron toxicity tolerance in crops: A viable alternative to soil amelioration. Crop Sci 48:854–865. https://doi.org/10.2135/cropsci2007.10.0539
Zeng J et al (2014) Comparative transcriptome profiling of two Tibetan wild barley genotypes in responses to low potassium. PLOS ONE 9:e100567. https://doi.org/10.1371/journal.pone.0100567
Zhang G, Li C (2016) Exploration and utilization of aluminum-tolerant barley germplasm. In: Exploration, identification and utilization of barley germplasm, pp 181–208
Zhang X, Davidson EA, Mauzerall DL, Searchinger TD, Dumas P, Shen Y (2015) Managing nitrogen for sustainable development. Nature 528:51–59. https://doi.org/10.1038/nature15743
Zheng BY, Chapman SC, Christopher JT, Frederiks TM, Chenu K (2015) Frost trends and their estimated impact on yield in the Australian wheatbelt. J Exp Bot 66:3611–3623. https://doi.org/10.1093/jxb/erv163
Zhou GF, Delhaize E, Zhou MX, Ryan PR (2013) The barley *MATE* gene, *HvAACT1*, increases citrate efflux and Al^{3+} tolerance when expressed in wheat and barley. Ann Bot 112:603–612. https://doi.org/10.1093/aob/mct135
Zhou GF, Pereira JF, Delhaize E, Zhou MX, Magalhaes JV, Ryan PR (2014) Enhancing the aluminium tolerance of barley by expressing the citrate transporter genes *SbMATE* and *FRD3*. J Exp Bot 65:2381–2390. https://doi.org/10.1093/jxb/eru121
Zhou G, Broughton S, Zhang XQ, Ma Y, Zhou M, Li C (2016) Genome-wide association mapping of acid soil resistance in barley (*Hordeum vulgare* L.). Front Plant Sci 7:406. https://doi.org/10.3389/fpls.2016.00406
Zohary D (1999) Monophyletic versus polyphyletic origin of the crops on which agriculture was founded in the near east. Genet Resour Crop Evol 46:133–142. https://doi.org/10.1023/a:1008692912820

Genomics-Based Barley Breeding

16

Kevin P. Smith, William Thomas, Lucia Gutierrez and Hazel Bull

Abstract

The creation of the first barley reference genome opens up new and improved avenues for the translation of genetics and genomics to application in plant breeding. Despite steady progress in barley breeding, there is still a great need for improved barley cultivars that provide raw materials for a wide array of products, are adapted to diverse growing conditions, and provide sustainable and profitable income for farmers. In this chapter, we investigate some of the major challenges in barley breeding and describe how our ever-increasing understanding of the barley genome has led to new methods and approaches to meet those challenges. There are several excellent reviews of barley breeding that detail progress in breeding over time (e.g., Friedt et al. 2010). Here, we will focus on how genetic tools and resources have affected progress so far and the prospects to utilize current and emerging genomic tools to sustain barley improvement. In particular, we explore the evolution of marker development for mapping, genomic approaches to selecting parent combinations and from within segregating breeding populations, exploiting diverse germplasm, tapping into genomic regions of limited recombination, and utilizing bioinformatic characterization of genetic load. Access to a complete reference genome is already providing breeders and geneticists information about candidate genes for QTL that are targets for marker-assisted selection, the physical order of genetically linked markers in regions explored by fine mapping, and the ultimate consensus map with which to compare mapping studies. Additional full genome sequences leading to a pan-genome for barley, refinement of the reference genome, and new tools to access, analyze, and utilize genomic information will foster further integration of genomics into barley improvement.

K. P. Smith (✉)
Department of Agronomy and Plant Genetics, University of Minnesota, Saint Paul, MN, USA
e-mail: smith376@umn.edu

W. Thomas
The James Hutton Institute, Dundee, Scotland, UK
e-mail: bill.thomas@hutton.ac.uk

L. Gutierrez
Department of Agronomy, University of Wisconsin, Madison, WI, USA
e-mail: gutierrezcha@wisc.edu

H. Bull
James Hutton Limited, Dundee, Scotland, UK
e-mail: hazel.bull@huttonltd.com

16.1 Successes in Barley Breeding Before a Reference Genome

The past 56 years has seen a steady rise in the world average yield per unit area of barley that is in the order of 0.025 t/year producing a world

N. Stein and G. J. Muehlbauer (eds.), *The Barley Genome*, Compendium of Plant Genomes,
https://doi.org/10.1007/978-3-319-92528-8_16

average that was in 2014 just under 3 t/ha with minima and maxima of 0.33 (Oman) and 9.2 t/ha (Belgium), respectively (http://www.fao.org/faostat/en/#data/QC). Arguably, much of this improvement in world barley yields is due to improved management, it is undeniable that improved varieties developed by breeding programs targeting specific macro-environments have also had a substantial impact. Riggs et al. (1981) compared 37 spring barley varieties released from the late nineteenth century up to 1980 and found that breeding had improved the yield potential largely by improving harvest index and concurrently reducing height. As the lines were all grown together in the same trial, the changes observed are truly genetic, but it could be argued that the introduction of semidwarf types with some more recent varieties carrying either the *ari-e*.GP or *sdw1* gene could have biased the findings. More recent studies of official trials in the UK showed that the rate of gain in a high yielding environment from 1982 to 2007 is approx. 0.07 and 0.06 t/ha per year for winter and spring barley, respectively. Specifically, 42 and 86% of that increase was due to improved genetics of winter and spring barley, respectively (Mackay et al. 2011). Analysis of a single breeding program in the Midwest U.S. documented a more modest gain of 0.015 t/ha per year in yield over a 40 year period for six-row spring malting barley (Condon et al. 2009). Much of this potential is not realized on farm, however, and there is evidence in some countries that a yield plateau has been reached. In the UK, this plateau is visible in the average wheat yields since 2000 but progress in on-farm yields still appears to be happening in barley with annual rates of increase of 0.019 and 0.013 t/ha for winter and spring barley, respectively.

How have these yield increases been achieved? From studies of UK barley, it appears that much of the yield increase originates from increases in thousand grain weight (Thomas et al. 2014), although the other two key yield components (grain and spike number) were not measured. The study conclusively demonstrated that heading date and crop maturity had not changed over the survey period while breeding had significantly reduced height, although the latter can largely be ascribed to the transition to cultivars carrying the *sdw1* semidwarfing gene (Thomas et al. 2014).

Breeding malting barley poses additional challenges, as gains in yield must be accompanied by maintaining or improving a suite of malting quality traits. For example, the gains in yield made in the US Midwestern spring six-row program paralleled a favorable increase in malt extract and an unfavorable increase in soluble protein (Condon et al. 2009). Breeding has also significantly increased the malt extract of the spring barley crop in the UK, although the increase in the winter barley crop was not significant and bridging that quality gap remains a major challenge for winter barley breeders (Thomas et al. 2014). Wort viscosity has significantly decreased over time in the UK winter barley crop but not in the spring barley crop where it has remained within acceptable bounds. By contrast, characters such as wort beta-glucan content, diastatic power, and fermentability show no significant trends with time and considerable variation so there has been no consistent breeding progress for these aspects of malting quality.

Breeding for disease resistance is also very important for barley; however, there are few studies that have evaluated progress in disease resistance over time. Rajala et al. (2017) provide some evidence of improvement over a 100-year span in the resistance of Nordic cultivars to net blotch (spot and net form) as well as scald but few other studies have been published. This is partly due to much resistance breeding work focusing on race-specific resistance leading to the so-called boom and bust cycles for powdery mildew resistance in barley. The deployment of the *mlo* resistance gene in Europe since the late 1970s has provided a durable source of resistance to powdery mildew that genomic tools have now characterized as a loss of function of a susceptibility factor (Piffanelli et al. 2004) that is still effective in current cultivars. Non-race-specific resistance remains an attractive breeding objective but has proved particularly difficult to select. The development of molecular markers has significantly improved selection for resistance to diseases that are either difficult to characterize or not present in all environments, e.g., selection for resistance to barley mild mosaic and barley yellow mosaic viruses in NW Europe (discussed in the following section).

To examine progress made in plant breeding, it is useful to consider a modification of the breeder's equation $R_y = (h^2S)/C$, where R_y is the response to selection per year, C is the length in years of the breeding cycle (from cross to use of the progeny in a subsequent cross), h^2 is the narrow sense heritability, and S is the selection intensity. Heritability is a function of the available genetic variation and the accuracy of selection. In addition, progress is measured as the response per unit time. Therefore, breeding progress is a function of (1) genetic variation, (2) selection accuracy, (3) selection intensity, and (4) breeding cycle time. As we consider the various tools, methods, and activities of breeders both past and present, it will be useful to frame them within these four components of gain per unit time.

Historically, the free exchange of germplasm among worldwide breeding programs has contributed to the genetic variation exploited by breeders and the subsequent genetic gains. As programs mature, breeders select primarily from within their own elite gene pools which over time may reduce genetic variation (Condon et al. 2008). Breeders will, however, use reciprocal crossing agreement for the early introduction of material from competitor programs, but these are generally sampling the same local elite gene pool. Thus, to improve the rate of genetic gain breeders must increase the accuracy and/or intensity of selection or reduce the breeding cycle time. Selection intensity is limited by available resources and the cost to select among individuals using phenotypic or genotypic selection. As the cost of phenotyping has generally remained constant or increased, the primary means to increase selection intensity are through exploiting high-throughput genotyping platforms (see section below). The accuracy of phenotyping has improved with developments in plot machinery that improve the accuracy of sowing and harvesting of trials combined with improvements in computing power that have facilitated the use of more flexible experimental designs and statistical procedures that can be used to account for nongenetic effects in field trials (Sarker and Singh 2015). For many traits, however, the contribution of genotype × environment interaction will be a significant component of heritability that will limit gains and breeders still have to conduct multisite trials that capture their range of target environments. Even with improvements in selection accuracy, there are diminishing returns with selection intensities below 10%.

Breeders have therefore paid particular attention to reducing the breeding cycle time as having the first of a new "type" of cultivar on the market is often more critical to market success than having a better cultivar that may be 1–2 years later to market. Commercial breeding programs in Europe have been particularly successful in this regard. Examining the pedigrees and year of first recommendation of cultivars on the UK Cereals & Oilseeds Recommended Lists (https://cereals.ahdb.org.uk/varieties/ahdb-recommended-lists.aspx) shows that some lines are being recrossed within 3 years (Fig. 16.1). For example, there were four years between the first recommendation of the spring barley Westminster and its progeny variety Concerto. Yet, subsequent to this the breeding interval has been reduced further with only 3 years between the first recommendation of Concerto and Chronicle (the selection of a Concerto/Quench cross). This short cycle time continues with only 3 years between Chronicle and its progeny Sienna (Chronicle/Genie). From this, it is clear that the breeding cycle in spring barley using a breeder's own material is now down to 3 years and can sometimes be just two (Westminster, Concerto, Chronicle, Genie, and Sienna are all varieties from the Limagrain breeding programs). However, these rapid turnarounds seem limited to material from a breeder's own program, this is illustrated in the pedigree of Chronicle, Quench (a variety from Syngenta) was first recommended 5 years prior to the recommendation of its progeny Chronicle, suggesting utilizing another breeder's variety within a cross adds another two

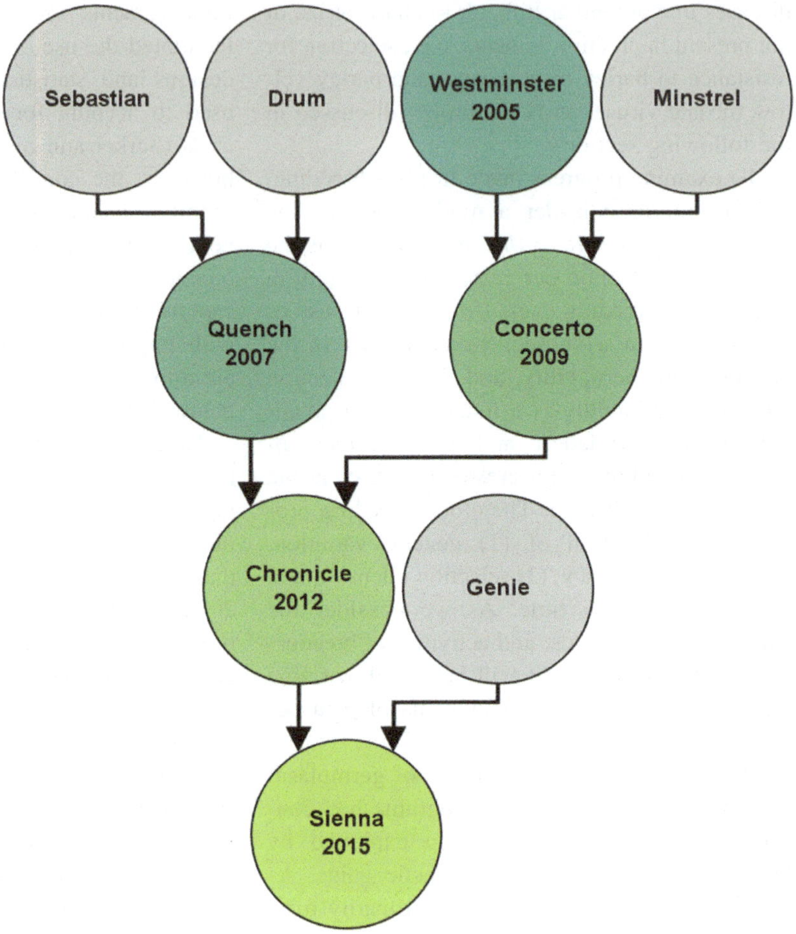

Fig. 16.1 Pedigree of some UK recommended spring barley cultivars. All the cultivars apart from Sebastian (Sejet) and Drum and Quench (Syngenta) are from the Limagrain program. Years in the circles are the first year that a cultivar was placed on the AHDB Cereals and Oilseeds recommended list for the UK. Colors progress from dark to light green to reflect older to newer first years of recommendation

years to the breeding cycle for a company while the potential of the variety is evaluated. This highlights the need for breeders to recognize and integrate valuable breeding material into their crossing programs at early generations so competitive advantage is maintained.

In addition to the basic pedigree breeding scheme, more rapid advancement to homozygosity has been developed through the application of doubled haploidy and single seed descent (SSD). More recently, SSD methodology has been combined with high-intensity lighting regimes to develop a technique called "speed breeding" where as many as 7 generations can be achieved in one year (Christopher et al. 2015). While this technique has been perfectly demonstrated with germplasm adapted to short-day growing, it remains to be seen if the same reduction can be achieved in germplasm adapted to long days, although multiple generations of <10 weeks were achieved in European spring barley under less intense lighting (Riggs and Hayter 1976).

Breeders have always been ready to adopt new technologies once they are convinced that it either improves the speed and/or the efficiency of selection or provides new sources of variation. This was seen in the 1950s with effort on mutation breeding to introduce semidwarf genes into cultivars. Later, breeders invested in doubled haploid production to speed up the production of inbred lines and improve the accuracy of selection. As molecular markers for key but "hard to measure" traits became available, breeders then adopted marker-assisted selection. Despite the substantial progress made through phenotypic selection and more recently with marker-assisted selection, there is still great need to accelerate breeding efforts to meet a wide array of agricultural challenges and improve our food system. Regardless of the specific challenge, improving breeding progress will require using **genetic variation**, applying **accurate selection**, imposing stringent **selection intensity** that is cost-effective, and reducing the **breeding cycle time**. Recent progress in barley genomics including access to a reference genome provides opportunities to improve existing breeding methods and develop novel approaches to barley improvement. In the following sections, we explore some of these opportunities and speculate how they may affect barley breeding.

16.2 Abundant and Reliable Markers

For molecular markers to be deployed cost-effectively in a breeding program they should increase the efficiency of selection over that of traditional phenotyping methods. For this to happen, markers should have reliable linkage with genes controlling specific traits of interest, exhibit a high level of polymorphism, be high throughput, occur at a high density in the genome, be relatively inexpensive, and be readily transferable among germplasm groups. In the past 25 years, barley researchers have actively applied different kinds of markers for the purpose of genetic mapping (Table 16.1). However, until relatively recently, there has been limited deployment of marker-assisted selection in barley breeding programs.

Isozymes, the earliest form of molecular marker, offered the possibility of using surrogate markers to help introgress disease resistance genes from exotic sources in a more precise manner and, more importantly, enable the introgression of recessive major genes without the need for progeny testing to establish phenotypes. However, there were insufficient isozyme markers and the levels of polymorphism were too low to make this approach widely applicable. Nevertheless, this approach was used to develop markers to select for the Barley Yellow Mosaic

Table 16.1 Characteristics of representative barley mapping populations from 1989 to 2016

Mapping population					Marker types						
Year	Name	Size	Num. of markers	Reference	Morph	Isozyme	RFLP	RAPD	AFLP	SSR	SNP
1989	Daikei A6636 × Tochikei 166	75	3	Konishi et al. (1989)		3					
1991	Igri × Franka	71	118	Graner et al. (1991)	2	2	114				
1993	Steptoe × Morex	150	295	Kleinhofs et al. (1993)	2	5	266	14			
2000	Lina × Canada Park (*H. spontaneum*)	86	347	Ramsay et al. (2000)					48	299	
2001	Oregon Wolfe	94	723	Costa et al. (2001)	12		87	5	586	33	
2003	Fredrickson × Stander	130	143	Mesfin et al. (2003)	1		85			57	
2003	Igri × Franka	65	105	Li et al. (2003)						105	
2005	Steptoe × Morex	150	370	Rostoks et al. (2005)			266	14			90
2009	Morex × Barke	93	1652	Close et al. (2009)							1536
2012	Morex × Barke	360	3973	Comadran et al. (2012)							3973
2016	Wild barley AB-NAM	796	4022	Nice et al. (2016)							4022

Virus (BaYMV)/Barley Mild Mosaic Virus (BaMMV) *rym4* resistance locus in winter barley breeding programs (Konishi et al. 1989). *rym4* still represents a major source of resistance to BaYMV and BaMMV in elite European winter barley germplasm.

Some of the earliest markers deployed were restriction fragment length polymorphisms (RFLP), used particularly in the early stage selection of traits such as malt extract QTL (Barr et al. 2000) and major pathogen resistance loci, e.g., cereal cyst nematode resistance (Kretschmer et al. 1997) and *rym4* BaMMV/BaYMV resistance (Graner and Bauer 1993). The gel-based screening methodology of RFLP and lack of sequence information made marker development and deployment slow and labor intensive, with consensus RFLP barley linkage maps consisting of approximately 587 genome-wide markers (Langridge et al. 1995), although polymorphism levels within each cross were in the order of 100–150 markers.

Random amplified polymorphic DNA (RAPD) markers are the result of the random amplification of genomic DNA in single 10 nucleotide primer PCR reactions. The pattern of the resulting random amplification is used to determine genotype. Like RFLP they are a gel-based screening method making them relatively slow and labor intensive. Additionally, due to the non-targeted nature of amplification, reproducibility of results can be low and therefore application limited; however, attempts were made to develop markers for *Rhynchosporium* resistance loci in barley (Barua et al. 1993).

Amplified fragment length polymorphisms (AFLPs) are a refinement of the RFLP technique, following digestion of genomic DNA with restriction enzymes; adaptor sequences are ligated to the fragments enabling targeted PCR amplification. Compared to other marker types, AFLPs were found to be more readily transferable across populations making them more amenable to use in breeding (Qi and Lindhout 1997). Despite these advantages, there appeared to be little use of AFLPs directly in breeding.

Simple sequence repeats (SSR) offered advantages over RFLP through their polyallelic nature, greater transferability between germplasm, higher genome density, and their semi-automated scoring through incorporation of fluorescently labeled primers (Ramsay et al. 2000). Several SSR have been readily deployed in breeding programs these include Bmac 29 & QLB1 for the selection of *rym4*/*rym5* BaYMV/BaMMV disease resistance alleles (Graner et al. 1999; Tyrka et al. 2008). While the linkage between the SSR marker and resistance remains robust, further work identifying the underlying gene means that diagnostic SNP markers can now be deployed for the selection of *rym4*/*rym5* resistance (Hofinger et al. 2011).

The development of SNP markers in tandem with improvements in genotyping technology has further revolutionized the accessibility of molecular markers to barley breeding programs. The Illumina BOPA1 and BOPA2 SNP platforms (Close et al. 2009), each comprising 1536 SNP identified through the Sanger sequencing of PCR amplicons and EST sequences, were the first opportunity for automated high-throughput multiplex SNP genotyping in barley. The subsequent development of high-density genetic linkage maps provided the first opportunity to undertake large-scale genome-wide comparative studies of elite cultivars to understand recombination patterns and associated haplotype selection blocks.

A genome-wide association study of the association genetics of UK elite barley (AGOUEB) collection in combination with morphological trait information used to ensure distinct, uniform, and stable (DUS) cultivars in the official UK varietal registration process (e.g., rachilla hair length) led to the development of a suite of SNP markers for the traceability of specific morphological DUS phenotypes within breeding programs (Cockram et al. 2010, 2012). As an extension to this the use of BOPA1 and BOPA2 markers in addition to morphological characterisation as part of the official registration process has been investigated, however the study concluded that 1:1 correspondence between genetic and phenotypic distance was not possible and raised questions as to what genetic distance threshold would be deemed appropriate to determine distinctness between varieties (Jones et al. 2013).

Since the development of the BOPA1 and BOPA2 SNP platform, advances in sequencing technologies have facilitated the development of higher density SNP platforms. The 9K iSelect platform included all the BOPA1 and BOPA2 SNP, 5000 SNPs developed from the RNA sequencing of radicle and embryo tissue of 10 spring barley cultivars, and additional SNP identified from resequencing studies of specific genes (Comadran et al. 2012). The 9K iSelect was superseded by the 50K iSelect, developed from the exome capture of 170 cultivars representing a diverse set of winter and spring cultivars and landraces (Bayer et al. 2017).

The higher SNP density has enabled the development of robust linkage maps with increased reliability in marker order. Nevertheless, the SNPs surveyed are a small subset of those existing in elite gene pools, let alone among wild barleys and it is a matter of chance as to when an individual SNP arose compared to a linked functional polymorphism. This means that the SNP most highly associated with the phenotype may not be the closest SNP to the functional polymorphism. In an association study of row type in barley, the closest BOPA1 SNP (11_20302) to the candidate gene (HvTB1) was not highly significantly associated with the phenotype whereas an adjacent SNP (11_20606) had the highest LOD score (Ramsay et al. 2011).

While the ability to map loci associated with traits has been greatly enhanced through high-density SNP genotyping platforms, the application of markers in breeding has been facilitated by low density, high throughput, and low-cost markers. In an attempt to bridge the gap in malting quality between the spring and winter barley crop, the higher marker density afforded by the 9K iSelect has been used for the tandem marker-assisted introgression of three consistently selected independent QTL associated with malting quality from spring germplasm into winter (JHI unpublished). The marker-assisted selection of multiple introgressions requires larger population sizes but their screening is facilitated by the development of cost-effective high-throughput single-marker genotyping systems such as the KASP marker system (LGC genomics) and automated plate-based DNA extraction technologies such as the QIAextract (QIAGEN).

Advancements in multiplex sequencing technology and subsequent reduction in cost have facilitated the use of genotyping by sequencing approaches to characterize germplasm in barley breeding programs. This approach combines SNP identification and genotyping into a single process; briefly, following genome complexity reduction and barcoding, multiple individuals are sequenced within a single sequencing reaction. Following sequencing, results are de-multiplexed and sequence aligned either using de novo assembly or more recently the reference genome (Poland et al. 2012a, b; Mascher et al. 2013c). Although in theory all SNPs distinguishing individuals in a population should be identifiable using this approach, in reality, it is often a trade-off between sequencing depth and cost. The deeper the sequencing, the fewer the sequencing artifacts, but the more expensive the genotyping. In a study, mapping drought tolerance loci within a population of *Hordeum spontaneum*/*Hordeum vulgare* inbred lines of the 41,554 SNP identified the following filtering for missing data and comparison with parental genotypes only 3744 SNP could be mapped (Honsdorf et al. 2014). While the information recovered is population specific, the relatively large-scale bioinformatics capability required to process GBS data currently makes this technology less amenable to situations where selection decisions need to be made relatively quickly in a breeding program. That said, efforts are ongoing to develop faster, more efficient GBS bioinformatics pipelines, which should accelerate deployment of this technology (Torkamaneh et al. 2017).

Genome-wide assays, such as SNP chips and GBS, have made parental characterization and QTL discovery much more routine. But while the cost per data point is small, the cost per sample is not when applied on the scale of a typical barley breeding program. Simple PCR assays facilitate the large-scale screening of specific alleles at key genetic loci in breeding populations. An example of this is the epiheterodendrin (Eph) marker that was deployed to select suitable varieties for use in the UK distilling industry. This marker screens for the non-production of the glycosidic nitrile epiheterodendrin, which is a cyanogenic glycoside precursor of the potential carcinogen ethyl carbamate (Hedley et al. 2005). A simple PCR diagnostic was developed to select for nonproducers of Epiheterodendrin and is being used to identify varieties suitable for use in distilling (Bringhurst 2015). Another example was the development of a cleaved amplified polymorphism (CAP) marker used to introgress the null allele at the *LOX-1* barley locus. Introgression of this allele from an Indian barley landrace into the elite cultivar CDC-Kendall through marker-assisted backcrossing improved beer flavor and head retention (Hirota et al. 2005, 2006a, b; Hoki et al. 2010). The resulting introgression line CDC-PolarStar was first registered in 2010 and occupied 0.93% of the 2.466 M/Ha of two-rowed malting barley area in Western Canada in 2016 (Canadian Grain Commission). In parallel, a joint project between Carlsberg and Heineken screened a batch of mutated seed for low or absence of LOX-1 activity and protected their intellectual property through separate patents (Skadhauge et al. 2010). The research program identified two mutants with no detectable activity and breeding programs have licensed the technology to introgress the character into adapted germplasm resulting in the release of a number of new spring cultivars with Chanson being added to the AHDB Cereals and Oilseeds list of Recommended Varieties for 2017.

In a few cases, where the genes underlying a specific locus have been cloned, "perfect" genetic markers exist enabling the direct selection of the phenotype of interest through genotyping. One such example is that of barley row-type, non-synonymous SNP and frameshift mutations within *HvHox1* cause the switch from two-rowed to six-rowed spike morphology within cultivated germplasm (Komatsuda et al. 2007). Therefore in a two-rowed by six-rowed cross, it is possible to reliably select for the specific row type of interest at the seedling stage. Additionally, genome-wide association mapping (GWAS) of barley row type has identified the historic consistent tandem selection of specific "six-rowed" alleles of the unlinked *HvTB1* gene with the six-rowed alleles at *HvHox1* (Ramsay et al. 2011). The reason for this tandem selection remains unclear, six-rowed alleles at *HvTB1* alone are insufficient to produce a six-rowed phenotype (Lundqvist and Lundqvist 1987), but the repeated selection for this combination suggests some selective advantage. The availability of "perfect" markers at both loci increases the intensity of selection within six-rowed breeding programs as all those lines without the desired six-rowed *HvHox1*, six-rowed *HvTB1* allele combination can be disregarded at the seedling stage.

The changing marker platforms and lack of overlap of markers among mapping populations and breeding populations severely limited the deployment of marker-based selection in breeding (Table 16.1). The reference genome provides a uniform platform to integrate results. The plethora of markers now available can all be anchored physically, enabling QTL identified in diverse studies to be overlaid, with the potential for improved confidence, resolution, and hence, selection accuracy. The ability to overlay QTL becomes very important when breeding programs need to go from understanding the locus that controls the trait to identifying a specific and informative marker that can be used for selection in their breeding program. In summary, the development of the reference genome will greatly facilitate the rapid translation of research findings to applied barley breeding.

16.3 Accurate Selection in Segregating Populations

Modern plant breeding activities consist of evaluating the genetic merit of lines by discerning genetic from environment and noise components (Bernardo 2010). Selection strategies in plant breeding could be grouped into three categories: traditional, marker-assisted (MAS; Tanksley 1983), and genomic selection (GS;

Meuwissen et al. 2001). Traditional plant breeding uses either phenotypic information per se or from relatives to evaluate their genetic value (Fehr 1991; Bernardo 2010). MAS, on the other hand, involves the identification of markers linked to genes or quantitative trait loci (QTL) for relevant traits, and then selecting individuals based on their marker scores (Tanksley 1993; Hospital and Charcosset 1997). Finally, GS involves the prediction of the genetic merit of individuals based on a large number of genome-wide marker scores and a statistical model (Meuwissen et al. 2001).

Compared to phenotypic selection, MAS has clear advantages for: (1) traits that are difficult to phenotype because they are either expensive to measure, time-consuming, or very complex; (2) traits that require specific environments or developmental stages; (3) during backcrossing with recessive alleles or to speed up the process by avoiding the need to progeny test; and (4) to pyramid multiple alleles (Xu and Crouch 2008). Evidence for the advantage of MAS schemes has been documented widely. This includes improving the pace and precision of backcross strategies (Tanksley 1983) using either foreground selection (Hospital and Charcosset 1997) or background selection (Visscher et al. 1996; Hospital and Charcosset 1997) to reduce linkage drag (Hospital 2005), increasing the rate of genetic gain compared to phenotypic selection (Eathington 2005; Crosbie et al. 2006; Ragot and Lee 2007), and successfully introgressing disease resistance or stress tolerance genes (Cregan et al. 1999; Cahil and Schmidt 2004; Johnson 2004; Niebur et al. 2004; Eathington 2005; Crosbie et al. 2006; Ragot and Lee 2007), or grain quality traits (Dubcovsky 2004). Marker-assisted selection can improve genetic gain by several means. First, MAS can improve selection accuracy for traits that are difficult to phenotype. For example, expensive traits such as malting quality are usually selected at the end of the breeding cycle, on small populations, and based on few or no replications. The use of markers can improve selection accuracy which should increase genetic gain (Gutierrez et al. 2011). Another example is when the phenotype is difficult to evaluate or environment specific such as complex adult plant disease resistance. The use of markers can improve selection accuracy in environments where the disease is not fully expressed or heritability is very low (Gutierrez et al. 2015). Second, the response to selection can be increased by choosing optimal parental combinations such as in gene pyramiding (Castro et al. 2003). Third, breeding cycles can be reduced if used for example for either foreground or background MAS with backcrossing. Finally, selection intensity can be increased when inexpensive markers are used instead of expensive phenotyping such as malting quality analyses. By using cheaper methods to evaluate the genetic merit of individuals, larger population sizes can be used, and therefore, greater selection intensities can be imposed without seriously compromising future genetic gain by narrowing the genetic diversity.

An MAS program requires the identification of genes or genomic regions associated with the trait(s) of interest (Paterson et al. 1988; Tanksley 1993), a validation step (Patterson et al. 1990), and ultimately deployment (Lande and Thompson 1990). The use of molecular markers in breeding programs has been explored since the 1980s (Stuber 1992). There are several strategies to identify quantitative trait loci (QTL) of relevant traits in plants including biparental population or traditional QTL mapping (Hayes et al. 1993) and GWAS (Jannink et al. 2001).

Traditional QTL mapping requires first the construction of balanced populations with known recombination history. Later, a statistical association between a molecular marker and the trait of interest is sought through linkage disequilibrium using either linear regression models (Haley and Knott 1992) or mixture distributions (Lander and Botstein 1989). Since all the recombination occurred within the limits of the experiment, the linkage disequilibrium is expected to be caused by physical linkage of the molecular marker and the QTL, and therefore, the location of the QTL can be inferred (Lander and Botstein 1989). Many QTL mapping studies have identified genomic regions associated with relevant quantitative traits in barley. Some of these studies include QTL for malting quality (see Szűcs et al.

2009 for a review); agronomic traits (Cuesta-Marcos et al. 2008, 2009); disease resistance for Fusarium head blight (de la Peña et al. 1999; Mesfin et al. 2003; Canci et al. 2003) and yellow rust (Castro et al. 2003); heading date (Sameri and Komatsuda 2004); photoperiod (Szűcs et al. 2006); vernalization (Dubcovsky et al. 2005) and winterhardiness (Hayes et al. 1993; Karsai et al. 1997, 2005; Pan et al. 1994; Francia et al. 2004).

GWAS is also based on a statistical association between the molecular marker and the trait of interest. However, since diverse populations without a known recombination history are used, the cause of linkage disequilibrium could be physical linkage and other causes such as selection, genetic drift, mutation, admixture, and population structure (Jannink et al. 2001). Therefore, controlling for population structure is crucial in GWAS (Pritchard et al. 2000). Advantages of GWAS over biparental QTL mapping include assessment of genetically diverse germplasm stocks; higher resolution mapping; effective use of historical data; avoidance of family-specific effects; and immediate applicability to cultivar development because the genetic background in which QTL are estimated is directly relevant for plant breeding (Kraakman et al. 2004). This strategy has been used successfully in barley for agronomic traits (Kraakman et al. 2004, 2006; Cuesta-Marcos et al. 2010; Locatelli et al. 2013; Pauli et al. 2014); flowering time (Stracke et al. 2009); winterhardiness (von Zitzewitz et al. 2011); malting quality (Beattie et al. 2010; Gutierrez et al. 2011; Mohammadi et al. 2014, 2015); disease resistance to Fusarium head blight (Massman et al. 2011), leaf rust (Gutierrez et al. 2015) stem rust (Zhou et al. 2014), spot blotch (Roy et al. 2010; Gutierrez et al. 2015), net blotch (Tamang et al. 2015), and stripe rust (Gutierrez et al. 2015). Furthermore, the development of cultivars with increased resistance to Fusarium head blight was successfully accomplished where QTL associated with Fusarium head blight resistance were identified (de la Peña et al. 1999; Mesfin et al. 2003; Canci et al. 2003) and MAS was initiated. GWAS studies were later performed (Hayes and Szűcs 2006), and marker–trait associations were validated and used for MAS (Navara and Smith 2014; Jefferies et al. 2003; Chutimanitsakun et al. 2013). Some reviews have also focused on different aspects of GWAS in barley (Cockram et al. 2008; Bradbury et al. 2011; Hayes and Szűcs 2006; Waugh et al. 2009; Wang et al. 2012; Muñoz-Amatriaín et al. 2014).

One of the main limitations of QTL studies is that the effect of some QTL can be overestimated due to insufficient population sizes (Beavis 1998). Furthermore, often the QTL that are identified have small effects and explain a small portion of the total variation (*missing heritability*, Manolio et al. 2009). Other limitations to QTL studies are: inaccurate positions of QTL with large confidence intervals (Melchinger et al. 1998) and sampling bias in QTL effects (Lande and Thompson 1990); poor phenotypic data quality (Kearsey and Farquhar 1998); genetic background interaction where markers identified in a particular background may not be effective in a different background (Langridge et al. 2001) due to small effect QTL (Charcosset and Moreau 2004) or epistasis (Lin et al. 2000); and QTL by environment interaction (Hayes et al. 1993; Romagosa et al. 1999; Mathews et al. 2008; Gutierrez et al. 2015). Another limitation specific to GWAS studies is properly accounting for population structure, especially in barley where different breeding objectives have been pursued between two- and six-row barley (Gutierrez et al. 2011). Furthermore, successful QTL introgression requires QTL validation (Stuber et al. 1999) using relevant parents, populations, and environments. A few QTL published in barley have been validated, including QTL for root architecture (Ahmad Naz et al. 2012), disease resistance (Canci et al. 2004; Castro et al. 2003; Yun et al. 2006; Navara and Smith 2014), and agronomic traits (Spaner et al. 1999; Muñoz-Amatriaín et al. 2008; Romagosa et al. 1999). All of these limitations make it challenging to use the QTL results in breeding programs. The few successful cases of QTL utilization have been for traits with simple inheritance or where the QTL explains a large proportion of the phenotypic variance (Massman et al. 2011). Marker-assisted backcrossing has successfully been used in barley to select for leaf (van Berloo et al. 2001) and stripe rust (Toojinda et al. 1998), and barley yellow dwarf virus (Jefferies et al. 2003) among others. Markers can also be used for combining alleles from several genes at the same time (i.e., gene pyramiding). This has been a successful strategy especially for stacking disease resistance genes in barley to obtain durable disease resistance to stripe rust (Castro et al. 2003), barley yellow mosaic virus (Okada et al. 2004; Werner et al. 2005), and Fusarium head blight (Navara and Smith 2014) among others. See the previous section for examples with other traits.

While MAS has been used to select for favorable alleles at specific loci, GS is used to select for quantitative traits using a large number of markers deployed to predict the performance of the individuals skipping the significance test for any specific marker (Meuwissen et al. 2001). The principle consists of developing a prediction model based on a large population thoroughly studied for both molecular and phenotypic information (i.e., the training population) and using the model to predict phenotypic performance in instances where phenotyping is not suitable (i.e., early generation testing, off-season nurseries, difficult traits to phenotype; Heffner et al. 2009). Genomic selection has the potential to increase gain in plant breeding programs (Heffner et al. 2010; Rutkoski et al. 2010; Lorenz 2013; Hayes et al. 2013). GS was first developed in animal breeding (Meuwissen et al. 2001) and has since been transferred into plant breeding (Heffner et al. 2009). There has been extensive work in evaluating GS model performance (e.g., Solberg et al. 2009; Crossa et al. 2010; Heffner et al. 2011; Heslot et al. 2012; de los Campos et al. 2013; Zhao et al. 2012; Storlie and Charmet 2013; Longin et al. 2015), optimization of the training population in terms of population size (e.g., Lorenzana and Bernardo 2009; Asoro et al. 2011; Lorenz et al. 2012; Riedelsheimer and Melchinger 2013), number of markers (e.g., Lorenzana and Bernardo 2009; Asoro et al. 2011; Heffner et al. 2011), and the relationship among individuals (e.g., Asoro et al. 2011; Rincent et al. 2012; Crossa et al. 2013; Isidro et al. 2014; Albrecht et al. 2014; Tayeh et al. 2015; Lorenz and Smith 2015; Akdemir et al. 2015), accounting for genotype by environment interaction (Resende et al. 2011; Heffner et al. 2011; Burgueño et al. 2012; Dawson et al. 2013; Jarquín et al. 2014; Lado et al. 2016) and strategies for implementing GS in breeding programs (Zhong et al. 2009; Burgueño et al. 2012; Crossa et al. 2010; Heffner et al. 2011; Crossa et al. 2011; Heslot et al. 2012; Lorenzana and Bernardo 2009; Poland et al. 2012a, b; Zhao et al. 2012; Longin et al. 2015; Hofheinz et al. 2012; He et al. 2016; Michel et al. 2016; Hoffstetter et al. 2016; Lado et al. 2017a, b). Several studies have shown larger genetic gains per unit time from GS than from phenotypic selection in plants (Combs and Bernardo 2013; Asoro et al. 2013; Rutkoski et al. 2015; Beyene et al. 2015; Krchov et al. 2015; Longin et al. 2015; Marulanda et al. 2016; Sallam and Smith 2016). Additionally, larger genetic gains from GS than from MAS in plants has also been extensively documented (e.g., Massman et al. 2013; Asoro et al. 2013; Rutkoski et al. 2015). Some of these studies have specifically optimized genomic prediction or genomic selection strategies in barley (Zhong et al. 2009; Iwata and Jannink 2011; Lorenz et al. 2012; Mohammadi et al. 2015; Lorenz and Smith 2015; Sallam and Smith 2016; Neyhart et al. 2017).

The barley reference genome will be a valuable resource to expand the application and improve the effectiveness of GS. Because GS relies on LD between the QTL and the marker, higher marker density increases the probability of finding a marker in perfect LD with the QTL (de los Campos et al. 2013). Therefore, GS can act directly on the causative mutations present in whole genome sequences (Meuwissen et al. 2016), and increases in prediction accuracy with sequence data have been found (Brøndum et al. 2015). Additionally, dense markers can improve the accuracy of the predictions by improving the estimation of the realized relationship matrix among individuals due to Mendelian sampling (Crossa et al. 2017). The use of whole genome sequence data becomes relevant especially when high population structure is found (Meuwissen et al. 2016) such as when including large training datasets from multiple breeding programs, the use of two-row and six-row or winter and spring breeding programs. The use of whole genome sequence data might require the use of nonlinear methods to account for the full variability present (Meuwissen et al. 2016).

With respect to the breeder's equation, GS can increase genetic gain per unit time in a number of ways. First, selection accuracy can be improved in situations where the phenotype is irrelevant such as greenhouse evaluations, off-season nurseries, or any situation where the heritability of the trait is zero (e.g., environments were diseases are not expressed). Second, the genetic variance can be maximized by choosing specific parental combinations instead of blindly crossing the best parents (see section below). Third, the breeding cycle can be reduced by the use of genomic selection in combination with other technologies such as doubled haploids, rapid cycle single seed descent methods in greenhouses or the use of off-season nurseries. Although the gain per cycle might not be improved, by increasing the number of cycles per year, a larger genetic gain can be achieved in shorter periods. Finally, selection intensity can also be increased where cheaper technology allows for evaluation of larger populations. Certainly, commercial breeders are evaluating the use of genomic prediction models in their programs (Schmidt et al. 2016) and, at the very least, they will be using them for the early selection of parents to reduce the breeding cycle to 2–3 years (see Sect. 16.1).

16.4 Selecting Optimal Parent Combinations

Selecting parent combinations that are likely to produce superior progeny is one of the most critical steps in plant breeding. For the most part, parent selection is based primarily on the performance per se of an individual and those individuals that are most superior (e.g., highest yielding or least diseased) are selected to be used as parents. The mean of the selfed progeny from a cross between two inbred parents is expected to be equal to the mean of the parents in the absence of epistasis (Bernardo 2010, p. 76). Selecting parents with superior performance focuses on creating progeny populations with favorable means. The use of this approach in "good by good" breeding has been successful in improving agronomic and malting quality traits in barley (e.g., Condon et al. 2009).

One consequence of selecting elite parents for crosses is that over time in a relatively closed breeding population the parents will become more genetically similar (Condon et al. 2008). When both parents have equally favorable means and are genetically similar, the expectation is that the progeny will exhibit low variation for the trait. In terms of the breeder's equation, we are reducing genetic variance. The only way to make progress in breeding is to generate progeny that exceed the value of the parents (i.e., transgressive segregation). Thus, it is desirable to design parent combinations that not only produce a progeny population with a high mean but also a high variance. The usefulness of a cross is a function of both the mean and the variance of a family (Schnell and Utz 1975). In a case where two crosses are expected to have the same mean but different variances, with all else being equal, we would expect the cross with the higher variance to produce more superior progeny (Fig. 16.2).

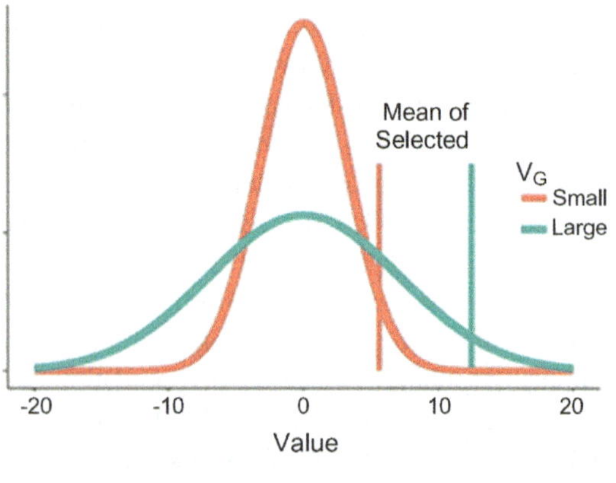

Fig. 16.2 Consider the distributions of two populations with the same mean value, but different variances (V_G). The mean of the superior progeny (i.e., top 10%; vertical lines) will be greater in a population with a larger variance (green) than in a population with a smaller variance (red). *Courtesy* J. Neyhart

It is easier to obtain a reliable estimate of the mean of a population than it is to obtain a reliable estimate of the variance but, from Fig. 16.2, it is clear that discriminating between populations on the basis of their genetic variances would greatly improve breeding progress. Several strategies for selecting parent combinations that should generate progeny populations with a large variance have therefore been evaluated in barley and other crops. While most studies report limited success predicting population variance, several were successful. For example, in oat, the coefficient of parentage as determined with trait data was effective at predicting progeny variance (Souza and Sorrells 1991). Other predictors that have been evaluated include the phenotypic difference between parents or various measures of the genetic distance between parents. In maize, the "distance" between parents as estimated by phenotype was much more predictive of progeny variance than using SSR or SNP (Hung et al. 2012) markers. In barley for two of five traits, the phenotypic distance between parents was significantly associated with genetic variance in the progeny while the genetic distance between parents as determined by markers was not (Kuczynska et al. 2007). Based on these examples and others, the phenotypic difference between parents is only modestly predictive of progeny variance but superior to genetic distance based on markers.

It is somewhat surprising that genetic distance is such a poor predictor of progeny variance. One explanation could be that genetic distance is derived from all (genome-wide) markers while the trait of interest is associated with only a subset of loci. Mohammadi et al. (2015) incorporated genetic marker information into an approach that uses marker effect estimates and simulated progeny to predict the progeny variance from parental combinations in barley. They used this simulation approach to predict the mean of the best 10% of the progeny from many simulated crosses and determined the extent to which the predicted mean of the population and the predicted variance of the population was associated with superior progeny mean. The population mean explained 82 and 88% of the superior progeny mean, for yield and mycotoxin content, respectively. However, adding the predicted population variance increased those correlations to nearly 100%. This procedure, called PopVar, was later validated using empirical disease data from barley breeding populations where it was more predictive of progeny variance than several methods using phenotypic difference and genetic distance between parents (Tiede et al. 2015). PopVar was also used to choose parental combinations for other traits in wheat revealing that modeling the variance was not very important for grain yield but was more important for grain quality traits (Lado et al. 2017a, b). The success of this approach depends on whether there are sufficient differences in variance among candidate crosses in order to select those predicted to have superior progeny and the reliability of marker effect estimates used in the simulation of progeny.

The availability of the first complete barley reference genome should offer possibilities to refine PopVar and other similar approaches (Mascher et al. 2017). Access to large numbers of informative markers may improve the ability to estimate marker effects that more completely capture the genetic variation in populations. Improved distribution of markers throughout the genome should improve the ability to capture recombination events in simulated populations. Lastly, knowledge of the genetic and physical positions of markers will provide more accurate descriptions of recombination in the genome that could be incorporated into methods that simulate progeny from parental crosses. Automated scripting of data analysis pipelines should enable breeders with trait and genotype data to more easily incorporate procedures that select potentially optimal parent combinations in addition to superior parents.

16.5 Efficiently Utilizing Exotic Germplasm for Novel Alleles

If a breeder cannot find sufficient genetic variation among elite lines within their breeding population, it may be necessary to utilize exotic germplasm to provide useful variation for crop improvement. In barley, breeders recognized this potential over 100 years ago when studying the inheritance of mildew resistance in crosses of cultivated barley with its wild progenitor, *H. spontaneum* (Biffen 1907). This resistance became known as the "spontaneum resistance" conditioned by the *Mla6* allele at the major Mla locus on chromosome 1H. For the next 70 years, researchers searched germplasm collections for additional sources of resistance, principally to powdery mildew, as described by (Wiberg 1974). Virtually all of these resistance sources came from unadapted sources and some were the mainstay of resistance breeding for the majority of the twentieth century. Most of the resistant alleles were dominant and were introduced through backcross or multi-parent mating designs where selection for the resistant phenotype could be conducted in the F1 generation provided there was a virulent form of mildew present. While these schemes proved efficient in introducing the desired resistance, they were far from precise and often resulted in the introgression of deleterious linked characters, often referred to as linkage drag, due to the relatively large size of the donor segment retained. This was highlighted by the finding that barley cultivars with the *Hordeum laevigatum* mildew resistance tended to have harder endosperms, higher grain beta-glucan content, and lower alpha-amylase levels than those without (Swanston 1987). The use of markers has also facilitated the rapid introgression of novel quality characters from exotic germplasm. A novel allele at the beta-amylase locus Bmy2 was identified in a collection of *H. spontaneum* lines and a marker developed to enable its transfer into adapted germplasm (Paris et al. 2002).

While there are a number of other examples of the use of exotic germplasm to identify novel alleles for barley improvement, these are largely for major genes with well characterized phenotypes that can either be used to directly screen for novel variants or to map the controlling gene and develop an efficient marker system for use in marker-assisted selection (MAS). The challenge is to identify beneficial sources of variation for quantitative traits. There are two possible approaches to this. One is to use diverse sources of exotic germplasm in a combined phenotyping and genotyping study and the other is to utilize DNA sequence information coupled with knowledge of biochemical pathways and/or localization of quantitative trait loci (QTL) to screen germplasm for DNA variants.

Several approaches have been devised to screen exotic germplasm phenotypically and genotypically and all require crossing and segregating population development. These include recombinant chromosome substitution lines, advanced backcross QTL population, and nested association mapping populations. All three of these approaches have been used to study *H. spontaneum* lines for barley improvement. Matus et al. (2003) report the generation of a set of RCSLs for barley that were subsequently used to identify genomic regions from a wild barley accession that improved tolerance to terminal drought stress (del Pozo et al. 2012). The first AB-QTL project reported in barley was a study of a population of DH lines derived from the BC1F2 of a cross between the variety Barke and the *H. spontaneum* accession HOR 11508 that found beneficial sources of variation from the wild barley that could be deployed in drought-prone environments (Talame et al. 2004). A number of other AB-QTL studies of populations derived from crosses between different *H. spontaneum* accessions and spring barley cultivars have been published since then with all reporting beneficial alleles from the exotic parent. For example, Sayed et al. (2012) reported alleles that increased proline accumulation and reduced leaf wilting derived from the wild barley accession ISR42-8 in an AB-QTL population derived from a cross between it and the spring barley Scarlett. The AB-QTL approach is, like the RCSL approach, limited to studying one or two exotic lines. The NAM method provides a means of assessing a slightly broader range of exotic germplasm as 10-30 accessions can be combined in one study. The brittle rachis of *H. spontaneum* presents a problem in using such accessions as direct parents in constructing NAM populations but can be overcome by combining the advanced backcross approach to eliminate or reduce the deleterious effects of exotic germplasm in a NAM crossing design to maximize the germplasm that can be usefully phenotyped for yield characters. A major advantage of this approach is that it results in more rapid decay of linkage disequilibrium than either AB-QTL or recombinant inbred line populations (Nice et al. 2016). This has enabled the localization of a QTL that improves yield under saline conditions by 30% to a 5 cM region on chromosome 2H that included several potential candidate genes (Saade et al. 2016).

The NAM approach still can only be applied to relatively few accessions from breeders' unadapted gene pools and the large-scale assessment of genebank accessions for beneficial alleles remains a challenge. Where potential candidate genes are known, screening large collections of lines can be done efficiently by pooling DNA samples and then screening for sequence variants. This approach has been applied to mutant populations using Targeting Induced Local Lesions IN Genomes (TILLING; McCallum et al. 2000). This approach has been used to develop the TILMORE populations of barley mutants (Talame et al. 2008) and can also be adapted to screen germplasm collections in the EcoTILLING method, which has been shown to identify sequence variants at two major powdery mildew resistance loci in barley (Mejlhede et al. 2006). The availability of extensive sequence data means that this process can now be accelerated and applied to large-scale screening of genetic resources through the deployment of exome capture arrays. An exome capture array has been developed for barley (Mascher et al. 2013b) and provides a means for the high throughput functional genotyping of collections of exotic barley germplasm to identify novel variants.

16.6 Exploiting Genomic Regions with Limited Recombination

Recombination is the major force behind the generation of genetic variation in breeding populations. From a breeder's perspective, more recombination is desirable in regions of the genome where unfavorable and favorable genes are physically linked. In these regions, recombination results in the independent assortment of alleles and the possibility to select individuals with fewer unfavorable linkages. On the other hand, less recombination is desirable in regions of the genome where blocks of favorable alleles at linked loci have been assembled through selection. Thus, knowledge of the patterns of recombination in the barley genome relative to genes that control important breeding traits would be incredibly useful to breeders.

The distribution of recombination events is not homogeneous throughout the genome with a predominantly distal location of chiasmata in barley (Higgins et al. 2012). This effect means that the relationship between the physical and genetic map varies along a chromosome with distances estimated as ranging from 0.3 to over 250 MB/cM on chromosome 3H in barley (Künzel and Waugh 2002). This trend is apparent in early barley linkage maps where markers are concentrated in the centromeric regions (e.g., Hayes et al. 1993). Using the physical positions of BOPA1 and BOPA2 SNP markers for chromosome 6H of the Steptoe × Morex population it is clear that there is dramatically reduced recombination for a large portion of the chromosome (Fig. 16.3). This can lead to association of unfavorable alleles at linked loci as recombination events are so rare that disrupting such linkage blocks rarely occurs. For example, the winter barley variety Igri carries a resistance

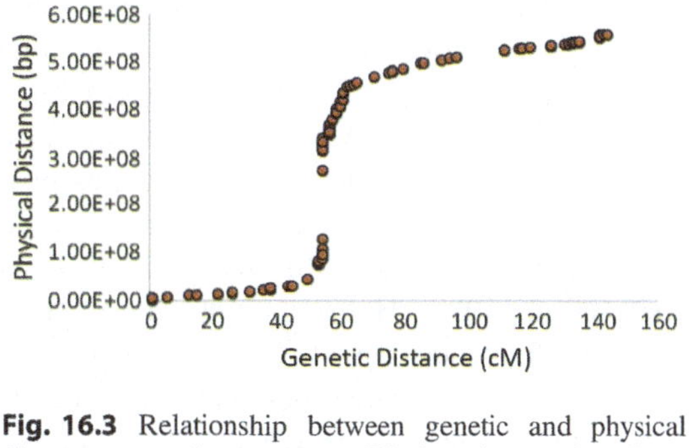

Fig. 16.3 Relationship between genetic and physical distance on chromosome 6H based on the Steptoe × Morex mapping populations using 109 BOPA1 and BOPA2 SNP markers

factor to net blotch (*Pyrenophora teres*) that is within 0.8 cM of the RFLP marker cMWG680 located in the centromeric region of chromosome 3H (Graner et al. 1996) yet the same marker co-segregates with a resistance locus for scald (*Rhynchosporium commune*) with Igri carrying the susceptible allele. This means that the expectation is that 1/125 progeny from a repulsion cross will carry resistant alleles to both diseases. Since the frequency of generating such a favorable recombinant is low, breeding must be done in two steps. First, one must assemble the favorable recombinant and then subsequently utilize that recombinant in breeding for other traits.

There are at least two ways that reduced recombination creates obstacles to progress in breeding. The first is the case of an unfavorable correlation between two traits. Breeders are always imposing selection on multiple traits in breeding populations and inevitably encounter two traits that are unfavorably correlated. There are several cases of correlated traits mapping to the same genomic region in barley. One of the notorious examples in barley involves FHB disease resistance which is unfavorably correlated with other agronomic traits. Two loci of major effect for FHB on chromosomes 2H and 6H have been identified in several populations including Gobernadora (Zhu et al. 1999), Chevron (de la Peña et al. 1999; Ma et al. 2000), Fredericksen (Mesfin et al. 2003), Zhedara (Dahleen et al. 2003), and CIho4196 (Horsley et al. 2006). The FHB resistance mechanism on chromosome 2H was characterized as early expression (i.e., before infection) of defense-related genes: cell reinforcement, PR proteins, and antimicrobial metabolites (Huang et al. 2016). This genomic region on 2H also maps with heading date (Steffenson 2003; Nduulu et al. 2007; Massman et al. 2011), where disease resistance is found on late heading individuals (Mesfin et al. 2003; de la Peña et al. 1999; Horsley et al. 2006). This unfavorable association has been documented as a series of closely linked loci, and recombinants were found when large population sizes and fine mapping with near-isogenic lines (NIL) were used (Nduulu et al. 2007). On the other hand, the 6H locus for FHB seems to be induced by the presence of the fungus (Huang et al. 2016) and maps together with several other agronomic traits. This region on chromosome 6H harbors a locus of major effect for GPC (Canci et al. 2003, 2004; de la Peña et al. 1999) with at least three alleles: low (Karl), moderate (Midwest six-row barley), and high (Chevron) GPC (Heidlebaugh et al. 2008; Jukanti et al. 2008; Jukanti and Fischer 2008; Mickelson et al. 2003). Regardless of the large effort to elucidate the genetic mechanisms of this unfavorable correlation between FHB and GPC, it remains largely unknown whether FHB and GPC on 6H are tightly linked loci or pleiotropic (Canci et al. 2003). Finally, this region on chromosome 6H is also associated with several agronomic traits including other disease resistances (Mascher et al. 2013a; Grewal et al. 2008; Richter et al. 1998; Cakir et al. 2003; Canci et al. 2004; Ma et al. 2000; Steffenson et al. 1996), heading date (Jukanti and Fischer 2008; Lacerenza et al. 2010), yield (Distelfeld et al. 2014), senescence, nitrogen storage and mobilization (Mickelson et al. 2003), lodging (Skov Kristensen et al. 2016), plant height, and morphological traits (Druka et al. 2011) as well as DON accumulation. It is believed that the association between senescence and nitrogen accumulation is pleiotropic (Jukanti and Fischer 2008). However, the association of FHB severity and DON accumulation are believed to be different loci tightly linked together because moderate FHB resistant lines have been found with high DON accumulation (Lamb et al. 2009) and similar FHB severity were found with segregation for DON accumulation (Smith et al. 2004). Other regions of tightly linked QTL or genes have been resolved by fine mapping with large populations and NILs for several malting quality traits on both chromosome 1H (malt extract, alpha-amylase, and diastatic power; Han et al. 1997) and 4H (malt extract alpha-amylase, diastatic power, and beta-glucan content; Gao et al. 2004). The genetic basis of trait correlation is either tight linkage of independent genes each controlling separate traits or pleiotropy. In most cases, it is very difficult to distinguish the two. If enough recombinants are screened to recover individuals in which the two traits are uncoupled, it is reasonable to draw the conclusion of linkage as the mechanism. However, the inability to recover such individuals is not sufficient evidence to conclude pleiotropy. To make this conclusion, one must demonstrate unambiguously that a single gene is responsible for both traits. This can be accomplished through gene silencing or direct gene modification of the gene suspected to control both traits.

A second way that reduced recombination creates obstacles to progress in breeding is through linkage of unfavorable alleles that control the same trait. If a series of closely linked genes with favorable and unfavorable allelic effects are always inherited as a unit then under simple additive model the effect of the inherited unit will be the sum of the effects. Thus, without recombination in this region, potentially useful alleles will go undiscovered in genetic mapping studies and would not contribute to predictions from genomic selection approaches. Thus, with the current methods of QTL mapping using low-resolution mapping, QTL in such regions would be undetectable. Exploring regions suspected to carry repulsion phase linkages by generating large numbers of recombinants in targeted regions may reveal the so-called fractionated QTL (Johnson et al. 2012). Discovering and exploring these regions could reveal useful genetic variation that is currently unexploited in breeding populations. Selecting for rare recombinants by screening large segregating populations or through targeted recombination may allow breeders to exploit this untapped resource (Bernardo 2017).

16.7 Improving Barley Through Targeting Deleterious Alleles

Barley improvement to date has relied on phenotypic information to impose direct selection or to develop marker prediction strategies for indirect selection. Strategies such as MAS and GS (discussed above) require phenotypic information, preferably from germplasm closely related to the target breeding population, to develop models (from single locus to genome-wide) that predict the phenotype of selection candidates. In the case of MAS, phenotypic information is associated with genotype using a mapping population to identify markers that are subsequently used for selection in segregating breeding populations. Genomic selection extends this approach by predicting the phenotype using all genome-wide marker effects. Both of these approaches require evaluating individuals for a phenotype in order to develop predictions that are based on DNA sequence.

The cumulative knowledge gained from studying gene function and crop genomes provides an opportunity to make inferences about the relative value of a gene based on its DNA sequence alone. Sequence variants that substantially alter the length or amino acid sequence of a protein from its ancestral state are more likely to have a negative effect on function and reduce fitness. These deleterious alleles would be expected to be eliminated from the population due to purifying selection. The effectiveness of purifying selection, in a self-pollinated species that is predominantly homozygous, is determined by the effective population size (N_e) and the relative disadvantage or selective coefficient (s) for one homozygous class over the other. If $N_e s$ is less than one, then variants are primarily subjected to genetic drift rather than selection. Thus, deleterious alleles (small values of s) will be subjected to selection and eventually purged from the population if N_e is large enough. However, if Ne is small as is often the case in breeding populations, deleterious alleles can persist in the population. In addition, regions of the genome with low effective recombination are also expected to accumulate deleterious mutations through a mechanism referred to as "Muller's ratchet" (Felsenstein 1974; Muller 1964). Persistent deleterious alleles segregating in breeding populations could be a new target for improving yield (i.e., fitness) in plant breeding (Morell et al. 2011).

Identification of putatively deleterious alleles based on sequence information alone is not limited to premature stop or nonsense mutations. Changes in DNA sequence that alter the amino

Table 16.2 Categories of SNPs from a diverse set of barley cultivars and the parents of a spring six-row breeding population at the University of Minnesota

Sample	Putatively deleterious[a]	Differences from Morex	Noncoding	Coding	Synonymous	Non-synonymous	Nonsense
Barke	1070	162,954	130,832	32,122	17,800	14,322	86
Bonus	882	135,540	108,762	26,778	15,184	11,594	65
Borwina	894	139,222	111,613	27,609	15,699	11,910	62
Bowman	998	130,335	105,575	24,760	13,645	11,115	77
Foma	1151	156,846	126,608	30,238	16,746	13,492	84
Gull	885	128,671	103,143	25,528	14,555	10,973	55
Harrington	1063	153,203	124,264	28,939	15,995	12,944	89
Haruna Nijo	1078	155,245	125,143	30,102	16,655	13,447	83
Igri	1077	161,224	130,100	31,124	17,275	13,849	98
Kindred	931	86,932	69,996	16,936	9458	7478	34
Morex[b]	700	5239	4602	637	253	384	4
Steptoe	1061	148,785	120,245	28,540	15,839	12,701	91
Vogelsanger Gold	910	146,303	117,557	28,746	16,168	12,578	70
FT11	1220	243,049	195,009	48,040	26,886	21,154	147
OUH602	1174	197,405	158,377	39,028	21,701	17,327	115
Breeding Parents[c]	659	146,279	111,488	34,791	17,764	17,027	242

[a]From supplemental Table 4 in Kono et al. (2016)
[b]Reference genome for SNP comparison
[c]Set of 21 parents that were founders of a new six-row spring breeding population developed at the University of Minnesota (Kono et al. unpublished)

acid sequence (non-synonymous) can modify the function of proteins and thus affect fitness. Bioinformatic approaches have been developed to determine which amino acid changes are likely to have deleterious effects based on DNA sequence conservation (Ng 2003; Adzhubei et al. 2010; Chun and Fay 2009). These approaches use sequence databases to determine whether a non-synonymous variant occurs in a region that is conserved across a phylogenetic taxon. If the variant is rare, particularly across related species, it is presumed to be deleterious. Beyond SNPs, other variants would include insertions and deletions that can also alter gene function. Thus, a variety of approaches can be used to assess sequence variants and predict the likelihood that they are deleterious.

Access to a complete reference genome and accessible resequencing technology enables the possibility to impose selection based entirely on the genotype with no phenotypic information. An alternative approach to barley improvement would be to identify unfavorable or deleterious alleles and reduce their frequency in the breeding population. Kono et al. (2016) characterized SNPs for a diverse set of 15 barley genotypes and found that 8.4–8.8% of the SNPs relative to Morex were non-synonymous (Table 16.2). They used the intersection of three methods based on sequence conservation to identify putatively deleterious mutations and found between 882 and 1220 per genome. This range excluded Morex as that was the reference genome used for comparison. Furthermore, ~40% of these putatively deleterious SNPs were private to an individual cultivar (Kono et al. 2016). Taken together, this indicates that there are a large number of potential deleterious alleles that could be reduced or eliminated through breeding.

The number of deleterious alleles carried by an individual determines the genetic load and is a predictor of the overall fitness. In an agronomic context, it can be argued that yield is analogous to fitness. In a spring six-row breeding population, SNPs categorized through bioinformatics as putatively deleterious explained a higher proportion of the variance for yield than coding SNPs, non-synonymous SNPs, and all SNPs (Kono et al. unpublished). Since genome-wide SNP breeding methods (i.e., GS) rely predominantly on common alleles, using information on rare putatively deleterious alleles could provide additional insight into breeding. It remains to be established how prevalent the differences in gene content and copy number observed between different maize genotypes (Swanson-Wagner et al. 2010) is in barley, but the availability of a high-quality reference genome will improve our ability to estimate such effects and predict whether or not these will contribute toward variation in deleterious alleles.

Assuming that deleterious alleles identified through bioinformatics have some predictive ability for yield, how could they be used in breeding? Since it appears that many putatively deleterious alleles are private, breeders could design crosses between elite performing lines that would maximize the segregation of deleterious alleles and select segregating individuals with minimal genetic load. Another possibility would be to first implement genomic selection and then impose selection for reduced genetic load. Of course, all of these approaches would depend on the genome distribution of deleterious alleles. Simulation studies show that deleterious alleles are more likely to accumulate near loci that are under positive selection (Hartfield and Otto 2011). This would suggest that large populations would be needed to generate sufficient recombination to break up the linkage between large effect favorable alleles and linked deleterious alleles. If such an approach seems impractical, then restoring function to deleterious alleles with larger effects through gene editing may be an alternate approach.

References

Adzhubei IA, Schmidt S, Peshkin L, Ramensky VE, Gerasimova A, Bork P, Kondrashov AS, Sunyaev SR (2010) A method and server for predicting damaging missense mutations. Nat Methods 7:248–249

Ahmad Naz A, Ehl A, Pillen K, Léon J (2012) Validation for root-related quantitative trait locus effects of wild origin in the cultivated background of barley (*Hordeum vulgare* L.). Plant Breed 3:392–398

Akdemir D, Sanchez JI, Jannink JL (2015) Optimization of genomic selection training populations with a genetic algorithm. Genet Sel Evol 47:38

Albrecht T, Auinger HJ, Wimmer V, Ogutu JO, Knaak C, Ouzunova M, Piepho H-P, Schön CC (2014) Genome-based prediction of maize hybrid performance across genetic groups, testers, locations, and years. Theor Appl Genet 127(6):1375–1386

Asoro FG, Newell MA, Beavis WD, Scott MP, Jannink JL (2011) Accuracy and training population design for genomic selection on quantitative traits in elite North American oats. Plant Genome 4:132

Asoro FG, Newell MA, Beavis WD, Scott MP, Tinker NA, Jannink JL (2013) Genomic, marker-assisted, and pedigree-BLUP selection methods for β-glucan concentration in elite oat. Crop Sci 53(5): 1894–1906

Barr AR, Jefferies SP, Warner P, Moody DB, Chalmers KJ, Langridge P (2000) Marker assisted selection in theory and practice. In: The 8th international barley genetics symposium Australia, 2000

Barua UM, Chalmers KJ, Hackett CA, Thomas WTB, Powell W, Waugh R (1993) Identification of RAPD markers linked to a *Rhynchosporium secalis* resistance locus in barley using near-isogenic lines and bulked segregant analysis. Heredity 71(2):177–184

Bayer MM, Rapazote-Flores P, Ganal M, Hedley PE, Macaulay M, Plieske J, Ramsay L, Russell J, Shaw PD, Thomas WTB, Waugh R (2017) Development and evaluation of a barley 50k iSelect SNP array. Front Plant Sci 8:1792

Beattie AD, Edney MJ, Scoles G, Rossnagel B (2010) Association mapping of malting quality data from Western Canadian two-row barley cooperative trials. Crop Sci 50:1649–1662. https://doi.org/10.2135/cropsci2009.06.0334

Beavis WD (1998) QTL analyses: power, precision, and accuracy. In: Paterson AH (ed) Molecular dissection of complex traits. CRC Press, New York, pp p145–p162

Bernardo R (2010) Breeding for quantitative traits in plants, 2nd edn. Stemma Press, Minnesota

Bernardo R (2017) Prospective targeted recombination and genetic gains for quantitative traits in maize. Plant Genome 10. https://doi.org/10.3835/plantgenome2016.11.0118

Beyene Y, Semagn K, Mugo S, Tarekegne A, Babu R, Meisel B, Sehabiague P et al (2015) Genetic gains in grain yield through genomic selection in eight bi-parental maize populations under drought stress. Crop Sci 55(1):154–163

Biffen R (1907) Studies in the inheritance of disease-resistance. J Agric Sci 2:109–128

Bradbury P, Parker T, Hamblin MT, Jannink JL (2011) Assessment of power and false discovery rate in genome-wide association studies using the BarleyCAP germplasm. Crop Sci 51:52–59

Bringhurst TA (2015) 125th anniversary review: barley research in relation to Scotch whisky production: a journey to new frontiers. J Inst Brew 121:1–18. https://doi.org/10.1002/jib.192

Brøndum RF, Su G, Janss L, Sahana G, Guldbrandtsen B, Boichard D, Lund MS (2015) Quantitative trait loci markers derived from whole genome sequence data increases the reliability of genomic prediction. J Dairy Sci 98:4107–4116

Burgueño J, de los Campos G, Weigel K, Crossa J (2012) Genomic prediction of breeding values when modeling genotype × environment interaction using pedigree and dense molecular markers. Crop Sci 52:707–719

Cahill DJ, Schmidt DH (2004) Use of marker-assisted selection in a product development breeding program. In: Fischer T (ed) New directions for a diverse planet. Proceedings of the 4th international crop science congress, Brisbane, QLD, Australia 26 Sept–1 Oct 2004

Cakir M, Gupta S, Platz GJ, Ablett GA, Loughman R, Emebiri LC, Poulsen D, Li CD, Lance RCM, Galwey NW, Jones MGK (2003) Mapping and validation of the genes for resistance to *Pyrenophora teres* f. teres in barley (*Hordeum vulgare* L.). Crop Pasture Sci 54:1369–1377

Canci PC, Nduulu LM, Dill-Macky R, Muehlbauer GJ, Rasmusson DC, Smith KP (2003) Genetic relationship between kernel discoloration and grain protein concentration in barley. Crop Sci 43(5):1671–1679

Canci PC, Nduulu LM, Muehlbauer GJ, Dill-Macky R, Rasmusson DC, Smith KP (2004) Validation of quantitative trait loci for Fusarium head blight and kernel discoloration in barley. Mol Breed 14:91–104. https://doi.org/10.1023/B:MOLB.0000037998.27661.58

Castro AJ, Capettini F, Corey AE, Filichkina T, Hayes PM, Kleinhofs A, Kudrna D, Richardson K, Sandoval-Islas S, Rossi C, Vivar H (2003) Mapping and pyramiding of qualitative and quantitative resistance to stripe rust in barley. Theor Appl Genet 107:922–930

Charcosset A, Moreau L (2004) Use of molecular markers for the development of new cultivars and the evaluation of genetic diversity. Euphytica 137(1):81–94

Christopher J, Richard C, Chenu K, Christopher M, Borrell A, Hickey L (2015) Integrating rapid phenotyping and speed breeding to improve stay green and root adaptation of wheat in changing, water-limited, Australian environments. In: Edwards D, Oldroyd G (eds) Agriculture and climate change - adapting crops to increased uncertainty (vol 29). Procedia Environmental Sciences, pp 175–176. https://doi.org/10.1016/j.proenv.2015.07.246

Chun S, Fay JC (2009) Identification of deleterious mutations within three human genomes. Genome Res 19:1553–1561

Chutimanitsakun Y, Cuesta-Marcos A, Chao S, Corey A, Filichkin T, Fisk S, Kolding M, Meints B, Ong YL, Rey JI, Ross AS (2013) Application of marker-assisted selection and genome-wide association scanning to the development of winter food barley germplasm resources. Plant Breed 132(6):563–570

Close TJ, Bhat P, Lonardi S, Wu Y, Rostoks N, Ramsay L, Druka A, Stein N, Svensson J, Wanamaker S, Bozdag S, Roose M, Moscou M, Chao S, Varshney R, Szucs P, Sato K, Hayes P, Matthews D, Kleinhofs A, Muehlbauer G, DeYoung J, Marshall D, Madishetty K, Fenton R, Condamine P, Graner A, Waugh R (2009) Development and implementation of high-throughput SNP genotyping in barley. BMC Genomics 10(1):582. https://doi.org/10.1186/1471-2164-10-582

Cockram J, White J, Leigh FJ, Lea VJ, Chiapparino E, Laurie DA, Mackay IJ, Powell W, O'Sullivan DM (2008) Association mapping of partitioning loci in barley. BMC Genet 9(1):16

Cockram J, White J, Zuluaga DL, Smith D, Comadran J, Macaulay M, Luo Z, Kearsey MJ, Werner P, Harrap D, Tapsell C, Liu H, Hedley PE, Stein N, Schulte D, Steuernagel B, Marshall DF, Thomas WTB, Ramsay L, Mackay I, Balding DJ, Consortium TA, Waugh R, O'Sullivan DM (2010) Genome-wide association mapping to candidate polymorphism resolution in the unsequenced barley genome. PNAS 107:21611–21616

Cockram J, Jones H, Norris C, O'Sullivan DM (2012) Evaluation of diagnostic molecular markers for DUS phenotypic assessment in the cereal crop, barley (*Hordeum vulgare* ssp. *vulgare* L.). Theor Appl Genet 125 (8):1735–1749. https://doi.org/10.1007/s00122-012-1950-3

Comadran J, Kilian B, Russell J, Ramsay L, Stein N, Ganal M, Shaw P, Bayer M, Thomas W, Marshall D, Hedley P, Tondelli A, Pecchioni N, Francia E, Korzun V, Walther A, Waugh R (2012) Natural variation in a homolog of *Antirrhinum CENTRORADIALIS* contributed to spring growth habit and environmental adaptation in cultivated barley. Nat Genet 44(12):1388–1392. https://doi.org/10.1038/ng.2447

Combs E, Bernardo R (2013) Genomewide selection to introgress semidwarf maize germplasm into US corn belt inbreds. Crop Sci 53:1427–1436

Condon F, Gustus C, Rasmusson DC, Smith KP (2008) Effect of advanced cycle breeding on genetic diversity in barley breeding germplasm. Crop Sci 48(3):1027–1036. https://doi.org/10.2135/cropsci2007.07.0415

Condon F, Rasmusson DC, Schiefelbein E, Velasquez G, Smith KP (2009) Effect of advanced cycle breeding on genetic gain and phenotypic diversity in barley breeding germplasm. Crop Sci 49:1751–1761. https://doi.org/10.2135/cropsci2008.10.0585

Costa JM, Corey A, Hayes PM, Jobet C, Kleinhofs A, Kopisch-Obusch A, Kramer SF, Kudrna D, Li M, Riera-Lizarazu O, Sato K, Szucs P, Toojinda T, Vales MI, Wolfe RI (2001) Molecular mapping of the Oregon Wolfe Barleys: a phenotypically polymorphic doubled-haploid population. Theor Appl Genet 103 (2–3):415–424

Cregan PB, Mudge J, Fickus EW, Danesh D, Denny R, Young ND (1999) Two simple sequence repeat markers to select for soybean cyst nematode resistance conditioned by the rhg1 locus. Theor Appl Genet 99:811–818

Crosbie TM, Eathington SR, Johnson GR, Edwards M, Reiter R, Stark S, Mohanty RG, Oyervides M, Buehler RE, Walker AK, Dobert R, Delannay X, Pershing JC, Hall MA, Lamkey KR (2006) Plant breeding: past, present, and future. In: Lamkey KR, Lee M (eds) Plant breeding: the Arnel R. Hallauer international symposium. Blackwell, Ames, pp 3–50

Crossa J, de los Campos G, Pérez P, Gianola D, Burgueño J, Araus JL, Makumbi D, Singh RP, Dreisigacker S, Yan J, Arief V, Banziger M, Braun H-J (2010) Prediction of genetic values of quantitative traits in plant breeding using pedigree and molecular markers. Genetics 186:713–724 https://doi.org/10.1534/genetics.110.118521

Crossa J, Pérez P, de los Campos G, Mahuku G, Dreisigacker S et al (2011) Genomic selection and prediction in plant breeding. J Crop Improv 25:239–261

Crossa J, Beyene Y, Kassa S, Pérez P, Hickey JM, Chen C, de los Campos G, Burgueño J, Windhausen VS, Buckler E, Jannink JL, Lopez Cruz MA, Babu R (2013) Genomic prediction in maize breeding populations with genotyping-by-sequencing. G3 3: 1903–1926

Crossa J, Pérez-Rodríguez P, Cuevas J, Montesinos-López O, Jarquín D, de los Campos G, Burgueño J, Camacho-González JM, Pérez-Elizalde S, Beyene Y, Dreisigacker S (2017) Genomic selection in plant breeding: methods, models, and perspectives. Trends Plant Sci. https://doi.org/10.1016/j.tplants.2017.08.011

Cuesta-Marcos A, Igartua E, Codesal P, Russell JR, Molina-Cano JL, Moralejo M, Szűcs P, Gracia MP, Lasa JM, Casas AM (2008) Heading date QTL in a spring × winter barley cross evaluated in Mediterranean environments. Mol Breed 21(4):455–471

Cuesta-Marcos A, Casas AM, Hayes PM, Gracia MP, Lasa JM, Ciudad F, Codesal P, Molina-Cano JL, Igartua E (2009) Yield QTL affected by heading date in Mediterranean grown barley. Plant Breed 128(1): 46–53

Cuesta-Marcos A, Szűcs P, Close TJ, Filichkin T, Muehlbauer GJ, Smith KP, Hayes PM (2010) Genome-wide SNPs and re-sequencing of growth habit and inflorescence genes in barley: implications for association mapping in germplasm arrays varying in size and structure. BMC Genom 11(1):707

Dahleen LS, Agrama HA, Horsley RD, Steffenson BJ, Schwarz PB, Mesfin A, Franckowiak JD (2003) Identification of QTLs associated with Fusarium head blight resistance in Zhedar 2 barley. Theor Appl Genet 108(1):95–104

Dawson JC, Endelman JB, Heslot N, Crossa J, Poland J, Dreisigacker S, Manès Y, Sorrells ME, Jannink J-L (2013) The use of unbalanced historical data for genomic selection in an international wheat breeding program. Field Crop Res 154:12–22

de la Peña RC, Smith KP, Capettini F, Muehlbauer GJ, Gallo-Meagher M, Dill-Macky R, Somers DA, Rasmusson DC (1999) Quantitative trait loci associated with resistance to Fusarium head blight and kernel discoloration in barley. Theor Appl Genet 99:561–569

de los Campos G, Vazquez AI, Fernando R, Klimentidis YC, Sorensen D (2013) Prediction of complex human traits using the genomic best linear unbiased predictor. PLoS Genet 9:e1003608

del Pozo A, Castillo D, Inostroza L, Matus I, Mendez AM, Morcuende R (2012) Physiological and yield responses of recombinant chromosome substitution lines of barley to terminal drought in a Mediterranean-type environment. Ann Appl Biol 160(2):157–167. https://doi.org/10.1111/j.1744-7348.2011.00528

Distelfeld A, Avni R, Fischer AM (2014) Senescence, nutrient remobilization, and yield in wheat and barley. J Exp Bot 65(14):3783–3798. https://doi.org/10.1093/jxb/ert477

Druka A, Franckowiak J, Lundqvist U, Bonar N, Alexander J, Houston K, Waugh R (2011) Genetic dissection of barley morphology and development. Plant Physiol 155(2):617–627. https://doi.org/10.1104/pp.110.166249

Dubcovsky J (2004) Marker-assisted selection in public breeding programs: the wheat experience. Crop Sci 44:1895–1898

Dubcovsky J, Chen C, Yan L (2005) Molecular characterization of the allelic variation at the VRN-H2 vernalization locus in barley. Mol Breed 15(4):395–407

Eathington SR (2005) Practical applications of molecular technology in the development of commercial maize hybrids. In: Proceedings of the 60th annual corn and sorghum seed research conference, 7–9 Dec 2005. American Seed Trade Association, Washington, DC

Fehr WR (1991) Principles of cultivar development. Vol. 1. Theory and technique. Macmillan Publishing Company, London

Felsenstein J (1974) The evolutionary advantage of recombination. Genetics 78(2):737–756

Francia E, Rizza F, Cattivelli L, Stanca AM, Galiba G, Toth B, Hayes PM, Skinner JS, Pecchioni N (2004) Two loci on chromosome 5H determine low-temperature tolerance in a 'Nure' (winter) × 'Tremois' (spring) barley map. Theor Appl Genet 108(4): 670–680

Friedt W, Horsley RD, Harvey BL, Poulsen DME, Lance RCM, Ceccarelli S, Grando S, Capettini F (2010) Barley breeding history, progress, objectives, and technology. In: Barley. Wiley-Blackwell, pp 160–220. https://doi.org/10.1002/9780470958636.ch8

Gao W, Clancy JA, Han F, Jones BL, Budde A, Wesenberg DM, Kleinhofs A, Ullrich SE (2004) Fine mapping of malting-quality QTL complex near the chromosome 4HS telomere in barley. Theor Appl Genet 109:750–760

Graner A, Bauer E (1993) RFLP mapping of the ym4 virus resistance gene in barley. Theor Appl Genet 86 (6):689–693

Graner A, Jahoor A, Schondelmaier J, Siedler H, Pillen K, Fischbeck G, Wenzel G, Herrmann RG (1991) Construction of an RFLP map of barley. Theor Appl Genet 83(2):250–256

Graner A, Foroughi-Wehr B, Tekauz A (1996) RFLP mapping of a gene in barley conferring resistance to net blotch (*Pyrenophora teres*). Euphytica 91:229–234

Graner A, Streng S, Kellermann A, Schiemann A, Bauer E, Waugh R, Pellio B, Ordon F (1999) Molecular mapping and genetic fine-structure of the rym5 locus encoding resistance to different strains of the Barley Yellow Mosaic Virus Complex. Theor Appl Genet 98(2):285–290. https://doi.org/10.1007/s001220051070

Grewal TS, Rossnagel BG, Pozniak CJ, Scoles GJ (2008) Mapping quantitative trait loci associated with barley net blotch resistance. Theor Appl Genet 116(4):529–539. https://doi.org/10.1007/s00122-007-0688-9

Gutierrez L, Cuesta-Marcos A, Castro AJ, von Zitzewitz J, Schmitt M, Hayes PM (2011) Association mapping of malting quality quantitative trait loci in winter barley: positive signals from small germplasm arrays. Plant Genome 4:256–272

Gutierrez L, German S, Pereyra S, Hayes PM, Perez CA, Capettini F, Locatelli A, Berberian NM, Falcioni E, Estrada R, Fros D, Gonza V, Altamirano H, Huerta-Espino J, Neyra E, Orjeda G, Sandoval-Islas S, Singh R, Turkington K, Castro AJ (2015) Multi-environment multi-QTL association mapping identifies disease resistance QTLs in barley germplasm from Latin America. Theor Appl Genet 128(3): 501–516. https://doi.org/10.1007/s00122-014-2448-y

Haley CS, Knott SA (1992) A simple regression method for mapping quantitative trait loci in line crosses using flanking markers. Heredity 69:315–324

Han F, Romagosa I, Ullrich SE, Jones BL, Hayes PM, Wesenberg DM (1997) Molecular marker-assisted selection for malting quality traits in barley. Mol Breed 3(6):427–437

Hartfield M, Otto SP (2011) Recombination and hitchhiking of deleterious alleles. Evolution 65:2421–2434. https://doi.org/10.1111/j.1558-5646.2011.01311.x Epub 2011 Apr 26

Hayes P, Szűcs P (2006) Disequilibrium and association in barley: thinking outside the glass. Proc Natl Acad Sci 103(49):18385–18386

Hayes PM, Liu BH, Knapp SJ, Chen F, Jones B, Blake T, Franckowiak J, Rasmusson D, Sorrells M, Ullrich SE, Wesenberg D, Kleinhofs A (1993) Quantitative trait locus effects and environmental interaction in a sample of North American barley germplasm. Theor Appl Genet 87:392–401

Hayes BJ, Cogan NOI, Pembleton LW, Goddard ME, Wang J, Spangenberg GC, Forster JW (2013) Prospects for genomic selection in forage plant species (OA Rognli, ed). Plant Breed 132:133–143

He S, Schulthess AW, Mirdita V et al (2016) Theor Appl Genet 129:641. https://doi.org/10.1007/s00122-015-2655-1

Hedley PE, Russell JR, Hein I, Booth A, Williamson S, Morris J, Lacomme C, Powell W (2005) Epiheterodendrin in malting barley: molecular evidence for cytochrome P450-mediated production. In: 2nd UK small grain cereals workshop, SCRI, Dundee, UK

Heffner EL, Sorrells ME, Jannink J-L (2009) Genomic selection for crop improvement. Crop Sci 49:1–12

Heffner EL, Lorenz AJ, Jannink J-L, Sorrells ME (2010) Plant breeding with genomic selection: gain per unit time and cost. Crop Sci 50:1681

Heffner EL, Jannink J-L, Sorrells ME (2011) Genomic selection accuracy using multifamily prediction models in a wheat breeding program. Plant Genome 4:65–75

Heidlebaugh NM, Trethewey BR, Jukanti AK, Parrott DL, Martin JM, Fischer AM (2008) Effects of a barley (*Hordeum vulgare*) chromosome 6 grain protein content locus on whole-plant nitrogen reallocation under two different fertilisation regimes. Funct Plant Biol 35(7):619–632

Heslot N, Yang H-P, Sorrells ME, Jannink J-L (2012) Genomic selection in plant breeding: a comparison of models. Crop Sci 52:146

Higgins JD, Perry RM, Barakate A, Ramsay L, Waugh R, Halpin C, Armstrong SJ, Franklin F (2012) Spatiotemporal asymmetry of the meiotic program underlies the predominantly distal distribution of meiotic crossovers in barley. Plant Cell 24:4096–4109

Hirota N, Kaneko T, Kuroda H, Kaneda H, Takashio M, Ito K, Takeda K (2005) Characterization of lipoxygenase-1 null mutants in barley. Theor Appl Genet 111(8):1580–1584. https://doi.org/10.1007/s00122-005-0088-y

Hirota N, Kaneko T, Ito K, Takeda K (2006a) Mapping a factor controlling the thermostability of seed lipoxygenase-1 in barley. Plant Breed 125(3):231–235

Hirota N, Kuroda H, Takoi K, Kaneko T, Kaneda H, Yoshida I, Takashio M, Ito K, Takeda K (2006b) Development of novel barley with improved beer foam and flavor stability-the impact of lipoxygenase-1-less barley in the brewing industry. Tech Quarterly-Master Brewers Assoc Am 43(2):131–135

Hoffstetter A, Cabrera A, Huang M, Sneller C (2016) Optimizing training population data and validation of genomic selection for economic traits in soft winter wheat. G3 6:2919–2928. https://doi.org/10.1534/g3.116.032532

Hofheinz N, Borchardt D, Weissleder K, Frisch M (2012) Genome-based prediction of test cross performance in two subsequent breeding cycles. Theor Appl Genet 125:1639–1645. https://doi.org/10.1007/s00122-012-1940-5

Hofinger BJ, Russell JR, Bass CG, Baldwin T, Dos Reis M, Hedley PE, Li Y, Macaulay M, Waugh R, Hammond-Kosack KE (2011) An exceptionally high nucleotide and haplotype diversity and a signature of positive selection for the eIF4E resistance gene in barley are revealed by allele mining and phylogenetic analyses of natural populations. Mol Ecol 20(17): 3653–3668

Hoki T, Saito W, Hirota N, Shirai M, Kihara M, Rossnagel BG, Ichikawa S (2010) The outcomes of joint breeding program in Western Canada-the breeding of LOX-1-less malting barley variety 'CDC-Polarstar'. In: 6th Canadian barley symposium proceedings, Saskatchewan, Canada, 2010

Honsdorf N, March T, Hecht A, Eglinton J, Pillen K (2014) Evaluation of juvenile drought stress tolerance and genotyping by sequencing with wild barley introgression lines. Mol Breed 34:1475. https://doi.org/10.1007/s11032-014-0131-2

Horsley RD, Schmierer D, Maier D, Kudrna D, Urrea CA, Steffenson BJ, Schwarz PB, Franckowiak JD, Green MJ, Zhang B, Kleinhofs A (2006) Identification of QTLs associated with Fusarium head blight resistance in barley accession CIho 4196. Crop Sci 46:145–156

Hospital F (2005) Selection in backcross programmes. Phil Trans R Soc B 360:1503–1511. https://doi.org/10.1098/rstb.2005.1670

Hospital F, Charcosset A (1997) Marker-assisted introgression of quantitative trait loci. Genetics 147(3): 1469–1485

Huang Y, Li L, Smith KP, Muehlbauer GJ (2016) Differential transcriptomic responses to *Fusarium graminearum* infection in two barley quantitative trait loci associated with *Fusarium* head blight resistance. BMC Genom 17(1):387

Hung H-Y, Browne C, Guill K, Coles N, Eller M, Garcia A, Lepak N, Melia-Hancock S, Oropeza-Rosas M, Salvo S, Upadyayula N, Buckler ES, Flint-Garcia S, McMullen MD, Rocheford TR, Holland JB (2012) The relationship between parental genetic or phenotypic divergence and progeny variation in the maize nested association mapping population. Heredity 108:490. https://doi.org/10.1038/hdy.2011.103

Isidro J, Jannink J-L, Akdemir D, Poland J, Heslot N, Sorrells ME (2014) Training set optimization under population structure in genomic selection. Theor Appl Genet 128:145–158

Iwata H, Jannink JL (2011) Accuracy of genomic selection prediction in barley breeding programs: a simulation study based on the real single nucleotide polymorphism data of barley breeding lines. Crop Sci 51(5):1915–1927

Jannink J-L, Bink MCAM, Jansen RC (2001) Using complex plant pedigrees to map valuable genes. Trends Plant Sci 6:337–342

Jarquín D, Crossa J, Lacaze X, Du Cheyron P, Daucourt J, Lorgeou J, Piraux F, Guerreiro L, Pérez P, Calus M, Burgueño J, de los Campos G (2014) A reaction norm model for genomic selection using high-dimensional genomic and environmental data. Theor Appl Genet 127:595–607

Jefferies SP, King BJ, Barr AR, Warner P, Logue SJ, Langridge P (2003) Marker-assisted backcross introgression of the Yd2 gene conferring resistance to barley yellow dwarf virus in barley. Plant Breed 122(1):52–56

Johnson GR (2004) Marker-assisted selection. Plant Breed Rev 24:293–301

Johnson EB, Haggard JE, St. Clair DA (2012) Fractionation, stability, and isolate-specificity of QTL for resistance to *Phytophthora infestans* in cultivated tomato (*Solanum lycopersicum*). G3 2(10):1145–1159. https://doi.org/10.1534/g3.112.003459

Jones H, Norris C, Smith D, Cockram J, Lee D, O'sullivan DM, Mackay Í (2013) Evaluation of the use of high-density SNP genotyping to implement UPOV Model 2 for DUS testing in barley. Theor Appl Genet 126(4):901

Jukanti AK, Fischer AM (2008) A high-grain protein content locus on barley (*Hordeum vulgare*) chromosome 6 is associated with increased flag leaf proteolysis and nitrogen remobilization. Physiol Plant 132 (4):426–439

Jukanti AK, Heidlebaugh NM, Parrott DL, Fischer IA, McInnerney K, Fischer AM (2008) Comparative transcriptome profiling of near-isogenic barley (*Hordeum vulgare*) lines differing in the allelic state of a major grain protein content locus identifies genes with possible roles in leaf senescence and nitrogen reallocation. New Phytol 177(2):333–349

Karsai I, Meszaros K, Hayes PM, Bedő Z (1997) Effects of loci on chromosomes 2 (2H) and 7 (5H) on developmental patterns in barley (*Hordeum vulgare* L.) under different photoperiod regimes. Theor Appl Genet 94(5):612–618

Karsai I, Szűcs P, Mészáros K, Filichkina T, Hayes PM, Skinner JS, Láng L, Bedő Z (2005) The Vrn-H2 locus is a major determinant of flowering time in a facultative × winter growth habit barley (*Hordeum vulgare* L.) mapping population. Theor Appl Genet 110(8):1458–1466

Kearsey MJ, Farquhar AGL (1998) QTL analysis in plants; where are we now? Heredity 80(2):137–142

Kleinhofs A, Kilian A, Maroof MAS, Biyashev RM, Hayes P, Chen FQ, Lapitan N, Fenwick A, Blake TK, Kanazin V, Ananiev E, Dahleen L, Kudrna D, Bollinger J, Knapp SJ, Liu B, Sorrells M, Heun M, Franckowiak JD, Hoffman D, Skadsen R, Steffenson BJ (1993) A molecular, isozyme and morphological map of the barley (*Hordeum vulgare*) genome. Theor Appl Genet 86(6):705–712

Komatsuda T, Pourkheirandish M, He CF, Azhaguvel P, Kanamori H, Perovic D, Stein N, Graner A, Wicker T, Tagiri A, Lundqvist U, Fujimura T, Matsuoka M, Matsumoto T, Yano M (2007) Six-rowed barley originated from a mutation in a homeodomain-leucine zipper I-class homeobox gene. Proc Natl Acad Sci USA 104:1424–1429

Konishi T, Kawada N, Yoshida H, Sohtome K (1989) Linkage relationship between two loci for the Barley Yellow Mosaic resistance of Mokusekko 3 and esterase isozymes in barley (*Hordeum vulgare* L.). Jpn J Breed 39(4):423–430. https://doi.org/10.1270/jsbbs1951.39.423

Kono TJY, Fu F, Mohammadi M, Hoffman PJ, Liu C, Stupar RM, Smith KP, Tiffin P, Fay JC, Morrell PL (2016) The role of deleterious substitutions in crop genomes. Mol Biol Evol 33:2307–2317. https://doi.org/10.1093/molbev/msw102

Kraakman AT, Niks RE, Van den Berg PM, Stam P, Van Eeuwijk FA (2004) Linkage disequilibrium mapping of yield and yield stability in modern spring barley cultivars. Genetics 168(1):435–446

Kraakman ATW, Martinez F, Mussiraliev B, van Eeuwijk FA, Niks RE (2006) Linkage disequilibrium mapping of morphological, resistance, and other agronomically relevant traits in modern spring barley cultivars. Mol Breed 17:41–58

Krchov L, Gordillo GA, Bernardo R (2015) Multienvironment validation of the effectiveness of phenotypic and genomewide selection within biparental maize populations. Crop Sci 55:1068–1075. https://doi.org/10.2135/cropsci2014.09.0608

Kretschmer JM, Chalmers KJ, Manning S, Karakousis A, Barr AR, Islam AKMR, Logue SJ, Choe YW, Barker SJ, Lance RCM, Langridge P (1997) RFLP mapping of the Ha 2 cereal cyst nematode resistance gene in barley. Theor Appl Genet 94(8):1060–1064. https://doi.org/10.1007/s001220050515

Kuczynska A, Surma M, Adamski T (2007) Methods to predict transgressive segregation in barley and other self-pollinated crops. J Appl Genet 48:321–328

Künzel G, Waugh R (2002) Integration of microsatellite markers into the translocation-based physical RFLP map of barley chromosome 3H. Theor Appl Genet 105:660–665

Lacerenza JA, Parrott DL, Fischer AM (2010) A major grain protein content locus on barley (*Hordeum vulgare* L.) chromosome 6 influences flowering time and sequential leaf senescence. J Exp Bot 61(11): 3137–3149. https://doi.org/10.1093/jxb/erq139

Lado B, Gonzalez-Barrios P, Quincke M, Silva P, Gutierrez L (2016) Modelling genotype by environment interaction for genomic selection with unbalanced data from a wheat (*Triticum aestivum* L.) breeding program. Crop Sci 56:1–15. https://doi.org/10.2135/cropsci2015.04.0207

Lado B, Battenfield S, Guzman C, Quincke M, Singh RP, Dreisigacker S, Peña J, Fritz A, Poland J, Gutiérrez L (2017a) Strategies to select crosses using genomic prediction in two wheat breeding programs. Plant Genome. https://doi.org/10.3835/plantgenome2016.12.0128

Lado B, Battenfield S, Guzman C, Quincke M, Singh RP, Dreisigacker S, Peña J, Fritz A, Poland J, Gutiérrez L (2017b) Strategies to select crosses using genomic prediction in two wheat breeding programs. Plant Genome. https://doi.org/10.3835/plantgenome2016.12.0128

Lamb KE, Gonzalez-Hernandez JL, Zhang B, Green M, Neate SM, Schwarz PB, Horsley RD (2009) Identification of QTL conferring resistance to Fusarium head blight resistance in the breeding line C93-3230-24. Crop Sci 49:1675–1680. https://doi.org/10.2135/cropsci2008.11.0642

Lande R, Thompson R (1990) Efficiency of marker-assisted selection in the improvement of quantitative traits. Genetics 124:743–756

Lander ES, Botstein D (1989) Mapping mendelian factors underlying quantitative traits using RFLP linkage maps. Genetics 121:185–199

Langridge P, Karakousis A, Collins N, Kretschmer J, Manning S (1995) A consensus linkage map of barley. Mol Breed 1(4):389–395

Langridge P, Lagudah ES, Holton TA, Appels R, Sharp PJ, Chalmers KJ (2001) Trends in genetic and genome analyses in wheat: a review. Crop Pasture Sci 52:1043–1077

Li JZ, Sjakste TG, Röder MS, Ganal MW (2003) Development and genetic mapping of 127 new microsatellite markers in barley. Theor Appl Genet 107(6):1021–1027. https://doi.org/10.1007/s00122-003-1345-6

Lin H, Yamamoto T, Sasaki T et al (2000) Characterization and detection of epistatic interactions of 3 QTLs, Hd1, Hd2, and Hd3, controlling heading date in rice using nearly isogenic lines. Theor Appl Genet 101:1021. https://doi.org/10.1007/s001220051576

Locatelli A, Cuesta-Marcos A, Gutierrez L, Hayes PM, Smith KP, Castro AJ (2013) Genome-wide association mapping of agronomic traits in relevant barley germplasm in Uruguay. Mol Breed 31(3):631–654

Longin CF, Mi X, Würschum T (2015) Genomic selection in wheat: optimum allocation of test resources and comparison of breeding strategies for line and hybrid breeding. Theor Appl Genet 128:1297–1306

Lorenz AJ (2013) Resource allocation for maximizing prediction accuracy and genetic gain of genomic selection in plant breeding: a simulation experiment. G3 3:481–491

Lorenz AAJ, Smith KP (2015) Adding genetically distant individuals to training populations reduces genomic prediction accuracy in barley. Crop Sci. https://doi.org/10.2135/cropsci2014.12.0827

Lorenz A, Smith KP, Jannink J-L (2012) Potential and optimization of genomic selection for Fusarium head blight resistance in six-row barley. Crop Sci 52:1609–1621

Lorenzana RE, Bernardo R (2009) Accuracy of genotypic value predictions for marker-based selection in biparental plant populations. Theor Appl Genet 120(1):151–161

Lundqvist U, Lundqvist A (1987) An intermedium gene present in a commercial 6-row variety of barley. Hereditas 107(2):131–135

Ma Z, Steffenson BJ, Prom LK, Lapitan NL (2000) Mapping of quantitative trait Loci for Fusarium head blight resistance in barley. Phytopathology 90 (10):1079–1088 https://doi.org/10.1094/PHYTO.2000.90.10.1079

Mackay I, Horwell A, Garner J, White J, McKee J, Philpott H (2011) Reanalyses of the historical series of UK variety trials to quantify the contributions of genetic and environmental factors to trends and variability in yield over time. Theor Appl Genet 122:225–238. https://doi.org/10.1007/s00122-010-1438-y

Manolio TA, Collins FS, Cox NJ, Goldstein DB, Hindorff LA, Hunter DJ, Visscher PM et al (2009) Finding the missing heritability of complex diseases. Nature 461(7265):747–753

Marulanda JJ, Mi X, Melchinger AE et al (2016) Theor Appl Genet 129:1901. https://doi.org/10.1007/s00122-016-2748-5

Mascher M, Muehlbauer GJ, Rokhsar DS, Chapman J, Schmutz J, Barry K, Waugh R (2013a) Anchoring and ordering NGS contig assemblies by population sequencing (POPSEQ). Plant J 76(4):718–727. https://doi.org/10.1111/tpj.12319

Mascher M, Richmond TA, Gerhardt DJ, Himmelbach A, Clissold L, Sampath D, Ayling S, Steuernagel B, Pfeifer M, D'Ascenzo M, Akhunov ED, Hedley PE, Gonzales AM, Morrell PL, Kilian B, Blattner FR, Scholz U, Mayer KFX, Flavell AJ, Muehlbauer GJ, Waugh R, Jeddeloh JA, Stein N (2013b) Barley whole exome capture: a tool for genomic research in the genus Hordeum and beyond. Plant J 76(3):494–505. https://doi.org/10.1111/tpj.12294

Mascher M, Wu S, Amand PS, Stein N, Poland J (2013c) Application of genotyping-by-sequencing on semiconductor sequencing platforms: a comparison of genetic and reference-based marker ordering in barley. PLoS ONE 8(10):e76925. https://doi.org/10.1371/journal.pone.0076925

Mascher M et al (2017) A chromosome conformation capture ordered sequence of the barley genome. Nature 544:427–433. https://doi.org/10.1038/nature22043

Massman J, Cooper B, Horsley R, Neate S, Dill-Macky R, Chao S, Dong Y, Schwarz P, Muehlbauer GJ, Smith KP (2011) Genome-wide association mapping of Fusarium head blight resistance in contemporary barley breeding germplasm. Mol Breed 27(4):439–454

Massman JM, Jung HJG, Bernardo R (2013) Genome-wide selection versus marker-assisted recurrent selection to improve grain yield and stover-quality traits for cellulosic ethanol in maize. Crop Sci 53:58–66

Mathews KL, Malosetti M, Chapman S, McIntyre L, Reynolds M, Shorter R, van Eeuwijk F (2008) Multi-environment QTL mixed models for drought stress adaptation in wheat. Theor Appl Genet 117(7):1077–1091

Matus I, Corey A, Filichkin T, Hayes PM, Vales MI, Kling J, Riera-Lizarazu O, Sato K, Powell W, Waugh R (2003) Development and characterization of recombinant chromosome substitution lines (RCSLs) using *Hordeum vulgare* subsp *spontaneum* as a source of donor alleles in a *Hordeum vulgare* subsp *vulgare* background. Genome 46(6):1010–1023

McCallum CM, Comai L, Greene EA, Henikoff S (2000) Targeting induced local lesions in genomes (TILLING) for plant functional genomics. Plant Physiol 123(2):439–442

Mejlhede N, Kyjovska Z, Backes G, Burhenne K, Rasmussen SK, Jahoor A (2006) EcoTILLING for the identification of allelic variation in the powdery mildew resistance genes mlo and Mla of barley. Plant Breed 125(5):461–467

Melchinger AE, Utz HF, Schön CC (1998) Quantitative trait locus (QTL) mapping using different testers and independent population samples in maize reveals low power of QTL detection and large bias in estimates of QTL effects. Genetics 149(1):383–403

Mesfin A, Smith KP, Dill-Macky R, Evans CK, Waugh R, Gustus CD, Muehlbauer GJ (2003) Quantitative trait loci for Fusarium head blight resistance in barley detected in a two-rowed by six-rowed population. Crop Sci 43:307–318

Meuwissen THE, Hayes BJ, Goddard ME (2001) Prediction of total genetic value using genome-wide dense marker maps. Genetics 157:1819–1829

Meuwissen T, Hayes B, Goddard M (2016) Genomic selection: a paradigm shift in animal breeding. Anim Front 6:6–14

Michel S, Ametz C, Gungor H, Epure D, Grausgruber H, Löschenberger F, Buerstmayr H (2016) Genomic selection across multiple breeding cycles in applied bread wheat breeding. Theor Appl Genet 129:1179–1189

Mickelson S, See D, Meyer FD, Garner JP, Foster CR, Blake TK, Fischer AM (2003) Mapping of QTL associated with nitrogen storage and remobilization in barley (*Hordeum vulgare* L.) leaves. J Exp Bot 54 (383):801–812. https://doi.org/10.1093/jxb/erg084

Mohammadi M, Endelman JB, Nair S, Chao S, Jones SS, Muehlbauer GJ, Ullrich SE, Baik BK, Wise ML, Smith KP (2014) Association mapping of grain hardness, polyphenol oxidase, total phenolics, amylose content, and β-glucan in US barley breeding germplasm. Mol Breed 34(3):1229–1243

Mohammadi M, Tiede T, Smith KP (2015) PopVar: a genomewide procedure for predicting genetic variance and correlated response in bi-parental breeding populations. Crop Sci 55:2068–2077

Morell PL, Buckler ES, Ross-Ibarra J (2011) Crop genomics: advances and applications. Nat Rev Genet 13:85–96

Muller HJ (1964) The relation of recombination to mutational advance. Mutat Res-Fundam Mol Mech 1:2–9

Muñoz-Amatriaín M, Castillo AM, Chen XW, Cistué L, Vallés MP (2008) Identification and validation of QTLs for green plant percentage in barley (*Hordeum vulgare* L.) anther culture. Mol Breed 22:119–129

Muñoz-Amatriaín M, Cuesta-Marcos A, Endelman J, Comadran J, Bonman J, Bockelman H, Chao S, Russell J, Waugh R, Hayes P, Muehlbauer G (2014)

The USDA collection of barley landraces and cultivars: genetic diversity, population structure, and potential for genome-wide association studies. PLoS One 14 9(4):394688
Navara S, Smith KP (2014) Using near-isogenic barley lines to validate deoxynivalenol (DON) QTL previously identified through association analysis. Theor Appl Genet 127(3):633–645
Nduulu LM, Mesfin A, Muehlbauer GJ, Smith KP (2007) Analysis of the chromosome 2(2H) region of barley associated with the correlated traits Fusarium head blight resistance and heading date. Theor Appl Genet 115:561–570
Neyhart JL, Tiede T, Lorenz AJ, Smith KP (2017) Evaluating methods of updating training data in long-term genomewide selection. G3 1499–1510. https://doi.org/10.1534/g3.117.040550
Ng PC (2003) SIFT: predicting amino acid changes that affect protein function. Nucleic Acids Res 31:3812–3814
Nice LM, Steffenson BJ, Brown-Guedira GL, Akhunov ED, Liu C, Kono TJY, Morrell PL, Blake TK, Horsley RD, Smith KP, Muehlbauer GJ (2016) Development and genetic characterization of an advanced backcross-nested association mapping (AB-NAM) population of wild × cultivated barley. Genetics 203(3):1453+. https://doi.org/10.1534/genetics.116.190736
Niebur WS, Rafalski JA, Smith OS, Cooper M (2004) Applications of genomics technologies to enhance rate of genetic progress for yield of maize within a commercial breeding program. In: Fischer T (ed) New directions for a diverse planet. Proceedings of 4th international crop science congress, Brisbane, QLD, Australia, 26 Sept–1 Oct 2004. The Regional Institute Ltd., Gosford, NSW, Australia
Okada Y, Kanatani R, Arai S, Ito K (2004) Interaction between barley yellow mosaic disease-resistance genes rym1 and rym5, in the response to BaYMV strains. Breed Sci 54(4):319–325
Pan A, Hayes PM, Chen F, Chen THH, Blake T, Wright S, Karsai I, Bedö Z (1994) Genetic analysis of the components of winterhardiness in barley (*Hordeum vulgare* L.). Theor Appl Genet 89(7):900–910
Paris M, Jones MGK, Eglinton JK (2002) Genotyping single nucleotide polymorphisms for selection of barley β-amylase alleles. Plant Mol Biol Report 20(2): 149–159. https://doi.org/10.1007/bf02799430
Paterson AH, Lander ES, Hewitt JD, Peterson S, Lincoln S, Tanksley SD (1988) Resolution of quantitative traits into Mendelian factors by using a complete linkage map of restriction fragment length polymorphisms. Nature 335(721726):6170
Paterson AH, DeVerna JW, Lanini B, Tanksley SD (1990) Fine mapping of quantitative trait loci using selected overlapping recombinant chromosomes, in an interspecies cross of tomato. Genetics 124:735–742
Pauli D, Muehlbauer GJ, Smith KP, Cooper B, Hole D, Obert DE, Ullrich SE, Blake TK (2014) Association mapping of agronomic QTLs in US spring barley breeding germplasm. Plant Genome 1 7(3)
Paulitz TC, Steffenson BJ (2010) Biotic stress. In: Ullrich SE (ed). Barley: disease problems and solutions. Barley: production, improvement, and uses. Wiley-Blackwell, Oxford, UK. https://doi.org/10.1002/9780470958636.ch11
Piffanelli P, Ramsay L, Waugh R, Benabdelmouna A, D'Hont A, Hollricher K, Jorgensen JH, Schulze-Lefert P, Panstruga R (2004) A barley cultivation-associated polymorphism conveys resistance to powdery mildew. Nature 430(7002):887–891
Poland JA, Brown PJ, Sorrells ME, Jannink J-L (2012a) Development of high-density genetic maps for barley and wheat using a novel two-enzyme genotyping-by-sequencing approach. PLoS ONE 7(2):e32253
Poland J, Endelman J, Dawson J, Rutkoski J, Wu S, Manes Y, Dreisigacker S, Crossa J, Sánchez-Villeda H, Sorrells M, Jannink J-L (2012b) Genomic selection in wheat breeding using genotyping-by-sequencing. Plant Genome 5:103–113. https://doi.org/10.3835/plantgenome2012.06.0006
Pritchard JK, Stephens M, Donnelly P (2000) Inference of population structure using multilocus genotype data. Genetics 155:945–959
Qi X, Lindhout P (1997) Development of AFLP markers in barley. MGG 254(3):330–336. https://doi.org/10.1007/s004380050423
Ragot M, Lee M (2007) Marker-assisted selection in maize: current status, potential, limitations and perspectives from the private and public sectors. In: Guimaraes EP et al (eds) Marker-assisted selection, current status and future perspectives in crops, livestock, forestry, and fish. FAO, Rome, pp 117–150
Rajala A, Peltonen-Sainio P, Jalli M, Jauhiainen L, Hannukkala A, Tenhola-Roininen T, Ramsay L, Manninen O (2017) One century of Nordic barley breeding: nitrogen use efficiency, agronomic traits and genetic diversity. J Agric Sci 155(4):582–598. https://doi.org/10.1017/s002185961600068x
Ramsay L, Macaulay M, Ivanissevich Sd, MacLean K, Cardle L, Fuller J, Edwards KJ, Tuvesson S, Morgante M, Massari A, Maestri E, Marmiroli N, Sjakste T, Ganal M, Powell W, Waugh R (2000) A simple sequence repeat-based linkage map of barley. Genetics 156(4):1997–2005
Ramsay L, Comadran J, Druka A, Marshall DF, Thomas WTB, Macaulay M, MacKenzie K, Simpson C, Fuller J, Bonar N, Hayes PM, Lundqvist U, Franckowiak JD, Close TJ, Muehlbauer GJ, Waugh R (2011) INTERMEDIUM-C, a modifier of lateral spikelet fertility in barley, is an ortholog of the maize domestication gene TEOSINTE BRANCHED 1. Nat Genet 43(2):169–172
Resende MFR, Muñoz P, Acosta JJ, Peter GF, Davis JM, Grattapaglia D, Resende MDV, Kirst M (2011) Accelerating the domestication of trees using genomic selection: accuracy of prediction models across ages and environments. New Phytol 193:617–624

Richter K, Schondelmaier J, Jung C (1998) Mapping of quantitative trait loci affecting Drechslera teres resistance in barley with molecular markers. TAG Theor Appl Genet 97(8):1225–1234. https://doi.org/10.1007/s001220051014

Riedelsheimer C, Melchinger AE (2013) Optimizing the allocation of resources for genomic selection in one breeding cycle. Theor Appl Genet 126:2835–2848. https://doi.org/10.1007/s00122-013-2175-9

Riggs T, Hayter A (1976) Practical aspects of the single seed descent method in barley breeding. In: Proceedings of the international barley genetics symposium

Riggs TJ, Hanson PR, Start ND, Miles DM, Morgan CL, Ford MA (1981) Comparison of spring barley varieties grown in England and Wales between 1880 and 1980. J Agric Sci 97(DEC):599–610

Rincent R, Laloë D, Nicolas S, Altmann T, Brunel D, Revilla P, Rodriguez VM et al (2012) Maximizing the reliability of genomic selection by optimizing the calibration set of reference individuals: comparison of methods in two diverse groups of maize inbreds (*Zea mays* L.). Genetics 192(2):715–728

Romagosa I, Han F, Ullrich SE, Hayes PM, Wesenberg DM (1999) Verification of yield QTL through realized molecular marker-assisted selection responses in a barley cross. Mol Breed 5:143–152

Rostoks N, Mudie S, Cardle L, Russell J, Ramsay L, Booth A, Svensson JT, Wanamaker SI, Walia H, Rodriguez EM, Hedley PE, Liu H, Morris J, Close TJ, Marshall DF, Waugh R (2005) Genome-wide SNP discovery and linkage analysis in barley based on genes responsive to abiotic stress. Mol Genet Genom 274(5):515–527. https://doi.org/10.1007/s00438-005-0046-z

Roy JK, Smith KP, Muehlbauer GJ, Chao S, Close TJ, Steffenson BJ (2010) Association mapping of spot blotch resistance in wild barley. Mol Breed 26:243–256

Rutkoski JE, Heffner EL, Sorrells ME (2010) Genomic selection for durable stem rust resistance in wheat. Euphytica 179:161–173

Rutkoski J, Singh RP, Huerta-Espino J, Bhavani S, Poland J, Jannink J-L, Sorrells ME (2015) Efficient use of historical data for genomic selection: a case study of stem rust resistance in wheat. Plant Genome 8. https://doi.org/10.3835/plantgenome2014.09.0046

Saade S, Maurer A, Shahid M, Oakey H, Schmockel SM, Negrao S, Pillen K, Tester M (2016) Yield-related salinity tolerance traits identified in a nested association mapping (NAM) population of wild barley. Sci Rep 6. https://doi.org/10.1038/srep32586

Sallam AH, Smith KP (2016) Genomic selection performs similarly to phenotypic selection in barley. Crop Sci 56:2871–2881. https://doi.org/10.2135/cropsci2015.09.0557

Sameri M, Komatsuda T (2004) Identification of quantitative trait loci (QTLs) controlling heading time in the population generated from a cross between oriental and occidental barley cultivars (*Hordeum vulgare* L.). Breed Sci 54(4):327–332

Sarker A, Singh M (2015) Improving breeding efficiency through application of appropriate experimental designs and analysis models: a case of lentil (*Lens culinaris* Medikus subsp. *culinaris*) yield trials. Field Crops Res 179:26–34. https://doi.org/10.1016/j.fcr.2015.04.007

Sayed MA, Schumann H, Pillen K, Naz AA, Leon J (2012) AB-QTL analysis reveals new alleles associated to proline accumulation and leaf wilting under drought stress conditions in barley (*Hordeum vulgare* L.). BMC Genet 13. https://doi.org/10.1186/1471-2156-13-61

Schmidt M, Kollers S, Maasberg-Prelle A, Großer J, Schinkel B, Tomerius A, Graner A, Korzun V (2016) Prediction of malting quality traits in barley based on genome-wide marker data to assess the potential of genomic selection. Theor Appl Genet 129:203–213

Schnell FW, Utz HF (1975) F1-leistung und elternwahl euphyder züchtung von selbstbefruchtern. In: Bericht über die Arbeitstagung der Vereinigung Österreichischer Pflanzenzüchter. BAL Gumpenstein, Gumpenstein, Austria

Skadhauge B, Lok F, Breddam K, Olsen O, Bech LM, Knudsen S (2010) Barley with reduced lipoxygenase activity. Google Patents

Skov Kristensen P, Dockter C, Lundqvist U, Lu Q, Gregersen PL, Thordal-Christensen H, Hansson M (2016) Genetic mapping of the barley lodging resistance locus *Erectoides-k*. Plant Breed 135(4):420–428. https://doi.org/10.1111/pbr.12377

Smith KP, Evans CK, Dill-Macky R, Gustus C, Xie W, Dong Y (2004) Host genetic effect on deoxynivalenol accumulation in Fusarium head blight of barley. Phytopathology 94(7):766–771. https://doi.org/10.1094/PHYTO.2004.94.7.766

Solberg TR, Sonesson AK, Woolliams JA, Meuwissen THE (2009) Reducing dimensionality for prediction of genome-wide breeding values. Genet Sel Evol 41:29

Souza E, Sorrells ME (1991) Prediction of progeny variation in oat from parental genetic relationships. Theor Appl Genet 82:233–241. https://doi.org/10.1007/BF00226219

Spaner D, Rossnagel BG, Legge WG, Scoles GJ, Eckstein PE, Penner GA, Tinker NA, Briggs KG, Falk DE, Afele JC, Hayes PM, Mather DE (1999) Verification of a quantitative trait locus affecting agronomic traits in two-row barley. Crop Sci 39:248–252

Steffenson BJ (2003) Fusarium head blight of barley: impact, epidemics, management, and strategies for identifying and utilizing genetic resistance. In: Leonard KJ, Bushnell WR (eds) Fusarium head blight of wheat and barley, pp 241–295

Steffenson BJ, Hayes PM, Kleinhofs A (1996) Genetics of seedling and adult plant resistance to net blotch (*Pyrenophora teres* f. *teres*) and spot blotch (*Cochliobolus sativus*) in barley. Theor Appl Genet 92(5):552–558. https://doi.org/10.1007/s001220050162

Storlie E, Charmet G (2013) Genomic selection accuracy using historical data generated in a wheat breeding program. Plant Genome 6. https://doi.org/10.3835/plantgenome2013.01.0001

Stracke S, Haseneyer G, Veyrieras J-B, Geiger HH, Sauer S, Graner A, Piepho H-P (2009) Association mapping reveals gene action and interactions in the determination of flowering time in barley. Theor Appl Genet 118:259–273

Stuber CW (1992) Biochemical and molecular markers in plant breeding. Plant Breed Rev 9:37–61

Stuber CW, Polacco M, Senior ML (1999) Synergy of empirical breeding, marker-assisted selection, and genomics to increase crop yield potential. Crop Sci 39(6):1571–1583

Swanson-Wagner RA, Eichten SR, Kumari S, Tiffin P, Stein JC, Ware D, Springer NM (2010) Pervasive gene content variation and copy number variation in maize and its undomesticated progenitor. Genome Res 20:1689–1699

Swanston JS (1987) The consequences, for malting quality, of *Hordeum laevigatum* as a source of mildew resistance in barley breeding. Ann Appl Biol 110(2): 351–355

Szűcs P, Blake VC, Bhat VR, Close TJ, Cuesta-Marcos A, Muehlbauer GJ, Ramsay LV, Waugh R, Hayes PM (2009) An integrated resource for barley linkage map and malting quality QTL alignment. Plant Genome 2:123–140

Szűcs P, Karsai I, von Zitzewitz J, Meszaros K, Cooper LLD, Gu YQ, Chen THH, Hayes PM, Skinner JS (2006) Positional relationships between photoperiod response QTL and photoreceptor and vernalization genes in barley. Theor Appl Genet 112:1277–1285

Talame V, Sanguineti MC, Chiapparino E, Bahri H, Ben Salem M, Forster BP, Ellis RP, Rhouma S, Zoumarou W, Waugh R, Tuberosa R (2004) Identification of Hordeum spontaneum QTL alleles improving field performance of barley grown under rainfed conditions. Ann Appl Biol 144(3):309–319

Talame V, Bovina R, Sanguineti MC, Tuberosa R, Lundqvist U, Salvi S (2008) TILLMore, a resource for the discovery of chemically induced mutants in barley. Plant Biotechnol J 6(5):477–485

Tamang P, Neupane A, Mamidi S, Friesen T, Brueggeman R (2015) Association mapping of seedling resistance to spot form net blotch in a worldwide collection of barley. Phytopathology 26 105(4):500–508

Tanksley D (1983) Molecular markers in plant breeding. Plant Mol Biol Rep 1:3–8

Tanksley SD (1993) QTL analysis of transgressive segregation in an interspecific tomato cross. Genetics 134(2):585–596

Tayeh N, Klein A, Le Paslier M-C, Jacquin F, Houtin H, Rond C, Chabert-Martinello M et al (2015) Genomic prediction in pea: effect of marker density and training population size and composition on prediction accuracy. Front Plant Sci 6:941. https://doi.org/10.3389/fpls.2015.00941

Thomas W, Comadran J, Ramsay L, Shaw P, Marshall D, Newton A, O'Sullivan D, Cockram J, Mackay I, Bayles R, White J, Kearsey M, Luo Z, Wang M, Tapsell C, Harrap D, Werner P, Klose S, Bury P, Wroth J, Argillier O, Habgood R, Glew M, Bochard A-M, Gymer P, Vequaud D, Christerson T, Allvin B, Davies N, Broadbent R, Brosnan J, Bringhurst T, Booer C, Waugh R (2014) Project report No. 528: association genetics of UK elite barley (AGOUEB). HGCA. http://cereals.ahdb.org.uk/media/337662/pr528-final-project-report.pdf

Tiede T, Kumar L, Mohamadi M, Smith KP (2015) Predicting genetic variance in bi-parental breeding populations is more accurate when explicitly modeling the segregation of informative genomewide markers. Mol Breed. https://doi.org/10.1007/s11032-015-0390-6

Toojinda T, Baird E, Booth A, Broers L, Hayes P, Powell W, Thomas W, Vivar H, Young G (1998) Introgression of quantitative trait loci (QTLs) determining stripe rust resistance in barley: an example of marker-assisted line development. Theor Appl Genet 96(1):123–131

Torkamaneh D, Laroche J, Bastien M, Abed A, Belzile F (2017) Fast-GBS: a new pipeline for the efficient and highly accurate calling of SNPs from genotyping-by-sequencing data. BMC Bioinform 18(1):5. https://doi.org/10.1186/s12859-016-1431-9

Tyrka M, Perovic D, Wardyńska A, Ordon F (2008) A new diagnostic SSR marker for selection of theRym4/Rym5 locus in barley breeding. J Appl Genet 49(2): 127–134. https://doi.org/10.1007/bf03195605

van Berloo R, Aalbers H, Werkman A, Niks RE (2001) Resistance QTL confirmed through development of QTL-NILs for barley leaf rust resistance. Mol Breed 8(3):187–195

Visscher PM, Haley CS, Knott SA (1996) Mapping QTLs for binary traits in backcross and F2 populations. Genet Res 68:55–63

von Zitzewitz J, Cuesta-Marcos A, Condon F, Castro AJ, Chao S, Corey A, Filichkin T, Fisk SP, Gutierrez L, Haggard K, Karsai I, Muehlbauer GJ, Smith KP, Veisz O, Hayes PM (2011) The genetics of winter-hardiness in barley: perspectives from genome-wide association mapping. Plant Genome 4(1):76–91

Wang M, Jiang N, Jia T, Leach L, Cockram J, Waugh R, Ramsay L, Thomas B, Luo Z (2012) Genome-wide association mapping of agronomic and morphologic traits in highly structured populations of barley cultivars. Theor Appl Genet 124(2):233–246

Waugh R, Jannink J-L, Muehlbauer GJ, Ramsay L (2009) The emergence of whole genome association scans in barley. Curr Opin Plant Biol 12:218–222

Werner K, Friedt W, Ordon F (2005) Strategies for pyramiding resistance genes against the barley yellow mosaic virus complex (BaMMV, BaYMV, BaYMV-2). Mol Breed 16(1):45–55

Wiberg A (1974) Genetical studies of spontaneous sources of resistance to powdery mildew in barley. Hereditas 77(1):89–148. https://doi.org/10.1111/j.1601-5223.1974.tb01357.x

Xu Y, Crouch JH (2008) Marker-assisted selection in plant breeding: from publications to practice. Crop Sci 48(2):391–407

Yun SJ, Gyenis L, Bossolini E, Hayes PM, Matus I, Smith KP, Steffenson BJ, Tuberosa R, Muelbauer GJ (2006) Validation of quantitative trait loci for multiple disease resistance in barley using advanced backcross lines developed with a wild barley. Crop Sci 46:1179–1186

Zhao Y, Gowda M, Liu W, Wurschum T, Maurer HP, Longin FH, Ranc N, Reif JC (2012) Accuracy of genomic selection in European maize elite breeding populations. Theor Appl Genet 124:769–776

Zhong S, Dekkers JC, Fernando RL, Jannink J-L (2009) Factors affecting accuracy from genomic selection in populations derived from multiple inbred lines: a barley case study. Genetics 182:355–364

Zhou H, Steffenson BJ, Muehlbauer G, Wanyera R, Njau P, Ndeda S (2014) Association mapping of stem rust race TTKSK resistance in US barley breeding germplasm. Theor Appl Genet 127(6):1293–1304

Zhu H, Gilchrist L, Hayes P, Kleinhofs A, Kudrna D, Liu Z, Prom L, Steffenson B, Toojinda T, Vivar H (1999) Does function follow form? Principal QTLs for Fusarium head blight (FHB) resistance are coincident with QTLs for inflorescence traits and plant height in a doubled-haploid population of barley. Theor Appl Genet 99(7):1221–1232

Barley Domestication, Adaptation and Population Genomics

17

Karl Schmid, Benjamin Kilian and Joanne Russell

Abstract

Wild and cultivated barley are characterized by a high level of genetic diversity and a pronounced geographic population structure. Numerous studies using a diversity of markers showed that the centre of diversity of both wild and cultivated barley is in the Western part of the Fertile Crescent where the species was presumably domesticated. Comparisons of geographic diversity patterns suggested additional centres of domestication, of which the Eastern part of the Fertile Crescent (Iran or Himalaya), are most strongly supported. In wild barley, the geographic distribution of genetic and phenotypic diversity largely follows a neutral isolation by distance pattern, but common-garden experiments and environmental association studies indicate that local adaptation by natural selection also had a significant influence on these patterns but, so far no strong candidate genes for local adaptation were identified. Cultivated barley landraces and elite material have a significantly reduced level of genetic diversity compared to wild barley which also shows significant geographic differentiation and evidence for local adaptation. Several major domestication genes have already been cloned and patterns of diversity largely confirm the hypotheses that these genes were exposed to strong domestication-related selection that caused a reduction of diversity in these genes. Ex situ genebank collections of wild and domesticated barley were used to define core collections that have been phenotyped and genotyped to facilitate allele mining and introgression into elite varieties. The future utilization of barley genetic diversity will be facilitated by a good reference genome. The rapid progress of sequencing technologies and modern breeding methods like genomic selection and genome editing will contribute to an efficient utilization of barley genetic diversity.

K. Schmid (✉)
Institute of Plant Breeding, Seed Science and Population Genetics, University of Hohenheim, Fruwirthstrasse 21, 70599 Stuttgart, Germany
e-mail: karl.schmid@uni-hohenheim.de

B. Kilian
Global Crop Diversity Trust, Platz der Vereinten Nationen 7, 53113 Bonn, Germany

J. Russell
The James Hutton Institute, Invergowrie, Dundee DD2 5D, UK

17.1 Introduction

Barley (*Hordeum vulgare* L. subsp. *vulgare*) is a founder crop of Old World agriculture (Zohary et al. 2012). Since its domestication from its closest wild relative *Hordeum vulgare* subsp. *spontaneum* (C. Koch) Thell. in the Fertile Crescent, about

N. Stein and G. J. Muehlbauer (eds.), *The Barley Genome*, Compendium of Plant Genomes,
https://doi.org/10.1007/978-3-319-92528-8_17

10,000 years ago, the geographic range of its cultivation area has expanded dramatically, especially within temperate regions of the Northern and Southern hemispheres. Domesticated barley became adapted to diverse local climatic conditions, which subsequently has led to a large diversity of both landraces and modern elite varieties. Historically, domesticated barley was used as a food crop for human consumption, which still continues in certain regions of Africa, Asia and Latin America (Fischbeck 2003), but its main use today is for animal feed and alcohol production from malted barley. In addition to being an economically important crop species, it is also a model species for cereal genomics, mainly because it is a predominantly self-pollinated and diploid species ($2n = 14$).

Modern barley varieties tend to have a narrow genetic basis compared to their wild ancestors due to genetic drift and high levels of inbreeding in breeding populations (Caldwell et al. 2005). While a well-defined genetic basis for yield-related traits is highly desirable, there are also traits that benefit from high levels of genetic diversity. These include resistance genes to allow coevolution with pathogens and genes involved in adaptation to local conditions and environmental stress factors such as extreme temperatures or periods of drought. Landraces and crop wild relatives (CWR) are recognized as valuable sources of novel genetic variants for introgression into modern varieties if they are resistant to pathogens. As for landraces, these should be reasonably adapted to the target environment of the crop. This insight has led to an important research paradigm in crop plant genetics: The search for agronomically important genes in CWR and landrace germplasm and the introgression of useful alleles into modern varieties by either marker-assisted selection (MAS) or genetic engineering (Tanksley and McCouch 1997; Zamir 2001; McCouch et al. 2013; Dempewolf et al. 2017). Considerable effort has been made to introgress desirable alleles from wild barley and landraces into breeding populations; the most successful have focused on disease resistance (Schmalenbach et al. 2008; Friedt et al. 2011), abiotic stress (Kalladan et al. 2013; Fan et al. 2015; Mora et al. 2016) and recently root architecture traits (Sayed et al. 2016). These studies have provided a catalogue of useful variation but with recent technological advances in sequencing and the application of genomic selection and genome editing, we envisage a step change that will lead to the deployment of this diversity for crop improvement.

For an efficient use of plant genetic resources, it is necessary to obtain a comprehensive collection of genetic resources that ideally captures most of the standing genetic variation, and to characterize those lines phenotypically and genetically. We review the current state of research on phenotypic and genetic variation in wild and domesticated barley from a population genetics perspective, and describe the various approaches to characterize this diversity in barley. Furthermore, we discuss the opportunities and problems of identifying agriculturally interesting genes that may be used to improve modern elite varieties.

17.2 Phenotypic Variation in Wild Barley

Wild barley occurs throughout the Eastern Mediterranean, Middle East and Central Asia up to the Himalayas (Harlan and Zohary 1966). Its core area of natural distribution is the Fertile Crescent, where it is frequently found in large local populations and exhibits the greatest ecological and morphological diversity. Israel and the West Bank are centres of barley diversity, where geographically and phenotypically distinct ecotypes were identified (Snow and Brody 1984) and confirmed by a phenotypic analysis of multiple traits (Hübner et al. 2012a). Elsewhere (e.g. the Himalayas), wild barley occurs in lower densities and is more dispersed. North African populations from Libya and Morocco were also reported (Molina-Cano et al. 1987, 1999; Orabi et al. 2007). Wild barley grows in diverse environments, which includes primary habitats (Hübner et al. 2012b), but it is also associated with sites disturbed by human activities, where it has a weedy and colonizing character. Since it can interbreed with domesticated barley, such weedy forms may constitute hybrids.

Phenotypic variation between genotypes from different regions (frequently called ecotypes) may represent neutral divergence between genetically separated populations or local adaptation to habitats that have different environments varying in precipitation levels, light intensities, soil characteristics or temperature (Hübner et al. 2012b; Bedada et al. 2014; Russell et al. 2014). To understand these complex environmental interactions, reciprocal common-garden experiments have been proposed as a means to compare fitness parameters of ecotypes from different geographical origins (Verhoeven et al. 2004). In both wild and domesticated barley, fitness differences between native and transplanted ecotypes were observed, suggesting that at least some proportion of the observed phenotypic variation is adaptive (Harlan and Martini 1938). In Israel, a strong temperature and precipitation gradient occurs within a short geographic distance, providing the means to studying local adaptation (Volis et al. 2001). A comparison of ecotypes from Mediterranean habitats in Northern Israel and xeric habitats in the Negev desert identified significant genetically determined fitness differences (Volis et al. 2001, 2002, 2010, 2011). Native ecotypes showed higher fitness values than transplanted ecotypes, and trait differences observed were consistent with previous hypotheses. Mediterranean ecotypes occur in a more predictable environment with respect to rainfall and produce larger seeds, whereas desert ecotypes produce more seeds with a lower germination rate, but have a higher sensitivity to environmental cues such as spring precipitation. Taken together, these studies indicate local adaptation of different ecotypes, which is caused by genetic differences that may represent useful variation for plant breeding.

17.3 Presence of Population Structure in Wild Barley

As expected for a predominately self-fertilizing species, wild barley shows a strong population structure at different geographic scales. An early DNA sequence survey of nine genes in 25 accessions representing the species range from Israel to the Himalayas partitioned genetic diversity into three geographic regions consisting of a Western population (Israel, Jordan, Syria), the Zagros mountain region (Eastern Turkey, Iraq, Northern Iran), and the Eastern population (Afghanistan, Turkmenistan, Tajikistan and Iran) (Morrell et al. 2003). Some loci showed a distinct geographic structure, which is defined as the restricted occurrence of certain haplotypes in one region. In contrast, other genes showed no evidence for such a structure, despite high levels of polymorphism at these loci, which suggests sufficient migration to move mutations across the species range whether or not the variants are adaptive or advantageous. The level of polymorphism in the Western (Levant) region was significantly higher than in the Zagros and Eastern regions, the former region is considered as the site of barley domestication. This large-scale population structure was essentially confirmed by most recent studies of species-wide diversity (Jakob et al. 2014; Russell et al. 2014, 2016; Pankin et al. 2017). Genetically distinct populations were also observed in the Western margin of the natural distribution range in North Africa (Molina-Cano et al. 1999).

Strong genetic differentiation also occurs within the major regions and at the geographically small scale, within a few hundred metres. Most of these studies have focused on the Levant region with its distinct physical geography and climatic variability, ranging from harsh desert to fertile Mediterranean environments (Volis et al. 2001, 2002, 2009; Hübner et al. 2009, 2012b). Hübner et al. (2009) systematically collected wild barleys, known as Barley1K, which reflect this variability. All of the phenotypic and genotypic data generated from this collection identify a strong and highly significant relationship between genetic and geographic distance, i.e. geographically close accessions tend to be genetically more similar than distant accessions (Hübner et al. 2009, 2012b; Bedada et al. 2014).

Of particular interest is the population structure on the small geographical scale. The natural habitat of wild barley in the Levant region is characterized by local microsites that differ environmentally over short distances. Probably the best known of these is 'Evolution Canyon' at Nahal Oren, Mt. Carmel, Israel (Nevo 1995). The canyon is oriented in an East-West direction, one slope exposed to the South and constitutes a hot and dry (xeric) environment; the opposite slope is exposed to the North and characterized by a more humid and cooler microclimate. This is accompanied by different vegetation types and suggests that species that occur on both slopes have adapted locally by natural selection to their respective conditions. Numerous studies using a variety of markers such as RAPDs (Owuor et al. 1999), SSRs (Huang et al. 2002; Nevo et al. 2005), ribosomal DNA (Gupta et al. 2004), BARE1 transposons (Kalendar et al. 2000), and DNA sequences of gene fragments (Bedada et al. 2014) detected significant genetic differentiation between opposing slopes of the Canyon. The level of genetic differentiation, G_{ST} (Nei 1973) was comparable to the differentiation of two populations with a distance of 100 km (Turpeinen et al. 2001). A similar differentiation between slopes was also observed at other Canyons in the region (Hübner et al. 2009). These studies highlight divergence at the microclimatic scale, not just in wild barley but in other organisms that survive on both slopes of this unusually diverse region (Nevo 2014).

Comparative range-wide studies, extending from North Africa, through the Fertile Crescent, and into Central Asia, have been used to address population structure and divergence in relation to domestication (reviewed by Dawson et al. (2015)). A strong and highly significant relationship between genetic and geographic distance may arise from isolation by distance (IBD), which is a neutral demographic process independent of selection and the predominately self-fertilizing nature of barley. Self-fertilization facilitates the coexistence of genetically differentiated lineages in close proximity that contribute to a strong geographic clustering in population structure inference methods like STRUCTURE (Ingvarsson and Eckert 2002). The IBD hypothesis was formally tested using SNP and SSR markers (Russell et al. 2014) and exome sequencing data (Russell et al. 2016). In both cases, the geographic pattern of genetic diversity shows strong IBD but was also consistent with possible Pleistocene refugia at the eastern Mediterranean and Central Asia because of a slightly increased similarity of accessions from these regions. At a more local level within the Levant, patterns of IBD are very strong and highly significant (Hübner et al. 2009). Patterns of diversity at different scales indicate that geographic distribution of genetic variation is largely determined by neutral processes.

In contrast, a highly geographically structured population may reflect differential adaptation to environmental variables. This hypothesis can be tested by searching for significant correlation of genetic diversity with environmental variables. Climate modelling showed a significant effect of climate on the geographic distribution of SNP and SSR alleles (Jakob et al. 2014; Russell et al. 2014). This was further confirmed with exome capture sequencing data from 91 wild barley *H. spontaneum* accessions and a subsequent association of genetic variation with 36 environmental variables reflecting their geographic origin (Russell et al. 2016). Generally, the association of SNP allele frequency distributions with environmental variables was low (correlation coefficient of 0.07 or less), but some SNPs and genomic regions showed highly significant associations that indicate the contribution of these regions to local adaptation. Similarly, on a smaller regional scale, Hübner et al. (2009) observed associations between environmental variables and genetic diversity within the Barley1K collection. Both studies recognize the role of environment in shaping genetic diversity. A principal component analysis coupled with an ordination analysis revealed that the environmental parameters elevation, midday temperature and rainfall explained a large proportion of the variance, and the geographic distribution of inferred genetic clusters aligns well with these gradients (Hübner et al. 2012b).

Taken together, these studies clearly indicate that wild barley genetic diversity shows a strong population structure that is, at least, partially influenced by local adaption, consistent with the results of common-garden experiments along environmental gradients. Wild barley not only harbours considerable novel variation but this diversity is functional and will be useful for future plant breeding efforts to a changing and challenging climate (Ellis et al. 2000; Dawson et al. 2015).

17.4 Candidate Genes for Local Adaptation

A significant (but weak) correlation of environmental variables and patterns of genetic diversity may allow the mapping of genes that contribute to local adaptation. To date, two studies have used a similar genome-wide and population genetics approach to scan for associations between environment and genes (Fang et al. 2014; Russell et al. 2016). Both used divergent statistics, based on allele frequencies (Bayenv2; Günther and Coop 2013) to identify SNPs that are associated with environmental variables, either treating individuals as a single population (Russell et al. 2016) or defining populations using ordination or clustering approaches (Fang et al. 2014). The former, using exome capture data, used an outlier approach to identify multiple SNPs, distributed throughout the genome, significantly associated with different environmental variables, a particularly strong footprint on chromosome 1H, with unknown function. The latter study also revealed associations with both temperature and precipitation variables for SNPs in previously identified genes on chromosome 2H and 5H, respectively. Importantly, many associations were clustered in a manner consistent with structural variants.

Given the large size of the barley genome (ca. 5.1 Gbp) and the potentially complex types of polymorphisms (e.g. insertion–deletion polymorphisms, regulatory versus protein coding variants, chromosomal inversions versus individual gene variants, strong effect versus weak effect polymorphisms; Weigel and Nordborg 2015), much adaptive genetic diversity will be missed in genomic scans of local adaptation until routine whole genome sequencing using single-molecule long-read sequencing becomes standard. In the future, genome resequencing and genomic linkage analysis coupled with suitable statistical analysis may provide much greater power to identify causal adaptive genetic variants (Vitti et al. 2013).

17.5 Effect of Domestication on Genome-Wide Polymorphisms

Archaeological findings indicate that barley was domesticated in the Levant region about 10,500 B.P. (Tanno and Willcox 2011; Zohary et al. 2012). Early genetic studies of barley domestication were interpreted as supporting a single domestication in the Eastern Fertile Crescent (e.g. Badr et al. 2000; Salamini et al. 2002). Other sites of domestication have been postulated, based on the diversity observed in landraces and wild barleys from these areas. For example, Morocco (Molina-Cano et al. 1999), Ethiopia (Orabi et al. 2007), and the Tibetan Plateau (Dai et al. 2012, 2014) were suggested as possible secondary sites of domestication. In the case of Morocco, Blattner and Badani Méndez (2001) suggested that this was an unlikely second site of domestication and the diversity could be attributed to migration and secondary introgression. Most authors now agree that more complex domestication events have occurred within the distributions of wild progenitors, but do not preclude additional domestication events elsewhere (Morrell and Clegg 2007; Brown et al. 2009; Russell et al. 2011; Cиván and Brown 2017). Because wild and domesticated barley occur in the same geographical locations, gene flow is possible, complicating the analysis of domestication and suggesting that a simple single event is insufficient to account for the complexity of the relationship between crops and their wild ancestors. Many studies have focused on comparing wild barley to cultivated accessions usually with varying sample sizes and unmatched collection sites. In general, all studies agree that wild barleys are more diverse than cultivated accessions with both nuclear and chloroplast markers (Clegg et al. 1984; Neale et al. 1988;

Maroof et al. 1990; Badr et al. 2000; Wei et al. 2005; Kilian et al. 2006; Morrell and Clegg 2007; Orabi et al. 2007; Backes et al. 2008; Jilal et al. 2008; Russell et al. 2011; Comadran et al. 2012; Dai et al. 2012; Hübner et al. 2012b; Poets et al. 2015a). Table 17.1 shows diversity estimates of several studies that are based on SNP genotyping, sequencing of individual genes, or resequencing of large numbers of exomes. The difference in estimated levels of genetic diversity between studies focusing on few genes and exome resequencing studies may result from a bias towards more polymorphic genes or more diverse samples than the resequencing studies.

Gene flow between wild and domesticated barley seems to be quite common. For this reason, several researchers have suggested caution when using ex situ genebank repository accession to address questions regarding domestication and gene flow. Jakob et al. (2014) found that recently collected accessions of wild barley showed an unambiguous geographic assignment in resequencing data whereas accessions from ex situ genebank repositories did not and exhibited footprints of introgression from domesticated barley that may have occurred during the ex situ regeneration of genebank material. A comparison of wild barley accessions from Israel with domesticated barley landraces using 1536 mapped SNPs of the Illumina Golden Gate Barley OPA1 (BOPA1) array (Close et al. 2009) also uncovered gene flow between both groups (Hübner et al. 2012b). In a matched sampling study across Syria and Jordan, using the same SNP array, Russell et al. (2011) observed secondary contact between wild and landrace barleys. More extensive sequence comparisons revealed a limited effect of domestication on genetic diversity of barley landraces (Morrell et al. 2013), which was confirmed by the analysis of 7842 SNP markers (Comadran et al. 2012) in a large sample of wild and domesticated barley accessions from the whole species range (Poets et al. 2015a). Patterns of allele sharing between wild and domesticated barley revealed significantly increased allele sharing with proximate (e.g. geographically close) wild barley populations in different regions of barley cultivation. Taken together, these analyses indicate that despite a clear differentiation of wild and domesticated barley in population cluster-type analyses (principal component analysis, PCA, and STRUCTURE; Falush et al. 2007), barley was likely domesticated in multiple regions with

Table 17.1 Comparison of selected genetic diversity estimates in wild and cultivated barley

Loci sequenced	Sample size, n		Diversity measure	Diversity level		Percentage	References
	H. spontaneum	*H. vulgare*		*H. spontaneum*	*H. vulgare*	Wild/Dom.	
7 genes (5616 bp)	25	20	θ_π	0.00765	0.00279	36	Kilian et al. (2006)
5 genes (ca. 2600 bp)	19	263	θ_π	0.0106	0.0085	80	Saisho and Purugganan (2007)
1135 SNPS	131	317	H	0.217	0.187	86	Russell et al. (2011)
7 genes (9296 bp)	25/45	36	θ_W	0.0069	0.0056	82	Morrell et al. (2013)
59.5 MB exome	91	137	θ_π	NA	NA	73	Russell et al. (2016)
13.8 MB exome	344	89	θ_π	0.00297	0.00153	52	Pankin et al. (2017)

Diversity measures H is heterozygosity. θ_W and θ_π are different estimators estimates of the population mutation parameter $\theta = 4N_e\mu$

subsequent exchange of genetic diversity between wild and cultivated genotypes. Recent archaeobotanical studies provide support for multiple and independent domestications of barley (Willcox 2005; Fuller et al. 2011a, b; Riehl et al. 2013). In particular, wild barley populations contributed more strongly to geographically close (proximal) landraces. A low level of genome-wide linkage disequilibrium (LD) and short tracts of identity by state (IBS) alleles indicate that this contribution does not originate from recent gene flow but reflects an ancient contribution that may date back to the beginning of barley cultivation. The analysis of genetic diversity by Poets et al. (2015a) also indicated that wild barley from the Southern Levant contributed most to domesticated barley, suggesting that barley was domesticated in this region, consistent with archaeological findings (Lev et al. 2005) and other studies (Russell et al. 2011).

17.6 Levels of Linkage Disequilibrium

Wild barley is a predominately self-fertilizing species and therefore expected to show a high level of linkage disequilibrium. However, LD decays to half the initial value within a gene, which is unexpected (Morrell et al. 2005). Parametric estimates of both mutation and recombination showed that recombination is as important for generating haplotypic diversity as mutation in barley populations in a set of 18 genes (Morrell et al. 2006). This study also showed evidence for gene conversion, which does generally not contribute to genetic map based estimates of recombination rate (Andolfatto and Nordborg 1998). For this reason, recombination rate in barley is likely underestimated. Linkage disequilibrium in cultivated barley populations is higher than in wild barley, and within cultivated barley higher in elite breeding populations than in landraces (Caldwell et al. 2005). The high level of LD within cultivated barley implies that significantly smaller numbers of markers than in wild barley are sufficient for QTL mapping in genome-wide association scan (GWAS) (Gawenda et al. 2015) or achieving high prediction accuracies in genomic prediction (Thorwarth et al. 2017). Genotyping of genetically diverse barley material representing worldwide diversity like the barley core collection (Muñoz-Amatriaín et al. 2014) or regional breeding material such as the Nordic countries (Bengtsson et al. 2017) revealed stronger differences in LD than in diversity that likely reflects different selection histories in geographic regions or traits.

17.7 Population Genomics of Domestication Genes

It has long been hypothesized that similar traits and genes may have been selected in all domesticated grasses, including seed size and dispersal, photoperiod, dormancy, flowering time, disease resistance and stress response (Paterson et al. 1995). Consequently, among other traits, wild cereals were transformed from photoperiod sensitive, small-seeded plants with seed shattering often owing to a disarticulating or brittle inflorescence into less photoperiod sensitive, nonshattering, large-seeded domesticates with either a summer or winter annual habit, which enabled humans to plant and harvest (Paterson et al. 1995). Several domestication genes were identified by QTL mapping and positional cloning in various crop species, of which the majority encodes transcription factors whose domesticated alleles are functional and lead to temporal or spatial changes in expression, or to a change in protein function (Doebley et al. 2006; Lenser and Theißen 2013). Sequence variation at or near some domestication genes shows a footprint of selection such as a reduced level of nucleotide variation, excess of rare polymorphisms or high linkage disequilibrium (LD), suggesting that advantageous alleles were rapidly fixed by artificial selection (Wang et al. 2005; Wright et al. 2005; Comadran et al. 2012). A good understanding of the genetic basis of domestication phenotypes (or Domestication Syndrome Factors, DSF) has important implications for plant breeding, since beneficial alleles can be identified in CWR and landraces and introgressed into modern materials (Ellis et al. 2000; von Korff et al. 2006).

The transition from wild to domesticated crops mainly entails changes to three morphological features that make the crop easier to harvest: seed size, ear rachis stiffness and the glume structure. Additional domestication- or improvement-related traits for barley are spike row number (two- versus six-rowed), caryopsis type (hulled versus naked), seed size, ear length, awn length, flowering times (early and late), reduced seed dormancy, vernalization requirement and photoperiod insensitivity (Salamini et al. 2002). Frequently, domestication traits are controlled by major effect genes that produce highly visible phenotypes, which facilitate their genetic and functional analysis. In barley, several major QTLs controlling these traits have been mapped (Pourkheirandish and Komatsuda 2007; Somers et al. 2007; Pasam et al. 2012), but so far only a few candidate genes have been identified (see Table 17.1 in Dawson et al. (2015) for genes cloned). One reason is that the resolution of basic QTL (biparental mapping) experiments is generally rather poor (5–20 cM), and for a positional cloning approach in barley a resolution of at least 0.05 cM is needed. This requires testing thousands of progeny for quantitative variation, which in most cases cannot be estimated reliably on the single plant level. Complementary methods like expression QTL (eQTL) mapping (Potokina et al. 2008; Chen et al. 2010; mainly disease-resistance genes), association studies (Long et al. 2013; Alqudah et al. 2014; Gawenda et al. 2015; Maurer et al. 2016), and selective sweep mapping (Comadran et al. 2012; Russell et al. 2016) are complementary and increasingly powerful methods for identifying genes controlling both qualitative and quantitative traits in barley.

Since only a small number of domestication-related genes have been cloned, little is known about the role of natural or artificial selection and their footprint on patterns of variation in these genes. A classical assumption is that genes, which contribute to the domestication syndrome and are therefore selected during domestication show a stronger footprint of selection (i.e. a reduction of genetic diversity) relative to nonselected genes, which are only affected by demographic domestication bottlenecks (Tanksley and McCouch 1997; Wright et al. 2005).

The best characterized domestication genes in barley are *Btr1* and *Btr2*, which both independently determine the brittleness of the rachis, and therefore control seed shattering. In both genes, recessive alleles produce the nonbrittle phenotype (Pourkheirandish et al. 2015). The fine mapping and positional cloning of the two genes required the analysis of two mapping populations consisting of more than 10,000 lines each. Sequence analyses of allelic variation at the *btr1* and *btr2* alleles suggested that the brittleness mutation originated twice in different regions in the Levant. Both nonbrittle alleles are monophyletic and originated by microdeletions that cause frameshifts in the protein coding regions. The selection of the *btr1* allele occurred in the Southern Levant region and selection of the *btr2* allele in the Northern Levant. Nonbrittle accessions from Central Asia all carried the *btr1* allele suggesting that it migrated there after domestication by human activities. At both loci, the respective nonbrittleness allele segregates at a high frequency in modern barley populations and show very low levels of polymorphisms and high LD, indicating a strong genetic bottleneck, potentially resulting from selection of these haplotypes during domestication. In comparison to wild barley, haplotype numbers at *Btr1* and *Btr2* are reduced to 7 and 20%, respectively, in domesticated barley, and a network analysis revealed that all domesticated barley haplotypes either belong to the *btr1* or *btr2* lineages, consistent two independent domestication events. More recently, PCR genotyping an independent barley landrace collection provided evidence for at least three independent origins of the stiff rachis phenotype in barley (Civáň and Brown 2017). This further supports a more protracted domestication scenario. Given these observations, a formal test of selection that accounts for the genome-wide effects of domestication needs to be conducted to differentiate between single and multiple domestication events.

There are two major types of domesticated barley that differ by their spike row number. Two-rowed spikes represent the ancestral condition found in wild barley. It is controlled by a dominant allele at the *Vrs1* locus, and the six-rowed type which is caused by a point mutation in this gene (Komatsuda et al. 2007). Six-rowed barley originated at least three times from either wild or cultivated two-rowed barley by independent loss-of-function mutations in *Vrs1*, which encodes an HD-ZIPI class homeobox protein. A phylogenetic analysis revealed that these mutations probably originated in the Western Mediterranean about 7000 to 6000 years BP and spread rapidly throughout the Western cultivation range of barley.

INTERMEDIUM-C (*INT-C*) influences lateral spike fertility by modifying the action of *Vrs1*. It is a homolog of maize *teosinte branched-1* (*tb1*), and a polymorphism survey in an association panel indicated that two different variants of the gene are associated with two-rowed (*int-c.b*) and six-rowed barley (*Int-c.a*) genotypes (Ramsay et al. 2011). A high proportion of non-synonymous polymorphism indicated that this allele was strongly selected in the six-rowed barley. It was not shown that these alleles are causal for the phenotypic differences and they may be linked to more distant causal polymorphisms. However, both alleles were also found in wild barley suggesting that the mutation causing the phenotypic differences predates domestication (Ramsay et al. 2011).

Selective sweep mapping for genes involved in adaptation to postdomestication environments in Europe, which are characterized by longer day length and different temperature regimes identified the barley homolog of the Antirrhinum gene *CENTRORADIALIS*, *HvCEN*, which controls flowering time and has been enriched in European spring and winter barley accessions (Comadran et al. 2012). Resequencing of *HvCEN* and the identification of haplotypes in a large collection of wild and domesticated barleys indicated an enrichment of the early flowering alleles in domesticated accessions, and showed that selected alleles predate domestication and therefore the phenotype in domestication is not due to new mutations. For example, haplotype III is present in wild barley and fixed in spring cultivars from Northern Europe, haplotype II is frequent in wild barley from the eastern Mediterranean and also in domesticated winter lines. The enrichment of different haplotypes in domesticated barley groups indicates strong diversifying selection.

The control of flowering time by photoperiod (day length) is an important trait in crops as it has a strong effect on yield. In barley, a recessive allele of the *Ppd-H1* gene (the day length nonresponsive allele *ppd-H1*) determines flowering time under long day conditions, which is an important adaptation to Northern latitudes. The causal variant was identified by genetic analysis (Turner 2005), and by sequencing, the gene in 72 wild barleys and 194 landraces (Jones et al. 2008) showed a typical domestication-related bottlenecked locus although two studies differ in the inferred causal variant at this gene. The frequency of the responsive *Ppd-H1* and nonresponsive *ppd-H1* alleles show a highly significant gradient along latitude in Europe, which indicates postdomestication selection. Both alleles originated in wild barley indicating that the nonresponsive allele is not a novel mutation that arose during the expansion in Europe. Furthermore, this allele likely originated in the Eastern part of the Fertile Crescent near the Zagros mountain, which is close to the postulated secondary centre of barley domestication (Morrell and Clegg 2007).

The genetic control of seed dormancy also shows a footprint of selection during barley domestication. Seed dormancy is an important trait for adaptation because it controls germination in response to environmental variation. In domesticated barley, synchronous germination by reducing seed dormancy is a selected trait; on the other hand in wet climates, a very short seed dormancy contributes to preharvest sprouting. QTL analyses of populations led to the identification of two genes controlling barley seed dormancy. *Qsd1* encodes an alanine aminotransferase (Sato et al. 2016), and *Qsd2* encodes a Mitogen-Activated Protein Kinase Kinase 3 (Nakamura et al. 2016). Resequencing of both genes in large panels of wild and domesticated barley revealed distinct evolutionary patterns of genetic diversity. At the *Qsd1* gene, the reduced dormancy allele *Qsd1* originated in the South Levant by mutation from the long dormancy allele *qsd1* which is present in wild barley. At the Qsd-2 locus, the dormant Az allele originated by a single non-synonymous mutation from a precursor allele that originated in the centre of the wild barley distribution range, but which is more frequent in East Asia. The dormant Az allele is prevalent in East Asia possibly in response to selection for preharvest sprouting because the rainy season and the harvest season for barley overlap in this region.

Taken together, only a few barley domestication genes have been identified and their polymorphism characterized in greater detail (Dawson et al. 2015). No genome-wide population genomic analyses to identify novel domestication genes by a systematic and statistically rigorous comparison of diversity have been carried out so far. Nevertheless, the available analyses indicate that natural variation used during and after domestication was segregating in wild barley populations and does not represent new postdomestication mutations. This observation is not unique to barley, a similar pattern has been found in maize, e.g. with the *teosinte branched 1* gene (Studer et al. 2011). For this reason, one may conclude that soft selective sweeps of standing variation (Hermisson and Pennings 2005) were more important for barley domestication than hard selective sweeps of new mutations. Another common pattern of both genome-wide and gene-specific comparisons of genetic diversity between wild and domesticated barley is the presence of a domestication bottleneck, which is not very strong and comparable to other crops like maize (Wright et al. 2005). A putative multiregional domestication and secondary gene flow may mask weak domestication-related selection or the footprints of soft selective sweeps (Gross and Olsen 2010). For this reason, a more rigid population genetic modelling required to uncover subtle signals of selection in both wild and domesticated barley populations by combining multiple tests of selection that also account for demographic history (Vitti et al. 2013).

17.8 Genetic Variation in Barley Landraces and Modern Cultivars

In addition to characterizing the genetic diversity of available resources, there are multiple applications for modern genotyping and sequencing technologies in barley pre-breeding as well as commercial breeding programmes, mainly in speed, efficiency and cost savings. This allows the definition of core collections for pre-breeding, genome-wide association studies and/or genomic selection. Current technologies available are SNP genotyping arrays, reduced representation sequencing methods such as genotyping by sequencing (GBS), and low coverage sequencing for rapid and efficient characterization.

Genebanks harbour tens of thousands of barley accessions that represent the diversity of domesticated barley (Knüpffer 2009; https://www.genesys-pgr.org). Several subsets of these collections were established and further characterized. Pasam et al. (2014) genotyped a large collection of 1485 spring barley landraces from the IPK genebank with 42 SSR markers, to determine the population structure and genetic diversity of this material. Subsequent studies using SNP arrays included smaller landrace collections and conducted environmental association studies to identify SNPs which are associated with climatic variables. Scandinavian landraces genotyped with 384 SNP markers (Moragues et al. 2010) were genetically distinct from Central European and Mediterranean landraces (Forsberg et al. 2014; Aslan et al. 2015). In addition, five SNP markers were identified that were significantly associated with day length and precipitation potentially involved in local adaptation (Aslan et al. 2015). Using the BOPA1 SNP array with 1536 markers, a sample of 65 worldwide domesticated barley accessions of the Israeli Plant Gene Bank (IPG) were genotyped and compared to wild barley accessions from Israel Hübner et al. (2012b) identified a number of admixed genotypes, which most likely originated from recent gene flow. Until recently, the iSelect genotyping array reflected most advanced

SNP array, the current generation with >7842 SNPs (Comadran et al. 2012) and has been used in a number of studies. Tondelli et al. (2013) genotyped 216 European spring two-rowed barley cultivars of genebank accessions representing old cultivars and modern elite varieties. The complete set was phenotyped for various quantitative traits across different European field sites, and with the increased marker density a GWAS, identified SNPs associated with plant height, lodging and necking. Several of these SNPs were located in genomic regions with previously characterized genes shown to influence these traits. Similarly, the increased density was sufficient to conduct tests of selection and to determine whether QTLs identified in GWAS overlap with selective sweep regions. In particular, the authors identified an overlap for two of three most significant QTL for plant height, exhibiting a footprint of local selection. The markers also provided the means to determine genomic regions with no genetic diversity detected at the markers surveyed in modern germplasm, likely a result of crossing elite varieties that were fixed by breeding, for the same key traits. The 9K iSelect SNP array was used to genotype 2417 barley accessions from the USDA National Small Grains Collection which holds over 30,000 barley accessions (Muñoz-Amatriaín et al. 2014). The genotyped collection consisted of 815 landraces and 781 cultivars or breeding lines representative of most of the diversity present in domesticated barley. Using the genotyping data, five populations were inferred that reflect the geographic origin and the row type (two- or six-rowed). Further analyses identified genomic regions strongly differentiated between the clusters, associated with high LD, which may reflect strong selection for major genes (row-type and flowering associated genes). Furthermore, genome-wide studies using multilocation phenotypic field trial data also found SNP markers located within or in close linkage to flowering time (e.g. *Ppd-H1*, *EPS2*) or row-type genes (*VRS1*, *VRS3*, *INT-C*). Finally, using allelic diversity, the authors identified small sets of highly diverse genotypes, providing mini core collections which are useful for more detailed and cost-effective phenotyping. The same array was also used to successfully identify candidate QTLs for quality traits like zinc and iron content in local landraces from Ethiopia and Eritrea (Mamo et al. 2014).

GBS provides another approach for an efficient survey of genetic diversity. In barley research, it has been applied to identify novel SNPs (Guo et al. 2014), establish genetic maps (Poland et al. 2012; Mascher et al. 2013), analyse segregation (Bélanger et al. 2016a, b), and map QTLs (Liu et al. 2014; Lin et al. 2015). GBS was also used to infer population structure in collections of barley genetic material (Pourkheirandish et al. 2015). Exome capture sequencing was also used for the characterization of genetic diversity in wild and landrace barleys (Russell et al. 2016). The analysis of 137 Eurasian and African landraces showed a high level of population structure. For the landraces, there was strong support for 14 populations, differentiated mainly by geography. Spatial autocorrelation indicated higher geographic-based structuring for the landraces compared to wild barley, most likely reflecting recent expansion of domesticated barley. The landraces had fewer private alleles than wild barleys, which is a consequence of the domestication process. The most strongly differentiated genomic regions between the two- and six-rowed barley were at the known row-type determining genes, *Vrs1* and *INT-C*, although there were other peaks differentiating row type on chromosomes 1H and 3H, which require further characterization. An analysis of selective sweeps identified two significant regions on chromosomes 1H and 3H, the SNP on 3H targets a member of the large pentatricopeptide repeat (PPR) family, whose diversity is also strongly reduced in comparison to wild barley. The most recent study by Pankin et al. (2017) found a larger number of selective sweep regions using a different test of selection. By comparing nucleotide diversity of wild and domesticated accession, 13 genomic regions harbouring 41 target genes were identified. Two candidate regions overlapped with the brittleness genes *Brt1* and *Brt2*, and the naked barley controlling gene *NUD*. However, in both cases, no significant SNP was directly located in the genes. Other significant regions targeted homologs of genes involved in light signalling, photoperiod, circadian clock, abscisic acid (ABA) and carbohydrate metabolism pathways. None of the genes have been functionally characterized in barley and thus they are candidates for further functional investigations.

Although large-scale patterns of genetic diversity in cultivated barley landraces are congruent, a comparison of different collections covering a similar geographic range that were

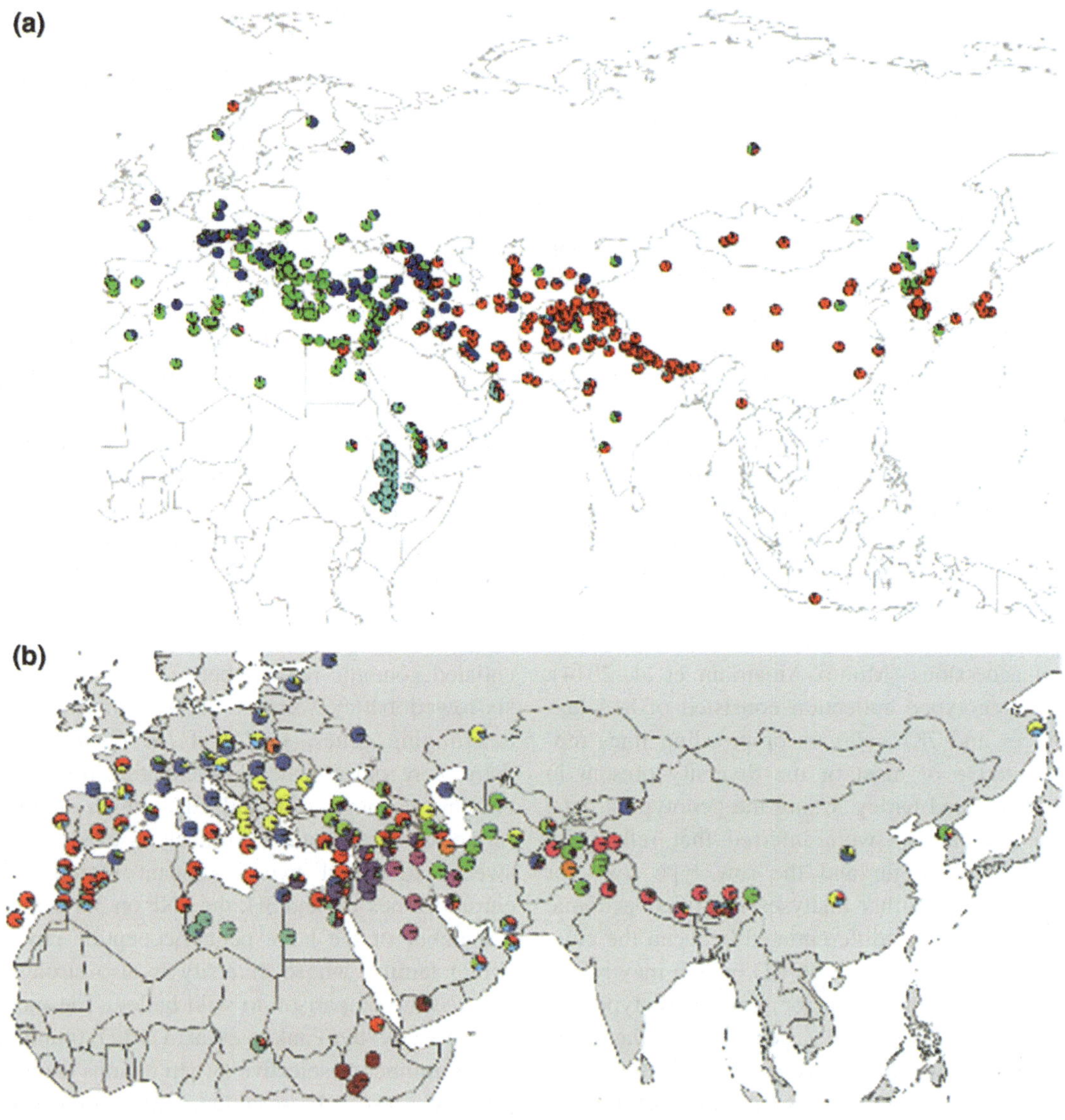

Fig. 17.1 Population genetic structure of cultivated barley landraces. **a** Population structure of barley landraces based on 6152 SNPs with $K = 4$ as optimal number of populations. Colours of pie diagrams indicate the proportion of assignment to each of the inferred clusters. *Source* Poets et al. (2015a). **b** Population structure of barley landraces based on exome capture sequencing with $K = 14$ as best-fitting number of populations. *Source* Russell et al. (2016). It should be noted that Poets et al. (2015a) limited the number of clusters to obtain better allele frequency estimates, whereas Russell et al. (2016) did not apply such a restriction, which explains the very different number of genetic clusters

genotyped with different methods (SNP genotyping versus exome capture sequencing) show some important differences like the number of populations that are most consistent with the data and the geographic distribution (Fig. 17.1). These results clearly indicate that large samples, a high density of sequence coverage, and explicit population genetic modelling are required for a robust inference of the genetic structure and demographic history of cultivated barley.

17.9 Monitoring Genetic Diversity Over Time

Current sequencing technologies allow genome-wide analyses of allele frequency changes in the genome, which can be used to study local adaptation over time in a changing environment or to track changes in diversity in breeding programmes. Allele frequency changes can be detected by *Fst*-based statistics, which are normally used for analysing spatial population differentiation. Thormann et al. (2017) monitored allele frequency changes in Jordanian wild barley accessions over 31 years, between 1981 and 2012 using 38 EST-derived SSR markers. The accessions collected in 2012 were unchanged with respect to genetic diversity but showed a greater spatial homogeneity. In contrast, a comparison of 10 wild barley populations collected in 1980 and 2008 in Israel exhibited significant changes in flowering time and SSR diversity, interpreted by the authors as a selection-driven response to climate change because recently collected accessions flowered significantly earlier and accessions grouped by collection year and less by geographic origin (Nevo et al. 2012).

Changes in allele frequency over time were also monitored in breeding populations. Poets et al. (2015b) analysed genotyping data of 3971 accessions from the North American Barley Coordinated Agricultural Project (CAP) that were genotyped with the BOPA 1 and 2 SNP arrays (2882 SNPs; Close et al. 2009). The material consisted of genotypes from diverse, mostly public, breeding programmes including two- and six-rowed as well as winter and spring types. Correspondingly, the average pairwise diversity in the material ranged from 0.15 to 0.19. The most prominent genetic differentiation between pools (measured as *Fst*) was not between row-type (0.23) or growth habit (winter versus spring) (0.17), but between breeding programmes (0.37). Marker density was sufficient to identify genomic regions with footprints of selection using outlier tests. These comprised major genes, such as flowering time response (*Ppd-H1*) and cold acclimatization (*Cbf4*) genes, as well as genes for spike type (*INT-C*). However, allele frequency differences at these loci were not particularly strong, which may reflect a high level of haplotype sharing that overlap with high/F_{ST}/values in some genomic regions indicative of strong selection. Finally, the analysis also revealed different levels of genetic drift within and gene flow between breeding populations, mirroring the breeding history of North American programmes.

17.10 Utilization of Diversity Studies for Allele Mining and Introgression

The analysis of population structure, demographic history and local adaptation of wild and landrace barleys is important for elucidating the evolutionary history of the crop. This knowledge can also be used to develop barley varieties that are resistant against pathogens and tolerate future climatic conditions by introgressing this variation into modern elite varieties. Because both wild and cultivated barley are fully interfertile, the former can be used in breeding programmes to introduce novel variation. The use of wild barleys has been limited because it brings undesirable traits such as grain shattering, uneven germination, stiff rachis and variation in vernalization requirements. Linkage drag of deleterious variation in regions with low recombination is also an issue. For this reason, wild barley is used in pre-breeding and crossed with suitable cultivated lines genotypes followed by rounds of backcrossing and selfing, which is subsequently used for conventional breeding (Ellis et al. 2000;

Fischbeck 2003; Dawson et al. 2015). Of particular interest is the introgression of disease-resistance genes (Nice et al. 2016), also from the secondary gene pool such as *Hordeum bulbosum* (Wendler et al. 2015).

In recent years, a number of populations were generated to map useful genes and facilitate their introgression into modern varieties (Zamir 2001; Dekkers et al. 2002). This includes chromosome substitution lines (RCSL) (Matus et al. 2003; Inostroza et al. 2008), nested association mapping (NAM) panels (Schnaithmann et al. 2014), advanced backcross lines (Pillen et al. 2003; Nice et al. 2016) and multiparent segregating populations (MAGIC) (Sannemann et al. 2015). The future development of introgression populations will further benefit from ongoing large-scale efforts to characterize collections of wild and barley both phenotypically (precision phenotyping methods) and genotypically (whole genome sequencing method identifying millions of SNPs). New mapping populations can therefore be designed to investigate complex traits. For traits that are mainly controlled by single genes, it may not be necessary to create segregating populations for introgression because novel, potentially beneficial alleles may be found by screening large collections of genetic resources (allele mining). The usefulness of this approach was demonstrated by the *eIF4E* resistance gene. Alleles at this gene provide resistance against the Barley Yellow Mosaic Virus (BYMC), and the two recessive alleles, *rym4* and *rym5*, were introgressed in most modern European varieties. To identify novel resistance alleles, the *eIF4E* gene was sequenced in a panel of 1090 barley landraces, which grouped into 47 haplotypes, showing significantly different frequencies in geographic regions and a footprint of selection potentially reflecting host–pathogen coevolution (Hofinger et al. 2011). A subsequent sequence survey and resistance tests of the same and the additional BYMV resistance gene *HvPDIL5* in cultivated and wild barley (Yang et al. 2016) revealed that the two genes showed different patterns of function and evolution. Alleles of *eIF4E* conferring resistance differed by non-synonymous and in-frame insertion/deletion polymorphisms in the protein coding sequence, whereas in the *HvPDIL5* gene resistance was provided by loss-of-function mutations. The analysis of selection revealed a strong footprint of adaptive evolution in domesticated barley, particularly in South East Asia, but no evidence for selection in wild barley. The new resistance alleles can now be introgressed into elite breeding lines by backcrossing or genome editing.

Another example of allele mining in barley is the search for new genes for the nonhost-specific resistance against powdery mildew (Spies et al. 2012). Since many new genes controlling agronomically interesting traits are identified and functionally characterized, an allele mining approach may hold great promise for the targeted improvement of elite varieties.

17.11 The Future of Genetic Diversity Studies in Barley

The study of genetic diversity is still limited by the large genome size of this species, which does not yet allow the complete sequencing of large numbers of individuals as in model plants (Alonso-Blanco et al. 2016) or crops with smaller genomes like rice (Li et al. 2014). For this reason, the analysis of demographic history and the identification of adaptive trait genes remains incomplete. This is evident by the fact that most studies utilizing genetic diversity identified the same well-characterized major domestication or adaptation genes, but the high proportion of missing genomic regions or the relatively small sample sizes that can be analysed using costly methods like exome capture sequencing reduce the power to identify novel genes or regulatory mutations. Genome technologies like single-molecule sequencing, chromosomal conformation capture and optical mapping will contribute to a better resolution of highly repetitive and large genomes as demonstrated with the most recent version of the barley reference genome (Mascher et al. 2017). In addition, the population genomic approaches of whole genome resequencing data will greatly facilitate the identification of novel genes useful for plant breeding (Shi and Lai 2015).

Plant breeding is currently undergoing another major transformation because of the increasing use of genomic selection and genome editing (Morrell et al. 2012; Hu and Lübberstedt 2015; Bevan et al. 2017). Genomic selection (GS) predicts the phenotype of individuals in a breeding population, based on constructing a quantitative genetic model of the breeding value using a training population that has been both phenotyped and genotyped (Schmid and Thorwarth 2014). GS will increase the selection gain per time unit and therefore accelerate the development of improved barley varieties. New methods for GS allow the inclusion of prior information like alleles with a positive effect on the trait that was identified by selective sweep mapping or GWAS, but it also enables selection against alleles that are deleterious (Morrell et al. 2012; Kono et al. 2016). GS also allows the breeder maintaining a high level of genetic diversity in breeding populations despite strong selection for desired traits. It may also be used for the targeted introduction of genetic diversity from CWR and landraces into genetically narrow elite breeding pools such as the German winter barleys (Thorwarth et al. 2017) by combining a comprehensive genomic characterization of genetic resources with GS.

The other major development is genome editing, which is the targeted induction of novel genetic diversity using systems like CRISPR/Cas9 and hold great promise for plant breeding (Schaart et al. 2016). Novel alleles that were identified by selection scans, allele mining or high-resolution GWAS can be introgressed into modern elite varieties directly by genome editing rather than backcrossing, thereby alleviating problems like linkage drag, lack of recombination or incompatible epistatic interactions (Østerberg et al. 2017). A bottleneck remains the requirement of genetic transformation and plant regeneration (Gurushidze et al. 2017; Kapusi et al. 2017). Nevertheless, the efficient utilization of genetic variation identified in wild relatives or landraces appears to be a realistic perspective for barley breeding.

References

Alonso-Blanco C, Andrade J, Becker C et al (2016) 1,135 genomes reveal the global pattern of polymorphism in *Arabidopsis thaliana*. Cell 166:481–491. https://doi.org/10.1016/j.cell.2016.05.063

Alqudah AM, Sharma R, Pasam RK et al (2014) Genetic dissection of photoperiod response based on GWAS of pre-anthesis phase duration in spring barley. PLoS One 9:e113120. https://doi.org/10.1371/journal.pone.0113120

Andolfatto P, Nordborg M (1998) The effect of gene conversion on intralocus associations. Genetics 148:1397–1399

Aslan S, Forsberg NEG, Hagenblad J, Leino MW (2015) Molecular genotyping of historical barley landraces reveals novel candidate regions for local adaption. Crop Sci 55:2766. https://doi.org/10.2135/cropsci2015.02.0119

Backes G, Orabi J, Wolday A et al (2008) High genetic diversity revealed in barley (*Hordeum vulgare*) collected from small-scale farmer's fields in Eritrea. Genet Res Crop Evol 56:85–97. https://doi.org/10.1007/s10722-008-9347-5

Badr A, M K, Sch R et al (2000) On the origin and domestication history of barley (*Hordeum vulgare*). Mol Biol Evol 17:499–510. https://doi.org/10.1093/oxfordjournals.molbev.a026330

Bedada G, Westerbergh A, Nevo E et al (2014) DNA sequence variation of wild barley *Hordeum spontaneum* (L.) across environmental gradients in Israel. Heredity 112:646–655. https://doi.org/10.1038/hdy.2014.2

Bélanger S, Clermont I, Esteves P, Belzile F (2016a) Extent and overlap of segregation distortion regions in 12 barley crosses determined via a Pool-GBS approach. Theor Appl Genet 129:1393–1404. https://doi.org/10.1007/s00122-016-2711-5

Bélanger S, Esteves P, Clermont I et al (2016b) Genotyping-by-sequencing on pooled samples and its use in measuring segregation bias during the course of androgenesis in barley. Plant Genome 9. https://doi.org/10.3835/plantgenome2014.10.0073

Bengtsson T, Manninen O, Jahoor A, Orabi J, PPP Barley Consortium (2017) Genetic diversity, population structure and linkage disequilibrium in Nordic spring barley (Hordeum vulgare L. subsp. vulgare). Genet Resour Crop Evol 64(8):2021–2033

Bevan MW, Uauy C, Wulff BB et al (2017) Genomic innovation for crop improvement. Nature 543:346–354

Blattner FR, Badani Méndez AG (2001) RAPD data do not support a second centre of barley domestication in Morocco. Genet Res Crop Evol 48:13–19. https://doi.org/10.1023/a:1011299021969

Brown TA, Jones MK, Powell W, Allaby RG (2009) The complex origins of domesticated crops in the Fertile Crescent. Trends Ecol Evol 24:103–109. https://doi.org/10.1016/j.tree.2008.09.008

Caldwell K, Russell J, Langridge P, Powell W (2005) Extreme population-dependent linkage disequilibrium detected in an inbreeding plant species, *Hordeum vulgare*. Genetics 172:557–567. https://doi.org/10.1534/genetics.104.038489

Chen X, Hackett CA, Niks RE et al (2010) An eQTL analysis of partial resistance to *Puccinia hordei* in barley. PLoS One 5:e8598. https://doi.org/10.1371/journal.pone.0008598

Civáň P, Brown TA (2017) A novel mutation conferring the nonbrittle phenotype of cultivated barley. New Phyt 214:468–472. https://doi.org/10.1111/nph.14377

Clegg M, Brown A, Whitfeld P (1984) Chloroplast DNA diversity in wild and cultivated barley: implications for genetic conservation. Genet Res 43:339–343

Close TJ, Bhat PR, Lonardi S et al (2009) Development and implementation of high-throughput SNP genotyping in barley. BMC Genomics 10:582. https://doi.org/10.1186/1471-2164-10-582

Comadran J, Kilian B, Russell J et al (2012) Natural variation in a homolog of *Antirrhinum CENTRORADIALIS* contributed to spring growth habit and environmental adaptation in cultivated barley. Nat Genet 44:1388–1392. https://doi.org/10.1038/ng.2447

Dai F, Nevo E, Wu D et al (2012) Tibet is one of the centers of domestication of cultivated barley. Proc Natl Acad Sci USA 109:16969–16973. https://doi.org/10.1073/pnas.1215265109

Dai F, Chen Z-H, Wang X et al (2014) Transcriptome profiling reveals mosaic genomic origins of modern cultivated barley. Proc Natl Acad Sci USA 111: 13403–13408. https://doi.org/10.1073/pnas.1414335111

Dawson IK, Russell J, Powell W et al (2015) Barley: a translational model for adaptation to climate change. New Phyt 206:913–931. https://doi.org/10.1111/nph.13266

Dekkers JC, Hospital F, others (2002) The use of molecular genetics in the improvement of agricultural populations. Nat Rev Genet 3:22–32

Dempewolf H, Baute G, Anderson J et al (2017) Past and future use of wild relatives in crop breeding. Crop Sci 57:1070. https://doi.org/10.2135/cropsci2016.10.0885

Doebley JF, Gaut BS, Smith BD (2006) The molecular genetics of crop domestication. Cell 127:1309–1321. https://doi.org/10.1016/j.cell.2006.12.006

Ellis R, Forster B, Robinson D et al (2000) Wild barley: a source of genes for crop improvement in the 21st century? J Exp Bot 51:9–17. https://doi.org/10.1093/jexbot/51.342.9

Falush D, Stephens M, Pritchard JK (2007) Inference of population structure using multilocus genotype data: dominant markers and null alleles. Mol Ecol Notes 7:574–578. https://doi.org/10.1111/j.1471-8286.2007.01758.x

Fan Y, Shabala S, Ma Y et al (2015) Using QTL mapping to investigate the relationships between abiotic stress tolerance (drought and salinity) and agronomic and physiological traits. BMC Genomics 16:43. https://doi.org/10.1186/s12864-015-1243-8

Fang Z, Gonzales AM, Clegg MT et al (2014) Two genomic regions contribute disproportionately to geographic differentiation in wild barley. Genes Genomes Genet 4:1193–1203

Fischbeck G (2003) Diversification through breeding. In: Developments in plant genetics and breeding. Elsevier, pp 29–52

Forsberg NEG, Russell J, Macaulay M et al (2014) Farmers without borders-genetic structuring in century old barley (*Hordeum vulgare*). Heredity 114:195–206. https://doi.org/10.1038/hdy.2014.83

Friedt W, Horsley RD, Harvey BL et al (2011) Barley breeding history, progress, objectives, and technology. In: Barley. Wiley-Blackwell, pp 160–220

Fuller DQ, Asouti E, Purugganan MD (2011a) Cultivation as slow evolutionary entanglement: comparative data on rate and sequence of domestication. Veg Hist Archaeobot 21:131–145. https://doi.org/10.1007/s00334-011-0329-8

Fuller DQ, Willcox G, Allaby RG (2011b) Cultivation and domestication had multiple origins: arguments against the core area hypothesis for the origins of agriculture in the near east. World Archaeol 43:628–652. https://doi.org/10.1080/00438243.2011.624747

Gawenda I, Thorwarth P, Günther T et al (2015) Genome-wide association studies in elite varieties of German winter barley using single-marker and haplotype-based methods. Plant Breed 134:28–39. https://doi.org/10.1111/pbr.12237

Gross BL, Olsen KM (2010) Genetic perspectives on crop domestication. Trends Plant Sci 15:529–537. https://doi.org/10.1016/j.tplants.2010.05.008

Günther T, Coop G (2013) Robust identification of local adaptation from allele frequencies. Genetics 195:205–220. https://doi.org/10.1534/genetics.113.152462

Guo G, Dondup D, Zhang L et al (2014) Identification of SNPs in barley (*Hordeum vulgare* L.) by deep sequencing of six reduced representation libraries. Crop J 2:419–425. https://doi.org/10.1016/j.cj.2014.06.008

Gupta PK, Sharma S, Kumar S et al (2004) Adaptive ribosomal DNA polymorphism in wild barley at a mosaic microsite, Newe Ya'ar in Israel. Plant Sci 166:1555–1563

Gurushidze M, Hiekel S, Otto I et al (2017) Site-directed mutagenesis in barley by expression of TALE nuclease in embryogenic pollen. In: Biotechnologies for plant mutation breeding. Springer, pp 113–128

Harlan HV, Martini ML (1938) The effect of natural selection in a mixture of barley varieties. J Agric Res 57(3)

Harlan J, Zohary D (1966) Distrbution of wild wheats and barley. Science 153:1074–1080

Hermisson J, Pennings PS (2005) Soft sweeps: molecular population genetics of adaptation from standing genetic variation. Genetics 169:2335–2352. https://doi.org/10.1534/genetics.104.036947

Hofinger BJ, Russel Joanne R, Bass CG et al (2011) An exceptionally high nucleotide and haplotype diversity and a signature of positive selection for the *eIF4E* resistance gene in barley are revealed by allele mining and phylogenetic analyses of natural populations. Mol

Ecol 17:3653–3668. https://doi.org/10.1111/j.1365-294x.2011.05201.x

Hu S, Lübberstedt T (2015) Getting the 'MOST' out of crop improvement. Trends Plant Sci 20:372–379. https://doi.org/10.1016/j.tplants.2015.03.002

Huang Q, Beharav A, Youchun U et al (2002) Mosaic microecological differential stess causes adaptive microsatellite divergence in wild barley, *Hordeum spontaneum*, at Neve Ya'ar, Israel. Genome 45:1216–1229

Hübner S, Höffken M, Oren E et al (2009) Strong correlation of wild barley (*Hordeum spontaneum*) population structure with temperature and precipitation variation. Mol Ecol 18:1523–1536. https://doi.org/10.1111/j.1365-294x.2009.04106.x

Hübner S, Bdolach E, Ein-Gedy S et al (2012a) Phenotypic landscapes: phenological patterns in wild and cultivated barley. J Evol Biol 26:163–174. https://doi.org/10.1111/jeb.12043

Hübner S, Günther T, Flavell A et al (2012b) Islands and streams: clusters and gene flow in wild barley populations from the Levant. Mol Ecol 21:1115–1129. https://doi.org/10.1111/j.1365-294x.2011.05434.x

Ingvarsson PK, Eckert C (2002) A metapopulation perspective on genetic diversity and differentiation in partially self-fertilizing plants. Evolution 56:2368. https://doi.org/10.1554/0014-3820(2002)056[2368:ampogd]2.0.co;2

Inostroza L, del Pozo A, Matus I et al (2008) Association mapping of plant height, yield, and yield stability in recombinant chromosome substitution lines (RCSLs) using *Hordeum vulgare* subsp. *spontaneum* as a source of donor alleles in a *Hordeum vulgare* subsp. *vulgare* background. Mol Breed 23:365–376. https://doi.org/10.1007/s11032-008-9239-6

Jakob SS, Rödder D, Engler JO et al (2014) Evolutionary history of wild barley (*Hordeum vulgare* subsp. *spontaneum*) analyzed using multilocus sequence data and paleodistribution modeling. Genome Biol Evol 6:685–702. https://doi.org/10.1093/gbe/evu047

Jilal A, Grando S, Henry RJ et al (2008) Genetic diversity of ICARDA's worldwide barley landrace collection. Genet Res Crop Evol 55:1221–1230. https://doi.org/10.1007/s10722-008-9322-1

Jones H, Leigh FJ, Mackay I et al (2008) Population-based resequencing reveals that the flowering time adaptation of cultivated barley originated east of the Fertile Crescent. Mol Biol Evol 25:2211–2219. https://doi.org/10.1093/molbev/msn167

Kalendar R, Tanskanen J, Immonen S et al (2000) Genome evolution of wild barley (*Hordeum spontaneum*) by BARE-1 retrotransposon dynamics in response to sharp microclimatic divergence. Proc Natl Acad Sci USA 97:6603–6607. https://doi.org/10.1073/pnas.110587497

Kalladan R, Worch S, Rolletschek H et al (2013) Identification of quantitative trait loci contributing to yield and seed quality parameters under terminal drought in barley advanced backcross lines. Mol Breed 32:71–90. https://doi.org/10.1007/s11032-013-9853-9

Kapusi E, Corcuera-Gómez M, Melnik S, Stoger E (2017) Heritable genomic fragment deletions and small indels in the putative engase gene induced by CRISPR/Cas9 in barley. Front Plant Sci 8:540. https://doi.org/10.3389/fpls.2017.00540

Kilian B, Ozkan H, Kohl J et al (2006) Haplotype structure at seven barley genes: relevance to gene pool bottlenecks, phylogeny of ear type and site of barley domestication. Mol Genet Genomics 276:230–241. https://doi.org/10.1007/s00438-006-0136-6

Knüpffer H (2009) Triticeae genetic resources in *ex situ* genebank collections. In: Genetics and genomics of the triticeae. Springer, pp 31–79

Komatsuda T, Pourkheirandish M, He C et al (2007) Six-rowed barley originated from a mutation in a homeodomain-leucine zipper I-class homeobox gene. Proc Natl Acad Sci USA 104:1424–1429. https://doi.org/10.1073/pnas.0608580104

Kono TJY, Fu F, Mohammadi M, Hoffman PJ, Liu C, Stupar RM, Smith KP, Tiffin P, Fay JC, Morrell PL (2016) The role of deleterious substitutions in crop genomes. Mol Biol Evol 33:2307–2317. https://doi.org/10.1093/molbev/msw102

Lenser T, Theißen G (2013) Molecular mechanisms involved in convergent crop domestication. Trends Plant Sci 18:704–714. https://doi.org/10.1016/j.tplants.2013.08.007

Lev E, Kislev ME, Bar-Yosef O (2005) Mousterian vegetal food in Kebara Cave, Mt. Carmel. J Archaeol Sci 32:475–484. https://doi.org/10.1016/j.jas.2004.11.006

Li J-Y, Wang J, Zeigler RS (2014) The 3,000 rice genomes project: new opportunities and challenges for future rice research. GigaScience 3:8. https://doi.org/10.1186/2047-217x-3-8

Lin M, Cai S, Wang S et al (2015) Genotyping-by-sequencing (GBS) identified SNP tightly linked to QTL for pre-harvest sprouting resistance. Theor Appl Genet 128:1385–1395. https://doi.org/10.1007/s00122-015-2513-1

Liu H, Bayer M, Druka A et al (2014) An evaluation of genotyping by sequencing (GBS) to map the *breviaristatum-e* (*ari-e*) locus in cultivated barley. BMC Genomics 15:104. https://doi.org/10.1186/1471-2164-15-104

Long NV, Dolstra O, Malosetti M et al (2013) Association mapping of salt tolerance in barley (*Hordeum vulgare* L.). Theor Appl Genet 126:2335–2351. https://doi.org/10.1007/s00122-013-2139-0

Mamo BE, Barber BL, Steffenson BJ (2014) Genome-wide association mapping of zinc and iron concentration in barley landraces from Ethiopia and Eritrea. J Cereal Sci 60:497–506. https://doi.org/10.1016/j.jcs.2014.08.007

Maroof MAS, Allard RW, Zhang QF (1990) Genetic diversity and ecogeographical differentiation among ribosomal DNA alleles in wild and cultivated barley. Proc Natl Acad Sci USA 87:8486–8490. https://doi.org/10.1073/pnas.87.21.8486

Mascher M, Wu S, Amand PS et al (2013) Application of genotyping-by-sequencing on semiconductor sequencing platforms: a comparison of genetic and reference-based marker ordering in barley. PLoS One 8: e76925. https://doi.org/10.1371/journal.pone.0076925

Mascher M, Gundlach H, Himmelbach A et al (2017) A chromosome conformation capture ordered sequence of the barley genome. Nature 544:427–433. https://doi.org/10.1038/nature22043

Matus I, Corey A, Filichkin T et al (2003) Development and characterization of recombinant chromosome substitution lines (RCSLs) using *Hordeum vulgare* subsp. *spontaneum* as a source of donor alleles in a *Hordeum vulgare* subsp. *vulgare* background. Genome 46:1010–1023. https://doi.org/10.1139/g03-080

Maurer A, Draba V, Pillen K (2016) Genomic dissection of plant development and its impact on thousand grain weight in barley through nested association mapping. J Exp Bot 67:2507–2518. https://doi.org/10.1093/jxb/erw070

McCouch S, Baute GJ, Bradeen J et al (2013) Agriculture: feeding the future. Nature 499:23–24

Molina-Cano JL, Fra-Mon P, Salcedo G et al (1987) Morocco as a possible domestication center for barley: biochemical and agromorphological evidence. Theor Appl Genet 73:531–536. https://doi.org/10.1007/bf00289190

Molina-Cano JL, Moralejo M, Igartua E, Romagosa I (1999) Further evidence supporting Morocco as a centre of origin of barley. Theor Appl Genet 98:913–918. https://doi.org/10.1007/s001220051150

Mora F, Quitral YA, Matus I et al (2016) SNP-based QTL mapping of 15 complex traits in barley under rain-fed and well-watered conditions by a mixed modeling approach. Front Plant Sci 7:909. https://doi.org/10.3389/fpls.2016.00909

Moragues M, Comadran J, Waugh R et al (2010) Effects of ascertainment bias and marker number on estimations of barley diversity from high-throughput SNP genotype data. Theor Appl Genet 120:1525–1534. https://doi.org/10.1007/s00122-010-1273-1

Morrell PL, Clegg MT (2007) Genetic evidence for a second domestication of barley (*Hordeum vulgare*) east of the Fertile Crescent. Proc Natl Acad Sci USA 104:3289–3294. https://doi.org/10.1073/pnas.0611377104

Morrell PL, Lundy KE, Clegg MT (2003) Distinct geographic patterns of genetic diversity are maintained in wild barley (*Hordeum vulgare* ssp. *spontaneum*) despite migration. Proc Natl Acad Sci USA 100: 10812–10817. https://doi.org/10.1073/pnas.1633708100

Morrell PL, Toleno DM, Lundy KE, Clegg MT (2005) Low levels of linkage disequilibrium in wild barley (*Hordeum vulgare* ssp. *spontaneum*) despite high rates of self-fertilization. Proc Natl Acad Sci USA 102:2442–2447. https://doi.org/10.1534/genetics.105.054502

Morrell PL, Toleno DM, Lundy KE, Clegg MT (2006) Estimating the contribution of mutation, recombination and gene conversion in the generation of haplotypic diversity. Genetics 173:1705–1723. https://doi.org/10.1534/genetics.105.054502

Morrell PL, Buckler ES, Ross-Ibarra J (2012) Crop genomics: advances and applications. Nat Rev Genet 13(2):85–96. https://doi.org/10.1038/nrg3097

Morrell PL, Gonzales AM, Meyer KK, Clegg MT (2013) Resequencing data indicate a modest effect of domestication on diversity in barley: a cultigen with multiple origins. J Heredity 105:253–264. https://doi.org/10.1093/jhered/est083

Muñoz-Amatriaín M, Cuesta-Marcos A, Endelman JB et al (2014) The USDA barley core collection: genetic diversity, population structure, and potential for genome-wide association studies. PLoS One 9: e94688. https://doi.org/10.1371/journal.pone.0094688

Nakamura S, Pourkheirandish M, Morishige H, Kubo Y, Nakamura M, Ichimura K, Seo S, Kanamori H, Wu J, Ando T, Hensel G, Sameri M, Stein N, Sato K, Matsumoto T, Yano M, Komatsuda T (2016) Mitogen-activated protein kinase kinase 3 regulates seed dormancy in barley. Curr Biol 26:775–781. https://doi.org/10.1016/j.cub.2016.01.024

Neale D, Saghai-Maroof M, Allard R et al (1988) Chloroplast DNA diversity in populations of wild and cultivated barley. Genetics 120:1105–1110

Nei M (1973) Analysis of gene diversity in subdivided populations. Proc Natl Acad Sci USA 70:3321–3323

Nevo E (1995) Asian, African and American biota meet at 'Evolution Canyon' Israel: local tests of global biodiversity and genetic diversity patterns. Proc Roy Soc B 262:149–155

Nevo E (2014) Evolution of wild barley at "Evolution Canyon": adaptation, speciation, possible domestication, and improvement

Nevo E, Beharav A, Meyer R et al (2005) Genomic microsatellite adaptive divergence of wild barley by microclimatic stress in 'Evolution Canyon', Israel. Biol J Linn Soc 84:205–224

Nevo E, Fu Y-B, Pavlicek T et al (2012) Evolution of wild cereals during 28 years of global warming in Israel. Proc Natl Acad Sci USA 109:3412–3415. https://doi.org/10.1073/pnas.1121411109

Nice LM, Steffenson BJ, Brown-Guedira GL et al (2016) Development and genetic characterization of an advanced backcross-nested association mapping (AB-NAM) population of wild × cultivated barley. Genetics 203:1453–1467. https://doi.org/10.1534/genetics.116.190736

Orabi J, Backes G, Wolday A et al (2007) The Horn of Africa as a centre of barley diversification and a potential domestication site. Theor Appl Genet 114:1117–1127. https://doi.org/10.1007/s00122-007-0505-5

Østerberg JT, Xiang W, Olsen LI et al (2017) Accelerating the domestication of new crops: feasibility and approaches. Trends Plant Sci 22:373–384. https://doi.org/10.1016/j.tplants.2017.01.004

Owuor ED, Fahima T, Beharav A et al (1999) RAPD divergence caused by microsite edaphic selection in wild barley. Genetica 105:177–192. https://doi.org/10.1023/A:1003781711908

Pankin A, Altmüller J, Becker C, von Korff M (2017) Targeted re-sequencing reveals the genetic signatures and the reticulate history of barley domestication. bioRxiv 070078

Pasam RK, Sharma R, Malosetti M et al (2012) Genome-wide association studies for agronomical traits in a world wide spring barley collection. BMC Plant Biol 12:16. https://doi.org/10.1186/1471-2229-12-16

Pasam RK, Sharma R, Walther A et al (2014) Genetic diversity and population structure in a legacy collection of spring barley landraces adapted to a wide range of climates. PLoS One 9:e116164. https://doi.org/10.1371/journal.pone.0116164

Paterson AH, Lin Y-R, Li Z et al (1995) Convergent domestication of cereal crops by independent mutations at corresponding genetic loci. Science 269:1714–1718. https://doi.org/10.1126/science.269.5231.1714

Pillen K, Zacharias A, Léon J (2003) Advanced backcross QTL analysis in barley (*Hordeum vulgare* L.). Theor Appl Genet 107:340–352. https://doi.org/10.1007/s00122-003-1253-9

Poets AM, Fang Z, Clegg MT, Morrell PL (2015a) Barley landraces are characterized by geographically heterogeneous genomic origins. Genome Biol 16:173. https://doi.org/10.1186/s13059-015-0712-3

Poets AM, Mohammadi M, Seth K et al (2015b) The effects of both recent and long-term selection and genetic drift are readily evident in North American barley breeding populations. Genes Genomes Genet 6:609–622. https://doi.org/10.1534/g3.115.024349

Poland JA, Brown PJ, Sorrells ME, Jannink J-L (2012) Development of high-density genetic maps for barley and wheat using a novel two-enzyme genotyping-by-sequencing approach. PLoS One 7:e32253. https://doi.org/10.1371/journal.pone.0032253

Potokina E, Druka A, Luo Z et al (2008) Gene expression quantitative trait locus analysis of 16 000 barley genes reveals a complex pattern of genome-wide transcriptional regulation. Plant J 53:90–101. https://doi.org/10.1111/j.1365-313x.2007.03315.x

Pourkheirandish M, Komatsuda T (2007) The importance of barley genetics and domestication in a global perspective. Annals Bot 100:999–1008. https://doi.org/10.1093/aob/mcm139

Pourkheirandish M, Hensel G, Kilian B et al (2015) Evolution of the grain dispersal system in barley. Cell 162:527–539. https://doi.org/10.1016/j.cell.2015.07.002

Ramsay L, Comadran J, Druka A et al (2011) *INTERMEDIUM-C*, a modifier of lateral spikelet fertility in barley, is an ortholog of the maize domestication gene *teosinte branched 1*. Nat Genet 43:169–172. https://doi.org/10.1038/ng.745

Riehl S, Zeidi M, Conard NJ (2013) Emergence of agriculture in the foothills of the Zagros mountains of Iran. Science 341:65–67. https://doi.org/10.1126/science.1236743

Russell J, Dawson IK, Flavell AJ et al (2011) Analysis of >1000 single nucleotide polymorphisms in geographically matched samples of landrace and wild barley indicates secondary contact and chromosome-level differences in diversity around domestication genes. New Phyt 191:564–578. https://doi.org/10.1111/j.1469-8137.2011.03704.x

Russell J, van Zonneveld M, Dawson IK et al (2014) Genetic diversity and ecological niche modelling of wild barley: refugia, large-scale post-LGM range expansion and limited mid-future climate threats? PLoS One 9:e86021. https://doi.org/10.1371/journal.pone.0086021

Russell J, Mascher M, Dawson IK et al (2016) Exome sequencing of geographically diverse barley landraces and wild relatives gives insights into environmental adaptation. Nat Genet 48:1024–1030. https://doi.org/10.1038/ng.3612

Saisho D, Purugganan MD (2007) Molecular phylogeography of domesticated barley traces expansion of agriculture in the old world. Genetics 177:1765–1776. https://doi.org/10.1534/genetics.107.079491

Salamini F, Özkan H, Brandolini A et al (2002) Genetics and geography of wild cereal domestication in the Near East. Nat Rev Genet 3:429–441

Sannemann W, Huang BE, Mathew B, Léon J (2015) Multi-parent advanced generation inter-cross in barley: high-resolution quantitative trait locus mapping for flowering time as a proof of concept. Mol Breed 35:86. https://doi.org/10.1007/s11032-015-0284-7

Sato K, Yamane M, Yamaji N, Kanamori H, Tagiri A, Schwerdt JG, Fincher GB, Matsumoto T, Takeda K, Komatsuda T (2016) Alanine aminotransferase controls seed dormancy in barley. Nat Commun 7:11625. https://doi.org/10.1038/ncomms11625

Sayed MA, Hamada A, Lèon J, Naz AA (2016) Genetic mapping reveals novel exotic QTL alleles for seminal root architecture in barley advanced backcross double haploid population. Euphytica 213:2. https://doi.org/10.1007/s10681-016-1809-9

Schaart JG, van de Wiel CC, Lotz LA, Smulders MJ (2016) Opportunities for products of new plant breeding techniques. Trends Plant Sci 21:438–449. https://doi.org/10.1016/j.tplants.2015.11.006

Schmalenbach I, Körber N, Pillen K (2008) Selecting a set of wild barley introgression lines and verification of QTL effects for resistance to powdery mildew and leaf rust. Theor Appl Genet 117:1093–1106. https://doi.org/10.1007/s00122-008-0847-7

Schmid KJ, Thorwarth P (2014) Genomic selection in barley breeding. In: Biotechnological approaches to barley improvement. Springer, pp 367–378

Schnaithmann F, Kopahnke D, Pillen K (2014) A first step toward the development of a barley NAM population and its utilization to detect QTLs conferring leaf rust seedling resistance. Theor Appl Genet 127:1513–1525. https://doi.org/10.1007/s00122-014-2315-x

Shi J, Lai J (2015) Patterns of genomic changes with crop domestication and breeding. Curr Opin Plant Biol 24:47–53. https://doi.org/10.1016/j.pbi.2015.01.008

Snow L, Brody T (1984) Genetic variation of *Hordeum spontaneum* in Israel, ecogeographical races detected by trait measurements. Plant Syst Evol 145:15–28

Somers D, Gyenis L, Yun S et al (2007) Genetic architecture of quantitative trait loci associated with morphological and agronomic trait differences in a wild by cultivated barley cross. Genome 50:714–723. https://doi.org/10.1139/g07-054

Spies A, Korzun V, Bayles R et al (2012) Allele mining in barley genetic resources reveals genes of race-non-specific powdery mildew resistance. Front Plant Sci 2. https://doi.org/10.3389/fpls.2011.00113

Studer A, Zhao Q, Ross-Ibarra J, Doebley J (2011) Identification of a functional transposon insertion in the maize domestication gene *tb1*. Nat Genet 43:1160–1163. https://doi.org/10.1038/ng.942

Tanksley SD, McCouch SR (1997) Seed banks and molecular maps: unlocking genetic potential from the wild. Science 277:1063–1066. https://doi.org/10.1126/science.277.5329.1063

Tanno K, Willcox G (2011) Distinguishing wild and domestic wheat and barley spikelets from early holocene sites in the Near East. Veg Hist Archaeobot 21:107–115. https://doi.org/10.1007/s00334-011-0316-0

Thormann I, Reeves P, Thumm S et al (2017) Changes in barley (*Hordeum vulgare* L. subsp. *vulgare*) genetic diversity and structure in Jordan over a period of 31 years. Plant Genet Resour. https://doi.org/10.1017/s1479262117000028

Thorwarth P, Ahlemeyer J, Bochard A-M et al (2017) Genomic prediction ability for yield-related traits in German winter barley elite material. Theor Appl Genet. https://doi.org/10.1007/s00122-017-2917-1

Tondelli A, Xu X, Moragues M et al (2013) Structural and temporal variation in genetic diversity of European spring two-row barley cultivars and association mapping of quantitative traits. Plant Genome. https://doi.org/10.3835/plantgenome2013.03.0007

Turner A (2005) The pseudo-response regulator *ppd-h1* provides adaptation to photoperiod in barley. Science 310:1031–1034. https://doi.org/10.1126/science.1117619

Turpeinen T, Tenhola T, Manninen O et al (2001) Microsatellite diversity associated with ecological factors in *Hordeum spontaneum* populations in Israel. Mol Ecol 10:1577–1591

Verhoeven KJ, Vanhala TK, Biere A, Nevo E, van Damme JM (2004) The genetic basis of adaptive population differentiation: a quantitative trait locus analysis of fitness traits in two wild barley populations from contrasting habitats. Evolution 58(2):270–283

Vitti JJ, Grossman SR, Sabeti PC (2013) Detecting natural selection in genomic data. Ann Rev Genet 47:97–120. https://doi.org/10.1146/annurev-genet-111212-133526

Volis S, Ward D, ad BYAIS, Zur V, Mendlinger S (2001) Tests for adaptive rapd variation in population genetic structure of wild barley, *Hordeum spontaneum* Koch. Biol J Linn Soc 4:289–303

Volis S, Mendlinger S, Turuspekov Y, Esnazarov U (2002) Phenotypic and allozyme variation in Mediterranean and desert populations of wild barley, *Hordeum spontaneum* Koch. Evolution 56:1403–1415

Volis S, Zaretsky M, Shulgina I (2009) Fine-scale spatial genetic structure in a predominantly selfing plant: role of seed and pollen dispersal. Heredity 105:384–393. https://doi.org/10.1038/hdy.2009.168

Volis S, Zaretsky M, Shulgina I (2010) Fine-scale spatial genetic structure in a predominantly selfing plant: role of seed and pollen dispersal. Heredity 105(4):384

Volis S, Shulgina I, Zaretsky M, Koren O (2011) Epistasis in natural populations of a predominantly selfing plant. Heredity 106(2):300

von Korff M, Wang H, Leon J, Pillen K (2006) AB-QTL analysis in spring barley: II. Detection of favourable exotic alleles for agronomic traits introgressed from wild barley (*H. vulgare* ssp. *spontaneum*). Theor Appl Genet 112:1221–1231. https://doi.org/10.1007/s00122-006-0223-4

Wang H, Nussbaum-Wagler T, Li B, Zhao Q, Vigouroux Y, Faller M, Bomblies K, Lukens L, Doebley JF (2005) The origin of the naked grains of maize. Nature 436:714–719. https://doi.org/10.1038/nature03863

Wei YM, Baum BR, Nevo E, Zheng YL (2005) Does domestication mimic speciation? 1. A population-genetic analysis of *Hordeum spontaneum* and *Hordeum vulgare* based on AFLP and evolutionary considerations. Can J Bot 83:1496–1512. https://doi.org/10.1139/b05-151

Weigel D, Nordborg M (2015) Population genomics for understanding adaptation in wild plant species. Ann Rev Genet 49:315–338. https://doi.org/10.1146/annurev-genet-120213-092110

Wendler N, Mascher M, Himmelbach A et al (2015) *Bulbosum* to go: a toolbox to utilize *Hordeum vulgare/bulbosum* introgressions for breeding and beyond. Mol Plant 8:1507–1519. https://doi.org/10.1016/j.molp.2015.05.004

Willcox G (2005) The distribution, natural habitats and availability of wild cereals in relation to their domestication in the near east: multiple events, multiple centres. Veg Hist Archaeobot 14:534–541. https://doi.org/10.1007/s00334-005-0075-x

Wright SI, Bi IV, Schroeder SG et al (2005) The effects of artificial selection on the maize genome. Science 308:1310–1314. https://doi.org/10.1126/science.1107891

Yang P, Habekuß A, Hofinger BJ et al (2016) Sequence diversification in recessive alleles of two host factor genes suggests adaptive selection for bymovirus resistance in cultivated barley from East Asia. Theor Appl Genet 130:331–344. https://doi.org/10.1007/s00122-016-2814-z

Zamir D (2001) Improving plant breeding with exotic genetic libraries. Nat Rev Genet 2:983–989

Zohary D, Hopf M, Weiss E (2012) Domestication of plants in the Old World: the origin and spread of domesticated plants in Southwest Asia, Europe, and the Mediterranean Basin. Oxford University Press, Oxford

The Genomes of the Secondary and Tertiary Gene Pools of Barley

18

Neele Wendler

Abstract

In plant breeding, crop wild relatives are usually a valuable genetic resource for crop improvement. Thereby, wild relatives of a plant are divided into three different gene pools, depending on how difficult it is to produce hybrids between a species and the crop itself. In barley (*Hordeum vulgare spp.* L), the primary gene pool consists of barley itself, landraces and *H. spontaneum*. These germplasms are easy to utilize and were frequently used for barley improvement. By contrast, species of the secondary and tertiary gene pool are much more difficult to access and have thus not yet been reported to be utilized for barley breeding. *H. bulbosum* is the only member within the secondary gene pool. A relatively large set of hybrids, substitution and introgression lines between barley and *H. bulbosum* have been produced and characterized on a physiological and molecular level. Recently, the latest molecular tools such as genotyping-by-sequencing and Exome Capture have been utilized to unlock these genetic resources. While about 31 highly diverse species belong to the tertiary gene pool, not much success has been reported so far to access them.

N. Wendler (✉)
KWS LOCHOW GMBH, KWS Cereals Biotechnology, Grimsehlstr. 31, 37574 Einbeck, Germany
e-mail: neele.wendler@kws.com

18.1 Introduction

Wild relatives of crop plants are often in the spotlight of plant breeders attention, due to the rich pool of desirable traits for crop improvement they contain. These traits may have been lost during domestication and breeding. Due to the gene pool concept, barley and its wild relatives are divided into three different gene pools (Harlan and de Wet 1971). The primary gene pool consists of cultivated barley (*Hordeum vulgare* L.), landraces and its direct ancestor *H. vulgare* L. spp. *spontaneum*. Crossing between species of the primary gene pool can be performed without barriers and the resulting hybrids are usually fertile and have good chromosome pairing (Harlan and de Wet 1971). The gene segregation of the hybrids is approximately normal, so genes can be transferred without obstacles (Harlan and de Wet 1971).

Members of the secondary gene pool belong to a different species than the primary pool. Crossing with primary gene pool crops invokes barriers. Therefore, offspring can be obtained at lower frequencies and may be weak, only partially fertile or infertile. If hybrids are formed between primary and secondary gene pools, chromosome pairing may be poor or even

N. Stein and G. J. Muehlbauer (eds.), *The Barley Genome*, Compendium of Plant Genomes,
https://doi.org/10.1007/978-3-319-92528-8_18

completely lacking (Harlan and de Wet 1971). For barley, *Hordeum bulbosum* L. is the only member within the secondary gene pool (von Bothmer et al. 1995).

H. bulbosum belongs to the genus *Hordeum* and is perennial (von Bothmer et al. 1995). Its geographical distribution is broad and involves the Mediterranean basin and the Fertile Crescent (von Bothmer et al. 1995). *H. bulbosum* has two propagation systems. It can propagate sexually from seeds and also vegetatively by producing bulbs. The latter form by swelling of the basal internodes and they were the basis to the species being named *H. bulbosum*. *H. bulbosum* occurs as diploid as well as auto-tetraploid cytotypes. The out-breeding species has a strong, two gene-based self-incompatibility system (Lundqvist 1962).

The tertiary gene pool is more distantly related to the primary gene pool and inter-pool crossing invokes strong barriers. The transfer of traits to the primary gene pool requires special techniques like embryo rescue, the fusion of protoplasts, chromosome doubling or the use of other species as a bridge for crossing (bridge crossing) (Harlan and de Wet 1971). There are about 31 species belonging to the tertiary gene pool of barley (von Bothmer et al. 1995). These include diploid, tetraploid, hexaploid, annual, perennial, self-pollinating, and out-breeding species (von Bothmer et al. 1995). Geographical the distribution of the tertiary gene pool species is large and the members are native to the northern as well as the southern hemisphere such as Central Asia, South Africa, Europe, the Middle East, North and South America (von Bothmer et al. 1995). Thus, species of the genus *Hordeum* are adapted to climates from subtropical to arctic and can be found on sea level at more than 4500 m above sea level (von Bothmer et al. 1995).

Due to strong crossability barriers and low chromosome pairing in hybrids with *H. vulgare*, all efforts to transfer traits of tertiary gene pool *Hordeum* species to *H. vulgare* have been unsuccessful (von Bothmer et al. 1983; von Bothmer and Jacobsen 1986). Protoplast fusion between barley and species of other genera e.g. *Elymus*, *Thinopyrum* and *Pseudoroegneria* (Kim et al. 2008), rice (Kisaka et al. 1998), rye (Fedak and Armstrong 1981) and of different families like carrot (*Daucus carota)* (Kisaka and Kameya 1998), soybean (*Glycine max)* (Kao et al. 1974) and tobacco (*Nicotiana tabacum)* (Somers et al. 1986) have been attempted and partially successful. However, the generated hybrids were mostly been sterile or died before maturity. Thus, to date, no introgression lines (IL) between species of the tertiary gene pool and barley have been obtained and so those species have not played a role for barley improvement.

18.2 *Hordeum bulbosum* as a Genetic Resource

While the primary gene pool of barley has been frequently used to integrate traits into the elite breeding material, the secondary and tertiary gene pools are much more difficult to access. Nonetheless, several decades of research have sought to unlock the potential of the secondary and partly also the tertiary gene pools of barley.

Several decades ago *H. bulbosum* was recognized as a valuable genetic resource for barley improvement, especially in respect to biotic stress resistances (Walther et al. 2000; Pickering et al. 1987; Ruge et al. 2003; Szigat and Szigat 1991). *H. bulbosum* is likely to comprise non-host resistances against several barley pathogens and is therefore thought to be a source for durable resistance (Johnston 2007). Also, perenniality and self-incompatibility are traits that may be of interest for breeders (Pickering and Johnston 2005). The formation of fertile hybrids—which is the first step for trait introgression, is hindered by strong crossability barriers between barley and *H. bulbosum*. If *H. vulgare* and *H. bulbosum* are cross-fertilized, chromosomes of *H. bulbosum* usually get eliminated during the first mitotic divisions, leaving behind a haploid barley embryo (Kasha et al. 1970). If the latter is treated with colchicine, the chromosomes of the haploid plant can be doubled and may recover fertile and completely homozygous plants (doubled haploid). After the discovery of this process, *H. bulbosum* has been frequently used for the production of doubled haploids in barley breeding programs (Kasha et al. 1970). Recently, this technique was replaced by the more efficient anther and microspore culture (Pickering and Devaux 1992).

The production of hybrids between barley and *H. bulbosum* became more efficient when it was found that the frequency of retained *H. bulbosum* chromosomes can be increased, if the genome ploidy ratio favored the *H. bulbosum* parent (Kasha and Sadasiva 1971).

Later on, it was discovered that the degradation of the *H. bulbosum* genome varies between genotypes (Pickering 1984). Furthermore, it was found that the frequency of retained *H. bulbosum* chromosomes can be increased when crossing and plant regeneration was performed at temperatures below 18 °C (Pickering 1984). In wide crosses, it is a common process that endosperms get degenerated after fertilization. The same is true in hybrids between barley and *H. bulbosum* (Pickering and Johnston 2005). Thus, it is necessary to apply embryo culture to regenerate *H. vulgare/H. bulbosum* hybrids.

Diploid hybrids that contain one haploid copy of the barley and *H. bulbosum* genome, are completely sterile, but fertility can be restored if the young plants are treated with colchicine (Pickering 2000). Tetraploid hybrids can be directly generated if the barley chromosomes are duplicated with colchicine before crossing with tetraploid *H. bulbosum* genotypes (Szigat and Pohler 1982). Furthermore, triploid hybrids are generated if diploid barley is crossed with tetraploid *H. bulbosum* (Pickering 1991b). These plants are partially fertile without colchicine treatment (Pickering 1991b). Since homeologous pairing occurs in hybrids between *H. vulgare* and *H. bulbosum* (Pickering 1991a), it is possible to obtain recombination between the two genomes. The latter is necessary to introgress and transfer chromosomal regions, and therefore traits from *H. bulbosum* to barley. The frequency of homeologous pairing was found to vary between the different chromosomes (Pickering 1991a) as well as the parental genotypes that are used (Thomas and Pickering 1985).

The first evidence that *H. bulbosum* chromatin was successfully introgressed into barley, was observed in 1982 (Szigat and Pohler 1982). However, it took about 10 more years and the emergence of techniques such as in situ hybridization and Southern hybridization to confirm the transfer of *H. bulbosum* chromatin at the molecular level (Pickering et al. 1995; Xu and Kasha 1992). Even though, crossing over between *H. vulgare* and *H. bulbosum* chromosomes is possible, the frequency of these events are very low (Pickering 1991a). If hybrids are either selfed or backcrossed, ILs and substitution lines may be obtained (Pickering et al. 1994; Pickering 2000). In some cases, the generated offspring will recover a diploid *H. vulgare* genome with either introgressed fragments of *H. bulbosum* chromosomes (ILs) or even substituted chromosomes (substitution lines).

Today, a relatively large number of ILs and substitution lines between barley and *H. bulbosum* have been generated. (Walther et al. 2000; Szigat and Szigat 1991; Ruge et al. 2003; Pickering et al. 1987; Johnston et al. 2009; Pickering et al. 1994). These introgressions and substitutions covered all barley chromosomes except for a chromosome 1H substitution. The identified introgressed segments of *H. bulbosum* were located toward the distal ends of the barley chromosomes (Johnston et al. 2009) (Fig. 18.1).

Phenotyping of the ILs revealed that a relatively large and diverse set of *H. bulbosum* traits have been carried over to barley, such as disease resistance against leaf rust (*Puccinia hordei*), barley mild mosaic virus (BaMMV), barley yellow mosaic virus (BaYMV-1,-2), barley yellow dwarf virus (BYDV), stem rust (*Puccinia graminis*), scald (*Rhynchosporium secalis*), powdery mildew and septoria speckled leaf blotch (*Septoria passerinii*) (Fetch et al. 2009; Pickering et al. 2004; Ruge et al. 2003; Scholz et al. 2009; Shtaya et al. 2007; Singh et al. 2004; Toubia-Rahme et al. 2003) (Fig. 18.1).

Initially, ILs were mainly identified by *H. bulbosum*-specific morphological characteristics, such as pubescent leaf sheath, overall growth habit, agronomic performance, or by screening for different disease resistance. However, this

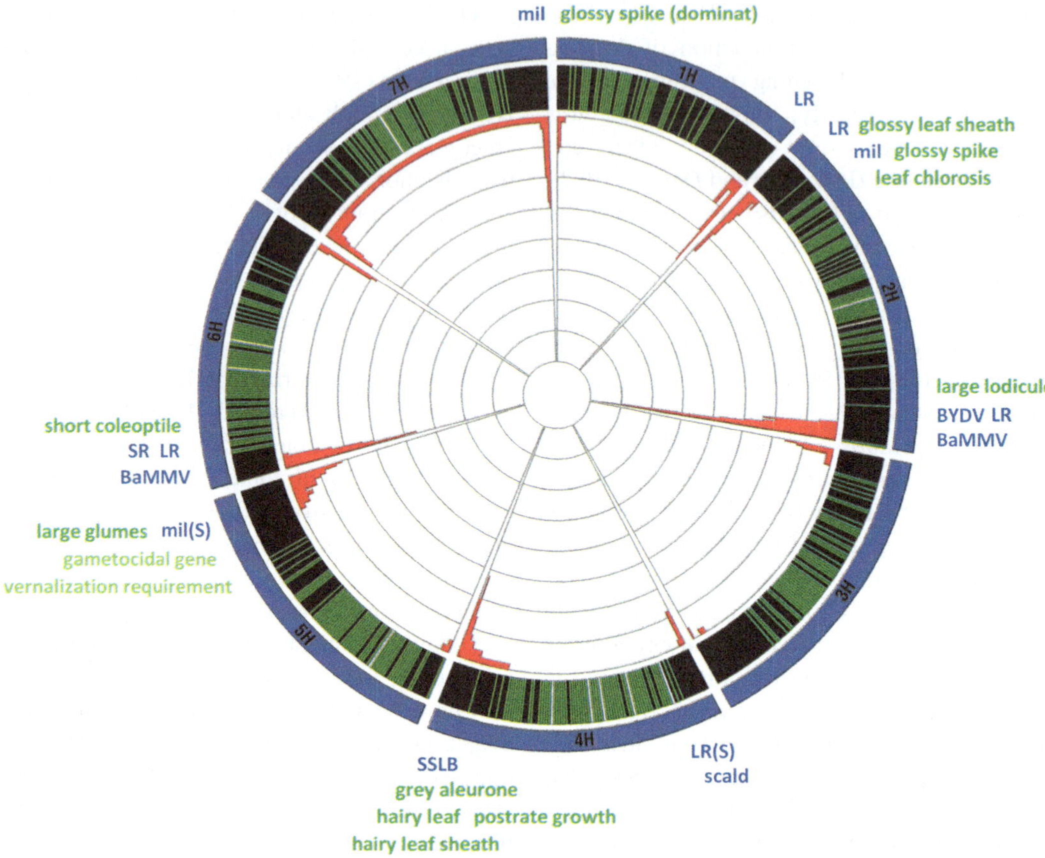

Fig. 18.1 Overview of the Distribution and Frequency of Detected Hb Segments and Hv/Hb SNPs along the Barley Genome. The figure includes a set of three concentric circles. The outer circle represents the seven barley chromosomes in a clockwise manner from the short to the long arms, scaled in 5 Mbp bins along the physical barley reference map (International Barley Sequencing Consortium 2012). Moving inwards, the second circle is a heat map of genotyping-by-sequencing Hv/Hb SNPs frequencies found between Hv cultivar "Emir" and Hb clone "A17/1" at a minimum fivefold sequence read coverage. The frequency is plotted in 5 Mbp bins along the physical barley reference (International Barley Sequencing Consortium 2012). The color code of these bins is set as: white, 0 SNPs/bin; dark green, 1–20 SNPs/bin; and black, >20 SNPs/bin. The innermost circle is a histogram of the frequency of Hb segments that were detected with genotyping-by-sequencing of the 146 ILs. The frequency is given in 5 Mbp bins along the barley genome and the gridlines of the y-axis are on a scale of 5. The three concentric circles were generated using Circos (Krzywinski et al. 2009). Outside the circles, Hb traits are listed that were uniquely attributed to Hb segments on individual chromosome arms of barley within the panel of 146 characterized ILs. Disease-associated traits are colored in blue, morphological traits are colored in dark green, and other traits are colored in light green. BaMMV, barley mild mosaic virus; BYDV, barley yellow dwarf virus; LR, leaf rust; mil, powdery mildew; (S), susceptible; scald, Rhynchosporium commune; SR, stem rust; SSLB, Septoria speckled leaf blotch. Figure and legend were published before in Wendler et al. 2015

method may have incorporated a selection slope towards particular positions of introgressed segments, which confer these morphological and disease resistance traits. Later on, the discovery of the pSc119.1 retrotransposon-based Polymerase Chain Reaction (PCR) assay allowed a fast and unbiased screen for *H. bulbosum* chromatin, since this method does not depend on the position of the introgressed segment (Johnston et al. 2009; Johnston and Pickering 2002).

Even though a large number of ILs are currently available, so far there is not a single report of a *H. bulbosum* trait that was released in a barley variety. The reason may be that the generations of an IL represent only the initial step towards trait utilization. Subsequently, an intensive breeding process as well as further research is necessary to unlock the *H. bulbosum* trait for barley breeding.

The main problem is constituted by negative linkage drag, which is usually associated with the *H. bulbosum* fragment (Stam and Zeven 1981; Brinkman and Frey 1977). Negative linkage drag is negative *H. bulbosum* characteristics that will cause yield losses. These yield limiting factors need to be eliminated by reducing the size of the initially transferred *H. bulbosum* segment (Pickering and Johnston 2005). The latter will need several rounds of selfing or back-crossing. Since the frequency of recombination events between barley and *H. bulbosum* chromatin is usually strongly reduced (Johnston et al. 2013; Ruge-Wehling et al. 2006), it is necessary to generate considerably larger populations to increase the possibility to find appropriate recombinants.

The success of this breeding process will also depend on the depths of the genetic characterization of the IL. To identify recombinants with reduced negative linkage drag, molecular markers are necessary to trace the *H. bulbosum* traits within the breeding population as well as the *H. bulbosum* chromatin surrounding the trait of interest. Therefore, all *H. bulbosum* segments must be localized within the barley genome and the precise dimension of each introgressed fragment must be determined. If the introgressed segment is genetically characterized, informative markers may be developed. Eventually, these markers will also allow mapping and cloning of a gene underlying the desired trait of *H. bulbosum*.

18.2.1 Genomics of *Hordeum bulbosum*

Barley and *H. bulbosum* have diverged approximately 6 million years ago and are therefore closely related (Jakob and Blattner 2006). On the basis of 136 molecular markers, it was found that the two species genomes are highly conserved and collinear (Salvo-Garrido et al. 2001). Recently, Fluorescence-in situ-hybridization of barley probes of chromosome 3H on *H. bulbosum* further approved collinearity between the species (Aliyeva-Schnorr et al. 2016). Therefore, the molecular tools and resources that have been successful in barley may be applied to characterize *H. bulbosum* and its ILs as well (Wendler et al. 2014, 2015).

Initially, the localization of *H. bulbosum* segments within the barley genome was performed by in situ hybridization (Johnston et al. 2009). This relatively low-resolution technique (Lukaszewski et al. 2005) allowed to define the chromosome arm that contains the *H. bulbosum* segment. With the emergence of an advanced barley reference sequence (International Barley Sequencing Consortium 2012; Mascher et al. 2017) and highly efficient NGS-based genotyping technologies such as genotyping-by-sequencing (GBS) and exome capture (Poland et al. 2012; Mascher et al. 2013b) the precision of the characterization of *H. bulbosum* ILs could be significantly improved. In two recent studies, GBS and exome capture were utilized for a detailed characterization of ILs between barley and *H. bulbosum* (Wendler et al. 2014, 2015). Thereby, re-sequencing of the ILs and the corresponding barley and *H. bulbosum* donor genotypes, allowed detection of *H. bulbosum*-specific polymorphisms within the ILs (Wendler et al. 2014, 2015). To order the detected polymorphisms, mapping and ordering of the sequencing reads were performed on the basis of the latest barley reference sequence and reference maps (Mascher et al. 2013a; International Barley Sequencing Consortium 2012; Wendler et al. 2014, 2015). Thus, the two techniques provide a fast and efficient way to detect and localize *H. bulbosum* segments within the barley genome.

In principle marker assays that were originally designed for barley polymorphisms can be applied to *H. bulbosum* as well (Wendler 2016). However, it is likely that due to an ascertainment bias (Lachance and Tishkoff 2013) the diversity of *H. bulbosum* will be significantly underestimated, even though *H. bulbosum* as a species itself is highly polymorphic (Jaffe et al. 2000).

In fact, it was found that pre-designed barley markers often fail to detect polymorphisms on *H. bulbosum* chromatin (Johnston 2007; Pickering and Johnston 2005).

As a globally applicable pool for *H. bulbosum* IL marker development, an integrated *H. vulgare/H. bulbosum* sequence resource was developed, which contains information on 112,847 conserved, interspecific SNPs between *H. vulgare* and *H. bulbosum* and about 35 Mbp of flanking sequence information (Wendler et al. 2015). The basis for this was exome capture re-sequencing data from 13 *H. vulgare* cultivars and 5 *H. bulbosum* accessions (Wendler et al. 2015).

It is likely that the genetic and physical positions provided by the barley reference in some cases may not reflect the true genomic positions *H. bulbosum*. Rearrangements between the genomes are expected, since the two species diverged several millions of years ago (Jakob and Blattner 2006). A high density genetic map of *H. bulbosum* as well as a detailed collinearity analysis to barley would be highly advantageous for the future.

References

Aliyeva-Schnorr L, Stein N, Houben A (2016) Collinearity of homoeologous group 3 chromosomes in the genus Hordeum and Secale cereale as revealed by 3H-derived FISH analysis. Chromosome Research:1–12. https://doi.org/10.1007/s10577-016-9518-8

Brinkman MA, Frey KJ (1977) Yield component analysis of oat isolines that produce different grain yields. Crop Sci 17:165–168

Fedak G, Armstrong KC (1981) Cytogenetics of the Trigeneric Hybrid, (*Hordeum vulgare X Triticum aestivum*) X *Secale cereale*. Theor Appl Genet 60 (4):215–219. https://doi.org/10.1007/bf02342541

Fetch T, Johnston PA, Pickering R (2009) Chromosomal location and inheritance of stem rust resistance transferred from *Hordeum bulbosum* into cultivated barley (*H. vulgare*). Phytopathology 99(4):339–343. https://doi.org/10.1094/phyto-99-4-0339

Harlan JR, de Wet JMJ (1971) Toward a rational classification of cultivated plants. Taxon 20:509–517

International Barley Sequencing Consortium (2012) A physical, genetic and functional sequence assembly of the barley genome. Nature 491(7426):711–716. http://www.nature.com/nature/journal/v491/n7426/abs/nature11543.html#supplementary-information

Jaffe B, Caligari PDS, Snape JW (2000) A skeletal linkage map of *Hordeum bulbosum* L. and comparative mapping with barley *(H. vulgare* L.). Euphytica 115(2):115–120. https://doi.org/10.1023/a:1003946329232

Jakob SS, Blattner FR (2006) A chloroplast genealogy of *Hordeum* (*Poaceae*): long-term persisting haplotypes, incomplete lineage sorting, regional extinction, and the consequences for phylogenetic inference. Molecular Biology and Evolution 23(8):1602–1612. https://doi.org/10.1093/molbev/msl018

Johnston PA (2007) Molecular characterisation of chromatin introgressed from *Hordeum bulbosum* L. into *Hordeum vulgare* L., University of Otago, Dunedin, New Zealand

Johnston PA, Niks RE, Meiyalaghan V, Blanchet E, Pickering R (2013) *Rph22*: mapping of a novel leaf rust resistance gene introgressed from the non-host *Hordeum bulbosum* L. into cultivated barley (*Hordeum vulgare* L.). Theor Appl Genet 126(6):1613–1625. https://doi.org/10.1007/s00122-013-2078-9

Johnston PA, Pickering RA (2002) PCR detection of *Hordeum bulbosum* introgressions in an *H. vulgare* background using a retrotransposon-like sequence. Theor Appl Genet 104(4):720–726. https://doi.org/10.1007/s00122-001-0791-2

Johnston PA, Timmerman-Vaughan GM, Farnden KJF, Pickering R (2009) Marker development and characterisation of *Hordeum bulbosum* introgression lines: a resource for barley improvement. Theor Appl Genet 118(8):1429–1437. https://doi.org/10.1007/s00122-009-0992-7

Kao KN, Constabe F, Michaylu MR, Gamborg OL (1974) Plant protoplast fusion and growth of intergeneric hybrid cells. Planta 120(3):215–227. https://doi.org/10.1007/bf00390290

Kasha KJ, Kao KN, Reinberg E (1970) Genetic control over chromosome stability in hybrids from interspecific *Hordeum* Crosses. Genetics 64(2):263–275

Kasha KJ, Sadasiva R (1971) Genome relationship between *Hordeum vulgare* L. and *H. bulbosum* L. Chromosoma 35(3):264–287. https://doi.org/10.1007/bf00326278

Kim NS, Fedak G, Han F, Cao W (2008) Cytogenetic analyses of intergeneric hybrids between barley and nine species of *Elymus*. Genome 51(11):897–904. https://doi.org/10.1139/g08-074

Kisaka H, Kameya T (1998) Cold and salt tolerance of somatic hybrid calli between barley *(Hordeum vulgare* L.) and carrot (*Daucus carota* L.). Breed Sci 48 (1):11–15

Kisaka H, Kisaka M, Kanno A, Kameya T (1998) Intergeneric somatic hybridization of rice (*Oryza sativa* L) and barley (*Hordeum vulgare* L) by protoplast fusion. Plant Cell Rep 17(5):362–367

Krzywinski M, Schein J, Birol I, Connors J, Gascoyne R, Horsman D, Jones SJ, and Marra MA (2009)Circos:

An information aesthetic for comparative genomics. Genome Res 19:1639–1645

Lachance J, Tishkoff SA (2013) SNP ascertainment bias in population genetic analyses: why it is important, and how to correct it. BioEssays: news and reviews in molecular, cellular and developmental biology 35(9). https://doi.org/10.1002/bies.201300014

Lukaszewski AJ, Lapinski B, Rybka K (2005) Limitations of in situ hybridization with total genomic DNA in routine screening for alien introgressions in wheat. Cytogenet Genome Res 109(1–3):373–377. https://doi.org/10.1159/000082422

Lundqvist A (1962) Self-incompatibility in diploid *Hordeum bulbosum* L. Hereditas 48(1–2):138–152. https://doi.org/10.1111/j.1601-5223.1962.tb01803.x

Mascher et al. (2017) A chromosome conformation capture ordered sequence of the barley genome. Nature 544:427–433. https://doi.org/10.1038/nature22043

Mascher M, Muehlbauer GJ, Rokhsar DS, Chapman J, Schmutz J, Barry K, Muñoz-Amatriain M, Close TJ, Wise RP, Schulman AH, Himmelbach A, Mayer KFX, Scholz U, Poland JA, Stein N, Waugh R (2013a) Anchoring and ordering NGS contig assemblies by population sequencing (POPSEQ). Plant J 76(4):718–727. https://doi.org/10.1111/tpj.12319

Mascher M, Richmond TA, Gerhardt DJ, Himmelbach A, Clissold L, Sampath D, Ayling S, Steuernagel B, Pfeifer M, D'Ascenzo M, Akhunov ED, Hedley PE, Gonzales AM, Morrell PL, Kilian B, Blattner FR, Scholz U, Mayer KFX, Flavell AJ, Muehlbauer GJ, Waugh R, Jeddeloh JA, Stein N (2013b) Barley whole exome capture: a tool for genomic research in the genus *Hordeum* and beyond. Plant J 76(3):494–505. https://doi.org/10.1111/tpj.12294

Pickering R (2000) Do the wild relatives of cultivated barley have a place in barley improvement? Barley genetics VIII: proceedings of the 8th international barley genetics symposium. Barley Genetics Conference, Adelaide, Australia

Pickering R, Johnston PA (2005) Recent progress in barley improvement using wild species of *Hordeum*. Cytogenet Genome Res 109(1–3):344–349. https://doi.org/10.1159/000082418

Pickering R, Niks RE, Johnston PA, Butler RC (2004) Importance of the secondary genepool in barley genetics and breeding II. Disease resistance, agronomic performance and quality. Czech J Genet Plant Breed 40(3):79–85

Pickering RA (1984) The influence of genotype and environment on chromosome elimination in crosses between *Hordeum vulgare* L. X *Hordeum bulbosum* L. Plant Sci Lett 34(1–2):153–164. https://doi.org/10.1016/0304-4211(84)90138-x

Pickering RA (1991a) Comparison of crossover frequencies in barley (*Hordeum vulgare*) and *H. vulgare* x *H. bulbosum* hybrids using a paracentric inversion. Genome Biol 34:666–673

Pickering RA (1991b) The production of fertile triploid hybrids from crosses between *Hordeum vulgare* L. ($2n = 4\times = 28$) and *H. bulbosum* L. ($2n = 2\times = 14$). Hereditas 114:227–236

Pickering RA, Devaux P (1992) Haploid production: approaches and use in plant breeding. Barley: genetics, biochemistry, molecular biology and biotechnology. CAB International, Wallingford, UK

Pickering RA, Hill AM, Michel M, Timmerman-Vaughan GM (1995) The transfer of a powdery mildew resistance gene from *Hordeum bulbosum* L. to barley (*H. vulgare* L.) chromosome 2 (2L). Theor Appl Genet 91(8):1288–1292

Pickering RA, Rennie WF, Cromey MG (1987) Disease resistant material available from the wide hybridization programme at DSIR. Barley Newsletter 31:248–250

Pickering RA, Timmerman GM, Cromey MG, Melz G (1994) Characterization of progeny from backcrosses of triploid hybrids between *Hordeum vulgare* L (2x) and *H. bulbosum* L. (4x) to *Hordeum vulgare*. Theor Appl Genet 88(3–4):460–464

Poland JA, Brown PJ, Sorrells ME, Jannink JL (2012) Development of high-density genetic maps for barley and wheat using a novel two-enzyme genotyping-by-sequencing approach. Plos One 7(2):e32253. https://doi.org/10.1371/journal.pone.0032253

Ruge-Wehling B, Linz A, Habekuss A, Wehling P (2006) Mapping of *Rym16*(Hb), the second soil-borne virus-resistance gene introgressed from *Hordeum bulbosum*. Theor Appl Genet 113(5):867–873. https://doi.org/10.1007/s00122-006-0345-8

Ruge B, Linz A, Pickering R, Proeseler G, Greif P, Wehling P (2003) Mapping of *Rym14* (Hb), a gene introgressed from *Hordeum bulbosum* and conferring resistance to BaMMV and BaYMV in barley. Theor Appl Genet 107(6):965–971. https://doi.org/10.1007/s00122-003-1339-4

Salvo-Garrido H, Laurie DA, Jaffé B, Snape JW (2001) An RFLP map of diploid *Hordeum bulbosum* L. and comparison with maps of barley (*H. vulgare* L.) and wheat (*Triticum aestivum* L.). Theor Appl Genet 103(6–7):869–880. https://doi.org/10.1007/s001220100707

Scholz M, Ruge-Wehling B, Habekuss A, Schrader O, Pendinen G, Fischer K, Wehling P (2009) *Ryd4* (Hb): a novel resistance gene introgressed from *Hordeum bulbosum* into barley and conferring complete and dominant resistance to the barley yellow dwarf virus. Theor Appl Genet 119(5):837–849. https://doi.org/10.1007/s00122-009-1093-3

Shtaya MJY, Sillero JC, Flath K, Pickering R, Rubiales D (2007) The resistance to leaf rust and powdery mildew of recombinant lines of barley (*Hordeum vulgare* L.) derived from *H. vulgare* x *H. bulbosum* crosses. Plant Breeding 126(3):259–267. https://doi.org/10.1111/j.1439-0523.2007.01328.x

Singh AK, Rossnagel BG, Scoles GJ, Pickering RA (2004) Identification of a quantitatively inherited source of *Hordeum bulbosum* derived scald resistance from barley line 926K2/11/1/5/1. Can J Plant Sci 84(3):935–938

Somers DA, Narayanan KR, Kleinhofs A, Cooperbland S, Cocking EC (1986) Immunological evidence for transfer of the barley nitrate reductase structural gene to nicotiana-tabacum by protoplast fusion. Mol General Genet 204(2):296–301. https://doi.org/10.1007/bf00425513

Stam P, Zeven AC (1981) The theoretical proportion of the donor genome in near-isogenic lines of self-fertilizers bred by backcrossing. Euphytica 30:227–238

Szigat G, Pohler W (1982) *Hordeum bulbosum* x *H. vulgare* hybrids and their backcrosses with cultivated barley. Cereal Res Commun 10:73–78

Szigat G, Szigat G (1991) Amphidiploid hybrids between *Hordeum vulgare* and *H. bulbosum*—basis for the development of new initial material for winter barley breeding. In: Winkel A (ed) Vortr Pflanzenzüchtg. EUCARPIA, Wageningen, Netherlands, pp 34–39

Thomas HM, Pickering RA (1985) The influence of parental genotype on the chromosome behaviour of *Hordeum vulgare* × *H. bulbosum* diploid hybrids. Theor Appl Genet 71:437–442

Toubia-Rahme H, Johnston PA, Pickering RA, Steffenson BJ (2003) Inheritance and chromosomal location of *Septoria passerinii* resistance introgressed from *Hordeum bulbosum* into *Hordeum vulgare*. Plant Breeding 122(5):405–409. https://doi.org/10.1046/j.1439-0523.2003.00850.x

von Bothmer R, Flink J, Jacoben N, Kotimäki M, Landström T (1983) Interspecific hybridization with cultivated barley (*Hordeum vulgare* L.). Hereditas 99:219–244

von Bothmer R, Jacobsen N (1986) Interspecific crosses in Hordeum (Poaceae). Plant Syst Evol 153(1/2): 49–64

von Bothmer R, Jacobsen N, Baden C, Jorgensen RB, Linde-Laursen I (1995) An ecogeographical study of the genus *Hordeum*. Systematic and ecogeographic studies on crop genepools, 2nd edn. International Plant Genetic Resources Institute, Rome, Italy

Walther U, Rapke H, Proeseler G, Szigat G (2000) *Hordeum bulbosum*—a new source of disease resistance—transfer of resistance to leaf rust and mosaic viruses from *H. bulbosum* into winter barley. Plant Breeding 119(3):215–218. https://doi.org/10.1046/j.1439-0523.2000.00475.x

Wendler N (2016) Unlocking the secondary gene pool of barley for breeding and research. Martin-Luther-Universität Halle-Wittenberg, Halle (Saale)

Wendler N, Mascher M, Himmelbach A, Johnston P, Pickering R, Stein N (2015) *Bulbosum* to go: a toolbox to utilize *Hordeum vulgare/bulbosum* introgressions for breeding and beyond. Mol Plant 128:1137–1149

Wendler N, Mascher M, Nöh C, Himmelbach A, Scholz U, Ruge-Wehling B, Stein N (2014) Unlocking the secondary gene-pool of barley with next-generation sequencing. Plant Biotechnol J 12(8): 1122–1131. https://doi.org/10.1111/pbi.12219

Xu J, Kasha KJ (1992) Transfer of a dominant gene for powdery mildew resistance and DNA from *Hordeum bulbosum* into cultivated barley (*Hordeum vulgare*). Theor Appl Genet 84(7–8):771–777

Barley Proteomics

19

Hans-Peter Mock, Christine Finnie, Katja Witzel
and Birte Svensson

Abstract

The present chapter summarizes the current status of proteome research on barley. The importance of barley as a model for cereals and as a major crop is reflected by a large number of publications using proteomics as an approach to address fundamental or applied research questions. Progress through technological developments in mass spectrometry, the central analytical technique in proteomics, forms the background methodology applied for protein or peptide separation and protein identification as outlined in the first section. The grain is of central relevance for the use of barley as a crop and seed biology is a central topic in plant science. Hence, a large number of studies focus on the grain proteome as well as the changes in proteome composition during grain maturation and germination. Separate sections cover research on abiotic and biotic stress defence responses. The next section is dedicated to subcellular proteomics, isolation of organelles or subcellular fractions being a powerful strategy to cope with the complexity of the plant proteome. Typical current analytical tools can cover only a small fraction of the complete proteome. With regard to the number of genes, any proteome is increased in complexity by a high number of post-translational modifications and many potential splicing variants. In addition, the dynamic range of the individual protein abundance covers many orders of magnitude exceeding the limits of current detection methods. Although this complexity of the proteome is demanding for in-depth protein analysis, information on, e.g. post-translational modifications cannot be derived from other approaches such as transcriptomics. These aspects and potential developments are addressed in the final section of our contribution.

H.-P. Mock (✉)
Department of Physiology and Cell Biology, Applied Biochemistry, Leibniz Institute of Plant Genetics and Crop Plant Research (IPK), Corrensstraße 3, 06466 Stadt Seeland, Germany
e-mail: mock@ipk-gatersleben.de

C. Finnie
Carlsberg Research Laboratory, JC Jacobsens Gade 4, 1799 Copenhagen V, Denmark

K. Witzel
Leibniz Institute of Vegetable and Ornamental Crops (IGZ), Theodor-Echtermeyer-Weg 1, 14979 Großbeeren, Germany

B. Svensson
Department of Biotechnology and Biomedicine, Enzyme and Protein Chemistry, Technical University Denmark, Søltofts Plads bldg. 224, 2800 Lyngby, Denmark

N. Stein and G. J. Muehlbauer (eds.), *The Barley Genome*, Compendium of Plant Genomes,
https://doi.org/10.1007/978-3-319-92528-8_19

19.1 The Basics: Technical Aspects of Current Proteome Techniques

Exploring the Barley proteome has been extensively carried out based on the two currently available proteomics strategies, (i) protein separation by two-dimensional gel electrophoresis in combination with protein identification by mass spectrometry, and (ii) proteolytic fragmentation of complex protein fractions, peptide separation and identification by combined liquid chromatography and mass spectrometry (Fig. 19.1):

(1) The classical technique for protein separation is two-dimensional gel electrophoresis (2-DE), which typically combines separation of proteins according to their isoelectric point (IEF; isoelectric focusing) in the first dimension followed in the second dimension by separation according to molecular mass by SDS-PAGE. After staining of proteins with Coomassie Blue, silver or fluorescent dyes, digital images of the protein pattern are analysed using the appropriate software. Protein visualization may also be performed prior to separation by labelling strategies, e.g. using Cy dyes in the DIGE (differential in gel electrophoresis) strategy, allowing separation of up to three different samples on the same gel. For further details of fluorescence labelling, the reader is referred to a review (Patton 2000). Image analysis and statistical evaluation of expression patterns enable selection of the most relevant spots for subsequent identification. Great progress in mass spectrometry for protein and peptide analysis was enabled by the introduction of instruments capable of ionizing macromolecules by laser ablation (MALDI; matrix-assisted laser desorption ionization) or electrospray ionization (ESI). This work,

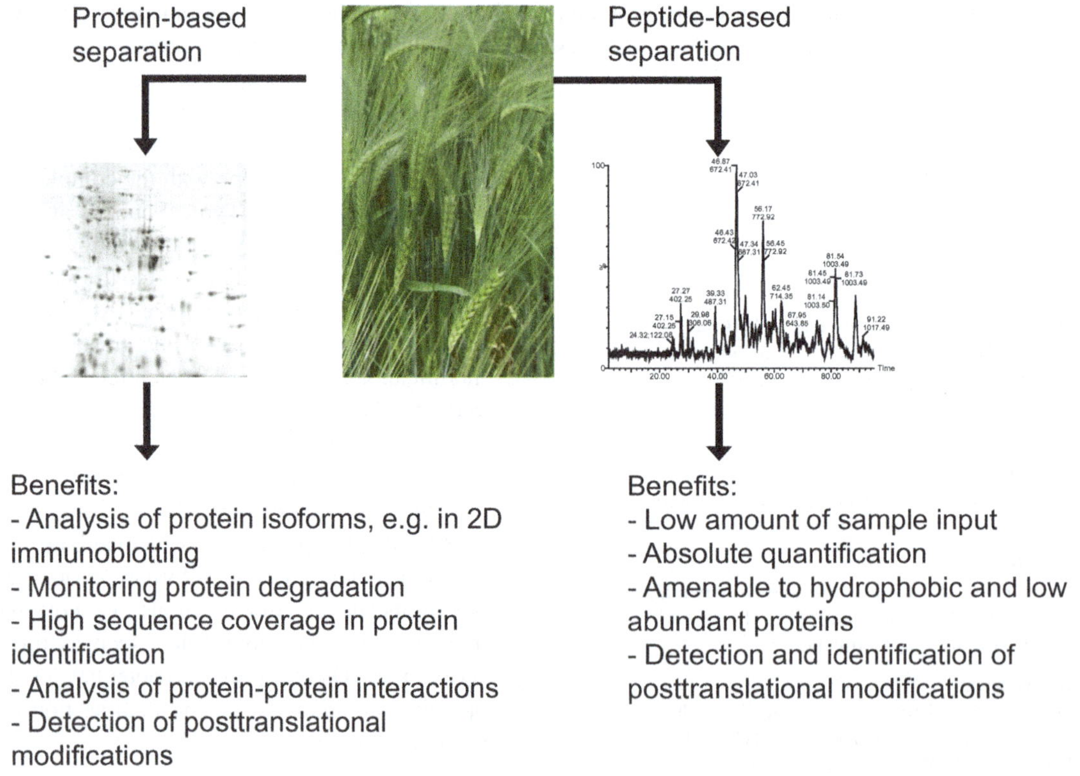

Fig. 19.1 Overview of the different strategies for proteomic analysis and highlighting their respective advantages

pioneered by Fenn and Tanaka, was awarded the Nobel prize in 2002 (see their Nobel lectures: (Fenn 2003; Tanaka 2003). Typically, the proteins to be identified are first digested using suitable proteases. The most frequently used protease for this purpose is trypsin, which cleaves peptide bonds after lysine and arginine residues. By this procedure, peptides are generated which are typical for each protein; the resulting 'peptide mass fingerprints' of proteins can be used to query genome databases for matching gene products. Peptide mass fingerprinting can be sufficient for identification of single proteins, or proteins in simple mixtures; however, for more complex samples, further information is needed. To this end, modern mass spectrometers capable of performing fragmentation of the peptides and MS analysis of the resulting fragments are used to provide additional sequence information. The increased availability of nucleotide sequence databases was a further prerequisite to allow identification of proteins based on mass spectrometric data sets.

(2) With the progress in mass spectrometry instrumentation, the standard approach for proteome analyses has shifted from 2-DE towards liquid chromatography (LC) for separation; with this approach, the proteome is analysed on the level of peptides resulting from digestion of complex protein extracts, again most commonly using trypsin (Matros et al. 2011). A major advantage of LC-MS-based proteomics is that much less protein is required per analysis. Thus, several hundred μg of protein are typically loaded onto one 2D gel, whereas proteolytic digests corresponding to a few μg or even less are sufficient to obtain a comparable coverage of the proteome using LC-MS. LC-MS is also preferable for the analysis of hydrophobic proteins, which are difficult to separate by 2-DE. On the other hand, 2-DE can be the method of choice to resolve closely related proteins: members of the same gene family, post-translationally modified forms of the same gene product; or protein degradation products. Targeted proteome analysis based on selected reaction monitoring (SRM) allows specific quantification of proteins of interest, even those not found in a non-targeted LC-MS approach (Picotti and Aebersold 2012). Triple quadrupole mass spectrometers are most often used for SRM assays, whereas other MS detectors such as TOF (time-of-flight) are used for non-targeted analysis. For a detailed background to mass spectrometry, the reader is referred to an excellent textbook (Gross 2011).

19.2 Early Proteomic Studies in Barley

Early work describing the barley proteome was performed in the group of Angelika Görg, one of the pioneers in establishing and improving protein analysis using 2-DE. Barley leaf extracts were separated by 2-DE followed by silver staining. The leaf proteome of 14 cultivars was compared and varieties could be distinguished based on characteristic spot patterns (Görg et al. 1988). Barley seed extracts were analysed and the application of immobilized pH gradients in the first dimension IEF resulted in a great improvement in resolution of cultivar-specific protein patterns of 55 winter and spring barley cultivars (Weiss et al. 1991). Around the same time, in one of the first studies to approach a subcellular proteome in barley, 2-DE was also applied to separate plasma membrane-enriched samples from barley roots (Hurkman and Tanaka 1986).

19.3 Grain Protein Composition, Germination and Malting

Seeds constitute the major propagation strategy for plants. In addition, mature seeds provide a high proportion of human and animal nutrition. Seed formation and seedling establishment are therefore prime topics in plant biology. Cereal grains are of particular importance for human nutrition. In the case of barley, the major economic value of grains is based on malting for production of beer, distilled spirits or vinegar. This is reflected by numerous proteomic studies on cereal grains. Barley is of interest as a crop, but also as a model for other cereals. The reader is referred to a review on grain proteomics (Finnie et al. 2011), a review on the analysis of malt proteins (Lastovickova and Bobalova 2012) and a review on the application of 'omics'-studies related to barley germination (Daneri-Castro et al. 2016).

The water-soluble seed proteome of the barley cultivars Golden Promise and Barke was studied using a 2-DE approach (Finnie et al. 2004; Ostergaard et al. 2004). Among several hundreds of spots detected, 16 spots differed when the gel images of the two cultivars were compared, of which 11 could be identified by mass spectrometry. The seed proteome patterns were further influenced by the cultivation conditions of the mother plants, i.e. when plants were either grown in the greenhouse or in the field, probably related to nitrogen availability during grain filling. Expression of transgenes (phosphinothricine N-acetyltransferase or green fluorescent protein, GFP) demonstrated that heterologously produced proteins could occur in several spots, indicating post-translational modifications. Apart from the foreign target protein, minor changes, if any, of the seed proteome were observed (Finnie et al. 2004). In addition to that, a protocol for the 2D separation of the basic protein fraction (p*I* 6-11) was established and applied to study grain and malt proteins in cv. Barke (Bak-Jensen et al. 2004).

To gain insight into the physiological processes during malting and brewing, and to identify proteins involved in foam formation in beer, Perrocheau et al (2005) compared the water-soluble proteome of grains with the heat-stable water-soluble proteins from barley, malt and beer. Cysteine-rich and highly stable proteins, such as serpin-like chymotrypsin inhibitors, amylase and bifunctional amylase/protease inhibitors and lipid transfer proteins, were recovered in the protein fraction of beer (Perrocheau et al. 2005).

The starch granule associated proteome was obtained by extraction from isolated granules followed by 2-DE (Borén et al. 2004). Enzymes involved in starch biosynthesis such as granule bound starch synthase I (GBSS I) were identified, together with hordeins B and D and other proteins. In total, 150 protein spots were observed and 74 were identified. Remarkably, 49 of these were annotated as GBSS I, indicating in vivo proteolysis (Borén et al. 2004).

Starch accumulation during grain development is tightly regulated. The group of 14-3-3 proteins possesses regulatory functions by binding to phosphorylated protein partners, while their distinct function is still unclear. The barley 14-3-3A protein was used as a bait to identify interacting proteins in developing grains by affinity chromatography. The resulting fraction was separated by 1-DE and 2-DE, and 54 proteins were identified (Alexander and Morris 2006). The largest group with a distinct function was involved in carbohydrate metabolism, including sucrose-synthetic enzymes and GBSS I.

The largest proportion of barley grain proteins are hordeins. These are storage proteins present in the alcohol-extractable protein fraction, with considerably low sequence diversity and biased amino acid compositions with low lysine content, making them of low nutritional quality. Their main relevance is in the malting process where they are broken down. Flodrová et al. (2012) used SDS-PAGE in combination with MALDI-TOF MS to monitor hordein profiles during malting and applied an isobaric peptide labelling strategy (iTRAQ) for quantification of C-hordein.

Seed germination involves a series of developmental processes exerted by different seed tissues. In cereals, a well-known example is the synthesis of gibberellic acid by the embryo which then acts a signal to trigger the synthesis of hydrolytic enzymes in the aleurone layer necessary to degrade the starchy endosperm. Using gibberellic acid-induced aleurone layers, insight into protein secretion processes leading to nutrient mobilization was obtained and novel glycosylation sites in aleurone layer proteins were identified (Barba-Espin et al. 2014). Changes in the proteome of the embryo, the aleurone and the endosperm during germination were followed in a kinetic manner at 8 time points until 72 h post imbibition (PI) using 2-DE (Bonsager et al. 2007). Early changes in the embryo proteome at 4 h PI were observed for desiccation stress-associated late embryogenesis abundant (LEA) proteins and ABA-induced proteins, an HSP70 fragment and a β-type proteasome subunit. These rapid changes in protein composition are consistent with the predisposition in the mature seeds of components relevant for the onset of germination. Several redox-related proteins differed in the pattern at the end of germination and onset of radicle elongation. Ascorbate peroxidase was observed only in the embryo, increasing in its abundance at 36 h PI (Bonsager et al. 2007).

Thioredoxins (Trx) are proteins involved in redox regulation of many cellular processes in most organisms including plants (Lozano et al. 1996). Trx reduce disulphide bonds in target proteins after which the reduced forms of Trx are regenerated by thioredoxin reductases in the cytosol, mitochondria (NADPH-dependent thioredoxin reductase NTR) or chloroplast (FAD-dependent thioredoxin reductase FTR). In plants, the cytosolic h-type thioredoxins exert a range of functions during seed germination, such as the inactivation of small proteinaceous inhibitors of proteolytic and amylolytic enzymes, the activation of hydrolytic enzymes and increasing the solubility of storage proteins (see Hägglund et al. 2013, 2016 for reviews). In barley grains, two thioredoxin h isoforms, Trxh1 and Trxh2, were identified using proteomics (Maeda et al. 2003). While Trxh2 was found mainly in the embryo, Trxh1 additionally accumulated in the endosperm and the aleurone layer. During germination, Trxh1 decreased in the aleurone and the endosperm. Decrease of Thrxh2 was also observed, but Trxh1 remained present in the embryo (Maeda et al. 2003). Potential target proteins of thioredoxins h were elucidated by applying different proteomic strategies. Incubation of protein extracts with thioredoxin leads to reduction of disulphide bonds in potential target proteins. The reduced cysteines then are amenable to labelling with thiol-reactive compounds like monobromobimane (Marx et al. 2003) or the fluorescent dye Cy5 (Maeda et al. 2004). After separation of proteins by 2-DE, additional spots are detectable relative to controls without thioredoxin treatment. A gel-free approach was later introduced making use of isotope-coded affinity tags (ICAT) for protein labelling. As the ICAT reagents contain a biotin tag, labelled proteins can be selectively enriched prior to mass spectrometrical analysis; for further technical details see (Hägglund et al. 2013). Using this approach, more than 100 putative targets of thioredoxin h could be identified in dissected embryos or proteins released from the aleurone layer in response to gibberellic acid (Hägglund et al. 2008, 2010).

Laser capture micro-dissection was used to obtain the nucellar projection and endosperm transfer cells of the developing barley grain (Kaspar et al. 2010b). Both tissues are the major transfer tissues for allocation of resources into the developing grain (Sreenivasulu et al. 2010). The minute amount of sample obtained was analysed by LC-MS. The nucellar projection tissue was enriched in stress defence components, whereas the tissues contained proteins in common related to assimilation, transport and mobilization of nutrients. Most of the identified proteins in the endosperm transfer cells were involved in the degradation of carbohydrates and in translation processes (Kaspar et al. 2010b).

Proteome patterns of grains at five developmental stages were compared using an LC-MS based approach (Kaspar-Schoenefeld et al. 2016). Three distinct developmentally related profiles were observed. Proteins of group I showed the highest accumulation in the middle phase of development, called the transition phase. Proteins identified were related to cell cycle regulation, protein synthesis, oxidative stress defence and energy production via photosynthesis. Accumulation of proteins in group II increased towards the end of the developmental stages investigated and was related to the storage phase. They comprised storage proteins and proteins involved in defence of storage reserves against pathogens. Group III proteins contained a mixture of peptides with both group I and II profiles and thus could not be allocated with certainty to either group. The mixed peptide pattern probably indicated the presence of protein isoforms. Most enzymes were detected in this group (Kaspar-Schoenefeld et al. 2016).

A shotgun proteomics approach was recently applied to compare the seed proteome of the two-rowed cultivar Conrad with the six-rowed cultivar Lacey. In total, 1168 unique proteins were identified, of which 20 differed in their abundance between the cultivars. Among this small set of differentially accumulated proteins were hordoindolines, which may contribute to the contrasting seed hardiness of the two cultivars (Mahalingam 2017).

To investigate the dependence of protein expression on the genotype, barley grain proteome maps have been subjected to genetical genomics approaches for quantitative trait loci (QTL) mapping (see Jansen and Nap 2001). Combining proteomic and genetic analyses allowed for the detection of the genetic localization of protein abundances represented by 48 spot variations (Finnie et al. 2009). First, grain protein maps as well as simple sequence repeat (SSR) markers of 18 barley cultivars were inspected and correlated. Grouping of cultivars with superior malting quality was closer when based on the grain proteome pattern as compared to genetic markers, indicating that the grain proteome is tightly connected to the malting phenotype. A doubled-haploid population of a malting barley and a feeding barley was then used to map protein expression pattern to genomic locations (Finnie et al. 2009).

Grains of 45 lines of a doubled-haploid population representing the complete wild barley genome within a modern breeding line background were assessed by 2-DE (Witzel et al. 2011). Expression profiles of 2718 protein spots were employed for QTL analysis using microsatellite and single nucleotide polymorphism markers. A total of 34 QTL for protein expression were detected, with some of them associated to markers previously assigned to agronomic QTL.

Wild barley introgression lines were used to identify proteins associated with two QTL for malting quality. Wild barley (*Hordeum vulgare* ssp. *spontaneum*) alleles at chromosome 1H reduced malting quality, whereas at chromosome 4H, malting quality was improved. 2-DE revealed 14 candidate proteins that could be involved in this trait (March et al. 2012).

19.4 Abiotic Stress Responses

Although this review focuses on barley, the reader is also referred to recent more general reviews on abiotic stress proteomics by Kosova and colleagues (Kosova et al. 2014, 2015) as well as by Johnova and colleagues (2016).

19.4.1 Drought Tolerance

A review on drought stress proteomics in leaves has recently been published (Wang et al. 2016).

Two contrasting barley varieties bred in the UK (Golden Promise) and Iraq (Basrah) were compared using DIGE (see above) (Wendelboe-Nelson and Morris 2012). Under the given experimental set-up for drought stress, the variety Basrah showed a higher relative water content in roots and shoots after 1 week of drought conditions. Both roots and leaves were analysed from controls and drought-exposed plants. Proteomic analysis revealed that the variety Basrah showed a constitutive high accumulation of proteins associated with the defence of reactive oxygen species (ROS) production and protein folding. Some of these proteins are further induced by drought. Photosynthetic enzymes showed a decreased abundance in this variety, consistent with reduced photosynthetic activity and ROS production. Assays for enzyme activities showed overall a good correlation with proteomic abundance data (Wendelboe-Nelson and Morris 2012). It is important to note that enzyme assays usually integrate the activities of several isoforms, if present, whereas proteomic analysis might only cover selected isoforms.

A set of barley genotypes from the Pakistani National Agriculture Research Centre in Islamabad was analysed for drought tolerance to select contrasting (three sensitive and three tolerant) accessions for subsequent detailed proteomic characterization (Kausar et al. 2013). Treatments with 20% polyethylene glycol or by withholding water were used separately to induce drought stress. The shoot proteome was analysed by 2-DE. Analysis of changes in proteome profiles pointed to the importance of plastidal metabolism as well as energy-related proteins in the adaptation of seedlings to drought (Kausar et al. 2013).

Egyptian barley landraces were used to assess drought stress responses in leaves (Ashoub et al. 2013). Chlorophyll fluorescence (Fv/Fm) was used as a parameter to select two contrasting genotypes which were analysed in more detail using again a DIGE proteomic approach. Conspicuous differences in the proteome composition were found for energy balance and chaperone proteins, which were more abundant in the drought-tolerant genotype #15141. However, higher abundance of proteins related to the formation of osmotically active compounds were found in the drought-sensitive genotype #15163 (Ashoub et al. 2013). In a follow-up study, responses to a combined drought and heat stress treatment revealed changes in barley proteins which were annotated with unknown functions (Ashoub et al. 2015).

Two genetically diverse barley genotypes, but both adopted to drought-prone environments, were studied with respect to responses to drought, high temperature or a combination of both (Rollins et al. 2013). Under the given experimental set-up, drought treatment resulted in a strong reduction of biomass and yield without causing profound changes in the proteome. In contrast, heat or combined heat and drought stress resulted in reduced photosynthesis and changes in the leaf proteome. In total, 99 protein spots showed changed volumes, 14 of them in a genotype-specific manner. Identified proteins covered roles in photosynthesis, detoxification, energy metabolism and protein biosynthesis. Variation in the stress responses of both adopted genotypes encourages further studies to pinpoint particular aspects of molecular responses with respect to breeding and fundamental research (Rollins et al. 2013).

Two barley genotypes contrasting in drought tolerance were compared using 2-DE as a proteomic technique (Chmielewska et al. 2016). Root and leaf proteomes of the two spring cultivars Maresi and Cam/B1/CI were analysed. Cam/B1/CI is a Syrian genotype whereas Maresi is of German origin. Although yield reduction (grain weight per plant) of the Syrian genotype under drought stress was less pronounced, Maresi had a higher yield under both control and stress conditions. Proteome analysis was complemented by additional analysis of a metabolome fraction using GC-MS. Distinct genotypic responses were observed in both roots and leaves for proteins as well as for metabolites. Despite distinct changes in metabolite and proteome pattern, the study also revealed changes common to both genotypes, and could not exclude a significant contribution to drought tolerance at this point (Chmielewska et al. 2016).

19.4.2 Salinity

The reader is also referred to our recent review on salinity stress proteomics (Witzel and Mock 2016).

Changes in the root proteome of the barley genotypes Steptoe and Morex contrasting in salinity tolerance were analysed using a 2-DE approach by comparing the protein pattern of controls or plants treated with either 100 or 150 mM NaCl (Witzel et al. 2009). Higher concentrations were not used in order to avoid induction of excessive cellular damage. Among others, two proteins related to the glutathione-based detoxification of ROS were prominent in the more tolerant genotype under the given experimental settings. Western blot analysis was used to validate the 2-DE result for catalase, which was shown to decrease under salinity in the more sensitive genotype Steptoe

(Witzel et al. 2009). In a follow-up study, using the same pair of contrasting genotypes, root samples were harvested at different time points after stress application (1, 4, 7 and 10 days). This experimental scheme allowed early and late responses to be distinguished (Witzel et al. 2014). In total, 74 proteins were identified that were either cultivar-specific or stress-responsive. Hierarchical cluster analysis grouped the protein expression profiles of the candidates into five clusters. A number of new protein candidates associated with salinity tolerance in barley were discovered, including numerous as yet uncharacterized proteins. Proteins related to detoxification pathways and terpenoid biosynthesis were detected in the early phase of the defence response, probably related to growth-regulating mechanisms and membrane stability via fine-tuning of phytohomormes and secondary metabolism. Proteomic data were confirmed by Western blotting for selected candidates and complemented by phytohormone analysis (Witzel et al. 2014). This refined study resulted in the identification of a range of additional candidates for further functional analysis.

The grain proteome of the parental lines of the Oregon Wolfe Barley mapping population was studied to identify proteins related to contrasting salinity tolerance during subsequent germination (Witzel et al. 2010). Two sensitive as well as two tolerant offspring lines were included in the study and helped to considerably narrow down the number of putative candidates associated with the seed proteome of this mapping population. Glucose/ribitol dehydrogenase which showed higher abundance in the tolerant genotypes was heterologously overexpressed and provided higher tolerance in a salt-sensitive yeast strain (Witzel et al. 2010).

Rasoulnia and colleagues performed a leaf proteome analysis of two genotypes contrasting in salinity tolerance (Rasoulnia et al. 2011). The salt-sensitive genotype L-527 showed a higher number of responsive proteins than the more tolerant Afzal.

The same genotypes were analysed in a follow-up study monitoring responses towards salinity after longer exposure (three weeks) (Fatehi et al. 2012).

The proteome of *H. spontaneum* is considered as a valuable resource to improve tolerance of barley breeding lines. The proteome response of *H. spontaneum* towards salinity has been explored after exposing seedlings at the four-leaf stage to 300 mM NaCl for 3 weeks (Fatehi et al. 2013). The proteome of leaf 4 was analysed by 2-DE and 29 protein spots showed differences in their volumes with respect to controls. Proteins with higher abundance under stress treatment included antioxidant proteins and the oxygen-evolving enhancer protein OEE2; other proteins identified covered a range of cellular processes (Fatehi et al. 2013).

The salt-tolerant barley cv.72 was compared with the Tibetan wild barley genotype XZ16 which has even superior salt tolerance (Wu et al. 2014). Major differences between the genotypes pointed to ion transporters as well as redox homeostasis as relevant cellular functions in the more tolerant XZ16 (Wu et al. 2014).

A multi-omics approach combining ion, metabolite and protein analysis has been applied to study changes in the shoot of two Tibetan barley genotypes (Shen et al. 2016). The more tolerant genotype XZ26 showed a lower shoot accumulation of Na, fewer changes in the proteome and in sugar metabolism (Shen et al. 2016). Time-course experiments involving ion analysis and proteomics were recently performed on another Tibetan wild barley genotype with higher salt stress tolerance (Shen et al. 2017).

Responses of the Syrian barley cultivar 'Tadmor' were compared with those of a halophytic wild barley species, *Hordeum marinum* (Marsalova et al. 2016). The salinity treatments led to an accumulation of dehydrin proteins isolated within a heat-soluble fraction. In *H. marinum*, dehydrin accumulation was lower than in Tadmor. Proteomic analysis was then performed with the non-heat-soluble fractions for control and a treatment with 300 mM NaCl. Proteins extracted in 8 M urea were separated by 1-D SDS-PAGE and lanes were cut into four separate fractions for each sample. Gel fractions were then digested and peptide profiles obtained by LC-MS analyses. Numerous changes in protein profiles associated with distinct molecular pathways were observed in both genotypes in response to salt stress treatments. A model summarizing major changes could be derived. The proteome analysis of the cultivar Tadmor revealed severe damage under salinity, suggested by the presence of apoptotic processes, protein degradation and a decrease in proteins related to energy metabolism. The proteome of *H. marinum* indicated a successful acclimation to high salinity as revealed by the identification of proteins associated with ROS scavenging, dehydration stress-responsive transcription factors and enhanced levels of proteins to support energy metabolism (Marsalova et al. 2016).

19.4.3 UV-B

The impact of UV-B on the proteome and flavonoid content was investigated in the leaf epidermal layer and mesophyll tissue of barley seedlings (Kaspar et al. 2010a). UV-B is a particular stress factor resulting in DNA and protein damage. The epidermal layer of leaves is thought to shield the mesophyll from UV-B thereby protecting the plant's photosynthetic capacity. 2-DE analysis revealed 11 UV-B responsive proteins in the epidermal layer, but only two in the mesophyll. LC-MS based proteomics showed further 15 responsive proteins in the epidermal tissue. Most of the responsive proteins were associated with oxidative and pathogen stress defence. The study highlights the significance of a spatially resolved analysis to decipher specific responses (Kaspar et al. 2010a).

19.4.4 Heavy Metal and Nutrient Stress

Heavy metal contamination imposes severe environmental threats. Cadmium is of particular significance for human nutrition, due to its high persistence. Breeding for varieties with lower Cd accumulation is necessary to reduce human intake of this heavy metal. Better understanding of Cd uptake and distribution mechanisms within the plant is a prerequisite for designing novel approaches to improve current crop genotypes including wheat, which is a major source of human Cd uptake. Analysis of the apoplastic proteins of barley plants exposed to Cd stress revealed the accumulation of pathogenesis-related (PR) proteins in barley leaves (Pos et al. 2011). Their precise role in Cd stress defence has not been fully elucidated. However, it is also known that Cd interacts with several signalling pathways in plants (Chmielowska-Bak et al. 2014). Vacuoles are a major component for the detoxification of heavy metals in plants (Sharma et al. 2016). Cadmium exposure performed at two different concentrations led to changes in the tonoplast proteome of barley leaf mesophyll (Schneider et al. 2009). A total of 56 tonoplast transporter proteins were identified, of which only a minor fraction showed differential accumulation under Cd stress relative to controls. Among the candidates was a natural resistance-associated macrophage protein, probably associated with a higher cytosolic iron concentration preventing an exchange with the toxic cadmium. In addition, a cation/proton exchanger protein, CAX1a was identified, which is likely to be involved in vacuolar transport of cadmium (Schneider et al. 2009).

For further reading on plant responses to heavy metal stress, the reader is referred to a recent review (Mustafa and Komatsu 2016).

Nitrogen is a major limiting factor for crop plants, requiring the addition of N fertilizers. Improvement of nitrogen use efficiency is needed to reduce soil and water pollution by unused N. N is taken up by most plants in the form of nitrate or ammonium. In a 2-DE-based study to analyse responses to N supply in barley shoots and roots (Moller et al. 2011), plants were grown in a hydroponic system allowing careful control of the nutrient supply. Control plants were grown in 5 mM nitrate for 33 days and compared to plants grown under N-deficiency (0.5 mM nitrate for 33 days) or exposed to short-term N starvation (28 days with 5 mM nitrate followed by 5 days with no N source). The effect on the shoot proteome of growth with ammonium was investigated in another group of plants grown for 28 days with 5 mM nitrate, followed by 5 days with 5 mM ammonium. The root and shoot proteomes responded differently to the treatments, and distinct but overlapping responses were observed to the three N regimes. Amino acid metabolism was strongly affected in the roots exposed to short-term N starvation but roots exposed to long-term nitrogen deficiency showed reduction in proteins involved in defence against pathogens. In both cases, proteins potentially involved in root growth were affected. Both short-term N starvation and growth on ammonium caused the appearance of proteolytic fragments of Rubisco in shoot proteomes, suggesting that this abundant protein is used as an immediate N source prior to metabolic reprogramming (Moller et al. 2011).

19.5 Biotic Interactions

The grain surface constitutes an interface for the interaction with the environment, including microbial communities. The grain surface proteome of two barley genotypes has been analysed by 2-DE and LC-MS approaches (Sultan et al. 2016). The grain surface was washed with a slightly acidic buffer system and efficiency verified by analysing the removal of surface-associated xylanase activity. A large number of plant proteins as well as bacterial and fungal proteins were identified by the proteomic analysis. The successful implementation of a method for analysis of the grain surface proteome will allow the highly dynamic nature of the plant-microbial interface to be explored (Sultan et al. 2016).

Infection of barley with *Fusarium graminearum* causes fusarium head blight and can lead to severe contamination of grains with mycotoxins, toxic secondary metabolites secreted by the fungus (Eggert and Pawelzik 2011). Early plant defence reactions were monitored in inoculated spikelets after 1, 2 and 3 days (Yang et al. 2010b). Several PR proteins accumulated in the plant tissue, along with increased abundance of glycolytic enzymes that might aid the fungal growth. The extent of *F. graminearum* infection and concomitant prominent proteolytic degradation of kernel proteins was found to be increased in plants grown with a lower fertilizer level (Yang et al. 2010a).

Interaction of the pathogen with naked barley was studied by analysing the proteome of grains at five developmental stages (Trumper et al. 2016). Proteome analysis revealed that in the early stages of grain development fungal infection affected primary metabolism and inhibition of proteases. Higher accumulation of thaumatin-like proteins was observed at all stages investigated, indicating their particular significance in the plant defence responses towards the pathogen (Trumper et al. 2016).

Powdery mildew infection was studied at different stages upon inoculation of barley with the obligate biotroph *Blumeria graminis* f. sp. *hordei* (Bindschedler et al. 2009). A shotgun approach based on SDS-PAGE separation of proteins coupled with LC-MS was used. In total, 827 proteins of the pathogen were identified and related to conidia, hyphae and haustoria. In addition to a deeper understanding of the infection process, the proteomic data supported the prediction of gene models (Bindschedler et al. 2009).

The interaction of powdery mildew pathogen *Blumeria graminis* f. sp. *hordei* includes the secretion of more than 500 candidate effectors. One of these Blumeria effector candidate (BEC) proteins, BEC1054, was studied by using pull-down assays to identify putative interacting proteins. BEC1054 encodes an RNAse-like effector protein. After incubation of the effector protein with barley protein extracts, putative interactors were identified by LC-MS. Negative controls were included to reduce false-positives from unspecific binding, and a yeast two-hybrid system was applied to confirm interactions revealed by the pull-down assay. Host target proteins of BEC1054 were a glutathione-S-transferase, a malate dehydrogenase and a PR5 protein isoform. These potential target proteins indicate a role for BEC1054 in compromising the defence response of the host (Pennington et al. 2016).

The root mutualistic fungus *Piriformospora indica* was shown to enhance plant stress tolerance under particular environmental conditions. Its potential impact on salt stress tolerance was studied using the barley cultivar 'Pallas' at two different concentrations of NaCl (Alikhani et al. 2013). Colonized plants showed increased potassium to sodium ratios, consistent with an improved salt stress tolerance. Leaf proteome analysis revealed a large number of proteins with modified abundance in response to the mutualistic interaction, covering a wide range of cellular functions (Alikhani et al. 2013).

Responses towards spot blotch disease caused by *Cochliobolus sativus* were compared between two barley genotypes with contrasting tolerance. Leaf tissue adjacent to infected areas was sampled and analysed by 2-DE and relevant spots identified by MALDI-MS. Proteome pattern changed in response to pathogen infection and between genotypes (Al-Daoude et al. 2015)

19.6 Subproteome Analysis

A comprehensive online compendium with data on the subcellular distribution of crop proteins, which is collated from previously published fluorescent tagging or mass spectrometry studies, as well as pre-computed subcellular predictions for barley, wheat, rice and maize proteomes, represents a valuable resource for investigating the proteome of cellular compartments (Hooper et al. 2016).

19.6.1 Plastids

A review on barley proteomics with an outline of the technical background as well as an emphasis on plastidal analysis has been provided (Petersen et al. 2013).

The proteome of plastids is characterized by numerous protein and pigment-protein complexes, which might vary in response to environmental conditions, developmental programs, and plant species (Zolla et al. 2002). For example, photosystem II monomers are formed by about 20 subunit proteins (Shen 2015). The absence of photosystem I in the barley *viridis zb63* mutant results in a reduced abundance of light-harvesting complex (LHC) proteins and this is governed at the post-transcriptional level by the plastoquinone redox state (Frigerio et al. 2007). A particular form of 2-DE has been applied to separate first the complexes and then resolve their proteins in the second dimension. This is achieved using blue-native PAGE in the first dimension, followed by denaturing SDS-PAGE in the second dimension. With this approach, the proteome of barley thylakoid membranes was mapped (Granvogl et al. 2006). Labelling of the thylakoid proteins with the fluorescent dye Cy3 proved a highly sensitive detection method for the separated proteins (Granvogl et al. 2006).

Etioplasts are plastids formed in dark-grown angiosperm seedlings, and lack thylakoid membranes and the photosynthetic complexes constituting photosystem I and II as well as the LHC I and II. Light triggers the transition of etioplasts into mature chloroplasts. The Eichacker lab showed that all low-molecular weight proteins of photosystem II accumulated in barley etioplasts (Ploscher et al. 2009). A comparison of the membrane protein complexes using the DIGE approach (see above) demonstrated that etioplasts and chloroplasts share a number of proteins associated with electron transport, chlorophyll and protein synthesis or fatty acid biosynthesis (Ploscher et al. 2011). The assembly of the photosystem complexes during de-etiolation has been studied in a kinetic manner by isolating plastids from dark-grown barley seedlings after exposure to light for different times (Shevela et al. 2016), combining plastidal proteomics with a range of functional approaches. During the de-etiolation process, the capability of photosystem II to split water preceded the assembly of the PS II-LHC II supercomplexes (Shevela et al. 2016).

19.6.2 Nucleus

A specific probe homologous to the barley satellite DNA was used to isolate chromatin associated with barley centromeres (Zeng and Jiang 2016). Proteins were isolated from the centromeric chromatin and both α- and β-cenH3 variants of the centromeric histone H3 were detected. In addition, several different variants of the histones H2A and H2B were also found in the MS analysis (Zeng and Jiang 2016).

19.6.3 Mitochondria

Respiratory supercomplexes were studied in mitochondria of Arabidopsis, potato, bean and barley (Eubel et al. 2003). Different electrophoretic techniques were applied for separation, namely blue-native PAGE and two-dimensional separation combining blue-native PAGE separation in the first dimension followed either by blue-native- or SDS-PAGE in the second dimension. The authors concluded that supercomplex formation between complexes I and III limits access of the alternative oxidase towards its substrate ubiquinol as a regulatory control of alternative respiration (Eubel et al. 2003).

The same technique has been applied to characterize mitochondrial and etioplastic protein complexes, leading to the identification of a mitochondrial acetyl-CoA carboxylase essential in plastidial fatty acid biosynthesis (Focke et al. 2003).

19.6.4 Vacuoles

A paper on tonoplast analysis in the context of cadmium detoxification has already been mentioned (Schneider et al. 2009). In the search for tonoplast sucrose transporters in barley, vacuoles from barley mesophyll were isolated and the tonoplast proteins were separated by SDS-PAGE (Endler et al. 2006). Around 100 proteins were identified, among them the sucrose transporter HvSUT2. Its localization was confirmed by transient expression of a HvSUT2-GFP fusion protein in Arabidopsis leaves and onion cells (Endler et al. 2006). A later study extended the analysis by focusing on phosphorylated tonoplast proteins (Endler et al. 2009). By enrichment of phosphorylated peptides by IMAC or titanium dioxide affinity chromatography, the authors were able to identify 65 phosphopeptides from 27 vacuolar membrane proteins.

19.6.5 Plasma Membrane

The plasma membrane from the aleurone layer of grains was prepared using two-phase partitioning followed by reversed-phase chromatography (Hynek et al. 2006). The plasma membrane fraction is prone to contamination due to its low abundance. The additional chromatographic step helped enrich for integral membrane proteins and two novel isoforms of H^+-ATPase and two proteins potentially involved in ion-channel regulation were found. In a follow-up study, the plasma membrane of germinating barley embryos was analysed (Hynek et al. 2009). Proteins eluting from the chromatographic column in a fraction containing H^+-ATPase as analysed by Western blotting were further separated by SDS-PAGE. Fourteen prominent bands in the gel were observed and in total, 61 protein sequences were retrieved by LC-MS/MS. With respect to the earlier publication (Hynek et al. 2006), it was concluded that despite common proteins, the plasma membrane fractions of aleurone and the embryo contain specific proteins corresponding to their functions during germination (Hynek et al. 2009).

19.7 Conclusions and Perspectives

Our review demonstrates that proteomics has already found many applications in barley research. Most of the cited studies were gel-based, but a shift towards LC-MS-based proteomics is apparent. Protein identification benefits from the advancement of nucleotide sequence information for barley and improved gene models. Compared with transcriptomics, the comprehensiveness of most proteome approaches is restricted for several technical reasons. Proteomic analysis has to cope with the higher chemical diversity of proteins based on an alphabet of 20 amino acids in comparison with nucleic acids which only rely on a four-letter alphabet, the four nucleotides.

The complexity of the proteome is further increased through a dynamic protein turnover, as shown for leaves of ^{15}N-labelled barley plants (Nelson et al. 2014), as well as a plethora of possible post-translational modifications; to date, more than 200 such modifications have been described, including phosphorylation, acetylation and glycosylation, to name a few. Potentially, a protein could be altered by one or several of these modifications simultaneously, resulting in a number of protein variants. Detection of these variants, which may each constitute only a small fraction of the particular protein, is challenging. Enrichment methods have been developed to capture peptides or proteins with particular modifications, for example, phosphorylation (Baginsky 2016). Many examples are known where phosphorylation/dephosphorylation modulates enzymatic activities or protein function, which allows dynamical adaptation of metabolism to the needs of cells under given environmental conditions. In addition to enrichment, particular protein variants might be quantified in targeted approaches (Osinalde et al. 2017). Although such analytical procedures often require optimization for particular applications, the study of post-translational modifications is a major area in which to apply proteomics, as such information cannot be derived from nucleotide-centred approaches. The information obtained is central to unravel the cellular regulatory mechanisms operating in a particular biological context. Another challenging issue in

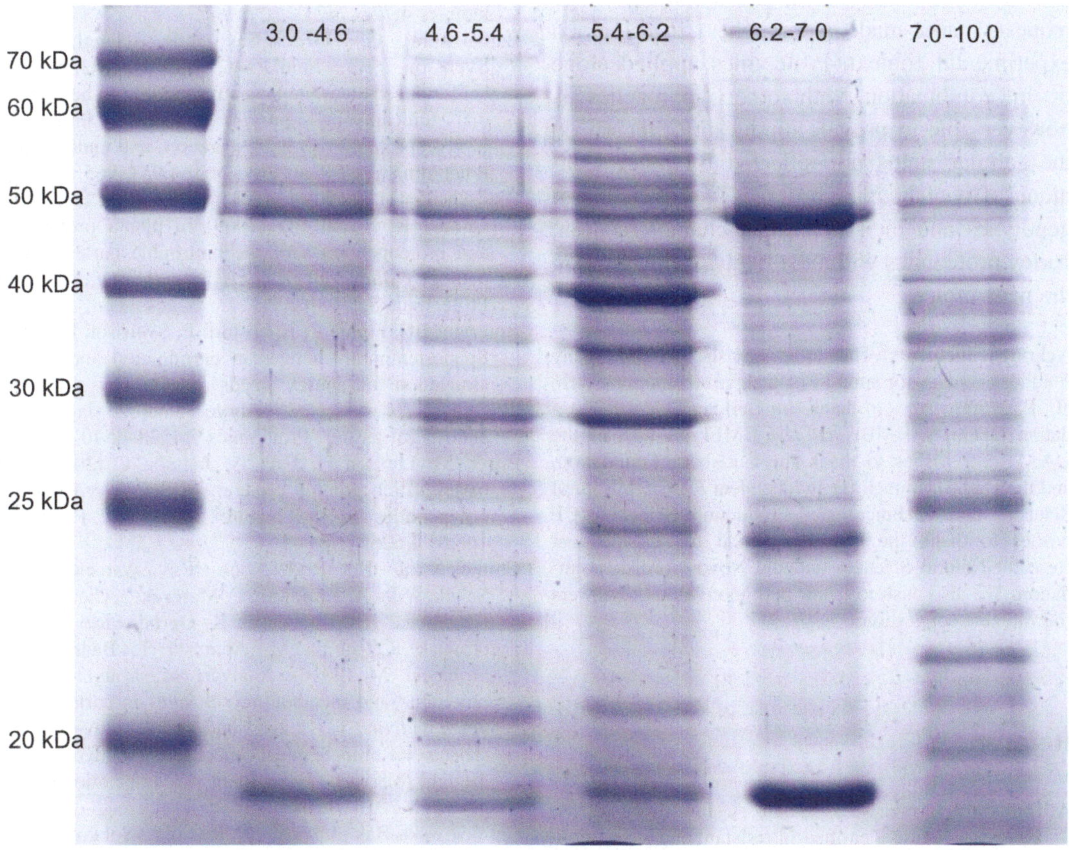

Fig. 19.2 Reduction of sample complexity achieved by in-solution isoelectric focusing for identification of low abundant proteins. Crude protein extract of barley leaf was separated into five fractions of different pH using a ZOOM IEF Fractionator (Invitrogen). The resulting fractions are suitable for downstream applications, such as 2-DE and LC-MS/MS. Subsequently, ten μg of protein was loaded onto SDS-PAGE and the gel was stained with colloidal Coomassie Blue

current proteomics is the highly dynamic range of protein abundance, estimated to cover up to ten orders of magnitude. With many of the current approaches, the full proteome of a complex tissue cannot be reached. Consistently, many of the gel-based studies summarized in our review reflect the most abundant proteins of cells, related to central metabolic functions. However, proteome coverage can be increased by cellular fractionation or by applying chemical pre-fractionation such as exemplified in Fig. 19.2. Using the so-called SWATH approach, an extension of selected reaction monitoring, it was possible to quantify yeast proteins down to 50 copies per cell (Picotti et al. 2009).

Collectively, the complexity of the proteome as reflected by the numerous post-translational modifications and its high dynamic range requests tailor-made approaches in a given experimental context. Proteomics applied alone or in combination with other approaches is however able to provide unique information on the cellular status not reflected by other techniques. Given the rapid progress in sequencing of genotypes and their functional characterization, barley proteomics will most likely also expand in the near future.

Acknowledgements The authors wish to thank many funding agencies for support of their proteomic research. H. -P. Mock wishes to thank the German Science Foundation DFG, the BMBF and the BMEL as well as the DAAD. BS wishes to thank the Carlsberg Foundation, and the Danish Council for Independent Research|Natural Sciences and Technical and Production Sciences. CF wishes to thank the Danish Council for Independent Research|Natural Sciences and the Novo Nordisk Foundation. We also wish to thank two anonymous reviewers for valuable suggestions.

References

Al-Daoude A, Arabi MIE, Shoaib A, Jawhar M (2015) Protein variation identified in resistant and susceptible barley genotypes infected with *Cochliobolus sativus*. Phyton-Annales Rei Botan 55:149–158

Alexander RD, Morris PC (2006) A proteomic analysis of 14-3-3 binding proteins from developing barley grains. Proteomics 6:1886–1896

Alikhani M, Khatabi B, Sepehri M, Nekouei MK, Mardi M, Salekdeh GH (2013) A proteomics approach to study the molecular basis of enhanced salt tolerance in barley (*Hordeum vulgare* L.) conferred by the root mutualistic fungus *Piriformospora indica*. Mol BioSyst 9:1498–1510

Ashoub A, Beckhaus T, Berberich T, Karas M, Bruggemann W (2013) Comparative analysis of barley leaf proteome as affected by drought stress. Planta 237:771–781

Ashoub A, Baeumlisberger M, Neupaertl M, Karas M, Bruggemann W (2015) Characterization of common and distinctive adjustments of wild barley leaf proteome under drought acclimation, heat stress and their combination. Plant Mol Biol 87:459–471

Baginsky S (2016) Protein phosphorylation in chloroplasts—a survey of phosphorylation targets. J Exp Bot 67:3873–3882

Bak-Jensen KS, Laugesen S, Roepstorff P, Svensson B (2004) Two-dimensional gel electrophoresis pattern (pH 6-11) and identification of water-soluble barley seed and malt proteins by mass spectrometry. Proteomics 4:728–742

Barba-Espin G, Dedvisitsakul P, Hagglund P, Svensson B, Finnie C (2014) Gibberellic acid-induced aleurone layers responding to heat shock or tunicamycin provide insight into the N-glycoproteome, protein secretion, and endoplasmic reticulum stress. Plant Physiol 164:951–965

Bindschedler LV, Burgis TA, Mills DJS, Ho JTC, Cramer R, Spanu PD (2009) In planta proteomics and proteogenomics of the biotrophic barley fungal pathogen *Blumeria graminis* f. sp. *hordei*. Mol Cell Proteomics 8:2368–2381

Bonsager BC, Finnie C, Roepstorff P, Svensson B (2007) Spatio-temporal changes in germination and radical elongation of barley seeds tracked by proteome analysis of dissected embryo, aleurone layer, and endosperm tissues. Proteomics 7:4528–4540

Borén M, Larsson H, Falk A, Jansson C (2004) The barley starch granule proteome—internalized granule polypeptides of the mature endosperm. Plant Sci 166:617–626

Chmielewska K, Rodziewicz P, Swarcewicz B, Sawikowska A, Krajewski P, Marczak L, Ciesiolka D, Kuczynska A, Mikolajczak K, Ogrodowicz P, Krystkowiak K, Surma M, Adamski T, Bednarek P, Stobiecki M (2016) Analysis of drought-induced proteomic and metabolomic changes in barley (*Hordeum vulgare* L.) leaves and roots unravels some aspects of biochemical mechanisms involved in drought tolerance. Front Plant Sci. https://doi.org/10.3389/fpls.2016.01108

Chmielowska-Bak J, Gzyl J, Rucinska-Sobkowiak R, Arasimowicz-Jelonek M, Deckert J (2014) The new insights into cadmium sensing. Front Plant Sci. https://doi.org/10.3389/fpls.2014.00245

Daneri-Castro SN, Svensson B, Roberts TH (2016) Barley germination: Spatio-temporal considerations for designing and interpreting 'omics' experiments. J Cereal Sci 70:29–37

Eggert K, Pawelzik E (2011) Proteome analysis of *Fusarium* head blight in grains of naked barley (*Hordeum vulgare* subsp. *nudum*). Proteomics 11:972–985

Endler A, Meyer S, Schelbert S, Schneider T, Weschke W, Peters SW, Keller F, Baginsky S, Martinoia E, Schmidt UG (2006) Identification of a vacuolar sucrose transporter in barley and arabidopsis mesophyll cells by a tonoplast proteomic approach. Plant Physiol 141:196–207

Endler A, Reiland S, Gerrits B, Schmidt UG, Baginsky S, Martinoia E (2009) In vivo phosphorylation sites of barley tonoplast proteins identified by a phosphoproteomic approach. Proteomics 9:310–321

Eubel H, Jansch L, Braun HP (2003) New insights into the respiratory chain of plant mitochondria. Supercomplexes and a unique composition of complex II. Plant Physiol 133:274–286

Fatehi F, Hosseinzadeh A, Alizadeh H, Brimavandi T, Struik PC (2012) The proteome response of salt-resistant and salt-sensitive barley genotypes to long-term salinity stress. Mol Biol Rep 39:6387–6397

Fatehi F, Hosseinzadeh A, Alizadeh H, Brimavandi T (2013) The proteome response of *Hordeum spontaneum* to salinity stress. Cereal Res Commun 41:78–87

Fenn JB (2003) Electrospray wings for molecular elephants (Nobel lecture). Angew Chem (International Edition) 42:3871–3894

Finnie C, Steenholdt T, Noguera OR, Knudsen S, Larsen J, Brinch-Pedersen H, Holm PB, Olsen O, Svensson B (2004) Environmental and transgene expression effects on the barley seed proteome. Phytochemistry 65:1619–1627

Finnie C, Bagge M, Steenholdt T, Ostergaard O, Bak-Jensen K, Backes G, Jensen A, Giese H, Larsen J, Roepstorff P, Svensson B (2009) Integration of the barley genetic and seed proteome maps for chromosome 1H, 2H, 3H, 5H and 7H. Funct Integr Genomics 9:135–143

Finnie C, Sultan A, Grasser KD (2011) From protein catalogues towards targeted proteomics approaches in cereal grains. Phytochemistry 72:1145–1153

Flodrová D, Salplachta J, Benkovska D, Bobalova J (2012) Application of proteomics to hordein screening in the malting process. Eur J Mass Spectrom 18:323–332

Focke M, Gieringer E, Schwan S, Jansch L, Binder S, Braun HP (2003) Fatty acid biosynthesis in mitochondria of grasses: malonyl-coenzyme A is generated by a mitochondrial-localized acetyl-coenzyme A carboxylase. Plant Physiol 133:875–884

Frigerio S, Campoli C, Zorzan S, Fantoni LI, Crosatti C, Drepper F, Haehnel W, Cattivelli L, Morosinotto T, Bassi R (2007) Photosynthetic antenna size in higher plants is controlled by the plastoquinone redox state at the post-transcriptional rather than transcriptional level. J Biol Chem 282:29457–29469

Görg A, Postel W, Domscheit A, Gunther S (1988) Two-dimensional electrophoresis with immobilized pH gradients of leaf proteins from barley (*Hordeum vulgare*)—method, reproducibility and genetic aspects. Electrophoresis 9:681–692

Granvogl B, Reisinger V, Eichacker LA (2006) Mapping the proteome of thylakoid membranes by de novo sequencing of intermembrane peptide domains. Proteomics 6:3681–3695

Gross JH (2011) Mass spectrometry: a textbook. Springer, Berlin

Hägglund P, Bunkenborg J, Maeda K, Svensson B (2008) Identification of thioredoxin disulfide targets using a quantitative proteomics approach based on isotope-coded affinity tags. J Proteome Res 7:5270–5276

Hägglund P, Bunkenborg J, Yang F, Harder LM, Finnie C, Svensson B (2010) Identification of thioredoxin target disulfides in proteins released from barley aleurone layers. J Proteomics 73:1133–1136

Hägglund P, Bjornberg O, Navrot N, Jensen JM, Maeda K, Kirkensgaard K, Shahpiri A, Sultan A, Bunkenborg J, Gubler F, Barrero JM, Henriksen A, Finnie C, Svensson B (2013) The barley grain thioredoxin system—an update. Front Plant Sci. https://doi.org/10.3389/fpls.2013.00151

Hägglund P, Finnie C, Yano H, Shahpiri A, Buchanan BB, Henriksen A, Svensson B (2016) Seed thioredoxin h. Biochim Biophys Acta-Proteins Proteomics 1864:974–982

Hooper CM, Castleden IR, Aryamanesh N, Jacoby RP, Millar AH (2016) Finding the subcellular location of barley, wheat, rice and maize proteins: the compendium of crop proteins with annotated locations (cropPAL). Plant Cell Physiol 57

Hurkman WJ, Tanaka CK (1986) Solubilization of plant membrane proteins for analysis by two-dimensional gel electrophoresis. Plant Physiol 81:802–806

Hynek R, Svensson B, Jensen ON, Barkholt V, Finnie C (2006) Enrichment and identification of integral membrane proteins from barley aleurone layers by reversed-phase chromatography, SDS-PAGE, and LC-MS/MS. J Proteome Res 5:3105–3113

Hynek R, Svensson B, Jensen ON, Barkholt V, Finnie C (2009) The plasma membrane proteome of germinating barley embryos. Proteomics 9:3787–3794

Jansen RC, Nap JP (2001) Genetical genomics: the added value from segregation. Trends Genet 17:388–391

Johnova P, Skalak J, Saiz-Fernandez I, Brzobohaty B (2016) Plant responses to ambient temperature fluctuations and water-limiting conditions: a proteome-wide perspective. Biochim Biophys Acta-Proteins Proteomics 1864:916–931

Kaspar S, Matros A, Mock HP (2010a) Proteome and flavonoid analysis reveals distinct responses of epidermal tissue and whole leaves upon UV-B radiation of barley (*Hordeum vulgare* L.) seedlings. J Proteome Res 9:2402–2411

Kaspar S, Weier D, Weschke W, Mock HP, Matros A (2010b) Protein analysis of laser capture micro-dissected tissues revealed cell-type specific biological functions in developing barley grains. Anal Bioanal Chem 398:2883–2893

Kaspar-Schoenefeld S, Merx K, Jozefowicz AM, Hartmann A, Seiffert U, Weschke W, Matros A, Mock HP (2016) Label-free proteome profiling reveals developmental-dependent patterns in young barley grains. J Proteomics 143:106–121

Kausar R, Arshad M, Shahzad A, Komatsu S (2013) Proteomics analysis of sensitive and tolerant barley genotypes under drought stress. Amino Acids 44:345–359

Kosova K, Vitamvas P, Prasil IT (2014) Proteomics of stress responses in wheat and barley-search for potential protein markers of stress tolerance. Front Plant Sci. https://doi.org/10.3389/fpls.2014.00711

Kosova K, Vitamvas P, Urban MO, Klima M, Roy A, Prasil IT (2015) Biological networks underlying abiotic stress tolerance in temperate crops-a proteomic perspective. Int J Mol Sci 16:20913–20942

Lastovickova M, Bobalova J (2012) MS based proteomic approaches for analysis of barley malt. J Cereal Sci 56:519–530

Lozano RM, Wong JH, Yee BC, Peters A, Kobrehel K, Buchanan BB (1996) New evidence for a role for thioredoxin h in germination and seedling development. Planta 200:100–106

Maeda K, Finnie C, Ostergaard O, Svensson B (2003) Identification, cloning and characterization of two thioredoxin H isoforms, HvTrxh1 and HvTrxh2, from the barley seed proteome. Eur J Biochem 270:2633–2643

Maeda K, Finnie C, Svensson B (2004) Cy5 maleimide labelling for sensitive detection of free thiols in native protein extracts: identification of seed proteins targeted by barley thioredoxin h isoforms. Biochem J 378:497–507

Mahalingam R (2017) Shotgun proteomics of the barley seed proteome. Bmc Genomics 18

March TJ, Richter D, Colby T, Harzen A, Schmidt J, Pillen K (2012) Identification of proteins associated with malting quality in a subset of wild barley introgression lines. Proteomics 12:2843–2851

Marsalova L, Vitamvas P, Hynek R, Prasil IT, Kosova K (2016) Proteomic response of *Hordeum vulgare* cv. Tadmor and *Hordeum marinum* to salinity stress: similarities and differences between a glycophyte and a halophyte. Front Plant Sci. https://doi.org/10.3389/fpls.2016.01154

Marx C, Wong JH, Buchanan BB (2003) Thioredoxin and germinating barley: targets and protein redox changes. Planta 216:454–460

Matros A, Kaspar S, Witzel K, Mock HP (2011) Recent progress in liquid chromatography-based separation and label-free quantitative plant proteomics. Phytochemistry 72:963–974

Moller ALB, Pedas P, Andersen B, Svensson B, Schjoerring JK, Finnie C (2011) Responses of barley root and shoot proteomes to long-term nitrogen deficiency, short-term nitrogen starvation and ammonium. Plant, Cell Environ 34:2024–2037

Mustafa G, Komatsu S (2016) Toxicity of heavy metals and metal-containing nanoparticles on plants. Biochim Biophys Acta-Proteins Proteomics 1864:932–944

Nelson CJ, Alexova R, Jacoby RP, Millar AH (2014) Proteins with high turnover rate in barley leaves estimated by proteome analysis combined with in planta isotope labeling. Plant Physiol 166:91–108

Osinalde N, Aloria K, Omaetxebarria MJ, Kratchmarova I (2017) Targeted mass spectrometry: an emerging powerful approach to unblock the bottleneck in phosphoproteomics. J Chromatogr B-Anal Technol Biomed Life Sci 1055:29–38

Ostergaard O, Finnie C, Laugesen S, Roepstorff P, Svensson B (2004) Proteome analysis of barley seeds: identification of major proteins from two-dimensional gels (pl 4-7). Proteomics 4:2437–2447

Patton WF (2000) A thousand points of light: the application of fluorescence detection technologies to two-dimensional gel electrophoresis and proteomics. Electrophoresis 21:1123–1144

Pennington HG, Gheorghe DM, Damerum A, Pliego C, Spanu PD, Cramer R, Bindschedler LV (2016) Interactions between the powdery mildew effector BEC1054 and barley proteins identify candidate host targets. J Proteome Res 15:826–839

Perrocheau L, Rogniaux H, Boivin P, Marion D (2005) Probing heat-stable water-soluble proteins from barley to malt and beer. Proteomics 5:2849–2858

Petersen J, Rogowska-Wrzesinska A, Jensen ON (2013) Functional proteomics of barley and barley chloroplasts—strategies, methods and perspectives. Front Plant Sci. https://doi.org/10.3389/fpls.2013.00052

Picotti P, Aebersold R (2012) Selected reaction monitoring-based proteomics: workflows, potential, pitfalls and future directions. Nat Methods 9:555–566

Picotti P, Bodenmiller B, Mueller LN, Domon B, Aebersold R (2009) Full dynamic range proteome analysis of *S. cerevisiae* by targeted proteomics. Cell 138:795–806

Ploscher M, Granvogl B, Zoryan M, Reisinger V, Eichacker LA (2009) Mass spectrometric characterization of membrane integral low molecular weight proteins from photosystem II in barley etioplasts. Proteomics 9:625–635

Ploscher M, Reisinger V, Eichacker LA (2011) Proteomic comparison of etioplast and chloroplast protein complexes. J Proteomics 74:1256–1265

Pos V, Hunyadi-Gulyas E, Caiazzo R, Jocsak I, Medzihradszky KF, Lukacs N (2011) Induction of pathogenesis-related proteins in intercellular fluid by cadmium stress in barley (*Hordeum vulgare* L.)—a proteomic analysis. Acta Aliment 40:164–175

Rasoulnia A, Bihamta MR, Peyghambari SA, Alizadeh H, Rahnama A (2011) Proteomic response of barley leaves to salinity. Mol Biol Rep 38:5055–5063

Rollins JA, Habte E, Templer SE, Colby T, Schmidt J, von Korff M (2013) Leaf proteome alterations in the context of physiological and morphological responses to drought and heat stress in barley (*Hordeum vulgare* L.). J Exp Bot 64:3201–3212

Schneider T, Schellenberg M, Meyer S, Keller F, Gehrig P, Riedel K, Lee Y, Eberl L, Martinoia E (2009) Quantitative detection of changes in the leaf-mesophyll tonoplast proteome in dependency of a cadmium exposure of barley (*Hordeum vulgare* L.) plants. Proteomics 9:2668–2677

Sharma SS, Dietz KJ, Mimura T (2016) Vacuolar compartmentalization as indispensable component of heavy metal detoxification in plants. Plant, Cell Environ 39:1112–1126

Shen JR (2015) The structure of photosystem II and the mechanism of water oxidation in photosynthesis. In Merchant SS (ed) Annual review of plant biology, vol 66, pp 23–48

Shen QF, Fu LB, Dai F, Jiang LX, Zhang GP, Wu DZ (2016) Multi-omics analysis reveals molecular mechanisms of shoot adaption to salt stress in Tibetan wild barley. Bmc Genomics 17

Shen QF, Fu LB, Qiu L, Xue F, Zhang GP, Wu DZ (2017) Time-course of ionic responses and proteomic analysis of a Tibetan wild barley at early stage under salt stress. Plant Growth Regul 81:11–21

Shevela D, Arnold J, Reisinger V, Berends HM, Kmiec K, Koroidov S, Bue AK, Messinger J, Eichacker LA (2016) Biogenesis of water splitting by photosystem II during de-etiolation of barley (*Hordeum vulgare* L.). Plant, Cell Environ 39:1524–1536

Sreenivasulu N, Borisjuk L, Junker BH, Mock HP, Rolletschek H, Seiffert U, Weschke W, Wobus U (2010) Barley grain development: toward an integrative view. Int Rev Cell Mol Biol 281:49–89

Sultan A, Andersen B, Svensson B, Finnie C (2016) Exploring the plant-microbe interface by profiling the surface-associated proteins of barley grains. J Proteome Res 15:1151–1167

Tanaka K (2003) The origin of macromolecule ionization by laser irradiation (Nobel lecture). Angew Chem (International Edition) 42:3860–3870

Trumper C, Paffenholz K, Smit I, Kossler P, Karlovsky P, Braun HP, Pawelzik E (2016) Identification of regulated proteins in naked barley grains (*Hordeum vulgare nudum*) after *Fusarium graminearum* infection at different grain ripening stages. J Proteomics 133:86–92

Wang XL, Cai XF, Xu CX, Wang QH, Dai SJ (2016) Drought-responsive mechanisms in plant leaves revealed by proteomics. Int J Mol Sci 17

Weiss W, Postel W, Gorg A (1991) Barley cultivar discrimination. 2. Sodium dodecyl sulfate-Polyacrylamide gel-electrophoresis and isoelectric-focusing with immobilized pH gradients. Electrophoresis 12:330–337

Wendelboe-Nelson C, Morris PC (2012) Proteins linked to drought tolerance revealed by DIGE analysis of drought resistant and susceptible barley varieties. Proteomics 12:3374–3385

Witzel K, Mock H-P (2016) A proteomic view of the cereal and vegetable crop response to salinity stress. In: Agricultural proteomics, vol 2. Springer, Berlin, pp 53–69

Witzel K, Weidner A, Surabhi GK, Borner A, Mock HP (2009) Salt stress-induced alterations in the root proteome of barley genotypes with contrasting response towards salinity. J Exp Bot 60:3545–3557

Witzel K, Weidner A, Surabhi GK, Varshney RK, Kunze G, Buck-Sorlin GH, Borner A, Mock HP (2010) Comparative analysis of the grain proteome fraction in barley genotypes with contrasting salinity tolerance during germination. Plant, Cell Environ 33:211–222

Witzel K, Pietsch C, Strickert M, Matros A, Roder M, Weschke W, Wobus U, Mock HP (2011) Mapping of quantitative trait loci associated with protein expression variation in barley grains. Mol Breeding 27:301–314

Witzel K, Matros A, Strickert M, Kaspar S, Peukert M, Mühling KH, Börner A, Mock H-P (2014) Salinity stress in roots of contrasting barley genotypes reveals time-distinct and genotype-specific patterns for defined proteins. Mol Plant 7:336–355

Wu D, Shen Q, Qiu L, Han Y, Ye L, Jabeen Z, Shu Q, Zhang G (2014) Identification of proteins associated with ion homeostasis and salt tolerance in barley. Proteomics 14:1381–1392

Yang F, Jensen JD, Spliid NH, Svensson B, Jacobsen S, Jorgensen LN, Jorgensen HJL, Collinge DB, Finnie C (2010a) Investigation of the effect of nitrogen on severity of Fusarium Head Blight in barley. J Proteomics 73:743–752

Yang F, Jensen JD, Svensson B, Jorgensen HJL, Collinge DB, Finnie C (2010b) Analysis of early events in the interaction between *Fusarium graminearum* and the susceptible barley (*Hordeum vulgare*) cultivar Scarlett. Proteomics 10:3748–3755

Zeng ZX, Jiang JM (2016) Isolation and proteomics analysis of barley centromeric chromatin using PICh. J Proteome Res 15:1875–1882

Zolla L, Rinalducci S, Timperio AM, Huber CG (2002) Proteomics of light-harvesting proteins in different plant species. Analysis and comparison by liquid chromatography-electrospray ionization mass spectrometry. Photosystem I. Plant Physiology 130:1938–1950

Organellar Genomes in Barley

20

Hirokazu Handa

Abstract

Similar to other higher plants, barley (*Hordeum vulgare*) contains the following three cell organelles that possess their own genomes: nucleus, mitochondrion, and chloroplast. In this chapter, the genome structures, genetic content, and functions of two cytoplasmic organelles, i.e., mitochondrion and chloroplast, in barley are discussed. The barley mitochondrial genome (mt genome) is 525,599 bp in size, which is 73 kb larger than that of wheat, and the gene content is well conserved among grass species; notably, the contents of intact protein-coding genes in barley are the same as those in wheat. However, the mt genome structure is markedly different among grass species, and rearrangements and fragmentations of homologous regions prevent the reconfiguration of evolutionary processes, even in the same Triticeae lineage, which includes barley and wheat. However, the genome structure and gene content of chloroplast genome (cp genome) are highly conserved among grass species. The cp genome in barley is 136,462 bp in size, and the quadripartite structures that are common in the cp genome of higher plants are conserved. Most sequences are collinear between wheat and barley, and the gene content and gene order in barley are identical to those in wheat. Chloroplasts and mitochondria are essential organelles, and the genes encoded in both organellar genomes are indispensable for plant cell survival. Several genetic interactions among the cell organelles, nucleus, mitochondrion, and chloroplast occur within a cell. In this chapter, these genetic interactions and outcomes, including cytoplasmic male sterility (CMS) and chloroplast dysfunction, are reviewed. These phenomena are interesting and important for the understanding of the physiological function of both cytoplasmic organelles and their potential use in plant breeding. We have only recently begun to understand these genetic interactions due to the publication of the complete genomes of the nucleus, mitochondrion, and chloroplast in barley.

20.1 Introduction

In plant cells, the following organelles possess their own genomes: nucleus, mitochondrion, and chloroplast.

H. Handa (✉)
Plant Genome Research Unit, Institute of Crop Science, National Agriculture and Food Research Organization (NARO), 2-1-2, Kan-non-dai, 305-8518 Tsukuba, Japan
e-mail: hirokazu@affrc.go.jp

N. Stein and G. J. Muehlbauer (eds.), *The Barley Genome*, Compendium of Plant Genomes,
https://doi.org/10.1007/978-3-319-92528-8_20

Most cp genomes have highly conserved structures with a quadripartite organization comprising a large single-copy (LSC) region, a small single-copy (SSC) region, and two copies of inverted repeats (IRs). The cp genomes range from 107 to 218 kb and commonly include approximately 120–130 genes (Daniell et al. 2016). However, the mt genomes in higher plants are much larger and widely differ in size compared to those in other organisms such as animals that possess circular DNA molecules of 15–17 kb in size. Plant mt genomes typically vary in size from 200 to 750 kb; however, in certain lineages, longer extensions have been observed that reach the megabase range. In contrast to the cp genome, plant mt genomes do not have a common structure. Despite their larger sizes, plant mt genomes contain fewer genes than mt genomes in other organisms, such as animals and fungi. The number of known genes typically ranges from 50 to 60, which is less than the number of genes in the cp genome (Gualberto et al. 2014).

Compared with the nuclear genome, the cp and mt genomes are smaller in size, and the numbers of encoded genes are lower. However, these two genomes are not less important than the nuclear genome. The chloroplast and mitochondrion are essential organelles in plant cells. The chloroplast conducts photosynthesis in the presence of sunlight, and the mitochondrion indirectly supplies energy to plant cells; the functions of both organelles are essential for the survival of plant cells. Moreover, the genes encoded in both organellar genomes are indispensable for photosynthesis and oxidative phosphorylation, which are required for producing ATP.

The cp genomes in major cereal crops, such as rice, wheat, and maize, were fully sequenced between the late 1980s and early 2000s (rice, Hiratsuka et al. 1989; maize, Maier et al. 1995; wheat, Ogihara et al. 2002), but the mt genome sequences were not revealed until the mid-2000s (rice, Notsu et al. 2002; maize, Clifton et al. 2004; wheat, Ogihara et al. 2005). Compared to these cereal crops, the sequencing of barley organellar genomes occurred relatively recently as follows: the cp genome was sequenced in 2007 (Saski et al. 2007), and the mt genome was sequenced in 2016 (Hisano et al. 2016). More recently, the nuclear genome sequence in barley has been described (Mascher et al. 2017). Currently, we can access the complete genome information of the nucleus, mitochondrion, and chloroplast in barley, enabling a complete understanding of the gene functions and gene interactions in this important crop.

In this chapter, the current understanding of the organellar genomes in barley, its relationship with other cereal crops and inter-organellar gene interactions are reviewed.

20.2 Mitochondrial Genome

The mitochondrion is a semiautonomous organelle that produces cellular ATP through oxidative phosphorylation, and this function is conserved in all eukaryotic cells, i.e., animal, fungus, and plant cells. In addition to its role as a supplier of cellular ATP, the plant mitochondrion is a key player in a variety of cellular processes, including development, fitness, and reproduction, because the autotrophic life style forces plant cells to synthesize all components. Furthermore, the plant mitochondrion plays roles in the responses to several stresses, particularly oxidative stress, under highly variable external conditions (Jacoby et al. 2012). Although the mitochondrial contribution to various cellular processes in plant cells depends on the cells' own genetic system, the mt genomes in higher plants exhibit several unique features that distinguish them from those in animals or fungi. Plant mt genomes are larger and show extensive structural rearrangements resulting from homologous recombination events, and there are specific modes of gene expression (e.g., cis- and trans-splicing and RNA editing) (Kubo and Newton 2008; Bonen 2008; Takenaka et al. 2008).

Almost no information was available regarding the organization and content of the mt genome in barley prior to 2016. However, the first mt genome sequence in barley was released by

Hisano et al. (2016), which enhanced our detailed understanding of its activity.

20.2.1 Genome Structures

The mitochondrial gene (mt gene) sequence in barley for the gene encoding the ATP synthase subunit 9 (*atp9*) was first reported in 1993 (Rish and Breiman 1993), while the first mt gene sequences in rice and wheat were reported in the early 80 s (*cox2* genes in both plants; rice, Kao et al. 1984 and wheat, Bonen et al. 1984). Subsequently, only a few mt genes were sequenced, as follows: *rps2* (Kubo et al. 2005), *Ψrps14* (Ong and Palmer 2006) and *rps7* (Byers et al. 2010). To date, information regarding the barley mt genes and genomes is insufficient to elucidate the mitochondrial function in barley.

Recently (2016), Hisano and colleagues assembled a 525,599 bp circular molecule for the mt genome in barley (AP017301, Hisano et al. 2016) using purified mtDNA samples and a next-generation sequencing (NGS) approach. The barley mt genome is 73 and 35 kb larger than that in wheat and rice (452,528 bp in wheat, AP008982 and 490,520 bp in Japonica rice, BA000029), respectively, and comparable in size to maize (569, 630 bp in the NB cytoplasm, AY506529) (Table 20.1). The size difference in the mt genomes among cereal crops or even among Triticeae species is not surprising because the mt genomes in higher plants differ widely in size, even among closely related species, and this size variation is independent of the size variation in their nuclear genomes. For example, the rapeseed mt genome is 221,853 bp in size, which is only two-thirds the size of that in *Arabidopsis* (366,924 bp), although both species are closely related and belong to the same family, i.e., Brassicaceae (Handa 2003).

The size variation in higher plant mt genomes reflects the presence of repeat sequences that are dispersed throughout the genome. Hisano et al. (2016) detected 6 large and 17 small repeat sequences in the barley mt genome, reaching 155 kb, while in the wheat mt genome, the repeated regions reached 86 kb (Ogihara et al. 2005). The size difference (approximately 70 kb) in the repeat regions between barley and wheat corresponds to the difference in the total mt genome sizes in both plants. These repeat sequences in the mt genome can function as intra- and/or intermolecular recombination sites, leading to further complex genome structures through rearrangements (Kubo and Newton 2008; Gualberto et al. 2014). Figure 20.1 shows the results of 2 BLAST analyses using MegaBLAST comparing the mt genome in barley with that in wheat or rice; these results indicate extensive rearrangements in the mt genome structures among these three cereal crops (AP017301, AP008982 and BA000029 for barley, wheat and rice, respectively). The rearrangements and fragmentations in the homologous regions are highly extensive and complicated, and reconfiguring the evolutionary process is impossible, even in the same Triticeae lineage, such as barley and wheat.

During successive genome rearrangements, certain sequences are captured by plant mt genomes, further contributing to the variations in the sizes of the plant mt genomes. The acquisition of cp genome sequences by mt genomes is one example. In the barley mt genome, 18 sequence regions of plastid origins were identified, totaling 19.8 kb in size, which is comparable to that in wheat (13.5 kb) (Hisano et al. 2016; Ogihara et al. 2005).

Another interesting feature of plant mt genomes is the presence of DNA sequences of unknown origin (Kubo and Newton 2008). Most sequences outside the well-conserved gene region, after accounting for chloroplast and nuclear DNA insertions, have unrecognizable origins, suggesting that these sequences have no apparent homology to any DNA sequences in the public databases. In total, 173,598 bp of DNA sequences in the barley mt genome are not homologous to the wheat mt genome ("not homologous" refers to the remaining sequences after subtracting the homologous sequences with an e-value = 0 by MegaBLAST), corresponding to 33% of the entire mt genome. Using these nonhomologous sequences as queries against the nonredundant nucleotide sequence database,

Table 20.1 Gene contents among mitochondrial genomes in grass species

Plant	Barley	Wheat	*T. timopheevi*	*Ae. Speltoides*	*L. perenne*	Maize	Sorghum	Rice (Japonica)	Rice (Indica)
Acc. No.	AP017301	AP008982	AP013106	AP013107	JX999996	AY506529	DQ984518	BA000029	DQ167399
Size (bp)	525,599	452,528	443,419	476,091	678,580	569,630	468,628	490,520	491,515
nad1	+	+	+	+	+	+	+	+	+
nad2	+	+	+	+	+	+	+	+	+
nad3	+ (2)	+	+	+	+	+	+	+	+
nad4	+	+	+	+	+	+	+	+	+
nad4L	+	+	+	+	+	+	+	+	+
nad5	+	+	+	+	+	+	+	+	+
nad6	+	+	+	+	+	+	+	+	+
nad7	+	+	+	+	+ (2)	+	+	+	+
nad9	+ (2)	+	+	+	+	+	+	+	+
cob	+	+	+ (2)	+	+	+	+ (2)	+	+
cox1	+	+	+	+	+	+	+	+	+
cox2	+	+	+	+	+	+	+	+	+
cox3	+	+	+	+	+	+	+	+	+
atp1	+	+	+	+	+	+ (2)	+ (2)	+	+
atp4	+	+	+	+	+	+	+	+	+
atp6	+	+ (2)	+	+	+	+	+	+	+
atp8	+	+ (2)	+	+	+	+	+	+	+
atp9	+	+	+	+	+	+	+	+	+
ccmB	+	+	+	+	+	+	+	+	+
ccmC	+ (2)	+	+	+	+	+	+ (2)	+	+
ccmFC	+	+	+	+	+	+	+	+	+
ccmFN	+	+	+	+	+	+	+	+	+
rpl2	Ψ	Ψ	Ψ	Ψ	−	−	−	+	+
rpl5	+ (2)	+	+	+	+ (2)	−	−	+	+

(continued)

Table 20.1 (continued)

Plant	Barley	Wheat	*T. timopheevi*	*Ae. Speltoides*	*L. perenne*	Maize	Sorghum	Rice (Japonica)	Rice (Indica)
rpl16	+	+	+	+	+	+	+	+	+
rps1	+	+	+	+	+	+	+	+	+
rps2	+ (2)	+	+	+	+	+ (2)	+	+	+
rps3	+	+	+	+	+	+	+	+	+
rps4	+	+	+	+	+	+	+	+	+
rps7	+	+	+	+	+ (2)	+	+ (2)	+	+
rps11	−	−	−	−	−	−	−	Ψ	Ψ
rps12	+	+	+	+	+	+	+	+	+
rps13	+	+	+	+	+	+	+	+	+
*rps14**	Ψ	Ψ	Ψ	Ψ	Ψ (2)	−	−	Ψ	Ψ (2)
rps19	−	Ψ	Ψ	Ψ	−	−	−	+	+
matR	+	+	+	+	+	+	+	+	+
mttB	+	+	+	+	+	+	+	+	+
rrn5	+ (3)	+ (3)	+ (3)	+ (4)	+ (2)	+	+	+	+ (2)
rrn18	+ (3)	+ (3)	+ (3)	+ (4)	+ (2)	+	+	+	+ (2)
rrn26	+ (2)	+ (2)	+ (2)	+ (2)	+ (2)	+	+	+	+ (2)

+, presence of the gene; −, absence of the gene; Ψ, pseudo gene; () gene copy numbers in parenthesis
*Ong and Plamer (2006)

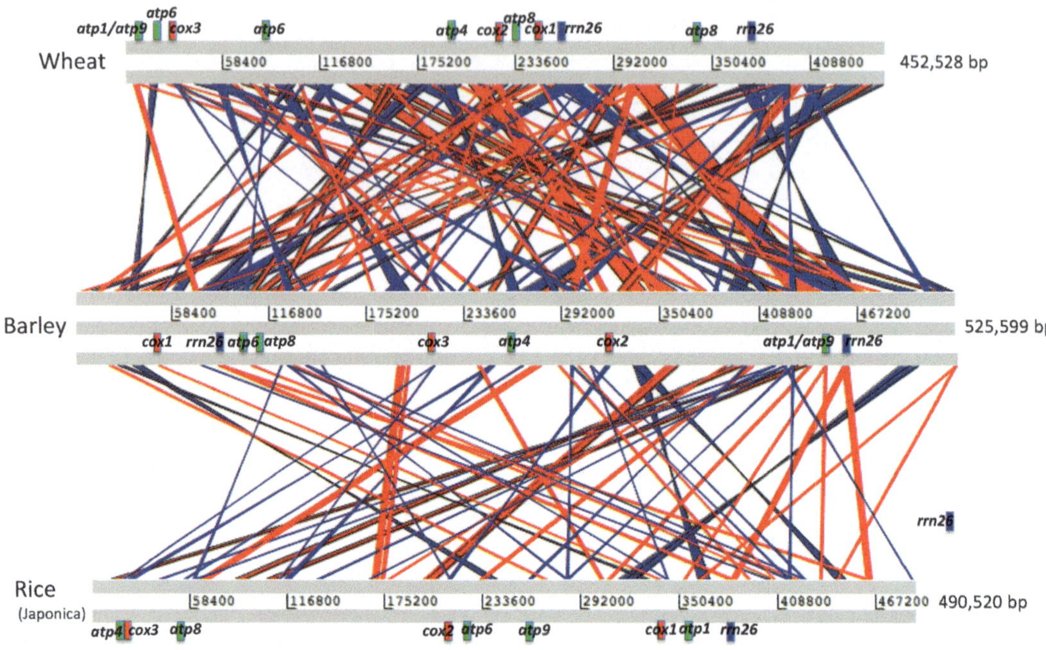

Fig. 20.1 Relative arrangement of homologous sequences in three mt genomes in cereal plants, i.e., barley (AP017301), wheat (AP008982), and rice (BA000029). Homologous regions were identified by performing 2 BLAST analyses using MegaBLAST for each pair of species. Red and blue colors represent the forward and reverse orientations, respectively, of the homologous sequences. The relative genomic positions of certain genes are indicated by a box; complex IV genes (red; *cox1*, *cox2*, *cox3*), complex V genes (green; *atp1*, *atp4*, *atp6*, *atp8*, and *atp9*) and *rrn26* (blue)

only three small sequence stretches were identified, which were homologous to the cp genome in barley cv. Morex (4637 bp, EF11541), the nuclear genome in wheat chromosome 3B (4455 bp, HG670306), and the mt genome in the wheat K-type cytoplasm (4170 bp, GU985444). The origins of the remaining 160 kb regions are completely unknown, and these sequences are specific to the barley mt genome. Identifying the functions of these sequences would be interesting; thus, further studies are needed to identify the functions of the open reading frames (ORFs) in these regions.

20.2.2 Gene Contents

Although the mt genomes in higher plants vary in size, the gene content in various plant species is similar. Typically, plant mt genomes contain 50–60 known genes (not considering copy number), and most genes are commonly conserved in most plant species (Gualberto et al. 2014). Hisano et al. (2016) identified 33 protein-coding genes, three ribosomal RNA genes and 16 transfer RNA genes in the barley mt genome. Table 20.1 summarizes the contents of the known protein-coding genes and ribosomal RNA genes in eight grass species, including barley. Overall, the gene content is well conserved among these grass species, particularly the genes encoding complex I (NADH dehydrogenase), complex III (cytochrome bc_1 complex), complex IV (cytochrome *c* oxidase), complex V (ATP synthase), and cytochrome *c* biogenesis (Table 20.1). Most differences in the gene content are observed among genes encoding ribosomal proteins. For example, the *rpl2* gene is intact in rice mitochondria but is a pseudogene in Triticeae lineages, including barley, and is lost in Panicoideae species and *L. perenne* (Table 20.1; Fig. 20.2). The *rps19* gene shows a similar

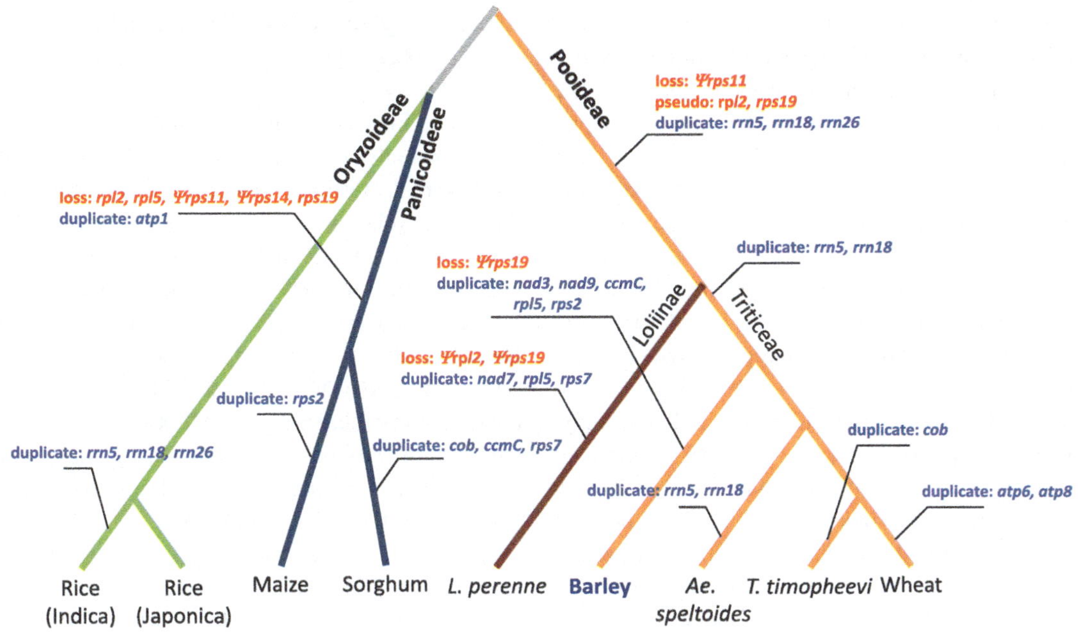

Fig. 20.2 Phylogenetic relationships and evolutionary events (gene loss, pseudogenization and copy multiplication) underlying the mt genomes in eight grass species (i.e., barley, wheat, *T. timopheevi*, *Ae. speltoides*, *L. perenne*, sorghum, maize, rice (Japonica) and rice (Indica)). Phylogenetic relationships are based on Fig. 4 in Hisano et al. (2016) with permission from Dr. Hisano

pattern; however, the barley mt genome lacks this gene even though *rps19* is pseudogenized in other Triticeae species. Most genes lost from, or pseudogenized in, mt genomes are likely transferred to the nuclear genome (Adams and Palmer 2003). Using the intact *rps19* gene sequence in the rice mt genome, two homologous sequences were identified. One sequence (HORVU4Hr1G044800) is located on chromosome 4H, and the other sequence (HORVU1Hr1G076060) is located on chromosome 1H. Although the nuclear copy on 4H is more similar to the rice mt *rps19* gene than that on 1H, the 4H copy lacks the N-termini signal sequence required for targeting to mitochondria. On the other hand, the 1H copy is similar to the EST sequence (BI959815) that encodes nuclear-transferred gene for mitochondrial RPS19 (Fallahi et al. 2005), although mismatches exist between the genomic and EST sequences, and further evaluation is needed. Currently, we can access the full sequence information of the barley nuclear genome; thus, it would be interesting to examine the gene transfer from the mitochondrion to the nucleus.

The gene content of the barley mt genome is similar to that of the wheat mt genome. The only difference observed between the protein-coding and ribosomal RNA genes is the *rps19* gene, which is missing from barley but is a pseudogene in wheat (Table 20.1). The barley and wheat mt genomes encode 18 and 17 genes for transfer RNAs, respectively, and 16 of these genes are common to both plants (Hisano et al. 2016; Ogihara et al. 2005). Two transfer RNA genes, i.e., *trnR* and *trnV*, are specific to barley, while *trnA* is missing from barley. Both tRNA genes, i.e., *trnR* and *trnV*, are located in chloroplast-derived sequences inserted in the barley mt genome, and these insertions are not detected in other Pooideae species (*Triticum-Aegilops* complex and *Lollium*). Although no data are available regarding the expression of these genes, determining whether these genes are expressed in barley mitochondria would be interesting. The similarity in the gene content between the barley and wheat mt genomes is unsurprising because of the closely related taxonomically position and sequence similarity in the common genes as shown in the phylogenetic analysis (Fig. 20.2 in this chapter and Fig. 4 in Hisano et al. (2016)). However, after including gene loss, acquisition, and pseudogenization in this analysis (Fig. 20.2), a different conclusion is obtained. For example, *rps19* is maintained in Oryzoideae, but is lost from Panicoideae and pseudogenized in Pooideae. Additionally, the pseudogenized copy of *rps19* is lost from two sub-lineages of Pooideae, Loliinae and the genus *Hordeum* of Triticeae, independently. However, in the *Triticum-Aegilops* complex of Triticeae, the mt genome maintains the *Ψrps19* gene. In total, four independent events occurred in one gene during the evolutionary process of cereal plants. The *rpl2* and *Ψrps11* genes show the same pattern as *rps19* as follows: two independent gene losses were identified in each gene in the grass species lineage. Higher plant mt genomes are extremely conserved with respect to coding sequences and somewhat conserved with respect to RNA editing, but the noncoding regions in plant mitochondrial DNA are extraordinarily dynamic with respect to structural changes, sequence acquisition and/or sequence loss (Handa 2003). Currently, no information is available regarding RNA editing in barley mitochondrial transcripts; however, extreme conservation of coding sequences has been observed in barley mt genes. By comparing the sequences of 24 protein-coding genes (all genes coding electron transport chain and cytochrome *c* biogenesis, *matR* and *mttB*, which were used to draw Fig. 20.2) in barley and wheat, 8 and 13 genes showed a perfect match and 99% identity, respectively, except for the *atp6* gene, whose N-terminal region diverged and showed no similarity between barley and wheat. However, this variation in the N-terminus of the *atp6* gene is often observed in plant mt genomes.

Notably, plant mitochondrial DNA evolves rapidly in structure, but slowly in sequence (Palmer and Herbon 1988), which also occurs in barley.

20.2.3 Cytoplasmic Male Sterility (CMS)

CMS is a maternally inherited trait that causes a defect in male reproductive organs but does not affect female fertility; this trait is important for the commercial production of hybrid seeds. CMS is caused by lesions or rearrangements in the mt genome, and nuclear-encoded "restorer of fertility (Rf)" genes can restore fertility (Hanson and Bentolila 2004). In barley, hybrid seed production using the CMS-Rf system has previously been commercialized (HYVIDO® from Syngenta, https://www.syngenta.co.uk/varieties), but the molecular mechanisms underlying the barley CMS-Rf system remain unknown.

Two male-sterile cytoplasms, i.e., msm1 and msm2, were identified by Ahokas (1979, 1982) in two accessions of wild barley (*H. vulgare* ssp. *spontaneum*) and a single dominant gene, i.e., *Rfm1a* in the *spontaneum* accession with the msm1 cytoplasm, can restore fertility induced by both male-sterile cytoplasms. To clone the *Rfm1* gene, high-resolution genetic linkage mapping and BAC physical mapping have narrowed the *Rfm1* locus on chromosome 6H to a sub-centimorgan genetic interval and a 208-kb physical interval (Ui et al. 2015). However, no studies have investigated the CMS-causal gene or the mt genome in barley. Notably, information from the mitochondrial side is necessary to fully understand the CMS-Rf system in barley.

Recently, Hisano and colleagues sequenced the mt genome in wild barley (*H. vulgare* ssp. *Spontaneum*, acc. H602) and Japanese cultivated barley (ssp. *vulgare*, cv. Haruna Nijo) (Hisano et al. 2016). Surprisingly, the mt genomes in these two accessions are identical in size and gene content, and the only differences observed between these genomes are three nucleotide changes. According to Hisano et al. (2016), accession H602 is not a suitable representative of wild barley, and we are unlikely to obtain any information regarding a CMS-inducing gene in barley from this accession.

Typically, marked differences are observed in the structures of the mt genomes between normal (non-CMS-inducing) and CMS-inducing cytoplasms. In maize, in addition to the two normal cytoplasms, i.e., NA and NB, the mt genomes in three CMS cytoplasms have been sequenced, i.e., CMS-C, CMS-S, and CMS-T (Allen et al. 2007). Their genome sizes ranged from 536 kb in CMS-T to 740 kb in CMS-C, and the structural organization dramatically varied, although the gene contents (except for the copy number variation) and sequences of known genes were highly conserved. Several unknown ORFs are specific to each cytoplasm, and certain ORFs are postulated to be causal genes for CMS, such as *T-urf13* in CMS-T (Dewey et al. 1987) and *orf355/orf77* in CMS-S (Zabala et al. 1997). In rice, Kazama and Toriyama (2016) sequenced the mt genome in the Boro-Taichung (BT)-type CMS cytoplasm but failed to assemble a circular DNA molecule similar to that in the normal cytoplasm (Notsu et al. 2002), indicating that structural differences exist between normal and CMS mt genomes. Cytoplasm-specific ORFs were also observed, including *orf79* in the BT mt genome, which is associated with CMS (Iwabuchi et al. 1993; Akagi et al. 1994).

To identify the CMS-causing gene in barley and elucidate the evolutionary relationship between wild and cultivated barley, more extensive sequencing studies investigating the mt genomes in wild-type accessions, including the CMS-inducing cytoplasm, are necessary.

20.3 Chloroplast Genome

The chloroplast is the metabolic center for photosynthesis and other cellular reactions, including the synthesis of cellular compounds (e.g., amino acids, nucleotides and fatty acids) and the activation of signaling pathways in response to environmental factors (e.g., heat, drought, and light) (Jensen and Leister 2014). The cp genome contains many key protein-coding genes that are involved in photosynthesis and other metabolic pathways.

Compared to studies investigating nuclear genomes, studies on the organization and contents of the cp genome in barley are limited. However, the first cp genome sequence was delineated in 2007 (Saski et al. 2007) using cv. Morex, which was also used as a material for the sequencing of nuclear genome. Subsequently, Middleton et al. (2014) sequenced the cp genome in cv. Barke.

20.3.1 Genome Organization and Gene Content in the Barley Chloroplast Genome

The barley cp genome is 136,462 bp (Saski et al. 2007) and comprises a single circular molecule with a quadripartite structure that is highly conserved among the cp genomes in higher plants. The barley cp genome includes two copies of an IR region of 21,579 bp that separates LSC and SSC regions of 80,600 and 12,704 bp, respectively. The barley cp genome contains 113 protein and ribosomal RNA genes, 18 of which are duplicated in the IR region, yielding a total of 131 genes. Furthermore, 30 distinct tRNA genes have been identified, 7 of which are duplicated in the IR region (Saski et al. 2007). Despite its smaller size, the cp genome contains more genes than the mt genome in barley as described above, which is also a common feature in higher plant organellar genomes.

20.3.2 Comparison of the Chloroplast Genomes in Barley and Other Grass Species and Within the Hordeum Species

In contrast to the mt genome, the cp genomes in higher plants have highly conserved structures and content organization (Daniell et al. 2016). These common features in the cp genomes are also observed in barley and related grass species. Initially, the size of the barley cp genome

(136,462 bp) was thought to be comparable to that of other grass cp genomes (wheat, 134,545 bp; japonica rice, 134,525 bp; sorghum, 140,754 bp; maize, 140,384 bp), and the sizes of the grass cp genomes are typically smaller than those in dicot plants (approximately 150,000 bp and more). Figure 20.3 shows a comparison of the genome structure in barley, wheat, and rice based on the results of 2 BLAST analyses using MegaBLAST. In contrast to the mt genome (see Fig. 20.1), the cp genomes in these three plants are highly homologous in their organizations. Most sequences, including genes and intergenic regions, are collinear between barley and wheat or rice, and no extensive rearrangements are observed, except for certain gaps in homology. The gene contents are highly conserved in barley and other grass species, and the lack of the *accD*, *ycf1*, and *ycf2* genes is a characteristic of the grass cp genomes, distinguishing these plants from dicot plants. In particular, barley and wheat have an identical gene content and gene order (Saski et al. 2007; Middleton et al. 2014).

Cp genomes are often used in the investigations of phylogenetic relationships among closely related plant species, such as those belonging to the same genus or family, because of the maternal inheritance and moderate speed of nucleotide substitution rate. In contrast, the mt genome is less informative due to the slower rate of nucleotide substitution, the presence of RNA editing, and highly divergent noncoding regions. Nishikawa et al. (2002) analyzed the molecular phylogeny of the genus *Hordeum* (46 individuals of 32 taxa) using the sequences of three cp genomic regions. Genetic relationships were observed among taxa, and the origin of polyploid species was demonstrated to be within the genus *Hordeum*. Jakob and Blattner (2006) extended this work and more precisely and accurately identified the phylogenetic relationships and speciation mechanisms in *Hordeum* by comparing 875 individuals covering 31 species of this genus. Jakob and Blattner (2006) revealed the phylogenetic and genealogical relationships of 88 chloroplast haplotypes and, by performing genealogical (network) analyses, identified the biogeographical patterns and indicated processes involved in speciation in *Hordeum*.

20.3.3 Barley Mutants for Chloroplast Biogenesis

Barley is a model plant in the molecular physiological study of chloroplast biogenesis. Several mutant lines of barley with chloroplast dysfunction are available, such as chloroplast mutator (*cpm*), *albostrians*, Okina-mugi, Saskatoon, striata, etc. (Greiner 2012).

The barley chloroplast mutator (*cpm*) induces several types of maternally inherited chlorophyll deficiencies and is controlled by a nuclear gene allele (Prina 1992; Prina et al. 2009). The characterization of cp genomes in several mutant lines induced by *cpm* revealed that *cpm* induced sequence polymorphisms in the cp genome, most of which are point mutations, such as single base pair transitions and insertions (Landau et al. 2016). Mutants induced by *cpm* are useful for examining the functions of genes encoded in the cp genome. For example, the *cpm* mutant lines for *infA*, which encodes translation initiation factor 1, and *ycf3*, that encodes a chaperone involved in PSI assembly, have been physiologically and molecular genetically characterized (Landau et al. 2007, 2009, 2011). These mutant lines possessed point mutations resulting in changes in the conserved amino acid residue or defects in the intron splicing of their target genes, which are the primary causes of the chlorophyll deficiencies observed in these mutants. Thus, molecular changes observed in the cp genome are valuable sources of chloroplast variability, and the *cpm* mutants are good experimental materials to investigate the mechanism of chloroplast biogenesis.

The *albostrians* mutation in barley is caused by a recessive nuclear allele (Hagemann and Scholz 1962) and is characterized by green, white, and striped leaves with identical genotypes. Plastids in white leaves and the white parts of striped leaves are due to ribosome deficiency and the lack of all plastid-encoded proteins;

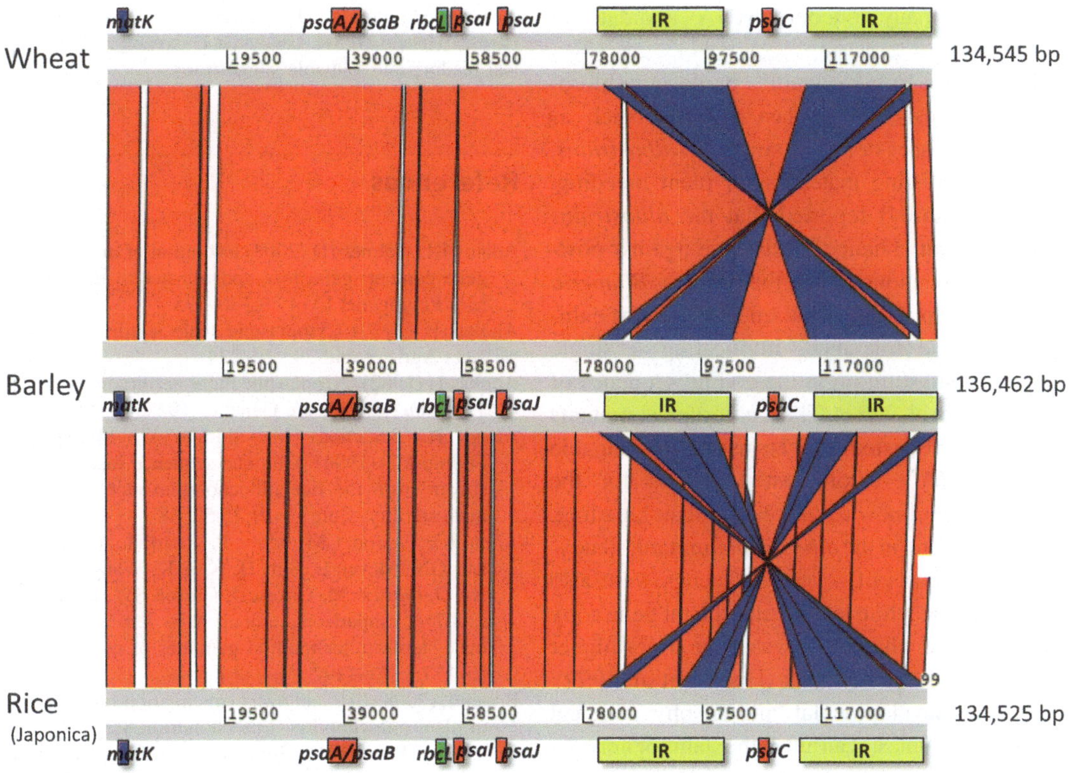

Fig. 20.3 Relative arrangement of homologous sequences in the three cp genomes in cereal plants, i.e., barley (EF115541), wheat (AB042240), and rice (X15901). Homologous regions were identified by performing 2 BLAST analyses using MegaBLAST for each pair of species. Red and blue colors represent the forward and reverse orientations, respectively, of homologous sequences. The relative genomic positions of certain genes and regions are indicated by a box; IRs (yellow), photosystem I genes (red; *psaA*, *psaB*, *psaC*, *psaI*, and *psaJ*), *rbcL* (green) and *matK* (blue)

thus, these plastids are photosynthetically inactive and have only rudimentary membrane systems (Hess et al. 1994a, b). However, in contrast to the *cpm* mutants, the DNA in the affected plastids exhibits a wild-type restriction pattern and is replicated to normal amounts, suggesting that the cp genome in the mutant is intact (Hess et al. 1993). However, the transcription of several photosynthesis-associated chloroplast and nuclear genes are extremely inhibited in the white parts, indicating the presence of signal pathways from chloroplast to nucleus that control nuclear gene expression and vice versa (Hess et al. 1994a). The inhibition of gene expression in the *albostrians* mutant reflects a lack of plastid-encoded plastid RNA polymerase (PEP), and the cp genome contains functional genes coding for the core subunits of PEP. A second polymerase, i.e., nuclear-encoded plastid RNA polymerase (NEP), has also been implicated in plastid transcription. Therefore, plastid transcription in the white parts of *albostrians* is exclusively achieved by NEP, while both PEP and NEP are involved in the transcription in the green part. Based on these observations, Zhelyazkova et al. (2012) obtained a genome-wide map of transcription start sites (TSSs) in the plastids of barley mature leaves. Numerous TSSs were obtained, suggesting that more promoters than genes are present in the cp genome, and most genes have both PEP and NEP promoters. Furthermore, many TSSs mapped to the intergenic regions and opposite strands of the annotated genes, which might indicate the potential existence of many noncoding RNA candidates in chloroplasts.

In addition to basic studies exploring chloroplast biogenesis, chloroplast mutants, such as *cpm*-derived ones or alleles similar to *albostrians*, may be interesting materials for plant breeding. Jain et al. (2004) reported that the *albostrians* mutant has an enhanced resistance against powdery mildew induced by the fungus *Blumeria graminis* f. sp. *hordei*, and an *cpm* mutant exhibits herbicide tolerance, implicating a single nucleotide substitution in the coding sequence of *psbA* (Rios et al. 2003). Thus, chloroplast mutants or the mutant system can be used as a valuable tool to modify chloroplast functions for the development of new resources for plant breeding.

Furthermore, in the *albostrians* mutant, steady-state RNA levels of several mt genes were also elevated in the white leaves, and the mt gene copy number in the white parts was three times higher than that in the green parts (Hedtke et al. 1999). Thus, the developmental and/or physiological status of chloroplasts can affect the status of not only nuclear but also mt genomes in gene copy number and/or gene expression, and inter-organellar crosstalk among these three organelles occurs frequently not only in the case of dysfunction but also in normal plant development.

20.4 Future Prospects

The coordinated gene expression among the three genomes present in the nucleus, mitochondrion, and chloroplast is highly important for normal development in plants. CMS caused by mt genome and chloroplast dysfunctions induced by nuclear-encoded genes are good examples of the close interactions among the genes encoded in the three genomes in barley. The identification of the signals involved in the coordination of the three organelles will increase our understanding of how inter-organellar crosstalk is achieved. Currently, the complete genomes of these three organelles in barley are available, enabling the clarification of their functions in barley cells and furthering the current understanding of the precise characteristics of this important crop.

Acknowledgements The author would like to thank Hiroshi Hisano (IPSR, Okayama University) for the critical reading and valuable suggestions.

References

Adams KL, Palmer JD (2003) Evolution of mitochondrial gene content: gene loss and transfer to the nucleus. Mol Phylogenet Evol 29:380–395

Ahokas H (1979) Cytoplasmic male sterility in barley. Acta Agric Scand 29:219–224

Ahokas H (1982) Cytoplasmic male sterility in barley. Xi. The *msm2* cytoplasm. Genetics 102:285–295

Akagi H, Sakamoto M, Shinjyo C, Shimada H, Fujimura T (1994) A unique sequence located downstream from the rice mitochondrial atp6 may cause male sterility. Curr Genet 25:52–58

Allen JO, Fauron CM, Minx P, Roark L, Oddiraju S, Lin GN, Meyer L, Sun H, Kim K, Wang C, Du F, Xu D, Gibson M, Cifrese J, Clifton SW, Newton KJ (2007) Comparisons among two fertile and three male-sterile mitochondrial genomes of maize. Genetics 177:1173–1192

Bonen L (2008) *Cis*- and *trans*-splicing of group II introns in plant mitochondria. Mitochondrion 8:26–34

Bonen L, Boer PH, Gray MW (1984) The wheat cytochrome oxidase subunit II gene has an intron insert and three radical amino acid changes relative to maize. EMBO J 3:2531–2536

Byers E, Rueger J, Bonen L (2010) Impact of genomic environment on mitochondrial *rps7* mRNA features in grasses. Mol Genet Genomics 284:207–216

Clifton SW, Minx P, Fauron CM, Gibson M, Allen JO, Sun H, Thompson M, Barbazuk WB, Kanuganti S, Tayloe C, Meyer L, Wilson RK, Newton KJ (2004) Sequence and comparative analysis of the maize NB mitochondrial genome. Plant Physiol 136:3486–3503

Daniell H, Lin C-S, Yu M, Chang W-J (2016) Chloroplast genomes: diversity, evolution, and applications in genetic engineering. Genome Biol 17:134

Dewey RE, Timothy DH, Levings CS (1987) A mitochondrial protein associated with cytoplasmic male sterility in the T cytoplasm of maize. Proc Natl Acad Sci USA 84:5374–5378

Fallahi M, Crosthwait J, Calixte S, Bonen L (2005) Fate of mitochondrially located S19 ribosomal protein genes after transfer of a functional copy to the nucleus in cereals. Mol Gen Genet 273:76–83

Greiner S (2012) Plastome mutants of higher plants. In: Bock R, Knoop V (eds) Genomics of chloroplasts and mitochondria, advances in photosynthesis and respiration, vol 35. Springer, The Netherlands, pp 237–266

Gualberto JM, Mileshina D, Wallet C, Niazi AK, Weber-Lotfi F, Dietrich A (2014) The plant mitochondrial genome: dynamics and maintenance. Biochimie 100:107–120

Hagemann R, Scholz F (1962) Ein Fall Gen-induzierter Mutationen des Plasmotyps bei Gerste. Der Züchter 32:50–59

Handa H (2003) The complete nucleotide sequence and RNA editing content of the mitochondrial genome of rapeseed (*Brassica napus* L.): comparative analysis of the mitochondrial genomes of rapeseed and Arabidopsis thaliana. Nucl Acids Res 31:5907–5916

Hanson M, Bentolila S (2004) Interactions of mitochondrial and nuclear genes that affect male gametophyte development. Plant Cell 16:S154–S169

Hedtke B, Wagner I, Börner T, Hess WR (1999) Inter-organellar crosstalk in higher plants: impaired chloroplast development affects mitochondrial gene and transcript levels. Plant J 19:635–643

Hess WR, Hoch B, Zeltz P, Hübschmann T, Kössel H, Börner T (1994a) Inefficient *rpl2* splicing in barley mutants with ribosome-deficient plastids. Plant Cell 6:1455–1465

Hess WR, Müller A, Nagy F, Börner T (1994b) Ribosome-deficient plastids affect transcription of light-induced nuclear genes: genetic evidence for a plastid-derived signal. Mol Gen Genet 242:305–312

Hess WR, Prombona A, Fieder B, Subramanian AR, Börner T (1993) Chloroplast *rps15* and the *rpoB/C1/C2* gene cluster are strongly transcribed in ribosome-deficient plastids: evidence for a functioning non-chloroplast-encoded RNA polymerase. EMBO J 12:563–571

Hisano H, Tsujimura M, Yoshida H, Terachi T, Sato K (2016) Mitochondrial genome sequences from wild and cultivated barley (*Hordeum vulgare*). BMC Genom 17:824

Hiratsuka J, Shimada H, Whittier R, Ishibashi T, Sakamoto M, Mori M, Kondo C, Honji Y, Sun CR, Meng BY, Li YQ, Kanno A, Nishizawa Y, Hirai A, Shinozaki K, Sugiura M (1989) The complete sequence of the rice (*Oryza sativa*) chloroplast genome: intermolecular recombination between distinct tRNA genes accounts for a major plastid DNA inversion during the evolution of the cereals. Mol Gen Genet 217:185–194

Iwabuchi M, Kyozuka J, Shimamoto K (1993) Processing followed by complete editing of an altered mitochondrial *atp6* RNA restores fertility of cytoplasmic male sterile rice. EMBO J 12:1437–1446

Jacoby RP, Li L, Huang S, Lee CP, Millar AH, Taylor NL (2012) Mitochondrial composition, function and stress response in plants. J Integr Plant Biol 54:887–906

Jain SK, Langen G, Hess W, Börner T, Hüchkelhoven R, Kogel K-H (2004) The white barley mutant *albostrians* shows enhanced resistance to the biotroph *Blumeria graminis* f. sp. *hordei*. Mol Plant-Microbe Interact 17:374–382

Jakob SS, Blattner FR (2006) A chloroplast genealogy of *Hordeum* (Poaceae): long-term persisting haplotypes, incomplete lineage sorting, regional extinction, and the consequences for phylogenetic inference. Mol Biol Evol 23:1602–1612

Jensen PE, Leister D (2014) Chloroplast evolution, structure and functions. F1000Prime Rep 6:40

Kao T-H, Moon E, Wu R (1984) Cytochrome oxidase subunit II gene of rice has an insertion sequence within the intron. Nucl Acids Res 12:7305–7315

Kazama T, Toriyama K (2016) Whole mitochondrial genome sequencing and re-examination of a cytoplasmic male sterility-associated gene in Boro-Taichung-type cytoplasmic male sterile rice. PLoS ONE 11:e0159379

Kubo N, Salomon B, Komatsuda T, von Bothmer R, Kadowaki K (2005) Structural and distributional variation of mitochondrial *rps2* genes in the tribe Triticeae (Poaceae). Theor Appl Genet 110:995–1002

Kubo T, Newton KJ (2008) Angiosperm mitochondrial genomes and mutations. Mitochondrion 8:5–14

Landau A, Díaz Paleo A, Civitillo R, Jaureguialzo M, Prina AR (2007) Two *infA* gene mutations independently originated from a mutator genotype in barley. J Hered 98:272–276

Landau A, Lencina F, Pacheco MG, Prina AR (2016) Plastome mutations and recombination events in barley chloroplast mutator seedlings. J Hered 107:266–273

Landau AM, Lokstein H, Scheller HV, Lainez V, Maldonado S, Prina AR (2009) A cytoplasmically inherited barley mutant is defective in photosystem I assembly due to a temperature-sensitive defect in *ycf3* splicing. Plant Physiol 151:1802–1811

Landau AM, Pacheco MG, Prina AR (2011) A second *infA* plastid gene point mutation shows a compensatory effect on the expression of the cytoplasmic line 2 (CL2) syndrome in barley. J Hered 102:633–639

Maier RM, Neckermann K, Igloi GL, Kossel H (1995) Complete sequence of the maize chloroplast genome: gene content, hotspots of divergence and fine tuning of genetic information by transcript editing. J Mol Biol 251:614–628

Mascher M, Gundlach H, Himmelbach A, Beier S, Twardziok SO, Wicker T, Radchuk V, Dockter C, Hedley PE, Russellet J et al (2017) A chromosome conformation capture ordered sequence of the barley genome. Nature 544:427–433

Middleton CP, Senerchia N, Stein N, Akhunov ED, Keller B, Wicker T (2014) Sequencing of chloroplast genomes from wheat, barley, rye and their relatives provides a detailed insight into the evolution of the Triticeae tribe. PLoS ONE 9:e85761

Nishikawa T, Salomon B, Komatsuda T, von Bothmer R, Kadowaki K (2002) Molecular phylogeny of the genus *Hordeum* using three chloroplast DNA sequences. Genome 45:1157–1166

Notsu Y, Masood S, Nishikawa T, Kubo N, Akiduki G, Nakazono M, Hirai A, Kadowaki K (2002) The complete sequence of the rice (*Oryza sativa* L.) mitochondrial genome: frequent DNA sequence acquisition and loss during the evolution of flowering plants. Mol Genet Genomics 268:434–445

Ogihara Y, Isono K, Kojima T, Endo A, Hanaoka M, Shiina T, Terachi T, Utsugi S, Murata M, Mori N, Takumi S, Ikeo K, Gojobori T, Murai R, Murai K, Matsuoka Y, Ohnishi Y, Tajiri H, Tsunewaki K

(2002) Structural features of a wheat plastome as revealed by complete sequencing of chloroplast DNA. Mol Genet Genomics 266:740–746

Ogihara Y, Yamazaki Y, Murai K, Kanno A, Terachi T, Shiina T, Miyashita N, Nasuda S, Nakamura C, Mori N, Takumi S, Murata M, Futo S, Tsunewaki K (2005) Structural dynamics of cereal mitochondrial genomes as revealed by complete nucleotide sequencing of the wheat mitochondrial genome. Nucleic Acids Res 33:6235–6250

Ong HC, Palmer JD (2006) Pervasive survival of expressed mitochondrial *rps14* pseudogenes in grasses and their relatives for 80 million years following three functional transfers to the nucleus. BMC Evol Biol 6:55

Palmer JD, Herbon LA (1988) Plant mitochondrial DNA evolves rapidly in structure, but slowly in sequence. J Mol Evol 28:87–97

Prina AR (1992) A mutator nuclear gene inducing a wide spectrum of cytoplasmically inherited chlorophyll deficiences in barley. Theor Appl Genet 85:245–251

Prina AR, Landau A, Colombo N, Jaureguialzo M, Arias MC, Rios RD, Pacheco MG (2009) Genetically unstable mutants as novel sources of genetic variability: the chloroplast mutator genotype in barley as a tool for exploring the plastid genome. In: Shu QY (ed) Induced plant mutations in the genomics era. FAO, Rome, pp 255–256

Rios RD, Saione H, Roberdo C, Acevedo A, Colombo N, Prina AR (2003) Isolation and molecular characterization of atrazine tolerant barley mutants. Theor Appl Genet 106:696–702

Rish SY, Breiman A (1993) RNA editing of the barley (*Hordeum vulgare*) mitochondrial ATP synthase subunit 9. Plant Mol Biol 22:711–714

Saski C, Lee SB, Fjellheim S, Guda C, Jansen RK, Luo H, Tomkins J, Rognli OA, Daniell H, Clarke JL (2007) Complete chloroplast genome sequences of *Hordeum vulgare*, *Sorghum bicolor* and *Agrostis stolonifera*, and comparative analyses with other grass genomes. Theor Appl Genet 115:571–590

Takenaka M, Verbitskiy D, van der Merwe JA, Zehrmann A, Brennicke A (2008) The process of RNA editing in plant mitochondria. Mitochondrion 8:35–46

Ui H, Sameri M, Pourkheirandish M, Chang M-C, Shimada H, Stein N, Komatsuda T, Handa H (2015) High-resolution genetic mapping and physical map construction for the fertility restorer Rfm1 locus in barley. Theor Appl Genet 128:283–290

Zabala G, Gabay-Laughnan S, Laughnan JR (1997) The nuclear gene *Rf3* affects the expression of the mitochondrial chimeric sequence R implicated in S-type male sterility in maize. Genetics 147:847–860

Zhelyazkova P, Sharma CM, Förstner KU, Liere K, Vogel J, Börner T (2012) The primary transcriptome of barley chloroplasts: numerous noncoding RNAs and the dominating role of the plastid-encoded RNA polymerase. Plant Cell 24:123–136

Databases and Tools for the Analysis of the Barley Genome

21

Sebastian Beier, Daniel M. Bolser, Uwe Scholz, Manuel Spannagl and Paul J. Kersey

Abstract

Ever-increasing advances seen in barley genome sequencing over the last years have enabled scientists to generate databases and tools specially designed in helping researchers and breeders. Both genomic as well as expressed sequences were obtained through various experimental setups ranging from BAC sequencing over Illumina iSelect 9k SNP chip to RNA-Seq to form heterogeneous datasets. Where possible datasets were cross-linked and enriched in information to build a basis for further research. Ensembl Plants, a web portal designed for exploring genomic data for various plant species, have been utilized to explore differences and similarities between barley and its related species. Furthermore, the barley genome explorer BARLEX was constructed to be the central repository and hub of genomic sequences of barley sequencing efforts. Powerful visualizations of interconnected BACs and other sequencing information enable to backtrack every position that makes up the barley reference sequence and help in understanding the connection to other datasets. Further tools utilizing other barley data are discussed and described for more specialized use cases. Last but not least a list of URLs is given for a comprehensive overview of barley-centric resources.

S. Beier · U. Scholz (✉)
Leibniz Institute of Plant Genetics and Crop Plant Research (IPK) Gatersleben, Corrensstr. 3, 06466 Seeland, Germany
e-mail: scholz@ipk-gatersleben.de

S. Beier
e-mail: beiers@ipk-gatersleben.de

D. M. Bolser · P. J. Kersey
European Molecular Biology Lab, The European Bioinformatics Institute, Wellcome Trust Genome Campus, CB10 1SD Hinxton, Cambridge, UK
e-mail: dbolser@ebi.ac.uk

P. J. Kersey
e-mail: pkersey@ebi.ac.uk

M. Spannagl
Plant Genome and Systems Biology (PGSB), Helmholtz Center Munich, 85764 Neuherberg, Germany
e-mail: manuel.spannagl@helmholtz-muenchen.de

Abbreviations

AA	Amino acid
AGP	A golden path (filetype)
AHRD	Automatic assignment of human-readable description
BAC	Bacterial artificial chromosome
BLAST	Basic local alignment search tool
CDS	Coding sequence
cDNA	Copy DNA
CSH	Cold Spring Harbor laboratory
DNA	Deoxyribonucleic acid
EBI	European Bioinformatics Institute
EI	Earlham Institute (formerly TGAC)

N. Stein and G. J. Muehlbauer (eds.), *The Barley Genome*, Compendium of Plant Genomes,
https://doi.org/10.1007/978-3-319-92528-8_21

EST	Expressed sequence tag
FPC	Fingerprinted contig
FTP	File transfer protocol
GO	Gene ontology
HC	High confidence
HTML	Hypertext markup language
IBSC	International Barley Genome Sequencing Consortium
IPK	Leibniz Institute of Plant Genetics and Crop Plant Research
Iso-Seq	Isoform sequencing
JHI	James Hutton Institute
LC	Low confidence
PGSB	Plant Genome and Systems Biology unit at the Helmholtz Center Munich
QTL	Quantitative trait loci
RNA	Ribonucleic acid
RNA-Seq	RNA sequencing
SNP	Single nucleotide polymorphism
TAIR	The Arabidopsis Information Resource
TCAP	Triticeae Coordinated Agricultural Project
TGAC	The Genome Analysis Centre
WGS	Whole genome shotgun
XML	Extensible markup language

21.1 Introduction

The molecular characterization of the barley genome has advanced rapidly in recent years, with the publication of successively more complete assemblies of the barley genome reference sequence (International Barley Genome Sequencing International Barley Genome Sequencing Consortium et al. 2012; Mascher et al. 2017). These assemblies are written as strings composed from a four characters alphabet, representing the succession of nucleotide bases that comprise a strand of DNA. But additional information is required to use these data for the understanding (and potential manipulation) of biological processes, and allow for the identification of functional elements, particularly the sequences of other individuals, populations and species—with patterns of conservation (and change) indicating likely functional roles (and their scope across the taxonomy). But other types of information are also critical—including expression data (directly indicating which elements of the genome are used, and when), and phenotypic observations (which typically result from the interaction of the genome and the environment). The resulting data sets are often very large, so to accelerate their search and interpretation, a number of resources have been developed, providing access to data and implementations of common algorithms for data analysis. Such tools allow researchers to quickly summarize what is known (and what, even if not previously explicitly stated, is easily inferable) and formulate new hypotheses (or, in the case of barley breeders, plan a breeding program), in full knowledge of the relevant facts.

The dominant paradigm for interactive exploration of genome-scale data is a software platform known as a 'genome browser', typically accessible via the World Wide Web. Genome browsers represent chromosomes (or chromosome fragments) as linear sequence (i.e. using the primary structure of the DNA polymer as the organizing principle), and allow users to zoom and scroll through each region of contiguous sequence (treating it as a one-dimensional map). Functional elements in the genome (e.g. genes, promoters) are marked, as are elements that code for functional features in molecules transcribed from the genome (e.g. splice sites, translational start sites, etc.). In addition to this positional information, additional views within the application provide non-positional information (e.g. names, functional descriptions, hyperlinks to other web resources representing the same object or related objects). As the reference barley genome assembly has become steadily more complete, it has become more suitable for dissemination through such a tool.

Any large map needs an index in order to be useful. Two sorts of index are commonly used in association with genome browsers (and other bioinformatics tools). First, there are tools that index metadata, associated with the sequence and annotation. These tools, often based on common text-search algorithms (e.g. Apache Lucene), but are more powerful if the data is structured, which allows the development of more advanced search tools and reduces the probability that interesting data is missed by accident, because it happens to have been described in an unexpected way. This observation has driven efforts by data generators and custodians to agree on common data formats, grammars and structured vocabularies to express this information. Second, the actual sequence can be indexed, allowing users to find portions of the genome that match (exactly or approximately) to a query sequence already of interest. As molecular sequences can be represented as character strings, essentially these are text-matching algorithms refined to better meet biological use cases (by favouring alignments that are more likely to indicate the evolution and function of the sequences), e.g. (Needleman and Wunsch 1970; Smith and Waterman 1981; Altschul et al. 1990). Either type of search can be used to find a region of interest within the genome, which users can then explore using the browser and associated functionality.

Other tools are also important for accessing genome data. Specialized tools provide the evidence supporting the construction of the genome assembly, e.g. the sequencing of BACs along a minimal tiling path or the creation procedure of pseudomolecules. Where contiguous sequence is unavailable, genetic information can provide a scaffold (and indeed, the language of the 'genetic map' precedes the development of software visualizations). Where sequence is missing from the genome, transcriptome data may be searched to locate missing genes. In addition, certain large-scale analyses may provide vital context for interpretation. Numerous algorithms apply the principles of sequence alignment to propose functions and histories for families of related genes, and to identify events that have had causative impact on phenotypes and traits such as gene sequence mutation, duplication and extinction, defunctionalization and neofunctionalization. Data sets linked to such algorithms are available as part of many resources, as well as tools that allow users to perform such analyses with a combination of private data and public reference resources.

Of particular relevance to the barley genome is the fact that many related species have also had their genomes sequenced, providing a wealth of interpretable context. These include other species from within the grass family (*Poaceae*) closer relatives from the *Triticeae* tribe, including other important crops. In addition, genomic data is also available from an increasing number of cultivated and wild barleys, which can (in the most part) be organized using the reference genome sequence as an anchor.

21.2 Overview of Barley Resources

21.2.1 Ensembl Plants and Gramene

Ensembl Plants (http://plants.ensembl.org) (Kersey et al. 2016) is a web portal offering access to genomic data for 45 different plant species, at the moment of this writing, including barley. Current and previous barley genome assemblies are available through the site and can be explored using a genome browser interface. The portal also contains several additional tools for exploring the data, and in addition, provision is made for accessing the data programmatically, in bulk, and through fast, specialized interfaces supporting frequent use cases. Ensembl Plants manages a joint data set and common interface tools with the Gramene database (http://www.gramene.org) (Tello-Ruiz et al. 2016), and the same data can be accessed through either site.

Available data types include assembled genome sequence, protein-coding and nonprotein-coding gene structures, protein domain analysis and functional classification, and comparative analysis amongst related genomes. The platform also supports genetic variation data, which can be linked to populations and phenotypes, or analysed on the fly. New genome assemblies are often originally made available first via a prerelease site (http://pre.plants.ensembl.org), before subsequent full integration into the resource. Older assembly versions remain available after release on an archive site (http://

archive.plants.ensembl.org). As of August 2017, the site currently includes two versions of the barley genome assembly, including an improved version of the IBSC_1.0 assembly (International Barley Genome Sequencing International Barley Genome Sequencing Consortium et al. 2012) (with additional information used to scaffold the contigs derived from population genomics (Ariyadasa et al. 2014; Mascher et al. 2013a) (located on the main site) and the new assembly (Mascher et al. 2017) (located on the pre-site). An assembly of the chloroplast genome (Middleton et al. 2014) has also been included in both assemblies. When the new assembly migrates into the main site, the previous assembly will remain visible in the archive.

Specific data for barley within Ensembl Plants includes mappings to the genome of the probe sets from Barley1 GeneChip array, the Agilent barley full-length cDNA array, and the barley PGRC1 10k A and B arrays. Transcriptomic data has also been aligned from 21 experiments in barley. Variation has been identified from five sources: the WGS survey sequencing of four cultivars (Barke, Bowman, Igri, Haruna Nijo and a wild barley (*H. spontaneum*)) (International Barley Genome Sequencing International Barley Genome Sequencing Consortium et al. 2012); RNA-Seq performed on the embryo tissues of 9 spring barley varieties (Barke, Betzes, Bowman, Derkado, Intro, Optic, Quench, Sergeant and Tocada) and Morex using Illumina HiSeq 2000 (International Barley Genome Sequencing International Barley Genome Sequencing Consortium et al. 2012); ~5 million variations from population sequencing of 90 Morex × Barke individuals (Mascher et al. 2013a); ~6 million variations from population sequencing of 84 Oregon Wolfe barley individuals (Mascher et al. 2013a); and SNPs from the Illumina iSelect 9k barley SNP chip. 2600 markers associated with the iSelect SNPs are also displayed (Comadran et al. 2012). The browser also provides access to alignments to the barley genome of transcriptomic data from related organisms, including RNA-Seq, EST and Unigene data sets derived from *Triticum aestivum*, and RNA-Seq assemblies of two subspecies of *Triticum monococcum*.

Within Ensembl Plants, barley protein sequences are analysed using the domain and family classification tool InterProScan (Finn et al. 2017), and from the domain assignments, functional annotation is derived, expressed in the language of the Gene Ontology (Gene Ontology Gene Ontology Consortium 2015). In addition, with each release, protein sequences from the entirety of Ensembl Plants are clustered, a tree is constructed from each cluster, and each tree is compared with the species history to infer gene duplication and deletion events and thus the existence of true orthologues and paralogues within each cluster (Vilella et al. 2009). In addition, whole genome alignments have been performed between barley and several closely related species, including *Aegilops tauschii*, *Brachypodium distachyon*, *Oryza sativa* (Japonica group), and *Triticum urartu*.

21.2.2 BARLEX—A Network-Based Genome Explorer for Barley Assembly Components

Genome projects usually combine a large amount of heterogeneous data types that in a final step are combined into a single resource. Although intermediate products of genome sequencing harbour untapped potential and evidence of biological interest, they are rarely reported in scientific literature. Often just a simple linear sequence of pseudomolecules is reported as the final product of a genome sequencing endeavour. For complex organisms like the barley genome, constructing a linear sequence is a considerable piece of work. To the author's knowledge, this has been the first genome to combine such a large diversity of data sources for a complex genome. Among the genome size, the high amount of repetitive content makes sequencing and assembling the barley genome a tremendous challenge. Short read sequencing alone cannot overcome this problem so the international barley sequencing consortium had to facilitate many different techniques and combine numerous data sets to finally arrive at the destination of presenting the first linear ordered reference sequence of barley. These resources cover physical (Ariyadasa et al. 2014) and genetic maps (Mascher et al. 2013a), several libraries of bacterial artificial chromosomes (Mascher et al. 2013a; Schulte et al. 2009), Sanger-sequenced BAC ends, chromosome conformation capture sequences, the gene centric draft genome based on whole genome shotgun sequencing (Schulte et al. 2009; Mascher et al. 2013a; International Barley Genome Sequencing International Barley Genome Sequencing Consortium et al. 2012), exome capture data (Mascher et al. 2013b) and expression profile based on RNA-Seq and Iso-Seq.

As a complement to the linear genome browser Ensembl Plants, a network-based view for the visualization of the genome increases the usage possibilities of the barley resources. With this network illustration, the high interconnectivity between different approaches, like BAC sequencing, genetic maps, exome capture data or other resources could be illustrated. The web-based application BARLEX (Colmsee et al. 2015) was developed to fulfil these requirements.

For this endeavour, a powerful Oracle database and the Oracle Application Express (APEX) software with the graphical plugin Cytoscape Web (Lopes et al. 2010) was combined. The plugin enabled us to visualize graph networks. Each edge and each node can be dragged around, giving the user the possibility to entangle complex graph structures within the given boundaries of the application and page. Tooltips and many interlinked pages based on sequencing effort can be accessed and navigated in an intuitive way.

The backbone of BARLEX is the minimal tiling path of BACs that was utilized to sequence the genome of barley. Each BAC clone has information based on its assembly such as the number of contigs, complete sequenced size of individual BAC assemblies and other key figures displayed in a tabular manner. In the process of fingerprinting, the whole collection of BACs so-called fingerprinted contigs were assembled with the utilization of restriction enzyme digestion (Soderlund et al. 1997). As a possibility to compare the sequencing effort with these physical contigs, sequenced BACs were screened for homology with megaBLAST (Zhang et al. 2000). With this concept, sequence clusters were formed and are visualized directly beside the physical contig for comparison.

Currently, the information of 87 075 BAC clones (consisting of minimal tiling path clones from all seven chromosomes and gene-bearing clones (Munoz-Amatriain et al. 2015) sequenced independently) are available in the BARLEX database. Two different gene sets were imported into BARLEX as well. The first published gene set based on several RNA-Seq experiments, cDNA libraries and the WGS assembly published by the International Barley Sequencing Consortium. The most recent gene set comprises of 39 734 high confidence and 41 949 low confidence genes. Additional links to Gene Ontology (Ashburner et al. 2000), InterPro (Finn et al. 2017) and Pfam (Finn et al. 2016) were also imported. Expression profiles for 16 developmental stages as well as exome capture targets have been integrated into BARLEX too.

Repeat annotation of mobile elements based on the pseudomolecule sequence has been integrated and gives a first overview over the composition of these elements of the complete barley genome. Sequences were screened against the REdat_9.7_Poaceae section of the PGSB transposon library (Spannagl et al. 2016) with vmatch (Kurtz 2003).

Inspecting of sequence complexity can also be done by looking at *K*-mer frequencies over the length of sequenced BAC contigs and is visualized insight BARLEX with the Kmasker software (Schmutzer et al. 2014). This particular visualization shows how the sequence composition is represented by short *K*-mers common to the barley genome. High abundance sequences will show a high occurrence in common *K*-mers while unique or low copy sequence will show a low occurrence of common *K*-mers.

Several entry points for BARLEX were set up. Fingerprinted contigs, sequence cluster, and BACs can be looked up with their names or their positions (genetic/physical). Likewise, a table of annotated genes can be searched with gene identifiers or the names of the sequence contig carrying them and is directly linked to the BAC information. All BLAST results underlying the links between BACs and contigs are also accessible as searchable tables with hyperlinks to information pages. Marker information from the 9k Illumina iSelect SNP chip were mapped to the reference sequence and can be accessed in BARLEX (Comadran et al. 2012).

Finally, direct links to the barley BLAST server and a portlet connecting BARLEX to the LAILAPS search engine (Lange et al. 2009) are embedded on the front page of BARLEX.

21.2.3 IPK Barley BLAST Server

For homology-based searches, several BLAST databases (Deng et al. 2007) have been set up at the IPK. They include the published draft assemblies of the barley cultivars Morex, Barke and Bowman (International Barley Genome Sequencing International Barley Genome Sequencing Consortium et al. 2012), as well as multiple special purpose databases (e.g. exome capture, BAC end sequences, full-length cDNA). Furthermore, the new pseudomolecule sequence and all associated resources are presented in this database as well. This includes the pseudomolecule sequence of the reference cultivar Morex, the annotated gene prediction (high- and low confidence gene sets) and the individual BAC assemblies used for the construction of the pseudomolecule sequence. These can be accessed via http://webblast.ipk-gatersleben.de/barley (Tables 21.1 and 21.2).

21.2.4 PGSB/MIPS PlantsDB Resources for Barley Genome Data

PGSB (Plant Genome and Systems Biology; formerly MIPS—Munich Institute for Protein Sequences) PlantsDB (Spannagl et al. 2016) is a database framework for the integration and analysis of plant genome data, with a major focus on comparative genomics and intuitive user navigation. PlantsDB is maintained over a decade now and stores and integrates data for individual (reference) genomes from both model and crop plants. Comparative genomics and analysis tools include CrowsNest, a viewer to explore conserved gene order (synteny), as well as RNASeqExpressionBrowser. The Repeat Element Database (pgsb-REdat) and the Repeat Element Catalog (pgsb-REcat) provide access to repetitive elements from various plant genomes, including a classification scheme. Data exchange with partners as well as integrated search functionality is facilitated using standards and technology developed within the transPLANT project.

PGSB PlantsDB can be accessed at http://pgsb.helmholtz-muenchen.de/plant/genomes.jsp.

21.2.5 PGSB PlantsDB Data Content and Access

PlantsDB currently stores and integrates genome data for the plant model organisms *Arabidopsis thaliana*, *Oryza sativa* (Rice), *Medicago truncatula*, *Zea mays* (Maize), *Solanum lycopersicum* (Tomato) and many more. To accommodate data and analysis results from the complex genomes of barley, wheat and rye, an instance specifically attributed to Triticeae genomes was generated within PlantsDB.

To access data interactively within PlantsDB, users navigate the database in a genome-oriented way. For every plant species, a dedicated database instance has been set up and can be accessed from the entries' page phylogenomic tree. Every genome database features the same basic navigation structure, with views and interfaces for many genetic elements such as genes, transposable elements, noncoding RNA, sequence contigs and many more. Genetic element reports, representing detailed information on particular genes, provide download options in HTML, XML or FASTA format. Cross-references were set up to access entries in external databases associated with the entry (e.g. links to the corresponding gene report in Ensembl Plants (Bolser et al. 2017) or BARLEX (Colmsee et al. 2015)). Other options to access data in PlantsDB include a generalized search function, a sequence

Table 21.1 Nucleotide databases relating to genome sequencing of barley

Dataset	No. of sequences	Complete size [bp]	Largest sequence [bp]	Mean sequence size [bp]	N50 [bp]
Barley CDS HC Mai2016	39,734	38,486,703	15,048	969	1404
Barley CDS LC Mai2016	41,545	16,162,413	7137	389	453
Barley representative Transcripts HC (including introns) Mai2016	39,734	59,562,803	19,746	1499	2068
Barley representative Transcripts LC (including introns) Mai2016	41,949	39,770,363	26,603	948	1362
Barley Genomic (start at 1st exon end of last exon) HC Genes Mai2016	39,734	238,841,754	820,598	6011	18,143
Barley Genomic (start at 1st exon end of last exon) LC Genes Mai2016	40,819	92,975,688	1059,687	2278	7865
Barley Pseudomolecules Masked Apr2016	8	4,833,791,107	–	–	–
Barley Pseudomolecules Aug2015	8	4,833,791,107	–	–	–
Barley Pseudomolecule Contigs Masked Apr2016	464,895	4,787,302,407	297,092	10,298	79,239
Barley Pseudomolecule Contigs Aug2015	464,895	4,787,302,407	297,092	10,298	79,239
Barley BAC Assemblies Aug2015	850,266	11,303,595,359	467,463	13,294	60,140
Assembly_WGSMorex	2,670,738	1,869,516,600	36,084	700	1425
Assembly_WGSBarke	2,742,077	2,018,168,672	38,386	736	1419
Assembly_WGSBowman	2,077,901	1,778,683,256	37,442	856	1986
454BacContigs	86,251	512,675,944	181,550	5,944	34,519
BacEndSequences	571,814	373,394,542	1004	653	723
HC_genes_CDS_Seq	26,159	31,892,184	14,874	1219	1539
LC_genes_CDS_Seq	53,220	16,996,719	5442	319	372
IlluminaBacContigs	2183	277,159,242	280,571	126,962	129,725
SortedChromosomes	20,478,866	7,361,557,118	772	359	419
Full-length cDNA	28,622	47,571,668	7384	1662	1897
ipk 206,633 barley ESTs	206,633	106,723,244	770	516	567
Exome Capture Regions 10x	168,097	48,359,102	7426	287	568
Exome Capture Regions 5x	207,448	67,701,329	9125	326	673
Barley Agilent Array	46,848	40,057,878	7364	855	1033

Table 21.2 Protein databases relating to genome sequencing of barley

Dataset	No. of proteins	Largest protein [aa]	Mean protein size [aa]
Barley HC Proteins Mai2016	39,734	5016	323
Barley LC Proteins Mai2016	41,545	2379	130
HC_genes_AA_Seq	26,159	4958	406
LC_genes_AA_Seq	53,220	1814	106

similarity search via BLAST and various download options. Besides gene IDs, the generalized search option operates on a set of pre-calculated functional gene descriptions, also referred to as 'human readable descriptions'. These descriptions either were derived from curated resources (such as from TAIR for Arabidopsis thaliana (Lamesch et al. 2012)) or computed using the AHRD ('Automatic assignment of human-readable descriptions') tool (Tomato Genome Tomato Genome Consortium 2012).

Beside information on individual genetic elements, more and more combinatorial and comparative queries become important to address more complex scientific questions. To address these more specific tasks, specialized tools and interfaces were developed or integrated in PlantsDB, including the synteny browser CrowsNest and the RNASeqExpressionBrowser, both outlined in more detail later.

21.2.6 Barley Genome Resources in PlantsDB

Genome resources in PlantsDB for barley mainly originated from the whole genome shotgun (WGS) analysis study and draft genome sequence published in 2012 (International Barley Genome Sequencing International Barley Genome Sequencing Consortium et al. 2012). PGSB has set up a database instance to integrate, manage and accommodate the resulting, often heterogeneous, data sets and provide users an entry point to search, browse, analyse and download that data in various ways and formats. With the emergence and availability of the barley reference genome sequence (expected in 2017), the PlantsDB barley database instance will be updated with the latest genome assembly, predicted gene calls and transposable elements as well as with associated data. PlantsDB views and tools that feature barley genome data include:

A. Access to Chromosome-arm sorted whole genome draft sequences (cultivars Morex, Barke and Bowman), generated by (International Barley Genome Sequencing International Barley Genome Sequencing Consortium et al. 2012). Gene predictions as well as functional annotation were performed by PGSB and data can be downloaded from PlantsDB (via structured FTP server at ftp://ftpmips.helmholtz-muenchen.de/plants/barley/public_data/). This download also provides access to expression data, physical and genetic maps plus their integration/anchoring, POPSEQ data (Mascher et al. 2013a), GenomeZipper data (see B.) and repeat annotation.

B. GenomeZipper data for barley: To overcome limitations introduced by often fragmented genome sequence assemblies and short contigs, virtually ordered gene maps were constructed for barley (Mayer et al. 2011), rye (Martis et al. 2013) and wheat (International Wheat Genome Sequencing International Wheat Genome Sequencing Consortium 2014). The GenomeZipper concepts (Mayer et al. 2011) hereby utilize the observation of long stretches of conserved gene order between many grass genomes ('synteny') (Moore et al. 1995), and integrate next generation sequencing data, chromosome sorting, array hybridization and fl-cDNAs, ESTs and genetic markers. That way, information-rich scaffolds of complex cereal crop genomes can be constructed from less complex, fully sequenced grass model species such as rice, sorghum and *Brachypodium distachyon*. To visualize GenomeZipper data and make all integrated data searchable to users, PlantsDB provides dedicated GenomeZipper interfaces. These provide access to all raw data/sequences anchored and Zipper results can be queried either via a visual map interface or via reference gene models from rice, Sorghum or *Brachypodium distachyon*. Download options include Excel- and/or CSV-formatted files (http://pgsb.helmholtz-muenchen.de/plant/barley/gz/download/index.jsp).

C. Barley gene predictions based on 2012 WGS draft genome sequence: all gene models predicted on the IBSC 2012 WGS draft genome sequence (International Barley Genome Sequencing International Barley Genome Sequencing Consortium et al. 2012) have been integrated into PlantsDB and are available for search by ID, keyword, functional description or sequence similarity via BLAST. Gene reports were set up for all barley genes, summarizing gene structure, sequence features, domains, GO terms, external links, confidence assignment and sequence downloads.

Table 21.3 Species with pairwise syntenic comparisons in CrowsNest

CrowsNest organism 1	CrowsNest organism 2
Brachypodium distachyon	*Oryza sativa*
Brachypodium distachyon	*Sorghum bicolor*
Oryza sativa	*Sorghum bicolor*
Hordeum vulgare	*Brachypodium distachyon*
Hordeum vulgare	*Aegilops tauschii*
Hordeum vulgare	*Oryza sativa*

21.2.7 CrowsNest—A Tool to Explore and Visualize Syntenic Relationships in Grasses

To analyse structural genome characteristics of plants such as conserved gene order between plant genomes, customized visualization tools are required. As a part of PlantsDB, CrowsNest was developed as a synteny viewer to facilitate comparisons on the basis of genetically and physically anchored genomes. Four different hierarchically ordered view levels visualize syntenic segments, orthologous and homologous gene pairs for selected plant genome comparisons (Table 21.3). CrowsNest allows the comparison of up to three genomic datasets (one target and one or two reference datasets). The CrowsNest tool can be accessed at http://pgsb.helmholtz-muenchen.de/plant/crowsNest/.

For barley, CrowsNest makes use of the integrated genome map and gene calls generated and published by IBSC in 2012 (International Barley Genome Sequencing International Barley Genome Sequencing Consortium et al. 2012). Figure 21.1 provides an example view comparing the gene order in barley (on the basis of the 2012 IBSC genome orientation) with the gene order in *Brachypodium distachyon*. This view corresponds to level 2 with visualization of individual chromosomes.

21.2.8 PGSB RNASeqExpression Browser to Explore and Analyse Barley Transcriptome Data

With the emergence and availability of multiple expression data sets for individual species, it has become increasingly important to both structure and visualize these data in the context of the genome data. To facilitate the analysis, communication and sharing of RNA-Seq transcriptome data, RNASeqExpressionBrowser (Nussbaumer et al. 2014) was developed as a web-based tool and integrated into PlantsDB (accessible at http://pgsb.helmholtz-muenchen.de/plant/RNASeqExpressionBrowser).

At date, the RNASeqExpressionBrowser stores the RNA-Seq expression data for two different barley experiments: (i) publically available RNA-Seq data from the IBSC 2012 publication (International Barley Genome Sequencing International Barley Genome Sequencing Consortium et al. 2012) (ii) expression data from a study related to Bacteria-triggered systemic immunity in barley (Dey et al. 2014). The RNASeqExpressionBrowser provides three different ways to access the processed RNA-Seq datasets: Wildcard-search based on keywords (such as gene identifier or domain annotation), a BLAST search against the corresponding

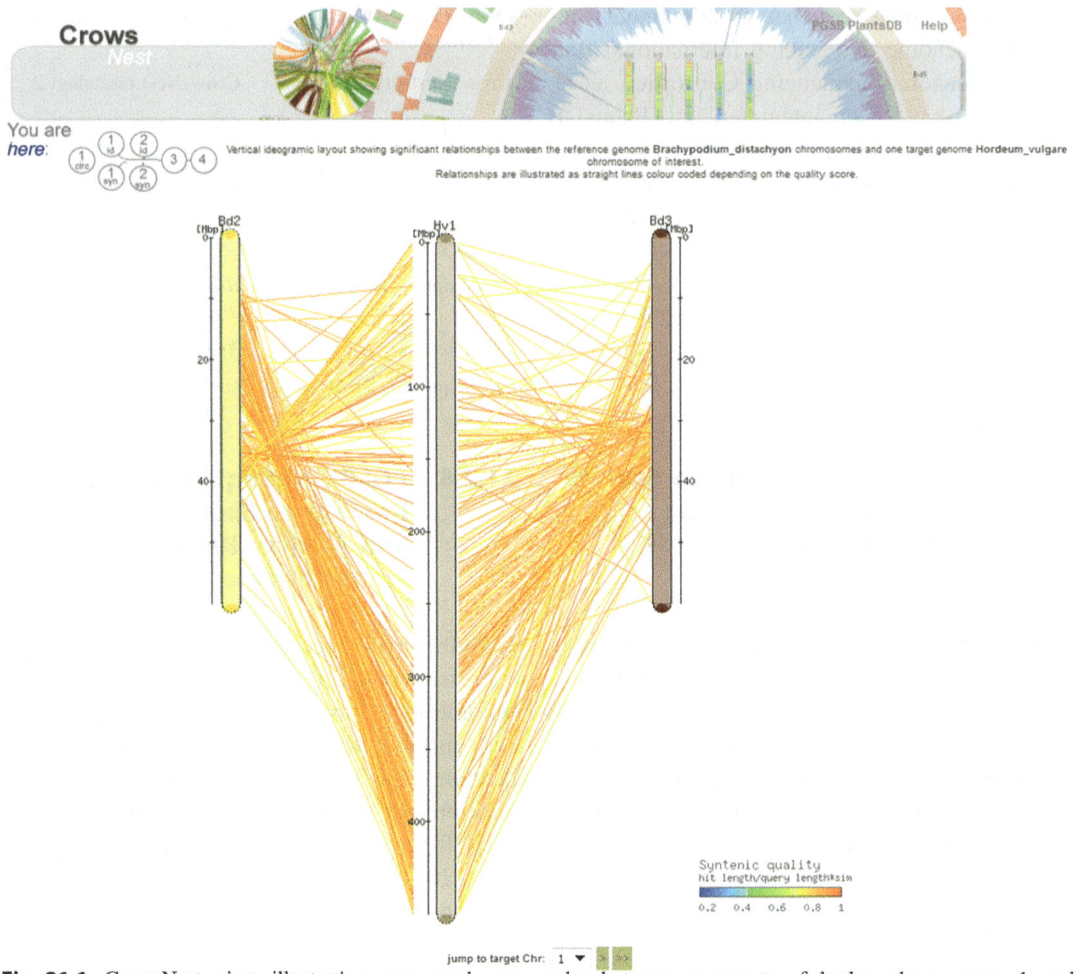

Fig. 21.1 CrowsNest view illustrating conserved gene order between segments of barley chromosome 1 and *Brachypodium distachyon* chromosomes 2 and 3. Additional (hierarchical) views can be accessed using the navigation bar in the top left corner

sequences, or a search based on a gene list. The RNASeqExpressionBrowser is also available as a portable stand-alone software platform enabling import and visualization of own RNA-Seq derived expression data.

21.2.9 PGSB Repeat Element Database (mips-REdat) and Catalog (mips-REcat)

The Repeat Element Database (pgsb-REdat) and the Repeat Element Catalog (pgsb-REcat) provide various interfaces for browsing and downloading repetitive elements. pgsb-REdat stores repetitive sequences from many different plant genomes/species which can be downloaded and used for, e.g. masking repetitive elements in gene prediction efforts. The mips-REcat repeat catalog provides a hierarchical classification scheme for repetitive elements with different levels, starting from main groups down to more specific repeat families. Both pgsb-REdat and pgsb-REcat assist users in the in silico detection of repeats and in the analysis of more complex and nested repeat insertions, e.g. present in the barley genome. pgsb-REdat and pgsb-REcat can be accessed via the PlantsDB tools section or directly at http://pgsb.helmholtz-muenchen.de/plant/recat/index.jsp.

21.2.10 Specialized Resources

GrainGenes (Carollo et al. 2005), a database for Triticeae and Avena, is a comprehensive resource for molecular and phenotypic information for wheat, barley, rye and other related species, including oat. The main data types provided for barley (https://wheat.pw.usda.gov/GG3/barley_blvd) are genetic maps, markers and germplasm information.

RNA-Seq data from the Morex cultivar can be searched at morexGenes (https://ics.hutton.ac.uk/morexGenes/), and epigenomic data can be accessed in the barley epigenome browser (https://ics.hutton.ac.uk/barley-epigenome/) (Baker et al. 2015). The Triticeae Toolbox (http://triticeaetoolbox.org) is a database schema enabling plant breeders and researchers to combine, visualize and interrogate the wealth of phenotype and genotype data generated by the Triticeae Coordinated Agricultural Project (TCAP) (Blake et al. 2016).

A database for barley genes and genetic stocks, including nomenclature for many genes, is available at NordGen (http://www.nordgen.org/bgs/index.php?pg=bgs_tables&m=loc).

21.3 Specific Use Cases for the Barley Genome

Use Case 1:

(International Barley Genome Sequencing International Barley Genome Sequencing Consortium et al. 2012; Yang et al. 2014) was able to identify two fingerprinted contigs (FPCs) containing flanking genetic markers for a gene that induces susceptibility to barley yellow mosaic virus (BaYMV). In order to detect the gene, 17 BACs were sequenced between these markers. The hypothesis was that both FPCs were overlapping and manual inspection showed that to be true. With this information (Yang et al. 2014) was able to locate a candidate gene.

By using only the tools available in BARLEX, we were able to reproduce these results (Fig. 21.2).

Use Case 2:

A general approach in finding your region of interest in the barley genome: You usually start with a protein of interest. On the BARLEX website, you paste the sequence into the BLAST field and hit the button 'go to BLAST search'. This will automatically open the BLAST tool. The output of this homology search is directly linked to the BARLEX system and will show you the sequence with the highest homology in the barley data set, usually a BAC sequence contig (or a gene). Using the hyperlinks will bring you to that specific BAC contig and its complexity plot (from Kmasker analysis (Schmutzer et al. 2014)). From here, there are two possible ways to obtain more information.

Either you follow the link to the parent BAC and then proceed in looking at its FPC and other neighbouring BACs from its cluster or you can try the AGP List and paste the BAC contig identifier into the 'Sequence Position -> AGP Position'-tool. If the BAC contig is part of the nonredundant sequence you will be able to find a corresponding position in the pseudomolecules and can use this for downstream analysis.

Use Case 3

Use a sequence to find a gene in barley and then its homologues in other *Triticeae* species; obtain functional information about the gene family, and identify differences in sequence likely to account for any divergence in function.

1. Go to http://plants.ensembl.org/index.html and click on 'HMMER' in the header. HMMER is an alternative sequence search tool to BLAST and is a quick and accurate way of matching protein sequences, that uses Hidden Markov Models to identify likely conserved domains (Finn et al. 2015).
2. Paste your sequence into the search box, and press 'return'. Figure 21.3 contains a query sequence used in this example.
3. After a short delay, a results page appears, showing the protein domains identified within your query sequence and below, a ranked list of protein sequence matches

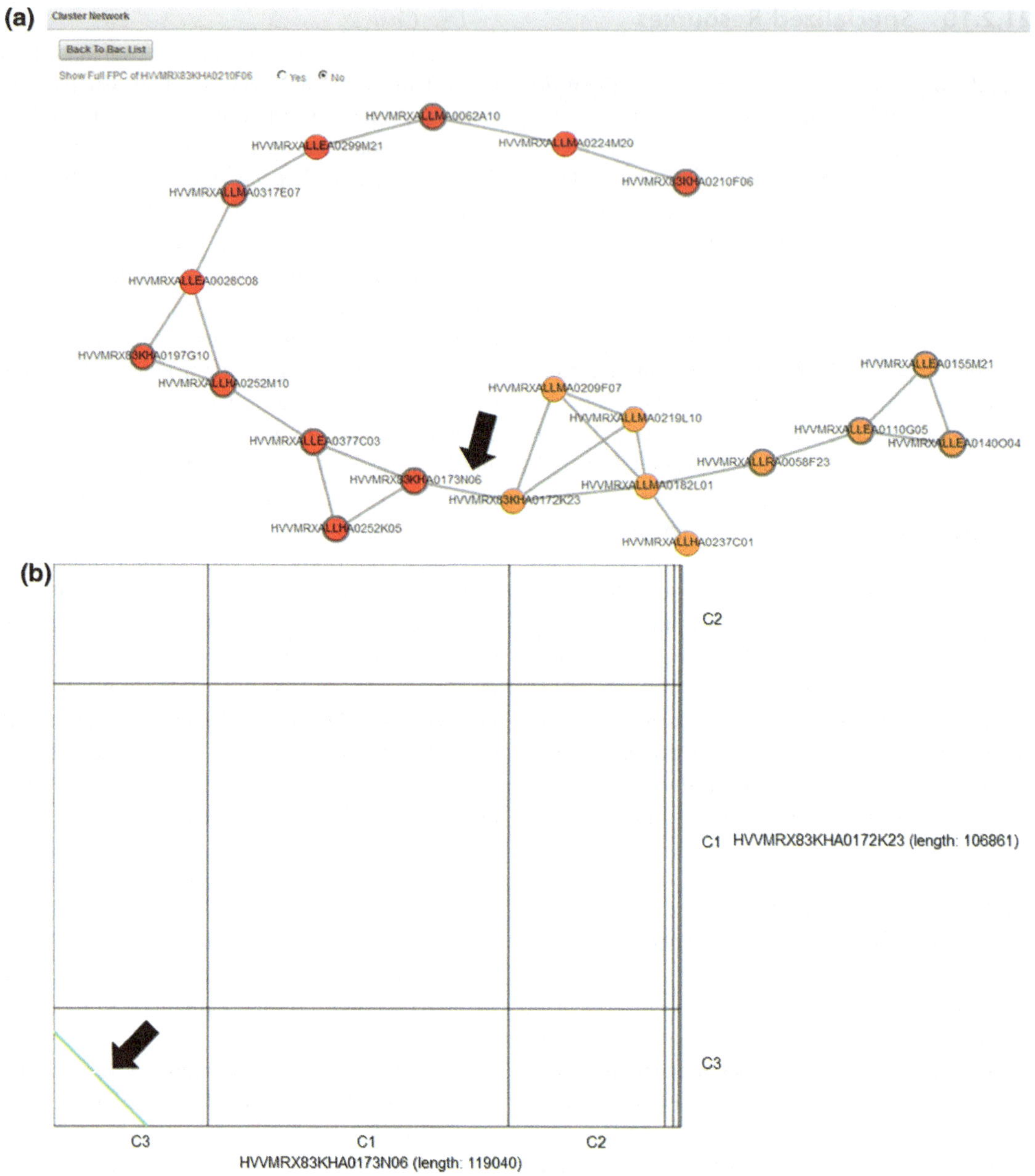

Fig. 21.2 Automatic determination of sequence overlaps between BACs (Yang et al. 2014). The automatic determination based on homology in the BARLEX system was able to reproduce and visualize the previously shown overlap between two FPCs (**a**). The Dotplot of the two connecting BACs showed a clear sequence shared by both clones represented by their identifiers: HVVMRX83kHA0173N06 and HVVMRX83kHA0172K23 (**b**)

VAAQLSARYIMGRHLPDKAIDLVDEACANVRVQLDSQPEEIDNLERKRIQLEVELHALEKEKDKASKARLVEVRKELDDLRD
KLQPLTMKYRKEKERIDEIRKLKQRREELQFTLQEAERRMDLARVADLKYGALQEIDAAIAKLEGETGENLMLTETVGPEQI
AEVVSRWTGIPVTRLGQNDKERLVGMADRLHTRVVGQTEAVNAVAEAVLRSRAGLGRPQQPTGSFLFLGPTGVGKTELAKAL
AEQLFDDENLLVRIDMSEYMEQHSVARLIGAPPGYVGHEEGGQLTEQVRRRPYSVILFDEVE

Fig. 21.3 Example sequence for use in use case

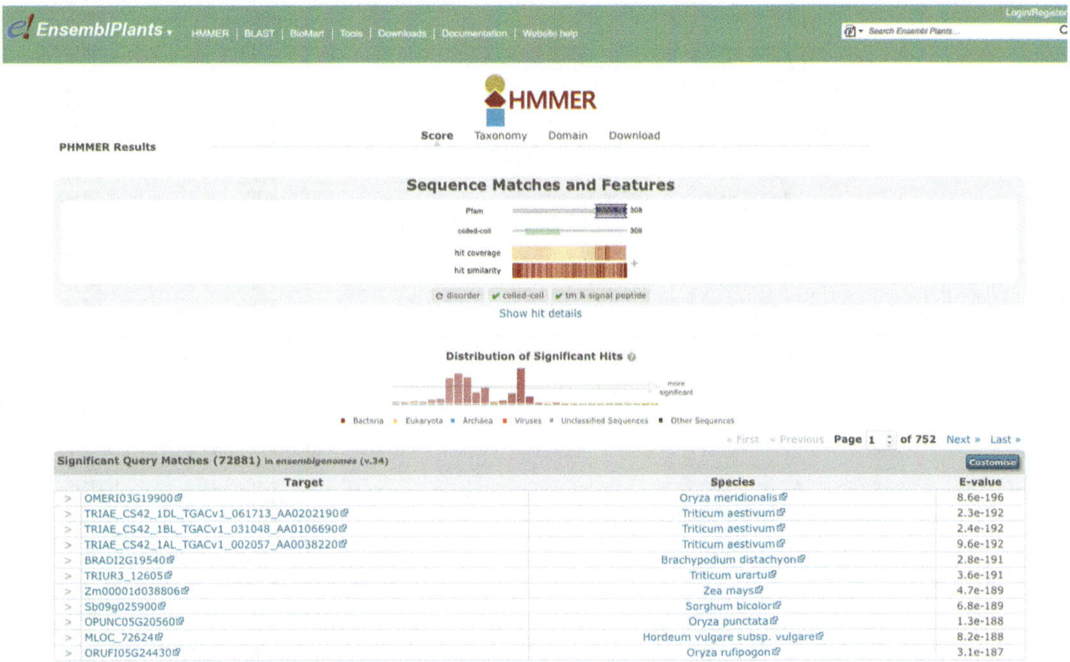

Fig. 21.4 Results panel from using the HMMER search in Ensembl Plants

from plant genomes (see Fig. 21.4). Select the best match from barley.

4. Clicking on the gene name ('MLOC_72624' if you used the example sequence, above) takes you to a page providing an overview about this gene. In the left-hand margin, there are various options to view additional information. Click on the link entitled 'Gene tree'.
5. The view shows an inferred evolutionary history of the gene family containing the selected gene. When (as in the example case) the family is large, the tree is shown in a contracted fashion. However, it can be expanded, in whole or in part. The selected gene is shown in red where portions of the tree are compressed shown by a solid triangle, the taxonomic range of the compressed portion of the tree is indicate to the right of the triangle. Meanwhile, each implied speciation or gene duplication event in the history of the family is shown by a small square box within the tree. Find the box corresponding to the node whose descendants contain the selected gene and other members of the *Poeceae* family. Hovering over the node reveals information about the event and options for analysis or visualization. Click on 'expand all-sub trees' to fully see the descendants of this node. The expanded tree is shown in Fig. 21.5.
6. Find the homologue of the gene in the well-annotated genome of *Oryza sativa japonica*. Click on it, and learn what you can about the gene's function.

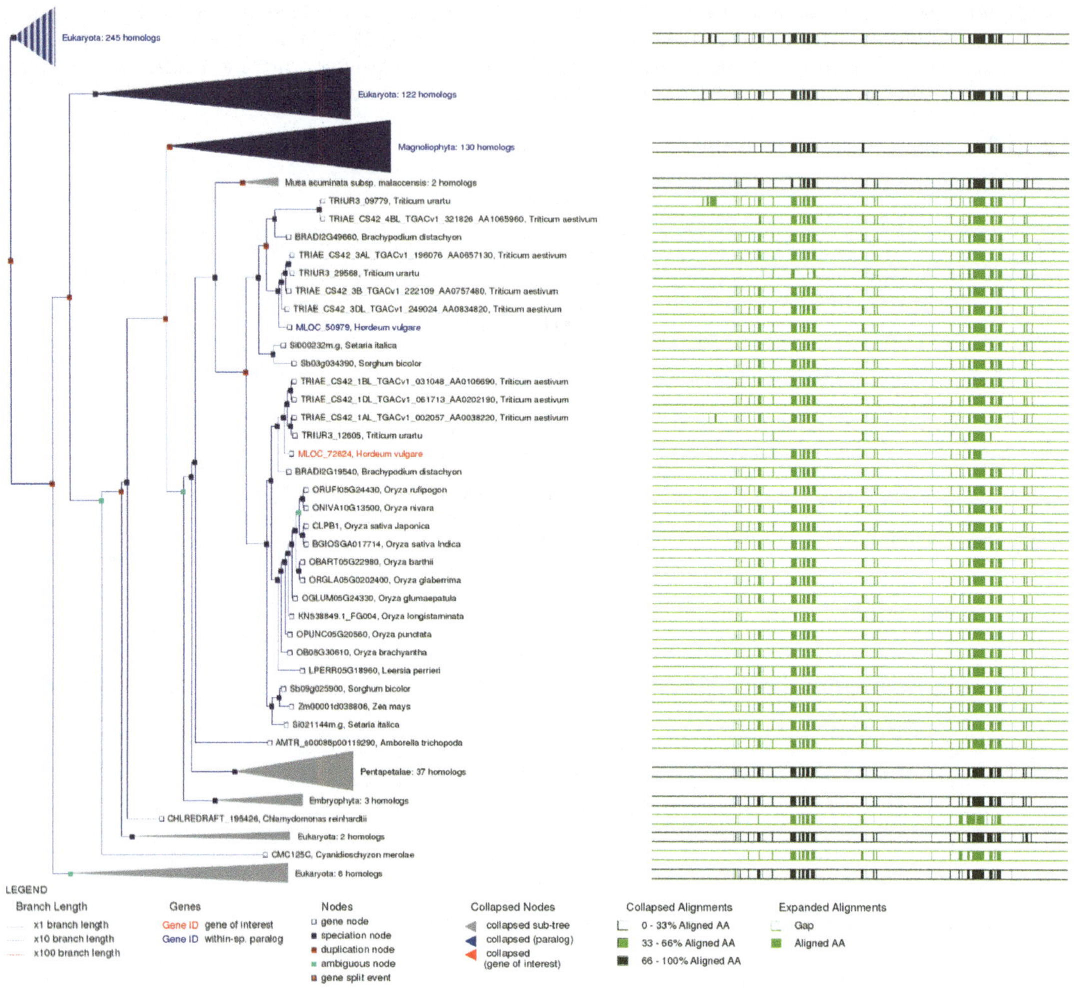

Fig. 21.5 Ensembl gene tree for barley gene MLOC_72624, expanded for the Poeceae (true grass) family

7. Now click on the node in the tree covering just the selected genes and its orthologues in the Triticeae. There are various options for viewing an alignment of these genes, including the interactive Wasabi alignment browser or export in various formats.
8. Notice that the tree indicates the selected gene has at least one paralogue, shown in blue. Click on the 'Paralogues' link in the left-hand menu to find out more.
9. Some of these paralogues are only distant family members. Identify the closest paralogue in the list and click on 'Alignment (protein)' in the 'Compare' column to see how the two paralogues have diverged.
10. Now click on 'Variant image' in the left-hand margin. This produces a graphic of known sequence polymorphisms in the barley population. Colours are used to indicate which variants are likely to disrupt the function of coding genes.
11. Click on 'Variant table', and use the button marked 'Consequences' to filter the contents of the table to see only missense variants.
12. Click on the variant ID to see more information about this variant.

21.4 Discussion: Future Perspectives

One limitation of the current barley genome and annotation is the relatively low availability of detailed functional information about barley genes. In the case of many genes, their function is unknown; their biochemical activity may be inferable from the presence of common domains in the proteins they encode, but their precise biological role is unknown. The availability of a high-quality reference genome sequence opens the way to large-scale genotypic-phenotypic analysis, and to the characterization of genes based on the molecular and morphological traits associated with variant forms. A major challenge is to incorporate this information in gene annotations, and find other appropriate ways to make the results of such analyses available to users.

Another source of information is from comparisons to genomes from other species. As genomes become more complete, orthology can be more reliably determined and information projected from well-studied sequences to others. A further benefit for barley could be provided from a standardization of naming across the Triticeae, providing an easy way to identify equivalent genes. This is made difficult by the complex evolutionary history of the tribe, and the existence of established, but inconsistent, naming conventions between barley and wheat.

We future developments in sequencing technology to drive further improvements in the reference assembly, but also in future sequencing of other strains of cultivated barley, wild barley and less studied relatives. High-quality assemblies will allow the construction of a definitive catalogue of genome diversity at all scales, including large-scale structural variation; and identify a wealth of usable genes that might be incorporated into elite lines. In an ideal model, this catalogue would be searchable, linked to gene banks and to repositories of phenotypic information, and available for on-demand search and genome-wide association analysis. But challenges remain in sequencing this material

Table 21.4 List of URLs for barley centric databases, tools, websites, and resources

Provider	URL	Short description
EBI	http://plants.ensembl.org	Ensembl Plants homepage
EBI/CSH	http://www.gramene.org	Gramene homepage
IPK	http://barlex.barleysequence.org	BARLEX homepage
IPK	http://webblast.ipk-gatersleben.de/barley	IPK barley web BLAST
PGSB	http://pgsb.helmholtz-muenchen.de/plant/genomes.jsp	PlantsDB
PGSB	http://pgsb.helmholtz-muenchen.de/plant/barley/gz/download/index.jsp	Barley GenomeZipper
PGSB	http://pgsb.helmholtz-muenchen.de/plant/crowsNest/	CrowsNest homepage
PGSB	http://pgsb.helmholtz-muenchen.de/plant/RNASeqExpressionBrowser	RNASeqExpressionBrowser homepage
PGSB	http://pgsb.helmholtz-muenchen.de/plant/recat/index.jsp	PGSB Repeat Element Database (REdat) and Catalog (REcat)
	https://wheat.pw.usda.gov/GG3/barley_blvd	GrainGenes Triticeae database
JHI	https://ics.hutton.ac.uk/morexGenes/	RNA-Seq data of barley cultivar Morex
JHI	https://ics.hutton.ac.uk/barley-epigenome/	Epigenome browser for barley
TCAP	http://triticeaetoolbox.org	TCAP homepage
NordGen	http://www.nordgen.org/bgs/index.php?pg=bgs_tables&m=loc	International Database for Barley Genes and Barley Genetic Stocks

and disseminating the information a number of broadly based international groups have formed an organization, DivSeek (http://divseek.org), to coordinate in this task for many crops. Phenotyping data, which can include molecular, laboratory, greenhouse or field data, presents further challenges owing to its diverse nature, the absence of generic repositories and the relative lack of standard ways of representing this information. A barley trait dictionary has been developed as part of the Crop Ontology (http://www.cropontology.org/ontology/CO_323/Barley%20Trait%20Dictionary) and is a first step towards being able to support the representation of barley phenotypes in a consistent fashion across multiple experiments. An overview of barley resources are summarized in Table 21.4.

References

Altschul SF, Gish W, Miller W, Myers EW, Lipman DJ (1990) Basic local alignment search tool. J Mol Biol 215(3):403–410. https://doi.org/10.1016/S0022-2836(05)80360-2

Ariyadasa R, Mascher M, Nussbaumer T, Schulte D, Frenkel Z, Poursarebani N, Zhou R, Steuernagel B, Gundlach H, Taudien S, Felder M, Platzer M, Himmelbach A, Schmutzer T, Hedley PE, Muehlbauer GJ, Scholz U, Korol A, Mayer KF, Waugh R, Langridge P, Graner A, Stein N (2014) A sequence-ready physical map of barley anchored genetically by two million single-nucleotide polymorphisms. Plant Physiol 164(1):412–423. https://doi.org/10.1104/pp.113.228213

Ashburner M, Ball CA, Blake JA, Botstein D, Butler H, Cherry JM, Davis AP, Dolinski K, Dwight SS, Eppig JT, Harris MA, Hill DP, Issel-Tarver L, Kasarskis A, Lewis S, Matese JC, Richardson JE, Ringwald M, Rubin GM, Sherlock G (2000) Gene ontology: tool for the unification of biology. The Gene Ontology Consortium. Nat Genet 25(1):25–29. https://doi.org/10.1038/75556

Baker K, Dhillon T, Colas I, Cook N, Milne I, Milne L, Bayer M, Flavell AJ (2015) Chromatin state analysis of the barley epigenome reveals a higher-order structure defined by H3K27me1 and H3K27me3 abundance. Plant J Cell Mol Biol 84(1):111–124. https://doi.org/10.1111/tpj.12963

Blake VC, Birkett C, Matthews DE, Hane DL, Bradbury P, Jannink JL (2016) The triticeae toolbox: combining phenotype and genotype data to advance small-grains breeding. Plant Genome 9(2). https://doi.org/10.3835/plantgenome2014.12.0099

Bolser DM, Staines DM, Perry E, Kersey PJ (2017) Ensembl plants: integrating tools for visualizing, mining, and analyzing plant genomic data. Methods Mol Biol 1533:1–31. https://doi.org/10.1007/978-1-4939-6658-5_1

Carollo V, Matthews DE, Lazo GR, Blake TK, Hummel DD, Lui N, Hane DL, Anderson OD (2005) GrainGenes 2.0. An improved resource for the small-grains community. Plant Physiol 139(2):643–651. https://doi.org/10.1104/pp.105.064485

Colmsee C, Beier S, Himmelbach A, Schmutzer T, Stein N, Scholz U, Mascher M (2015) BARLEX—the barley draft genome explorer. Mol Plant 8(6):964–966. https://doi.org/10.1016/j.molp.2015.03.009

Comadran J, Kilian B, Russell J, Ramsay L, Stein N, Ganal M, Shaw P, Bayer M, Thomas W, Marshall D, Hedley P, Tondelli A, Pecchioni N, Francia E, Korzun V, Walther A, Waugh R (2012) Natural variation in a homolog of Antirrhinum CENTRORADIALIS contributed to spring growth habit and environmental adaptation in cultivated barley. Nat Genet 44(12):1388–1392. https://doi.org/10.1038/ng.2447

Deng W, Nickle DC, Learn GH, Maust B, Mullins JI (2007) ViroBLAST: a stand-alone BLAST web server for flexible queries of multiple databases and user's datasets. Bioinformatics 23(17):2334–2336. https://doi.org/10.1093/bioinformatics/btm331

Dey S, Wenig M, Langen G, Sharma S, Kugler KG, Knappe C, Hause B, Bichlmeier M, Babaeizad V, Imani J (2014) Bacteria-triggered systemic immunity in barley is associated with WRKY and ETHYLENE RESPONSIVE FACTORs but not with salicylic acid. Plant Physiol 166(4):2133–2151

Finn RD, Attwood TK, Babbitt PC, Bateman A, Bork P, Bridge AJ, Chang HY, Dosztanyi Z, El-Gebali S, Fraser M, Gough J, Haft D, Holliday GL, Huang H, Huang X, Letunic I, Lopez R, Lu S, Marchler-Bauer A, Mi H, Mistry J, Natale DA, Necci M, Nuka G, Orengo CA, Park Y, Pesseat S, Piovesan D, Potter SC, Rawlings ND, Redaschi N, Richardson L, Rivoire C, Sangrador-Vegas A, Sigrist C, Sillitoe I, Smithers B, Squizzato S, Sutton G, Thanki N, Thomas PD, Tosatto SC, Wu CH, Xenarios I, Yeh LS, Young SY, Mitchell AL (2017) InterPro in 2017-beyond protein family and domain annotations. Nucleic Acids Res 45(D1):D190–D199. https://doi.org/10.1093/nar/gkw1107

Finn RD, Clements J, Arndt W, Miller BL, Wheeler TJ, Schreiber F, Bateman A, Eddy SR (2015) HMMER web server: 2015 update. Nucleic Acids Res 43(W1):W30–38. https://doi.org/10.1093/nar/gkv397

Finn RD, Coggill P, Eberhardt RY, Eddy SR, Mistry J, Mitchell AL, Potter SC, Punta M, Qureshi M, Sangrador-Vegas A, Salazar GA, Tate J, Bateman A (2016) The Pfam protein families database: towards a more sustainable future. Nucleic Acids Res 44(D1):D279–285. https://doi.org/10.1093/nar/gkv1344

Gene Ontology Consortium (2015) Gene ontology consortium: going forward. Nucleic Acids Res 43(D1): D1049–D1056

International Barley Genome Sequencing Consortium, Mayer KF, Waugh R, Brown JW, Schulman A, Langridge P, Platzer M, Fincher GB, Muehlbauer GJ, Sato K, Close TJ, Wise RP, Stein N (2012) A physical, genetic and functional sequence assembly of the barley genome. Nature 491(7426):711–716. https://doi.org/10.1038/nature11543

International Wheat Genome Sequencing Consortium (2014) A chromosome-based draft sequence of the hexaploid bread wheat (Triticum aestivum) genome. Science 345(6194):1251788. https://doi.org/10.1126/science.1251788

Kersey PJ, Allen JE, Armean I, Boddu S, Bolt BJ, Carvalho-Silva D, Christensen M, Davis P, Falin LJ, Grabmueller C, Humphrey J, Kerhornou A, Khobova J, Aranganathan NK, Langridge N, Lowy E, McDowall MD, Maheswari U, Nuhn M, Ong CK, Overduin B, Paulini M, Pedro H, Perry E, Spudich G, Tapanari E, Walts B, Williams G, Tello-Ruiz M, Stein J, Wei S, Ware D, Bolser DM, Howe KL, Kulesha E, Lawson D, Maslen G, Staines DM (2016) Ensembl Genomes 2016: more genomes, more complexity. Nucleic Acids Res 44 (D1):D574–580. https://doi.org/10.1093/nar/gkv1209

Kurtz S (2003) The Vmatch large scale sequence analysis software. Ref Type: Computer Program:4–12

Lamesch P, Berardini TZ, Li D, Swarbreck D, Wilks C, Sasidharan R, Muller R, Dreher K, Alexander DL, Garcia-Hernandez M, Karthikeyan AS, Lee CH, Nelson WD, Ploetz L, Singh S, Wensel A, Huala E (2012) The Arabidopsis Information Resource (TAIR): improved gene annotation and new tools. Nucleic Acids Res 40 (Database issue):D1202–1210. https://doi.org/10.1093/nar/gkr1090

Lange M, Spies K, Colmsee C, Flemming S, Klapperstück M, Scholz U (2009) The LAILAPS search engine: a feature model for relevance ranking in life science databases. J Integr Bioinform 7(3):183–200

Lopes CT, Franz M, Kazi F, Donaldson SL, Morris Q, Bader GD (2010) Cytoscape Web: an interactive web-based network browser. Bioinformatics 26 (18):2347–2348. https://doi.org/10.1093/bioinformatics/btq430

Martis MM, Zhou R, Haseneyer G, Schmutzer T, Vrana J, Kubalakova M, Konig S, Kugler KG, Scholz U, Hackauf B, Korzun V, Schon CC, Dolezel J, Bauer E, Mayer KF, Stein N (2013) Reticulate evolution of the rye genome. Plant Cell 25(10):3685–3698. https://doi.org/10.1105/tpc.113.114553

Mascher M, Gundlach H, Himmelbach A, Beier S, Twardziok SO, Wicker T, Radchuk V, Dockter C, Hedley P, Russel J, Bayer M, Ramsay L, Liu H, Haberer G, Zhang X-Q, Zhang Q, Barrero RA, Li L, Taudien S, Groth M, Felder M, Hastie AR, Simkova H, Stankova H, Vrana J, Chan S, Munoz-Amatriain M, Ounit R, Wanamaker S, Bolser DM, Colmsee C, Schmutzer T, Aliyeva-Schnorr L, Grasso S, Tanskanen J, Chailyan A, Sampath D, Heavens D, Clissold L, Cao S, Chapman B, Dai F, Han Y, Li H, Li X, Lin C, McCooke JK, Tan C, Wang P, Wang S, Yin S, Zhou G, Poland JA, Bellgard MI, Borisjuk L, Houben A, Dolezel J, Ayling S, Lonardi S, Kersey PJ, Langridge P, Muehlbauer G, Clark MD, Caccamo M, Schulman AH, Mayer KFX, Platzer M, Close TJ, Scholz U, Hansson M, Zhang G, Braumann I, Spannagl M, Li C, Waugh R, Stein N (2017) A chromosome conformation capture ordered sequence of the barley genome. Nature 544(7651):427

Mascher M, Muehlbauer GJ, Rokhsar DS, Chapman J, Schmutz J, Barry K, Munoz-Amatriain M, Close TJ, Wise RP, Schulman AH, Himmelbach A, Mayer KF, Scholz U, Poland JA, Stein N, Waugh R (2013a) Anchoring and ordering NGS contig assemblies by population sequencing (POPSEQ). Plant J Cell Mole Biol 76(4):718–727. https://doi.org/10.1111/tpj.12319

Mascher M, Richmond TA, Gerhardt DJ, Himmelbach A, Clissold L, Sampath D, Ayling S, Steuernagel B, Pfeifer M, D'Ascenzo M, Akhunov ED, Hedley PE, Gonzales AM, Morrell PL, Kilian B, Blattner FR, Scholz U, Mayer KF, Flavell AJ, Muehlbauer GJ, Waugh R, Jeddeloh JA, Stein N (2013b) Barley whole exome capture: a tool for genomic research in the genus Hordeum and beyond. Plant J Cell Mole Biol 76 (3):494–505. https://doi.org/10.1111/tpj.12294

Mayer KF, Martis M, Hedley PE, Simkova H, Liu H, Morris JA, Steuernagel B, Taudien S, Roessner S, Gundlach H, Kubalakova M, Suchankova P, Murat F, Felder M, Nussbaumer T, Graner A, Salse J, Endo T, Sakai H, Tanaka T, Itoh T, Sato K, Platzer M, Matsumoto T, Scholz U, Dolezel J, Waugh R, Stein N (2011) Unlocking the barley genome by chromosomal and comparative genomics. Plant Cell 23(4):1249–1263. https://doi.org/10.1105/tpc.110.082537

Middleton CP, Senerchia N, Stein N, Akhunov ED, Keller B, Wicker T, Kilian B (2014) Sequencing of chloroplast genomes from wheat, barley, rye and their relatives provides a detailed insight into the evolution of the triticeae tribe. PLoS ONE 9(3):e85761. https://doi.org/10.1371/journal.pone.0085761

Moore G, Devos KM, Wang Z, Gale MD (1995) Cereal genome evolution. Grasses, line up and form a circle. Curr Biol CB 5(7):737–739

Munoz-Amatriain M, Lonardi S, Luo M, Madishetty K, Svensson JT, Moscou MJ, Wanamaker S, Jiang T, Kleinhofs A, Muehlbauer GJ, Wise RP, Stein N, Ma Y, Rodriguez E, Kudrna D, Bhat PR, Chao S, Condamine P, Heinen S, Resnik J, Wing R, Witt HN, Alpert M, Beccuti M, Bozdag S, Cordero F, Mirebrahim H, Ounit R, Wu Y, You F, Zheng J, Simkova H, Dolezel J, Grimwood J, Schmutz J, Duma D, Altschmied L, Blake T, Bregitzer P, Cooper L, Dilbirligi M, Falk A, Feiz L, Graner A, Gustafson P, Hayes PM, Lemaux P, Mammadov J, Close TJ (2015) Sequencing of 15 622 gene-bearing BACs clarifies the gene-dense regions of the barley genome. Plant J Cell Mole Biol 84(1):216–227. https://doi.org/10.1111/tpj.12959

Needleman SB, Wunsch CD (1970) A general method applicable to the search for similarities in the amino acid sequence of two proteins. J Mol Biol 48(3):443–453

Nussbaumer T, Kugler KG, Bader KC, Sharma S, Seidel M, Mayer KFX (2014) RNASeqExpression Browser-a web interface to browse and visualize high-throughput expression data. Bioinformatics 30(17): 2519–2520. https://doi.org/10.1093/bioinformatics/btu334

Schmutzer T, Ma L, Pousarebani N, Bull F, Stein N, Houben A, Scholz U (2014) Kmasker–a tool for in silico prediction of single-copy FISH probes for the large-genome species Hordeum vulgare. Cytogenet Genome Res 142(1):66–78. https://doi.org/10.1159/000356460

Schulte D, Close TJ, Graner A, Langridge P, Matsumoto T, Muehlbauer G, Sato K, Schulman AH, Waugh R, Wise RP, Stein N (2009) The international barley sequencing consortium–at the threshold of efficient access to the barley genome. Plant Physiol 149(1): 142–147. https://doi.org/10.1104/pp.108.128967

Smith TF, Waterman MS (1981) Identification of common molecular subsequences. J Mol Biol 147(1):195–197

Soderlund C, Longden I, Mott R (1997) FPC: a system for building contigs from restriction fingerprinted clones. Comput Appl Biosci CABIOS 13(5):523–535

Spannagl M, Nussbaumer T, Bader KC, Martis MM, Seidel M, Kugler KG, Gundlach H, Mayer KF (2016) PGSB PlantsDB: updates to the database framework for comparative plant genome research. Nucleic Acids Res 44(D1):D1141–1147. https://doi.org/10.1093/nar/gkv1130

Tello-Ruiz MK, Stein J, Wei S, Youens-Clark K, Jaiswal P, Ware D (2016) Gramene: a resource for comparative analysis of plants genomes and pathways. Methods Mol Biol 1374:141–163. https://doi.org/10.1007/978-1-4939-3167-5_7

Tomato Genome Consortium (2012) The tomato genome sequence provides insights into fleshy fruit evolution. Nature 485(7400):635–641

Vilella AJ, Severin J, Ureta-Vidal A, Heng L, Durbin R, Birney E (2009) EnsemblCompara GeneTrees: complete, duplication-aware phylogenetic trees in vertebrates. Genome Res 19(2):327–335. https://doi.org/10.1101/gr.073585.107

Yang P, Lupken T, Habekuss A, Hensel G, Steuernagel B, Kilian B, Ariyadasa R, Himmelbach A, Kumlehn J, Scholz U, Ordon F, Stein N (2014) Protein Disulfide Isomerase Like 5-1 is a susceptibility factor to plant viruses. Proc Natl Acad Sci USA 111(6):2104–2109. https://doi.org/10.1073/pnas.1320362111

Zhang Z, Schwartz S, Wagner L, Miller W (2000) A greedy algorithm for aligning DNA sequences. J Comput Biol 7(1–2):203–214. https://doi.org/10.1089/10665270050081478

The manufacturer's authorised representative in the EU is Springer Nature Customer Service Centre GmbH, Europaplatz 3, 69115 Heidelberg, Germany. If you have any concerns regarding our products, please contact ProductSafety@springernature.com

Printed and bound by CPI Group (UK) Ltd, Croydon, CR0 4YY

30/09/2026

02225655-0020